USA in Space
Third Edition

USA in Space
Third Edition

Volume 3

Space Shuttle Mission STS 51-A—X-20 Dyna-Soar

Appendices

Indexes

1377–2014

Edited by

Russell R. Tobias
David G. Fisher

SALEM PRESS, INC.
Pasadena, California • Hackensack, New Jersey

Editor in Chief: Dawn P. Dawson
Editorial Director: Christina J. Moose
Acquisitions Editor: Mark Rehn
Research Supervisor: Jeffry Jensen
Manuscript Editor: Anna A. Moore
Production Editor: Kathy Hix
Design and Graphics: James Hutson
Editorial Assistant: Dana Garey
Layout: Eddie Murillo
Photograph Editor: Cynthia Beres

Library of Congress Cataloging-in-Publication Data

USA in space / edited by Russell R. Tobias and David G. Fisher.— 3rd ed.
    p. cm.
    Includes bibliographical references and index.
    ISBN-10: 1-58765-259-5 (set : alk. paper)
    ISBN-10: 1-58765-262-5 (vol. 3 : alk. paper)
    ISBN-13: 978-1-58765-259-2 (set : alk. paper)
    ISBN-13: 978-1-58765-262-2 (vol. 3 : alk. paper)
    [etc.]
    1. Astronautics—United States. I. Title: United States of America in space. II. Tobias, Russell R. III. Fisher, David G.
    TL789.8.U5U83 2006
    629.40973—dc22

                   2005030756

First Printing

PRINTED IN THE UNITED STATES OF AMERICA

Table of Contents

List of Illustrations

Complete List of Contents

Volume 1

Volume 2

Volume 3

Category List

List of Categories

Aerospace Agencies

National Aeronautics and Space
Administration, 973
National Commission on Space, 989
Space Task Group, 1620
United Space Alliance, 1760
United States Space Command, 1765

Communications Satellites

Amateur Radio Satellites, 6
Applications Technology Satellites, 174
Intelsat Communications Satellites, 569
Mobile Satellite System, 967
Telecommunications Satellites: Maritime,
1692
Telecommunications Satellites: Military, 1698
Telecommunications Satellites: Passive Relay,
1704
Telecommunications Satellites: Private and
Commercial, 1710
Tracking and Data-Relay Communications
Satellites, 1748

Earth Observation

Dynamics Explorers, 292
Earth Observing System Satellites, 303
Explorers: Air Density, 332
Explorers: Atmosphere, 346
Explorers: Ionosphere, 353
Geodetic Satellites, 481
Global Atmospheric Research Program,
497
Heat Capacity Mapping Mission, 519
Landsat 1, 2, and 3, 697
Landsat 4 and 5, 704
Landsat 7, 709
Mission to Planet Earth, 961
Seasat, 1159

Expendable Launch Vehicles

Atlas Launch Vehicles, 202-208
Delta Launch Vehicles, 284
Launch Vehicles, 719
Saturn Launch Vehicles, 1142
Soyuz Launch Vehicle, 1230
Titan Launch Vehicles, 1741

USA in Space
Third Edition

Space Shuttle Mission STS 51-A

Date: November 8 to November 16, 1984
Type of mission: Piloted spaceflight

STS 51-A was the first space retrieval mission. Using techniques devised on the spot, the crew captured and returned to Earth two satellites that otherwise would have been useless forever. The impressive success of this flight was an excellent demonstration of the value of people working in space.

Key Figures

Frederick H. "Rick" Hauck (b. 1941), STS 51-A commander
David M. Walker (1944-2001), STS 51-A pilot
Joseph P. Allen (b. 1937), STS 51-A mission specialist
Anna L. Fisher (b. 1949), STS 51-A mission specialist
Dale A. Gardner (b. 1948), STS 51-A mission specialist

Summary of the Mission

When five astronauts were assigned to STS 51-A in late 1983, they expected their mission to be a routine cargo-hauling trip into space. Yet with the first flight of 1984, STS 41-B, it began to appear as though STS 51-A might become something more unusual. The crew of STS 41-B saw perfect deployments of their payload: two communications satellites, Indonesia's Palapa B2 (Indonesian for "fruit of the effort") and Western Union's Westar 6. In both cases, however, the rocket motors designed to propel them from *Challenger* to their orbital destinations 35,800 kilometers above Earth shut down prematurely. The two satellites were left in useless orbits about one thousand kilometers high. During 1984, the crew of STS 51-A prepared for the retrieval of these two satellites. Meant for orbits out of reach of the space shuttle, they had not been designed for recapture, but, through creative thinking and clever design by engineers and the astronauts, an apparently workable solution was devised. While plans were being formulated and equipment manufactured, ground controllers gradually lowered the orbits of the two satellites to an altitude the space shuttle orbiter could reach. On November 8, 1984, *Discovery* began its second trip to space on time at 12:15:00 Coordinated Universal Time (UTC, or 7:15 A.M. eastern standard time). Frederick H. "Rick" Hauck commanded this fourteenth mission of the Space Transportation System (STS). His pilot was David M. Walker, and the mission specialists were Joseph P. Allen, Anna L. Fisher, and Dale A. Gardner. As soon as the *Discovery* was settled into its orbit 302 kilometers above Earth, they began preparing for their job in space.

The task of rendezvousing with an orbiting spacecraft is very complicated. In addition to getting to the right point in space to meet the target, it is essential to be in the same orbit as the target. Many fine adjustments are required to make the entire orbits match and not simply intersect at one point. Matching orbits ensures that the two spacecraft stay together once they meet. It would require a total of forty-four carefully timed burns of *Discovery*'s rocket engines over several days to bring it to its first quarry and still allow it to be in the correct positions at the correct times to deploy its payload of two communications satellites.

Telesat of Canada owned the first satellite the astronauts were to deploy. It was the third Canadian communications satellite taken into space by the space shuttle. This one was designated Anik D-2 (Inuit for "brother") or Telesat-H. After confirming that the satellite was in good condition, the crew deployed it on schedule at 21:04 UTC on November 9 and fired *Discovery*'s engines to move away to a safe distance. This separation burn served the additional purpose of contributing to the intricate orbital dance required for the rendezvous. Forty-five minutes later, a timer on the Anik fired its rocket engine, and, unlike the satellites being pursued on this mission, it went smoothly to its targeted orbit.

The second satellite on board was to be leased to the United States Navy. Thus, the manufacturer (Hughes Aircraft Company) called it Leasat 1, while the Navy referred to it as Syncom IV-1. It was released from the payload bay on November 10 at 12:56 UTC (7:56 A.M.) and successfully attained its intended orbit.

After completing the delivery of the two satellites, the crew could devote full attention to the retrieval plans. While *Discovery* continued to stalk Palapa on November 10, Allen and Gardner tested the space suits they would use for their extravehicular activities (EVAs, or spacewalks). On November 11, the orbital chase progressed.

As Hauck and Fisher completed the rendezvous with Palapa on November 12, Walker helped Allen and Gardner into their bulky space suits. By the time the two spacewalkers had exited the air lock and entered *Discovery*'s payload bay, the orbiter had closed to within 9 meters of Palapa. Allen donned the Manned Maneuvering Unit (MMU). With its small thrusters, this backpack allowed him to "fly" in space. Gardner helped him attach the 2-meter-long "stinger" to the front of his MMU. The stinger was a pole with extendable fingers like the ribs of an umbrella, and it projected in front of Allen like a lance in front of a knight. It would be used to dock with Palapa.

Palapa was a cylinder 2.7 meters high and 2.1 meters in diameter. At one end was the nozzle of the spent rocket engine, and the other end held the fragile antennae that would have been put into service had the satellite reached its intended orbit. The astronauts would have liked to capture the satellite with the orbiter's 15-meter mechanical arm (the Remote Manipulator System, or RMS), but the satellite's smooth exterior presented nothing for the arm to snag. It was the job of the EVA astronauts to attach something for the RMS to hold and then to stop the satellite from rotating at 1.5 revolutions per minute so the RMS could capture it.

After confirming that all of his equipment was working as intended, Allen waited until an orbital sunrise and used his MMU to travel to the nozzle end of Palapa. He slowly approached the satellite and inserted the stinger rod into the nozzle of the rocket engine. With a control lever, he opened the rod's fingers inside the engine. The force of the ex-

Space shuttle mission STS 51-A was the first satellite retrieval mission. Two communications satellites had failed to reach their desired Earth orbits and had to be returned to Earth. Here one of the astronauts can be seen spinning with the satellite after reaching it by means of the Manned Maneuvering Unit. (NASA CORE/Lorain County JVS)

tended prongs against the interior of the satellite provided a connection between the astronaut and the spacecraft. He had successfully docked with the errant satellite.

At that point, Allen began rotating with the satellite. By using the gyroscopic stabilization system in his MMU, he was able to stop both himself and Palapa from spinning. Fisher then was able to guide the RMS to a fixture on the side of the stinger. *Discovery* finally had control over Palapa. To secure the satellite in the payload bay, it would be necessary to connect the nozzle end to a cradle in the bay. The antennae made the other end too fragile. Therefore, the next goal was to attach a temporary fixture (the Antenna Bridge Structure, or ABS) over the antennae so the RMS could hold it by that end. Allen and Gardner then would connect mounting brackets to the nozzle end that would allow the satellite to attach to the cradle waiting in the payload bay, and the RMS would lower it into place.

While Allen relaxed in his position at one end of the satellite (still connected to it with the stinger), Fisher positioned the entire assembly so that Gardner could attach the ABS to the other end. To his surprise and frustration, a small protrusion on the spacecraft prevented the ABS from fitting properly. He struggled but was unable to make the critical connection. Without it, the RMS would not be able to place Palapa into its cradle.

After some quick discussion, the crew, with the concurrence of mission controllers, chose a backup procedure that had been practiced briefly before the flight. With Walker orchestrating the operations from inside *Discovery*, Allen detached himself from the stinger and returned the MMU to its station. He positioned himself in a foot restraint attached to the side of the payload bay, and Fisher used the RMS to hand him the 597-kilogram Palapa. He took it by the antenna end and held it while Gardner worked at the other end.

In orbit, objects have essentially no weight. They do, however, still have mass. Allen found that he was able to control the huge mass of the satellite through slow, careful movements. He was in an un-

comfortable position, with the satellite above his head, but he endured long enough for his coworker to complete the necessary tasks. For one orbit of Earth, lasting ninety minutes, he held the satellite while Gardner removed the stinger from the nozzle and attached the bracket for the cradle. This chore had been planned for the two crew members working together, but because Allen was now replacing the function of the ABS, Gardner had to do this task alone. The two men then managed to position the satellite in its holder and lock it into the payload bay.

The astronauts collected their tools and finished the EVA six hours after it had begun. Palapa, despite their difficulties, was secured in the payload bay. Yet there was no time for celebrating. Westar was in orbit more than 1,100 kilometers ahead. The crew would spend the next day chasing it and recharging their space suits in preparation for capturing that satellite. Meanwhile, they were faced with the question of whether there might be an interfering protrusion on Westar as there had been on Palapa. If there were, the ABS would not fit again. With Palapa berthed in the bay, the working space would be more crowded, and it had been fatiguing for Allen to keep the satellite positioned for Gardner. The astronauts and ground controllers devised a new plan.

On November 14, with Westar rotating only 9 meters from Gardner's home in space, he flew his MMU/stinger combination to the nozzle end. He docked with it and stopped its rotation just as Allen had done two days earlier. Instead of guiding the RMS to the grapple target on the stinger, Fisher used it to hold a foot restraint. Allen fixed himself in it as if he was on a cherry picker, and Fisher positioned him so he could grab the antenna end of Westar. While he held it, Gardner removed and stowed his MMU. With the RMS holding Allen and Allen holding Westar, Gardner attached the mounting bracket. Allen was in a more comfortable position than he had been in when he had held Palapa, and Gardner had the benefit of his experience as he attached the mounting bracket by himself once again. The EVA ended 5 hours and 43

An STS 51-A astronaut locks onto a stranded satellite. The maneuvering of massive objects in the weightless environment of Earth orbit was one of the accomplishments of this mission. (NASA CORE/Lorain County JVS)

minutes after it had begun, with both satellites safely in the payload bay awaiting return to Earth.

The next day, the proud crew prepared for the return to Earth, held a press conference, and frequently looked into the payload bay to assure themselves they had indeed captured the satellites. At 10:55 UTC on November 16, while above the Indian Ocean, the crew fired *Discovery*'s engines for 184 seconds to bring them back to Earth. Thirteen minutes after sunrise, at 11:59:56 UTC, the orbiter touched down at Kennedy Space Center. *Discovery* had begun its 5.3-million-kilometer journey only 5 kilometers away with two satellites to deploy and a hopeful crew. It ended with two satellites retrieved and a jubilant crew.

Contributions

The two communications satellites *Discovery* carried into space attained their intended orbital po-

sitions. At the orbital altitude of both satellites (35,800 kilometers), it takes twenty-four hours to circle the globe. The orbital speed then matches the rotational speed of Earth, and each satellite appears stationary in the sky. Many communications satellites are placed in such geosynchronous orbits, where they can act as relay antennae in continuous view of widely separated ground stations.

The Anik D-2 was stored in orbit for almost two years before being used. This allowed Telesat to launch it while launch costs were still relatively low, and it was available in space should an unexpectedly early need to use it have arisen. It was brought into service on November 1, 1986, and formed part of a network of communications satellites serving Canada. It is located above the equator at longitude 110.5° west. The Leasat assumed duties immediately after it was checked out in orbit and continued to function as expected, aiding in communications for the Navy. It is stationed in orbit over the Atlantic Ocean at longitude 15° west.

Shortly after *Discovery* reached orbit, the crew activated the diffusive mixing of organic solutions experiment. It ran throughout the flight with little attention required from the astronauts. The experiment was designed to investigate the growth of crystals from organic compounds in the near weightlessness of orbit. The failure of some valves prevented some of the chemicals from crystallizing, but other samples produced hundreds of crystals. As scientists had hoped, the crystals grown in space, without the distorting effects of gravity, were larger and purer than those produced in Earth-based laboratories. The detailed analysis of their properties is expected to shed light on the complex process of crystal formation and aid in producing better crystals both on Earth and in space (when space manufacturing becomes a reality).

The knowledge that astronauts could manipulate extremely massive objects in space was an unexpected benefit of the mission. If everything had worked according to plan, this need would never have arisen. After Allen held the satellites and was able to position them wherever Gardner asked, it was realized that this ability could be applied in many other situations.

Context

That ability to handle large, massive objects in orbit did indeed prove very useful for NASA. Relying on the results of STS 51-A, the crew of STS 51-I successfully captured the malfunctioning Leasat 3 deployed on STS 51-D. The confidence and experience gained from STS 51-A made it possible to develop a plan for having the astronauts hold and move this 6,890-kilogram satellite. Again devising procedures during the mission, three spacewalking astronauts on STS-49 manually cap-

tured and maneuvered the 4,064-kilogram Intelsat VI F-3. Astronauts on STS-61 were able to accurately position nearly 300-kilogram instruments in their EVA work to repair and upgrade the Hubble Space Telescope. The direct handling of massive objects, as demonstrated first on STS 51-A, is expected to permit great flexibility in future space retrieval and construction tasks.

The failure of Palapa and Westar to reach their targeted orbits after their deployment on STS 41-B, although not NASA's fault, did tarnish the agency's image. The recovery of the satellites restored the "can-do" reputation NASA valued. It also was another example of the value of having people in space, as the recovery was clearly beyond the realm of remotely controlled machines.

The Indonesian government bought Palapa to help tie together its many isolated regions. Westar

After the satellites' retrieval, astronaut Dale Gardner suggests an alternative approach to dealing with the recovered spacecraft. (NASA)

was built for Western Union's use in its commercial work. Both satellites belonged to insurance underwriters after their rocket failures. Insurers had paid $180 million to the original owners, and, by paying a total of $10.5 million for the recovery operation, they hoped to resell the satellites and recoup much of their losses. The failures of these satellites occurred in a year in which spacecraft insurers paid out almost $300 million in claims, and there was great concern over the enormity of these payments. The retrieval helped alleviate that problem.

The recovery agreements had specified that the two salvaged satellites would be relaunched with the Space Transportation System. After the reevaluation of the United States' space policy in the wake of the *Challenger* (STS 51-L) accident, it was decided that commercial communications satellites would no longer be taken into space on these human flights. Yet there were not enough expendable launch vehicles in the nation's inventory to loft all of the satellites that were ready, and the unexpected hiatus in the capability of the United States to launch satellites caused long delays in the relaunching of Palapa and Westar. Westar 6 was eventually purchased by the Asia Satellite Telecommunications Co., Ltd., and was refurbished and launched as AsiaSat 1 on April 7, 1990. This launch,

using the People's Republic of China's Long March 3 launch vehicle, marked the first time that a spacecraft made in the United States was launched on a Chinese launch vehicle. The refurbished Palapa B2, renamed Palapa B2R, was launched on a Delta 6925 six days later and joined two other Palapa communications satellites that were already in orbit.

The commander of this challenging mission, Frederick H. "Rick" Hauck, gained a great deal of experience from the complexity of the flight. His next trip into space was as commander of the very successful STS-26, the first flight after the failed STS 51-L. After the *Challenger* was destroyed and the crew died on STS 51-L, the STS underwent many changes, and greater emphasis was placed on missions requiring people in space rather than routine cargo-hauling trips. STS 51-A continues to shine as a stellar example both of the kind of mission that requires people in space and of the ability of NASA to accomplish impressive and valuable tasks in space.

See also: Manned Maneuvering Unit; Solar Maximum Mission; Space Shuttle Flights, 1984; Space Shuttle Mission STS 41-C; Space Shuttle Flights, January-June, 1985; Space Shuttle Flights, July-December, 1985.

Further Reading

Allen, Joseph P., and Russell Martin. *Entering Space: An Astronaut's Odyssey.* Rev. ed. New York: Stewart, Tabori and Chang, 1984. This exquisite book was written by one of the STS 51-A mission specialists and includes his firsthand account of the flight. Suitable for all audiences, it describes the experiences of spaceflight, both the routine chores of living and working and the excitement and drama of being in space. It presents more than two hundred color photographs displaying the beautiful views available in space as well as the activities performed by astronauts.

Cooper, Henry S. F., Jr. *Before Liftoff: The Making of a Space Shuttle Crew.* Baltimore: Johns Hopkins University Press, 1987. The long process of training astronauts for a flight on the space shuttle is described by following the crew of STS 41-G, the flight immediately before STS 51-A. Although astronauts train for a specific mission, the book contains a great deal of interesting insight into the difficulty and importance of the preflight training for any mission. All readers will gain an appreciation of the challenges that precede a space shuttle flight and how hard the astronauts have to work in order to make their mission look so easy and smooth.

Furniss, Tim. *Manned Spaceflight Log.* Rev. ed. London: Jane's Publishing Company, 1986. With a description of every human mission into space through Soyuz T-15 in March, 1986, this book is entertainingly written and should be enjoyed by general audiences. It provides the essential facts from each flight and allows the reader to understand any spaceflight in the context of humankind's efforts to explore and work in space.

Harland, David M. *The Space Shuttle: Roles, Missions, and Accomplishments.* Hoboken, N.J.: John Wiley, 1998. *The Space Shuttle* is written thematically, rather than purely chronologically. Topics include shuttle operations and payloads, weightlessness, materials processing, exploration, Spacelabs and free-flyers, and the shuttle's role in the International Space Station.

Harrington, Philip S. *The Space Shuttle: A Photographic History.* San Francisco, Calif.: Brown Trout, 2003. With one hundred full-color photographs by Roger Ressmeyer and others and with text by popular astronomy writer Phil Harrington, this book tells the story of the space shuttle program from 1972 to 2003. Its beautiful photographs allow the general reader to survey the history of the space shuttle program and be uplifted by the pioneering spirit of one of humanity's grandest enterprises.

Jenkins, Dennis R. *Space Shuttle: The History of the National Space Transportation System: The First 100 Missions.* Stillwater, Minn.: Voyageur Press, 2001. This is a concisely written technical reference account of the space shuttle and its ancestors, the aerodynamic lifting bodies. It details some of the advantages and inherent disadvantages of using a reusable space vehicle. Each of the vehicles is illustrated by line drawings. The book follows the space shuttle from its original concepts and briefly chronicles its first one hundred flights.

Joëls, Kerry Mark, and Gregory P. Kennedy. *The Space Shuttle Operator's Manual.* Designed by David Larkin. Rev. ed. New York: Ballantine Books, 1988. This book contains a wealth of information on space shuttle systems and flight procedures. It is written as a manual for imaginary crew members on a generic mission and will be appreciated by anyone interested in how the astronauts fly the orbiter, deploy satellites, conduct spacewalks, and live in space. Contains many drawings and some photographs of equipment.

Powers, Robert M. *Shuttle: The World's First Spaceship.* Harrisburg, Pa.: Stackpole Books, 1979. Despite having been written before the first flight of the Space Transportation System, this book contains excellent descriptions of the space shuttle systems and the types of missions conducted. There are very good explanations of why and how the environment of space is used for a variety of scientific and technological applications. The book includes many paintings and drawings, a glossary, and an index.

Slayton, Donald K., with Michael Cassutt. *Deke! U.S. Manned Space: From Mercury to the Shuttle.* New York: Forge, 1995. This is the autobiography of the last of the Mercury astronauts to fly in space. After being grounded from flying in Project Mercury for what turned out to be a minor heart murmur, Slayton was appointed head of the Astronaut Office. Later, he commanded the Apollo-Soyuz flight in 1975. He stayed with NASA through the STS-2 flight. Although he was no longer an active participant in NASA's activities, he gives some insight into the behind-the-scenes adventures of the shuttle program.

Marc D. Rayman

Space Shuttle Flights, January-June, 1985

Date: January 24 to June 24, 1985
Type of mission: Piloted spaceflight

STS 51-C was the first shuttle mission entirely dedicated to a classified U.S. Department of Defense payload. Mission 51-D was the first flight to have an unscheduled spacewalk and the first to carry a nonpilot, nonscientist astronaut, Senator Jake Garn of Utah. STS 51-B carried the Spacelab 3 science payload into orbit, including rats and monkeys as well as human beings. STS 51-G was one of the most successful of the series. Four satellites were released, one was retrieved, and many important technological and scientific investigations were completed.

Key Figures

Thomas K. "Ken" Mattingly II (b. 1936), STS 51-C commander
Loren J. Shriver (b. 1944), STS 51-C pilot
Ellison S. Onizuka (1946-1986), STS 51-C mission specialist
James F. Buchli (b. 1945), STS 51-C mission specialist
Gary E. Payton, STS 51-C payload specialist
Karol J. Bobko (b. 1937), STS 51-D commander
Donald E. Williams (b. 1942), STS 51-D pilot
Margaret Rhea Seddon (b. 1947), STS 51-D mission specialist
Jeffrey A. Hoffman (b. 1944), STS 51-D mission specialist
S. David Griggs (1939-1989), STS 51-D mission specialist
Charles D. Walker (b. 1948), STS 51-D payload specialist
Edwin Jacob "Jake" Garn (b. 1932), U.S. senator and STS 51-D payload specialist
Robert F. Overmyer (1936-1996), STS-51B commander
Frederick D. Gregory (b. 1941), STS-51B pilot
Don L. Lind (b. 1930), STS-51B mission specialist
Norman E. Thagard (b. 1943), STS-51B mission specialist
William E. Thornton (b. 1929), STS-51B mission specialist
Lodewijk van den Berg, STS-51B payload specialist
Taylor G. Wang, STS-51B payload specialist
Daniel C. Brandenstein (b. 1943), STS 51-G commander
John O. Creighton (b. 1943), STS 51-G pilot
John M. Fabian (b. 1939), STS 51-G mission specialist
Steven R. Nagel (b. 1946), STS 51-G mission specialist
Shannon W. Lucid (b. 1943), STS 51-G mission specialist
Patrick Baudry, STS 51-G payload specialist
Sultan Salman A. al-Saud, STS 51-G payload specialist

Summary of the Missions

According to the National Aeronautics and Space Act of 1958, "The aeronautical and space activities of the United States shall be conducted so as to contribute . . . to the expansion of human knowledge of phenomena in the atmosphere and space. The [National Aeronautics and Space] Administration shall provide for the widest practicable and appropriate dissemination of information concerning its activities and the results thereof."

Unlike its Soviet counterpart, NASA opted to make public all of its activities relating to piloted spaceflight. This would allow the budding agency a chance to showcase its greatest accomplishments, as well as to lay open to criticism its shortcomings.

The chief personnel within NASA understood politics and budgets. They knew that the amount of money they would be allocated by the U.S. Congress would, for the most part, be dictated by the will of the American public. If the citizens believed that a project was worthwhile (for example, landing an American on the Moon before the Russians), NASA would be given support to accomplish the goal. From Mercury-Redstone 3, the first piloted spaceflight, in May, 1961, through the spectacular retrieval of two stranded satellites by the STS 51-A crew in November, 1984, every piloted journey into space was duly publicized. Information about the crew and mission was presented well in advance of the launch, and every aspect of flight, from liftoff to landing, was broadcast live around the world.

This open policy came to an end, at least temporarily, for the fifteenth flight of the space shuttle, STS 51-C, the first flight of a dedicated Department of Defense (DoD) payload. Because the DoD was in the habit of deploying secret satellites, the flight of 51-C would be shrouded in mystery. At least, that was the plan.

The flight that became STS 51-C was originally manifested as STS-10, to be flown in December, 1983. Its major payload would be a classified satellite, launched into geosynchronous orbit (that is, an orbit in which the satellite's velocity and altitude make it appear to hover over one spot on Earth's surface) by the Inertial Upper Stage (IUS). The IUS is a two-stage solid-propellant payload booster that can be taken into low-Earth orbit by either an expendable launch vehicle, such as the Titan, or the space shuttle.

The military crew for STS-10, announced in October, 1982, included veteran Thomas K. "Ken" Mattingly II, who had flown previously as Command Module pilot of Apollo 16 and commander of STS-4. He would be the commander of this flight. His pilot would be Loren J. Shriver, and the mission specialists would be Ellison S. Onizuka and James F. Buchli. The payload specialist, a U.S. Air Force piloted spaceflight engineer, would be named at a later date.

Following the failure of the IUS on STS-6 to place the first Tracking and Data-Relay Satellite (TDRS-A) into geosynchronous orbit, the Air

This photograph taken during the 51-B space shuttle mission in April, 1985, shows the San Andreas Fault. Los Angeles is located near the bottom right-hand corner. (NASA CORE/Lorain County JVS)

Force asked NASA to delay the launch of STS-10 until the problem had been resolved.

At the beginning of 1984, the crew for STS-10 (which now included Payload Specialist Gary Payton) and its payload were designated STS 41-E under the new numbering system (where the "4" represented Fiscal Year 1984, the "1" Kennedy Space Center [KSC], and "E" the fifth mission of the fiscal year). It was scheduled for launch on July 7, 1984, aboard *Challenger*.

When *Challenger* returned from its STS 41-G mission, loose tiles were found all over its skin. Nearly four thousand of them had to be replaced before the spacecraft could be flown again. NASA decided to use *Discovery*, just back from STS 51-A, for the now-designated STS 51-C. This would push the launch back to January, 1985.

NASA announced that the launch of STS 51-C would take place between 18:15 and 21:15 Coordinated Universal Time (UTC, or 1:15 and 4:15 P.M., eastern standard time), on January 23, 1985; the administration would not reveal the exact launch time until the countdown resumed after its last scheduled hold at nine minutes before launch. No information about the military payload would be released. On December 18, *The Washington Post* revealed that the payload for STS 51-C was an electronic monitoring satellite, the advanced signal intelligence satellite (SigInt). Although many in the DoD came close to calling the announcement an act of treason, *The Washington Post* had not actually given any details about the satellite.

On January 23, the mission was delayed for twenty-four hours because of what NASA called "extreme weather conditions." This was the first time a piloted spaceflight had been postponed because of cold weather. At 16:40 UTC (11:40 A.M., eastern standard time), on January 24, the crew transfer van carrying the astronauts arrived at Launch Complex 39A. The crew usually arrived two and one-half hours prior to launch, but that mark passed and the public countdown clocks remained blank. Suddenly, at 19:41 UTC, a NASA launch commentator announced, "T minus nine minutes and counting."

Discovery was launched at 19:50:00 UTC (2:50 P.M.), 1 hour and 35 minutes into the 3-hour launch window. Two problems involving orbiter systems caused the delays, but each was corrected on its own. All systems performed normally during launch, and the orbiter was placed into an orbit of 332 by 341 kilometers. NASA announced only that the shuttle was in orbit and that everything was going well. Status reports were given every eight hours thereafter.

Although not confirmed officially, at approximately 12:00 UTC on January 25, the SigInt satellite was deployed along with its IUS. The IUS fired about forty-five minutes later and placed the satellite into its proper orbit. Later, the DoD announced that the IUS had been deployed and had successfully completed its mission.

At 17:23 UTC on January 27, NASA announced that *Discovery* would be landing at 21:23 UTC the next day. Prior to the flight, NASA had indicated that it would announce the landing time sixteen hours before it was to take place. The deorbit burn took place at 20:18 UTC over the Indian Ocean, and *Discovery* began its long glide back to the KSC. At 21:23:23 UTC, its main landing gear touched down 792.48 meters beyond the threshold on runway 15, ending the mission after three days, 1 hour, 33 minutes, and 23 seconds. There was no postflight interview of the astronauts, who were whisked away for debriefing.

The mission of STS 51-D, *Discovery*'s fourth flight, got off to a shaky start but clearly demonstrated the overall flexibility of the space shuttle program. Originally, the crew of STS 51-D was supposed to fly STS 51-E, on *Challenger*, in early March, 1985. Only six days before launch, severe problems were discovered with the primary payload: the second Tracking and Data-Relay Satellite (TDRS-B). The flight, already delayed five times previously, was finally canceled outright. In order to keep to schedule as much as possible, a revised STS 51-D mission was quickly assembled, to be launched a mere six weeks later. The new payload would be the Telesat 1 (Anik C) satellite from STS 51-E, and the Hughes Communications Syncom IV-3 Navy communications satellite originally scheduled for STS 51-D.

All but one of the original crew members would fly: French Payload Specialist Patrick Baudry was replaced by Charles D. Walker from McDonnell Douglas as a result of payload requirements. This was the second flight for Walker, his first having been mission 41-D, about ten months earlier. The commander, Karol "Bo" Bobko, would be making his second flight as well. He first flew as the pilot on STS-6, the maiden voyage of *Challenger.* Navy Commander Donald E. Williams, the pilot for STS 51-D, was making his first trip into space. The mission specialists chosen for the flight were also space rookies: Margaret Rhea Seddon, M.D.; Navy Captain S. David Griggs; and Jeffrey A. Hoffman, Ph.D.

One of the other notable elements of this flight was that it would be the first mission to have a nonscientist, nonpilot crew member, Senator Edwin Jacob "Jake" Garn of Utah. As the chairperson of the Housing and Urban Development and Independent Agencies Subcommittee, Senator Garn had control over funding for the space program. Much was made of the senator's participation, considered by some to be a public relations stunt, but he participated in valuable medical experiments, particularly in the area of motion sickness. Motion sickness plagues most astronauts, causing many to have violent fits of nausea lasting up to two days. Because of the costs and short duration of the missions, the last thing needed is a sick crew, so extreme care was taken to avoid the malady. Because Garn had no active role in the payload deployment or piloting of the spacecraft, he was the ideal candidate to get sick deliberately, offering the crew physicians important data without jeopardizing the mission.

Discovery got off the ground on April 12, 1985, at 13:59:05 UTC, four years to the day after STS-1. The crew successfully launched the Telesat com-

This view of southern Florida was taken during the 51-C space shuttle mission in January, 1985, showing smoke plumes from grass fires in the Everglades. (NASA CORE/Lorain County JVS)

munications satellite at about nine and a half hours into the mission.

Early on the second day, the Syncom IV-3 satellite was deployed successfully, using the Frisbee technique. Normally the satellites were seated vertically in the shuttle's payload bay, spun up like tops (the technique used for the Telesat), and shot straight out. The Syncom was much larger, however, being a squat cylinder 5 meters in diameter, 4 meters high, and weighing about 6,800 kilograms. It was therefore stowed on its side and practically "rolled" out of the spacecraft such that it resembled a giant Frisbee being thrown. While the physical deployment was on time and apparently correct, the "omni" antenna on top of the satellite did not erect itself, as it was supposed to do about a minute after release. Furthermore, the kick motor used to boost the spacecraft into its final orbit did not fire after deployment.

Immediately a special analysis team was formed; it decided that the on/off lever on the side of the satellite had not been switched to "on" at the time of deployment. This lever was supposed to activate an internal sequencing timer that in turn would deploy the antenna and fire the engine. Several options were discussed, the main ones calling for an unplanned extravehicular activity (EVA) to throw the switch manually. Two EVA crewmen would attach snaring devices on the end of the Remote Manipulator System (RMS) arm. Afterward, the snares would be used in an attempt to trip the lever once the astronauts were safely inside.

Because of the change in plans, the mission was extended by two days, while ground crews put the finishing touches on the EVA schedule and designed the snaring devices. The capsule communicator (CapCom) radioed up descriptions of the snares; one was termed a "flyswatter" and the other a "lacrosse stick," for the objects they resembled. The crew constructed them out of tape, plastic pages from their onboard documents, and "swizzle sticks" used by the crew to reach distant switches while strapped into their seats. Dr. Seddon used her onboard surgical equipment to construct the devices, assisted by Senator Garn.

The flyswatter consisted of a long plastic sheet with three large square holes in a vertical row, looking much like a small ladder with four rungs a few centimeters wide. The objective would be to snag the lever with one of the rungs; if the flyswatter missed or broke a rung, the lever would likely hit a second. A single stick would attach it to the RMS. The lacrosse stick looked much like the swatter, except that it had a loop of wire across the top and was supported by two sticks on either side so as to give it slightly more strength.

Although most missions do not include spacewalks, flight rules dictate that there always be two crew members trained for emergency operations. Such emergencies might include repair of the delicate tiles, releasing a satellite stuck in the payload bay, or manually cranking the payload-bay doors closed. On this mission, astronauts Hoffman and Griggs were the two trained in EVA operations. Seddon would assist them in manipulating the RMS as required. The EVA took place on the fifth day of the flight and lasted slightly more than three

hours. Once the EVA was completed, preparations were begun for the following day's rendezvous with Syncom and snaring of the switch.

Visual sighting first took place while the Syncom was about 70 kilometers away in the early morning hours of April 17. At a distance of 15 meters, the crew could clearly see the separation lever and that it was apparently thrown as it should have been originally. This meant either that there was an internal failure of the timer or that the switch needed only a tiny bit of motion to trip it, because the internal microswitches were activated in only the last few degrees of movement.

At five days, 13 minutes into the flight, Seddon was given the go-ahead to snare the lever. She slowly moved the arm toward the center of the satellite, exactly where the lever would be when it passed by the arm. At one time the lever could be seen slicing one of the rungs. Another time, the arm appeared to bump up against the Syncom. Six minutes later the attempts were halted and the crew reported that they got "a hard physical contact on at least two occasions." With that, *Discovery* separated from the satellite for the last time and maneuvered to an attitude from which the crew could observe the firing of the kick motor. No such firing was seen, and the satellite was left in a 300-by-416-kilometer orbit for the possibility of a rescue mission on another flight.

Landing took place on Friday morning, April 19, 1985, at the KSC. As if enough had not happened during the flight, the landing was plagued by more troubles. First, the weather delayed reentry by one revolution. Next, during landing, crosswinds gusting to 15 knots blew the vehicle off the centerline of the runway. Bobko tried to correct this with "moderate" braking. As a result, both right brakes locked and the inboard tire blew out, rendering all tires unusable. Later investigation revealed that a tile had fallen off during ascent, exposing bare aluminum to the heat of reentry. More than 120 other tiles were damaged and required replacement.

The STS 51-B mission, carrying the Spacelab 3 scientific payload, was the seventeenth flight in the shuttle series. The primary areas of investigation for this mission were the life sciences and microgravity (the weightless environment and its effects on humans, animals, and properties of matter).

The STS 51-B commander was Robert F. Overmyer, who had previously flown on the fifth shuttle mission as pilot in 1982. His pilot on STS 51-B was to be Frederick D. Gregory, who was on his first shuttle mission. Three NASA mission specialists were named: Don L. Lind, a physicist, who was making his first flight; Norman E. Thagard, M.D., who had flown previously on STS-7; and William E. Thornton, M.D., who was on his second flight, having flown on STS-8, and who, at age fifty-six, was the oldest person to date to have flown in space. The two payload specialists for the mission were selected from a group representing the scientific interests of the mission: Taylor G. Wang, a fluid mechanics expert and a principal investigator in one of the Spacelab 3 investigations, and Lodewijk van den Berg, a materials science expert. During the mission, these crew members were to work in twelve-hour shifts: The "gold" (day) shift would consist of Overmyer, Lind, Thornton, and Wang. The "silver" (night) shift would be Gregory, Thagard, and van den Berg. Though no spacewalk (EVA) was planned, Gregory and Thagard were trained as contingency EVA crew members.

The payload and area of investigation for Spacelab 3 had been finalized several years before launch, and as a result of development difficulties with Spacelab 2, as well as several launch delays within the program, the manifest placed the launch of Spacelab 3 (which had been designated the first "operational" flight of the Spacelab series) before the Spacelab 2 flight. The countdown for 51-B proceeded without major incident, and its life science payload—two monkeys and twenty-four rats—was installed in the holding facility twenty-four hours before launch, allowing the animals to become accustomed to their new home. At 16:02:18 UTC (12:02 P.M. eastern standard time), *Challenger* lifted off and followed a nominal ascent profile to orbit.

The crew's first major activity was to deploy two small satellites from Get-Away Special (GAS) canisters in the payload bay. In order to ensure successful deployment and efficient battery power, it had been decided to deploy these satellites on flight day 1. The Northern Utah State Satellite (Nusat) was deployed by spring ejection 4 hours and 14 minutes into the mission; the Global Low-Orbiting Message Relay (GLOMR) satellite, which was to have been deployed some 14 minutes later, did not eject from its canister, though the GAS door did open. It was returned to Earth for reassignment to a later flight.

Challenger was then maneuvered into a gravity-gradient stabilization attitude with the tail toward Earth and *Challenger*'s port wing pointing in the direction of flight. The rest of the first day of the flight was spent in activating Spacelab and its experiments. For the next few days, the crew of STS 51-B worked hard to gather as much information as possible from their experiments.

Difficulties with the French Very Wide Field Camera (VWFC) were encountered some eight hours into the flight, when Lind had trouble using Spacelab's air lock in connection with the camera. A bent handle in the air lock forced ground controllers to decide against using the system for the rest of the flight, and the VWFC had to be abandoned with almost no data obtained.

On the whole, the crew members kept to their flight plan well, but only with much effort. Problems with the toilet system plagued them, and van den Berg experienced difficulties with his crystal-growing experiments during the second day of the flight. The animals, however, adapted well to their strange new environment. As time in orbit increased, thoughts of extending the mission were aired, a move that would enable the crew to complete research that had been slowed by the need to repair faulty equipment. Van den Berg had to shut down his experiment while he tried to repair its equipment, which he succeeded in doing during flight day 3, and Thagard had to repair the urine-monitoring system experiment and the Atmospheric Trace Molecules Spectroscopy experiment (ATMOS). These efforts were worth the trouble. The ATMOS Experiment yielded some spectacular results despite operating for only three days and obtaining only 19 of 60 planned data takes. Throughout his twelve-hour shifts in the Spacelab, Lind was able to obtain excellent photographs and record precise visual descriptions of auroral displays, which had been of scientific interest to him for years, especially after a period of intense solar activity on April 30.

A fine demonstration of the value of the human presence in space appeared in the untiring efforts of Wang, whose drop dynamics fluid mechanics experiment failed to work early in the flight because of an electrical malfunction. Determined not to return home unsuccessful, Wang almost completely rewired the machine himself, finally getting the experiment to function after several days' work. Another example of the importance of the "human factor" occurred when one of the two monkeys refused to eat for several days, apparently suffering from a bout of space sickness; Thornton was able to encourage the animal to eat by hand-feeding him, probably saving his life.

The last full day in space was spent in stowing experiments and equipment in preparation for reentry into Earth's atmosphere. *Challenger* touched down on runway 17 at Edwards Air Force Base after an uneventful descent at 16:12:04 UTC (9:12:04 A.M., Pacific daylight time), May 6, 1985, and rolled to a wheel stop 47 seconds later. The shuttle had traveled more than 4 million kilometers in seven days, 9 minutes, and 46 seconds, and had completed 110 orbits of Earth.

Despite a lightning strike on the 24-meter-tall lightning mast the night before, the fifth flight of *Discovery* on mission STS 51-G began with a perfect liftoff at 11:33:00 UTC (7:33 A.M., eastern daylight time), on June 17, 1985. Some 8 minutes and 46 seconds after leaving KSC's Launch Complex 39A, *Discovery* was in orbit.

Upon reaching the desired orbit, the crew began preparing for a seven-day mission that would see the release of a record four satellites, the retrieval of one of them, and a variety of experiments ranging from studies of human biology to measurements of the emissions of distant galaxies. The crew

of NASA astronauts included the commander, Daniel C. Brandenstein; the pilot, John O. Creighton; and Mission Specialists John M. Fabian, Shannon W. Lucid, and Steven R. Nagel. While the crew was preparing the orbiter for its stay in space, Patrick Baudry began a series of measurements to study the flow of blood in his body. A payload specialist, Baudry was a French space traveler who had served as backup for his countryman Jean-Loup Chrétien on Soyuz T-6. The French Echocardiograph Experiment (FEE) that Baudry was using was very similar to the equipment Chrétien had used on Salyut 7.

Both Patrick Baudry and the other payload specialist, Sultan Salman Abdulaziz al-Saud, were on board to conduct specific experiments from scientists in their countries. Al-Saud's was sponsored by Saudi Arabia. The twenty-two-member Arab Satellite Communications Organization was hiring NASA to deploy a satellite from the space shuttle. NASA policy allowed customers to include a passenger in such cases; neither Baudry nor al-Saud had any responsibilities directly related to the principal goals of the flight.

The first important goal in orbit was to release the Morelos-A communications satellite, which belonged to Mexico. Morelos was ejected into space only 8 hours and 5 minutes after *Discovery* itself was launched. *Discovery* fired its maneuvering rockets to move away from the free-floating satellite, and after 45 minutes a timer on Morelos activated its rocket engine to propel it to the desired orbit.

Although the Arabsat 1B release was scheduled for the next day, that satellite demanded attention when ground controllers received a signal indicating that one of its solar panels (used to convert sunlight into electricity for the satellite) had partially opened under the sunshield. The crew could see no evidence of this defect, so the next day the Remote Manipulator System (RMS) arm was used to get a closer view of the satellite. A camera at the "wrist" of the RMS allowed a detailed inspection of the solar array, and it was determined that the indicator was faulty. The array was correctly tucked against the side of the satellite. The release of the Arabsat took place on schedule and without difficulty on the second day of the mission, at 13:36 UTC. The successful deployment of AT&T's communications satellite Telstar 3D occurred at 11:20 UTC on June 19 and completed the delivery of communications satellites to orbit.

With the three satellites on their way, the crew had time to turn to other activities. One of these was the High-Precision Tracking Experiment (HPTE), an unclassified test in collaboration with the Strategic Defense Initiative Organization (SDIO). The purpose was to evaluate various schemes for training a laser on an Earth target over a period of time. Earth's turbulent atmosphere makes this very difficult. The plan called for the astronauts to install a special 22-centimeter mirror in one of the windows. When a low-power laser at Mount Haleakala in Maui, Hawaii, was directed toward *Discovery*, the mirror would send some of the light back to the ground station, which could then measure how accurately the laser had "hit" the orbiter.

The astronauts entered the location of the ground-based laser into the computers so that the orbiter could orient itself correctly to ensure that the mirror would reflect the laser back to the ground test facility. Shortly before the test, the astronauts reported that the window was not pointing toward Hawaii, but was pointing into space instead. The astronauts were able to see the blue-green light from the laser (it covered an area on the spacecraft about 9 meters in diameter), but the mirror was not in a position to return any of the light. It was soon determined that the location of the laser transmitter had been entered into the computer incorrectly. Thus the computer miscalculated the laser's location. By the time this error was discovered, it was too late to correct it.

Before another attempt to perform the tracking experiment was made, the last satellite release had to be undertaken. The Shuttle-Pointed Autonomous Research Tool for Astronomy (SPARTAN) was designed to conduct observations independently of the space shuttle for as long as two days. This first test flight had instruments for measuring the x rays emanating from the center of the Milky Way galaxy and from the galaxies clustered in a

group known as Perseus. Shannon W. Lucid used the RMS to remove SPARTAN 1 from its cradle in the payload bay and point it so that its own systems would be directed toward the Sun and the star Vega. This orientation would allow the satellite to establish and maintain its own direction in space so that it could aim its instruments at the preprogrammed targets. The satellite, about the size of a telephone booth, was gently released at 16:03 UTC on June 20. In order to minimize the cost and complexity of the SPARTAN, no communications equipment was built into it, so, to inform the crew that all of its systems had passed a self-test; the free-flying spacecraft performed a "pirouette." Satisfied that everything was working as planned, the astronauts directed *Discovery* away from the satellite to allow SPARTAN's observations to proceed without interference from the orbiter. The distance between the two craft would slowly grow to more than 190 kilometers.

On June 21, the laser tracking experiment was performed successfully. The next day, the crew members began a twenty-hour, thirteen-orbit chase to recover the SPARTAN. When they were fewer than 50 meters away from the satellite, they could see that it was not in the correct orientation. It was stable, but the fixture that the RMS needed to attach itself to was not on the side facing *Discovery*. The orbiter closed to within 10 meters of the craft, and John M. Fabian manipulated the RMS to reach behind the SPARTAN and pluck it from orbit. He then returned the satellite to its resting place in the payload bay so that scientists and engineers on Earth could evaluate both the data from its astronomical observations and the overall performance of the new, reusable satellite.

On June 24, Commander Brandenstein landed *Discovery* on the dry lake bed at Edwards Air Force Base. The spacecraft rolled for 40 seconds before stopping, the final part of its 4.7-million-kilometer journey marred only by the left main landing gear digging 15 centimeters into the ground as *Discovery* came to a halt. The orbiter incurred minimal damage, however, and STS 51-G concluded with the same level of success it had demonstrated since it began.

This view taken during the STS 51-B mission is of the Atlantic Ocean. The large quantities of ice seen here act like a cap on the ocean. It prevents the exchange of heat between the ocean and lower levels of the atmosphere, thus helping to cool Earth. (NASA CORE/Lorain County JVS)

Contributions

Although neither the U.S. Air Force nor NASA gave specific information about the military payloads flown aboard Department of Defense missions, it is generally believed that each satellite was placed into geosynchronous orbit and performed as designed. The IUS, which had proved troublesome on the STS-6 mission, worked as planned. Other experiments were flown aboard the flights, and some information about them has been released.

One of the experiments on mission STS 51-C, known as the aggregation of red blood cells experiment, was flown in two canisters in the middeck of the orbiter. A computer controlled the experiment, which involved the passing of eight donors' blood between two glass sheets so that cameras and a digital data system could assess how the blood aggregated in the near-weightless conditions of orbital spaceflight. Because blood aggregates differently in persons with disease from the way it aggregates in those who are free from disease, researchers can understand better how blood circulates by viewing its activities under controlled conditions.

Many of the shuttle missions are primarily targeted toward delivery of satellites; some have few if any major experiments. That was the case with STS 51-D. Vital experience was gained in on-the-spot planning for the unscheduled EVA, verifying that the current systems of training and mission planning did work.

On board the shuttle were two Get-Away Special canisters (GAS cans, used for low-cost experimental units requiring little crew intervention). G035 was an experiment to test the surface tension, viscosity of fluids, and solids and alloy furnaces. A malfunction prevented G471, the capillary pump loop priming experiment, from being conducted.

Payload Specialist Charles D. Walker operated the McDonnell Douglas Continuous Flow Electrophoresis System (CFES), which was used to develop techniques required for generating a pure pharmaceutical material. McDonnell Douglas reported that "all samples were processed and no contamination of the product was observed." What exactly the product was, however, was kept secret.

On the lighter side was the Toys in Space project. This served to demonstrate the behavior of more than thirty miniature mechanical systems in zero gravity. Films were taken of the crew that would later be used in classroom science discussions. Common toys, such as Slinkies, paper airplanes, jacks, yo-yos, and paddleballs were used.

The scientific objectives of the STS 51-B Spacelab 3 mission could be divided into four areas: materials processing, environmental observations, fluid mechanics, and life sciences. Out of fifteen experiments, the crew gained useful data from fourteen, resulting in 250 billion bits of scientific data. Mission Scientist George Fichiti described preliminary data from the Spacelab 3 experiments as excellent.

In the materials science experiments, a mercury iodide crystal about the size of a sugar cube was grown from a seed crystal in the vapor crystal growth system over a period of 104 hours, using a vapor transport technique. A fluid transport technique was used to grow two triglycerine sulfate crystals in the fluid experiment system. The vapor crystal growth and fluid experiment systems provided the first opportunity to observe in detail crystal growth in a microgravity environment and to determine the difference between growth on Earth and in orbit.

A total of 102 hours of geophysical fluid flow cell operations were completed on the flight to understand convection on the Sun, in planetary atmospheres, in Earth's oceans, and in basic fluid physics. The first experimental data on the behavior of a free-floating fluid in a microgravity environment were obtained in the drop dynamics module. For the first time, both solid and liquid samples were acoustically positioned and maneuvered in weightlessness; drops of varying sizes and viscosities were formed, rotated, and oscillated. In addition, studies were made on compound drops (drops formed within drops).

The ATMOS experiment obtained nineteen sequences of more than 150 independent atmospheric spectra, each containing more than 100,000 individual measurements used to analyze Earth's stratosphere and mesosphere at altitudes between 10 and 150 kilometers. In addition, high-resolution infrared spectra of the Sun were obtained during five calibrations and provided some surprising evidence about the molecular constituents of the Sun.

Mounted outside the pressurized module, the Ionization of Solar and Galactic Cosmic-Ray Heavy Nuclei experiment (IONS) recorded data on high-energy particles emitted from the Sun and other, more distant galactic sources. The life science experiments revealed that both the rats and the monkeys could successfully adjust to spaceflight conditions, although one of the monkeys did suffer from space adaptation syndrome, recovering in a manner similar to human recovery.

The flight of SPARTAN 1 during STS 51-G yielded important scientific and technical results. The primary objective of mapping the x-ray emissions from the central region of the Milky Way and the complete Perseus cluster of galaxies was achieved.

The French experiments on the adaptation to spaceflight and the readaptation to normal gravity completed all the planned tests. The results of Baudry's measurements of the cardiovascular system and other systems in the body added to a small but growing database on the effects of gravity (or its absence) on living organisms.

One of the other experiments conducted throughout the mission was designed to shed light on the process of convection in the melting of certain materials. Operated automatically while the astronauts slept (so that their movements about the orbiter would not cause disturbances in the sensitive experiment), it melted samples for later study to examine to what extent the absence of gravity reduced convection. The postflight analysis revealed the surprising fact that convection plays a very important role in solidification even when gravitational effects are greatly reduced.

Context

STS 51-C, which had been delayed in getting off the ground because of cold weather, landed a year and a day before the next flight to be seriously affected by the cold, STS 51-L. Ellison S. Onizuka would be on that tragic mission, too. After that cold morning in January, 1986, the space shuttle program would forever change. Never again would a flight be attempted in freezing weather.

The Department of Defense would fly another "secret" mission aboard the space shuttle—STS 51-J in October, 1985. After the *Challenger* accident in January, 1986, the military opted to puts its launch efforts back into expendable launch vehicles.

The flexibility of the space shuttle system was decisively demonstrated in the first unplanned EVA on mission 51-D. The ability of ground personnel, combined with the contingency training of the flight crew, made for a successful operation even though the satellite was not activated. The space program was seen to have reached a new level of maturity. Senator Garn's presence served to emphasize that space travel was becoming increasingly routine.

STS 51-I was launched on August 27, 1985, to "hot-wire" the Syncom deployed by the *Discovery* crew on STS 51-D. (The physical design of the Syncom precluded any return to Earth.) Four days later, astronauts James "Ox" Van Hoften and William F. Fisher left *Discovery* to snare the satellite manually and return it the payload bay. There Fisher installed the Hughes-designed Spun Bypass Unit to bypass the failed sequencer. Once the unit was installed, Fisher threw four switches, bringing the $90 million machine back to life. Using handles that he had installed in the side of Syncom, Van Hoften spun it up by hand to two revolutions per minute and "launched" it back into its own orbit.

The success of this flight, along with STS 51-D, deflated the arguments of those who claimed that piloted spaceflight was a mere luxury and that robots could do all that was necessary.

The flight of STS 51-B provided scientists with their first real opportunity to investigate the science of fluid mechanics on an American piloted spaceflight since the Skylab missions of 1973 and 1974. The growing of crystals in space has significant applications in the world of microelectronics, military systems, telescopes, cameras, and infrared monitors. The important link between the crew and ground teams was demonstrated in the work of Wang, who for the first time operated and repaired his own equipment in space, recovering the experimental facility and achieving almost all preflight goals, despite a late start in the mission.

Results from the round-the-clock operations on STS 51-B provided vast amounts of new data to complement the results obtained prior to the mission. For example, the ATMOS Experiment gathered more data on one flight than previously obtained in decades of similar research with balloon-borne high-altitude sorties. The flying of principal investigators allowed them to refine their own hardware and to experience at first hand the operation of their experiments. What they learned was of great importance in the development of follow-up experiments. An understanding of the effects of spaceflight on hardware and the difficulties and advantages of the microgravity environment allowed Wang and van den Berg to make plans for more efficient and productive experimental hardware for later flights and use by other crews.

The three communications satellites sent into orbit by *Discovery*'s crew on STS 51-G reached their appointed slots in geosynchronous orbit. The

Morelos satellite helps bring educational television, commercial programs, telephone and facsimile, and data transmissions to virtually every area in Mexico. SPARTAN's flight demonstrated the capabilities of a new family of free-flying spacecraft.

The high-precision tracking experiment provided experimenters with about 2.5 minutes (more than twice as long as needed) of data on the performance of different tracking techniques. It was the first Strategic Defense Initiative Organization experiment conducted with the space shuttle, and more were planned to follow.

See also: Extravehicular Activity; Get-Away Special Experiments; Launch Vehicles; Launch Vehicles: Reusable; Materials Processing in Space; Space Shuttle: Microgravity Laboratories and Payloads; Space Shuttle Mission STS 51-A; Space Shuttle Flights, July-December, 1985; Spacelab Program.

Further Reading

Furniss, Tim. *Manned Spaceflight Log.* Rev. ed. London: Jane's Publishing Company, 1986. This reference provides a nontechnical overview of piloted spaceflight since its beginning. Covers the flights of American astronauts, Soviet cosmonauts, and space travelers from other nations who have flown with them. Accounts of each flight are accompanied by black-and-white photographs.

_____. *Space Shuttle Log.* London: Jane's Publishing Company, 1986. From STS-1 through STS 61-A, the space shuttles are covered in depth. Furniss discusses the design concepts for the Space Transportation System and provides a concise account of each shuttle flight. Includes black-and-white photographs from the missions.

Gurney, Gene, and Jeff Forte. *Space Shuttle Log: The First Twenty-Five Flights.* Blue Ridge Summit, Pa.: TAB Books, 1988. A collection of chapters summarizing the first twenty-five missions of the Space Transportation System program, covering the operational flight time of April, 1981, to January, 1986, the first five years of the flight program. Each entry, from STS-1 through STS-25, lists information on crews, payloads, launch preparations, launches, orbital operations, and landing phases. A mission summary and text are accompanied by a selection of black-and-white photographs for each mission.

Harland, David M. *The Space Shuttle: Roles, Missions, and Accomplishments.* Hoboken, N.J.: John Wiley, 1998. *The Space Shuttle* is written thematically, rather than purely chronologically. Topics include shuttle operations and payloads, weightlessness, materials processing, exploration, Spacelabs and free-flyers, and the shuttle's role in the International Space Station.

Harrington, Philip S. *The Space Shuttle: A Photographic History.* San Francisco, Calif.: Brown Trout, 2003. With one hundred full-color photographs by Roger Ressmeyer and others and with text by popular astronomy writer Phil Harrington, this book tells the story of the space shuttle program from 1972 to 2003. Its beautiful photographs allow the general reader to survey the history of the space shuttle program and be uplifted by the pioneering spirit of one of humanity's grandest enterprises.

Jenkins, Dennis R. *Space Shuttle: The History of the National Space Transportation System: The First 100 Missions.* Stillwater, Minn.: Voyageur Press, 2001. This is a concisely written technical reference account of the space shuttle and its ancestors, the aerodynamic lifting bodies. It details some of the advantages and inherent disadvantages of using a reusable space vehicle. Each of the vehicles is illustrated by line drawings. The book follows the space shuttle from its original concepts and briefly chronicles its first one hundred flights.

Kerrod, Robin. *Space Shuttle*. New York: Gallery Books, 1984. Most valuable for its beautiful color photographs of the space shuttle, this volume conveys the essence of the Space Transportation System vehicles and the people who fly them. Highlights of the first dozen missions are presented, as well as a fanciful look at the future of space exploration.

National Aeronautics and Space Administration. *Space Shuttle Mission Press Kits*. http://www-pao.ksc.nasa.gov/kscpao/presskit/presskit.htm. Provides detailed preflight information about each of the space shuttle missions. Accessed March, 2005.

_____. *Spacelab 3*. NASA EP-203. Washington, D.C.: Government Printing Office, 1984. This NASA publication provides a preflight overview of the 51-B mission and the scientific investigations planned for Spacelab 3. Background chapters on the flight crew, mission planning, and each of the major areas of scientific research are included, along with diagrams and color photographs, providing a handy summary of preflight objectives.

Otto, Dixon P. *On Orbit: Bringing on the Space Shuttle*. Athens, Ohio: Main Stage Publications, 1986. Early designs of the space shuttle are discussed. Also presents an account of each of the first twenty-five flights, including the crew, the payloads, and the flight objectives. Illustrated in black and white.

Slayton, Donald K., with Michael Cassutt. *Deke! U.S. Manned Space: From Mercury to the Shuttle*. New York: Forge, 1995. This is the autobiography of the last of the Mercury astronauts to fly in space. After being grounded from flying in Project Mercury for what turned out to be a minor heart murmur, Slayton was appointed head of the Astronaut Office. Later, he commanded the Apollo-Soyuz flight in 1975. He stayed with NASA through the STS-2 flight. Although he was no longer an active participant in NASA's activities, he gives some insight into the behind-the-scenes adventures of the shuttle program.

Smith, Melvyn. *An Illustrated History of Space Shuttle: X-15 to Orbiter*. Newbury Park, Calif.: Haynes Publications, 1986. A concise overview of the space shuttle and the experimental aircraft that led to its design. Spanning the period from 1959 through 1985, the book was written for the general reader. Illustrated with many photographs of the early lifting bodies, which help to show how the shuttle orbiter came to look as it does today. Arranged chronologically.

Wilson, Andrew. *Space Shuttle Story*. New York: Crescent Books, 1986. Traces the history of the space shuttle from the early days of rocketry to the *Challenger* accident. Furnished with more than one hundred color photographs, this volume provides little detail but emphasizes the men and women who fly the space planes to and from orbit.

Yenne, Bill. *Space Shuttle*. New York: Gallery Books, 1986. A large-format picture book that covers the shuttle in the most general fashion, from construction to operations. Sections are devoted to history, manufacture, and mission profiles. A brief flight log is included in the back.

Russell R. Tobias (STS 51-C) Michael Smithwick (STS 51-D)
David J. Shayler (STS 51-B) Marc D. Rayman (STS 51-G)

Space Shuttle Flights, July-December, 1985

Date: July 29 to December 3, 1985
Type of mission: Piloted spaceflight

During STS 51-F, astronauts conducted experiments in solar and space plasma physics and astrophysics. STS 51-I deployed three communications satellites, and crew members performed inflight maintenance on a malfunctioning satellite. STS 51-J, the maiden flight of Atlantis, featured the deployment of two military communications satellites during the second classified American piloted spaceflight. STS 61-A was the first American spaceflight to have its control, once the shuttle was in orbit, centered outside the United States. Spacelab D-1 was an international mission dedicated to various scientific and technological investigations. During STS 61-B, astronauts launched three communications satellites and practiced assembling large structures in space, in preparation for the assembly of space stations.

Key Figures

Charles Gordon Fullerton (b. 1936), STS 51-F commander
Roy D. Bridges, Jr. (b. 1943), STS 51-F pilot
F. Story Musgrave (b. 1935), STS 51-F mission specialist
Anthony W. England (b. 1942), STS 51-F mission specialist
Karl G. Henize (1926-1993), STS 51-F mission specialist
Loren W. Acton, STS 51-F payload specialist
John-David F. Bartoe, STS 51-F payload specialist
Joseph H. Engle (b. 1932), STS 51-I commander
Richard O. Covey (b. 1946), STS 51-I pilot
James D. A. "Ox" Van Hoften (b. 1944), STS 51-I mission specialist
John M. "Mike" Lounge (b. 1946), STS 51-I mission specialist
William F. Fisher (b. 1946), STS 51-I mission specialist
Karol J. Bobko (b. 1937), STS 51-J commander
Ronald J. Grabe (b. 1945), STS 51-J pilot
Robert L. Stewart (b. 1942), STS 51-J mission specialist
David C. Hilmers (b. 1950), STS 51-J mission specialist
William A. Pales, STS 51-J mission specialist
Henry W. Hartsfield, Jr. (b. 1933), STS 61-A commander
Steven R. Nagel (b. 1946), USAF, STS 61-A pilot
James F. Buchli (b. 1945), USMC, STS 61-A mission specialist
Guion S. Bluford, Jr. (b. 1942), STS 61-A mission specialist
Bonnie J. Dunbar (b. 1949), STS 61-A mission specialist
Reinhard Furrer (1940-1995), STS 61-A payload specialist
Ernst W. Messerschmid (b. 1945), STS 61-A payload specialist

Wubbo J. Ockels (b. 1946), STS 61-A payload specialist
Brewster H. Shaw, Jr. (b. 1945), STS 61-B commander
Bryan D. O'Connor (b. 1946), STS 61-B pilot
Mary L. Cleave (b. 1947), STS 61-B mission specialist
Sherwood C. Spring (b. 1944), STS 61-B mission specialist
Jerry L. Ross (b. 1948), STS 61-B mission specialist
Rudolfo Neri Vela (b. 1952), STS 61-B payload specialist
Charles D. Walker (b. 1948), STS 61-B payload specialist

Summary of the Missions

The nineteenth space shuttle mission returned important data on the Sun, the stars, and the space environment with an advanced array of sophisticated instruments. The mission also demonstrated the Instrument Pointing System (IPS), designed as part of the Spacelab science system for the shuttle. Although the flight itself was designated STS 51-F, it is best known by its primary payload, Spacelab 2. Its scientific objectives were to scan the sky and to analyze the near-space environment around Earth.

The first attempt to launch *Challenger* on mission 51-F was aborted on the launch pad. Apparently, an engine had been slow in starting only three seconds before liftoff on July 12, 1985. The launch was reset for July 29.

After a ninety-minute delay caused by flight computer problems, *Challenger* lifted off at 21:00:00 Coordinated Universal Time (UTC, or 5:00 P.M., eastern daylight time). Ascent was normal until 5 minutes and 45 seconds after liftoff, when *Challenger*'s computer automatically shut down the center engine. That forced the crew to implement a procedure known as "abort to orbit." The remaining two engines would use the rest of the propellant to burn one minute longer in order to achieve orbit. Although called an "abort," this maneuver actually allows the mission to proceed by placing the spacecraft in orbit, albeit lower than the one that had been planned. *Challenger*'s engines were, in fact, performing as designed, but a pair of temperature sensors on a turbopump gave erroneous high readings and the computer deactivated the engine. A flight controller saw that all other engine readings were normal. When sensors on a second engine functioned the same way, the crew was in-

structed to override the computer's automatic command to deactivate this second engine.

Challenger was successfully inserted into an orbit 322 kilometers (instead of 385 kilometers) high. Because energy was conserved during the flight, Mission Control could extend the mission by one day to allow the Solar Science Team to recover part of the experiment time lost when the instrument pointing system was malfunctioning. The mission ended at 19:45:26 UTC (12:45 P.M. Pacific daylight time), on August 6, when *Challenger* landed at Edwards Air Force Base in California.

Instruments in the Spacelab 2 payload were designed to return data in the fields of astrophysics and solar and space plasma physics; also on board were experiments in the areas of atmospheric physics, technology development, and the life sciences. These were assembled on three U-shaped Spacelab pallets and on a special structure carried in *Challenger*'s payload bay.

The solar and atmospheric physics instruments were attached to an IPS designed to aim telescopes at the Sun, the stars, Earth, and other targets as the shuttle flew through space. Spacelab 2 had been delayed for several years because of problems in developing this highly sophisticated system.

Four solar instruments were mounted on the forward pallet of the instrument pointing system: the High-Resolution Telescope and Spectrograph (HRTS), the Solar Optical Universal Polarimeter (SOUP), the Coronal Helium Abundance Experiment (CHASE), and the Solar Ultraviolet Spectral Irradiance Monitor (SUSIM). Only the first three were considered to be true solar physics instruments, because the SUSIM was designed to support studies of Earth's atmosphere. The SUSIM had flown earlier, on the STS mission in 1982.

The HRTS and the SOUP were complementary instruments designed to return data on active regions of the visible surface of the Sun. It was hoped that the SOUP would help produce photographs of individual magnetic field activities within granules (convective eddies rising to the Sun's surface) during their 5- to 20-minute lives and within supergranules during their 20- to 40-hour lives. The HRTS would aid scientists in their study of the outer layers of the solar atmosphere, especially the transitional region between the chromosphere and the corona.

Astrophysics instruments were mounted on pallets and structures through the remainder of the payload bay. These were the X-Ray Telescope (XRT); the small, helium-cooled Infrared Telescope (IRT); and the elemental composition and energy spectra of the Cosmic-Ray Nuclei (CRN) in-

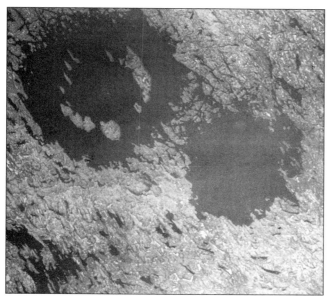

This photo of two impact craters located in northwestern Quebec was made during the 61-A space shuttle mission in early November, 1985. Both craters are now filled with water to make lakes. Most of the Earth's ancient craters have been deformed by volcanism, erosion, and mountain-building processes. (NASA CORE/Lorain County JVS)

strument. The XRT actually consisted of two telescopes of similar design; each used a pinhole mask to project images of the sky in high-energy x rays. The IRT used liquid helium to cool the detectors at the focal plane of a 15-centimeter telescope so that cold objects in the 1- to 120-micron wavelength range, such as stellar nurseries and nebulae, could be observed. The telescope itself scanned at right angles to the shuttle's line of flight to construct images line by line. The CRN, also called the "Chicago egg" because of its shape and its origin (the University of Chicago), was the largest cosmic-ray instrument placed in orbit. Its design allowed for detection of extremely heavy cosmic rays at energies between 400 and 4,000 gigaelectron volts. (One electron volt of energy is equivalent to 1.6×10^{-19} joules.) Cosmic rays are not actually rays but atomic and subatomic particles released after the explosion of stars and other such violent events.

Two plasma physics instruments shared pallet space with the IRT: the ejectable Plasma Diagnostics Package (PDP) and the Vehicle-Charging and Potential Experiment (VCAP). Both the PDP and the VCAP had been flown on STS in 1982. The PDP carried several instruments in an ejectable package that the shuttle was to circulate in order to help scientists determine the effects of large vehicles on the space environment. The VCAP, carrying complementary instruments, included an electron gun to probe the plasma environment, while the PDP measured responses. A third investigation, designed to return data on plasma depletion for ionospheric and radio astronomical studies, had no special equipment aboard *Challenger.* Instead, it used the shuttle's thrusters to burn "holes" in the ionosphere.

One technology experiment, whose subject was the properties of superfluid helium in zero gravity, was carried on the pallet. Superfluid helium (He II), in which electrical resistance disappears, behaves according to the laws of quantum mechanics. Two life science experiments were carried in the shuttle cabin to measure the effects of weightlessness on life; the first focused on the normal cycles in human bones, and the second focused on the production of lignin (a tough cellulose) in mung bean and pine seedlings.

The robot arm retrieved the PDP on the third mission day. It was then released so the shuttle could retreat to a point a few kilometers away and maneuver around it for six hours. Some of these activities required expert flying by the crew to place the shuttle and the PDP on the same magnetic field lines. Four of the eight in-orbit rocket firings planned for the plasma depletion experiments were canceled because of the fuel expenditure made during ascent.

STS 51-I was, perhaps, the most ambitious flight of the year. Plans for the mission included the deployment of three communications satellites. The first two were smaller HS76 satellites, while the third was the rather large Syncom IV-4 (Synchronous Communication Satellite IV-4) set to roll out of *Discovery*'s payload bay on the third day of the mission. However, the portion of the flight destined to get the most attention was the on-orbit repair of Leasat 3 (also known as Syncom IV). The satellite, a near duplicate of Syncom IV-4, had been deposited into orbit by the crew of STS 51-D four months earlier. The apparent failure of a triggering latch to activate resulted in the failure of the satellite's booster rocket to fire. Two astronauts on this flight would patch the electronics and send Syncom on its way.

On August 27, 1985, *Discovery* and its five-member crew were launched at 10:58:01 UTC (6:58:01 A.M., eastern daylight time) from Launch Complex 39A. The ascent was normal, and *Discovery* was placed in a nearly circular orbit 306 kilometers above Earth.

The commander of the flight was Air Force Colonel Joseph H. Engle. He made his first trip into space as commander of STS in 1981. Air Force Lieutenant Colonel and Pilot Richard O. Covey was making his first flight into space. The mission specialists on the flight were James D. A. Van

Hoften, Ph.D., John M. "Mike" Lounge, and William F. Fisher, M.D. The only space-experienced mission specialist was Van Hoften, who flew on STS 41-C and participated in the Solar Maximum Mission (SMM) satellite repairs.

Six and a half hours after liftoff, the Aussat (Australian Satellite) was deployed; it later achieved its proper geosynchronous orbit with the aid of the Payload Assist Module, Delta class (PAM-D). A second satellite, American Satellite Company 1 (ASC), was deployed five hours later. The third satellite, Syncom IV-4, which is leased to the Department of Defense by its builder, Hughes Aircraft, was deployed as scheduled on flight day three. All three achieved geosynchronous orbit and became operational.

Discovery rendezvoused with the ailing Syncom IV on day five of the mission, and the Remote Manipulator System (RMS) arm grappled it. It was then lowered into position in the payload bay for the repairs. Astronauts Van Hoften and Fisher performed two spacewalks (on days five and six) for a total of 11 hours and 27 minutes. During this time they replaced parts in the satellite needed to fire the satellite's booster rocket. After the activation lever was repaired, Van Hoften grabbed the satellite and gave it a gentle spin while releasing it. Eventually, commands were sent to Leasat, and it rode its booster rocket to the proper geosynchronous orbit.

The Physical Vapor Transport of Organic Solids (PVTOS) experiment operated during the sleep periods on the mission. *Discovery* landed at Edwards Air Force Base in California, at 13:15:43 UTC (6:16 A.M., Pacific daylight time), September 3, 1985. The mission lasted seven days, 2 hours, and 18 minutes.

Atlantis, the fourth of NASA's shuttle orbiters, was almost identical to its sister ship *Discovery.* It was named for the Woods Hole Oceanographic Institute research ship used from 1930 to 1966, which was the first American-operated vessel designed especially for oceanic research.

The commander of STS 51-J was Karol J. Bobko, a veteran of two previous shuttle flights. He was the pilot of *Challenger* for STS-6 and the commander of *Discovery* for STS 51-D. The remainder of the crew included Pilot Ronald J. Grabe, Mission Specialists David C. Hilmers and Robert L. Stewart, and Department of Defense (DoD) Payload Specialist William A. Pales. Stewart was the only other veteran, having test-flown the Manned Maneuvering Unit (MMU) on STS 41-B.

The security surrounding STS 51-J was tighter than that surrounding STS 51-C; the DoD was not going to tolerate the leaking of any information regarding the payload. NASA imposed the same restrictions about flight details that it had on 51-C. The cargo for the flight was reported to be two military communications satellites known as Defense Satellite Communications System (DSCS).

The twenty-first space shuttle mission was launched at 15:15:30 UTC (11:15:30 A.M., eastern daylight time), on October 3, 1985, with very little notice. All systems worked as planned during the launch phase, and *Atlantis* was placed in a shuttle record-high orbit of 469 by 476 kilometers. Almost immediately, its secret payload was deployed. NASA remained silent about the progress of the flight until twenty-four hours before the planned landing. STS 51-J ended successfully on October 7, 1985, as *Atlantis* touched down on runway 23 at Edwards Air Force Base. The mission had lasted four days, 1 hour, 44 minutes, and 38 seconds.

The year 1985 was busy for shuttle operations, with nine piloted missions flown. By the time STS 61-A took to the air, the media interest in reporting shuttle missions had dropped considerably; thus, this mission carrying the Spacelab D-1 payload was notable not only for the largest crew ever launched into space by one vehicle but also for the low-key coverage the general media devoted to the science mission.

Shuttle veteran Henry W. Hartsfield, Jr., who had flown on STS-4 in 1982 and STS 41-D in 1984, commanded the crew. His pilot was Steven R. Nagel, who had flown on STS 51-G in 1985. Mission specialists for STS 61-A included Bonnie J. Dunbar, who had degrees in ceramic and biomedical engineering and who was on her first spaceflight;

James F. Buchli, who had flown on STS 51-C earlier in 1985; and Guion S. Bluford, Jr., the first African American in space, who had flown on STS-8 in 1983. In addition, three rookie European payload specialists were members of the crew: Ernst W. Messerschmid and Reinhard Furrer, both from West Germany, and Wubbo J. Ockels, a Dutch national from the European Space Agency (ESA).

As with all Spacelab missions, the crew members were to alternate in twelve-hour shifts to operate the experiments in the Spacelab module. The blue shift was led by Nagel with Dunbar and Furrer; the red shift was led by Buchli with Bluford and Messerschmid. Hartsfield and Ockels were not assigned to a team and worked with either team as required. Buchli and Dunbar trained as contingency extravehicular activity (EVA) crew members.

Despite a few minor problems, the countdown for STS 61-A was one of the smoothest and most trouble-free of the program to date. With eight astronauts aboard, *Challenger* left the pad exactly on time, at 17:00:00 UTC (noon, eastern standard time), on October 30, 1985. All stages of the ascent were nominal. Once orbit had been achieved, the payload bay doors were opened and a week of orbital science experiments began for the crew.

Three hours after launch, the crew floated into Spacelab to activate the experiments and equipment. Two hours later, this task was completed, and the control of payload operations was transferred from NASA Mission Control in Johnson Space Center, Houston, Texas, to the West German Space Operations Center in Oberpfaffenhofen, near Munich, a facility operated by the Deutsche Forschungs-und Versuchsanstalt für Luft- und Raumfahrt (German Federal Ministry of Research and Technology).

As soon as Spacelab was activated, the crew split into its two shifts, one team settling down for a sleep period, the other beginning the round-the-clock work in the science module. Once orbital operations had begun, the coverage of the mission began to decrease, an indication of how routine shuttle flights had become. Meanwhile, the crew successfully deployed the Global Low-Orbiting Message Relay (GLOMR) satellite, which had failed to deploy on STS 51-B five months earlier. During the week in space, seventy-three experiments out of a projected seventy-six were successfully activated, and their data were recorded.

Several of the experiments were conducted as forerunners to a planned Spacelab D-1 and the U.S. International Space Station, then planned for the early 1990's. One experiment by Ockels investigated a new sleep restraint designed to alleviate the sensation of floating in space and therefore disturbing sleep cycles. Tubes in the restraint were inflated to apply pressure to the body. More important was a demonstration of the capability of a crew of eight to work together in the confined environment of the Spacelab module and shuttle flight and middeck areas.

After a normal descent, *Challenger* landed on Runway 17 at Edwards Air Force Base just before 17:44:53 UTC (9:45 A.M., Pacific standard time), then began its rollout down the runway to a wheel stop. During the 2,560-meter rollout, Hartsfield completed the last experiment of the flight with the nosewheel steering test. Crews of several previous missions had experienced difficulty in controlling the orbiter during the runway rollout, and instrumentation had been fitted to *Challenger* for this flight to investigate the nosewheel steering inputs the pilot conducts during the rollout. Hartsfield successfully moved *Challenger* first to the left of the central line, then to the right, before moving back to the central line for a wheel stop.

Challenger had logged seven days and 45 minutes in space and traveled more than 4 million kilometers in 111 orbits, landing during the 112th. Despite the signs of wear on its thermal protection system that were noticed after the landing of STS 61-A, mission planners were looking forward to seeing *Challenger* fly at least five more times in 1986.

STS 61-B began with the liftoff, at 00:29:00 UTC, November 27, 1985 (7:29

P.M., eastern standard time, November 26, 1985), of the orbiter *Atlantis*. During the week-long flight, the second one for *Atlantis*, the crew deployed three communications satellites and demonstrated the techniques needed for building large structures in space. The three satellites were the American Satcom 2, the Mexican Morelos-B, and the Australian Aussat.

Brewster H. Shaw, Jr., commanded the mission. Bryan D. O'Connor was the pilot. Sherwood C. Spring, Jerry L. Ross, and Mary L. Cleave were mission specialists on the flight. McDonnell Douglas engineer Charles D. Walker and Mexican engineer Rudolfo Neri Vela, payload specialists, accompanied them. Vela was present to observe the deployment of the Morelos communications satellite, to operate several Mexican-built medical experiments, and to photograph areas of Mexico from space.

In addition to the three satellites, *Atlantis* carried several payloads in its crew compartment.

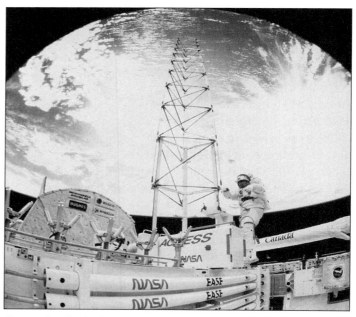

Space shuttle STS 61-B mission crew members Jerry Ross and Sherwood Spring assemble the Experimental Assembly of Structures in EVA (EASE). This first-ever demonstration of zero-gravity construction was designed to validate International Space Station assembly concepts. (NASA)

Among these was the Continuous Flow Electrophoresis System (CFES), a commercial payload built by McDonnell Douglas. Electrophoresis is a process for separating cells using weak electrical charges. All living cells have a small negative charge on their surfaces; different types of cells have different charges, and in solution it is possible to separate them because of these differences. On Earth, such separation is difficult because the charges are extremely small and gravity causes sedimentation and convection. In the microgravity environment of orbital flight, however, such separation can be more easily accomplished. STS 61-B was the seventh flight to carry the CFES. Charles D. Walker, the payload specialist selected by McDonnell Douglas to operate the experiment, was on his third flight into space with the payload. Walker also operated a handheld protein growth experiment, a device to study the feasibility of crystallizing enzymes, hormones, and other proteins. Again, trying to crystallize such materials on Earth is extremely difficult. Nevertheless, successful crystallization permits the study of their three-dimensional atomic structure—important knowledge for enhancing or inhibiting certain functions of the proteins in the development of improved pharmaceuticals.

Other payloads in the crew compartment included the Diffusion Mixing of Organic Solution (DMOS), an experiment built by the 3M Corporation to try to grow large organic crystals for optical and electrical uses. Scientists at 3M wanted to see if larger, perfect crystals could be grown in space. Such crystals could be used for optical switches and computers that process information with light rather than electricity. This device also had cells to observe the mixing of fluids in weightlessness.

In the payload bay, *Atlantis* carried several attached payloads. One of these was an IMAX camera. IMAX, a Canadian large-format camera, had flown in the crew compartment on three previous space shuttle missions. The footage collected from these was used for a film first shown at the National Air and Space Museum in Washington, D.C. *Atlantis* also carried a small, self-contained payload, or Get-Away Special (GAS) experiment, for

Telesat of Canada. The result of a national competition among high school students in Canada, this experiment sought to fabricate better mirrors than those made on Earth by placing gold, silver, and aluminum coatings on quartz plates.

Atlantis flew a direct-ascent trajectory into its 352-kilometer-high initial orbit. The first satellite deployment of the mission, the Morelos 2, came only seven hours after launch, at 7:57 UTC. About forty-five minutes later, the Payload Assist Module (PAM) attached to the satellite fired and moved it on a trajectory toward its eventual geosynchronous orbit. Early the next morning, at 01:20 UTC, the crew released Aussat. The third and final satellite release of the mission, Satcom 2, occurred on Thursday, November 28.

After the satellites were deployed, the crew prepared for the next major mission activities, a pair of extravehicular activities (EVAs). During the spacewalks, Spring and Ross would build a 13.7-meter-tall tower and a 3.7-meter-wide tetrahedron in *Atlantis*'s payload bay. These two structures represented the culmination of nearly a decade of research on large space structures.

The tower, designed and fabricated by engineers at the Langley Research Center in Hampton, Virginia, was called Assembly Concept for Construction of Erectable Space Structures, or ACCESS. The other structure, developed jointly by the Marshall Space Flight Center and the Massachusetts Institute of Technology, was named Experimental Assembly of Structures in Extravehicular Activity, or EASE.

Each structure required a different assembly technique. EASE was a geometric structure resembling an inverted pyramid and consisted of a few large beams and connecting nodes. ACCESS, by comparison, was a high-rise tower consisting of many small struts and nodes. ACCESS could be assembled from a fixed workstation in the payload bay, while EASE required the astronauts to move about the structure during assembly. Both were anchored on a special pallet that bridged the payload bay.

The first EVA began at 21:45 UTC on November 29. For five and a half hours, Spring and Ross practiced assembling the structures. They first built the ACCESS tower. During preflight underwater simulations, they had taken an average of 58 minutes to build the tower. For the orbital EVA, mission planners had allotted two hours for the task. After only fifty-five minutes, however, they were finished. Spring and Ross disassembled the tower. They stowed the ACCESS components away and began working with EASE.

The pair assembled and disassembled the tetrahedron eight times during the first EVA. For the first four times, Ross acted as low man, handing beams to Spring and later putting them back in their storage rack. After the third assembly, Spring indicated that his hands were tired and his fingers were getting numb. By the fourth time, fatigue was beginning to set in, so Spring and Ross traded places for the remaining assemblies. Ross was supposed to perform only two assemblies from the upper position, but the process went fast enough that he had time for four. The first EVA provided fundamental information on space construction.

The second EVA, on December 1, explored specific space station assembly tasks and evaluated the use of the shuttle's RMS arm in these operations. For the arm tests, the astronauts attached a portable foot restraint to the arm. Spring and Ross assembled the ACCESS tower. Then, while Ross stood on the foot-restraint platform, Cleave moved the arm to various work locations from inside *Atlantis*. While attached to the arm, Ross assembled one bay of the ACCESS tower's struts and nodes. Then, as Cleave maneuvered him along the tower, he attached a simulated electrical cable to its length, demonstrating a common space station assembly task. For the next test, Spring released the tower from its assembly jig in the payload bay, and Ross then maneuvered it by hand, demonstrating manual movement and positioning of large space structures. After Ross put the ACCESS tower back in its jig, he and Spring exchanged positions on the end of the arm. Spring then practiced removing and replacing tower struts. He also moved the 13.7-meter tower by hand.

The astronauts then turned their attention to the EASE pyramid and assembled it while Spring was on the end of the arm, a different assembly technique from the one used during the first EVA. Ross then exchanged positions with Spring and tried moving the completed tetrahedron by hand. Following the conclusion of these tests, the astronauts disassembled the structure, stowed the components, and reentered *Atlantis*'s air lock. The EVA had lasted six and a half hours.

On December 3, the crew brought *Atlantis* back to Earth. Mission Commander Shaw landed *Atlantis* at Edwards Air Force Base, California, at 21:33:49 UTC (1:33 P.M., Pacific standard time). After the landing, Shaw applied only light braking, allowing the orbiter to roll 3,279 meters.

Contributions

Although most of the mission objectives were achieved, the scientific results from Spacelab 2 were not all useful. When the payload bay liner reflected more sunlight than expected, overheating fogged many of the HRTS images. Nevertheless, scientists have called some of the images remarkable. They have discovered jets or explosive events in the solar corona, possibly where the solar magnetic field is perpendicular to the surface. Magnetic activity may also have been observed in superspicules, jets of hot gas rising to 15,000 kilometers above the solar surface and lasting three to five minutes. The HRTS also observed smaller spicules that match known spicules observed for some time in white light.

The CHASE instrument did not return useful data on helium ratios because of time constraints and because of internal instrument problems. Although the SOUP film was shot quickly, the experiment yielded more than six thousand striking, high-resolution images showing unusual evolution of granular structures. As the SOUP helped scientists discover, granules explode or break into bright rings that fade or they are destroyed by interaction with granules that have exploded. Also, the granules were found to be absent where solar magnetic activity is most intense.

The twin XRTs produced images of the center of the galaxy and other stellar objects in a broader energy range than had previously been observed. The IRT provided useful data at short and long infrared wavelengths, including images of the center of the galaxy. Data collected in the mid-range, though, were useless. The CRN detected millions of low-energy events and some ten thousand high-energy events of interest. A few registered as high as 10 teraelectron volts, and a gamma-ray burst was detected by the secondary particles it created when it hit the detector shell.

Results from the plasma experiments were substantial. The most striking was the opening of a "hole" in the ionosphere through which the radio telescope at Hobart, Tasmania, could observe stars on wavelengths that normally are reflected by the electrified upper atmosphere. The "hole" was actually an area depleted of electrons and ions and thus transparent to those wavelengths for a few minutes, until the ionosphere regenerated itself. A similar "hole" was generated over New England and persisted for only fifteen minutes. The plasma wake left by the shuttle was found to be complex and turbulent. Thruster firings, water dumps, gas leaks, and other emissions from the shuttle generated a large cloud of neutral gas that expanded around the vehicle and altered the ionosphere. Water ions not normally present at the shuttle's orbiting altitude were detected in large quantities to a distance of several hundred meters from the shuttle, especially in the plasma wake. The PDP fly-around activity placed the shuttle and PDP on the same magnetic field line four times and showed that electrostatic noise (first detected on STS) extends far "downstream" from the shuttle but only a short distance "upstream."

Jerry Ross during STS 61-B after deploying the truss-like Assembly Concept for Construction of Erectable Space Structures (ACCESS) with fellow astronaut Sherwood Spring. His feet are secured by a restraint on the Remote Manipulator System arm. (NASA)

STS 51-I successfully delivered its trio of satellites to their deployment spots and, once again, demonstrated the role that astronauts play in ensuring mission success. A stuck sunshield covering the Aussat satellite would have resulted in an aborted launch had a crew not been there to evaluate and correct the problem. More important, astronauts were required to accomplish the capture, repair, and redeployment of the Syncom IV.

The on-site service of Syncom IV utilized EVA techniques developed on previous spacewalks. The astronauts were able to modify their procedures when required. The experience gained on the flight would be put to use on future repair missions and for the construction of the international space station.

A NASA experiment on STS 51-J called Bios, designed to study the damage to biological materials from high-energy cosmic rays, was positioned in the middeck area of *Atlantis*. A solid-state dosimeter was used to survey the interior of the orbiter to identify the areas most likely to be affected by such radiation. The experiments help to ensure that astronauts (and other biological passengers) are not exposed to potentially hazardous radiation from the Sun and other celestial bodies. In addition, the data are useful in the design of the Space Station and future space vehicles.

The Spacelab D-1/STS 61-A mission flew a scientific package consisting of seventy-six investigations involving fluid physics, solidification, biology, medicine, space-time interaction, GLOMR satellite deployment, flight test maneuvers during descent, and the steering wheel test during rollout. In the fields of biological and life sciences, the ground-based scientists were able to evaluate their data almost in real time, to confirm their findings and determine the success or failure of each experiment. The samples from the materials science experiments were evaluated over a longer period of several months after the conclusion of the mission.

From the materials science double rack with the isothermal heating facility, mirror heating facility, gradient heating facility with its quenching device, and fluid physics module, a total of 75 to 125 percent positive runs and stored data flows were obtained. From the process chamber with the holographic interferometric apparatus and the Marangini convection experiment in an open boat, a recorded level of 90 percent positive runs was achieved. Following early operational difficulties, from the Material Science Experiment Double Rack for Experiment Modules and Apparatus (MEDEA) payload element carrying the monoellipsoid heating facility and the gradient furnace with a quenching device, a 110 percent success rate was achieved during the flight. From the life sciences experiments, success levels of 95 and 100 percent were recorded for the investigations. In addition, the vestibular sled, which was flown for the first time on this flight, achieved 120 percent test-run success.

Among the navigation experiments, the clock synchronization and one-way distance measurement experiments were 100 percent successful.

In all, Spacelab D-1 provided a vast wealth of scientific data from experiments and investigations that had been years in the making; these data made planning for the D-1 mission much easier. The success of the data gathering of Spacelab D-1 was even more remarkable since only 40 percent transmission time could be achieved from the spacecraft to Earth because only one Tracking and Data-Relay Satellite was operational instead of the planned two.

The STS 61-B mission introduced a new satellite upper stage, the PAM-D2. All three satellites used the PAM, as had many of the satellites released during previous shuttle missions. The 61-B satellites, however, used an improved, more powerful version of the motor. PAM-D2 could propel satellites weighing 1,900 kilograms to geosynchronous orbit, while earlier PAM-D motors had only a 1,270-kilogram capacity. In fact, Satcom 2, with a weight of 1,860 kilograms, was the heaviest payload ever propelled by a PAM.

Nevertheless, the most significant results of the mission were from the two EVAs. During twelve hours outside *Atlantis*, Spring and Ross had demonstrated many of the techniques needed to build space stations and other large space structures. With ACCESS, they showed how to construct a long, thin tower structure from a fixed workstation with an assembly jig. EASE, on the other hand, required one of the crew to be free-floating as they assembled the large, pyramid-shaped structure.

During the second EVA, Spring, Ross, and Cleave showed that astronauts inside the spacecraft could work in concert with astronauts outside the vehicle. Cleave maneuvered the two EVA astronauts on the RMS arm along the lengths of both EASE and ACCESS. She positioned her fellow crew members precisely at predetermined work locations. Working with Cleave, Ross attached a length of rope along the tower. This demonstration showed that it was possible to construct a structural framework in space, then route cables and electrical leads along it. Ross and Spring also showed that a space suit-clad astronaut could move large structures manually and could position them precisely by hand.

Context

Like Spacelabs 1 and 3, Spacelab 2 advanced space science in several areas and demonstrated that the shuttle/Spacelab combination is an effective platform for conducting space science experiments. Observations made with the cluster of solar instruments marked the first time since Skylab's Apollo Telescope Mount in 1973-1974 that a piloted solar observatory had been operated in space.

Despite the problems with the IPS, the data from the HRTS and SOUP telescopes were outstanding. Results from both experiments revealed details of solar activity that had been suspected but unobserved because Earth's atmosphere blocks the view. The images from the SOUP would provide a better understanding of the evolution and importance of granules (discovered only two centuries ago) in transporting energy to the solar surface. The HRTS images of superspicules show that a particular spicule phenomenon is larger than previously believed and plays a greater role in the transport of mass and energy from the solar surface into the transitional region where temperatures rise rapidly. A raster survey of 25 percent of the solar disk would help establish global properties of the fine structures of the solar surface. The SUSIM provided measurements of the solar ultraviolet output, with an accuracy of 6 to 10 percent. The SUSIM flown on Spacelab 2 was the beginning of a long-term program to collect data with respect to the influence of the Sun on the terrestrial environment, including the ozone hole. The value of these data will become known as SUSIM-type instruments are reflown over the next few decades. X-ray and infrared images of the skies, and cosmic-ray data, are adding to scientists' understanding of celestial objects.

The shuttle's utility as a scientific platform was demonstrated with mixed results. The abort-to-orbit

maneuver demonstrated a need for performance margins for experiments and for less intense mission planning. After Spacelab 2, infrared observations aboard the shuttle were perceived as risky at best, based on the IRT results. The plasma experiments, however, provided a wealth of data, as scientists were able to disturb the plasma environment in a controlled way and observe the effects with a nearby craft. Although the scientific results were generally good, the operation of the setup would need refining.

The capability to perform on-orbit repairs to the shuttle, its payloads, and previously deployed payloads had been demonstrated on flights prior to 51-I. The Syncom IV repair mission built on this experience and refined the techniques that would be used in later flights. It also gave NASA confidence in the ability of astronauts to work with large orbiting objects, carrying large amounts of explosive propellants. If a space station was in NASA's future, it would have to be constructed by astronauts working in close proximity to extremely large and potentially dangerous objects.

Atlantis came back from the STS 51-J mission virtually unscathed, a far cry from the various damaged parts and nonworking systems the others brought back. More than anything, experience gained from the previous twenty missions contributed to the success of *Atlantis*. Major changes included improved construction materials and electronic hardware, lighter thermal protection systems, and provisions for carrying a Centaur high-energy booster. The Centaur, a liquid-propellant vehicle, was the upper stage of the Atlas and Delta launch vehicles. It was to be used to boost several large deep-space probes from the shuttle, including the Galileo probe to Jupiter. Like the orbiter's main engines, Centaur used liquid hydrogen and liquid oxygen for propellants. After the *Challenger* accident, the Centaur program was canceled, because the vehicle was considered too hazardous to be carried in the shuttle's payload bay.

The flight of Spacelab D-1 on STS 61-A provided a valuable link for the Americans between the early biomedical and materials experiments carried out on the earlier Apollo and Skylab missions in the 1970's and the planned flights on the International Space Station. The Spacelab D-1/STS 61-A mission continued the scientific investigations carried out on the Spacelabs 1 and 3 long module missions in 1983 and then in 1985.

From the human point of view, the flight of a crew of mixed sexes, races, and nationalities pointed the way to international cooperation on the Space Station and talks on the need of a united program of exploration of Mars in the next century. The effective use of the confined habitable quarters of the shuttle during D allowed spacecraft designers to determine the most efficient, pleasing, and functional interior designs of the Space Station and future spacecraft. The Russians have gone a long way toward spacecraft habitability with their Salyut and Mir series. As crews increase in size and missions increase in duration, with the added complications of mixed sexes, religions, and races, suitable internal designs of spacecraft are an important element in planning. Spacelab D-1 represented a milestone in this ongoing aspect of the space program.

STS 61-B was the twenty-third flight of the space shuttle and the fifty-fourth American piloted spaceflight. It came at a time when NASA managers were selecting the configuration for the International Space Station, which they planned for the mid-1990's. Validating the concept of EVA construction by inflight experience was important as they made their decisions.

The EVAs were the culmination of nearly a decade's development and testing. During this period of hardware development, personnel aboard shuttle missions were demonstrating an amazing capability for EVA operations. The first space shuttle EVA was performed during STS-6 in April, 1983. One year later, astronauts repaired the ailing SMM satellite during the STS 41-C mission. In November, 1984, two communications satellites that had been placed in incorrect orbits were retrieved and returned to Earth for refurbishment and relaunch. In August, 1985, the Leasat 3 satellite, which failed just after being released from the orbiter *Discovery* some four months earlier, was jump-started in space. Thus, by the time the EASE and ACCESS experiments flew on *Atlantis*, American astronauts had considerable experience inon-orbit satellite servicing and retrieval. The STS 61-B mission added experience with in-space construction.

See also: Extravehicular Activity; Materials Processing in Space; Solar Maximum Mission; Space Shuttle; Space Shuttle: Life Science Laboratories; Space Shuttle: Microgravity Laboratories and Payloads; Space Shuttle Flights, January-June, 1985; Space Shuttle Mission STS 51-I; Spacelab Program.

Further Reading

Furniss, Tim. *Manned Spaceflight Log.* Rev. ed. London: Jane's Publishing Company, 1986. This updated version of the 1983 first edition covers in a minihistory the world's piloted spaceflights in chronological order, from Yuri A. Gagarin's historic first flight in April, 1961, to the *Challenger* accident and the triumph of Mir in 1986, twenty-five years later. Presented in launch order, 115 piloted spaceflights from the United States and the Soviet Union are described, along with thirteen X5 Astro flights of the American research aircraft of the 1960's.

_____. *Space Shuttle Log.* London: Jane's Publishing Company, 1986. A collection of highly readable reports on the first twenty-two shuttle flights, from April, 1981, to the *Challenger* mission of October/November, 1985. The text provides a useful mission summary and data on each flight in sequence, as well as a collection of biographical sketches of shuttle astronauts up to 1985. Suitable for general readers.

Gurney, Gene, and Jeff Forte. *Space Shuttle Log: The First Twenty-Five Flights.* Blue Ridge Summit, Pa.: TAB Books, 1988. A collection of chapters summarizing the first twenty-five missions of the space shuttle program, covering the operational flights from April, 1981, to the loss of *Challenger* in January, 1986. The entry on each mission covers data on crew and payload and flight records; the main text describes launch preparations and experiments and investigations on each flight. A collection of black-and-white photographs from all the missions accompanies the text.

Harland, David M. *The Space Shuttle: Roles, Missions, and Accomplishments*. Hoboken, N.J.: John Wiley, 1998. *The Space Shuttle* is written thematically, rather than purely chronologically. Topics include shuttle operations and payloads, weightlessness, materials processing, exploration, Spacelabs and free-flyers, and the shuttle's role in the International Space Station.

Harrington, Philip S. *The Space Shuttle: A Photographic History*. San Francisco, Calif.: Brown Trout, 2003. With one hundred full-color photographs by Roger Ressmeyer and others and with text by popular astronomy writer Phil Harrington, this book tells the story of the space shuttle program from 1972 to 2003. Its beautiful photographs allow the general reader to survey the history of the space shuttle program and be uplifted by the pioneering spirit of one of humanity's grandest enterprises.

Jenkins, Dennis R. *Space Shuttle: The History of the National Space Transportation System: The First 100 Missions*. Stillwater, Minn.: Voyageur Press, 2001. This is a concisely written technical reference account of the space shuttle and its ancestors, the aerodynamic lifting bodies. It details some of the advantages and inherent disadvantages of using a reusable space vehicle. Each of the vehicles is illustrated by line-drawings with important features pointed out with lines and text. The book follows the space shuttle from its original concepts and briefly chronicles its first one hundred flights.

Kerrod, Robin. *Space Shuttle*. New York: Gallery Books, 1984. Most valuable for its beautiful color photographs of the space shuttle, this volume conveys the essence of the Space Transportation System vehicles and the people who fly them. Highlights of the first dozen missions are presented, as well as a fanciful look at the future of space exploration.

National Aeronautics and Space Administration. *Marshall Space Flight Center. Spacelab 2*. NASA EP17. Washington, D.C.: Government Printing Office, 1985. This educational publication describes the instruments aboard Spacelab 2 and the planned scientific experiments. Written for reporters covering the mission.

_____. *Space Shuttle Mission Press Kits*. http://www-pao.ksc.nasa.gov/kscpao/presskit/presskit.htm. Provides detailed preflight information about each of the space shuttle missions. Accessed March, 2005.

Otto, Dixon P. *On Orbit: Bringing on the Space Shuttle*. Athens, Ohio: Main Stage Publications, 1986. Early designs of the space shuttle are discussed. Also presents an account of each of the first twenty-five flights, including the crew, the payloads, and the flight objectives. Illustrated in black-and-white.

Shayler, David J. *Shuttle Challenger: Aviation Fact File*. London: Salamander Books, 1987. A book devoted to the career and achievements of space shuttle orbiter OV-099, *Challenger*. This comprehensive text covers the role of *Challenger* in the shuttle program, the construction and components of the vehicle, and its missions—including an account of the STS 51-L accident and summaries of all the astronauts and payloads *Challenger* carried into space on its ten missions. A selection of tables logging accumulated time and hardware data completes the work. Includes a selection of color photographs. A large-format book, this commemorative work on *Challenger* is aimed at a general readership.

Slayton, Donald K., with Michael Cassutt. *Deke! U.S. Manned Space: From Mercury to the Shuttle.* New York: Forge, 1995. This is the autobiography of the last of the Mercury astronauts to fly in space. After being grounded from flying in Project Mercury for what turned out to be a minor heart murmur, Slayton was appointed head of the Astronaut Office. Later, he commanded the Apollo-Soyuz flight in 1975. He stayed with NASA through the STS-2 flight. Although he was no longer an active participant in NASA's activities, he gives some insight into the behind-the-scenes adventures of the shuttle program.

Smith, David H., and Thornton L. Page. "Spacelab 2: Science in Orbit." *Sky and Telescope* 72 (November, 1986): 438-445. An extensive survey of scientific results from the Spacelab 2 mission. Well written and well illustrated with color photographs and charts. For the educated reader interested in astronomy.

Smith, Melvyn. *An Illustrated History of Space Shuttle: X5 to Orbiter.* Newbury Park, Calif.: Haynes Publications, 1986. A concise overview of the space shuttle and the experimental aircraft that led to its design. Spanning the period from 1959 through 1985, the book was written for the general reader. Illustrated with many photographs of the early lifting bodies, which help to show how the shuttle orbiter came to look as it does today. Arranged chronologically.

Wilson, Andrew. *Space Shuttle Story.* New York: Crescent Books, 1986. Traces the history of the space shuttle from the early days of rocketry to the *Challenger* accident. Furnished with more than one hundred color photographs, this volume provides little detail but emphasizes the men and women who fly the space planes to and from orbit.

Yenne, Bill. *The Astronauts.* New York: Exeter Books, 1986. Presents an overview of the Soviet and American space programs and tells of the international passengers carried on various missions. Illustrated with several hundred photographs taken in both countries.

Dave Dooling (STS 51-F) Dennis Chamberland (STS 51-I)
Russell R. Tobias (STS 51-J) David J. Shayler (STS 61-A)

Space Shuttle Mission STS 51-I

Date: August 27 to September 3, 1985
Type of mission: Piloted spaceflight

STS 51-I was the twentieth flight of the United States' space shuttle program. In addition to deploying three communications satellites, crew members captured a malfunctioning satellite launched by a previous shuttle, repaired it successfully, and returned it to orbit. It was the second such in-orbit satellite repair in history.

Key Figures

Joseph H. Engle (b. 1932), STS 51-I commander
Richard O. Covey (b. 1946), STS 51-I pilot
James D. A. "Ox" Van Hoften (b. 1944), STS 51-I mission specialist
John M. "Mike" Lounge (b. 1946), STS 51-I mission specialist
William F. Fisher (b. 1946), STS 51-I mission specialist

Summary of the Mission

STS mission 51-I marked the sixth spaceflight of the orbiter *Discovery*. The seven-day mission proved the extraordinary versatility of the space shuttle as a crewed delivery and repair platform.

The first launch date scheduled for mission 51-I was August 24, 1985, from Kennedy Space Center's Launch Complex 39A. The launch was rescheduled for the next day because of thunderstorms. During the second attempt, on August 25, 1985, an onboard computer malfunctioned during the countdown. National Aeronautics and Space Administration (NASA) engineers described the malfunction as a "GPC 5 byte fault." GPC 5 is general purpose computer number 5, located inside the space shuttle orbiter. This computer contained backup flight system software essential to the shuttle's launch. Engineers reinitialized the software after the error was found, but the error appeared again only 11 minutes later. The flight was postponed again, this time for two days, so GPC 5 could be removed and replaced.

The third countdown, on August 27, 1985, proceeded smoothly. *Discovery* and its five-member crew were launched without significant delays at 10:58:01 Coordinated Universal Time (UTC, or 6:58:01 A.M. eastern daylight time), from Launch Complex 39A. The launch proceeded normally in all respects. The solid-fueled rocket boosters separated without any problems 2 minutes and 1 second after liftoff. The three main engines were shut down 6 minutes and 27 seconds later, 18 seconds before the large External Tank was jettisoned. (It later burned in the atmosphere over the Indian Ocean.)

For many missions it is necessary to fire the orbital maneuvering system (OMS) rocket engines twice to refine the orbital parameters. In this mission, however, the ship was flown in an ascent mode called a "direct insertion ascent trajectory," which precluded the necessity to initiate the first scheduled OMS firing. Exactly 40 minutes and 21 seconds after liftoff, the OMS engines were fired for 3 minutes and 3 seconds, placing the orbiter in a nearly circular orbit approximately 306 by 306 kilometers above Earth. At 14:41 UTC on August 27, 1985, *Discovery* was safely in its proper orbit.

Immediately following the OMS burn, the orbiter's payload bay doors were opened, exposing the satellites in the bay. At 2 hours and 2 minutes after liftoff, the sunshield covering the Australian satellite (Aussat) in the payload bay was commanded to open so that the satellite's systems could be checked prior to being deployed from *Discovery*. The sunshield, however, did not fully open. It was determined that the shield was probably binding on an antenna bracket located on top of the satellite. Two hours after the problem was discovered, ground engineers authorized the crew to use the Remote Manipulator System (RMS), sometimes called the robot arm or the Canadarm, to help push the sunshield open and expose the satellite.

During these operations, the RMS "elbow joint" did not respond to computer commands. Fortunately, the arm had a backup system, and that backup system was used for the remainder of the mission. Yet the failure caused the cancellation of some operations involving use of the RMS-mounted video cameras, such as the monitoring of satellite engine burns and a wastewater dump from the orbiter.

Aussat was finally deployed at 6.5 hours into the mission. Three days later, the satellite reached its station in orbit 35,800 kilometers above Earth, propelled by an engine attached to it called the Payload Assist Module, Delta class (PAM-D). Aussat is used to provide communications relay services for Australia and its offshore islands. A second satellite, American Satellite ASC-1, was deployed at 11 hours and 9 minutes into the mission. Its deployment was successful, and ASC 1 reached its orbit on August 31. ASC-1 is a communications satellite for American business and government agencies.

The second day on board *Discovery* was much more relaxed than the first. It was spent performing experiments with an experimental package called PVTOS, for Physical Vapor Transport of Organic Solids, a package sponsored by the 3M Company and designed to collect data from chemistry experiments conducted in the weightless environment of space. The second day was also used to check out the third satellite still in the payload bay and prepare it for deployment on the third day.

The third satellite, Synchronous Communication Satellite IV-4 (Syncom IV-4, also called Leasat 4), was deployed as scheduled on August 30. It reached geosynchronous orbit, at 35,800 kilometers in altitude, successfully. At such an altitude, satellites rotate at the same speed as Earth so that they appear to remain stationary in the sky. Unfortunately, for unknown reasons, all communications with the satellite were later lost.

Meanwhile, *Discovery* was effecting orbital corrections to rendezvous with Leasat 3, which had been launched by STS 51-D (also from *Discovery*) some four and one-half months earlier. Leasat 3's booster rocket, which would have placed it in a high, geosynchronous orbit, had failed. The *Discovery* crew planned to maneuver the malfunctioning satellite into the payload bay and repair it.

The fourth day in space was spent preparing for the encounter with Leasat 3. Two rendezvous maneuvers were performed while the crew members tested their extravehicular mobility units (EMUs, or space suits). That included charging their batteries and checking out the Remote Power Unit (RPU), whose batteries had been charged on the second day. They would use the RPU to repair the satellite.

The following day, *Discovery* maneuvered to within a few meters of the ailing satellite. Crew members William F. Fisher, a physician, and James D. A. "Ox" Van Hoften, a researcher, exited the orbiter by way of the air lock into the payload bay. With the help of the RMS and some muscle, they captured the satellite and locked it into place in the payload bay to begin the long task of repairing it. Fisher and Van Hoften's extravehicular activity (EVA) set a record for the longest spacewalk of the shuttle program: 7 hours and 10 minutes. During that time, Fisher and Van Hoften worked to repair the satellite by replacing the parts needed to fire the satellite's booster rocket.

The sixth day was used to finish the repair work on the satellite in an EVA lasting 4 hours and 20 minutes. The crew members reentered *Discovery*, and Leasat 3 was deployed. Days later, ground controllers successfully fired its troublesome booster rocket; the satellite attained the proper geosynchronous orbit and began normal service.

The seventh day of the flight was used to prepare for reentry. During this time, the crew members pressurized the cabin to sea level pressure. They tested the forward thrusters and the flight control systems using an auxiliary power unit (APU), which provides the flight control systems with power during reentry. They also dumped wastewater into space and stowed all loose items in the cabin for reentry.

Early on the eighth day of flight, the payload bay doors were closed. The crew fired the OMS engines for 4 minutes and 9 seconds to reduce the orbiter's speed. Thirty minutes later, *Discovery* had descended from 305 kilometers to 126 kilometers, where it encountered the "atmospheric interface," or the upper, relatively dense portion of Earth's atmosphere. *Discovery* reentered Earth's atmosphere at twenty-five times the speed of sound (Mach 25). Thirty minutes after reaching the atmosphere, the spacecraft touched down at Edwards Air Force Base in California, at 13:15:43 UTC (6:15 A.M. Pacific standard time), September 3, 1985. The mission had lasted seven days, 2 hours, and 18 minutes.

Contributions

The success of STS 51-I underscores many of the general aims of the United States' space shuttle program. The system delivered multiple large satellites to orbit. As a crewed system, it was able to correct relatively simple payload problems in space (the failure of the Aussat sunshield to open, for example) that probably would have resulted in the loss of a robotic payload. Its own systematic problem (the failure of an RMS mode) was overcome because crew members were able to evaluate the situation. Also, perhaps most important, *Discovery* was able to perform an in-orbit repair.

The RMS system was used in a unique way on this mission, to assist in the capture and stowing of an in-orbit satellite. Because the RMS required the help of a backup system, the crew expanded the knowledge of RMS capabilities and just how far the RMS could be pushed beyond its design.

The capture and repair of Leasat 3 incorporated a body of knowledge into the Space Transportation System that would be used in future repair missions and even eventual space construction. In the weightlessness of space, the crew members were able to maneuver the massive satellite and its attached PAM-D booster, weighing thousands of kilograms on Earth, into the spacecraft's restraints in the cargo bay. They were then able to anchor it into place for the repair work, releasing it later for boosting into its final orbit. Using the crew's experience, researchers would be better equipped to design tools for the most efficient methods of construction in weightlessness.

The repair provided valuable knowledge about bypassing complex electronic systems with alternate circuits and externally modified systems. It required assessing the problem from ground telemetry, working up a probable scenario of the circuitry involved, and designing a system to bypass the troubled circuits. Ground researchers had accomplished this over the span of a few months, and the astronauts were trained and sent into orbit to effect the repairs. All these activities provided a baseline of experience and knowledge that could be used repeatedly as a successful example of how such in-orbit repair missions could be effected in the future.

The seven-hour EVA set a very important precedent for work activities in space. It proved the ability of people to work in space for long periods and established that space suits and life-support systems are functional under extremely rigorous conditions.

The PVTOS experiment provided knowledge of the transport of organic solids by vaporizing organic materials in what were called "reactor cells" within the experimental package. Data were obtained and stored in a special computer storage system that was a part of the PVTOS package itself. The exact data and parameters obtained were returned to the 3M Company as proprietary information; the data concerned chemical reactions that can be performed only in weightlessness.

Context

Prior to this *Discovery* mission, deployment of satellites from the shuttle system had become, for all practical purposes, commonplace. Twenty-four satellites had been deployed from the shuttle on previous missions. All deployments had been successful; yet, after leaving the shuttle payload bay, several had malfunctioned in orbit—which was not the fault of the shuttle delivery system.

The in-orbit repair of satellites, however, was not common at all. Although the single previous attempt, STS 41-C, had been successful, such repair missions incorporated many uncertainties. For example, the exact cause of the malfunction of the spacecraft could only be narrowed down to a list of possibilities, because the system could not be examined directly. Then the satellite engineers were required to manufacture a solution to cover the entire range of possibilities, plan how the crew could install these in space, and assist the mission planners in training the astronauts to execute the repairs. The pilot and commander of the shuttle, meanwhile, were required to train in rendezvous maneuvers in simulators while other mission specialists trained in the use of the RMS, which would help maneuver the satellite into the payload bay. All of these were mere contingencies; the spectrum of the training program would also have to cover any unplanned events.

NASA engineers had been designing a system whereby an Orbital Maneuvering Vehicle (OMV) would ascend to higher orbits, retrieve malfunctioning spacecraft, and transport them down for repair in lower orbits. Yet, that system was not available to mission 51-I.

As STS 51-I flew, the Soviet Union was at least two and one-half years away from the first launch of its shuttle system, and the Europeans had hardly released word of their planned shuttle-type system, Hermes. Hence, the United States was the only nation at the time to have such capability of launching and repairing spacecraft in orbit from a crewed vehicle.

Unfortunately, the United States' shuttle system would fly only four more times before the tragic loss of the orbiter *Challenger.* Not only would the United States lose its lead in the operation of shuttle-type systems, but the program itself also would emerge fundamentally changed after that paralyzing tragedy.

In September, 1988, *Discovery* flew the first shuttle flight in the post-*Challenger* era. STS-26 proved the capabilities of the shuttle system, but nonmilitary satellites would no longer be launched from or rescued by an orbiter's crew.

See also: Electronic Intelligence Satellites; Get-Away Special Experiments; Manned Maneuvering Unit; Materials Processing in Space; Space Shuttle; Space Shuttle Mission STS 51-A; Space Shuttle Flights, July-December, 1985; Space Shuttle Mission STS 61-C; Space Shuttle Mission STS 51-L; Space Shuttle Activity, 1986-1988; Spacelab Program.

Further Reading

Clarke, Arthur C. *Ascent to Orbit.* New York: John Wiley and Sons, 1984. This work is a compilation of many of Clarke's works from his early material (1930's) to his work of the mid-1980's. It is most effective in presenting the "history of conception" of space systems, from communications satellites to the distant future of space exploration. A mixture of technical and purely entertaining essays that can be appreciated by most readers with any interest in the space sciences. Illustrated.

Harland, David M. *The Space Shuttle: Roles, Missions, and Accomplishments.* Hoboken, N.J.: John Wiley, 1998. *The Space Shuttle* is written thematically, rather than purely chronologically. Topics include shuttle operations and payloads, weightlessness, materials processing, exploration, Spacelabs and free-flyers, and the shuttle's role in the International Space Station.

Harrington, Philip S. *The Space Shuttle: A Photographic History*. San Francisco, Calif.: Brown Trout, 2003. With one hundred full-color photographs by Roger Ressmeyer and others and with text by popular astronomy writer Phil Harrington, this book tells the story of the space shuttle program from 1972 to 2003. Its beautiful photographs allow the general reader to survey the history of the space shuttle program and be uplifted by the pioneering spirit of one of humanity's grandest enterprises.

Jenkins, Dennis R. *Space Shuttle: The History of the National Space Transportation System: The First 100 Missions*. Stillwater, Minn.: Voyageur Press, 2001. This is a concisely written technical reference account of the space shuttle and its ancestors, the aerodynamic lifting bodies. It details some of the advantages and inherent disadvantages of using a reusable space vehicle. Each of the vehicles is illustrated by line drawings. The book follows the space shuttle from its original concepts and briefly chronicles its first one hundred flights.

Joëls, Kerry Mark, and Gregory P. Kennedy. *The Space Shuttle Operator's Manual*. Designed by David Larkin. Rev. ed. New York: Ballantine Books, 1988. This book contains a wealth of information on space shuttle systems and flight procedures. It is written as a manual for imaginary crew members on a generic mission and will be appreciated by anyone interested in how the astronauts fly the orbiter, deploy satellites, conduct spacewalks, and live in space. Contains many drawings and some photographs of equipment.

Nova: Adventures in Science. Reading, Mass.: Addison-Wesley, 1982. A collection of essays and photographs from the public television series *Nova*. Includes several essays on space exploration, with details on the United States' space shuttle system. The book discusses the role of science in daily life, making it an especially valuable tool for referencing the space program and the shuttle's link with everyday existence. It is aimed toward the general reader.

O'Neill, Gerard K. *2081: A Hopeful View of the Human Future*. New York: Simon & Schuster, 1981. Princeton physicist, founder of the Space Studies Institute, and "father" of the space colony, Gerard K. O'Neill has pieced together a thoughtful look at the year 2081. The book speculates on future shuttle systems in an insightful way that reveals the grand vision of today's missions. Illustrated and directed toward the general audience with an interest in the future and in space exploration.

Shayler, David J. *Walking in Space: Development of Space Walking Techniques*. Chichester, England: Springer-Praxis, 2003. Shayler provides a comprehensive overview and analysis of EVA techniques, drawing on original documentation, personal interviews with astronauts with experience in EVAs, and accounts by those involved in suit design and EVA planning and operations.

Slayton, Donald K., with Michael Cassutt. *Deke! U.S. Manned Space: From Mercury to the Shuttle*. New York: Forge, 1995. This is the autobiography of the last of the Mercury astronauts to fly in space. After being grounded from flying in Project Mercury for what turned out to be a minor heart murmur, Slayton was appointed head of the Astronaut Office. Later, he commanded the Apollo-Soyuz flight in 1975. He stayed with NASA through the STS-2 flight. Although he was no longer an active participant in NASA's activities, he gives some insight into the behind-the-scenes adventures of the shuttle program.

Dennis Chamberland

Space Shuttle Mission STS 61-C

Date: January 11 to January 18, 1986
Type of mission: Piloted spaceflight

During STS 61-C, the twenty-fourth flight of the space shuttle, astronauts aboard Columbia *launched a commercial communications satellite, tested a new payload carrier system, and photographed Halley's comet.*

Key Figures

Robert L. "Hoot" Gibson (b. 1946), STS 61-C mission commander

Charles F. Bolden, Jr. (b. 1946), STS 61-C pilot

George D. "Pinky" Nelson (b. 1950), STS 61-C mission specialist

Steven A. Hawley (b. 1951), STS 61-C mission specialist

Franklin R. Chang-Díaz (b. 1950), STS 61-C mission specialist

Robert J. Cenker, STS 61-C payload specialist

Clarens William "Bill" Nelson (b. 1950), Republican representative from Florida and STS 61-C payload specialist

Summary of the Mission

On January 11, 1986, Space Transportation System (STS) mission 61-C began with the predawn liftoff of the space shuttle *Columbia.* The 11:55:00 Coordinated Universal Time (UTC, or 6:55 A.M., eastern standard time) launch was the twenty-fourth space shuttle mission and the seventeenth flight for *Columbia.* In its payload bay, *Columbia* carried the RCA Satcom KU-1, thirteen Get-Away Special (GAS) canisters, the Hitchhiker payload carrier, a materials science laboratory, and an infrared imaging experiment. Inside the crew compartment, the astronauts operated the initial blood-storage experiment, the Comet Halley Active Monitoring Program, and three Shuttle Student Involvement Program experiments. *Columbia* also carried special flight instrumentation to determine more precisely orbiter aerodynamic and reentry heating characteristics.

Veteran Astronaut Robert L. "Hoot" Gibson commanded *Columbia.* This was his second trip into space. The STS 61-C pilot was Charles F. Bolden, Jr., making his first spaceflight. The crew also included three mission specialists: Franklin R. Chang-Díaz, Steven A. Hawley, and George D. "Pinky" Nelson. RCA Engineer Robert J. Cenker and Congressman Clarens William "Bill" Nelson (R-Florida) served as payload specialists.

STS 61-C was the first mission for *Columbia* since the STS 9 flight in late 1983. Following that mission, National Aeronautics and Space Administration (NASA) managers returned *Columbia* to Rockwell International in Palmdale, California, for an eighteen-month overhaul. The hundreds of changes made to the first orbiter to fly in space included updating its navigation system, adding a cylindrical housing to its vertical stabilizer, and building a new nose cap to house the Shuttle Entry Air Data System (SEADS). For its first flight in two years, *Columbia* also carried instrumentation to sample air at its surface in the upper atmosphere and pressure transducers on the top and bottom sides of the wings to determine wing loading during ascent and reentry. *Columbia* originally had 90 wing load sensors. During its overhaul, engineers added 200 more.

The RCA Satcom KU-1 was the only deployable payload aboard *Columbia* for the STS 61-C mission. The satellite cost $50 million, and RCA paid NASA $14.2 million to launch it from the shuttle. It was a Ku-band communications satellite to provide voice, television, facsimile, and data services to commercial customers throughout the forty-eight contiguous U.S. states. Like all earlier communications satellites carried aboard space shuttles, the RCA Satcom was attached to a booster motor that would propel it from low-Earth orbit to a higher, geosynchronous orbit. The motor attached to the Satcom was called the Payload Assist Module D2 (PAM-D2). Satellites that orbit at an altitude of 36,000 kilometers are in a geosynchronous orbit; that is, at that altitude, it takes twenty-four hours to complete one orbit. Thus, a satellite orbiting 36,000 kilometers above the equator will remain fixed in space with respect to an observer on the ground. Ground stations can receive broadcasts from geosynchronous satellites with fixed antennae.

This was the second of three planned vehicles for the RCA American Domestic Satellite System. The first RCA Satcom had been launched during the STS 61-B mission. RCA Satcom is a version of the RCA 4000 series of three-axis stabilized satellites. It carries sixteen operational transponders and six spares, each with an output of 45 watts. These are powerful enough to permit ground stations to receive their transmissions with antennae as small as 1 meter in diameter. The RCA Satcom system can provide direct-to-home television program distribution and television service to hotels, apartment houses, and other large institutions.

Early in the space shuttle program, NASA created the Get-Away Special (GAS) program. Get-Away Specials are small, self-contained payloads, carried in the orbiter's payload bay, that may be flown in space at a cost of as little as $3,000. GAS experiments must be entirely self-contained. That is, they must have their own power and data-recording systems. All an astronaut will normally do with a GAS experiment is turn it on and off during the flight. They are flown on a space-available basis and are accessible to private individuals, foreign governments, and corporations. NASA provides standardized GAS containers for mounting in the payload bay. The containers are about 85 centimeters tall and 50 centimeters wide.

STS 61-C was the maiden flight of a new piece of GAS support equipment: the GAS bridge. The GAS bridge was an aluminum structure that spanned the width of the shuttle's cargo bay and could accommodate up to twelve GAS canisters. A thirteenth GAS canister was attached to the inside wall of the cargo bay near the GAS bridge. It contained instrumentation to measure the environment of the bridge during launch and landing. Prior to this flight, all GAS canisters had been attached to the inside wall of the cargo bay. By the time *Columbia* flew the first space shuttle mission, NASA had sold more than two hundred GAS reservations. At the time of the STS 61-C flight, the backlog was even greater. Engineers at the Goddard Space Flight Center in Greenbelt, Maryland, devised the GAS bridge as a means of carrying large numbers of GAS payloads on individual shuttle missions to reduce the backlog.

Hitchhiker, a new payload carrier system, was also tested on STS 61-C. Like the GAS experiments, Hitchhiker was devised as a method of providing researchers with rapid and economical access to space. Hitchhiker can support scientific, technological, and commercial payloads. It has limited instrument pointing and data processing capabilities. Developed as a payload-of-opportunity carrier, Hitchhiker uses cargo space remaining after the space shuttle's primary payload has been accommodated. Unlike the autonomous GAS canisters, Hitchhiker payloads are connected to the orbiter's communications and power systems. Communications with the payload are provided through a Payload Operations Control Center at Goddard, enabling real-time customer interaction and control.

NASA developed two separate Hitchhiker systems. Engineers at Goddard developed Hitchhiker-G, the type flown on STS 61-C. It was mounted on the front wall of the orbiter payload bay and could accommodate up to four experiments with a combined weight of up to 340 kilograms. The other system, Hitchhiker-M, was developed at the Marshall Space Flight Center in Huntsville, Alabama. It was a structure similar to the GAS bridge that could carry payloads heavier than those carried by Hitchhiker-G. Hitchhiker was created to support payloads too large for GAS and too small for the Spacelab carrier.

U.S. Congressman Clarens William "Bill" Nelson accompanied the crew as a payload specialist and congressional observer. He was the chairperson of the House of Representatives' Space Science and Applications Subcommittee. Nelson represented the Eleventh Congressional District in Florida. During the mission, he operated the handheld protein crystal growth experiment. This experiment sought to use the weightless environment of space to produce protein crystals of sufficient size and quality to allow their nature and structure to be analyzed. Nelson also participated in detailed studies for NASA's Biomedical Research Institute. These studies provided additional data on the effects of spaceflight on the human body.

The flight of STS 61-C was delayed four times before *Columbia* was launched on January 11, 1986. During the early part of the ascent, cockpit instruments indicated that one of *Columbia*'s engines had a helium leak. The situation was serious enough to threaten a shutdown of one of the orbiter's three main engines. If such a malfunction had occurred at that point in the flight, the mission would have been aborted. Pilot Bolden took immediate action to correct the problem, and the mission proceeded according to schedule. Thirty seconds after liftoff, *Columbia* entered the area of maximum dynamic pressure, or max Q. (This pressure is the product of air density times velocity squared.) After reaching a maximum during the first minute of flight, the aerodynamic forces on the vehicle decreased as the shuttle climbed higher.

The ascent profile flown by *Columbia* was deliberately selected to place greater stresses than ever before on the vehicle. Because actual stresses on the craft are greater than what is predicted based on wind-tunnel testing and other experiments,

launch profiles that were less stressful (and therefore less capable of testing the shuttle's full payload capability) had been flown on past missions. STS 61-C was one of three flights planned to collect data that would explain the difference between actual flight results and wind-tunnel predictions. Once acquired, the new information could lead to a relaxation of ascent load constraints, allowing space shuttles to carry heavier payloads.

After the main engines finished their nine-minute burn, and after two firings by the orbital maneuvering system (OMS) engines, *Columbia* was in a 323-kilometer-high circular orbit. Nine hours after launch, during the seventh orbit, the crew opened the sunshield, which protected the Satcom in the cargo bay, and released the satellite from its launch cradle. Forty-five minutes later, the PAM-D2 motor fired and placed the satellite into a highly elliptical orbit which took it to 36,000 kilometers. On January 15, another rocket motor contained in Satcom fired and circularized the orbit at geosynchronous altitude.

During their first day in space, the crew also activated the Material Science Laboratory 2 (MSL-2). MSL-2 comprised three experiments in the cargo bay to study the behavior of materials in microgravity. Two of the experiments studied how melted materials solidify; the third observed liquid behavior in zero gravity. MSL-2 experiments continued throughout the mission.

The astronauts observed Halley's comet on the second day of the mission. The equipment used for this experiment, called the Comet Halley Active Monitoring Program, included a 35-millimeter handheld camera system provided by the University of Colorado. For this experiment, crew members photographed the comet using standard filters to obtain images and spectra. Unfortunately, an intensifier that boosted the light-gathering power of the camera malfunctioned, so the experiment returned only very limited results.

Cenker, the RCA payload specialist, operated another payload, an infrared imaging experiment. Developed by RCA, this was one of two experiments aboard STS 61-C that supported the Strategic Defense Initiative (SDI) and future space-borne surveillance systems. This experiment was mounted on the aft wall of the cargo bay. As he operated the setup, Cenker observed aircraft to measure their infrared signatures. The exact location and types of aircraft were classified. Cenker also used the system, which has possible applications for civilian remote-sensing systems, to observe such unclassified targets as cities and volcanoes.

Hitchhiker carried the other SDI-related payload. Developed by the U.S. Air Force, it was called the Particle-Analysis Camera System (PACS). It comprised two 35-millimeter cameras and a strobe light to take photographs every 120 seconds, recording the amount and type of floating debris surrounding the space shuttle orbiter.

The STS 61-C mission was scheduled to land at the Kennedy Space Center (KSC) in Florida. This was the first landing scheduled for KSC since the STS 51-D mission in April, 1985. During that flight's landing, the orbiter *Discovery*'s right main landing tire experienced a blowout. After that landing, orbiters had landed on the dry lake bed of the Dryden Flight Research Center until mission 61-B. The STS 61-B flight concluded with a landing on the paved runway at Edwards Air Force Base in California. Following this successful landing, NASA managers opted for the shuttle to land at KSC. The weather at KSC, however, prevented a landing there on either January 16 or 17, so *Columbia* returned to Edwards on January 18 instead. The orbiter touched down at 13:58:51 UTC (9:59 P.M. Pacific standard time) on Runway 22. The duration of the mission was six days, 2 hours, and 4 minutes. During reentry and atmospheric flight, infrared sensors in a housing at the top of *Columbia*'s stabilizer measured the temperatures on the craft's upper surfaces. Other instruments in *Columbia* studied the composition of the upper atmosphere and provided precise measurements of the craft's flight attitude.

Contributions

The six-day STS 61-C mission demonstrated the utility and versatility of the space shuttle. On a sin-

gle mission, NASA flew a diverse group of payloads. Major crew activities during the mission included deploying one commercial satellite, testing new payload support equipment, evaluating new space-based Earth imaging systems, and conducting materials processing experiments in space. *Columbia* also carried instrumentation that provided high-resolution infrared images of the top of the orbiter's left wing to create detailed maps of aerodynamic heating during reentry.

The one deployable payload was the RCA Satcom KU-1. This was the second of three Ku-band communications satellites. Most previous communications had operated in the C-band frequency range, which can interfere with terrestrial microwave systems. Because the Ku-band frequencies are not shared with microwave traffic, antennae served by the RCA Satcoms can be located inside major metropolitan areas. In addition, most C-band satellite transponders emit a signal strength of 12 to 30 watts. The RCA Satcom transponders transmit 45 watts of power. This makes direct reception from the satellites possible with antennae of less than 1 meter in diameter.

The Hitchhiker payload system, first demonstrated on this mission, promised to provide researchers with low-cost and rapid access to space. From its inception, the system was designed for simplicity and economy. It incorporated such features as standardized interfaces with orbiter systems and reusability to reduce hardware costs. In addition, with the introduction of the Hitchhiker, NASA reduced the level of paperwork and documentation normally required for shuttle payloads.

Another new piece of payload support hardware tested on this flight was the GAS bridge. Engineers at Goddard devised the GAS program as a means of providing researchers access to space at the lowest possible cost. Through this program, individuals and organizations, both public and private from all countries, have an opportunity to send experiments into space aboard the space shuttle.

This was also the second flight to have a congressional observer as a payload specialist. The first such flight of an elected official was the STS 51-D mission in April, 1985. On that flight, Senator Edwin Jacob "Jake" Garn (R-Utah) flew aboard *Discovery*. Providing flight opportunities for appropriate congressional leaders gave them firsthand experience with spaceflight that they could use when evaluating proposed programs. In addition, they provided NASA physicians with an opportunity to evaluate the effects of spaceflight on individuals who were not career astronauts.

Context

The STS 61-C mission was the twenty-fourth flight of the Space Transportation System Program and the last space shuttle mission before the loss of *Challenger* in January, 1986. STS 61-C demonstrated the flexibility of shuttle payload scheduling by mixing deployable and attached payloads on *Columbia*.

Columbia was the first space shuttle orbiter to orbit Earth. On April 12, 1981, *Columbia* lifted off for the first time. Seven months later, it made its second voyage into space, becoming the first piloted spacecraft to be reused. *Columbia* made four more spaceflights after that, then was temporarily removed from service for an overhaul. As the first operational orbiter, *Columbia* did not have many of the refinements that were built into subsequent vehicles. These included a "heads-up" display for the commander and pilot to use during landing, improvements in the thermal protection system, and structural changes. During its eighteen-month stay at the Rockwell International plant in Palmdale,

California, engineers made these and hundreds of other modifications to *Columbia*.

While these modifications were under way, the other three orbiters made fourteen flights into space. Payloads included commercial satellites, classified Department of Defense experiments, scientific laboratories, and research satellites. The flight of STS 61-C combined many of these types of payloads into a single mission.

Representative Bill Nelson ran an unsuccessful campaign as the Democratic candidate for governor of Florida in 1990. From 1995 to 2000, he was the treasurer and insurance commissioner of Florida. In 2000, Nelson won the election for Florida senator, beating then-Representative Bill McCollum. He went on to serve on the Senate Committee on Commerce, Science, and Transportation, which conducted hearings on the shuttle *Columbia* accident in 2003. On February 1, 2003, *Columbia* had broken up approximately 16 minutes before landing, during reentry over Texas, en route to the Kennedy Space Center. The STS-107 crew members, who were returning home after a successful sixteen-day scientific research mission, were all killed.

See also: Asteroid and Comet Exploration; Get-Away Special Experiments; Launch Vehicles; Launch Vehicles: Reusable; Materials Processing in Space; Space Shuttle; Space Shuttle Flights, July-December, 1985; Space Shuttle Mission STS 51-I; Space Shuttle Mission STS 51-L; Space Shuttle Activity, 1986-1988; Spacelab Program.

Further Reading

Couvalt, Craig. "Delays in *Columbia* Mission Complicate Shuttle Scheduling." *Aviation Week and Space Technology* 124 (January 20, 1986): 20-22. This article provides an overview of the launch and early flight results of the STS 61-C mission. It also discusses the impact of this mission's launch delays on the flight schedule planned prior to the loss of *Challenger*.

Harland, David M. *The Space Shuttle: Roles, Missions, and Accomplishments.* Hoboken, N.J.: John Wiley, 1998. *The Space Shuttle* is written thematically, rather than purely chronologically. Topics include shuttle operations and payloads, weightlessness, materials processing, exploration, Spacelabs and free-flyers, and the shuttle's role in the International Space Station.

Harrington, Philip S. *The Space Shuttle: A Photographic History.* San Francisco, Calif.: Brown Trout, 2003. With one hundred full-color photographs by Roger Ressmeyer and others and with text by popular astronomy writer Phil Harrington, this book tells the story of the space shuttle program from 1972 to 2003. Its beautiful photographs allow the general reader to survey the history of the space shuttle program and be uplifted by the pioneering spirit of one of humanity's grandest enterprises.

Jenkins, Dennis R. *Space Shuttle: The History of the National Space Transportation System: The First 100 Missions.* Stillwater, Minn.: Voyageur Press, 2001. This is a concisely written technical reference account of the space shuttle and its ancestors, the aerodynamic lifting bodies. It details some of the advantages and inherent disadvantages of using a reusable space vehicle. Each of the vehicles is illustrated by line drawings. The book follows the space shuttle from its original concepts and briefly chronicles its first one hundred flights.

Microgravity Science and Applications Division, Office of Space Science and Applications. *Microgravity: A New Tool for Basic and Applied Research in Space.* NASA EP-212. Washington, D.C.: Government Printing Office, 1984. Written for a general audience, this document provides an overview of NASA's space materials processing programs.

Nelson, C. William. "Ascent." *Final Frontier* 1 (July/August, 1988): 18-21, 57. In this article, Congressman Nelson provides a firsthand account of the first 8.5 minutes of the STS 61-C mission.

Nelson, C. William, with Jamie Buckingham. *Mission.* New York: Harcourt Brace Jovanovich, 1988. This is a personal account by Congressman Bill Nelson of his flight aboard STS 61-C.

Gregory P. Kennedy

Space Shuttle Mission STS 51-L

Date: January 28, 1986

. *Type of mission:* Piloted spaceflight

STS 51-L, the twenty-fifth mission of the U.S. space shuttle, would have launched the second Tracking and Data-Relay Satellite and a scientific mission called SPARTAN-Halley into Earth orbit. In addition, Teacher-in-Space S. Christa McAuliffe would have broadcast a series of lessons to schoolchildren throughout America. However, STS 51-L broke apart only 73 seconds after launch, killing its crew and completely destroying the orbiter Challenger *and its satellite cargo. In the wake of the STS 51-L accident, the U.S. space program was severely set back, and the shuttle did not fly again for almost three years.*

Key Figures

Francis R. "Dick" Scobee (1939-1986), STS 51-L commander

Michael J. Smith (1945-1986), STS 51-L pilot

Ellison S. Onizuka (1946-1986), STS 51-L mission specialist

Ronald E. McNair (1950-1986), STS 51-L mission specialist

Judith A. Resnik (1949-1986), STS 51-L mission specialist

Gregory B. Jarvis (1944-1986), STS 51-L payload specialist

S. Christa McAuliffe (1948-1986), STS 51-L payload specialist and the first teacher in space

William P. Rogers (1913-2001), the chairperson of the Presidential Commission on the Space Shuttle *Challenger* Accident

Summary of the Mission

Preparations for the Space Transportation System's (STS) twenty-fifth mission, STS 51-L, began more than eighteen months before launch. When the flight crew was originally selected, on January 27, 1985, 51-L's launch was scheduled for the summer of 1985. Delays and a series of cargo changes, however, postponed the flight to mid-January, 1986. Because of these delays, both the detailed flight planning process and the crew's training were interrupted.

The major payloads carried on STS 51-L were the second National Aeronautics and Space Administration (NASA) Tracking and Data-Relay Satellite (TDRS) and the SPARTAN-Halley comet research observatory. In addition to these payloads, several small experiments were carried in the crew cabin, and the flight was to include the Teacher-in-Space activities of S. Christa McAuliffe.

According to preflight planning, mission 51-L was to last six days. During this time the crew would launch the TDRS satellite, activate and launch SPARTAN, conduct astronomical and medical experiments, recover SPARTAN from orbit, and broadcast lessons to students on the ground.

The planned 1986 shuttle launch schedule was very tight, and several very high priority missions were to take place in the early part of the year. Within NASA, plans were discussed to skip 51-L if the launch date slipped beyond February 1 and proceed with the rest of the schedule. The purpose of this move would have been to clear the pad for the next launch (an important mission scheduled for March) and to begin readying *Challenger* for its planned launch of an international mission to explore Jupiter and the Sun.

An afternoon launch was originally planned for

51-L. Although scientists leading the SPARTAN project argued for retaining this time for scientific reasons, NASA mission planners insisted on changing the liftoff to mid-morning. NASA's reasoning for a morning launch was based on safety concerns. Were the vehicle to suffer an "engine-out" during its ascent from the Kennedy Space Center (KSC), it would have to glide to an emergency landing site at Casablanca on the west coast of Africa. Casablanca's runway was not equipped with lighting for night landings. It was decided, therefore, that the shuttle would be launched in the morning, local time, so that there would still be light in Casablanca, 6,400 kilometers to the east.

The countdown for STS 51-L began on January 24, but weather forecasts caused the launch to be postponed to January 27. The crew spent the extra time reviewing flight plans and watching the Super Bowl football game from their quarters. During this period, pressures within NASA to launch 51-L mounted. The SPARTAN satellite required a launch before January 31. With flights of even higher priority on NASA's schedule, the prospect of canceling 51-L became greater. Every effort was made to make sure the shuttle would be ready on the twenty-seventh, when the weather cleared.

On January 27, the vehicle was fueled and the crew was aboard when the ground crew reported a problem with the exterior handle on the orbiter's hatch. While the countdown was halted, winds at the KSC return-to-launch-site landing strip increased and exceeded the allowable velocity for a

The STS 51-L crew, who lost their lives in the tragic accident on January 28, 1986 (from left): Ellison S. Onizuka, Mike Smith, Sharon Christa McAuliffe, Dick Scobee, Greg Jarvis, Ron McNair, and Judy Resnik. (NASA)

landing. When the launch window closed, NASA was forced to reschedule the launch once again, this time for January 28.

During the evening of January 27 and the early morning hours of the next day, a series of meetings were held among intermediate-level managers from NASA's Marshall Space Flight Center and the shuttle program's major industrial contractors, Rockwell, Morton Thiokol, and Martin Marietta. Marshall had final responsibility for the solid-fueled rocket boosters (SRBs). The purpose of these meetings was to assess the status of the launch. Such meetings take place during every countdown.

During these prelaunch meetings, concerns were expressed about the possible effects of a cold weather front approaching the Cape. Some engineers from Morton Thiokol, the manufacturer, expressed three concerns. In particular, Thiokol engineers Roger Boisjoly and Alan McDonald thought it possible that the booster's O-rings, used to seal the joints between its segments, might become stiff in the cold. Once stiffened, these O-rings (which are supposed to be resilient) would not act to seal the SRBs. Without a good seal, exhaust would leak from the one of the joints (termed "blow-by"), rather than exiting normally through the nozzle. The booster would be likely to explode or rupture—ending the mission in catastrophe. In one conversation, McDonald went so far as to say that if anything happened, he would not want to have to explain it to a board of inquiry. McDonald noted that no shuttle had ever been launched at a temperature below 12° Celsius. Even at that temperature the SRBs had experienced some exhaust blow-by.

The possibility of severe blow-by raised concerns and was the reason for many of the meetings that were held that night. Unfortunately, no actual tests of the boosters had ever been made at low temperatures; therefore, there was no clear case for what would happen. Thiokol's engineers recommended the launch be delayed until later in the day, or perhaps until January 29. NASA managers, however, feeling increasing pressure to launch, pressed for a firm decision from Thiokol. In testimony to the

Rogers Commission (convened to discover the cause of the accident), NASA managers stated that the probability of an O-ring failure was believed to be low because a second O-ring backed up each primary O-ring for increased protection, in case blow-by did occur.

Perhaps sensing the impatience of some NASA officials, Thiokol managers overruled their own engineers and signed a waiver form, stating that the SRBs were safe for launch. Without such a signature, the launch could not have occurred.

During the final hours before the launch of STS 51-L, all of *Challenger*'s mechanical and electrical systems were checked. The crew was awakened at 11:00 Coordinated Universal Time (UTC, or 6:00 A.M. eastern standard time) and ate breakfast. At 13:36 UTC they arrived at the pad and boarded the shuttle. Because there was a buildup of ice on the launch pad and there had been some delays overnight in fueling the huge External Tank, the launch was pushed back first from 14:48 UTC, to 15:38 UTC, and then to 16:38 UTC. These intermittent delays were unusual in the STS program. Several of the onboard scientific experiments required launch times earlier than 15:00 UTC on any given day in order to have the right lighting conditions in orbit. These experiments were sacrificed in order to get the flight launched that morning. The countdown proceeded.

As the final few minutes passed, Pilot Mike Smith powered up *Challenger*'s auxiliary power units (APUs) for flight. Only two minutes before launch, Commander Francis R. "Dick" Scobee called to crew members McNair, Jarvis, and McAuliffe on the shuttle's lower passenger deck, "Two minutes, downstairs; you got a watch running down there?"

At launch minus thirty seconds, *Challenger*'s onboard computers took control. First, the orbiter's three powerful main engines were pressurized, and then thousands of electronic checks were performed to verify the engines were ready to start. At minus 6.6 seconds, the main engines were ignited, one at a time, about a second apart. When 90 percent of flight-level thrust was reached on all

Challenger *exploded about 73 seconds after launch of STS 51-L. All seven crew members, including the first school-teacher in space, were killed.* (NASA)

three, the command was sent to ignite the SRBs. Both SRBs ignited simultaneously at exactly 16:38:00.010 UTC, and the vehicle rose from the pad. On board, Astronaut Judith A. Resnik exclaimed, "All right," as *Challenger* began its long-awaited push to orbit. In the launch control center three miles away, Thiokol engineers Boisjoly and McDonald relaxed a bit—apparently the O-rings had held. The ambient air temperature was approximately 2° Celsius, 10° Celsius colder than that of any previous shuttle launch.

Less than a half-second after the SRBs ignited, the first of eight small but ominous puffs of black smoke swirled from one of the lower joints in *Challenger*'s right booster. These puffs were not obvious to onlookers (no one is allowed within 5 kilometers

of the launch pad) but were recorded by cameras filming the launch. Later analysis by technical experts working with the Rogers Commission revealed that a primary O-ring had failed to seal in the right SRB and that its backup O-ring had failed as well. Blow-by had occurred. Engineering analyses have since indicated that either propellant residue or O-ring soot plugged this leak about two and one-half seconds into the flight.

For nearly a minute, the ascent went as planned. *Challenger* rolled to put itself on the proper flight path, thousands of electronic checks of onboard systems showed everything performing "nominally" (according to plan), and the vehicle properly throttled back its engines when aerodynamic forces increased. *Challenger* had entered the area of maximum dynamic pressure, or max Q. (Max Q is the product of air density times velocity squared.)

Approximately fifty-nine seconds into the flight, trouble began. A review of film from ground cameras recording the launch detected flames coming from the side of the right SRB closest to the External Tank (ET). *Challenger* was being buffeted by a combination of high-altitude winds and the aerodynamic stresses of max Q. In response the vehicle flexed slightly and began to steer its engines to counteract the wind. These forces probably reopened the hole in the right SRB caused by the blow-by at ignition. Over the next five seconds the plume of flame grew and grew. By sixty-four seconds into the flight, a gaping hole was formed in the casing of the SRB. The thrust escaping through this hole exerted a force of 45,000 kilograms on the shuttle, greater than the thrust of many jetliners. *Challenger*'s computers interpreted this force as unusually strong winds. To counteract this side-force, the shuttle automatically swung its engines slightly to the left. Inside the cockpit, the crew was jolted around by a combination of actual wind gusts, engine steering, and the thrust escaping from the breeched SRB. Pilot Mike Smith remarked, "Looks like we've got a lot of wind here today."

Diverted by the stream of air flowing past the vehicle, flames shooting from the hole wrapped themselves around the base of the External Tank. At 72 seconds into the flight, the attachment strut between the SRB and the External Tank either tore or burned loose. Eventually, the second lower support strut broke and the SRB flailed helplessly, attached to the ET only by its forward strut. Weakened by the flames, the base of the External Tank fell away, releasing the contents of the hydrogen tank located there. This produced an upward thrust that pushed the tank into the structural intertank section. In the final second of flight, computers on board *Challenger* detected a fuel line break and shut down each of the shuttle's three main engines. A moment later, the SRB slammed into and tore off *Challenger*'s right wing, then careened into the upper section of the External Tank. This effectively destroyed the giant orange-colored tank. No longer attached to the tank, the SRBs flew away. Simultaneously, the cockpit voice recorder taped the first and last indication that anyone on board knew of the serious trouble—Pilot Michael J. Smith said, "Uh oh." With no support from the External Tank and engulfed in the explosive burn of the propellants, *Challenger* began to break apart from the aerodynamic loads. At an altitude of 14.6 kilometers and a speed of Mach 2 (twice the speed of sound), several large pieces of *Challenger* emerged from the fireball.

On the ground, some spectators realized that the SRBs had separated too early. Others, unfamiliar with shuttle launches, thought this was the normal staging of SRBs. Soon, however, it was clear to all that the shuttle was nowhere to be seen in a widening fireball, and that the SRBs were wildly spinning off on their own, still under thrust.

At Mission Control in Houston, telemetry signals suddenly stopped. At first, having seen no indications of trouble during the launch, flight controllers believed that either the tracking station or the shuttle's radios had failed. Within seconds, however, radar tracking detected hundreds of pieces of debris following *Challenger*'s trajectory. Noting this, flight dynamics officer Brian Perry, a veteran shuttle flight controller, confirmed the tracking data and reported the incident to the flight director.

The aerodynamic breakup that destroyed *Challenger* (NASA's second and, at the time, most experienced space shuttle) also destroyed the two satellites carried in its payload bay. Its crew, Commander Francis R. "Dick" Scobee; Pilot Michael J. Smith; Mission Specialists Ellison S. Onizuka, Judith A. Resnik, and Ronald E. McNair; Payload Specialist Gregory B. Jarvis; and spaceflight participant S. Christa McAuliffe, were all killed.

In the days that followed the *Challenger* accident the nation mourned. President Ronald W. Reagan eulogized the crew both in a nationally televised speech and at a memorial ceremony at NASA's Johnson Space Center in Houston, Texas.

Within hours of the accident that destroyed *Challenger*, calls were made for an official investigation. At NASA, Dr. Jesse W. Moore, the official in charge of the shuttle program, set up a task force to carry out a technical investigation of the cause, or causes, of the accident. Moore's all-NASA team impounded all data relating to the flight and initiated a salvage effort to recover as much of the wreckage as possible from the ocean. The wreckage would provide physical evidence that would be available to help pinpoint the accident's cause.

There were calls for a non-NASA investigation. Such an investigation, it was said, would more likely be freer of bias than any investigation carried out by NASA. Heeding these calls on February 3, 1986, President Reagan appointed a group of thirteen distinguished engineers, test pilots, and scientists to investigate the *Challenger* accident. This group was officially known as the Presidential Commission on the Space Shuttle *Challenger* Accident. The commission's chairperson was William P. Rogers, a former secretary of state, former U.S. attor-ney general, and accomplished lawyer. Other members of the commission included Neil A. Armstrong, the first human to walk on the Moon; Sally K. Ride, an astrophysicist and the first American woman in space; and Richard Feynman, a physicist and Nobel Prize winner. Like the Apollo 1 investigating committee, the Rogers Commission was charged with carrying out a full assessment of all aspects of the accident and the shuttle program; unlike the Apollo investigation, the *Challenger* inquiry was performed publicly, by a basically non-NASA group.

The Rogers Commission took testimony from more than 160 individuals involved in the shuttle program and *Challenger*'s last flight. More than twelve thousand pages of sworn testimony were taken, and more than sixty-three hundred documents relating to the accident were reviewed. More than six thousand engineers, scientists, technicians, and other individuals participated in the commission's work. On June 6, 1986, the Rogers Commission released its final report, fixing the immediate cause of the accident as well as discussing the contributing factors that had led to the decision to launch *Challenger* on January 28, 1986. The commission's report also made recommendations to improve the design of the space shuttle and to prevent future accidents.

Using facts uncovered by the Rogers Commission, as well as supporting evidence and eyewitness accounts of the accident and the salvaged wreckage, it is possible to reconstruct the flight of STS 51-L and the crucial events that led to its ill-fated launch.

Context
The Rogers Commission made a methodical study of all the events leading up to the flight of STS 51-L.

The solid rocket booster's forward skirt, recovered after the accident. (NASA)

Also evaluated were flight records radioed to the ground, debris recovered from the ocean, and films of the flight taken by long-range cameras. Many individuals were interviewed, and a great number of technical studies were performed to test theories concerning the in-flight events of January 28.

The commission considered many things that could have caused *Challenger*'s destruction. Possible causes that were investigated included a failure of the main engines, a rupture of the huge external fuel tank, a problem in one of the payload rockets (such as the ignition of the TDRS upper stage), a failure in one of the SRBs, premature ignition of the shuttle's emergency destruct system, and sabotage. As the evidence mounted, many of the possible causes were eliminated from the list. By early February, only weeks after the launch, the investigators were already focusing their entire attention on the right SRB. Much of the reason for this early narrowing of the possibilities came about because films from automatic cameras developed after the flight clearly showed black smoke seeping from SRB joints at ignition. The films also depicted bright flames jetting from the rocket casings about fifty-eight seconds after launch. Engineers and technicians working for the commission considered propellant cracks, cracks in the rocket motor case, and O-ring seal problems as possible causes of the SRB failure.

In its final report, the commission pinpointed the cause of the accident and made several recommendations for improvements in the shuttle and its management. From a technical standpoint, the cause of the accident was quite clear. The commission found that the cause of the accident was a failure of the O-ring pressure seal of the right solid rocket motor. In reaching this conclusion, many possible SRB failure modes had been evaluated. Once the O-ring was identified as the cause, the commission went on to determine what specifically caused the O-ring to fail.

Had the O-ring been improperly installed or tested? Had sand or water got into the O-ring joint to prevent it from sealing? Had the cold been to blame? Had the elastic putty used in the O-ring joint failed to seal? Again, more tests were performed, and the flight data and debris were reanalyzed. The commission did not, however, draw a definite conclusion about this cause of the accident. Too much of the evidence had been destroyed in the explosion. Although it was certain that the right SRB had experienced a failure in one of its joints, it was possible that one or more of the above causes were to blame. The commission did, however, conclude that the SRB design was prone to certain failures, including the one that destroyed *Challenger.*

The Rogers Commission's findings went far beyond a determination of the immediate cause of the accident. The commission also concluded that there had been "serious flaws in the decision-making process" leading to 51-L's launch. In particular, it concluded that exceptions to established rules had been granted "at the expense of flight safety" and that Morton Thiokol's management "reversed its position and recommended launch contrary to the views of its engineers in order to accommodate a major customer."

It was found that previous ground tests and blow-by problems experienced on past flights should have alerted NASA and Thiokol to the serious deficiencies in the SRBs. The commission also found that the Marshall Space Flight Center had not properly passed evidence of SRB problems up the chain of command within the shuttle program but had instead "attempted to resolve them internally." The commission stated in its report to the president that this kind of management "is altogether at odds with successful flight missions."

After analyzing the cause of the *Challenger* accident, the Rogers Commission made a number of recommendations to NASA. These recommendations fell into several categories, including the design of the SRBs and shuttle management. The goal of the recommendations was to improve the reliability of the entire shuttle.

The commission recommended that the SRBs be redesigned and recertified to solve the numerous problems inherent in their O-ring joints. The redesign specifically called for an SRB that in future flights would "be insensitive to" environmental factors, including the cold and rain as well as "assembly procedures." Further, the commission called for a design that would have joints as strong as the rocket casings themselves. To verify the integrity of the new design, the commission recommended testing full-size boosters before the new SRBs were committed to actual flight. These tests began in the summer of 1987.

The commission also made specific recommendations relating to other potential problem areas in the shuttle. They insisted that the shuttle's brakes be improved (a long history of brake problems had occurred over many flights) and that a re-evaluation of crew abort and escape mechanisms be undertaken to determine if launch and landing problems could be made "more survivable." Finally, the commission insisted that the rate of shuttle flights be controlled to maximize safety. Such a policy had not been implemented in the past, the commission said.

Beyond technical matters, the Rogers Commission also recommended a number of sweeping changes in the shuttle program's management structure. These were designed to prevent the problems that led to a "flawed decision-making process" regarding the launch of 51-L. These specific recommendations included the establishment of a safety panel, with broad powers, reporting directly to the manager of the shuttle program and the establishment of the Office of Safety, Reliability, and Quality Assurance within NASA reporting directly to the NASA administrator, with broad powers to investigate and demand solutions to safety-related issues.

Additionally, it was recommended that a full review take place of all critical safety items in the space shuttle before the next flight. Finally, the commission recommended that astronauts be more fully involved in the shuttle program's management. This recommendation came in response to the anger expressed by some astronauts during the investigation that they, who were at greatest risk in each flight, had not been informed of the O-ring

blow-by and erosion problems prior to the accident. In response to this call, NASA placed senior astronauts—Robert L. Crippen, Jr., Sally K. Ride, Paul J. Weitz, and others—in key advisory roles.

The *Challenger* accident brought the United States space program to a halt. Within months of the accident, two robotic launchers failed as well. With no way of launching satellites until either these rockets or the shuttle was recertified for flight, both NASA and the military were "pinned down." New research in space could not be conducted. Replacement military and weather satellites could not be launched. Planned space missions stagnated, awaiting the availability of a launcher. More than a dozen scientific payloads were canceled, and seventy more were delayed for years. Space planners were forced to buy dozens of expendable Titan and Delta rockets to supplement the grounded shuttle program.

Slightly more than seventeen years after *Challenger,* the shuttle program suffered its second major inflight accident. On February 1, 2003, following a nominal mission, *Columbia*'s deorbit burn occurred at 13:15 UTC for a planned landing on KSC Runway 33. At approximately 13:52 UTC, *Columbia* was at 68 kilometers in altitude, crossing over the coast of California and entering Roll Reversal number one. It was traveling at Mach 20.9 at Mission Elapsed Time of fifteen days, 22 hours, 17 minutes, and 50 seconds. At approximately 14:00 UTC (early morning in the western and central United States), communication with the crew and loss of data occurred. The vehicle broke up while traveling at 20,200 kilometers per hour (Mach 18.3) at an altitude of 63 kilometers over east central Texas, resulting in the loss of both vehicle and crew.

See also: Cape Canaveral and the Kennedy Space Center; Escape from Piloted Spacecraft; Ethnic and Gender Diversity in the Space Program; National Aeronautics and Space Administration; Space Centers, Spaceports, and Launch Sites; Space Shuttle; Space Shuttle Mission STS 61-C; Space Shuttle Activity, 1986-1988; Space Shuttle Mission STS-26.

Further Reading

Durant, Frederick C., III, ed. *Between Sputnik and the Shuttle: New Perspectives on American Astronautics, 1957-1980.* San Diego: Univelt, 1981. A comprehensive history of American piloted spaceflight from 1957 to 1981. This book contains an excellent description of the origins of the Space Transportation System program.

Harland, David M. *The Space Shuttle: Roles, Missions, and Accomplishments.* Hoboken, N.J.: John Wiley, 1998. *The Space Shuttle* is written thematically, rather than purely chronologically. Topics include shuttle operations and payloads, weightlessness, materials processing, exploration, Spacelabs and free-flyers, and the shuttle's role in the International Space Station.

Harrington, Philip S. *The Space Shuttle: A Photographic History.* San Francisco, Calif.: Brown Trout, 2003. With one hundred full-color photographs by Roger Ressmeyer and others and with text by popular astronomy writer Phil Harrington, this book tells the story of the space shuttle program from 1972 to 2003. Its beautiful photographs allow the general reader to survey the history of the space shuttle program and be uplifted by the pioneering spirit of one of humanity's grandest enterprises.

Jenkins, Dennis R. *Space Shuttle: The History of the National Space Transportation System: The First 100 Missions.* Stillwater, Minn.: Voyageur Press, 2001. This is a concisely written technical reference account of the space shuttle and its ancestors, the aerodynamic lifting bodies. It details some of the advantages and inherent disadvantages of using a reusable space vehicle. Each of the vehicles is illustrated by line drawings. The book follows the space shuttle from its original concepts and briefly chronicles its first one hundred flights.

Lewis, Richard S. *Challenger: The Final Voyage.* New York: Columbia University Press, 1988. A factual account of the *Challenger* accident, this book relates the events of January 28, 1986. A popular version of the Rogers Commission Report.

Report of the Presidential Commission on the Space Shuttle Challenger Accident. Washington, D.C.: Government Printing Office, 1986. Contains the full text of the official report of the Rogers Commission. Technical in its content, this volume details both the immediate and root causes of the STS 51-L accident.

Scobee Rodgers, June. *Silver Linings: Triumph of the Challenger 7.* Macon, Ga.: Peake Road, 1996. Ten years after she watched her husband's space shuttle explode into flames, June Scobee Rodgers, wife of *Challenger* Commander Francis R. "Dick" Scobee, tells the story of the tragedy that changed her life. In *Silver Linings: Triumph of the Challenger 7,* June recounts her personal journey through intimate pictures and words, revealing how she found joy in the midst of her sorrow and triumph in the *Challenger* tragedy.

Shayler, David J. *Disasters and Accidents in Manned Spaceflight.* Chichester, England: Springer-Praxis, 2000. The author examines the challenges that face all crews as they prepare and execute their missions. The book covers all aspects of crewed spaceflight—training, launch to space, survival in space, and return from space—and follows with a series of case histories that chronicle the major incidents in each of those categories over the past forty years. The sixth section looks at the International Space Station and how it is planned to prevent major incidents.

Slayton, Donald K., with Michael Cassutt. *Deke! U.S. Manned Space: From Mercury to the Shuttle.* New York: Forge, 1995. This is the autobiography of the last of the Mercury astronauts to fly in space. After being grounded from flying in Project Mercury for what turned out to be a minor heart murmur, Slayton was appointed head of the Astronaut Office. Later, he commanded the Apollo-Soyuz flight in 1975. He stayed with NASA through the STS-2 flight. Although he was no longer an active participant in NASA's activities, he gives some insight into the behind-the-scenes adventures of the shuttle program.

Stern, S. Alan. *The U.S. Space Program After Challenger.* New York: Franklin Watts, 1987. A detailed look at the *Challenger* accident and its ramifications for the future of the United States space program.

Vaughan, Diane. *The Challenger Launch Decision: Risky Technology, Culture, and Deviance at NASA.* Chicago: University of Chicago Press, 1997. In this book, Vaughan re-creates the steps leading up to the decision to launch *Challenger* on the STS 51-L mission, contradicting conventional interpretations to prove that what occurred at NASA was not skullduggery or misconduct but a disastrous mistake.

Washington Post Editorial Staff. *Challengers.* New York: Simon & Schuster, 1986. A touching and personal series of biographies of each of the seven crew members of the ill-fated STS 51-L mission.

Alan Stern

Space Shuttle Activity, 1986-1988

Date: January 1, 1986, to December 31, 1988
Type of mission: Piloted spaceflight

Following nine shuttle flights in 1985, fifteen launches were scheduled for 1986 as the United States moved toward its aim of eventually having twenty-four launches annually. After the Challenger *accident early in 1986, which killed all on board, the entire space program was reevaluated, causing the cancellation of launches planned for 1986, 1987, and early 1988.*

Key Figures

Robert L. "Hoot" Gibson (b. 1946), STS 61-C mission commander

Charles F. Bolden, Jr. (b. 1946), STS 61-C pilot

George D. "Pinky" Nelson (b. 1950), STS 61-C mission specialist

Steven A. Hawley (b. 1951), STS 61-C mission specialist

Franklin R. Chang-Díaz (b. 1950), STS 61-C mission specialist

Robert J. Cenker, STS 61-C payload specialist

Clarens William "Bill" Nelson (b. 1950), Republican representative from Florida and STS 61-C payload specialist

Francis R. "Dick" Scobee (1939-1986), STS 51-L commander

Michael J. Smith (1945-1986), STS 51-L pilot

Ellison S. Onizuka (1946-1986), STS 51-L mission specialist

Ronald E. McNair (1950-1986), STS 51-L mission specialist

Judith A. Resnik (1949-1986), STS 51-L mission specialist

Gregory B. Jarvis (1944-1986), STS 51-L payload specialist

S. Christa McAuliffe (1948-1986), STS 51-L payload specialist and the first teacher in space

William P. Rogers (1913-2001), chairperson of the Presidential Commission on the Space Shuttle *Challenger* Accident

Frederick H. "Rick" Hauck (b. 1941), STS-26 commander

Richard O. Covey (b. 1946), STS-26 pilot

George D. "Pinky" Nelson (b. 1950), STS-26 mission specialist

David C. Hilmers (b. 1950), STS-26 mission specialist

John M. "Mike" Lounge (b. 1946), STS-26 mission specialist

Robert L. "Hoot" Gibson (b. 1946), STS-27 commander

Guy S. Gardner (b. 1948), STS-27 pilot

Richard M. Mullane (b. 1945), STS-27 mission specialist

Jerry L. Ross (b. 1948), STS-27 mission specialist

William M. Shepherd (b. 1949), STS-27 mission specialist

Summary of the Missions

The space shuttle was envisioned as the work-horse that would fly regularly into space to put satellites into orbit, to repair or return to Earth malfunctioning satellites, to conduct research of various sorts, and to serve as a foundation for large construction projects in space. Designed to carry a crew of seven, various space shuttles made twenty-four successful flights between April, 1981, and January, 1986. Although no venture into space is routine, the success rate of previous shuttles gave officials of the National Aeronautics and Space Administration (NASA) considerable confidence that they could move toward the agency's eventual aim of launching twenty-four satellites every year.

In order to achieve this end, it was necessary to adhere as closely as possible to a schedule that involved regular launches and fast turnarounds once a shuttle had returned to Earth. James M. Beggs, NASA's head since 1981, strove relentlessly to make sure that the agency's ambitious goals were met.

Immediately prior to the fateful launch of *Challenger* (flight 51-L) on January 28, 1986, NASA was facing difficulties. Launch schedules that had been set had to be altered when the shuttle *Columbia* (STS 61-C), originally scheduled to be launched on December 18, 1985, could not be launched as planned because of inclement weather. It was finally launched successfully from Florida's Kennedy Space Center (KSC) on January 12, 1986, after several delays. When it was scheduled to return to Earth six days later, after having deployed a communications satellite, performed numerous experiments, and photographed Halley's comet, bad weather prevented its landing at the KSC. It touched down instead at Edwards Air Force Base in California's Mojave Desert and was returned to the KSC atop a specially modified Boeing 747.

Challenger's launch was scheduled for January 22, 1986, but *Columbia*'s late return forced NASA to reschedule that flight to January 26. Bad weather forced another delay to January 27. That launch date was pushed back to January 28 because of a problem with the orbiter's external hatch handle.

Meanwhile, the pressure to launch *Challenger* was building because further exploratory missions were scheduled for May. The Galileo probe had to be launched precisely when Earth was aligned with Jupiter. If the May launch were missed, Galileo could not begin its three-and-a-half-year flight to Jupiter until June, 1987, the next date at which the proper Earth-Jupiter alignment would occur. A launch in March, 1986, was scheduled to put into low-Earth orbit various instruments to observe Halley's comet. A launch scheduled for September was to put the twelve-ton Hubble Space Telescope into Earth orbit. A substantial delay in any launch would necessarily affect other launches that were scheduled.

The morning of January 28 was unusually cold for Florida. It was a clear day, but icicles had formed on *Challenger*'s gantries. The liftoff proceeded as usual and seemed routine as *Challenger* streaked into the sky spewing behind it a characteristic trail of white smoke. Sixty seconds after launch, however, the white cloud that appeared behind *Challenger* was tinged with orange.

The shuttle came crashing down, its crew members dead, including the mission commander, Francis R. "Dick" Scobee, the shuttle's pilot, Michael J. Smith, and S. Christa McAuliffe, the first schoolteacher in space, who had hoped to speak with schoolchildren live from *Challenger* in an effort to stimulate them to give serious thought to studying science. The public had been particularly observant of this mission because of the human interest that McAuliffe's presence lent to it. They witnessed its tragic ending on their television sets. Grief and outrage swept the nation following the accident.

The cause of the accident was attributed to the failure of a rubber seal, called an O-ring, to hold at the joint of the right-hand solid-fueled rocket booster (SRB), where it had been installed. Low temperatures had caused the O-ring to become stiff and to give way. This caused hot rocket exhaust to shoot out of the side of the booster and cut into the base of the hydrogen fuel tank located in the bottom of the External Tank (ET). Its SRB mounting supports burned through, causing the booster to break loose and punch a hole in the oxygen tank at the top of the ET. With the structural integrity of the External Tank gone, the orbiter broke apart under heavy aerodynamic loads.

The *Challenger* accident put future launches on hold while a presidential commission headed by William P. Rogers investigated both the accident and the administration of NASA. No shuttles were launched from January 28, 1986, until September 29, 1988, when *Discovery* (STS-26) was sent aloft to put a communications satellite into geosynchronous orbit. This launch was followed on December 2, 1988, by the launching of *Atlantis* (STS-27), which carried a Department of Defense payload into outer space.

It was not until mid-October, 1989, that Galileo was launched from *Atlantis* (STS-34) on its mission to Jupiter. Ulysses was finally launched from *Discovery* (STS-41) in October, 1990, on its mission to orbit the Sun.

Contributions

The *Challenger* accident brought to light many problems that had plagued NASA. It also helped explain some problems in President Ronald W. Reagan's administration, which was relatively uninformed about the technical intricacies of space ventures. In fact, at the time of the accident, the post of presidential science adviser was vacant.

A president who had limited understanding of space policy now appointed a committee to investigate the accident. At its end, NASA canceled future launches until the committee had presented its findings. This committee was composed mainly of political bureaucrats rather than of the scientists who could best assess the status of the space program following the *Challenger* accident.

Soon reports of some astronauts' fears about the shuttles began to surface. There were serious concerns about the strength of their landing gear and about their braking system. Some critics pointed to weaknesses in the shuttles' main engines. There was an almost universal concern among the astro-

nauts about the lack of an escape system that could be used in case of emergencies, although it was generally conceded that no escape system could have saved the lives of *Challenger*'s crew.

The technicians at NASA learned a great deal from the *Challenger* accident. In the two and a half years during which there were no further launches, every effort was made to address the safety problems that the accident had pinpointed. Because the basic problem in this case was the failure of an O-ring to hold at a joint, the joints on all the rocket boosters were made stronger and an additional O-ring was used on joints to offer greater protection from an O-ring failure.

The Reaction Control Systems rockets were redesigned, as were the auxiliary power units. The landing gear about which many astronauts were deeply concerned was now strengthened, and the braking system was improved substantially. The main engines were enhanced, and the navigation equipment was provided with more backups.

Perhaps the most significant safety improvement was the addition in each orbiter of a parachute escape system that could save crew members if a crash landing seemed inevitable, although it could not save astronauts if a breakup similar to that experienced by *Challenger* should occur. This system permitted the entry hatch to be jettisoned by an explosive charge in case of an emergency. Once the entry hatch was dislodged, a strong metal rod would be installed in the crew compartment to curve out from the hatch opening. Crew members would attach a hook to the rod and slide out and away from the shuttle, before falling freely through the air. It was calculated that they would be at least 9 meters (30 feet) from the space vehicle by the time they reached the end of the rod. Once in free fall, they would deploy their parachutes and drift safely to a landing.

Following the *Challenger* accident, the safety of the astronauts became NASA's highest priority. Whereas economy had been a high priority at NASA during the early 1980's, the agency now eschewed economy in favor of safety. Questions were posed about the need to send humans into outer space if smart machines could gather important data instead. It was also suggested that a crew of seven was too large to assure safety and that smaller crews, between two and four astronauts, could do the job and also would have a better chance of escaping from the vehicle if there were an emergency. Using a smaller flight crew, shuttles could be equipped with escape systems like the ones used during the development of the B-70.

NASA's main task after the accident was to get shuttles into space again as soon as was possible without compromising safety. The agency worked assiduously to get the program back on track and to resume flying as quickly as was prudent.

Concerns were also voiced about the danger of carrying explosive propellants in the shuttles' cargo bay. Liquid hydrogen and liquid oxygen were being used as propellants for the boosters of deep-space probes such as Galileo and Ulysses. Indeed, it now became clear that launching planetary missions from the shuttle severely compromised the safety of the crew.

Context

The space program of the United States involved a spirited competition with the Soviet Union. Because it became a matter of national pride to prevail in the Space Race, the United States was frequently in a dead heat with the Soviets for ascendancy in space.

All was not well at NASA in 1985. When James M. Beggs's deputy, Dr. Hans Mark, left the agency, President Ronald W. Reagan made a political appointment, naming William R. Graham, Jr., to replace Mark. An outraged Beggs ignored Graham, refusing to have any communication with him.

On December 2, 1985, a federal grand jury indicted Beggs on spurious criminal fraud charges that were dismissed by the court two years later. Beggs refused to resign as director of NASA, in-

stead taking a leave of absence during which he came to his office every day. Graham, who had been on the job for only nine days and who had little interest in civilian space activities, was named acting director of NASA when Beggs went on leave.

The agency had been somewhat demoralized because the Reagan administration was not very forthcoming in support of it, although as a matter of public policy, the president had to appear publicly supportive. Funding was a perennial problem during the Reagan administration. Many thought that the president valued impressive displays that brought favorable publicity over the serious scientific work that space travel demanded.

As NASA became increasingly bureaucratized, the pressure to move quickly was constant and corners were cut to keep launches on schedule to whatever extent was possible. The pressure to launch Challenger in late January of 1986 was much greater because of the lineup of other spaceflights that had to be accomplished within a narrow time frame or delayed for more than a year.

Had the pressure to launch been less great, the January 28 launch of Challenger might have been delayed yet another time. Although the air temperature at KSC at the time of launching was near freezing, the day was deceptively clear and bright. The Rogers Commission's report on the accident pointed to significant mismanagement at NASA and a lack of sound judgment in unquestioningly following bureaucratic process and launching Challenger in spite of the cold temperatures.

See also: Asteroid and Comet Exploration; Astronauts and the U.S. Astronaut Program; Cape Canaveral and the Kennedy Space Center; Escape from Piloted Spacecraft; Funding Procedures of Space Programs; Space Shuttle; Space Shuttle Mission STS 61-C; Space Shuttle Mission STS 51-L; Space Shuttle Mission STS-26; Space Shuttle Flights, 1989.

Further Reading

Burrows, William E. *Exploring Space: Voyages in the Solar System and Beyond.* New York: Random House, 1990. Burrows addresses the *Challenger* accident extensively and directly, analyzing intelligently its effect on the space program and on NASA. In his chapter "Galileo: The Perils of Pauline," he reviews the agonizing attempts to get the Galileo space probe into orbit, a project that extended over a dozen years and was plagued by lack of sufficient funding. The *Challenger* accident robbed Galileo of its projected May, 1986, launch, necessitating a wait of more than three years before Jupiter was in a proper position for the launch to take place.

Damon, Thomas D. *Introduction to Space: The Science of Spaceflight.* 2d ed. Malabar, Fla.: Krieger Publishing Company, 1995. Damon's tables, which list all the shuttle flights from 1981 to 1994, are particularly useful. They provide in succinct form information about the ends that each flight was meant to accomplish. Damon has a comprehensive view of the space program and of NASA. He tells his story well. His book is rich in illustrative material, including photographs and graphic representations of various elements of the shuttles.

Harland, David M. *The Space Shuttle: Roles, Missions, and Accomplishments.* Hoboken, N.J.: John Wiley, 1998. *The Space Shuttle* is written thematically, rather than purely chronologically. Topics include shuttle operations and payloads, weightlessness, materials processing, exploration, Spacelabs and free-flyers, and the shuttle's role in the International Space Station.

Harrington, Philip S. *The Space Shuttle: A Photographic History.* San Francisco, Calif.: Brown Trout, 2003. With one hundred full-color photographs by Roger Ressmeyer and others and with text by popular astronomy writer Phil Harrington, this book tells the story of the space shuttle program from 1972 to 2003. Its beautiful photographs allow the general reader to survey the history of the space shuttle program and be uplifted by the pioneering spirit of one of humanity's grandest enterprises.

Heppenheimer, T. A. *Countdown: A History of Space Flight.* New York: John Wiley, 1997. A detailed historical narrative of the human conquest of space. Heppenheimer traces the development of piloted flight through the military rocketry programs of the era preceding World War II. Covers both the American and the Soviet attempts to place vehicles, spacecraft, and humans into the hostile environment of space. More than a dozen pages are devoted to bibliographic references.

Jenkins, Dennis R. *Space Shuttle: The History of the National Space Transportation System: The First 100 Missions.* Stillwater, Minn.: Voyageur Press, 2001. This is a concisely written technical reference account of the space shuttle and its ancestors, the aerodynamic lifting bodies. It details some of the advantages and inherent disadvantages of using a reusable space vehicle. Each of the vehicles is illustrated by line drawings. The book follows the space shuttle from its original concepts and briefly chronicles its first one hundred flights.

Murray, Bruce. *Journey into Space: The First Three Decades of Space Exploration.* New York: W. W. Norton, 1989. Murray discusses the politics of the space program better than any other writer to date. He shows how an astounding lack of technical knowledge during Ronald W. Reagan's presidency presented clear and present dangers to astronauts. Reagan isolated himself from the scientists who might have helped him understand the benefits to be gained from developing a full-blown, well-financed space program.

National Aeronautics and Space Administration. *Space Shuttle Mission Press Kits.* http://www-pao.ksc.nasa.gov/kscpao/presskit/presskit.htm. Provides detailed preflight information about each of the space shuttle missions. Accessed March, 2005.

Shayler, David J. *Disasters and Accidents in Manned Spaceflight.* Chichester, England: Springer-Praxis, 2000. The author examines the challenges that face all crews as they prepare and execute their missions. The book covers all aspects of crewed spaceflight—training, launch to space, survival in space, and return from space—and follows with a series of case histories that chronicle the major incidents in each of those categories over the past forty years.

Stott, Carole, and Steve Gorton. *Space Exploration.* New York: Alfred A. Knopf, 1997. The authors present a useful review of the gap in the space program that followed the *Challenger* accident. They review the changes in the United States space program and in NASA following the loss of *Challenger,* demonstrating how much was learned from that tragic occurrence. The book is well written and excellently documented.

Vaughan, Diane. *The Challenger Launch Decision: Risky Technology, Culture, and Deviance at NASA.* Chicago: University of Chicago Press, 1996. Vaughan presents the most comprehensive account to date of the unsettled situation at NASA immediately prior to the ill-fated *Challenger* launch in January, 1986. She explores the politics surrounding the decision to launch when conditions clearly indicated that a postponement, although not politically expedient, was appropriate.

Vogt, Gregory. *The Space Shuttle: Missions in Space.* Brookfield, Conn.: Millbrook Press, 1991. This thin volume packs a great deal of information into a limited space. The writing is clear, the information accurate. Vogt's target audience is adolescents, but his book is sufficiently important to warrant reading by adults interested in the topic. The author is especially informative in dealing with improvements made in shuttles after the loss of *Challenger.*

R. Baird Shuman

Space Shuttle Mission STS-26

Date: September 29 to October 3, 1988
Type of mission: Piloted spaceflight

Space Transportation System 26 (STS-26) was the first shuttle flight following the destruction of the space shuttle Challenger *thirty-two months before. STS-26 successfully returned the United States' piloted space program to an active flight status. During the flight, a communications satellite vital to the U.S. space program was launched into geosynchronous orbit.*

Key Figures

Frederick H. "Rick" Hauck (b. 1941), STS-26 commander
Richard O. Covey (b. 1946), STS-26 pilot
George D. "Pinky" Nelson (b. 1950), STS-26 mission specialist
David C. Hilmers (b. 1950), STS-26 mission specialist
John M. "Mike" Lounge (b. 1946), STS-26 mission specialist

Summary of the Mission

Space shuttle mission 26 was the first U.S. piloted spaceflight following the loss of the space shuttle *Challenger* and its seven-member crew on January 28, 1986. Because of the protracted recovery time of thirty-two months, STS-26 became widely regarded as America's return to space. In fact, the National Aeronautics and Space Administration (NASA) officially designated the mission "Return to Flight."

Not since the fatal launch pad fire of Apollo 1 on January 27, 1967 (which killed astronauts Virgil I. "Gus" Grissom, Edward White, and Roger B. Chaffee), had there been such an extensive reworking of an American spacecraft. Following the loss of *Challenger*, a thirteen-member investigative commission was appointed by President Ronald W. Reagan and headed by former Secretary of State William P. Rogers. Called the Rogers Commission, the panel issued its report to the president on June 9, 1986. More than four hundred changes in a $2.4-billion program were advocated to improve the shuttle and help ensure the safety of future flights. These changes included a complete redesign of the O-ring system connecting the solid-fueled rocket booster segments, the system that was blamed for the *Challenger* accident.

Problems encountered after the initial launch date was set for February, 1988, caused a series of delays. Space planners wanted to be certain that the tests of the solid-fueled rocket boosters were completed successfully. Finally, on July 4, 1988, after the redesigned joints had been approved, the assembled vehicle was rolled from the vehicle assembly building at the Kennedy Space Center (KSC) to Launch Complex 39B. *Discovery* was poised for launch.

Even with the shuttle positioned on the pad, NASA officials waited before setting another launch date, ostensibly to evaluate the assembled system further and to receive the results from yet more booster tests. The media proved especially critical of the space agency in the wake of the *Challenger* accident and these frequently shifting launch dates. After the final flight readiness review panel met, the launch date was set for Thursday, September 29, 1988.

The media gave as much attention to the upcoming launch as to any other space launch in American history, including the Apollo Moon flights. NASA had instituted a new launch management system to prevent a recurrence of what the Rogers Commission had called "a flawed decision-making process" for shuttle launches. Heading a launch management team was active Astronaut Robert L. Crippen, Jr., pilot of the first shuttle, who would make the final launch decision. In the glare of world attention, NASA and the U.S. space program could ill afford a problematic launch or mission.

The countdown proceeded smoothly on September 29. Many predictions prior to the launch held that the new and untested safety and management systems were so bulky that there would likely be days of delays, holds, and reconsiderations before STS-26 could be launched. These predictions proved unfounded. After a delay of only one hour and thirty-eight minutes, caused by high-altitude winds that were lighter than predicted, *Discovery* was launched from Pad 39B at 15:37:00.009 Coordinated Universal Time (UTC, or 11:37 A.M. eastern daylight time).

The launch was normal in every respect. The redesigned solid-fueled rocket boosters were jettisoned on schedule and dropped by parachute into the ocean off the coast of Florida. Subsequent inspection proved that they had come through the flight in pristine condition. The shuttle's three main engines performed well enough to preclude an initial burn of the orbital maneuvering system (OMS) engines to achieve stable orbit. Later, an OMS burn

was performed in order to place *Discovery* in a 348-kilometer orbit above Earth.

Aside from the principal function of proving viability of the extensively reworked shuttle system, the four-day mission was to include eleven scientific experiments and the deployment of a vital NASA communications satellite. The satellite, a $100 million Tracking and Data-Relay Satellite (TDRS), was to replace the one lost on *Challenger*. With a mass of 2,268 kilograms, this satellite, the TDRS-C, was one of the largest communications satellites ever launched; it was so massive that it could fit only in the shuttle's payload bay. The

Space shuttle mission STS-26, launching on September 29, 1988: The shuttle returns to flight following the Challenger *tragedy of January, 1986.* (NASA)

TDRS was successfully deployed by a spring-loaded platform from *Discovery*'s payload bay six hours into the mission. After the astronauts maneuvered the orbiter to a safe 72 kilometers away, TDRS-3 (as it was now called) automatically boosted itself 35,800 kilometers above Earth into geosynchronous orbit. (Satellites in geosynchronous orbit travel at the same speed as Earth rotates and thus appear to remain at a fixed point in the sky.)

The purpose of the TDRS system is to provide a multisatellite communications constellation for NASA missions. The TDRS system enables nearly constant communications between Earth and other orbiting spacecraft such as the orbiter fleet, the International Space Station, and the Hubble Space Telescope.

One of the first problems the crew encountered on STS-26 was the partial failure of a cooling device called a flash evaporator, which is used to cool the crew cabin and the equipment. During liftoff, the evaporator became frozen with ice and operated at less than its optimal capacity. To assist in melting the ice, ground engineers allowed the temperature in *Discovery* to rise to 29° Celsius for the first two days of the mission. The evaporator would malfunction again briefly during reentry, but it never endangered the crew or *Discovery*'s systems.

The Ku-band antenna, used to communicate with Earth-based stations, malfunctioned on the first day of the mission. The antenna failed to align itself properly for broadcast of signals to Earth, and it wobbled erratically on its mount. On the second day, the malfunctioning antenna was finally stowed back in the shuttle's payload bay; consequently, the flow of communications between *Discovery* and the ground for the remainder of the flight was reduced.

Eleven scientific experiments were carried on board the shuttle. Protein crystal growth experiments were conducted for use in the development of complex protein molecules that could be utilized in medicines and other chemical solutions. One of the crystals being examined was critical to the study of the acquired immunodeficiency syndrome (AIDS) virus and how it replicates. Two experiments investigated medicinal properties unique to weightless space. Two other experiments were designed to investigate molten metal resolidification and crystallization for development of stronger metal alloys. Also on board was an experiment involving crystal growth on a semipermeable membrane (the results could ultimately help reduce medical X-radiation doses). Yet another was designed to produce thin films of organic material and study their properties.

Two meteorological investigations were conducted on STS-26. One photographed lightning and the other the glow of Earth's horizon near sunrise and sunset. This last experiment was remarkably similar to one of the first scientific investigations ever conducted by humans in space.

On October 2, 1988, *Discovery*'s crew fired the OMS engines for two minutes and fifty seconds in the deorbital maneuver. Forty minutes later, *Discovery* reentered Earth's atmosphere. Twenty-one minutes after reentry, the shuttle and its crew landed safely at Edwards Air Force Base in the high desert of Southern California at 15:37:11.468 UTC (11:37 A.M. eastern daylight time). The duration of mission 26 had been four days and one hour.

Contributions

The primary goal of mission 26 was to prove the safety of the redesigned shuttle system and return American piloted spaceflight to an active status. The successful return of the crew with all major mission objectives met fulfilled that goal. Precise details of the success of each redesigned element would take many weeks of detailed analysis, but all major redesigned components were proved sound.

The newly designed O-ring system appeared to perform flawlessly. According to engineers who examined the system after it was recovered from the ocean following the launch, the boosters appeared in better condition than any recovered from any other mission.

The three main engines had undergone forty significant design changes in a $100 million program that included strengthening of the engine's high-speed pump components. These compo-

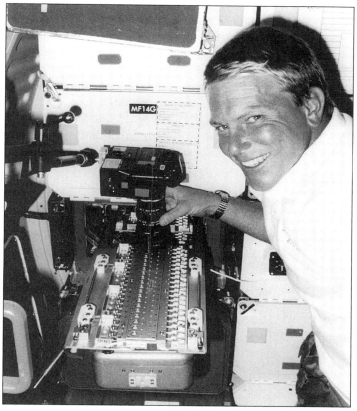

Mission Specialist George D. Nelson photographs the results of a Protein Crystal Growth (PCG) experiment on STS-26. (NASA)

sary redesigns gave NASA the confidence to incorporate future changes as they may be required in the shuttle program. One manifestation of this newfound capability was the redesign of a drogue chute braking system to be installed on all orbiters to absorb touchdown energies further and enhance the safety of landings.

Although overshadowed by the mission's main goal of returning the U.S. space program to flight status, the scientific experiments aboard STS-26 represented one of the primary reasons the space shuttle program was first conceived. The extensive study of protein crystal growth provided data for vital chemical studies that may have far-reaching implications. Crystal growth on membranes may one day reduce dosages of medical X-radiation, which account for the single largest source of ionizing radiation exposure for the American public. The experiment designed to study the AIDS virus also proved valuable. Information gathered from the experiment may lead to the development of medications.

nents also appeared to have performed well, although one engine apparently experienced an oxidizer leak near the end of the main combustion chamber.

The orbiter's tires, brakes, and nose-wheel steering system had all undergone extensive modifications. These systems absorb the monumental energies of landing when the shuttle orbiter touches down at speeds of more than 350 kilometers per hour. *Discovery*'s landing speed was said to have been 381 kilometers per hour. The vehicle required only 2,337 meters to stop, indicating a very good braking (energy absorption) efficiency when compared with that of past missions.

The performance of these critical systems indicates the viability of these redesigns. The ability of space planners to identify and produce the neces-

Most important, STS-26 proved the ability of the space agency to recover from a catastrophe and unprecedented tragedy in space to fly again. The system emerged from the flight of 26 fundamentally changed, with a better spacecraft and a baseline of knowledge that, unfortunately, might not have been obtained under other circumstances. From those hard-won lessons, the space program gained a clearer vision that would guide the United States into the next century.

Context

Mission STS-26 flew as the twenty-sixth space shuttle mission, the fifty-seventh United States piloted mission, and the twenty-second piloted mission of the operational shuttle system. During the thirty-two-month period of rebuilding, the Rogers

Commission and the world media had heavily criticized NASA for a lack of proper management and a poor decision-making process. The space agency worked ceaselessly during this long hiatus to regain the confidence of the American public, the confidence it had enjoyed since its formation by Congress in 1958.

NASA took various measures to help ensure a successful mission. The agency reviewed the cause of the *Challenger* accident and corrected the flawed O-ring system. It reviewed all shuttle systems, identifying and modifying any that could cause future problems.

Space planners at NASA also restructured the agency's management system, strengthening safety, quality, and reliability by making improvements in training and personnel. Finally, NASA reduced the planned flight rate and removed altogether the somewhat artificial dependence on civil payload manifests that had resulted in an overloaded system.

The crew selected to fly the mission was an all-veteran crew; NASA wanted to maximize the experience available on this important return to space. Frederick H. "Rick" Hauck and George D. "Pinky" Nelson had both flown twice before, and Richard O. Covey, David C. Hilmers, and John M. "Mike" Lounge had flown once before—making the STS-26 crew one of the most experienced ever to fly on a shuttle mission.

The space agency and the nation approached the launch with an apprehension that attracted much attention. The shuttle's systems were tested and retested as though *Discovery* had never flown before. The orbiter sat on the launch pad undergoing tests some thirteen weeks before launch, the longest launch pad preparation period since the earliest shuttle launches. Clearly, the launch was approached with maximum caution.

Just before the launch, crowds gathered at Kennedy Space Center. The atmosphere was reminiscent of that preceding the early Apollo Moon launches. Unlike the previous launch of the tragic mission STS 51-L, which was attended by relatively few and virtually dismissed by the press as "routine," the mission of 26 was anything but ordinary.

More than at any other time in history, the fate of the Western world's future in space was on the line. If STS-26 had failed, the piloted space effort would have suffered another catastrophic setback, perhaps giving way to critics' calls for the abandonment of the piloted program in favor of mechanical, robotic exploration. With the string of ongoing successes by the Soviets, another U.S. failure would have meant abdicating superiority in space exploration. The United States had only three shuttle orbiters left in its active fleet. The loss of another of these multibillion-dollar national resources would have been disastrous.

The success of STS-26 held an implicit significance for continuing piloted spaceflight. Without question, it was the mission that had to succeed. Succeed it did, brilliantly restoring the United States to its competitive position in space. The tragedy of mission 51-L and all it entailed had been replaced by triumph. All space projects were suddenly possible again: the Hubble Space Telescope, Magellan to Venus, Galileo to Jupiter, and the International Space Station. The United States had regained its foothold in space.

As of 2005, there were ten TDRS satellites in the constellation. TDRS-3 was in orbit at 84.8° east longitude. TDRS-8 and TDRS-10 were the primary satellites, with TDRS-9 serving as a spare.

See also: Funding Procedures of Space Programs; Space Shuttle; Space Shuttle: Microgravity Laboratories and Payloads; Space Shuttle Mission STS 51-L; Space Shuttle Activity, 1986-1988; Spacelab Program; Telecommunications Satellites: Military; Telecommunications Satellites: Private and Commercial; Tracking and Data-Relay Communications Satellites.

Further Reading

Baker, David. *The History of Manned Space Flight*. New York: Crown, 1982. This book offers a precise chronology of the history of piloted spaceflight, heavily oriented toward the

United States' effort. It is a detailed work that chronicles the U.S. piloted space effort from its beginning to the start of the space shuttle program. Includes analyses of U.S. spaceflights, beginning with the Mercury missions of the 1960's.

DeWaard, E. John, and Nancy DeWaard. *History of NASA: America's Voyage to the Stars.* New York: Exeter Books, 1984. A colorful pictorial essay on the history of NASA. It does not go into great detail on every flight but does include many color photographs of the various U.S. piloted programs.

Harland, David M. *The Space Shuttle: Roles, Missions, and Accomplishments.* Hoboken, N.J.: John Wiley, 1998. *The Space Shuttle* is written thematically, rather than purely chronologically. Topics include shuttle operations and payloads, weightlessness, materials processing, exploration, Spacelabs and free-flyers, and the shuttle's role in the International Space Station.

Harrington, Philip S. *The Space Shuttle: A Photographic History.* San Francisco, Calif.: Brown Trout, 2003. With one hundred full-color photographs by Roger Ressmeyer and others and with text by popular astronomy writer Phil Harrington, this book tells the story of the space shuttle program from 1972 to 2003. Its beautiful photographs allow the general reader to survey the history of the space shuttle program and be uplifted by the pioneering spirit of one of humanity's grandest enterprises.

Jenkins, Dennis R. *Space Shuttle: The History of the National Space Transportation System: The First 100 Missions.* Stillwater, Minn.: Voyageur Press, 2001. This is a concisely written technical reference account of the space shuttle and its ancestors, the aerodynamic lifting bodies. It details some of the advantages and inherent disadvantages of using a reusable space vehicle. Each of the vehicles is illustrated by line drawings. The book follows the space shuttle from its original concepts and briefly chronicles its first one hundred flights.

Shayler, David J. *Disasters and Accidents in Manned Spaceflight.* Chichester, England: Springer-Praxis, 2000. The author examines the challenges that face all crews as they prepare and execute their missions. The book covers all aspects of crewed spaceflight—training, launch to space, survival in space, and return from space—and follows with a series of case histories that chronicle the major incidents in each of those categories over the past forty years.

Slayton, Donald K., with Michael Cassutt. *Deke! U.S. Manned Space: From Mercury to the Shuttle.* New York: Forge, 1995. This is the autobiography of the last of the Mercury astronauts to fly in space. After being grounded from flying in Project Mercury for what turned out to be a minor heart murmur, Slayton was appointed head of the Astronaut Office. Later, he commanded the Apollo-Soyuz flight in 1975. He stayed with NASA through the STS-2 flight. Although he was no longer an active participant in NASA's activities, he gives some insight into the behind-the-scenes adventures of the shuttle program.

Dennis Chamberland

Space Shuttle Flights, 1989

Date: March 13 to November 27, 1989
Type of mission: Piloted spaceflight

Space shuttle crews orbited the Earth for more than four weeks during 1989 as NASA successfully launched and landed five space shuttle missions. Highlights of the 1989 space shuttle flights included the boosting of the Magellan/Venus and the Galileo/Jupiter spacecraft on their respective missions to explore the solar system.

Key Figures

James C. Fletcher (1919-1991), Space Shuttle Program administrator
Dale DeHaven Myers (b. 1922), Space Shuttle Program deputy administrator
Michael L. Coats (b. 1946), STS-29 commander
John E. Blaha (b. 1942), STS-29 and STS-33 pilot
Robert C. Springer (b. 1942), STS-29 mission specialist
James F. Buchli (b. 1945), STS-29 mission specialist
James P. Bagian (b. 1952), STS-29 mission specialist
David M. Walker (1944-2001), STS-30 commander
Ronald J. Grabe (b. 1945), STS-30 pilot
Mark C. Lee (b. 1952), STS-30 mission specialist
Norman E. Thagard (b. 1943), STS-30 mission specialist
Mary L. Cleave (b. 1947), STS-30 mission specialist
Brewster H. Shaw, Jr. (b. 1945), STS-28 commander
Richard N. "Dick" Richards (b. 1946), STS-28 pilot
James C. Adamson (b. 1946), STS-28 mission specialist
David C. Leestma (b. 1949), STS-28 mission specialist
Mark N. Brown (b. 1951), STS-28 mission specialist
Donald E. Williams (b. 1942), STS-34 commander
Michael J. McCulley (b. 1943), STS-34 pilot
Shannon W. Lucid (b. 1943), STS-34 mission specialist
Franklin R. Chang-Díaz (b. 1950), STS-34 mission specialist
Ellen S. Baker (b. 1953), STS-34 mission specialist
Frederick D. Gregory (b. 1941), STS-33 commander
Manley Lanier "Sonny" Carter, Jr. (1947-1991), STS-33 mission specialist
F. Story Musgrave (b. 1935), STS-33 mission specialist
Kathryn C. Thornton (b. 1952), STS-33 mission specialist

Summary of the Missions

After the successful conclusion of the STS-26 mission (September, 1988), the orbiter *Discovery* was returned from the Dryden Research Facility of the Kennedy Space Center for postflight inspection and reconstruction, a process that included the replacement of its three main engines. After

1444

the turn-around activities were completed, *Discovery* was transferred from the Orbiter Processing Facility (OPF) to the Vehicle Assembly Building (VAB), where it was mated to an expendable External Tank (ET) and two solid-fueled rocket boosters (SRBs). On February 3, the assembled space shuttle orbiter, mounted on the 47-meter-high ET and 45.46-meter SRBs, was rolled out of the VAB aboard a mobile launcher platform for the 7-kilometer journey to Launch Complex 39B. The payload for Mission STS-29 was installed, and the launch preparation countdown proceeded. On March 13, 1989, after a 1 hour, 50 minute delay owing to ground fog and upper winds, the 116,282-kilogram (256,357-pound) *Discovery* and cargo were launched at 9:57 A.M. eastern standard time.

The primary objective of STS-29 was to place a Tracking and Data-Relay Satellite (TDRS-D; renamed TDRS-4 after deployment) into orbit. The TDRS-4, which was attached to a solid-propellant Boeing/U.S. Air Force Inertial Upper Stage (IUS), was deployed about six hours after liftoff. The IUS boosters were then fired, sending the TDRS communication satellite into its proper geosynchronous orbit. TDRS-4 was the third of its kind to be deployed from a shuttle. The successful deployment of TDRS-4 completed the constellation of on-orbit satellites for NASA's advanced space communication system.

During the remainder of this five-day mission, the crew—commanded by Michael L. Coats—conducted chromosome and plant cell division experiments, studied the growth of protein crystals, performed two student-involvement program experiments, and tested a space station heat pipe advanced radiator. After completing seventy-nine orbits and traveling approximately 3.2 million kilometers, the crew returned on March 18 with beautiful pictures of the Earth that were photographed with a handheld IMAX camera. The IMAX camera system consists of the camera, magazine, lens, power filter, and power cable. A tape recorder and two microphones are used for audio recording.

The primary objective of the second mission of 1989, STS-30, was to deploy the Magellan Venus-exploration spacecraft. This mission received significant national media coverage, not only because the Magellan Project was the first planetary science mission attempted since 1978 but also because it was to be the first time that an interplanetary probe was to be deployed from a space shuttle. The proposed April 28 launch was scrubbed just 31 seconds before liftoff (T minus 31 seconds) owing to a problem with the liquid hydrogen recirculation pump. After repairs were made, the launch date was rescheduled for May 4. Cloud cover and high winds almost forced another postponement. Liftoff took place, however, at 18:46:59.011 Coordinated Universal Time (UTC, or 2:47 P.M. eastern daylight time), about 5 minutes before the allowed 64-minute window opening expired. This was the fourth flight for *Atlantis*, an orbiter that flew its inaugural flight in 1985.

The five-member crew on this historic mission included Commander David M. Walker, Pilot Ronald J. Grabe, and Mission Specialists Norman E. Thagard, Mary L. Cleave, and Mark C. Lee. Six hours and 14 minutes into flight, the crew deployed from the shuttle's cargo bay the Magellan/ Venus radar mapper spacecraft along with its solid-fuel IUS booster. Shortly thereafter, the IUS boosters were fired, and the Magellan spacecraft began its fifteen-month journey around the Sun and to Venus. Before the completion of the Magellan mission in October, 1994, the Magellan space probe would radar map 98 percent of the planet's surface, and collect and transmit to NASA scientists high-resolution gravity data of Venus.

The four-day STS-30 mission produced another dramatic, yet unplanned moment. While in flight, one of the five General Purpose Computers (GPC) failed and had to be replaced with an onboard computer hardware spare. This was the first time a GPC was switched while a shuttle spacecraft was in orbit. Fortunately, the new computer was fully functional. After completing its sixty-four planned orbits, the *Atlantis* landed safely at Edwards Air Force Base in California on May 8.

On August 8, the first of two 1989 classified Department of Defense shuttle missions was launched

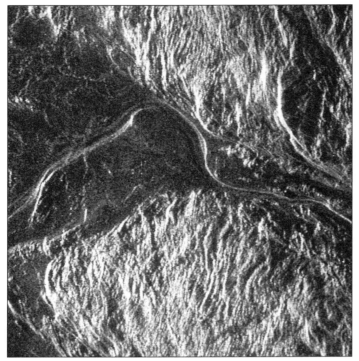

The Magellan probe radar mapped 98 percent of Venus's surface. This photo shows a sinuous volcanic channel. (NASA CORE/ Lorain County JVS)

from Pad 39B at the Kennedy Space Center. This was the first mission in more than three and one-half years for the space shuttle *Columbia*, America's original orbiter, which flew its first flight in 1981. This mission, STS-28, lasted five days, 1 hour, and landed without incident on August 13. Three months later, the second classified mission of 1989 (and the fifth shuttle mission dedicated to the Department of Defense) was scheduled for the orbiter *Discovery*. The original November 20 launch date for this mission was delayed in order to allow time to replace suspect integrated electronics assemblies on the twin solid rocket boosters. At 00:23:29.986 UTC on November 23 (7:23 P.M. eastern standard time on November 22) the thirty-second U.S. space shuttle mission was launched for orbit, mission STS-33. This was only the third shuttle mission launched at night, and the first night launching since the return of shuttle flights

following the *Challenger* accident in 1986. The five-day mission for the Department of Defense completed seventy-nine orbits, traveled 3.4 million kilometers, and ended without complication on November 27.

Perhaps the most dramatic of the 1989 shuttle flights was the STS-34 mission. This mission was twice rescheduled, once owing to faulty main engine controller on the number two main engine, and the second time owing to weather conditions that scientists believed would prevent a return-to-launch-site landing at the Kennedy Space Center's Shuttle Landing Facility. Following these delays, the five-member crew—Donald E. Williams, commander; Michael J. McCulley, pilot; and Franklin R. Chang-Díaz, Shannon W. Lucid, and Ellen S. Baker, mission specialists—was launched into orbit aboard the *Atlantis* at 16:53:40.020 UTC (12:54 P.M. eastern daylight time) on October 18. The primary task of the crew was to deploy the Galileo/Jupiter spacecraft and its attached IUS booster. This objective was accomplished six and one-half hours into flight. The IUS motors attached to the planetary orbiter and probe were fired, and Galileo was propelled on an interesting and historic trajectory to Jupiter.

Originally, the Galileo spacecraft had been designed for a direct flight to Jupiter, which would take about two and one-half years. Changes in the launch system after the *Challenger* accident, however, prevented this direct trajectory and forced engineers to design a new interplanetary flight path using several gravity-assisted swing-bys, once past Venus and twice around Earth. This Venus-Earth-Earth Gravity Assist trajectory (known as VEEGA) would send the Galileo orbiter and probe on a six-year, 3.7-billion-kilometer voyage to the mysterious Jupiter. The eight objectives of the sophisticated and ambitious Galileo Project include:

(1) to determine the temperature and pressure structure of Jupiter's atmosphere; (2) to determine the chemical composition of Jupiter; (3) to determine how many cloud layers exist, to find their location, and to characterize the cloud particles as to size and number density; (4) to measure the amount of helium relative to hydrogen on Jupiter to high accuracy; (5) to measure the winds in Jupiter's atmosphere and to determine how deep in the atmosphere the winds exist; (6) to measure how sunlight and energy coming from the deep interior are distributed in Jupiter's atmosphere; (7) to detect lightning if it occurs, to measure how energetic it is, and to observe the frequency of occurrence; and (8) to measure the characteristics of energetic protons and electrons trapped in Jupiter's magnetic field within a few Jovian radii from the planet.

After deploying the Galileo spacecraft, the STS-34 mission crew spent the remaining four days of their flight operating secondary payloads, conducting experiments, and taking IMAX photographs of the Earth and space. This flight, which set in motion the historic Galileo mission, ended with the successful landing of *Atlantis* on October 23 at Edwards Air Force Base.

A sixth mission, STS-32, was originally scheduled for 1989. The anticipated December 18, 1989, launch date for the mission, however, had to be postponed in order to allow time to complete and verify modifications to Pad A, a launch pad that had not been used since January, 1986, for the STS 61-C mission. STS-32 ultimately was launched on January 9, 1990.

Contributions

In 1989, for the first time, NASA used shuttle orbiters to deploy interplanetary spacecraft designed to explore our solar system. Mission STS-30 launched Magellan on a trajectory to Venus, the planet often called "the Earth's sister planet" because of its similar size and distance from the sun. Between August, 1990, and October, 1994, the Magellan spacecraft orbited Venus. Because a dense, opaque atmosphere clouds Venus, conventional

Liftoff of the STS-34 shuttle Atlantis *carrying the Galileo probe, destined for Jupiter.* (NASA CORE/Lorain County JVS)

optical cameras could not be used to image its surface. The Magellan, however, used a sophisticated imaging radar that provided scientists with the most highly detailed maps of Venus ever captured. It also made global maps of Venus's gravity field. Craters depicted in the radar images tell scientists that Venus's surface is relatively young. Although Venus, like the Earth, was formed about 4.6 billion years ago, its surface was re-formed by widespread volcanic eruptions only about 500 million years ago. Venus's present harsh environment has persisted since this resurfacing. Nothing was found to suggest the presence of oceans or lakes at any time in the planet's history. Nor does Venus appear to have plate tectonics—movements of crustal masses that result in earthquakes and continental drift. Unlike the Earth, Venus seems to lack an "asthenosphere," a buffer layer between the planet's mantle and crust. As a result, gravity fields on Venus are more affected by surface topography than they are on the Earth.

Mission STS-34, like STS-30, contributed to the success of a NASA robotic planetary mission. The Galileo orbiter and probe, launched aboard space shuttle *Atlantis* on October 18, 1989, carried a total of sixteen scientific instruments. Knowledge gained and to be gained from this mission is voluminous. While en route, Galileo flew within 110,000 kilometers of the Moon, obtaining multispectral lunar images and data that will be useful in comparing Earth's Moon with the Jovian satellites. Data from this lunar flyby suggest that the Moon has been more volcanically active and that the far side of the Moon has a thicker crust than researchers previously thought. Galileo also became the first spacecraft to fly closely by two asteroids, Gaspra and Ida. Data from these encounters confirmed and photographed a small moon, later named Dactyl, orbiting around Ida. In July, 1994, Galileo was the only observer in position to obtain images of the impact of fragments from Comet Shoemaker-Levy 9 on the far side of Jupiter. In December, 1995, Galileo became the first spacecraft to enter orbit around one of the outer planets of the solar system. On December 7, 1995, the Galileo probe successfully entered the atmosphere of Jupiter. A radio link between the probe and the orbiter lasted for about fifty-seven minutes during the probe's descent. Data sent from the probe to the orbiter were stored in the orbiter's computer memory and its tape recorder. These data were transmitted back to the Earth in the period from January, 1996, through May, 1996. The orbiter then began an intensive study from orbit of Jupiter's moons, magnetic field, radiation belts, and atmosphere. The Galileo spacecraft's fourteen-year odyssey came to an end on September 21, 2003, when the spacecraft passed into Jupiter's shadow, then disintegrated in the planet's dense atmosphere at 18:57 UTC. The spacecraft was purposely put on a collision course with Jupiter because the onboard propellant was nearly depleted and to eliminate any chance of an unwanted impact between the spacecraft and Jupiter's moon Europa, which Galileo discovered is likely to have a subsurface ocean.

Context

Two of the five completed missions of 1989 were classified Department of Defense flights, and little information is publicly available on these missions. Of the three remaining missions, however, two were highly publicized historic flights that initiated a new age of planetary space exploration. The Magellan and Galileo missions, both launched in 1989 from the shuttle orbiter *Atlantis*, have provided and will continue to provide scientists with a greater understanding of the composition and origins of the solar system. Galileo's probe of Jupiter's physical and orbital properties, in particular, offers spectacular opportunities to unlock the mysteries of the cosmos, because Jupiter and its four largest moons, which range in size from the diameter of the Earth's Moon to the size of the planet Mercury, are in some ways analogous to a mini-solar system. By observing this miniature solar system, Galileo transmitted new clues about how the Sun and the planets formed and about how they continue to interact and evolve.

Before 1989, the space shuttle—which takes off like a rocket, operates in orbit as a spacecraft, lands on Earth like an airplane, and then is refitted to fly again, perhaps for as many as eighty missions—had proven its effectiveness as a deployer and retriever of satellites and as a platform laboratory for conducting space research. In 1989, Missions STS-30 and STS-34 demonstrated the utility of the shuttle in another way. By serving as a base for the launching of the Magellan and Galileo space probes, the piloted space shuttle *Atlantis* helped inaugurate a new era of robotic interplanetary exploration.

As of 2005, there were ten TDRS satellites in the constellation. TDRS-4 is at 41.5° west longitude. TDRS-8 and TDRS-10 are the primary satellites, with TDRS-9 serving as a spare.

See also: Extravehicular Activity; Galileo: Jupiter; Get-Away Special Experiments; Magellan: Venus; Planetary Exploration; Space Shuttle Activity, 1986-1988; Space Shuttle Flights, 1990; Telecommunications Satellites: Military; Tracking and Data-Relay Communications Satellites.

Further Reading

Embury, Barbara, with Thomas D. Crouch. *The Dream Is Alive.* New York: Harper and Row, 1990. An interesting account of the shuttle program intended for juvenile readers.

Harland, David M. *The Space Shuttle: Roles, Missions, and Accomplishments.* Hoboken, N.J.: John Wiley, 1998. *The Space Shuttle* is written thematically, rather than purely chronologically. Topics include shuttle operations and payloads, weightlessness, materials processing, exploration, Spacelabs and free-flyers, and the shuttle's role in the International Space Station.

Harrington, Philip S. *The Space Shuttle: A Photographic History.* San Francisco, Calif.: Brown Trout, 2003. With one hundred full-color photographs by Roger Ressmeyer and others and with text by popular astronomy writer Phil Harrington, this book tells the story of the space shuttle program from 1972 to 2003. Its beautiful photographs allow the general reader to survey the history of the space shuttle program and be uplifted by the pioneering spirit of one of humanity's grandest enterprises.

Jenkins, Dennis R. *Space Shuttle: The History of the National Space Transportation System: The First 100 Missions.* Stillwater, Minn.: Voyageur Press, 2001. This is a concisely written technical reference account of the space shuttle and its ancestors, the aerodynamic lifting bodies. It details some of the advantages and inherent disadvantages of using a reusable space vehicle. Each of the vehicles is illustrated by line drawings. The book follows the space shuttle from its original concepts and briefly chronicles its first one hundred flights.

Lewis, Richard S. *The Voyages of Columbia: The First True Spaceship.* New York: Columbia University Press, 1984. An excellent description of space shuttle *Columbia* and its early missions. This early work, however, does not cover the *Columbia* missions of 1989.

National Aeronautics and Space Administration. *NASA Space Shuttle Mission STS-29 Press Kit.* Washington, D.C.: Author, 1989. A thirty-page NASA press release detailing, among other things, the objectives, countdown milestones, major activities, and landing and postlanding operations of the twenty-eighth space shuttle mission. This release, available in most research libraries on microfilm, is also available on the Internet at location http://www.ksc.nasa.gov/shuttle missions/sts-29-press-kit.txt. For similar descriptions of the two other nonclassified missions of 1989, see NASA space shuttle mission STS-30 press kit (April, 1989) and NASA space shuttle mission STS-34 press kit (October, 1989). Other pertinent information concerning 1989 shuttle flights can be accessed on the Internet through the NASA space shuttle launches homepage located at http://www.ksc.nasa.gov/shuttle/missions/missions.html. Accessed March, 2005.

_____. *Space Shuttle Mission Press Kits.* http://www-pao.ksc.nasa.gov/kscpao/presskit/presskit.htm. Provides detailed preflight information about each of the space shuttle missions. Accessed March, 2005.

Slayton, Donald K., with Michael Cassutt. *Deke! U.S. Manned Space: From Mercury to the Shuttle.* New York: Forge, 1995. This is the autobiography of the last of the Mercury astronauts to fly in space. After being grounded from flying in Project Mercury for what turned out to be a minor heart murmur, Slayton was appointed head of the Astronaut Office. Later, he commanded the Apollo-Soyuz flight in 1975. He stayed with NASA through the STS-2 flight. Although he was no longer an active participant in NASA's activities, he gives some insight into the behind-the-scenes adventures of the shuttle program.

United States Congress. Senate Committee on Appropriations. *Space Shuttle and Galileo Mission: Hearing Before a Subcommittee on Appropriations.* Washington, D.C.: Government Printing Office, 1980. For a look into the politics of space in general, and into the process of funding for the Galileo mission in particular, see this 132-page document. Students of both the space program and the U.S. government will enjoy comparing this early document with the 1988 fact sheet prepared by the General Accounting Office for the chair of the Subcommittee on Science, Technology, and Space. This thirty-two-page document is entitled *Space Exploration: Cost, Schedule and Performance of NASA's Galileo Mission to Jupiter.*

"Variable Phenomena in Jovian Planetary Systems." *Journal of Geophysical Research* 98 (October 25, 1993): 18,727-18,876. A technical update that discusses some of the recent discoveries and theories concerning Jupiter and its moons. Intended for scientific audiences.

Young, Carolyn, ed. *The Magellan Venus Explorer's Guide.* Pasadena, Calif.: NASA, Jet Propulsion Laboratory, 1990. This 197-page document produced by the California Institute of Technology provides a detailed description of the Magellan spacecraft, the Venus probes, and the space-based radar system used in this historic mission.

Terry D. Bilhartz

Space Shuttle Flights, 1990

Date: January 9 to December 11, 1990
Type of mission: Piloted spaceflight

NASA flew six space shuttle flights in 1990, missions that totaled more than thirty-eight days in space. Among the highlights of 1990 was mission STS-31, the flight that deployed the Hubble Space Telescope into an orbit from which it began collecting data on objects up to 14 billion light-years away.

Key Figures

Daniel C. Brandenstein (b. 1943), STS-32 commander
James D. Wetherbee (b. 1952), STS-32 pilot
Bonnie J. Dunbar (b. 1949), STS-32 mission specialist
Marsha S. Ivins (b. 1951), STS-32 mission specialist
G. David Low (b. 1956), STS-32 mission specialist
John O. Creighton (b. 1943), STS-36 commander
John H. Casper (b. 1943), STS-36 pilot
David C. Hilmers (b. 1950), STS-36 mission specialist
Richard M. Mullane (b. 1945), STS-36 mission specialist
Pierre J. Thuot (b. 1955), STS-36 mission specialist
Loren J. Shriver (b. 1944), STS-31 commander
Charles F. Bolden, Jr. (b. 1946), STS-31 pilot
Bruce M. McCandless II (b. 1937), STS-31 mission specialist
Steven A. Hawley (b. 1951), STS-31 mission specialist
Kathryn D. Sullivan (b. 1951), STS-31 mission specialist
Richard N. "Dick" Richards (b. 1946), STS-41 commander
Robert D. Cabana (b. 1949), STS-41 pilot
Bruce E. Melnick (b. 1949), STS-41 mission specialist
William M. Shepherd (b. 1949), STS-41 mission specialist
Thomas D. "Tom" Akers (b. 1951), STS-41 mission specialist
Richard O. Covey (b. 1946), STS-38 commander
Frank L. Culbertson, Jr. (b. 1949), STS-38 pilot
Charles D. "Sam" Gemar (b. 1955), STS-38 mission specialist
Robert C. Springer (b. 1942), STS-38 mission specialist
Carl J. Meade (b. 1950), STS-38 mission specialist
Vance DeVoe Brand (b. 1931), STS-35 commander
Guy S. Gardner (b. 1948), STS-35 pilot
Jeffrey A. Hoffman (b. 1944), STS-35 mission specialist
John M. "Mike" Lounge (b. 1946), STS-35 mission specialist
Robert Allen Ridley Parker (b. 1936), STS-35 mission specialist
Samuel T. Durrance (b. 1943), STS-35 payload specialist
Ronald A. Parise (b. 1951), STS-35 payload specialist

Summary of the Missions

The December 18, 1989, scheduled launch of Mission STS-32 was postponed to complete modifications to Launch Complex 39A, a launch site that had not been used since the January, 1986, launch of STS 61-C. The rescheduled launch date of January 8, 1990, also was scrubbed due to poor weather conditions. Finally, on January 9, 12:35:00.017 Coordinated Universal Time (UTC, or 7:35 A.M. eastern standard time), the space shuttle *Columbia* was launched on what would be an eleven-day, 172-orbit mission.

The crew on this thirty-third space shuttle flight performed a variety of deployment and retrieval tasks. Mission STS-32, for instance, deployed a communication satellite (Syncom IV-F5, later called Leasat 5) that was designed to provide worldwide UHF communications between ships, planes, and fixed facilities on Earth. This 6-meter-long, 7,700-kilogram satellite with UHF and omni-directional antennae was launched using the unique "Frisbee" or rollout method of deployment. On the fourth day of the flight, mission STS-32 retrieved from orbit the Long Duration Exposure Facility (LDEF). The LDEF was a twelve-sided, open grid structure, 9 meters long and 4.2 meters in diameter, which, since its deployment in April, 1984, had been exposing a variety of materials to the harsh space environment. Among the fifty-seven experiments conducted in the experiment trays of the LDEF between 1984 and 1990 were an interstellar gas experiment (designed to provide insight into the formation of the Milky Way galaxy by capturing and analyzing its interstellar gas atoms), a cosmic radiation experiment (designed to investigate the evolution of the heavier elements in our galaxy), and a micrometeoroid experiment (designed to increase our understanding of the processes involved in the evolution of the solar system).

In addition to deploying and retrieving payloads, the crew of STS-32 conducted a number of experiments from the laboratories of the space shuttle. Mission Specialist Marsha S. Ivins, for example, operated an American Flight Echocardiograph (AFRE), a medical ultrasonic imaging sys-

tem that, when attached to the skin of another crew member, provided in-flight measurements of the size and functioning of the astronaut's heart. Ivins, along with Mission Specialists Bonnie J. Dunbar and G. David Low, also conducted numerous Protein Crystal Growth (PCG) experiments. Because protein crystals grown in space are purer and larger than crystals produced on Earth, these experiments performed in microgravity provide scientists with a better understanding of the three-dimensional structure of proteins, information needed for the development of new drugs to combat cancer, acquired immunodeficiency syndrome (AIDS), and many other diseases.

After nearly eleven days in orbit, the longest space shuttle flight to date, STS-32 ended at 09:35:33.15 UTC (1:35 A.M. Pacific standard time), on January 20. It made the third night landing in shuttle flight history.

The second shuttle flight of 1990, STS-36, was the thirty-fourth U.S. shuttle mission and the sixth devoted to Department of Defense concerns. This flight, which originally was scheduled for launch on February 22, was postponed several times: first, because of Commander John O. Creighton's illness; second, because of a malfunction of the range safety computer; and third, because of unacceptable weather conditions. STS-36 marked the first time since the Apollo 13 mission of 1970 that a piloted space launch was affected by the illness of a crew member. After these delays, however, on February 28, at 07:50:22.000 UTC (2:50 A.M. eastern standard time), the five-member crew was launched aboard the space shuttle *Atlantis* in NASA's fourth nighttime launch. This classified mission ended after sixty-nine orbits with a successful landing on March 4.

The thirty-fifth space shuttle flight, STS-31, placed into orbit the Hubble Space Telescope (HST), a 2.4-meter reflecting telescope that was created in a cooperative effort by the European Space Agency (ESA) and NASA. The first attempted launch of this historic mission was scrubbed four minutes before the scheduled liftoff on April 10 owing to a faulty valve in the auxiliary power unit. Two weeks later, however, *Discovery* and its crew were launched into orbit. Liftoff took place on April 24 at 12:33:50.99 UTC (8:34 A.M. eastern daylight time).

Filling most of the payload bay area of *Discovery* was the HST, the largest space-based observatory ever built. On the second day of the five-day mission, this 11,000-kilogram, railroad-car-sized observatory (13.3 meters long, 4.3 meters in diameter) was deployed into a low-Earth (600-kilometer) orbit, a perch from which it would be able to image objects up to 14 billion light-years away. The later discovery of a spherical aberration on the lens of the original Wide Field/Planetary Camera (WFPC1) of the HST raised initial concerns about how the Hubble project was functioning. In December, 1993, however, the crew of another shuttle flight, STS-61, obviated the effects of the spherical aberration when it replaced the damaged WFPC1 with an optically corrected spare WFPC2 and installed corrective mirrors for the primary mirror.

In addition to delivering the HST, the STS-31 crew conducted a variety of middeck experiments, including protein crystal growth and polymer membrane processing studies. On April 29, after completing 76 orbits and traveling more than 3.2 million kilometers (2 million miles), the crew landed safely on Runway 22 of Edwards Air Force Base, California, at 13:49:57.25 UTC (6:50 A.M. Pacific daylight time).

After the completion of the successful STS-31 mission, NASA experienced six months of frustration before another shuttle was launched. The next two planned 1990 shuttle flights were the *Columbia*/STS-35 mission and the *Atlantis*/STS-38 mission. Launch dates for these flights originally were set, respectively, for May and July. Owing to a series of hardware problems, however, neither of these missions would be launched before the scheduled October 6 launching of the *Discovery*/ STS-41 mission.

NASA's frustration began with the forced scrubbing of STS-35 from its original May 16 launch in

order to change a faulty Freon coolant loop proportional valve in *Columbia*'s coolant system. Two weeks later a second launch was canceled during tanking owing to a hydrogen leak in the External Tank/orbiter disconnect assembly. Because repairs could not be made at the pad, *Columbia* was returned to the Vehicle Assembly Building (VAB), where it was demated from its External Tanks. It was then transferred to the Orbiter Processing Facility (OPF), where it was fitted with new umbilical hardware.

With STS-35 temporarily on hold, the *Atlantis* crew of STS-38 prepared for its scheduled July launch. The hydrogen leak on *Columbia*, however, prompted NASA officials to conduct three precautionary tanking tests on *Atlantis*. When these tests—performed June 29, July 13, and July 25— confirmed a hydrogen fuel leak, *Atlantis* also was returned to the VAB and to the OPF for demating and repair. Unfortunately, on the day that the *Columbia*/ STS-35 was being transferred back to the pad for another launch preparation, *Atlantis*, while parked outside the VAB, was caught in a hailstorm and suffered additional damage to its tiles.

Meanwhile, on August 9, the *Columbia*/STS-35 was rolled out to Pad A for a second time. Two days before the rescheduled September 1 launch, an avionics box on the payload equipment malfunctioned and was replaced and retested, forcing another postponement to September 6. During tanking for this launch, however, a high concentration of hydrogen was detected in the orbiter's aft compartment. NASA managers concluded that *Columbia* had experienced separate hydrogen leaks from the beginning: one in the umbilical assembly, which subsequently had been replaced, and another in the aft compartment, which had resurfaced. Three hydrogen recirculation pumps in the aft compartment were then replaced and retested, and the launch date was reset for September 18. When the leak in the aft compartment resurfaced again during tanking, NASA decided to put this mission on hold until a special tiger team assigned by the space shuttle director would resolve the problem. *Columbia* again was transferred from Pad A to Pad B to make room for the *Atlantis/ STS-36*. On October 9, the *Columbia* orbiter was moved yet again, this time to the VAB to protect it from tropical storm Klaus.

Problems with the *Columbia* and *Atlantis* orbiters, however, did not delay the scheduled October 6 launching of *Discovery/STS-41* mission. The 11:47:14.938 UTC (7:47 A.M. eastern daylight time) liftoff occurred just twelve minutes after the two-and-one-half hour launch window opened at 11:35 UTC. This mission's payload launch mass of 117,750 kilograms made this the heaviest payload to date.

The primary objective of STS-41 was to send the Ulysses spacecraft on a five-year probe to explore the polar regions of the Sun. On the first day of the mission, the Ulysses was released from the cargo

The firing room at Kennedy Space Center during the launch of space shuttle Discovery *on April 24, 1990. This mission (STS-31) deployed the Hubble Space Telescope.* (NASA)

bay. Afterward, its two-stage Inertial Upper Stage (IUS) rockets were fired, thus sending the European Space Agency-built Ulysses spacecraft on a sixteen-month voyage to Jupiter. To reach the never-before-seen polar regions of the Sun, the Ulysses had to use the massive gravity of Jupiter to swing it up and perpendicular to the plane of the orbits of the Earth and all the other planets.

Secondary objectives of mission STS-41 included studying the effects of atomic oxygen wear on solar panels; recording radiation levels in orbit; conducting a chromosome and plant cell division experiment designed to measure growth patterns of plant roots in microgravity; performing a physiological systems experiment designed to investigate how microgravity affects bone calcium, body mass, and immune cell functions; and completing a solid surface combustion experiment engineered to study flames in microgravity. Crew members on STS-41 also tested a new voice-actuated command system for future shuttle flights. After four days and sixty-five orbits, *Discovery* landed on October 10, 13:57:19 UTC (6:57 A.M. Pacific daylight time) at Edwards Air Force Base.

The *Atlantis/STS-38* mission, which suffered yet another delay in October owing to damages caused by a falling platform beam during hoisting operations, finally received a favorable flight readiness review (FRR) and was rescheduled with a launch date of November 9. Payload problems forced one more postponement, but at 23:48:15.006 UTC (6:48 P.M. eastern standard time) on November 15, the long-delayed mission was launched into orbit. This scheduled four-day flight, a classified Department of Defense mission, was extended for an additional day owing to unacceptable crosswinds at the original planned landing site at Edwards. Continued adverse conditions led to the decision to shift to the Kennedy Space Center landing site. The *Atlantis* landing, which took place at 21:42:46 UTC (4:43 P.M. eastern standard time) on November 20, was the first shuttle landing at Kennedy since April, 1985.

STS-35, the other long-delayed 1990 flight, finally received the go-ahead for a launch on December 2. The 06:49:01.0218 UTC (1:40 A.M. eastern standard time) launch of America's thirty-eighth shuttle mission was NASA's sixth night launch. The primary objective of this mission was to conduct round-the-clock observations of the celestial sphere. Working two twelve-hour shifts, the seven-member crew operated two sophisticated instruments: ASTRO, an ultraviolet astronomy observatory capable of making precise measurements of objects such as planets, stars, and galaxies in relatively small fields of view; and BBXRT, a broad band x-ray telescope designed to measure directly the amount of energy, in electron volts, of each x ray detected from targets such as active galaxies, clusters of galaxies, supernova remnants, and stars. This mission, which had numerous problems before liftoff, also suffered in-flight troubles when the loss of the data display units used for pointing the telescopes impacted the crew-aiming procedures. The problem was partially solved by allowing ground teams at Marshall Space Flight Center to aim the ultraviolet telescopes. The crew on this troubled mission also experienced difficulties dumping wastewater, owing to a clogged drain, a problem managed by using spare containers for the waste. Fittingly, the mission was cut short by one day due to impending bad weather at the primary landing site. *Columbia* and crew landed safely at Edwards Air Force Base at 05:54:09 UTC on December 11 (9:54 P.M. Pacific standard time, December 10).

Contributions

Two shuttle missions in 1990, as in 1989, were classified Department of Defense flights, and for these missions, little information has been released. However, an impressive array of scientific intelligence gathered from the four nonclassified 1990 shuttle flights has been disseminated.

STS-32 retrieved from space the LDEF, a facility that had been orbiting Earth since 1984 and at the time of retrieval was within one month of reentering the Earth's atmosphere. During the LDEF's 32,422 Earth orbits, it collected in its eighty-six trays a great variety of data about the space environment and its effects on various materials. Technical knowledge gained from these data led NASA management to conclude that (1) space environments are hostile to spacecraft materials and coatings, (2) synergistic effects of all aspects of the low-Earth-orbit environment must be considered in spacecraft design, (3) contamination should be a very significant consideration in design, (4) the pre-LDEF knowledge of space environmental effects on materials had major flaws, and (5) LDEF knowledge has forced the revision of environment-related test and qualification procedures. One of the many consequences of the LDEF findings was the decision to change the International Space Station radiator design from Teflon, a material that showed substantial deterioration due to atomic oxygen, to a Z-93 ceramic paint, a coating that proved to be very stable in the space environment.

In contrast to STS-32, a mission that retrieved scientific data that were many years in the collecting, the STS-31 and STS-41 missions launched into space vehicles that would be collecting and relaying information about the boundaries of the universe for years to come. For example, the Hubble Space Telescope (HST), which was deployed on STS-31, has enabled measurement of stars 30 million light-years away, and galaxies 100 million light-years away, thus extending the volume of space able to be surveyed to 100 times greater than that of the most advanced ground telescopes. By imaging the distant heavens, scientists were given a chance to determine more accurately the Hubble constant (a calculation of the rate at which the universe is expanding) and the deceleration parameter (a measure of whether the distant galaxies are receding at a slower rate than nearby, newer galaxies). Moreover, by studying the motion of distant galaxies, the HST began to collect data enabling astronomers to infer the masses of galaxies and, from these data, to compute the mass of the universe as a whole. Such data can provide insights that will help scientists address some of the grandest questions the human mind has ever pondered: How big is the universe? How old is it? Will it expand forever? At what rate is it expanding? How did structure (galaxies) arise from a fireball (Big Bang)?

The Ulysses spacecraft, launched from STS-41, also provided new measurements of and insights into the Sun. Some of the early findings taken from this probe suggest that in the Sun's polar region, the solar wind (a very hot, ionized flow of gases and energetic particles emanating from the Sun) flows at a very high and steady velocity of about 750 kilometers per second, twice the speed of solar winds at the Sun's lower latitudes. Prior to the Ulysses observations, scientists expected a continuous increase of velocity toward the poles. Moreover, many models of the solar magnetic field used prior to Ulysses assumed that the solar magnetic field was similar to that of a dipole. (For a dipole, the field strength over the poles is twice the strength it is over the equator.) Ulysses observations found, however, that the amount of outward magnetic flux in the solar wind did not vary greatly with latitude, thus indicating the importance of pressure forces near the Sun for evenly distributing the magnetic flux. Such technical knowledge of the Sun is of practical value for engineers because changes in the solar wind pressures and related magnetic disturbances impinge on Earth's magnetosphere, often causing severe disruption of radio communications. At a meeting in Paris on February 12, 2004, ESA's Science Program Committee unanimously approved a proposal to continue operating the highly successful Ulysses spacecraft until March, 2008.

The final flight of 1990, STS-35, was perhaps the most disappointing mission of the year. The in-flight problems with ASTRO and BBXRT forced NASA management to abandon the entire preplanned, optimized mission time line for a replanned, day-by-day schedule. The result was a lower-efficiency mission. In the end, BBXRT achieved a disappointing total of 185,000 seconds of observation time on cosmic x-ray sources. Even so, science teams at Marshall and Goddard Space Flight Centers estimated

that nearly 70 percent of the planned science data were obtained on this mission.

Context

For many Americans, 1990 appeared to be a time of stupendous developments and rapid change. Americans greeted the year still celebrating the extraordinary opening of the Berlin Wall in late 1989. Later in the spring of 1990, U.S. President George Bush and Soviet Premier Mikhail Sergeyevich Gorbachev signed a major trade agreement and accords on reducing strategic nuclear arsenals and ending the production of chemical weapons. By the year's end, the leaders of the two formerly hostile powers issued a joint statement condemning Iraq's invasion of Kuwait, agreed on the final settlement with respect to German reunification, and signed the Charter of Paris declaring an end to the military division of Europe. As Americans celebrated the ending of the Cold War, they also began to focus more of their attention on domestic concerns, in particular on the federal budget deficit. The 1990's would become a time when Americans would look for ways to reduce the federal government. In this new political environment, even the cherished U.S. space program would not be immune to increased budgetary scrutiny.

Unfortunately for space enthusiasts, during this time of national reevaluation NASA did not enjoy a banner year. While there was no incident similar to the 1986 STS 51-L (*Challenger*) accident, the shuttle missions of 1990 experienced a number of flight postponements, in-flight troubles, and scientific disappointments. Even the excitement over the April deployment of the Hubble Space Telescope—a highly sophisticated piece of human engineering with the capacity to explore the very boundaries of the universe—dwindled in late June when it was learned that the Hubble had a serious design flaw in one of its mirrors. This flaw, which would be rectified by an HST servicing crew on a 1993 shuttle flight, received more national media attention than the original HST deployment itself. The long-delayed STS-35 mission, and then the less than desired scientific results obtained from it, brought this frustrating year to a close.

In hindsight, however, 1990 was a better year for NASA than it appeared to many Americans at the time. During 1990, the United States safely launched and landed six shuttle crews, conducted thousands of hours of microgravity observations and experiments, and launched the Hubble and the Ulysses projects, robotic missions that ulti-mately would prove to be among the most success-ful space initiatives in U.S. aviation history.

See also: Hubble Space Telescope; Hubble Space Telescope: Science; Hubble Space Tele-scope: Servicing Missions; International Ultravio-let Explorer; Planetary Exploration; Space Shuttle Flights, 1989; Space Shuttle Flights, 1991; Tele-scopes: Air and Space; Ulysses: Solar Polar Mission.

Further Reading

Chaisson, Eric. *The Hubble Wars: Astrophysics Meets Astropolitics in the Two-Billion-Dollar Struggle over the Hubble Space Telescope.* New York: HarperCollins, 1994. An entertaining 386-page treatise on the politics of space exploration.

Collins, Carolyn. *Hubble Vision: Astronomy with the Hubble Space Telescope.* 2d ed. New York: Cambridge University Press, 1995. A scholarly study on space astronomy and the Hubble program that will interest general readers as well as academics.

Embury, Barbara, with Thomas D. Crouch. *The Dream Is Alive.* New York: Harper and Row, 1990. An interesting account of the shuttle program intended for juvenile readers.

Harland, David M. *The Space Shuttle: Roles, Missions, and Accomplishments.* Hoboken, N.J.: John Wiley, 1998. *The Space Shuttle* is written thematically, rather than purely chronologi-cally. Topics include shuttle operations and payloads, weightlessness, materials process-ing, exploration, Spacelabs and free-flyers, and the shuttle's role in the International Space Station.

Harrington, Philip S. *The Space Shuttle: A Photographic History.* San Francisco, Calif.: Brown Trout, 2003. With one hundred full-color photographs by Roger Ressmeyer and others and with text by popular astronomy writer Phil Harrington, this book tells the story of the space shuttle program from 1972 to 2003. Its beautiful photographs allow the gen-eral reader to survey the history of the space shuttle program and be uplifted by the pio-neering spirit of one of humanity's grandest enterprises.

Jenkins, Dennis R. *Space Shuttle: The History of the National Space Transportation System: The First 100 Missions.* Stillwater, Minn.: Voyageur Press, 2001. This is a concisely written tech-nical reference account of the space shuttle and its ancestors, the aerodynamic lifting bodies. It details some of the advantages and inherent disadvantages of using a reusable space vehicle. Each of the vehicles is illustrated by line drawings. The book follows the space shuttle from its original concepts and briefly chronicles its first one hundred flights.

Kerrod, Robin. *Hubble: The Mirror on the Universe.* Richmond Hill, Ont.: Firefly Books Ltd., 2003. The book covers the observable universe in six sections: "Stars in the Firmament," "Stellar Death and Destruction," "Gregarious Galaxies," "The Expansive Universe," "So-lar Systems," and "The Heavenly Wanderers." Clear and concise text explains the fasci-nating history of astronomy and the development of the HST. *Hubble* transports readers to the planets of our solar system and on to galaxies millions—even billions—of light years away. These dramatic, unforgettable images will bring into sharp focus how the universe is unfolding in new and astonishing ways.

National Aeronautics and Space Administration. *NASA Space Shuttle Mission STS-32 Press Kit.* Washington, D.C.: Author, 1989. Because few history monographs of shuttle missions since 1990 have been published, among the better sources of information on these missions are the official NASA press releases. These releases, generally prepared before scheduled liftoff, detail the objectives and planned time lines of the missions. The releases may be viewed on microfilm, or electronically on the Internet. For the other nonclassified missions of 1990, see *NASA Space Shuttle Mission STS-31 Press Kit* (April, 1990), *NASA Space Shuttle Mission STS-41 Press Kit* (October, 1990), and *NASA Space Shuttle Mission STS-35 Press Kit* (December, 1990).

_____. *Space Shuttle Mission Press Kits.* http://www-pao.ksc.nasa.gov/kscpao/presskit/presskit.htm. Provides detailed preflight information about each of the space shuttle missions. Accessed March, 2005.

National Aeronautics and Space Administration, Jet Propulsion Laboratory. *Ulysses: A Voyage to the Sun.* Washington, D.C.: Government Printing Office, 1986. An early report prepared by the Jet Propulsion Laboratory that outlines the objectives of the Ulysses project.

Page, D. Edgar, and Edward J. Smith. "Reflecting on the Findings of the Ulysses Spacecraft." *Eros* 76 (July 25, 1995): 297-302. A technical update on the progress of the mission.

Slayton, Donald K., with Michael Cassutt. *Deke! U.S. Manned Space: From Mercury to the Shuttle.* New York: Forge, 1995. This is the autobiography of the last of the Mercury astronauts to fly in space. After being grounded from flying in Project Mercury for what turned out to be a minor heart murmur, Slayton was appointed head of the Astronaut Office. Later, he commanded the Apollo-Soyuz flight in 1975. He stayed with NASA through the STS-2 flight. Although he was no longer an active participant in NASA's activities, he gives some insight into the behind-the-scenes adventures of the shuttle program.

Terry D. Bilhartz

Space Shuttle Flights, 1991

Date: April 5 to December 1, 1991
Type of mission: Piloted spaceflight

Three space shuttle missions were flown in 1991 by the shuttle Atlantis, *two more by* Discovery, *and one by* Columbia. *Satellites to monitor the upper atmosphere, relay data from shuttles to the ground, and monitor missile launchings and nuclear explosions were deployed. Shuttle astronauts conducted experiments to determine the physiological effects of spaceflight and demonstrate techniques for construction of the International Space Station.*

Key Figures

Steven R. Nagel (b. 1946), STS-37 commander
Kenneth D. Cameron (b. 1949), STS-37 pilot
Linda M. Godwin (b. 1952), STS-37 mission specialist
Jerry L. Ross (b. 1948), STS-37 mission specialist
Jerome "Jay" Apt (b. 1949), STS-37 mission specialist
Michael L. Coats (b. 1946), STS-39 commander
L. Blaine Hammond, Jr. (b. 1952), STS-39 pilot
Gregory J. Harbaugh (b. 1956), STS-39 mission specialist
Donald R. "Don" McMonagle (b. 1952), STS-39 mission specialist
Guion S. Bluford, Jr. (b. 1942), STS-39 mission specialist
Charles Lacy Veach (1944-1995), STS-39 mission specialist
Richard J. Hieb (b. 1955), STS-39 mission specialist
Bryan D. O'Connor (b. 1946), STS-40 commander
Sidney M. Gutierrez (b. 1951), STS-40 pilot
James P. Bagian (b. 1952), STS-40 mission specialist
Tamara E. Jernigan (b. 1959), STS-40 mission specialist
Margaret Rhea Seddon (b. 1947), STS-40 mission specialist
Francis A. "Drew" Gaffney (b. 1946), STS-40 payload specialist
Millie W. Hughes-Fulford (b. 1945), STS-40 payload specialist
John E. Blaha (b. 1942), STS-43 commander
Michael A. Baker (b. 1953), STS-43 pilot
Shannon W. Lucid (b. 1943), STS-43 mission specialist
G. David Low (b. 1956), STS-43 mission specialist
James C. Adamson (b. 1946), STS-43 mission specialist
John O. Creighton (b. 1943), STS-48 commander
Kenneth S. Reightler, Jr. (b. 1951), STS-48 pilot
Charles D. "Sam" Gemar (b. 1955), STS-48 mission specialist
James F. Buchli (b. 1945), STS-48 mission specialist
Mark N. Brown (b. 1951), STS-48 mission specialist

Frederick D. Gregory (b. 1941), STS-44 commander
Terence T. "Tom" Henricks (b. 1952), STS-44 pilot
James S. Voss (b. 1949), STS-44 mission specialist
F. Story Musgrave (b. 1935), STS-44 mission specialist
Mario Runco, Jr. (b. 1952), STS-44 mission specialist
Thomas P. Hennen (b. 1952), STS-44 payload specialist

Summary of the Missions

The U.S. piloted space program experienced a delay in early 1991, when cracked hinges were found on all three of the shuttle orbiters, *Atlantis*, *Discovery*, and *Columbia*. These hinges allowed a hatch, which covers the entry port for hoses from the shuttle's external fuel tank, to close during re-entry into the atmosphere. If this hatch did not close completely, the shuttle could burn up during reentry. The largest cracks were on the shuttle *Discovery*, and its launching, planned for February 28, was postponed for two months to permit repairs. Only one hairline crack was detected on *Atlantis*, so it was able to make the first shuttle flight of 1991.

The space shuttle *Atlantis* lifted off from the NASA Kennedy Space Center (KSC), Florida, at 14:22:44.988 Coordinated Universal Time (UTC, or 9:23 A.M. eastern standard time) on April 5, 1991, under the command of Air Force Colonel Steven Nagel, who was making his third trip into space. The STS-37 mission was piloted by Marine Corps Lieutenant Colonel Kenneth D. Cameron, who was making his first spaceflight. *Atlantis* also carried three mission specialists, Air Force Lieutenant Colonel Jerry L. Ross, who was making his third spaceflight, and Jerome "Jay" Apt and Linda M. Godwin, both on their first shuttle flight.

On April 7, the crew deployed the 15,400-kilogram Compton Gamma Ray Observatory (GRO), the heaviest payload ever carried aboard the shuttle, into a nearly circular orbit 451 kilometers (280 miles) above the Earth's surface. The GRO was the second of four large astronomical observatories NASA planned to launch in its Great Observatories Program. The first was the Hubble Space Telescope, launched in 1990. Unlike the Hubble, which observes astronomical objects in visible light, the GRO detects very high energy electromagnetic radiation, called gamma rays. While the GRO was still attached to the robot arm used to remove it from the shuttle's cargo bay, the 4.9-meter (16-foot) antenna that it would use to communicate with Earth failed to unfold despite repeated commands transmitted from the NASA ground stations. Astronauts Ross and Apt, in a four-and-a-half-hour emergency spacewalk, were able to free the antenna.

Astronauts Ross and Apt conducted an even longer spacewalk, lasting six and one-half hours, on April 8. They assembled a 14-meter-long (47-foot-long) rail in the cargo bay, and then tested prototypes of three carts, one manually propelled, one driven by an electric motor, and one using a mechanical pump; the latter was planned for use in the construction of NASA's Space Station *Freedom*.

After extending its flight for one day because of high winds in the landing area, *Atlantis* landed at Edwards Air Force Base in California at 13:55:29 UTC (6:55 A.M. Pacific daylight time) on April 11, 1991.

The space shuttle *Discovery* lifted off from KSC in Florida at 11:33:14.018 UTC (7:33 A.M. eastern daylight time) on April 28, 1991, on a military mission to test systems developed for the Strategic Defense Initiative program. Navy Captain Michael L. Coats, who was making his third shuttle flight, commanded the STS-39 mission, and Air Force Lieutenant Colonel L. Blaine Hammond, Jr., on his first flight on the shuttle, piloted *Discovery* into orbit. This was the first military mission of the space shuttle that was not cloaked in secrecy. On prior military flights, the press was not even given advance notice of the launch or landing times. However, in a budget-cutting measure, the secret launch control room at KSC and the secret flight control room at the Lyndon B. Johnson Space Center in Houston, Texas, had been closed in late 1990.

On April 29, the crew opened the cover on the Cryogenic Infrared Radiance Instrument for Shuttle (CIRRIS) instrument, designed to monitor the infrared radiation emitted by natural processes such as an aurora. This experiment was designed to provide the information needed to construct instruments to distinguish between the heat emitted by enemy missiles and that emitted by naturally occurring processes.

On May 1, the crew released the $94 million shuttle pallet satellite, Shuttle Pallet Satellite-II (SPAS-II), into space. Sensors on this satellite mon-

itored the emission from the rocket thrusters used to reposition the shuttle and compared this with emissions from vapor clouds sprayed from gas canisters aboard the shuttle. The objective was to determine how to distinguish actual rocket exhaust from vapor clouds released by enemy decoys.

On May 4, the crew overcame a problem with the data recorders on a third experiment, designed to distinguish between natural and human-induced x-ray sources. The objective was to monitor compliance with nuclear test ban treaties, because nuclear explosions release a burst of x rays.

Discovery had been scheduled to land at Edwards Air Force Base. However, high winds at the Califor-

Endeavour *arrives at Kennedy Space Center on May 7, 1991, riding a Boeing 747.* (NASA)

nia site forced it to divert to KSC in Florida, where the shuttle landed at 18:54:41 UTC (2:55 P.M. eastern daylight time) on May 6.

The space shuttle *Columbia* had been scheduled for a flight on May 21, 1991. However, this launching was postponed when *Columbia* developed three problems: electronic circuits linking the booster rockets to the orbiter failed, one of the five computers on the orbiter failed, and cracked welds were found in sensors that monitored the temperature of the liquid hydrogen fuel lines. The weld problem, which had not been detected in earlier inspections, was the most serious. If a sensor broke free, it could be sucked into the engine's pumps, causing an explosion. Further inspection found six cracks in welds on *Columbia*, as well as cracks in welds on *Atlantis* and *Discovery*.

Following the replacement of three temperature sensors, *Columbia* lifted off from KSC at 13:24:51.008 UTC (9:25 A.M. eastern daylight time), on June 5, 1991, under the command of Marine Corps Colonel Bryan D. O'Connor. The purpose of the nine-day STS-40 mission was to study the physiological effects of spaceflight on living organisms. The Spacelab module, a 6.7-meter-long by 4.9-meter-diameter pressurized cylinder, was carried aloft in *Columbia*'s cargo bay. It contained some of the world's most sophisticated medical instruments, as well as 29 rats and 2,748 jellyfish. The jellyfish have gravity sensing mechanisms similar to those in the human ear, and the *Columbia* scientists studied how the jellyfish oriented themselves and adapted to the weightless environment. The rats were dissected after the flight to see how their bodies had changed in response to weightlessness. *Columbia*'s flight ended on June 14, 1991, with a landing at Edwards Air Force Base at 15:39:10.9 UTC (8:39 A.M. Pacific daylight time).

The shuttle *Atlantis*, under the command of Air Force Colonel John E. Blaha and piloted by Navy Lieutenant Commander Michael A. Baker, lifted off on mission STS-43 from KSC at 15:01:59.986 UTC (11:02 A.M. eastern daylight time), on August 2, 1991. *Atlantis* carried a Tracking and Data-Relay Satellite (TDRS) in its cargo bay. The 2,130-kilogram

TDRS, built by TRW, is a communications satellite designed to relay data from the space shuttle and other Earth-orbiting satellites to ground stations. Because radio signals travel in straight lines, NASA had previously employed a worldwide network of ground stations to maintain communications with spacecraft in orbit. The use of TDRS relays allowed NASA to close many of these ground stations. TDRS-5, deployed six hours after liftoff, joined three others, placed in orbit previously.

The second purpose of this mission was to evaluate equipment being considered for use on the planned Space Station. The crew tested a fiber-optic communications link, new cooling systems, and a method to minimize the effects of weightlessness by creating a low-pressure force on the lower body. The crew, all but one of whom had flown previously, observed that the Earth was surrounded by an unusually thick haze that they speculated might have been caused either by the eruption of Mount Pinatubo in the Philippines in June or by residue from the oil well fires set in Kuwait at the end of the Gulf War. *Atlantis* landed at KSC at 12:22:25.00 UTC (8:23 A.M. eastern daylight time), on August 11, 1991. This was the first scheduled landing of a shuttle in Florida since 1986.

The shuttle *Discovery*, commanded by Navy Captain John O. Creighton and piloted by Navy Commander Kenneth S. Reightler, Jr., lifted off from KSC at 23:11:03.970 UTC (7:11 P.M. eastern daylight time), on September 12, 1991. STS-48 marked the beginning of NASA's Mission to Planet Earth, a decade-long effort to monitor the Earth's environment from space.

The Upper Atmosphere Research Satellite (UARS), weighing 14,500 pounds and costing $740 million, was released from the cargo bay on September 15. The UARS carried ten instruments to monitor the chemistry, winds, and heat distribution of the upper atmosphere of the Earth. One major objective was to study the ozone layer, which shields the Earth's surface from the ultraviolet rays that are harmful to biological organisms. Bad weather in Florida kept *Discovery* from making the first night landing on the KSC runway. The shuttle was diverted to Edwards Air Force Base, where it landed at 07:37:42 UTC (12:38 A.M. Pacific daylight time), on September 18.

The shuttle *Atlantis*, commanded by Air Force Colonel Frederick D. Gregory and piloted by Air Force Lieutenant Colonel Terence T. "Tom" Henricks, lifted off from KSC at 23:44:00.006 UTC (6:44 P.M. eastern standard time) on November 24, 1991. STS-44 was a military mission during which the crew observed ground sites to determine if human reconnaissance from space would be useful in times of world crises. They used binoculars, a small telescope, and a digital camera to observe twenty-five U.S. military sites around the globe.

The crew of *Atlantis* also deployed a 5,200-pound spy satellite containing a large infrared telescope to detect missile launches and nuclear explosions. Earlier in 1991, similar satellites had helped track Scud missiles launched by Iraq during the Gulf War. One of the three inertial measurement units, which monitor the velocity and orientation of the shuttle in space, failed during the flight, forcing *Atlantis* to return to Earth three days earlier than scheduled. *Atlantis* landed at Edwards Air Force Base at 22:32:58.77 UTC (2:35 P.M. Pacific standard time) on December 1.

Contributions

The Compton Gamma Ray Observatory satellite was launched by the shuttle *Atlantis* on April 7, 1991. It provided scientists with the opportunity to study some of the most violent astronomical events in the universe. Gamma rays, which are very high energy radiation, do not penetrate the Earth's atmosphere, and thus cannot be studied from the ground. Gamma-ray detectors flown on balloons and smaller satellites had provided some information on these violent events, but the sensors on the GRO were ten to twenty times more sensitive than any previously flown, allowing detection of sources at greater distances from the Earth.

Vela satellites, whose primary mission is to detect nuclear explosions on Earth, had detected bursts of gamma rays from space. One burst, detected on March 5, 1979, lasted only one five-thousandth of a second but gave off energy at a rate greater than the visible emission from the entire Milky Way galaxy. The GRO mapped the distribution of these gamma-ray bursts across the sky and found they were apparently randomly distributed. This indicates to astronomers that these sources are not located within our galaxy, and that they must be even more energetic than previously assumed because they are located at large distances from the Earth. Compton was safely deorbited and reentered the Earth's atmosphere on June 4, 2000.

Since the beginning of piloted spaceflight, experts in space medicine had recognized that the human body reacted adversely to weightlessness. Studies of the physiological effects of weightlessness were a major objective of the long-duration missions on the U.S. Skylab space station, in the 1970's. The nine-day flight of *Columbia* carried three mission specialists, Dr. James P. Bagian, a surgeon, Dr. Francis A. "Drew" Gaffney, a cardiologist, and Millie W. Hughes-Fulford, a cell biologist, to investigate these effects. The other *Columbia* astronauts were examined, and body fluid samples were collected throughout the flight to monitor the onset of the body's reaction to weightlessness.

On Earth, fluids accumulate in the legs due to gravity, but in weightlessness these fluids are redistributed more evenly throughout the body. The resulting increase in fluids in the torso causes the body to expel the perceived excess fluids. This fluid loss was observed to cause an increase in the heart rate. As the exposure to weightlessness continues, bones lose calcium and muscles atrophy. The physiological measurements on the *Columbia* crew provided data to support future long-duration space missions, including the Space Station and a human mission to Mars.

On its November flight, *Atlantis* crew members tested a rowing machine and a treadmill designed to determine if some of the physiological effects of weightlessness could be overcome by exercise.

Context

Mission to Planet Earth is NASA's contribution to the U.S. Global Change Research Program, an interagency effort to understand global change and the impact of human activity on the planet. This program is designed to distinguish changes caused by human activity from those resulting from natural processes. For example, both industrial processing and natural activity, such as volcanoes, alter the chemical composition of the atmosphere. The Nimbus 7 weather satellite carried some instruments to monitor the upper atmosphere, but the UARS, launched by the shuttle *Discovery* in September, 1991, was the first satellite completely dedicated to this effort. The UARS monitored the day/night variation in the abundances of chemical species important in maintaining the ozone layer as well as measuring the abundances of nitrous oxides and halocarbons, important in maintaining the nitrogen and the chlorine cycles, in the upper atmosphere.

Each of the last two shuttle missions of 1991 had to take evasive action to avoid space debris. On September 15, *Discovery* maneuvered to avoid flying within 2.7 kilometers (1.7 miles) of the upper stage of the rocket that had placed Kosmos 955 into orbit in September, 1977. NASA officials indicated this was the first time one satellite had to maneuver to avoid a close approach with another. Because of the catastrophic consequences of high-speed collisions, NASA flight rules require that the shuttle avoid coming within 8 kilometers (5 miles) of orbiting space debris. On November 28, the shuttle *Atlantis* maneuvered to avoid coming within 2.4 kilometers of part of the booster rocket of Kosmos 851, launched in 1976. These two incidents dramatically illustrated the increasing hazard to future spaceflight of human-made debris orbiting the Earth.

In April, 1991, the space shuttle *Endeavour*, the replacement for the shuttle *Challenger*, lost in January, 1986, was unveiled at the Palmdale, California, facility of Rockwell International Corporation. *Endeavour* was built using many spare parts left over from earlier shuttles, but improved systems allow *Endeavour* to stay in orbit for up to twenty-eight days. Although NASA sought permission to build an additional shuttle, Vice President Dan Quayle, in his role as chair of the National Space Council, announced on July 24, 1991, that no additional space shuttles would be built. In the 1970's, the shuttle was envisioned as America's only access to space, but Quayle indicated that the United States would shift to robotic rockets to launch future commercial and military satellites. Quayle said, "The space shuttle, with its precious human lives, is just too valuable to use on missions that don't need its unique capabilities."

As of 2005, there were ten TDRS satellites in the constellation. TDRS-5 was at 171.4° west longitude. TDRS-8 and TDRS-10 were the primary satellites, with TDRS-9 serving as a spare.

See also: Compton Gamma Ray Observatory; National AeroSpace Plane; Pegasus Launch Vehicles; Space Shuttle: Life Science Laboratories; Space Shuttle: Microgravity Laboratories and Payloads; Space Shuttle Flights, 1990; Space Shuttle Flights, 1992; Tracking and Data-Relay Communications Satellites; Ulysses: Solar Polar Mission; Upper Atmosphere Research Satellite.

Further Reading

Covault, Craig. "USAF/SDI Space Shuttle Flight to Execute Complex Maneuvers." *Aviation Week and Space Technology* 134 (March 4, 1991): 46-69. A well-illustrated account of the April, 1991, flight of the shuttle *Discovery*, describing the instruments and experiments on this military mission.

Harland, David M. *The Space Shuttle: Roles, Missions, and Accomplishments.* Hoboken, N.J.: John Wiley, 1998. *The Space Shuttle* is written thematically, rather than purely chronologically. Topics include shuttle operations and payloads, weightlessness, materials processing, exploration, Spacelabs and free-flyers, and the shuttle's role in the International Space Station.

Harrington, Philip S. *The Space Shuttle: A Photographic History*. San Francisco, Calif.: Brown Trout, 2003. With one hundred full-color photographs by Roger Ressmeyer and others and with text by popular astronomy writer Phil Harrington, this book tells the story of the space shuttle program from 1972 to 2003. Its beautiful photographs allow the general reader to survey the history of the space shuttle program and be uplifted by the pioneering spirit of one of humanity's grandest enterprises.

Jenkins, Dennis R. *Space Shuttle: The History of the National Space Transportation System: The First 100 Missions*. Stillwater, Minn.: Voyageur Press, 2001. This is a concisely written technical reference account of the space shuttle and its ancestors, the aerodynamic lifting bodies. It details some of the advantages and inherent disadvantages of using a reusable space vehicle. Each of the vehicles is illustrated by line drawings. The book follows the space shuttle from its original concepts and briefly chronicles its first one hundred flights.

National Aeronautics and Space Administration. *Space Shuttle Mission Press Kits*. http://www-pao.ksc.nasa.gov/kscpao/presskit/presskit.htm. Provides detailed preflight information about each of the space shuttle missions. Accessed March, 2005.

Schuiling, R. L., and S. Young. "Life Sciences Get Important New Data from Spacelab Mission." *Spaceflight* 33 (December, 1991): 336-341. Description of the June, 1991, flight of *Columbia*, including a thorough account of the biomedical results obtained on this mission.

_____. "STS-37 Mission Report: Astronauts Give GRO a Helping Hand." *Spaceflight* 33 (June, 1991): 194-205. A detailed, well-illustrated account of the April, 1991, flight of *Discovery*, including a description of the Compton Gamma Ray Observatory and its scientific objectives.

Slayton, Donald K., with Michael Cassutt. *Deke! U.S. Manned Space: From Mercury to the Shuttle*. New York: Forge, 1995. This is the autobiography of the last of the Mercury astronauts to fly in space. After being grounded from flying in Project Mercury for what turned out to be a minor heart murmur, Slayton was appointed head of the Astronaut Office. Later, he commanded the Apollo-Soyuz flight in 1975. He stayed with NASA through the STS-2 flight. Although he was no longer an active participant in NASA's activities, he gives some insight into the behind-the-scenes adventures of the shuttle program.

George J. Flynn

Space Shuttle Flights, 1992

Date: January 22 to December 9, 1992
Type of mission: Piloted spaceflight

Eight space shuttle missions were flown in 1992. Several new piloted spaceflight records were set during these missions, including the longest shuttle mission to date and the most spacewalks in one mission until 1993. The fourth space shuttle orbiter, Endeavour, *was put into service, replacing* Challenger, *which had been destroyed during a launch accident in 1986.*

Key Figures

Eugene F. Kranz (b. 1933), director, Mission Operations, Johnson Space Center
Robert B. Sieck, launch director, Kennedy Space Center
Ronald J. Grabe (b. 1945), STS-42 commander
Stephen S. Oswald (b. 1951), STS-42 pilot
Norman E. Thagard (b. 1943), STS-42 mission specialist
William F. Readdy (b. 1952), STS-42 mission specialist
David C. Hilmers (b. 1950), STS-42 mission specialist
Roberta L. Bondar, STS-42 payload specialist
Ulf Merbold (b. 1941), STS-42 payload specialist
Charles F. Bolden, Jr. (b. 1946), STS-45 commander
Brian Duffy (b. 1953), STS-45 pilot
Kathryn D. Sullivan (b. 1951), STS-45 mission specialist
David C. Leestma (b. 1949), STS-45 mission specialist
C. Michael Foale (b. 1957), STS-45 mission specialist
Dirk D. Frimout (b. 1941), STS-45 payload specialist
Byron K. Lichtenberg (b. 1948), STS-45 payload specialist
Daniel C. Brandenstein (b. 1943), STS-49 commander
Kevin P. Chilton (b. 1954), STS-49 pilot
Richard J. Hieb (b. 1955), STS-49 mission specialist
Bruce E. Melnick (b. 1949), STS-49 mission specialist
Pierre J. Thuot (b. 1955), STS-49 mission specialist
Kathryn C. Thornton (b. 1952), STS-49 mission specialist
Thomas D. "Tom" Akers (b. 1951), STS-49 mission specialist
Richard N. "Dick" Richards (b. 1946), STS-50 commander
Kenneth D. Bowersox (b. 1956), STS-50 pilot
Bonnie J. Dunbar (b. 1949), STS-50 mission specialist
Ellen S. Baker (b. 1953), STS-50 mission specialist
Carl J. Meade (b. 1950), STS-50 mission specialist
Lawrence J. DeLucas, STS-50 payload specialist
Eugene H. Trinh (b. 1958), STS-50 payload specialist

Loren J. Shriver (b. 1944), STS-46 commander
Andrew M. Allen (b. 1955), STS-46 pilot
Claude Nicollier (b. 1944), STS-46 mission specialist
Marsha S. Ivins (b. 1951), STS-46 mission specialist
Jeffrey A. Hoffman (b. 1944), STS-46 mission specialist
Franklin R. Chang-Díaz (b. 1950), STS-46 mission specialist
Franco Malerba (b. 1946), STS-46 payload specialist
Robert L. "Hoot" Gibson (b. 1946), STS-47 commander
Curtis L. Brown, Jr. (b. 1956), STS-47 pilot
Mark C. Lee (b. 1952), STS-47 mission specialist
Jerome "Jay" Apt (b. 1949), STS-47 mission specialist
N. Jan Davis (b. 1953), STS-47 mission specialist
Mae C. Jemison (b. 1956), STS-47 mission specialist
Mamoru Mohri (b. 1948), STS-47 payload specialist
James D. Wetherbee (b. 1952), STS-52 commander
Michael A. Baker (b. 1953), STS-52 pilot
Charles Lacy Veach (1944-1995), STS-52 mission specialist
William M. Shepherd (b. 1949), STS-52 mission specialist
Tamara E. Jernigan (b. 1959), STS-52 mission specialist
Steven G. MacLean (b. 1954), STS-52 payload specialist
David M. Walker (1944-2001), STS-53 commander
Robert D. Cabana (b. 1949), STS-53 pilot
Guion S. Bluford, Jr. (b. 1942), STS-53 mission specialist
James S. Voss (b. 1949), STS-53 mission specialist
Michael R. "Rich" Clifford (b. 1952), STS-53 mission specialist

Summary of the Missions

The year 1992 was dubbed International Space Year. Eight space shuttle missions were flown during the year. The orbiter *Endeavour* (OV-105) was added to the fleet of active orbiter vehicles, joining *Columbia* (OV-102), *Discovery* (OV-103), and *Atlantis* (OV-104). Each of the four shuttles flew two missions during the year. Most of the missions during the year were science missions. In addition to the science missions, satellite deployment and retrieval missions were also flown, and the final flight of the year was a classified military mission.

The first space shuttle mission of the year was STS-42. The orbiter *Discovery* was launched January 22, 1992, at 14:52:32.992 Coordinated Universal Time (UTC, or 9:53 A.M. eastern standard time) from Launch Complex 39A, Kennedy Space Center (KSC). The mission was primarily a scientific mission with the International Microgravity Laboratory-1 (IML-1) flying in the shuttle's cargo bay. Landing was postponed for one day to allow for additional scientific investigations. The landing finally occurred at 16:07:18.44 UTC (8:07 A.M. Pacific standard time) on Runway 22, Edwards Air Force Base, in California.

The International Microgravity Laboratory-1 was designed to investigate the effects of weightlessness on living organisms and materials processing techniques. The term "microgravity" is used rather than the older terms "zero gravity" or "weightlessness" since all objects have some gravity, and thus the experiments experienced almost no gravity but were not really done in zero-gravity conditions.

In addition to scientific experiments, astronauts also took film footage of the STS-42 mission with

IMAX cameras for a film to be made in commemoration of the International Space Year. This film was later shown in Omniplex theaters around the country.

The second space shuttle mission of the year, STS-45, began with the launch of the space shuttle *Atlantis* from Launch Complex 39A at KSC at 13:13:39.991 UTC (8:14 A.M. eastern standard time) on March 24, 1992. The launch was one day late due to a possible fuel leak, which had been detected during the filling of the shuttle's external fuel tank. This mission, like the first space shuttle mission of the year, was primarily a scientific mission. The primary payload of the *Atlantis* was the Atmospheric Laboratory for Applications and Science (ATLAS-1), which allowed astronauts to study the Earth's atmosphere from above. STS-45 ended with *Atlantis* landing on April 2, 1992, at 11:23:05 UTC (6:23 A.M.) on Runway 33 at KSC.

STS-45 was an international mission, with instruments in the ATLAS-1 investigations being provided by institutions and scientific agencies in seven different nations. ATLAS-1 allowed space scientists to study the composition and dynamics of the Earth's upper atmosphere from a unique perspective, from above. In addition to providing data for other ongoing satellite studies of the atmosphere, ATLAS-1 was designed to provide a set of comparison data for the later ATLAS missions, scheduled for space shuttle flights over the next few years.

The third space shuttle mission of 1992, STS-49, was the maiden voyage of the space shuttle *Endeavour.* Launch was from Pad 39B on May 7, 1992, at 23:40:00.019 UTC (7:40 P.M. eastern daylight time). The primary goal of this mission was the rescue of the Intelsat VI (F-3) satellite, which had been stranded in a useless orbit due to the malfunction of its Titan launch vehicle. Difficulties with the satellite rescue portions of the mission caused mission controllers to extend the mission two extra days to complete all of the flight's objectives. Landing occurred at 20:57:38 UTC (1:58 P.M. Pacific daylight time), on Runway 22 at Edwards Air Force Base, in California.

The rescue of the Intelsat VI satellite proved to be much more difficult than had been anticipated. The satellite had not been designed to be retrieved in space and thus did not have any good places for the astronauts to grab it with the shuttle's robot arm, the Remote Manipulator System (RMS). Difficulties in attaching a capture bar to the satellite for the RMS to grab forced a new, more daring plan to capture the satellite. Astronauts built a bridge structure across the shuttle's cargo bay, and three astronauts, Pierre J. Thuot, Richard J. Hieb, and Thomas D. "Tom" Akers, stood on the bridge and manually grasped the satellite to stabilize it so that the capture bar could be attached. A new kick motor was attached to the satellite to push it into a useful orbit, and after several unsuccessful attempts, the satellite was finally redeployed.

The fourth mission of the year, STS-50, began at 16:12:22.997 UTC (12:12 P.M., eastern daylight time) on June 25, 1992, with the launch of the space shuttle *Columbia* from Pad 39A. STS-50 was another science mission, studying the effects of microgravity on materials processing and on living organisms. The primary payload was the United States Microgravity Laboratory-1 (USML-1). The USML-1 experiments were similar to those performed in IML-1 in the STS-42 mission earlier in the year. The orbiter *Columbia* had been extensively refurbished during the year since its last flight, enabling STS-50 to be an extended duration mission of more than thirteen days, marking the longest space shuttle mission to that time. Landing was at KSC on Runway 33 at 11:42:27 UTC (7:42 A.M.) on July 9, 1992.

Several problems developed with the experimental equipment; however, almost all the mission objectives were met. A more serious problem developed on the second day of the flight when the new system installed on board the *Columbia* to remove carbon dioxide gas from the air on board the orbiter failed. A backup system consisting of lithium hydroxide canisters was used to absorb the carbon dioxide while the astronauts worked to repair the primary system. The regenerative carbon dioxide removal system was repaired by the sixth day of the

The United States Microgravity Laboratory (USML) in the space shuttle Columbia's *payload bay during the STS-50 mission.* (NASA)

USML-1 missions flown aboard the space shuttle, but the investigations last for longer durations than a single space shuttle flight. The STS-57 mission retrieved EURECA nearly one year later.

The TSS-1 mission objectives were never met. TSS-1 was to have been deployed on a 19-kilometer-long tether. As the satellite and its tether swept through space, scientists would have been able to study the Earth's magnetic field. The tether was to have been used as a test of generating electric power from the Earth's magnetic field, perhaps for a future orbiting space station. Because of a technical problem (a protruding bolt), the tether could be released to only about 260 meters. A reflight of the tether system (TSS-1R) was flown in February, 1996, on STS-75.

On September 12, 1992, the space shuttle *Endeavour* lifted off from Pad 39B at 14:23:00.010 UTC (10:23 A.M. eastern daylight time) to begin the sixth space shuttle mission of the year, STS-47. This mission was another science mission, studying the effects of microgravity. The primary payload was Spacelab-J, the first United States and Japan joint space shuttle mission. The mission was extended one day to allow for additional data to be gathered. Landing was on Runway 33 at KSC at 12:53:23.0 UTC (8:53 A.M. eastern daylight time) on September 20, 1992.

Astronauts conducted forty-three materials science and life science experiments during the Spacelab-J mission. This mission had an unusually large number of life science experiments. Among the firsts of this mission was the Frog Embryology Experiment, which marked the first study of an animal's ovulation, fertilization, and development under microgravity conditions. Astronauts also studied the effectiveness of using biofeedback techniques to control motion sickness, which commonly affects space travelers.

The seventh space shuttle mission of 1992 was STS-52. The space shuttle *Columbia* began the mis-

mission, and the mission was able to continue as planned.

The fifth mission of the year was STS-46. The space shuttle *Atlantis* lifted off from Pad 39B at 13:56:48.011 UTC (9:57 A.M. eastern daylight time) on July 31, 1992. This mission was both a science and a satellite deployment mission, requiring the deployment of two satellites. The European Retrievable Carrier (EURECA) was deployed by STS-46 to be retrieved later in another shuttle mission. Another satellite scheduled for deployment was the Tethered Satellite System (TSS-1). Difficulties with deployment of both primary payloads forced an extension of the mission of one day. Landing was August 8, 1992, at 9:12 A.M. eastern daylight time on Runway 33 at KSC.

The first part of the mission involved deployment of the EURECA satellite. This satellite was designed to investigate the effects of microgravity and space radiation on materials and living organisms. This is a similar mission to the IML-1 and

sion October 22, 1992, at 17:09:39.007 UTC (1:10 P.M. eastern daylight time) from Pad 39B at KSC. The mission was a joint science mission and satellite deployment mission. Many of the science experiments were performed as part of the United States Microgravity Payload-1 (USMP-1) in the cargo bay. STS-52 also deployed the Laser Geodynamic Satellite II (LAGEOS 2). Landing was at KSC on Runway 33 at 14:05:52 UTC (9:06 A.M. eastern standard time) on November 1, 1992.

The USMP-1 payload performed quite well. USMP-1 allowed scientists on the ground an opportunity to interact with their experiments in orbit using a new system for relaying telemetry from the experiments to the ground for immediate analysis. This allowed scientists to issue commands to the USMP-1 investigations while the experiments were under way and after initial data had already been examined.

LAGEOS 2 was deployed on the second day of the mission. A rocket motor on the satellite lifted it from the space shuttle's low-Earth orbit to a higher circular orbit. The LAGEOS 2 satellite replaced the LAGEOS 1 satellite launched in 1976 on a Delta rocket. These satellites were designed to use laser-ranging systems to study motions of Earth's crust, tides, and Earth's rotation.

The eighth, and final, space shuttle mission of the year was a Department of Defense mission, STS-53. The mission began with the launch of *Discovery* from Launch Complex 39A at KSC on December 2, 1992, at 13:23:59.993 UTC (8:24 A.M. eastern standard time). The primary mission was the deployment of a classified military payload. Landing was at 20:43:47 UTC (12:44 P.M. Pacific standard time), on Runway 22 at Edwards Air Force Base, in California, on December 9, 1992.

In addition to the classified primary payload, STS-53 astronauts experimented with a new system developed to determine the positions of targets on the Earth from orbit. The system is called the Handheld Earth-oriented, Real-time, Cooperative, User-friendly, Location-targeting and Environmental System (HERCULES). HERCULES allows astronauts to point a camera at a feature on Earth,

take its picture, and determine its latitude and longitude to an accuracy of less than 4 kilometers. The crew was unable to deploy the Orbital Debris Radar Calibration System (ODERACS) because of payload equipment malfunction.

Contributions

The materials science microgravity experiments performed in IML-1 (STS-42), USML-1 (STS-50), Spacelab-J (STS-47), and USMP-1 (STS-52) allowed scientists and engineers to learn how crystals grow in space. All crystals grown on Earth experience gravitational forces, and these forces are all in one direction during the formation of the crystal. Crystals are free to grow with minimal external influences under microgravity conditions, such as aboard the shuttle. These crystals tend to be more pure and contain fewer defects than those grown on Earth. Crystals also tend to grow larger in orbit than they do on the Earth. The EURECA satellite also had materials science experiments aboard, but the results of these experiments would not become known until the satellite was retrieved during the STS-57 mission in 1993.

In addition to crystal growth studies, one of the major investigations of the USMP-1 payload was the study of the lambda point transition of helium. This transition is the change of state that occurs in helium at 2.17 kelvins (degrees above absolute zero). At this temperature, liquid helium changes from an ordinary liquid to a superfluid. Superfluids have properties different from ordinary liquids, including the ability to flow without viscosity. While this transition of helium from normal liquid to superfluid had been studied, it had never been studied in microgravity.

Numerous life science experiments were performed during almost every space shuttle mission. Every mission involved study of the astronauts themselves as they adapted to microgravity conditions. Some surprises indicated that seeds tended to sprout sooner and plants grow more rapidly in space than on Earth. This finding was particularly apparent during the IML-1 mission. The Spacelab-J mission studied the growth of insects in orbit. Experiments with frogs provided the first studies of the conception and early developmental stages of animals under microgravity conditions. These studies allow scientists to learn in part what development is due to genetics and what development is environmental.

The ATLAS-1 mission (STS-45) provided valuable data on the Earth's atmosphere. Many of these investigations were designed to establish a baseline from which to make comparisons for other atmospheric studies. Ozone levels in the upper atmosphere were studied along with other measurements of atmospheric composition. ATLAS-1 discovered residual aerosols from the huge volcanic eruption in 1991 of Mount Pinatubo in the Philippines.

The STS-49 mission to rescue a stranded satellite provided valuable experience in securing and servicing a satellite from the space shuttle. This experience was put to use over a year later during the STS-61 mission to service and repair the Hubble Space Telescope.

Not all the knowledge gained by the space shuttle flights of 1992 was gained by experiments in orbit. Some useful engineering data was gathered in the landing of the orbiter. *Endeavour* and the newly refurbished *Columbia* employed drag chutes upon landing. These chutes allowed for shorter stopping distances on landing, thus adding an extra margin of safety. Unfortunately, these chutes tended to sometimes cause the orbiter to veer slightly to the side of the runway centerline during landing. This may have been due to crosswinds at landing.

Context

The space shuttle flights of 1992 continued the saga of piloted spaceflight with the setting of several new records. Among those records were a record number of four spacewalks, or extravehicular activities (EVAs). Two of the EVAs of the STS-49 mission became the longest EVAs conducted by astronauts, with one lasting 8 hours, 29 minutes and the other lasting 7 hours, 45 minutes. Both of these activities surpassed the previous record of 7 hours

and 37 minutes held by Apollo 17 astronauts on the surface of the Moon.

In addition to long spacewalks, *Columbia* spent more than thirteen days in space. This marked a new endurance record for piloted spacecraft. Only space station missions, such as Skylab or the Russian space station, Mir, have lasted longer than the STS-50 mission. The importance of long-duration missions is that they allow time for additional experiments or for space construction activities, such as would be needed to build a space station in orbit.

One of the chief rationales for the space shuttle program given in the 1970's was that such a system would allow astronauts to deliver satellites to orbit or to retrieve satellites already in orbit. The space shuttle deployed EURECA and a defense payload. Attempts to deploy TSS-1 failed, yet valuable information was gained from the attempt.

Much more important, the STS-49 mission proved the importance of the space shuttle system. By retrieving and replacing the motor of the Intelsat VI satellite, STS-49 saved millions of dollars that would have been lost replacing the satellite with another using yet another expendable launch vehicle. This rescue mission, in fact, required human ingenuity and could not have been done with a robotic spacecraft.

The microgravity experiments were vital to our understanding of how materials and living organisms adapt to gravity or the lack of gravity. Future humans may live in space, and it is important to know how long-term exposure to the lack of gravity or to the radiation in space affects living organisms. The plant studies are vital to any attempt that may be made to try to grow food in space. The extended missions allowed further studies to be made than would have been possible with shorter-duration flights, yet they also pointed to the need for a permanent piloted space station to conduct even longer-duration scientific investigations. The materials science investigations are essential to our understanding of crystal structure. Most of modern electronics depends upon solid-state technology, and this technology utilizes the crystal structure of semiconductors. Better understanding of crystal and crystal growth helps in the development of better computer chips. The higher purity and greater perfection of crystals grown in space would tend to indicate that space-based manufacturing of these crystals may be in our future.

The ATLAS-1 mission was the first of ten scheduled ATLAS missions. These missions provide data on the Earth's atmosphere from above, a perspective impossible for ground-based scientists.

See also: Atmospheric Laboratory for Applications and Science; Compton Gamma Ray Observatory; Funding Procedures of Space Programs; National AeroSpace Plane; Space Shuttle Flights, 1991; Space Shuttle Mission STS-49; Space Shuttle Flights, 1993; Strategic Defense Initiative; Tethered Satellite System; Ulysses: Solar Polar Mission; Upper Atmosphere Research Satellite.

Further Reading

Branley, Franklyn M. *From Sputnik to Space Shuttles.* New York: Thomas Y. Crowell, 1986. This is a very good juvenile book on the development of space exploration, from robotic satellites to the modern space shuttle. The book emphasizes the uses of the space shuttle more than the shuttle itself.

DeWaard, E. John, and Nancy DeWaard. *History of NASA: America's Voyage to the Stars.* New York: Exeter Books, 1988. This book is intended for the general audience. The last chapter of the book covers the beginnings of the space shuttle program. The need for a system such as the space shuttle is discussed. The book ends with the explosion that destroyed *Challenger* in 1986.

Harland, David M. *The Space Shuttle: Roles, Missions, and Accomplishments.* Hoboken, N.J.: John Wiley, 1998. *The Space Shuttle* is written thematically, rather than purely chronologically. Topics include shuttle operations and payloads, weightlessness, materials processing, exploration, Spacelabs and free-flyers, and the shuttle's role in the International Space Station.

Harrington, Philip S. *The Space Shuttle: A Photographic History.* San Francisco, Calif.: Brown Trout, 2003. With one hundred full-color photographs by Roger Ressmeyer and others and with text by popular astronomy writer Phil Harrington, this book tells the story of the space shuttle program from 1972 to 2003. Its beautiful photographs allow the general reader to survey the history of the space shuttle program and be uplifted by the pioneering spirit of one of humanity's grandest enterprises.

Jenkins, Dennis R. *Space Shuttle: The History of the National Space Transportation System: The First 100 Missions.* Stillwater, Minn.: Voyageur Press, 2001. This is a concisely written technical reference account of the space shuttle and its ancestors, the aerodynamic lifting bodies. It details some of the advantages and inherent disadvantages of using a reusable space vehicle. Each of the vehicles is illustrated by line drawings. The book follows the space shuttle from its original concepts and briefly chronicles its first one hundred flights.

Joëls, Kerry Mark, and Gregory P. Kennedy. *The Space Shuttle Operator's Manual.* Designed by David Larkin. Rev. ed. New York: Ballantine Books, 1988. This book contains a wealth of information on space shuttle systems and flight procedures. It is written as a manual for imaginary crew members on a generic mission and will be appreciated by anyone interested in how the astronauts fly the orbiter, deploy satellites, conduct spacewalks, and live in space. Contains many drawings and some photographs of equipment.

National Aeronautics and Space Administration. *Space Shuttle Mission Press Kits.* http://www-pao.ksc.nasa.gov/kscpao/presskit/presskit.htm. Provides detailed preflight information about each of the space shuttle missions. Accessed March, 2005.

Shapland, David, and Michael Rycroft. *Spacelab: Research in Earth Orbit.* New York: Cambridge University Press, 1984. Written in the early days of the space shuttle program, this general-audience-level book describes the types of scientific missions anticipated for the space shuttle. The book describes a spacelab, which is a living/working module that is flown in the space shuttle's cargo bay during science missions.

Slayton, Donald K., with Michael Cassutt. *Deke! U.S. Manned Space: From Mercury to the Shuttle.* New York: Forge, 1995. This is the autobiography of the last of the Mercury astronauts to fly in space. After being grounded from flying in Project Mercury for what turned out to be a minor heart murmur, Slayton was appointed head of the Astronaut Office. Later, he commanded the Apollo-Soyuz flight in 1975. He stayed with NASA through the STS-2 flight. Although he was no longer an active participant in NASA's activities, he gives some insight into the behind-the-scenes adventures of the shuttle program.

Torres, George J. *Space Shuttle: A Quantum Leap.* Novato, Calif.: Presidio Press, 1986. This general-audience-level book traces the development of piloted spaceflight, culminating with the space shuttle. Included are brief descriptions on many different types of shuttle missions. The book was written at the time of the *Challenger* accident and does not include any of the missions that occurred after the resumption of the shuttle program.

Raymond D. Benge, Jr.

Space Shuttle Mission STS-49

Date: May 7 to May 16, 1992
Type of mission: Piloted spaceflight

The newest space shuttle orbiter, Endeavour, *was built as a replacement for* Challenger, *lost during a launch accident in 1986. STS-49 was the maiden voyage of* Endeavour.

Key Figures

Daniel C. Brandenstein (b. 1943), STS-49 commander
Kevin P. Chilton (b. 1954), STS-49 pilot
Richard J. Hieb (b. 1955), STS-49 mission specialist
Bruce E. Melnick (b. 1949), STS-49 mission specialist
Pierre J. Thuot (b. 1955), STS-49 mission specialist
Kathryn C. Thornton (b. 1952), STS-49 mission specialist
Thomas D. "Tom" Akers (b. 1951), STS-49 mission specialist

Summary of the Mission

Every orbiter underwent an on-the-pad test firing of its main engines prior to first launch. *Endeavour's* Flight Readiness Firing (FRF), a twenty-two-second engine burn, was performed on April 6, 1992. At the postfiring press conference, the launch team announced it was quite pleased with the test firing, and it reported no significant problems were encountered.

Endeavour's first flight was entrusted to Commander Daniel C. Brandenstein and Pilot Kevin P. Chilton. The rest of the STS-49 crew complement included five mission specialists: Richard J. Hieb, Bruce E. Melnick, Pierre J. Thuot, Kathryn C. Thornton, and Thomas D. "Tom" Akers. With the exception of Chilton, each crew member already had shuttle flight experience. Brandenstein flew as pilot on STS-8 and as commander of both STS 51-G and STS-32. Melnick flew previously on STS-41, Hieb on STS-39, Thornton on STS-33, and Thuot on STS-36.

Although not an STS-49 payload, the primary focus of STS-49 centered on the Intelsat VI satellite. (Intelsat is an acronym for International Tele-

communications Satellite Consortium, a group of 124 member nations formed in the second half of the 1960's.) Intelsat VI was launched on March 14, 1990, atop a commercial Titan launch vehicle. A malfunction had left Intelsat VI attached to a spent Titan second stage. As a result, the satellite's perigee kick motor (PKM) was unable to fire to push the satellite toward geosynchronous orbit. In order to save Intelsat, controllers jettisoned its PKM and, using only small thrusters, managed to place the satellite into a stable orbit from which it might later be rescued.

STS-49 would include three back-to-back extravehicular activities (EVAs), a first for shuttle operations. During the first EVA, astronauts would attempt to attach a new PKM to Intelsat VI. This new PKM, developed by United Technologies, weighed 10,454 kilograms and was 235 centimeters in maximum diameter and 323 centimeters long. During the second and third EVAs, two teams of astronauts would demonstrate and assess several methods of space station assembly and maintenance. During the Assembly of Space Station by EVA Methods

(ASEM) exercises, five different crew rescue device prototypes (a bi-stem pole, inflatable pole, crew propulsive device, telescopic pole, and astrorope) would be evaluated.

The National Aeronautics and Space Administration (NASA) had sponsored a contest among schoolchildren throughout the country to name the new orbiter. Representatives from the school submitting the winning entry, *Endeavour*, were present at the Kennedy Space Center (KSC) to witness launch on the evening of May 7, 1992. Problems with the master events controller and weather conditions at a transatlantic abort landing strip delayed the countdown. *Endeavour* lifted off at 23:40:00.019 Coordinated Universal Time (UTC, or 7:40 P.M. eastern daylight time) on May 7, quickly rolling to the proper heading for a 28.35 degree-inclination orbit. *Endeavour* passed through cloud patches before, during, and after the point of maximum aerodynamic pressure on the vehicle. After the crew received the go-at-throttle-up call from Mission Control, skies over KSC were clear enough to provide an unusually splendid view of solid rocket booster separation. *Endeavour* entered orbit 9 minutes and 27 seconds after liftoff. The new orbiter was in space on the anniversary of its delivery to KSC.

Intelsat VI was 560 kilometers above Earth when *Endeavour* lifted off. *Endeavour* entered a 328-kilometer-high orbit, 13,500 kilometers away from the satellite. Controllers at Intelsat's Washington, D.C., control center began a series of maneuvers to actively engage their satellite in the rendezvous phase. Intelsat VI would perform four thruster firings over the next two days to lower its orbit closer to that of *Endeavour*.

The astronauts prepared for the Intelsat capture EVA. First they lowered the cabin pressure to 70,326 pascals to minimize the amount of time EVA astronauts would have to prebreathe pure oxygen before leaving the safety of *Endeavour*'s air lock. Second, the EVA teams checked out all of their space suits, making sure they were ready to support the planned EVA. Third, the Remote Manipulator System (RMS) was unberthed and commanded

through a set of preprogrammed maneuvers to verify it could execute required motions during the EVA.

Late in the evening of May 9, the astronauts were remote participants in activities being held at the Peabody Hotel in Orlando, Florida. Five of the original Mercury astronauts were being honored at the Give Kids the World annual banquet. The crew spoke to John H. Glenn, Jr., Donald K. "Deke" Slayton, L. Gordon Cooper, Jr., M. Scott Carpenter, and Alan Shepard, and also with U.S. Representative James Bacchus (D-Florida).

Thornton received special Mother's Day (May 10) greetings from her three daughters. The wake-up music they selected for their mother was from the film version of *Winnie the Pooh*. *Endeavour* was in a 305-by-300-kilometer orbit when the crew awoke.

The astronauts got their first glimpse of Intelsat VI around 16:30 UTC, when the satellite was still a little more than 75 kilometers away. Thuot and Hieb exited the air lock at 20:42 UTC, when Intelsat was still 270 meters from *Endeavour*. Brandenstein moved cautiously closer to Intelsat at a 0.3-meter-per-second closing rate. Thuot positioned himself on the RMS arm's end effector. He was moved close to the satellite by Melnick, who operated the RMS from inside *Endeavour* at its aft flight station. Thuot reached out with a specially designed capture bar to grasp the satellite and stop its rotation. Both the astronauts and Mission Control were surprised when Thuot failed on his first attempt. Interaction dynamics between the capture bar and Intelsat proved to be different from what Thuot had expected, based upon his training experiences. He was amazed at how reactive the massive satellite was to his application of relatively small forces. The satellite began to wobble. Thuot's second capture attempt also failed and resulted in an even greater coning motion, the cone angle estimated to be as much as 45°.

EVA 1 was terminated and Hieb and Thuot were recalled into *Endeavour*'s payload bay. Mission Control ordered the flight crew to execute a separation maneuver that would put 37 kilometers between *Endeavour* and Intelsat. During the night, Intelsat

Mission Specialists Thomas D. Akers (back) and Kathryn C. Thornton in the payload bay of Endeavour, *working on the Multipurpose Experiment Support Structures (MPESS) while in Earth orbit on STS-49 in May, 1992.* (NASA)

controllers stabilized the satellite for a second EVA attempt to capture the errant satellite.

The astronauts were awakened around 13:30 UTC on May 11. The EVA team exited the air lock around 21:10 UTC to begin preparations for another try at Intelsat VI capture. Prior to going after the satellite itself, Thuot attached himself to the RMS end effector and allowed Melnick to maneuver him to the side of the orbiter. There, Thuot practiced bumping the capture bar against *Endeavour* in a test of the dynamics of the bar and capture procedure.

Thuot was nearly perfectly positioned for his second capture attempt. At 23:18 UTC, he inserted part of the capture bar into the center of the satellite's underside and began to fire the latches that should have rigidified the capture bar to the satellite. Instead of firing properly, Thuot's efforts gave the satellite a small rate away from him and induced further wobbling.

The astronauts spent the next several minutes discussing what had gone wrong during this latest capture attempt and watched the satellite's motion to ascertain how best to next approach Intelsat. It was decided to try again using the original method of capture. The crew pulled Thuot back after this third failed capture attempt and then maneuvered him in even closer on the next try. Mission Control asked the astronauts about the possibility of what might have seemed a rather desperate plan— Thuot grabbing the satellite by hand.

Try as he might, Thuot's efforts continued to fail. EVA 2 was ordered terminated, having lasted 5 hours, 30 minutes. Thuot attempted a total of five captures before he induced unacceptable wobbles into Intelsat VI's motion.

NASA management took a hard look at a three-person EVA, with astronauts hand-grasping the large satellite. During a brief televised conference, Brandenstein pushed gently for approval of such a move. Just before midnight UTC, Mission Control approved the plan.

EVA preparations on May 13 were uneventful. After two previous EVAs, the astronauts had become quite proficient in preparing their space suits and other EVA equipment, and running through the checklist for depressurization and payload bay entry.

Once outside again, Hieb put himself into his assigned foot restraint. He held the capture bar nearby. Thuot assisted Akers into his foot restraint. From inside the orbiter, Melnick moved Thuot on the RMS arm to a position 120° away from each of his fellow spacewalkers. Hieb was stationed on the starboard longeron, and Akers was in the middle of the payload bay on the ASEM strut.

Brandenstein had to level *Endeavour* and the three spacewalkers relative to Intelsat's base. The astronauts grabbed Intelsat at its solar drum positioners, solid devices that would not be damaged by human contact, and slowly nulled the satellite's rotation. The astronauts spent nearly half an hour getting comfortable holding the huge satellite before going ahead with the capture bar installation. Akers and Thuot secured the capture bar's left end to Intelsat's aft ring. With the satellite attached to the capture bar, Melnick was able to grapple the bar's fixture and maneuver Intelsat. He moved Intelsat over to the PKM's location for attachment. Latches were thrown, and all indications suggested a hard mating between Intelsat and its new PKM. This marked the first time astronauts attached a live rocket motor to an orbiting satellite.

EVA 3 ran into the early morning hours of May 14. Thornton and Melnick began preparations for Intelsat deployment. The two astronauts would command deployment from the aft flight deck. Astronauts in the air lock would stand ready to return to the payload bay if an anomaly prevented deployment of Intelsat. After commands failed twice, the third time was the charm. Intelsat was deployed at 04:53 UTC, and it slowly departed *Endeavour*'s bay with a slow rotation about its symmetry axis provided by a spring-loaded deployment mechanism referred to as a super-zip. Down in Mission Control, the flight team applauded. Intelsat was deployed near orbital apogee. *Endeavour* then executed a separation maneuver that would place the orbiter a safe distance away from the satellite by the time of its PKM burn.

When it was all over, EVA 3 set a number of impressive records. It was the longest spacewalk ever—8 hours and 29 minutes—the first to include three space-suited astronauts, and the one hundredth spacewalk in the history of piloted spaceflight.

Mission Control informed the astronauts that STS-49 would be extended a full day in order to

provide them an ample rest period following upcoming EVA 4 activities and to give them plenty of time to prepare for reentry and landing operations.

Intelsat controllers, at their headquarters in Washington, D.C., commanded a sequencer to ignite the satellite's new PKM at 17:25 UTC on May 14. The burn was nominal and the astronauts had the added bonus of viewing it.

A problem threatened the viability of EVA 4—failure of Thornton's space suit display electronics. However, telemetry was being sent down to Mission Control, so the EVA was given a go to proceed under the provision that Mission Control would advise Thornton on her suit performance. Akers and Thornton left the air lock at 21:30 UTC on May 14 to perform abbreviated ASEM work, combining the most important aspects of the original plans for EVA 2 and EVA 3 into a single spacewalk. EVA 4 ended after Thornton and Akers stowed the ASEM hardware. Its duration was 7 hours and 45 minutes.

May 15 provided a leisurely day to prepare to end the mission. One thruster developed a minor leak following a hot-fire test, but that ceased after thermal restabilization. A news conference was held from orbit. The astronauts were told that President Bush had called NASA Administrator Daniel S. Goldin to congratulate the agency on its successful capture/redeploy of Intelsat VI.

The crew's final sleep period began just prior to 5:00 UTC on May 16. They were awakened at 12:40 UTC to prepare *Endeavour* for deorbit and landing. Runway 22 at Edwards Air Force Base was the selected landing site.

Deorbit was performed an hour before scheduled landing. Entry was nominal, and *Endeavour* touched down on the concrete runway at 20:57:38 UTC, ending the eight-day, 21-hour, 17-minute, and 38-second mission. *Endeavour*'s rollout took only 58 seconds under the braking effect of the red, white, and blue drag chute deployed from the base of *Endeavour*'s vertical stabilizer immediately following nosewheel touchdown.

Intelsat controllers executed an interim orbital maneuver on May 17, sending the satellite out of its

PKM-provided 36,800-by-312-kilometer transfer orbit, leaving it in a circular geosynchronous one. After a week, the satellite was on station ready for initial preoperational testing.

Contributions

STS-49 carried the latest version of the Protein Crystal Growth (PCG) facility, an experiment that had flown on twelve previous shuttle missions. On STS-49, the PCG was used to grow protein crystals of bovine insulin from solution. The Air Force Maui Optical System (AMOS) was tested again on STS-49. An electro-optical facility on the island Maui recorded orbiter signatures during orbital maneuvering System (OMS) and RCS thruster firings. Another part of the ambitious STS-49 flight plan was a space shuttle demonstration of the Assembly of Station by EVA Methods (ASEM).

The ASEM Truss had been partially constructed to support the EVA 3 Intelsat grab. Thornton and Akers completed that 4.5-by-4.5-by-4.5-meter pyramid-shaped truss assembly. ASEM work involving evaluation of astronaut large mass handling capability was eliminated from EVA 4. Thornton and Akers also skipped the part of the ASEM work devoted to evaluating the orbiter's nose as an assembly station for Space Station Freedom's components. Some evaluations of crew self-rescue aids were performed.

Context

Endeavour, orbiter OV-105, was authorized in late 1986 as the replacement for *Challenger* (OV-099), lost on mission STS 51-L during a launch accident. Rockwell International had to reassemble its space shuttle orbiter production line at the Palmdale plant. *Endeavour* was completed ahead of schedule and slightly under budget, a first for a shuttle orbiter. *Endeavour*'s design included technology updates and the ability to remain in orbit for up to twenty-eight days. *Endeavour* was rolled out of Palmdale plant 42 on April 25, 1991, to the theme from the motion picture *2001: A Space Odyssey* as thousands of spectators cheered. *Endeavour* had cost $1.9 billion and took 2.5 million manufac-

turing hours to complete after assembly began in August, 1987. *Endeavour* assembly flow benefited heavily from lessons learned during construction of previous shuttle orbiters. For example, *Columbia* required 55 hours per tile to complete installation of the 32,000 Thermal Protection System (TPS) tiles. *Endeavour*'s 24,000 TPS tiles were installed with only 11 hours per tile required. *Endeavour* was also 4,100 kilograms lighter than *Columbia* (the first production orbiter).

Endeavour began its maiden voyage on the specified launch date, May 7, 1992, performing with fewer anomalies than any other shuttle orbiter experienced during its first mission. The flight crew, through extraordinary efforts, was able to fulfill all STS-49 mission objectives and set a few spaceflight records in the process: first three-person EVA and first shuttle mission to support four EVAs.

STS-49 was Goldin's first mission as the new NASA administrator. Following STS-49, he formed a group of experts to investigate NASA pricing policies. Intelsat would easily generate more than $1 billion in revenue as a result of its salvage on STS-49, yet the Intelsat consortium paid NASA only $93 million in compensation for services rendered. It had also paid $46 million for the new PKM, but the total cost to Intelsat hardly covered the total expense of NASA training and flight operations for the STS-49 Intelsat rescue.

Through mid-2002, Intelsat was operational at 35,792 kilometers above its orbital position at 24.5° west longitude over the Atlantic Ocean. Its replacement, Intelsat 905, launched on time at 06:44 UTC on June 5, 2002. It arrived in its proper orbit after taking a nearly 21-minute ride atop an Ariane 4 rocket. Intelsat 905 assumed responsibility for providing television service to much of the eastern United States, South America, Europe, Africa, and parts of the Middle East.

See also: Astronauts and the U.S. Astronaut Program; Funding Procedures of Space Programs; Launch Vehicles; Launch Vehicles: Reusable; Space Shuttle; Space Shuttle: Microgravity Laboratories and Payloads; Space Shuttle Flights, 1992; Space Shuttle Flights, 1993.

Further Reading

"*Endeavour*'s Intelsat Rescue Sets EVA, Rendezvous Records." *Aviation Week and Space Technology*, May 18, 1992, 22-24. Describes the rendezvous sequence that led up to capture of Intelsat VI, and the EVA techniques used to attach a new perigee kick motor to the satellite.

"EVAs to Influence Development of Space Station Hardware." *Aviation Week and Space Technology*, May 18, 1992, 25-26. Describes lessons learned during the STS-49 EVAs that are applicable to construction of a space station in low-Earth orbit.

Harland, David M. *The Space Shuttle: Roles, Missions, and Accomplishments*. Hoboken, N.J.: John Wiley, 1998. *The Space Shuttle* is written thematically, rather than purely chronologically. Topics include shuttle operations and payloads, weightlessness, materials processing, exploration, Spacelabs and free-flyers, and the shuttle's role in the International Space Station.

Harrington, Philip S. *The Space Shuttle: A Photographic History*. San Francisco, Calif.: Brown Trout, 2003. With one hundred full-color photographs by Roger Ressmeyer and others and with text by popular astronomy writer Phil Harrington, this book tells the story of the space shuttle program from 1972 to 2003. Its beautiful photographs allow the general reader to survey the history of the space shuttle program and be uplifted by the pioneering spirit of one of humanity's grandest enterprises.

Jenkins, Dennis R. *Space Shuttle: The History of the National Space Transportation System: The First 100 Missions*. Stillwater, Minn.: Voyageur Press, 2001. This is a concisely written technical reference account of the space shuttle and its ancestors, the aerodynamic lifting bodies. It details some of the advantages and inherent disadvantages of using a reusable space vehicle. Each of the vehicles is illustrated by line drawings. The book follows the space shuttle from its original concepts and briefly chronicles its first one hundred flights.

"Mission Control Saved Intelsat Rescue from Software, Checklist Problems." *Aviation Week and Space Technology*, May 25, 1992, 78-79. Describes problems encountered during the Intelsat rescue, and also the first entry and landing of *Endeavour*.

National Aeronautics and Space Administration. *STS-49 Press Kit*. Washington, D.C.: Author, 1992. Fully documents facts and figures concerning the flight activities, *Endeavour* space shuttle, and Intelsat satellite.

Slayton, Donald K., with Michael Cassutt. *Deke! U.S. Manned Space: From Mercury to the Shuttle*. New York: Forge, 1995. This is the autobiography of the last of the Mercury astronauts to fly in space. After being grounded from flying in Project Mercury for what turned out to be a minor heart murmur, Slayton was appointed head of the Astronaut Office. Later, he commanded the Apollo-Soyuz flight in 1975. He stayed with NASA through the STS-2 flight. Although he was no longer an active participant in NASA's activities, he gives some insight into the behind-the-scenes adventures of the shuttle program.

David G. Fisher

Space Shuttle Flights, 1993

Date: January 13 to December 13, 1993
Type of mission: Piloted spaceflight

Seven space shuttle missions were flown in 1993. Several of the missions were science missions. There was also a mission to retrieve a satellite left in space a year earlier by another space shuttle as well as missions to deploy new satellites. The crown jewel of the 1993 space shuttle missions, though, was the service and repair mission of the Hubble Space Telescope in December by the STS-61 mission.

Key Figures

Eugene F. Kranz (b. 1933), director, Mission Operations, Johnson Space Center
Robert B. Sieck, launch director, Kennedy Space Center
John H. Casper (b. 1943), STS-54 commander
Donald R. "Don" McMonagle (b. 1952), STS-54 pilot
Mario Runco, Jr. (b. 1952), STS-54 mission specialist
Gregory J. Harbaugh (b. 1956), STS-54 mission specialist
Susan J. Helms (b. 1958), STS-54 mission specialist
Kenneth D. Cameron (b. 1949), STS-56 commander
Stephen S. Oswald (b. 1951), STS-56 pilot
C. Michael Foale (b. 1957), STS-56 mission specialist
Kenneth D. Cockrell (b. 1950), STS-56 mission specialist
Ellen Ochoa (b. 1958), STS-56 mission specialist
Steven R. Nagel (b. 1946), STS-55 commander
Terence T. "Tom" Henricks (b. 1952), STS-55 pilot
Jerry L. Ross (b. 1948), STS-55 mission specialist
Charles J. Precourt (b. 1955), STS-55 mission specialist
Bernard A. Harris, Jr. (b. 1956), STS-55 mission specialist
Ulrich Walter, STS-55 payload specialist
Hans Wilhelm Schlegel (b. 1951), STS-55 payload specialist
Ronald J. Grabe (b. 1945), STS-57 commander
Brian Duffy (b. 1953), STS-57 pilot
G. David Low (b. 1956), STS-57 mission specialist
Nancy J. Sherlock, STS-57 mission specialist
Peter J. K. "Jeff" Wisoff (b. 1958), STS-57 mission specialist
Janice E. Voss (b. 1956), STS-57 mission specialist
Frank L. Culbertson, Jr. (b. 1949), STS-51 commander
William F. Readdy (b. 1952), STS-51 pilot
James H. Newman (b. 1956), STS-51 mission specialist
Daniel W. Bursch (b. 1957), STS-51 mission specialist
Carl E. Walz (b. 1955), STS-51 mission specialist

John E. Blaha (b. 1942), STS-58 commander
Richard A. Searfoss (b. 1956), STS-58 pilot
Margaret Rhea Seddon (b. 1947), STS-58 mission specialist
William S. "Bill" McArthur, Jr. (b. 1951), STS-58 mission specialist
David A. Wolf (b. 1956), STS-58 mission specialist
Shannon W. Lucid (b. 1943), STS-58 mission specialist
Martin J. Fettman (b. 1956), STS-58 payload specialist
Richard O. Covey (b. 1946), STS-61 commander
Kenneth D. Bowersox (b. 1956), STS-61 pilot
Kathryn C. Thornton (b. 1952), STS-61 mission specialist
Claude Nicollier (b. 1944), STS-61 mission specialist
Jeffrey A. Hoffman (b. 1944), STS-61 mission specialist
F. Story Musgrave (b. 1935), STS-61 mission specialist
Thomas D. "Tom" Akers (b. 1951), STS-61 mission specialist

Summary of the Missions

Seven space shuttle missions were flown during 1993. Three of the four active space shuttle orbiters were involved in these missions. *Endeavour* (OV-105) flew three missions, while *Columbia* (OV-102) and *Discovery* (OV-103) each flew two missions. *Atlantis* (OV-104) spent most of the year being refurbished and did not fly again until 1994. All the missions involved scientific investigations; however, only three of the missions were dedicated to science only. Three missions had as a major objective satellite deployment or retrieval. The final mission of the year was the long-awaited Hubble Space Telescope servicing mission.

The first space shuttle mission of 1993 was designated STS-54. This mission began with the launch of the orbiter *Endeavour* from Launch Complex 39B at the Kennedy Space Center (KSC) on January 13, 1993, at 13:59:29.989 Coordinated Universal Time (UTC, or 8:59 A.M. eastern standard time). The primary objective of this mission was the deployment of the fifth Tracking and Data-Relay Satellite (TDRS-F). The *Endeavour* landed on January 19, 1993, on Runway 33 at KSC at 13:37:47 UTC (8:38 A.M. eastern standard time).

The TDR satellites relay telemetry from space shuttles, the International Space Station, the Hubble Space Telescope, and other satellites to controllers on the ground. The TDR satellites are located in geosynchronous orbits, orbits that are so high

above the surface of the Earth that it takes exactly one day to orbit the Earth. This means that satellites in geosynchronous orbits seem to always stay above one part of the Earth.

TDRS-F was deployed on the first day of the STS-54 mission. Because the space shuttle cannot fly high enough to put the satellite into a geosynchronous orbit, the satellite must be boosted into the proper orbit by a small rocket engine called the Inertial Upper Stage (IUS) booster. The IUS successfully boosted TDRS-F into its proper orbit, at which time it was renamed TDRS-6.

Crew members aboard *Endeavour* spent the remainder of the mission performing various microgravity experiments on board the orbiter, and on the final full day of the mission they spent several hours practicing extravehicular activities (EVAs) in the space shuttle's cargo bay. These activities were designed to give astronauts practice at maneuvering in weightlessness and moving heavy objects, tasks that might be necessary in repairing satellites in orbit or assembling components of a space station.

The second mission of 1993, STS-56, began at 05:28:59.986 UTC (1:29 A.M. eastern daylight time), on April 8, 1993, with the launch of the space shuttle *Discovery* from Pad 39B at KSC. Launch had initially been scheduled for April 6, 1993; however, the launch had been canceled only eleven seconds from liftoff due to an indication that there may have been a problem with a fuel valve. This was only the sixth night launch of a space shuttle. This mission was primarily a science mission studying the Earth's atmosphere from above. Landing was on Runway 33 at KSC at 11:37:19 UTC (7:37 A.M. eastern daylight time), on April 17, 1993. The landing had been delayed one day due to bad weather in Florida.

The primary payload of this mission was instrumentation for the Atmospheric Laboratory for Applications and Science-2 (ATLAS-2). The equipment was located on a Spacelab pallet in the cargo bay of the *Discovery.* A Spacelab pallet is a self-contained laboratory that can be loaded into a space shuttle. Astronauts climb into the Spacelab from the main cabin of the space shuttle to perform the experiments or make the measurements necessary for the mission.

ATLAS-2 was designed to study the Earth's atmosphere from above and to investigate changes in the Earth's atmosphere resulting from human activity and from interactions between the Sun and the atmosphere. Among the studies undertaken were studies of ozone depletion in the atmosphere, the level of pollutants in the atmosphere, and the composition of the atmosphere. The data from ATLAS-2 were compared to data collected from the ATLAS-1 mission flown one year earlier aboard the STS-45 mission.

Astronauts aboard *Discovery* also performed several astronomy, materials science, and life science experiments during the mission. In addition to the science aspects of the mission, astronauts communicated with schoolchildren around the world by shortwave radio as they passed overhead. They also made radio contact with the Russian space station Mir using the same amateur radio equipment.

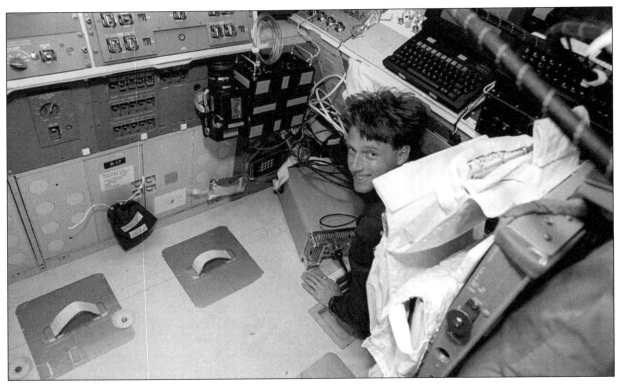

Space shuttle Discovery *(STS-56) Mission Specialist Michael Foale working in the Atmospheric Laboratory for Applications and Science (ATLAS-2).* (NASA)

The third space shuttle mission of the year was also a science mission. Designated STS-55, this mission began with the launch of *Columbia* at 14:50:00.017 UTC (10:50 A.M. eastern daylight time) on April 26, 1993, from Launch Complex 39A at KSC. *Columbia* carried into orbit a German Spacelab module called Spacelab D-2. Landing was on May 6, 1993, at 14:29:59 UTC (7:30 A.M. Pacific daylight time) on Runway 22 of Edwards Air Force Base in California.

Many of the experiments conducted in the Spacelab D-2 were extensions on the earlier German Spacelab D-1, which flew aboard the space shuttle mission 61-A in 1985. These experiments investigated the effects of microgravity on fluids, materials processing, and living organisms. While in orbit, the space shuttle experiences almost no gravitational effects, hence the term microgravity. Most scientists prefer the term "microgravity" to "zero gravity" in recognition of the fact that all mass has gravity, and thus the experiments in orbit, while away from the strong gravitational forces experienced on the Earth's surface, still experience tiny gravitational forces.

Among the materials science experiments, astronauts studied the effects of microgravity on crystal growth. Some of the life science experiments involved the astronauts themselves. The human body loses certain fluids and chemicals during spaceflight, and part of the experiments involved injections of saline solutions as an experiment investigating the body's response to direct replacement of fluid loss.

STS-57, the fourth space shuttle mission of the year, began June 21, 1993, with the launch of *Endeavour* from Launch Complex 39A at KSC at 13:07:21.989 UTC (9:07 A.M. eastern daylight time). In addition to biomedical and materials science experiments, a primary goal of the STS-57 was the retrieval of the European Retrievable Carrier (EURECA), which had been left in orbit during the STS-46 mission one year earlier. Landing was

The Tracking and Data Relay Satellite-F quickly moves away from the space shuttle Endeavour *during STS-54.* (NASA)

delayed two days due to bad weather in Florida. Landing finally occurred at 12:52:16 UTC (8:52 A.M. eastern daylight time), on Runway 33 at KSC. The EURECA satellite had been left in orbit a year earlier in order to study long-term effects of microgravity and the radiation in space on materials and living organisms. The original plan for retrieval of EURECA was for the satellite to be captured by *Endeavour*'s robot arm, called the Remote Manipulator System (RMS), and for EURECA's antennae to be retracted using power from the space shuttle. A misaligned electrical connector on the RMS prevented EURECA from getting power from *Endeavour* to retract its antenna. Astronauts were forced to don space suits and manually close the antennae after the satellite had been pulled into the shuttle's cargo bay.

During the rest of the mission, astronauts studied crystal growth and performed other materials science experiments. Astronauts also practiced manipulating equipment that might one day be needed in the construction of an orbiting space station. Astronauts also studied the effects of extended weightlessness on human posture.

The fifth space shuttle mission of the year was STS-51. This mission had numerous delays in launching. The mission was nearly two months late in launching, after twice having been stopped in the countdown mere seconds before launch. A further delay was caused by fears that the August 11, 1993, Perseid meteor shower might prove hazardous to spaceflight. Launch finally occurred at 11:45:00.006 UTC (7:45 A.M. eastern daylight time), September 12, 1993, when the space shuttle *Discovery* lifted off from Pad 39B at KSC. The primary objective of the STS-51 mission was the deployment of the Advanced Communications Technology Satellite (ACTS). Landing was at KSC, on Runway 15, at 07:56:11 UTC (3:56 A.M. eastern daylight time) on September 22, 1993. This marked the first nighttime landing of a space shuttle at KSC.

The deployment of ACTS was delayed through communication problems between *Discovery* and mission controllers on the ground. An accident that occurred during deployment caused minor damage to the interior of *Discovery*'s cargo bay. After deploying ACTS, astronauts performed various biomedical experiments and practiced EVAs that would be needed later in the year to repair the Hubble Space Telescope (HST).

The sixth space shuttle mission of the year, STS-58, was dedicated to life sciences research. The space shuttle *Columbia* lifted off at 14:53:10.0099 UTC (10:53 A.M. eastern daylight time) on October 18, 1993, from Pad 39B at KSC. Landing was at Edwards Air Force Base in California on Runway 22 at 15:05:42 UTC (7:06 A.M. Pacific standard time) on November 1, 1993.

Most of the scientific work done on the STS-58 mission was biomedical research. The long duration of this mission permitted scientists to study the physiological adjustments that occur during weightlessness. Several experiments were performed to attempt to counter the negative effects that occur to astronauts' bodies in microgravity. In addition to studies involving the astronauts, forty-eight rats were also taken into space on board *Columbia*.

The seventh and final space shuttle mission of 1993 was the long-awaited repair and servicing mission for the Hubble Space Telescope (HST). This mission, designated STS-61, involved a night launch and a night landing for the space shuttle *Endeavour*, both rare events for space shuttle missions. Launch was from Pad 39B at KSC on December 2, 1993, at 09:26:59.983 UTC (4:27 A.M. eastern standard time). Landing was on December 13, 1993, at 05:25:37 UTC (12:26 A.M. eastern standard time) on Runway 33 at KSC.

The HST had been placed into orbit three years earlier with the anticipation that it would eventually need a servicing mission to update its instruments and to replace any equipment on board that had failed during the intervening time. With this in mind, most of the equipment on board HST had been designed to be removed and replaced while the satellite was in orbit. Perhaps the most publicized repair of the mission was the Corrective Optics Space Telescope Axial Replacement (COSTAR) system. COSTAR was designed as an optical correction for a flaw that had been discovered in the telescope's primary mirror. This flaw prevented the telescope from properly focusing light, thus reducing its effectiveness for some of its scientific objectives. While correcting the telescope's optics got most of the press attention, this was only one of many tasks performed by the astronauts of the STS-61 mission.

Endeavour spent two days carefully maneuvering so it could safely grab hold of HST with its RMS arm. The astronauts had to be very careful not to damage the telescope. During the first EVA, astronauts changed two of HST's six gyroscopes, which keep the telescope pointed in the desired direction. The second EVA was to replace HST's solar panels. The original solar panels had flexed whenever the space telescope passed into or out of the Earth's shadow. This flexing caused the telescope to lose a stable fix on any object that it tried to view at that time. The new solar panels were designed not to flex when undergoing large temperature changes. The original plan had been to safely stow the old solar panels in *Endeavour*'s cargo bay and re-

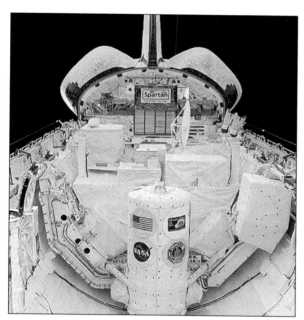

Atmospheric Laboratory for Applications and Science-2 (ATLAS-2) in the cargo bay of the shuttle Discovery. (NASA)

turn them to Earth; however, one of the solar arrays had become twisted and could not be stowed aboard the shuttle. This array was thrown overboard to eventually burn up in the Earth's atmosphere as its orbit spirals down toward the Earth.

The third EVA had two objectives. Astronauts changed out HST's original Wide Field/Planetary Camera (WFPC) with a new and more sophisticated Wide Field/Planetary Camera 2 (WFPC2). After installation of WFPC2, astronauts changed two magnetometers on the HST. The magnetometers sense the Earth's magnetic field, allowing the space telescope to orient itself relative to the Earth's magnetic field.

The fourth EVA also had two separate objectives. First, astronauts replaced HST's high-speed photometer with COSTAR. The high-speed photometer had been one of the least-used instruments on the HST, so it was sacrificed to allow installation of COSTAR to correct the flaw in the telescope's primary mirror. A second objective of this EVA was the upgrade of HST's onboard com-

puters. A fifth EVA replaced various other electronic components aboard HST. Then HST was released back into its orbit, thus successfully ending one of the most sophisticated space shuttle missions of all time.

Contributions

The ATLAS-2 investigations of STS-56 studied the composition of the Earth's atmosphere. Among the factors studied were the impact of both human-induced pollution and variations in the Sun's energy output on the composition of the upper atmosphere. ATLAS-2 measured ozone concentrations in the upper atmosphere and concentrations of greenhouse gases in the lower atmosphere. While the information gathered by ATLAS-2 is important in its own right, it is even more important when compared with that from ATLAS-1, which flew aboard STS-45 one year earlier, and ATLAS missions to be flown later. Comparisons between the measurements made by all the different ATLAS missions allow scientists to study changes in the atmosphere.

The crew of STS-58 took numerous infrared photographs of wildfires that were burning out of control in Southern California during the mission. These photographs were made available to scientists after the mission was over in the hope that some of the information in the photographs would help scientists understand the phenomena that led to those fires.

The EURECA satellite recovered by STS-57 permitted scientists to study the long-term impact on materials and organisms of exposure to microgravity and radiation in space. Spacelab D-2 of the STS-55 mission studied similar effects but for a much shorter period of time. Some of these data are still being evaluated, but some findings are available. For example, crystals grown in orbit tend to be larger and more perfect than those grown on Earth.

Astronauts aboard STS-59 studied changes in human physiology that occur in microgravity. This information is being used to design better living and working areas on board future shuttle missions. The STS-58 mission had numerous life science findings. Astronauts studied bone tissue loss that occurs in astronauts who spend extended periods of time in microgravity. Calcium appears not to be maintained in bones as efficiently in microgravity as in the comparatively heavy gravity of Earth. Several different experiments were performed to try to minimize calcium loss in astronauts. Astronauts also kept logs of fluid and food intake as well as physical activity in an attempt to determine the nutritional and energy needs of astronauts in microgravity conditions. The nutrition needs of humans in microgravity differ from those of humans under normal gravity. For the first few days of the mission, astronauts also performed experiments aimed at trying to determine the mechanism for space motion sickness, a form of motion sickness that has plagued astronauts since the early days of piloted spaceflight.

Throughout several missions during the year, astronauts conducted EVAs that were designed to provide familiarity and practice with moving around and manipulating tools and materials under microgravity conditions from within the confines of a space suit. The experience gained during these EVAs would be put into use during later HST servicing missions, as well as later satellite retrieval and repair missions and the construction of the International Space Station.

Context

The scientific investigations conducted during the year indicated the importance of space-based research. The atmospheric studies of ATLAS-2 could be done only from above the atmosphere. The studies of the atmosphere from above complement the studies done on Earth. Photographs and measurements of Earth from above give a much different perspective from those done on the Earth's surface. ATLAS-2 is the second of a series of ten ATLAS missions scheduled to study changes in the Earth's atmosphere over a decade.

The nearly weightless conditions aboard a spacecraft in orbit permit studies of materials and organisms under microgravity conditions. Material science studies often center on the study of crystal growth in orbit. These sorts of studies are very important for future technology. Much of modern technology is reliant upon semiconductors. Semiconductor devices may be manufactured with greater precision and accuracy under microgravity conditions.

Long-duration exposure to the microgravity conditions of space has demonstrated detrimental effects on the human body. Because astronauts will be exposed to long durations of microgravity in proposed missions to Mars or proposed space stations, it is important for scientists to understand the effects of microgravity on astronauts. To accomplish these studies, the space shuttle *Columbia*'s STS-58 mission was the fourth-longest piloted spaceflight in NASA's history.

The space shuttle demonstrated its versatility during the satellite deployment and retrieval missions of 1993. Only the space shuttle could have retrieved EURECA from orbit. Furthermore, the HST servicing mission also demonstrated the value of a piloted space vehicle. The space telescope could not have been repaired or upgraded without the space shuttle's unique capabilities.

The HST servicing and repair mission was perhaps the most complicated space shuttle mission flown to date. To service the space telescope, astronauts were required to perform five EVAs during the mission. This was a new record for most EVAs in a piloted spaceflight.

As of 2005 there were ten TDRS satellites in the constellation. TDRS-6 was located at 46.8° west longitude. TDRS-8 and TDRS-10 were the primary satellites, with TDRS-9 serving as a spare.

See also: Hubble Space Telescope; Hubble Space Telescope: Science; Hubble Space Telescope: Servicing Missions; Space Shuttle Flights, 1992; Space Shuttle Mission STS-49; Space Shuttle Flights, 1994; Spaceflight Tracking and Data Network; Strategic Defense Initiative; Telescopes: Air and Space; Tracking and Data-Relay Communications Satellites.

Further Reading

Branley, Franklyn M. *From Sputnik to Space Shuttles.* New York: Thomas Y. Crowell, 1986. This is a very good juvenile book on the development of space exploration, from robotic satellites to the modern space shuttle. The book emphasizes the uses of the space shuttle more than the shuttle itself.

Bruning, David. "Hubble: Better than New." *Astronomy* 22 (April, 1994): 44-49. This article has little information but lots of pictures taken with the newly repaired Hubble Space Telescope. It gives an idea of how much improved images from the space telescope will be with the new corrective optics installed.

DeWaard, E. John, and Nancy DeWaard. *History of NASA: America's Voyage to the Stars.* New York: Exeter Books, 1988. This book is intended for the general audience. The last chapter of the book covers the beginnings of the space shuttle program. The need for a system such as the space shuttle is discussed. The book ends with the explosion that destroyed *Challenger* in 1986.

Harland, David M. *The Space Shuttle: Roles, Missions, and Accomplishments.* Hoboken, N.J.: John Wiley, 1998. *The Space Shuttle* is written thematically, rather than purely chronologically. Topics include shuttle operations and payloads, weightlessness, materials processing, exploration, Spacelabs and free-flyers, and the shuttle's role in the International Space Station.

Harrington, Philip S. *The Space Shuttle: A Photographic History.* San Francisco, Calif.: Brown Trout, 2003. With one hundred full-color photographs by Roger Ressmeyer and others and with text by popular astronomy writer Phil Harrington, this book tells the story of the space shuttle program from 1972 to 2003. Its beautiful photographs allow the general reader to survey the history of the space shuttle program and be uplifted by the pioneering spirit of one of humanity's grandest enterprises.

Hoffman, Jeffrey A. "How We'll Fix the Hubble Space Telescope: An Astronaut's Anticipations." *Sky and Telescope* 21 (December, 1993): 23-29. Although this article was written prior to the space telescope servicing mission, the mission went so smoothly that there were virtually no deviations from the plans outlined in the article. The article, written by one of the astronauts to fly on the mission, gives a detailed account of the various activities involved in fixing the Hubble Space Telescope.

Jenkins, Dennis R. *Space Shuttle: The History of the National Space Transportation System: The First 100 Missions.* Stillwater, Minn.: Voyageur Press, 2001. This is a concisely written technical reference account of the space shuttle and its ancestors, the aerodynamic lifting bodies. It details some of the advantages and inherent disadvantages of using a reusable space vehicle. Each of the vehicles is illustrated by line drawings. The book follows the space shuttle from its original concepts and briefly chronicles its first one hundred flights.

Joëls, Kerry Mark, and Gregory P. Kennedy. *The Space Shuttle Operator's Manual.* Designed by David Larkin. Rev. ed. New York: Ballantine Books, 1988. This book contains a wealth of information on space shuttle systems and flight procedures. It is written as a manual for imaginary crew members on a generic mission and will be appreciated by anyone interested in how the astronauts fly the orbiter, deploy satellites, conduct spacewalks, and live in space. Contains many drawings and some photographs of equipment.

Kerrod, Robin. *Hubble: The Mirror on the Universe.* Richmond Hill, Ont.: Firefly Books Ltd., 2003. The author uses hundreds of the latest, most spectacular images from the HST to illustrate a comprehensive astronomy reference. The book explains how new discoveries are revising scientific understanding of the universe.

National Aeronautics and Space Administration. *Space Shuttle Mission Press Kits.* http://www-pao.ksc.nasa.gov/kscpao/presskit/presskit.htm. Provides detailed preflight information about each of the space shuttle missions. Accessed March, 2005.

Shapland, David, and Michael Rycroft. *Spacelab: Research in Earth Orbit.* New York: Cambridge University Press, 1984. Written in the early days of the space shuttle program, this general-audience-level book describes the types of scientific missions anticipated for the space shuttle. The book describes a Spacelab, which is a living/working module that is flown in the space shuttle's cargo bay during science missions.

Slayton, Donald K., with Michael Cassutt. *Deke! U.S. Manned Space: From Mercury to the Shuttle.* New York: Forge, 1995. This is the autobiography of the last of the Mercury astronauts to fly in space. After being grounded from flying in Project Mercury for what turned out to be a minor heart murmur, Slayton was appointed head of the Astronaut Office. Later, he commanded the Apollo-Soyuz flight in 1975. He stayed with NASA through the STS-2 flight. Although he was no longer an active participant in NASA's activities, he gives some insight into the behind-the-scenes adventures of the shuttle program.

Torres, George J. *Space Shuttle: A Quantum Leap.* Novato, Calif.: Presidio Press, 1986. This general-audience-level book traces the development of piloted spaceflight, culminating with the space shuttle. Included are brief descriptions of many different types of shuttle missions. The book was written at the time of the *Challenger* accident and thus does not include any of the missions that occurred after the resumption of the shuttle program.

Raymond D. Benge, Jr.

Space Shuttle Mission STS-61

Date: December 2 to December 13, 1993
Type of mission: Piloted spaceflight

The Hubble Space Telescope (HST), first in a series of NASA's Great Observatories, was launched in 1990 with a serious deficiency in its primary mirror. STS-61 was responsible for an impressive demonstration of the value of piloted spaceflight—the orbital repair of HST.

Key Figures

Richard O. Covey (b. 1946), STS-61 commander
Kenneth D. Bowersox (b. 1956), STS-61 pilot
Kathryn C. Thornton (b. 1952), STS-61 mission specialist
Claude Nicollier (b. 1944), STS-61 mission specialist
Jeffrey A. Hoffman (b. 1944), STS-61 mission specialist
F. Story Musgrave (b. 1935), STS-61 mission specialist
Thomas D. "Tom" Akers (b. 1951), STS-61 mission specialist
Edward Weiler, HST project scientist
Daniel S. Goldin (b. 1940), NASA administrator
Lyman Spitzer, Jr. (1914-1997), scientist who first proposed orbiting a Large Space Telescope
Edwin P. Hubble (1889-1953), astronomer for whom the HST is named

Summary of the Mission

On April 23, 1993, NASA issued a press release concerning STS-61/Hubble Space Telescope (HST) mission operations. NASA was concerned, as the list of required repairs continued growing at an alarming rate during the second quarter of 1993. Thought of lowering the scope of this first HST repair mission was not well received. Considering the bad press surrounding HST's problems, shuttle managers decided it would be preferable to attempt a more ambitious schedule and possibly not complete everything, than to set their sights lower and find the repair work easier than imagined.

Extravehicular Activity 1 (EVA 1) would consist principally of getting set up for repairs to follow, establishing work stations and unpacking tools, and replacing two Rate-Sensing Units (RSU). EVA 2's primary task would be removing and stowing the old solar arrays and putting new ones in their place. EVA 3 would be dedicated to installing the Wide Field/Planetary Camera 2 (WFPC2). EVA 4 was dedicated to installing the Corrective Optics Space Telescope Axial Replacement (COSTAR). EVA 5 would include working on areas of HST that were not originally designed to be EVA-friendly. A second magnetometer would be installed, and a coprocessor module added to the telescope's main computer.

NASA's cost estimates for the HST repair effort, a $692 million expenditure, included $429 million for STS-61, $12 million for new solar arrays from ESA, $86.3 million to characterize the HST primary mirror's spherical aberration, $49.9 million for COSTAR, and $23.8 million for WFPC2.

Within press circles, STS-61 was touted as a make-or-break mission for NASA. Public support ran somewhat on the cynical side prior to launch but quickly turned to a sense of excitement and sincere hope that the shuttle *Endeavour* astronauts would succeed in their repair efforts. Dr. Edward Weiler cautioned the press not to expect a guaranteed success, emphasizing this was a risky and complex mission.

Weather was excellent on December 2, and *Endeavour* lifted off in total darkness at 09:26:59.983 Coordinated Universal Time (UTC, or 4:27 A.M. eastern standard time). Solid-fueled rocket booster (SRB) exhaust quickly turned darkness into brilliant light as the vehicle punched off the pad and climbed along the proper flight azimuth. SRB separation was clean and first-stage performance nominal. Portions of ascent were visible for several hundred miles along the eastern sea coast. *Endeavour* flew a direct ascent trajectory to the targeted 493-by-344-kilometer orbit. Later, orbital maneuvering system (OMS) engines were fired to alter the orbit and initiate rendezvous. That burn, called OMS-2, was commanded when *Endeavour* was 9,440 kilometers behind HST. Over the next two days a sequence of small thruster firings was performed to close in slowly on HST.

The crew consisted of Commander Richard O. Covey, Pilot Kenneth D. Bowersox, Payload Commander F. Story Musgrave, and Mission Specialists Kathryn C. Thornton, Claude Nicollier, Jeffrey A. Hoffman, and Thomas D. "Tom" Akers. The astronauts were awakened at 00:00 UTC to begin their third day in space and to prepare for the final phase of rendezvous and subsequent grapple and capture operations. Maneuvers over the course of the next eight hours, each designed to slow approach speed, ultimately brought *Endeavour* and HST to within grappling range for the Remote Manipulator System (RMS) end effector. By December 4, *Endeavour* had narrowed the gap between HST and itself from 9,600 kilometers at liftoff to only 12.8 kilometers by terminal phase initiation.

The final portion of the rendezvous was flown very slowly. Claude Nicollier raised the RMS arm high above the payload bay. Four midcourse corrections were performed before HST sat 11 meters from the Orbiter's payload bay. The RMS was moved so its end effector camera's field of view was centered on the HST star tracker. *Endeavour*'s Reaction Control System (RCS) thrusters were turned off so as not to interfere with grapple operations. Nicollier grabbed HST at 08:46:56 UTC on December 4 without inducing the slightest shaking motion in the telescope. Nicollier very slowly inched HST down for berthing on the Flight Support System (FSS) in *Endeavour*'s payload bay.

The crew reported seeing more than just a bow in the starboard solar array. It had suffered a kink attributed to the thermally induced flexing that the array experienced on each orbit and to an anomalous event that occurred about eighteen months before the repair flight, the exact nature of which was unknown.

F. Story Musgrave and Jeffrey A. Hoffman started the first EVA about an hour earlier than scheduled by stepping into the cargo bay at 03:46 UTC on December 5. They began by unpacking tools, safety tethers, and work platforms. While Hoffman and Musgrave worked in the payload bay, Akers acted as choreographer, watching from the aft station inside *Endeavour* during EVA 1. Covey and Bowersox were responsible for maintaining or changing *Endeavour*'s attitude. Thornton documented the EVA using various cameras.

About one hour into EVA 1, Hoffman was ready to open up HST's aft shroud area to work inside at the RSU replacement. RSU replacement provided no surprises, as EVA 1 continued into December 5. Hoffman undid the latches, pulled out the units, handed them to Musgrave, and put in new ones.

The first difficulty of the mission surfaced as Hoffman could not simultaneously close both aft shroud doors at the top, bottom, and middle bolts. In order not to lose the advantage that had been gained from the ease of RSU replacement work, Hoffman was told to move on to change-out of the Electronic Assemblies (ECUs) for the RSUs. The astronauts then performed some work at the Solar Array Carrier (SAC), tasks that would help EVA 2 work begin quickly. Hoffman closed out both

ECUs and shut the Bay 10 door on HST at 08:10 UTC, moving on to replacing eight fuses about four and a half hours into the EVA.

The astronauts installed a portable foot restraint on the right side of HST so that Musgrave could assist Hoffman in closing the aft shroud doors. It was decided to use a bungee cord between latch handles to maintain even tension. At 10:30 UTC, Hoffman and Musgrave finally got the door bolts engaged. To retract the solar arrays, HST was rotated and pivoted on the FSS to the proper position as EVA 1 concluded after 7 hours, 54 minutes; it had been scheduled for only 6 hours.

For EVA 2, Thornton was fixed to the RMS foot restraint while Akers floated freely. The air-lock hatch was opened at 03:30 UTC on December 6. Before the EVA team exited the air lock, Mission Control rolled up the solar array that was slated for return to Earth. The damaged one would be discarded in orbit since it could not be retracted.

Thornton placed a handhold on the damaged solar array so that she could handle the array once it was disconnected from HST. Nicollier moved Thornton to the optimum position for holding the solar array once Akers completed the disconnection. Nicollier moved Thornton and the solar array high above the payload bay at 04:40 UTC to await orbital sunrise. Meanwhile, Akers started working on extracting the new solar arrays from their protective covers. Thornton simply let go of the array at 04:48 UTC. Nicollier pulled her quickly down into the bay away from the discarded array. The crew at the aft flight deck commanded the orbiter to move away from the solar array. When *Endeavour*'s exhaust hit the array, it picked up twisting and flapping motions. One large RCS firing clobbered

November 15, 1993: Launch of Endeavour *from Kennedy Space Center's Pad 39A, on its way to repair the Hubble Space Telescope.* (NASA)

the array with its shock wave. The array picked up a spin, a tumble, and a twist.

Thornton and Akers now turned their attention to the new solar arrays, removing them from protective carriers near the front of the payload bay and installing one in place of the discarded array before tackling the other array's replacement. Thornton and Akers completed all work associated with the first array by 06:00 UTC.

By 06:30 UTC, the discarded solar array was 18 kilometers away from *Endeavour*. Eventually, it was destroyed during its reentry. The first new solar array was in place and properly mated to HST. Inside the shuttle, the astronauts rotated HST 180° to provide Thornton and Akers access to the second solar array. EVA 2 concluded after Thornton and Akers did some EVA 3 prep work, setting up for WFPC removal and replacement.

EVA 3 began at 03:45 UTC on December 7. By the time Musgrave looked deep inside the WFPC cavity, nearly two hours of the EVA had passed. Hoffman put WFPC in its temporary stow position on *Endeavour*'s left side, extending over the payload bay longeron. WFPC2 installation was done even more slowly than WFPC removal. By 06:30 UTC WFPC2 was inserted. Hoffman tightened bolts and latches, and attached a grounding strap as Musgrave looked in to observe indicator lights from the fine guidance sensor bay; each light signaled the desired configuration.

For magnetometer attachment, both spacewalkers loaded an RMS caddy with tools by 08:00 UTC and were lifted up together to the work site. A small piece of multilayer thermal insulation got away from the astronauts. It moved away at a few feet per second off to the left side of *Endeavour*. This was no significant concern, as *Endeavour* would reboost HST away from the current shuttle's orbit after HST repair operations were finished.

EVA 4 began at 03:15 UTC on December 8. The first EVA 4 task to perform was removing the telephone booth-sized High Speed Photometer (HSP) from its aft shroud bay. HSP had the same mass-handling characteristics as COSTAR and was precisely the same size and shape. Thus it was a good

surrogate on which to practice before COSTAR insertion.

Shortly after 05:00 UTC, Akers returned to the open cavity from which HSP had been removed. Fifteen minutes later, he was in position, ready to assist Thornton in moving COSTAR into proper position. COSTAR installation was completed by 06:00 UTC. Thornton and Akers then released HSP from its temporary parking position. Nicollier moved Thornton back over to the forward part of the payload bay, putting HSP in COSTAR's protective enclosure.

The next job was the attachment of a new computer coprocessor. This was mounted right on top of HST's existing IBM-compatible 286 computer, using the threaded holes of previously used handles. Akers chuckled at how easy it was to make bolt contacts and electrical connector attachments. Coprocessor work was completed by 08:40 UTC, approximately 5 hours and 20 minutes into EVA 4. After preparing the payload bay for the next day's work, EVA 4 ended at 10:05:52 UTC. It had lasted 6 hours, 50 minutes, and 52 seconds.

EVA 5 began at 03:30 UTC on December 9. EVA 5's first task, change-out of the solar array drive electronics, was completed halfway through *Endeavour*'s revolution 104. The spacewalkers turned their attention next to getting the solar arrays down along the telescope's side. After that, the astronauts worked on the Goddard High Resolution Spectrometer (GHRS). A cabling kit bypassed a short circuit that had disabled one of GHRS's power supplies and essentially cut its capability in half. Installation of the redundancy kit was completed at 08:15 UTC.

The last task the astronauts undertook was to remove a small protective cover that they had placed over the telescope's low-gain antenna. HST was ready for redeployment except for unfurling the solar arrays and deployment of the high-gain antenna.

Hoffman entered the air lock, and Musgrave remained on the RMS while the first solar array deployment was commanded. Both arrays blossomed to full extension without kinks in the bi-stem. Once the arrays were deployed, HST had to be rotated to the position required for proper thermal control and even solar illumination until release. EVA 5 ended at 10:51 UTC after 7 hours, 21 minutes.

Difficulty encountered in initial attempts to communicate with the telescope's computer system delayed HST release. Nicollier grappled HST at 07:44 UTC on December 10 and then raised it off the berthing equipment. He held HST high above the bay before releasing the telescope.

Nicollier commanded the RMS end effector grapple wires to retract at 10:26:47 UTC. HST's attitude was stable as it began to fly freely again after nearly a week secured in *Endeavour*'s payload bay. HST immediately acquired the Sun and properly oriented its solar arrays. Communications through the Tracking and Data-Relay Satellite (TDRS) network was established, and HST controllers commanded the aperture door to reopen.

HST was released into an orbit 589 kilometers above the Earth's surface. Covey fired small thrusters twice to slowly separate away from HST without impinging thruster exhaust on the telescope.

On December 12, the crew gathered in *Endeavour*'s middeck and talked with reporters for over an hour. The crew began a sleep period at 14:00 UTC. They were awakened for the final time of the mission at 22:00 UTC. Weather in the KSC area was expected to degrade during the early morning hours of December 13, so Mission Control decided to bring the shuttle home one orbit early.

A double sonic boom signaled *Endeavour*'s return to the skies over Florida. Huge spotlights on the runway were turned off so that Covey could see just the landing lights. *Endeavour* flew behind the Vehicle Assembly Building on final approach and touched down on runway 33 at 05:25:37 UTC (12:26 A.M. eastern standard time) December 13, 1993.

Contributions

The scientific goals of the Hubble Space Telescope included measurement of Cepheid variable light curves for stars in galaxies at least 50 million light-years away in an attempt to improve the determination of the expansion rate of the universe; examination of gravitational influences and signatures of massive black holes in both normal and active galactic cores; determination of the age of globular clusters; and imaging faint, distant objects.

Many of these lofty goals could not be accomplished using the telescope in the state in which it was launched. After STS-61, those goals were finally within reach. WFPC2 was designed to detect objects over one hundred times fainter than those observable with ground-based telescopes, and with a tenfold improvement in spatial resolution. The unit would restore HST to its prelaunch expectations for imaging of distant, faint objects.

Final HST science operations were concluded during the early morning hours of December 3, 1993. WFPC and HSP solar system observations were made before closing HST's aperture door. For HSP, these would be its last observations. HSP would be replaced by the COSTAR corrective optics package and be returned to Earth, its final disposition uncertain.

STS-61 astronaut activities were largely devoted to HST repairs, of course. However, there were several science and technology payloads. The PILOT system, first flown on STS-58, designed to test shuttle piloting skills after a long exposure to weightlessness, was manifested on STS-61. Also, the Air Force Maui Optical System (AMOS) facility on Hawaii examined shuttle airglow, water dumps, and thruster firings when *Endeavour* passed over the island of Maui.

Five weeks after the first servicing mission, NASA held a news conference and a science briefing to discuss preliminary findings after testing the repaired Hubble. HST Scientist Edward Weiler proclaimed that Hubble was fixed beyond the Science Team's wildest expectations.

Prior to the repair mission, only about 12 percent of a star's light converged in its central image. The original requirement for HST was to have 60 percent of the light in the central image. Thus, HST suffered from a serious spherical aberration. After the repair, it was determined that the amount of light was at least 70 percent and possibly more. The diffraction limit of HST's optics was 85 percent of the light in the central image. As a result of STS-61 repair work, HST was in even better shape than had been called for in the original design specifications. In terms of a more terrestrial analog, the diffraction capability of HST after the repair mission was equivalent to detecting the light of a firefly in Tokyo, Japan, from Washington, D.C., and being able to clearly resolve two fireflies at that distance if those fireflies were no less than 3 meters apart.

Context

During the January 13, 1994, conferences, NASA Administrator Daniel S. Goldin formally declared the Hubble servicing mission successful:

> This is phase two of a fabulous, two-part success story. The world watched in wonder [last month] as the astronauts performed an unprecedented and incredibly smooth series of spacewalks. Now, we see the real fruits of their work and that of the entire NASA team. Men and women all across this agency committed themselves to this effort. They never wavered in their belief that the Hubble Space Telescope is a true international treasure.

The valiant actions of the well-trained and well-equipped STS-61 crew demonstrated the value of sending humans into space, by adapting to the un-expected and performing tasks not possible with machines. STS-61 partially restored the tarnished public image of NASA after a series of major problems such as the loss of the Mars Observer, failure to properly deploy the antenna on Galileo, the original flaws in HST's optics, and a series of nagging delays in launching space shuttle missions. With HST repaired, NASA had a world-class observatory uniquely qualified to address fundamental astrophysical issues.

In response to budgetary cuts in the wake of the STS-107 accident, NASA eliminated all space shuttle flights not directly supporting the International Space Station. This included scheduled servicing missions to the Hubble Space Telescope. NASA had planned to visit Hubble one last time in 2006 to change out instruments and replace its gyroscopes with the intent of keeping the telescope in service until at least 2011, when its heir apparent, the James E. Webb Space Telescope, is expected to launch. Scrapping the final servicing mission raises the likelihood that Hubble will fail before Webb is on orbit.

See also: Hubble Space Telescope; Hubble Space Telescope: Science; Hubble Space Telescope: Servicing Missions; National Aeronautics and Space Administration; Planetary Exploration; Space Shuttle Flights, 1993; Space Shuttle Flights, 1994; Space Shuttle Mission STS-63; Space Shuttle Mission STS-71/Mir Primary Expedition Eighteen.

Further Reading

Harland, David M. *The Space Shuttle: Roles, Missions, and Accomplishments.* Hoboken, N.J.: John Wiley, 1998. *The Space Shuttle* is written thematically, rather than purely chronologically. Topics include shuttle operations and payloads, weightlessness, materials processing, exploration, Spacelabs and free-flyers, and the shuttle's role in the International Space Station.

Harrington, Philip S. *The Space Shuttle: A Photographic History.* San Francisco, Calif.: Brown Trout, 2003. With one hundred full-color photographs by Roger Ressmeyer and others and with text by popular astronomy writer Phil Harrington, this book tells the story of the space shuttle program from 1972 to 2003. Its beautiful photographs allow the general reader to survey the history of the space shuttle program and be uplifted by the pioneering spirit of one of humanity's grandest enterprises.

Jenkins, Dennis R. *Space Shuttle: The History of the National Space Transportation System: The First 100 Missions.* Stillwater, Minn.: Voyageur Press, 2001. This is a concisely written technical reference account of the space shuttle and its ancestors, the aerodynamic lifting bodies. It details some of the advantages and inherent disadvantages of using a reusable space vehicle. Each of the vehicles is illustrated by line drawings. The book follows the space shuttle from its original concepts and briefly chronicles its first one hundred flights.

Kerrod, Robin. *Hubble: The Mirror on the Universe.* Richmond Hill, Ont.: Firefly Books Ltd., 2003. The author uses hundreds of the latest, most spectacular images from the HST to illustrate a comprehensive astronomy reference. He explains how new discoveries are revising scientific understanding of the universe. The book covers the observable universe in six sections: "Stars in the Firmament," "Stellar Death and Destruction," "Gregarious Galaxies," "The Expansive Universe," "Solar Systems," and "The Heavenly Wanderers." Clear and concise text explains the history of astronomy and the development of the HST.

National Aeronautics and Space Administration. *Exploring the Universe with the Hubble Space Telescope.* NP-126. Washington, D.C.: Superintendent of Documents, 1990. A prelaunch NASA publication that illustrates the design, construction, and mission of the Hubble Telescope. Amply illustrated. Provides historical context for the Hubble Space Telescope.

_____. *Hubble Space Telescope: Media Reference Guide.* Sunnyvale, Calif.: Lockheed Missiles and Space Company, 1990. A thorough reference guide to the Hubble Space Telescope prepared for NASA by one of the HST prime contractors. Filled with diagrams and charts describing all aspects of the HST program.

_____. *The Space Telescope.* NASA SP-392. Washington, D.C.: Superintendent of Documents, 1976. A collection of scientific works describing the capabilities of HST and the potential for advancing our understanding of the universe. Not for the casual reader.

Slayton, Donald K., with Michael Cassutt. *Deke! U.S. Manned Space: From Mercury to the Shuttle.* New York: Forge, 1995. This is the autobiography of the last of the Mercury astronauts to fly in space. After being grounded from flying in Project Mercury for what turned out to be a minor heart murmur, Slayton was appointed head of the Astronaut Office. Later, he commanded the Apollo-Soyuz flight in 1975. He stayed with NASA through the STS-2 flight. Although he was no longer an active participant in NASA's activities, he gives some insight into the behind-the-scenes adventures of the shuttle program.

Smith, Robert W. *The Space Telescope: A Study of NASA Science, Technology, and Politics.* New York: Cambridge University Press, 1989. A prelaunch tour of the history of the development of HST. Includes internal NASA decision-making processes, reactions of the scientific community, the role of politics in bringing HST to realization, and the interaction of contractors with government agencies.

David G. Fisher

Space Shuttle Flights, 1994

Date: February 3 to November 14, 1994
Type of mission: Piloted spaceflight

Seven space shuttle missions were flown in 1994. Unlike most years in the space shuttle program, the missions of this year were all dedicated science missions. Astronauts collected some of the best data ever on the Earth's atmosphere and conducted numerous experiments measuring the effects of microgravity on materials and living organisms.

Key Figures

Charles F. Bolden, Jr. (b. 1946), STS-60 commander
Kenneth S. Reightler, Jr. (b. 1951), STS-60 pilot
N. Jan Davis (b. 1953), STS-60 mission specialist
Ronald M. Sega (b. 1952), STS-60 mission specialist
Franklin R. Chang-Díaz (b. 1950), STS-60 mission specialist
Sergei Konstantinovich Krikalev (b. 1958), STS-60 mission specialist
John H. Casper (b. 1943), STS-62 commander
Andrew M. Allen (b. 1955), STS-62 pilot
Pierre J. Thuot (b. 1955), STS-62 mission specialist
Charles D. "Sam" Gemar (b. 1955), STS-62 mission specialist
Marsha S. Ivins (b. 1951), STS-62 mission specialist
Sidney M. Gutierrez (b. 1951), STS-59 commander
Kevin P. Chilton (b. 1954), STS-59 pilot
Jerome "Jay" Apt (b. 1949), STS-59 mission specialist
Michael R. "Rich" Clifford (b. 1952), STS-59 mission specialist
Linda M. Godwin (b. 1952), STS-59 mission specialist
Thomas D. "Tom" Jones (b. 1955), STS-59 mission specialist
Robert D. Cabana (b. 1949), STS-65 commander
James D. Halsell, Jr. (b. 1956), STS-65 pilot
Richard J. Hieb (b. 1955), STS-65 mission specialist
Carl E. Walz (b. 1955), STS-65 mission specialist
Leroy Chiao (b. 1960), STS-65 mission specialist
Donald A. Thomas (b. 1955), STS-65 mission specialist
Chiaki Mukai (b. 1952), STS-65 payload specialist
Richard N. "Dick" Richards (b. 1946), STS-64 commander
L. Blaine Hammond, Jr. (b. 1952), STS-64 pilot
Jerry M. Linenger (b. 1955), STS-64 mission specialist
Susan J. Helms (b. 1958), STS-64 mission specialist
Carl J. Meade (b. 1950), STS-64 mission specialist
Mark C. Lee (b. 1952), STS-64 mission specialist

Michael A. Baker (b. 1953), STS-68 commander
Terrence W. "Terry" Wilcutt (b. 1949), STS-68 pilot
Steven L. Smith (b. 1958), STS-68 mission specialist
Daniel W. Bursch (b. 1957), STS-68 mission specialist
Peter J. K. "Jeff" Wisoff (b. 1958), STS-68 mission specialist
Thomas D. "Tom" Jones (b. 1955), STS-68 mission specialist
Donald R. "Don" McMonagle (b. 1952), STS-66 commander
Curtis L. Brown, Jr. (b. 1956), STS-66 pilot
Ellen Ochoa (b. 1958), STS-66 mission specialist
Joseph R. Tanner (b. 1950), STS-66 mission specialist
Jean-François Clervoy (b. 1958), STS-66 mission specialist
Scott E. Parazynski (b. 1961), STS-66 mission specialist

Summary of the Missions

Seven space shuttle missions were flown during 1994, with all four active orbiters participating in missions. The orbiters *Columbia* (OV-102), *Discovery* (OV-103), and *Endeavour* (OV-105) each flew two missions. The orbiter *Atlantis* (OV-104) flew only one mission, near the end of the year. All the space shuttle missions of 1994 were dedicated science missions. Among the new records set during the year were the longest space shuttle mission and the first Russian cosmonaut to fly aboard an American spacecraft.

The first space shuttle flight of 1994 was the STS-60 mission. It began on February 3, 1994, with the launch of the *Discovery* from the Kennedy Space Center's (KSC) Launch Complex 39A at 12:10:00.000 Coordinated Universal Time (UTC, or 7:10 A.M. eastern standard time). The primary payloads were SPACEHAB 2 and the Wake Shield Facility (WSF). Landing was on February 11, 1994, at 19:19:22 UTC (2:19 P.M. eastern standard time) on Runway 15 at KSC.

The SPACEHAB 2 was a pressurized laboratory module which occupied part of the *Discovery*'s cargo bay. Astronauts working in the SPACEHAB module performed various experiments measuring the effects of microgravity on materials processing. While many of the experiments involved research that could one day lead to improved manufacturing of semiconductor crystals, several experiments were conducted that may lead to new techniques for pharmaceutical or biological processing.

The other major payload for STS-60 was the WSF-1 (Wake Shield Facility-1). The WSF was designed to conduct additional materials science experiments. Many materials science investigations must be performed in a vacuum to avoid contamination of the samples. WSF was designed to take advantage of the fact that space offers an environment that is much more nearly a total vacuum than can be produced on Earth. To further minimize contamination, WSF was designed to operate autonomously from the shuttle at a distance of several miles. Unfortunately, numerous difficulties over several days prevented astronauts from being able to deploy the WSF. Many of the planned experiments were conducted in the cargo bay of the space shuttle, though this environment was less than ideal.

The second space shuttle mission of the year was STS-62. The mission began with the launch of the *Columbia* from Pad 39B at KSC on March 4, 1994, at 13:53:00.009 UTC (8:53 A.M. eastern standard time). The primary payload for STS-62 was the United States Microgravity Payload-2 (USMP-2). Landing occurred on March 18, 1994, at 13:09:41 UTC (8:10 A.M. eastern standard time) on Runway 33 at KSC.

USMP-2 investigations studied materials and biological systems under the influence of microgravity. In orbit, the experiments are not subject to the same gravitational forces that they would be on the Earth's surface. This allows scientists to study systems as they would behave without the influence of gravity. Though the experiments in orbit can be performed with almost no gravitational influences, the term "microgravity" is used instead of "zero gravity" because tiny gravitational forces do still play a part and in fact can never be truly eliminated.

Among the materials science investigations, astronauts studied crystal growth and fabrication techniques for semiconductors. Astronauts also performed experiments designed to observe the element xenon at its critical point. A fluid's critical point is the combination of temperature and pressure at which the fluid is simultaneously a liquid and a gas. Biological investigations included experiments designed to minimize the muscle atrophy that astronauts experience after extended periods in space.

The third space shuttle mission of 1994, STS-59, was a mission dedicated to studying the Earth from space. The mission began with the launch of the orbiter *Endeavour* from KSC Launch Complex 39A on April 9, 1994, at 11:05:00.020 UTC (7:05 A.M. eastern daylight time). *Endeavour* landed eleven days later on April 20, 1994, at 16:54:30 UTC (12:55 P.M. eastern daylight time) on Runway 22 at KSC.

One of the major payloads of STS-59 was the Space Radar Laboratory-1 (SRL-1). A major com-

In October, 1994, at Edwards Air Force Base, Endeavour *rests after its return from the STS-68 mission, while* Columbia *takes a ride to nearby Palmdale for inspection.* (NASA-DFRC)

ponent of SRL was an imaging radar system designed for detailed investigations of surface features on the Earth. The radar could image details of the surface of the Earth through clouds, vegetation, or even very dry soil. This ability provides scientists the ability to study parts of the Earth that would otherwise be inaccessible.

Another portion of the mission included the Measurement of Air Pollution from Satellites (MAPS) system. MAPS took data on carbon monoxide concentrations over the Earth's surface. Among the sites studied were areas of the Earth scarred by forest fires in an attempt to measure how well the Earth's atmosphere rids itself of greenhouse gases. Greenhouse gases are gases that hold in heat from the Sun, thus raising the temperature of the Earth.

The fourth space shuttle mission of the year was an extended duration mission with the orbiter *Columbia*. The mission, STS-65, began with liftoff at 16:43:00.0013 UTC (12:43 A.M. eastern daylight time) on July 8, 1994, from Launch Complex 39A of KSC. *Columbia* carried the second International Microgravity Laboratory-2 (IML-2) into orbit. Landing was at 10:38:00 UTC (6:38 A.M. eastern daylight time) on July 23, 1994, on Runway 33 at KSC.

IML-2 was an international cooperative effort of scientists from six different nations. The scientific investigations centered on studying materials and biological systems under microgravity conditions. The materials science investigations generally focused on crystal growth and semiconductor manufacturing experiments. The life science investigations were targeted at investigating how gravity affects living organisms.

Among the experiments performed on board *Columbia* during STS-65 was an experiment designed to study crystal growth of proteins in an attempt to learn more about the structure of those proteins. Many of the experiments involved studies of human physiological changes in microgravity, with the astronauts themselves acting as test subjects. Critical point investigations similar to those of the STS-62 mission were also performed. *Colum-*

bia also carried into orbit numerous aquatic animals, such as jellyfish, goldfish, and sea urchins, to see how these animals react to microgravity. Plant studies were performed in a centrifuge to determine the lowest levels of gravity that begin to affect plant roots.

The fifth space shuttle mission of 1994 was STS-64. The mission began at 22:22:54.982 UTC (6:23 P.M. eastern daylight time) on September 9, 1994, with the launch of the space shuttle *Discovery* from Pad 39B at KSC. The mission had a variety of basic science objectives. Landing was on September 20, 1994, at Edwards Air Force Base, in California, on Runway 04 at 21:12:52 UTC (2:13 P.M. Pacific daylight time).

A major component of the mission was the use of a laser system to study particles in the Earth's atmosphere. This information can be used to determine human impact on the atmosphere. These measurements of the atmosphere from above were done in conjunction with ground-based observations of the atmosphere.

A separate experiment was conducted using the SPARTAN-201 satellite. This satellite was designed to measure solar activity and to make measurements of the Sun's corona. The satellite was deployed on September 13, 1994, and it moved slowly away from *Discovery*, far enough to avoid interference the space shuttle may have had with its measurements of the Sun. The satellite was retrieved on September 15, 1994.

Another component of the STS-64 mission was an untethered spacewalk. During this activity, astronauts tested equipment that was designed as a rescue aid for future astronauts who accidentally become untethered during a spacewalk and float away from their spacecraft or space station.

The sixth space shuttle mission of the year was STS-68. This mission was a continuation of the studies done by STS-59 earlier in the year, with the space shuttle *Endeavour* once again carrying SRL-2 and the MAPS into orbit. Launch occurred at 11:16:00.011 UTC (7:16 A.M. eastern daylight time) on September 30, 1994, from Launch Complex 39A at KSC. Landing was on October 11, 1994, at 17:02:08 UTC (10:02 A.M. Pacific daylight time) at Edwards Air Force Base, in California, on Runway 22.

STS-68 continued the radar observations of the Earth that the STS-59 mission had started in April, 1994. The system made radar images of volcanoes that had erupted during the interval between the two missions. Radar images of the Sahara and the North Atlantic were made in an attempt to understand climate changes on Earth.

The MAPS system measured carbon monoxide levels around the world from above. This information is important in our understanding of human interactions with our atmosphere. Furthermore, MAPS measured carbon monoxide levels in the area of several deliberately set small forest fires. These fires had been planned prior to the space shuttle mission as part of forest management strategies and were not set solely for the benefit of the STS-68 mission.

The seventh, and final, space shuttle mission of 1994 emphasized atmospheric studies. This mission, STS-66, began on November 3, 1994, at 16:59:43.004 UTC (seventeen seconds before noon) with the launch of the space shuttle *Atlantis* from Pad 39B at KSC. The primary payload was the Atmospheric Laboratory for Applications and Sciences-3 (ATLAS-3). Landing was at Edwards Air Force Base, in California, on Runway 22 at 15:33:45 UTC (7:34 A.M. Pacific standard time) on November 14, 1994.

ATLAS-3 was the third in a series of ten planned ATLAS missions to study the Earth's atmosphere from above. Astronauts aboard *Atlantis* made numerous measurements of atmospheric composition during the mission. A major goal of the mission was to further the understanding of the chemical reactions involving ozone in the upper atmosphere.

Contributions

All the space shuttle missions of 1994 were dedicated science missions. As a consequence, a wealth of scientific data was recorded over the course of the year.

Numerous materials science experiments were performed under microgravity conditions. These experiments allowed scientists to study the formation and growth of crystals without the influence of gravity. Eliminating the influence of gravity allows crystals to grow larger and more perfectly than those grown on Earth. By studying these crystals, scientists can learn more about the atoms and molecules themselves that make up the materials being studied. The critical point experiments permitted scientists to study critical points of materials, such as xenon, which are difficult or impossible to study on Earth due to the convection effects that are augmented by the larger gravitational effects on the Earth's surface.

Biological investigations studied the growth and development of animals and plants under microgravity. Several of these investigations utilized the astronauts as test subjects to study how the human body adapts to microgravity. Among the findings was the fact that calcium loss from bones, a major problem in long-duration spaceflights, can be lessened through physical exercise. Bone tissue replenishes calcium in bones under normal circumstances on Earth. However, under microgravity, bone tissue is much less efficient at replacing the calcium in bones. The experiments done during STS-65 indicated that compressing and stressing the bone tissues through physical exercise can minimize some of the detrimental effects of microgravity.

The two SRL missions, STS-59 and STS-68, provided valuable data on volcanic activity and fault motion. Furthermore, dry river beds under the sands of the Sahara Desert indicated that the northern portions of Africa were at one time much wetter than they are today. This information may be useful in understanding climate changes on Earth.

Atmospheric studies were an important aspect of the space shuttle missions of 1994. MAPS made measurements of carbon monoxide around the world. By flying on two missions during the year, MAPS was able to make some preliminary estimates of seasonal changes in carbon monoxide levels. This information is important in tracking greenhouse gas concentrations on Earth.

The ATLAS-3 was the third in a series of ten ATLAS missions. By comparing the findings of each of the ATLAS missions, scientists hope to understand how the Earth's atmosphere reacts to both human and solar activity. ATLAS-3 measurements were some of the best of the series. Excellent measurements were made of solar activity. Another major emphasis of the mission was measurements of concentrations of ozone in the upper atmosphere and measurements of other chemicals that interact with ozone. Scientists are using these measurements to gain a better understanding of ozone depletion in the upper atmosphere.

Context

Unlike most years, in which several space shuttle missions involved transporting satellites to or from orbit, the space shuttle missions of 1994 were all dedicated science missions. The science missions scheduled for the space shuttle could not have been done on the Earth, and thus had to be done in orbit. In accomplishing the objectives for the year, the space shuttle program set at least two new records. Sergei Konstantinovich Krikalev was the first Russian cosmonaut to fly aboard a U.S. space shuttle. The space shuttle *Columbia* set a new record for the longest space shuttle mission with STS-65 at fourteen days, 17 hours, and 55 minutes.

Materials science experiments, such as those done on STS-60, STS-62, and STS-65, helped scientists understand crystal growth. Crystal structure is fundamental to modern semiconductor electronics, and a better understanding of crystal structure is vital in the search for ways of building better semiconductors. Information learned about proteins on STS-62 and STS-65 may yield new drugs to help fight diseases on Earth.

The life science experiments performed during several missions throughout the year yielded information about the effects of microgravity on plants and animals. This information is essential to reduce the detrimental effects of microgravity on astronauts during long-duration spaceflights, such as a space station or a mission to Mars. Furthermore, some of the techniques used by astronauts to prevent calcium loss in their bones may also aid people on Earth suffering from certain bone disorders, such as osteoporosis.

The two SRL missions provided valuable data for geologists to study. This method of using radar to study a planetary surface is very similar to that used by the Magellan Spacecraft at the planet Venus, where dense clouds prevent the spacecraft from visually studying the planet's surface.

The atmospheric investigations of the year were among the most sophisticated performed on Earth's atmosphere. The MAPS data provided scientists with a tool for measuring the effects of greenhouse gases in the atmosphere. ATLAS-3 studied ozone depletion, among other things. Both ozone depletion and the buildup of greenhouse gases have been postulated to be the result of interactions between humans and the atmosphere of our planet. These missions help to understand the effects of pollution on the atmosphere.

See also: Atmospheric Laboratory for Applications and Science; Materials Processing in Space; Pegasus Launch Vehicles; Skylab Program; Space Shuttle: Life Science Laboratories; Space Shuttle: Microgravity Laboratories and Payloads; Space Shuttle: Radar Imaging Laboratories; Space Shuttle Flights, 1993; Space Shuttle Flights, 1995; SPACEHAB.

Further Reading

Branley, Franklyn M. *From Sputnik to Space Shuttles.* New York: Thomas Y. Crowell, 1986. This is a very good juvenile book on the development of space exploration, from robotic satellites to the modern space shuttle. The book emphasizes the uses of the space shuttle more than the shuttle itself.

DeWaard, E. John, and Nancy DeWaard. *History of NASA: America's Voyage to the Stars.* New York: Exeter Books, 1988. This book is intended for the general audience. The last chapter of the book covers the beginnings of the space shuttle program. The need for a system such as the space shuttle is discussed. The book ends with the explosion that destroyed the *Challenger* in 1986.

Harland, David M. *The Space Shuttle: Roles, Missions, and Accomplishments.* Hoboken, N.J.: John Wiley, 1998. *The Space Shuttle* is written thematically, rather than purely chronologically. Topics include shuttle operations and payloads, weightlessness, materials processing, exploration, Spacelabs and free-flyers, and the shuttle's role in the International Space Station.

Harrington, Philip S. *The Space Shuttle: A Photographic History.* San Francisco, Calif.: Brown Trout, 2003. With one hundred full-color photographs by Roger Ressmeyer and others and with text by popular astronomy writer Phil Harrington, this book tells the story of the space shuttle program from 1972 to 2003. Its beautiful photographs allow the general reader to survey the history of the space shuttle program and be uplifted by the pioneering spirit of one of humanity's grandest enterprises.

Jenkins, Dennis R. *Space Shuttle: The History of the National Space Transportation System: The First 100 Missions.* Stillwater, Minn.: Voyageur Press, 2001. This is a concisely written technical reference account of the space shuttle and its ancestors, the aerodynamic lifting bodies. It details some of the advantages and inherent disadvantages of using a reusable space vehicle. Each of the vehicles is illustrated by line drawings. The book follows the space shuttle from its original concepts and briefly chronicles its first one hundred flights.

Joëls, Kerry Mark, and Gregory P. Kennedy. *The Space Shuttle Operator's Manual.* Designed by David Larkin. Rev. ed. New York: Ballantine Books, 1988. This book contains a wealth of information on space shuttle systems and flight procedures. It is written as a manual for imaginary crew members on a generic mission and will be appreciated by anyone interested in how the astronauts fly the orbiter, deploy satellites, conduct spacewalks, and live in space. Contains many drawings and some photographs of equipment.

National Aeronautics and Space Administration. *Space Shuttle Mission Press Kits.* http://www-pao.ksc.nasa.gov/kscpao/presskit/presskit.htm. Provides detailed preflight information about each of the space shuttle missions. Accessed March, 2005.

Shapland, David, and Michael Rycroft. *Spacelab: Research in Earth Orbit.* New York: Cambridge University Press, 1984. Written in the early days of the space shuttle program, this general audience level book describes the types of scientific missions anticipated for the space shuttle. The book describes a Spacelab, which is a living/working module that is flown in the space shuttle's cargo bay during science missions.

Torres, George J. *Space Shuttle: A Quantum Leap.* Novato, Calif.: Presidio Press, 1986. This general audience level book traces the development of piloted spaceflight, culminating with the space shuttle. Included are brief descriptions of many different types of shuttle missions. The book was written at the time of the *Challenger* accident and thus does not include any of the missions that occurred after the resumption of the shuttle program.

Raymond D. Benge, Jr.

Space Shuttle Flights, 1995

Date: February 3 to November 20, 1995
Type of mission: Piloted spaceflight

The space shuttle flights of 1995 included seven missions, three of which involved a rendezvous with the Russian space station Mir. Ultraviolet data were collected and experiments performed under reduced gravity. Astronauts engaged in extended spacewalks to test the thermal performance of space suits.

Key Figures

James D. Wetherbee (b. 1952), STS-63 commander
Eileen M. Collins (b. 1956), STS-63 pilot
Bernard A. Harris, Jr. (b. 1956), STS-63 mission specialist
C. Michael Foale (b. 1957), STS-63 mission specialist
Janice E. Voss (b. 1956), STS-63 mission specialist
Vladimir Georgievich Titov (b. 1947), STS-63 mission specialist (Russian Space Agency, RSA)
Stephen S. Oswald (b. 1951), STS-67 commander
William G. Gregory (b. 1957), STS-67 pilot
John M. Grunsfeld (b. 1958), STS-67 mission specialist
Wendy B. Lawrence (b. 1959), STS-67 mission specialist
Tamara E. Jernigan (b. 1959), STS-67 mission specialist
Ronald A. Parise (b. 1951), STS-67 payload specialist
Samuel T. Durrance (b. 1943), STS-67 payload specialist
Vladimir Nikolaevich Dezhurov (b. 1962), Mir Permanent Crew 18 commander
Gennady Mikhailovich Strekalov (b. 1940), Mir Permanent Crew 18 flight engineer
Norman E. Thagard (b. 1943), Mir Permanent Crew 18 astronaut-investigator
Robert L. "Hoot" Gibson (b. 1946), STS-71 commander
Charles J. Precourt (b. 1955), STS-71 pilot
Ellen S. Baker (b. 1953), STS-71 mission specialist
Gregory J. Harbaugh (b. 1956), STS-71 mission specialist
Bonnie J. Dunbar (b. 1949), STS-71 mission specialist
Anatoly Yakovlevich Solovyev (b. 1948), STS-71 payload specialist, Mir Permanent Crew 19 commander
Nikolai Mikhailovich Budarin (b. 1953), STS-71 payload specialist, Mir Permanent Crew 19 flight engineer
Terence T. "Tom" Henricks (b. 1952), STS-70 commander
Kevin R. Kregel (b. 1956), STS-70 pilot
Donald A. Thomas (b. 1955), STS-70 mission specialist
Nancy J. Currie (b. 1958), STS-70 mission specialist
Mary Ellen Weber (b. 1962), STS-70 mission specialist
David M. Walker (1944-2001), STS-69 commander

Kenneth D. Cockrell (b. 1950), STS-69 pilot

James S. Voss (b. 1949), STS-69 mission specialist

James H. Newman (b. 1956), STS-69 mission specialist

Michael L. Gernhardt (b. 1956), STS-69 mission specialist

Kenneth D. Bowersox (b. 1956), STS-73 commander

Kent V. Rominger (b. 1956), STS-73 pilot

Catherine G. "Cady" Coleman (b. 1960), STS-73 mission specialist

Michael E. Lopez-Alegria (b. 1958), STS-73 mission specialist

Kathryn C. Thornton (b. 1952), STS-73 mission specialist

Fred W. Leslie (b. 1951), STS-73 payload specialist

Albert Sacco, Jr. (b. 1949), STS-73 payload specialist

Kenneth D. Cameron (b. 1949), STS-74 commander

James D. Halsell, Jr. (b. 1956), STS-74 pilot

Chris A. Hadfield (b. 1959), STS-74 mission specialist

Jerry L. Ross (b. 1948), STS-74 mission specialist

William S. "Bill" McArthur, Jr. (b. 1951), STS-74 mission specialist

Summary of the Missions

STS-63 with the *Discovery* shuttle orbiter entered an orbit to catch up with the Russian space station Mir after launch on February 3, 1995, at 17:22:03.994 Coordinated Universal Time (UTC, or 12:22 P.M. eastern standard time). On February 6, after careful maneuvering by Navy Commander James D. Wetherbee and Lieutenant Colonel Eileen M. Collins, *Discovery* approached Mir to within 11.2 meters, the two largest spacecraft ever to rendezvous in orbit.

Training for the historic meeting was aided with a laptop computer program that simulated the actual flight paths of both spacecraft. When an orbit of 340 kilometers was reached, *Discovery* approached Mir from below, closing in from ahead but along Mir's flight path. From 700 meters away, Wetherbee and Collins used manual control of the shuttle's thrusters to ease into a target located on Mir's module docking port. Live television cameras from both spacecraft documented the flight. The entire fly-around lasted one-half of an orbit or forty-five minutes.

Throughout the flight, the crew of *Discovery* performed a variety of experiments. Vladimir Georgievich Titov, a Russian cosmonaut on *Discovery*, used the orbiter's Remote Manipulator System to lift and position the imaging spectrograph to measure the glow produced by atomic oxygen striking the spacecraft. The spectrograph was also used to image the interstellar medium.

A major objective of the mission was to gather data on space suit comfort in extreme cold. Two astronauts performed extended spacewalks for 4.5 hours to provide space suit temperature data. Activities involved moving a 1,300 kilogram satellite with a special tool manipulated by hand. STS-63 ended with the landing of *Discovery* at the Kennedy Space Center (KSC) Shuttle Landing Facility on February 11, 1995, at 11:50:19 UTC (6:50 A.M. eastern standard time).

STS-67 with the *Endeavour* orbiter was launched on March 2, 1995, at 06:38:12.989 UTC (1:38 A.M. eastern daylight time). The primary objective of the mission was to operate the ultraviolet astronomy instruments, mapping out major uncharted regions of space that are usually blocked by ozone layers in the atmosphere. Scientists expected to detect the diffuse hydrogen and helium gas composing the intergalactic medium believed to be uniformly distributed after the Big Bang.

A key to the success of that task was ensuring that the Instrument Pointing System (IPS) functioned properly. The IPS used improved software that allowed astronauts and scientists on the ground to lock on to certain target stars closer to the Earth's limb, permitting longer observations of those stars and sources with three major instruments. The crew of *Endeavour* carried out detailed round-the-clock observations for the first days of the mission in two teams. One team, including Mission Commander Stephen S. Oswald, flew the orbiter while the others were involved with the experiments.

During the mission, the astronauts conducted more than forty hours of experiments with the middeck active control experiment. This package is designed to detect and reduce uncontrolled vibrations that can occur on a free-floating surface. Programs based on preflight configurations provided the information to control vibrations and maintain platform stability. Data from this experiment were expected to result in the development of control systems that would eliminate vibrations that interfere with the operations of structures on satellites and other space vehicles. *Endeavour* touched down on Edwards Air Force Base Runway 22 on March 18, 1995, at 21:47:01 UTC (1:47 P.M. Pacific standard time), setting an orbiter endurance record of sixteen days, 15 hours, 8 minutes, and 48 seconds.

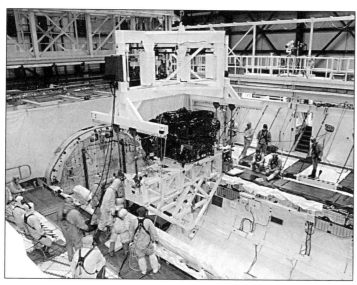

The Shuttle Pointed Autonomous Research Tool for Astronomy-201 (SPARTAN-201) is lowered into the payload bay of the space shuttle Endeavour *for the STS-69 mission.* (NASA)

The launch of STS-71 with the *Atlantis* orbiter occurred on June 27, 1995, at 19:32:18.988 UTC (3:32 P.M. eastern standard time) and marked the one hundredth American piloted spaceflight. The primary objective of the mission was to rendezvous and dock with the Mir 18 space station, exchanging crew members. All techniques for the close approach were pioneered on STS-63 except for one major difference. On the earlier mission, the *Discovery* orbiter approached Mir from the front along Mir's flight path. This time the *Atlantis* closed toward the space station from underneath directly along a line termed the "R-bar." This maneuver was easier to accomplish because gravity would naturally act as a brake, conserving fuel from *Atlantis*'s thrusters.

The new docking procedure was initiated by the concern that *Atlantis* might cause damage to Mir's solar panels and because the use of gravity would reduce the need for *Atlantis*'s crew to use the thrusters to lessen the approach speed. The actual docking was achieved on June 29 at an altitude of 340 kilometers and was assisted by television cameras, including one placed on the centerline of the docking system. The docking was the result of more than two years of preparation by flight controllers and resulted in new communication and procedures. *Atlantis* remained docked to the Mir Complex for five days of joint experiments. On July 6, 1995, at 14:54:36 UTC (10:55 A.M. eastern daylight time) the orbiter glided to a landing at the KSC.

The launch of STS-70, commanded by Terence T. "Tom" Henricks, originally scheduled for liftoff on June 8, did not launch until July 13, seven days after Mission 71 had landed. The delay was due to the discovery of 205 holes in the urethane foam insulation on the vehicle's external fuel tank. Several woodpeckers seeking a nest apparently created the holes. The delay cost the agency more than $2 million, including $100,000 to roll the shuttle to

The Tracking and Data Relay Satellite-G (TDRS-G) and its Inertial Upper Stage (IUS) are loaded into a payload cannister. This communications satellite was the primary payload for the STS-70 mission. (NASA)

the Vehicle Assembly Building and return it to the pad.

The eight-day mission of the *Discovery* orbiter began at 13:41:55.020 UTC (9:42 A.M. eastern daylight time) July 13, 1995, and featured the new Block 1 shuttle main engine. The new design, which included a dual duct hot-gas manifold and a redesigned main injector, was expected to provide greater reliability and performance. A second major objective was the orbital launching of the Tracking and Data-Relay Satellite-G (TDRS-G). TDRS-7 (as it was now called) was positioned in a geosynchronous orbit complementing five similar satellites already in orbit.

Discovery's crew also tested a camera system designed to take high-resolution images of ship tracks for the purpose of analysis of wave formations. The crew also performed a number of biological and materials experiments, including one involving the determination of the effects of microgravity on physiological changes in rats. The flight concluded at 12:02:01.99 UTC (8:02 A.M. eastern daylight time) on July 22, 1995, on Runway 33 at KSC's Shuttle Landing Facility.

The eleven-day flight of STS-69 with the *Endeavour* orbiter, launched on September 7, 1995, at 15:08:59.995 UTC (11:09 A.M. eastern daylight time) and commanded by David M. Walker, carried a variety of experiments, including the Wake Shield Facility-2 (WSF-2), a saucer-shaped satellite designed to fly free from the shuttle. The WSF allowed growth of thin films in the almost perfect vacuum produced by the wake of the satellite. This has an application to the semiconductor industry and the chemical growth of films for electrical instruments. To ensure a clean environment, the WSF is allowed to trail the *Endeavour* by as much as 65 kilometers.

Onboard payloads investigated the effects of microgravity or reduced gravity on cell changes and on the repeated melting and freezing of fluoride salts. This experiment, conducted by the National Institutes of Health, investigated the loss of bone mass during spaceflight. Other experiments determined the gravity-sensing mechanisms within mammalian cells and the effect of microgravity on neuromuscular disorders. The thermal energy storage experiment was designed to understand the long-term behavior of the spaces of a lithium fluoride salt produced in a repeated environment of melting and refreezing in microgravity. This salt is used to store thermal energy in solar-powered designs. *Endeavour* landed on September 18, 1995, at 11:37:56 UTC (7:38 A.M. eastern daylight time) at KSC.

STS-73 with the *Columbia* orbiter was launched on October 20, 1995, at 13:53:00.013 UTC (9:53 A.M. eastern daylight time) with Kenneth D. Bowersox as commander. The fifteen-day mission's objectives included deployment of the second microgravity laboratory, United States Microgravity Laboratory-2 (USMP-2), providing insights into the role of gravity in theoretical models of fluid physics, crystal growth, plant growth, and drop encapsulation. A particle dispersion experiment was placed on board to confirm the behavior of dust and particles in space.

The Geophysical Fluid Flow Cell (GFFC) experiment modeled planetary atmospheres and simulated fluid flow within the Earth's mantle. The planet Jupiter's atmosphere was modeled on several of the runs. On the final session, a lower rotation similar to that found in the fluids of the Earth's mantle was simulated. All these runs were performed by the crew members through the payload general support computer, with the status reported to ground investigators.

With a device called the glovebox, mission specialists monitored protein crystal growth exposed to a constantly changing gravity vector. The changing gravity was achieved by periodic changes in the vernier thrusters on board the spacecraft and was expected to cause unusual structural changes. The glovebox was used to study how the behavior and movement of fluids was affected by the attitude and shape of the container. In another container, five small potatoes were placed to study the feasibility of growing edible plants in space.

Drop encapsulation was successfully accomplished in the Drop Physics Module, a device that uses sound waves to levitate and maneuver liquid drops for close study. The coalescence of two drops into one was achieved, as was the formation of a chemical membrane between the two drops. Drop fissioning, the spinning of one drop until it splits into two, was accomplished on this mission.

The particle dispersion experiment was designed to confirm a theory about the behavior of dust and particle clouds. The experiment tested the theory that attraction occurs in the dust clouds of space due to static electrical charges. Variables such as particle size, cloud density, and type of material (volcanic material, quartz, and copper) were all tested. *Columbia*'s mission came to a successful

conclusion at the KSC on November 5, 1995, at 11:45:21.4 UTC (6:45 A.M. eastern standard time).

The final shuttle mission of 1995 occurred with the launch of STS-74 and the *Atlantis* orbiter on November 12 at 12:30:43.013 UTC (7:31 A.M. eastern standard time). The crew, with Colonel Kenneth D. Cameron as commander, utilized the experience of STS-71, which achieved the first shuttle docking with the Mir space station. The mission objectives included transporting food, water, and experiments up to the space station, retrieving test samples and hardware, and deploying and installing a new docking module and solar arrays. Both Russian and U.S. crew members as well as other officials were able to work closely on the demanding task of assembling a new space station.

The new 41-kilogram docking module was installed on Mir on November 14. The docking module was deployed from the cargo bay of *Atlantis* by means of a 15-meter Remote Manipulator System arm. The docking module, which would link the *Atlantis* to Mir, replaced the older Kristall Module, enabling future shuttles to dock with adequate clearance. Two sets of solar arrays were attached to the docking module and were deployed by the cosmonauts on Mir by means of extravehicular activities (EVAs), or spacewalks. *Atlantis* ended its historic mission with a landing at the KSC on November 20, 1995, at 17:01:29 UTC (12:01 P.M. eastern standard time).

Contributions

The joint missions of Mir and the shuttle on STS-63, STS-71, and STS-74 led to the development of new techniques for rendezvous and ultimately docking in orbit. STS-71 pioneered the "R-bar" approach and docking method of closing in on the space station from below and using gravity to act as a brake, conserving fuel from the thruster engines. These missions developed the ODS or Orbital Docking System, which was an automated system with a capture ring that facilitated docking by removing residual motion upon contact of the spacecraft.

STS-74 delivered a new docking module to Mir that was installed in space with a Remote Manipula-

The crew of STS-74 look through the windows of Atlantis *after returning from the Mir Space Station.* (NASA)

tor arm. The Remote Manipulator was operated by Mission Specialist Chris A. Hadfield. When docking was completed, the shuttle and Mir formed the largest combined spacecraft ever attempted, with a total mass of more than 226,000 kilograms. This was considered a major step in the assembly of an international space station.

The handshake between Robert "Hoot" Gibson and Vladimir Nikolaevich Dezhurov across the *Atlantis* and Mir docking tunnel on STS-71 was the first international spacecraft linkup since the U.S. Apollo and Soviet Soyuz flight on July 17, 1975. Three of the Mir crew, including Dezhurov, returned to Earth on *Atlantis* July 7 after almost four months in space. While in orbit, the *Atlantis* crew transferred 76 kilograms of water to Mir and used *Atlantis* to raise Mir's cabin pressure by 11 percent. STS-74 offloaded 970 kilograms of food, water, and equipment with supplies of nitrogen and oxygen to Mir. In turn, 371 kilograms of equipment and sam-

ples were taken on board *Atlantis* for the return trip. All these achievements demonstrated the feasibility of servicing an operating space station.

The extended spacewalks performed by STS-63 specialists C. Michael Foale and Bernard A. Harris, Jr. provided valuable data on space-suit comfort under extreme cold. It was discovered that the astronauts' fingertips became too cold for them to work in space. As a result, space suits were modified with heater coils placed in each fingertip of their gloves. Temperature sensors were also added to the astronauts' boots. Space suits were redesigned, allowing the coolant to flow around the body without interrupting flow to the suit electronics and other components.

The second flight of the Wake Shield Facility on STS-69 to test materials in the near vacuum created by the satellite's wake demonstrated the ability to grow exotic and ultra-pure thin films in space. The wake created on the top side of the satellite produced a vacuum several orders of magnitude greater than could be reached on Earth. The satellite also carried an experiment that measured how much electrical fields caused by ionized particles around the spacecraft interfered with communications.

The microgravity of the shuttle's environment led to successful results in a number of experiments. Scientists observed various heat-driven flow patterns in the GFFC experiment when the initial conditions were changed. These findings appear to substantiate theoretical mathematical models of planetary and solar fluid flow. The glovebox experiment showed differences in the way fluids adhered to container walls, indicating that there are additional physical factors involved not predicted by current theory.

The ultraviolet measurements taken by the crew of STS-67 included the first observations of an active volcano on Jupiter's moon Io. The first measurements of radiation and polarization were made on Nova Aquilae, a binary star system. The largest telescope locked on to distant quasars and measured the abundance of hydrogen and helium in the intergalactic medium.

Context

The space shuttle flights of 1995 demonstrated the importance of international cooperation in achieving the major goals of space exploration, including building the International Space Station. Russia and the United States were able to coordinate their space programs precisely not only to communicate and rendezvous in space but also to build jointly a docking module that would serve future missions. Both of these nations were able to use their separate facilities to train astronauts and cosmonauts for common missions.

Crew members on these flights were able to deploy and retrieve a remote satellite with capabilities for growing exotic films in the nearly perfect vacuum created in a wake. The high-quality chemical growth produced under these conditions will have applications for the development of semiconductors and other high-tech electrical instruments.

Observations of fluid flow in microgravity are expected to help scientists better understand the behavior of fluid flows on the Earth, as well as movements of water in the oceans and circulation within the atmosphere. The glovebox experiment showed how surfaces form in low gravity, and these insights will help in the design of fluid systems used in space, including those utilizing fuel.

Results from ultraviolet experiments indicated the advantages of using orbital telescopes for observing wavelengths that would otherwise be filtered out by the Earth's atmosphere. A nova star was imaged for the first time with the 3.7-meter-long telescope. A clearer image of the white hot gas pulled off a white dwarf star by a normal star in this binary star system was obtained than is possible from ground-based telescopes. Research based on these data may provide clues about the triggering mechanism for the resulting thermonuclear fusion process.

The abundance of hydrogen relative to helium in the intergalactic medium was also detected with ultraviolet telescopes, and data gathered will be used to confirm or rule out the quantities of helium predicted by the Big Bang theory. The medium of hydrogen and helium is thought to have formed in the first few minutes after the Big Bang, the epic event that astronomers generally believe created the universe. Data collected from ultraviolet measurements are being compiled into a galactic atlas that may prove an important key in unlocking some of the mysteries of the universe.

There are ten TDRS satellites in the constellation. TDRS-7 is at 150.8° west longitude. TDRS-8 and TDRS-10 are the primary satellites, with TDRS-9 serving as a spare.

See also: Astronauts and the U.S. Astronaut Program; Cooperation in Space: U.S. and Russian; Get-Away Special Experiments; International Space Station: Crew Return Vehicles; Russia's Mir Space Station; Space Shuttle: Microgravity Laboratories and Payloads; Space Shuttle Flights, 1994; Space Shuttle Mission STS-63; Space Shuttle Flights, 1996; Space Suit Development.

Further Reading

Asker, James R. "Mission 63 to Test U.S.-Russian Teaming." *Aviation Week and Space Technology*, January 30, 1995, 36-38. An excellent source of information with detailed coverage on all the current spaceflights. The shuttle flights of this manuscript are covered in considerable depth in subsequent issues. James Asker has done a commendable job keeping the reader informed of developments in all phases of the space program. Available at most public libraries.

Gore, Rick. "When the Space Shuttle Finally Flies." *National Geographic*, March, 1981. The development of the space shuttle is documented in this thirty-page article. There are abundant color photographs and diagrams featuring the cockpit interior, an exploded view of the shuttle's interior, and the power plants. Photographs show astronauts practicing at the underwater facility in Huntsville, Alabama. The first test flights in California are also described.

Harland, David M. *The Space Shuttle: Roles, Missions, and Accomplishments*. Hoboken, N.J.: John Wiley, 1998. *The Space Shuttle* is written thematically, rather than purely chronologically. Topics include shuttle operations and payloads, weightlessness, materials processing, exploration, Spacelabs and free-flyers, and the shuttle's role in the International Space Station.

Harrington, Philip S. *The Space Shuttle: A Photographic History*. San Francisco, Calif.: Brown Trout, 2003. With one hundred full-color photographs by Roger Ressmeyer and others and with text by popular astronomy writer Phil Harrington, this book tells the story of the space shuttle program from 1972 to 2003. Its beautiful photographs allow the general reader to survey the history of the space shuttle program and be uplifted by the pioneering spirit of one of humanity's grandest enterprises.

Jenkins, Dennis R. *Space Shuttle: The History of the National Space Transportation System: The First 100 Missions*. Stillwater, Minn.: Voyageur Press, 2001. This is a concisely written technical reference account of the space shuttle and its ancestors, the aerodynamic lifting bodies. It details some of the advantages and inherent disadvantages of using a reusable space vehicle. Each of the vehicles is illustrated by line drawings. The book follows the space shuttle from its original concepts and briefly chronicles its first one hundred flights.

Kerrod, Robin. *The Illustrated History of NASA*. New York: Gallery Books, 1987. A superb reference of color plates depicting advancements in the Space Age. Chapter 5 documents the development and first missions of the shuttles *Challenger* and *Columbia*. Chapter 6 is profusely illustrated with the flights of *Atlantis* and *Discovery*. A chronology of NASA's first twenty-five years covers significant events by year from 1960 to 1983. A fitting epilogue describes the last flight of *Challenger* on January 28, 1986.

National Aeronautics and Space Administration. *Space Shuttle Mission Press Kits*. http://www-pao.ksc.nasa.gov/kscpao/presskit/presskit.htm. Provides detailed preflight information about each of the space shuttle missions. Accessed March, 2005.

Powers, Robert M. *Shuttle: The World's First Spaceship*. Harrisburg, Pa.: Stackpole Books, 1979. A history of the development of the shuttle detailing how the design evolved from earlier rocket configurations. Excellent diagrams and photographs include cockpit detail and component dimensions. Specifications are included in the appendices. A glossary of NASA abbreviations and jargon is provided for the reader.

Slayton, Donald K., with Michael Cassutt. *Deke! U.S. Manned Space: From Mercury to the Shuttle.* New York: Forge, 1995. This is the autobiography of the last of the Mercury astronauts to fly in space. After being grounded from flying in Project Mercury for what turned out to be a minor heart murmur, Slayton was appointed head of the Astronaut Office. Later, he commanded the Apollo-Soyuz flight in 1975. He stayed with NASA through the STS-2 flight. Although he was no longer an active participant in NASA's activities, he gives some insight into the behind-the-scenes adventures of the shuttle program.

Time-Life Books. *Life in Space.* Boston: Little, Brown, 1984. Three chapters are devoted entirely to the shuttle, with the third chapter containing color photographs of six shuttle missions. A description is given of the various mission specialists and their functions. The arrangement and composition of the more than thirty thousand ceramic tiles covering the shuttle's outer skin is discussed. Cutaway diagrams reveal internal structure, and a full-page diagram depicts twenty-one positions of the shuttle's trajectory.

_____. *Outbound.* Richmond, Va.: Author, 1989. A photographic history of human exploration in space. Photographs show controllers at the Johnson Space Center monitoring the fourth flight of the shuttle *Challenger* in February, 1984. A good reference for information on the suits worn by the shuttle crew. Color diagrams reveal the interior details of the suit life-support and power systems.

_____. *Spacefarers.* Richmond, Va.: Author, 1989. Depicts the shuttle as a delivery vehicle for the construction of the Space Station. Diagrams show how the elements of the Space Station will fill the shuttle's cargo bay. Additional color illustrations display the Space Station under construction and the use of the mobile transporter that will carry the astronauts and their supplies to the work area.

Michael L. Broyles

Space Shuttle Mission STS-63

Date: February 3 to February 11, 1995
Type of mission: Piloted spaceflight

STS-63 began an important phase of cooperation between NASA and the Russian Space Agency (RSA), in which techniques for joint operations between the NASA space shuttle and the Russian Mir Space Station would demonstrate the feasibility of construction of an International Space Station.

Key Figures

James D. Wetherbee (b. 1952), STS-63 commander

Eileen M. Collins (b. 1956), STS-63 pilot

Bernard A. Harris, Jr. (b. 1956), STS-63 mission specialist

C. Michael Foale (b. 1957), STS-63 mission specialist

Janice E. Voss (b. 1956), STS-63 mission specialist

Vladimir Georgievich Titov (b. 1947), STS-63 mission specialist

Alexander Viktorenko, Soyuz TM-20/Mir 17 commander

Elena V. Kondakova (b. 1957), Soyuz TM-20/Mir 17 flight engineer

Valery Polyakov, Soyuz TM-18/Mir 15 physician-cosmonaut, who spent over a year on Mir

Summary of the Mission

NASA announced the crew for STS-63 and its primary focus on September 9, 1993. That crew included one important and historic distinction— the first female shuttle pilot. Selected as STS-63 commander was Navy Commander James D. Wetherbee. His pilot was Air Force Major Eileen M. Collins. Mission specialists included C. Michael Foale, Janice E. Voss, physician Bernard A. Harris, Jr., and Russian Air Force Colonel Cosmonaut Vladimir Georgievich Titov. STS-63's primary payloads were SPACEHAB 3 and the SPARTAN-204 astronomical research platform for studying the solar wind.

Weather conditions for launch were perfect. *Discovery* lifted off at 17:22:03.994 Coordinated Universal Time (UTC, or 12:22 A.M. eastern standard time) on February 3, 1995, into clear starlit skies. The vehicle appeared somewhat sluggish at solid-fueled rocket booster (SRB) ignition, but performed nominally. SRB separation was clean. The second-stage performance was nominal and steered *Discovery* into the proper insertion point at main engine cutoff. *Discovery* had returned to space on its twentieth mission.

Once in orbit, it was soon noticed that *Discovery* had a leaking Reaction Control System (RCS) thruster on its right orbital maneuvering system (OMS) pod. Flight rules for STS-63 required that *Discovery* have all of its aft-firing thrusters up and running nominally before it could approach to within the final 300 meters of Mir. Mission Control ordered Wetherbee to change *Discovery*'s attitude so that the offending thruster was facing the Sun in order to allow it to warm up for several hours. The thruster was losing about 1 kilogram of propellant per hour. This was certainly a manageable loss, one with no impact on the ability of the orbiter to remain in space for the duration of the scheduled mission. However, the gas could contaminate structures on Mir if it was to impinge upon them, structures such as the solar panels and high-optical-quality glass portholes.

Shortly after 05:00 UTC on February 4, the astronauts grappled the SPARTAN-204 satellite with the Remote Manipulator System (RMS) arm's end effector and lifted the payload out of *Discovery*'s cargo bay for a shuttle glow measurement. The payload was held about 11 meters above the bay for several hours and was then returned to its berthing position. Titov was primarily responsible for RMS operation.

The crew transmitted to Mission Control televised pictures of the leaking RCS thruster. Streams of fuel coming out of the thruster cone were plainly visible. Wetherbee and Collins performed some work to stop the leak, but their attempts failed. Russian engineers were reluctant to allow *Discovery* to get closer than 121 meters to Mir. Flight directors in Houston were decidedly reluctant not to fly to within the 11 meters previously agreed upon.

On February 6 the crew received its wake-up call at 05:21 UTC. For the first time in nearly twenty years, piloted American and Russian spacecraft were about to fly in formation. When the crew arose, they were still "no go" for the close approach. There were still six hours to go until completion of the rendezvous, but the astronauts already saw Mir in the predawn skies several hundred kilometers away. The astronauts transmitted televised images of Mir, the space station appearing only as a bright dot. Titov first made contact with his comrades on Mir, calling from *Discovery* at 15:00 UTC. The two spacecraft were still about 160 kilometers apart.

The RSA eventually approved the close approach to Mir at 15:25 UTC. *Discovery* would keep the leaking thruster and its companion on the same manifold shut down during the close approach, and a new combination of thrusters would be used instead. The orbiter's nose pointed forward, its payload bay to Mir. Mir Commander Alexander Viktorenko reported seeing thruster firings from his vantage point, noting that the thruster plumes did not affect Mir's solar arrays. Yuri Polyakov aboard Mir reported he could see Wetherbee through the aft overhead window of *Discovery*. Weth-

erbee responded with a wave back to the cosmonaut.

Wetherbee assumed manual control of the remainder of the approach when *Discovery* was only 606 meters away from Mir. He switched *Discovery* to the low-Z mode when 310 meters from Mir. This prevented thrusters from firing up against Mir structures. *Discovery* moved down a 16° cone, eventually approaching close to the Kristall Module of Mir at less than 3 centimeters per second.

The astronaut crew waited for a final "go" to get within 11.5 meters of the Mir docking port as the two vehicles passed over the South Atlantic Ocean. Foale continued to work as a flight engineer at the aft station, and *Discovery* remained 121 meters in front of Mir while final preparations were made prior to getting a "go" to proceed in further.

Polyakov, the current world spaceflight-endurance holder—going for eighteen months, well past a year already—could see flames from thruster firings as Wetherbee moved in on Mir. None of the thrusters he used fired up. The cosmonauts on Mir asked Titov to describe the condition of Mir from his vantage point on *Discovery*, specifically looking to note changes since the time he spent a year on board the space station. The shuttle crew saw no motion of Mir's solar arrays in response to *Discovery*'s thruster firings. The Mir cosmonauts agreed with that assessment.

Wetherbee began to brake *Discovery*'s approach and held the two vehicles in separation at a distance of 10 meters. The two vehicles were flying over the mid-Pacific Ocean in conditions of daylight on *Discovery*'s fifty-seventh orbit of the mission. Viktorenko noted that the cosmonauts inside Mir never felt any movement of Mir from the activities of *Discovery* during the close approach phase. After data refinement, it was announced that the closest distance between the top of SPACEHAB and the Kristall Module on Mir was 11.2 meters, a separation achieved at 18:20 UTC.

Discovery backed away to a 121-meter separation distance at 20:26 UTC and began a fly-around of the space station. Live television was downlinked from inside Mir showing Dr. Polyakov as *Discovery* crossed directly above the space station at a distance of 144 meters. To complete the fly-around, *Discovery* flew to a position directly behind Mir and then flew underneath the space station and returned to the point from which it started the fly-around.

Mir was in a so-called T-configuration. It would not be in this particular arrangement when *Atlantis* would attempt a docking during STS-71. Plans were in place to launch another module, Spektr, from Kazakhstan prior to the STS-71 docking mission. Kristall would be moved to another axial port on Mir to make room for Spektr. STS-71 would still make its docking to the Kristall Module, regardless of its location on Mir.

The final separation burn was performed by *Discovery* at 21:13 UTC. *Discovery* was directly above Mir at the time. The burn pushed *Discovery* behind, away, and down from Mir, entering a lower orbit that would speed up the orbiter and eventually put it ahead of Mir.

The flight was hardly over, although this aspect was by and large the most exciting part, something of which Flight Director Phil Engelauf reminded the press. He called the joint flight a very invigorating exercise that had required a great deal of work.

On February 7, Titov grappled the SPARTAN-204 satellite again and raised it high above the payload bay as *Discovery* passed over Brazil at an altitude of 384 kilometers. This time, he let the satellite go, and *Discovery* backed away from the payload so that it could begin to make its far ultraviolet measurements autonomously over the next several days. By the end of the day, Mir and *Discovery* were about 260 kilometers apart, continuing to separate further.

Extravehicular activity (EVA) preparations filled much of February 8, as did experiment work with SPACEHAB investigations. On February 9, Voss used the RMS arm to grapple the SPARTAN-204 satellite at 11:53 UTC, as *Discovery* flew over the Aleutian Islands. For most of its autonomous flight, the free-flyer had been about 68 kilometers ahead of *Discovery*.

By 12:00 UTC, the SPACEHAB hatch was open and the EVA was about to start. The two spacewalkers were both making their first excursion outside a spacecraft in orbit. When he exited the air lock, Harris became the first African American to perform an EVA.

The EVA astronauts wore warm thermal undergarments, sporting modifications made for the STS-61 Hubble Space Telescope (HST) first servicing mission EVA work. Cooling water was stopped short of the suit arms, and thermal mittens were available. When cold, each man could pull his fingers back out of the glove and into the palm to make a fist and restore circulation and warmth.

When Foale and Harris reported just how cold they eventually got (the temperature in Foale's glove went well under 4.5° Celsius), Collins joked with them, suggesting that they were in a deep freeze. Because of the cold, the orbiter's attitude was eventually changed to point its payload bay to Earth for reflected light to warm up the spacewalkers. However, that was not done until most thermal testing and mass-handling exercises were accomplished. Further, because of the cold, the spacewalk was terminated about twenty-five minutes early.

With Titov at the RMS controls, Harris and Foale were maneuvered to a position high above the payload bay. This test lasted twenty minutes, with both men doing very little except serving as guinea pigs for the temperature exercise. Both men got uncomfortably cold in the fingers.

Special handholds were attached to the SPARTAN-204 payload to facilitate its manipulation. By 14:10 UTC, the astronauts were working on SPARTAN, evaluating how easily it could be maneuvered manually about the payload bay. Harris held the 1,273-kilogram satellite while on the RMS arm, judging how well he could handle such a bulk. He was moving something almost three times as massive as had been moved about on the HST first servicing mission (STS-61).

A thermal cube had been attached to the RMS arm to record temperatures encountered during the EVA. Before the EVA concluded at 16:30 UTC, the astronauts demated that thermal cube and stowed it for return to Earth.

On February 10, Wetherbee and Collins thoroughly checked out *Discovery*'s flight control system. Thrusters were test-fired, and the computer software used on entry was verified. The GLO Experiment was allowed to observe the effect of the steering jets' firings on the orbiter's airglow phenomenon.

The astronauts had one final look at Mir. At 18:35 UTC, the maneuver was performed that allowed the crew to see Mir near the Earth's horizon at a distance of over 1,360 kilometers. To the crew, and the payload-bay television cameras, Mir appeared only as a small flashing light.

On February 11, *Discovery* was given a "go" for a nominal deorbit. As the vehicle entered the upper atmosphere, however, crosswinds at the Kennedy Space Center (KSC) Shuttle Landing Facility (SLF) picked up in intensity. Wetherbee would have to steer on runway 15 during touchdown and rollout as a result.

About 250 visitors gathered at a viewing site to watch *Discovery* glide out of the postdawn skies at KSC to what could only be described as a picture-perfect touchdown at 11:50:19 UTC (6:50 A.M. eastern standard time).

Contributions

The STS-63 flight plan was packed with diverse mission activities and objectives, the most exciting of which was undoubtedly the fly-around of the Russian space station Mir. In addition, *Discovery* carried the third SPACEHAB module with twenty individual experiments, a Hitchhiker payload holding low-cost shuttle experiments, and the SPARTAN-204 free-flyer to study stellar emissions in the far ultraviolet (UV) range. In addition, as part of preparations for the upcoming shuttle-Mir missions and space station assembly flights, a five-hour spacewalk by Foale and Harris tested modifications to the shuttle EVA suit. They also practiced handling large objects in weightlessness.

The primary instrument on SPARTAN-204 was the Far Ultraviolet Imaging Spectrograph (FUVIS).

The objective of FUVIS was to observe astronomical (and shuttle-created) sources of far ultraviolet (UV) radiation, looking for information on the composition, physical and chemical properties, and distribution of far UV-emitting materials in the interstellar medium. Some of the objects that SPARTAN observed included the Barnard Loop, North America nebula, Cygnus Loop, comets, and galactic structures outside the Milky Way. The principal investigator for FUVIS was Dr. George R. Carruthers at the Naval Research Laboratory.

SPACEHAB 3 contained twenty experiments sponsored by the NASA Office of Space Access and Technology, NASA Office of Life and Microgravity Sciences and Applications, and the Department of Defense (DoD). Many of these experiments had flown before on the shuttle, several having flown on numerous occasions.

One of the experiments that the crew performed on February 5 in the SPACEHAB module was the Solid Surface Combustion Experiment. Pieces of Plexiglas were burned in a weightless environment, a part of a continuing experiment program that had flown on several previous shuttle missions. Another experiment conducted in SPACEHAB was the Charlotte robotic device. This experiment was set up by the astronauts, then left to run by its own devices. Cables were strung for the robot to maneuver across.

Context

Discovery touched down at KSC, having completed 129 orbits of the Earth and traveled a total of 3,348,480 kilometers over the course of an eight-day, 6-hour, and 28-minute flight, one of the most exciting and activity-packed shuttle missions, one that laid the groundwork for an even more exciting shuttle mission: STS-71, the first shuttle-Mir complex docking.

Collins was notified during flight that she was now in line for a shuttle commander position. Often a shuttle pilot flies twice in the pilot's seat before being given a command assignment. There were no details released about any specific flight assignment for the first woman shuttle pilot.

Down on Earth, the events in space between NASA and the Russians sparked debate concerning the merits and even wisdom of attempting joint operations, based largely on two concerns: that the Russians were once a bitter rival, and that the Russian economy and government are both unpredictably volatile and unstable. With threats of further cuts in the NASA budget from Congress, NASA Administrator Daniel S. Goldin affirmed his commitment to reduce the $3.2 billion annual shuttle operating budget. Goldin stated that the budget could not hide behind issues of safety but stressed that safety would not be compromised, either.

The Republican-controlled House of Representatives raised an old issue: privatization of the shuttle fleet. Consideration was given to plans to turn the shuttle maintenance and operational aspects of flight over to a commercial concern for profit and release NASA to perform purely research and development tasks.

Shuttle-Mir would consist of eleven space shuttle flights and seven astronaut residencies on the Russian space station Mir. Mir's journey would end on March 23, 2001, as Mir reentered the Earth's atmosphere near Nadi, Fiji, and fell into the South Pacific.

See also: Russia's Mir Space Station; Space Shuttle; Space Shuttle Flights, 1995; Space Shuttle-Mir: Joint Missions; Space Shuttle Mission STS-71/Mir Primary Expedition Eighteen; SPACEHAB.

Further Reading

Hall, Rex, and David J. Shayler. *Soyuz: A Universal Spacecraft.* Chichester, England: Springer-Praxis, 2003. The authors review the development and operations of the reliable Soyuz family of spacecraft, including lesser-known military and unmanned versions. While most works on Soviet/Russian space operations focus on space station activities, the story of the Soyuz spacecraft has been largely neglected.

Harland, David M. *The Space Shuttle: Roles, Missions, and Accomplishments.* Hoboken, N.J.: John Wiley, 1998. Topics include shuttle operations and payloads, weightlessness, materials processing, exploration, Spacelabs and free-flyers, and the shuttle's role in the International Space Station.

Harrington, Philip S. *The Space Shuttle: A Photographic History.* San Francisco, Calif.: Brown Trout, 2003. With one hundred full-color photographs by Roger Ressmeyer and others and with text by popular astronomy writer Phil Harrington, this book tells the story of the space shuttle program from 1972 to 2003. Its beautiful photographs allow the general reader to survey the history of the space shuttle program and be uplifted by the pioneering spirit of one of humanity's grandest enterprises.

Jenkins, Dennis R. *Space Shuttle: The History of the National Space Transportation System: The First 100 Missions.* Stillwater, Minn.: Voyageur Press, 2001. This is a concisely written technical reference account of the space shuttle and its ancestors, the aerodynamic lifting bodies. It details some of the advantages and inherent disadvantages of using a reusable space vehicle. Each of the vehicles is illustrated by line-drawings with important features pointed out with lines and text. The book follows the space shuttle from its original concepts and briefly chronicles its first one hundred flights.

National Aeronautics and Space Administration. *STS-63 Press Kit.* Washington, D.C.: Author, 1995. Fully describes the Mir Space Station, shuttle orbiter *Discovery,* and the flight plan for this first shuttle fly-around of the Russian space station.

Oberg, James. *Star-Crossed Orbits: Inside the U.S.-Russian Space Alliance.* New York: McGraw-Hill, 2001. In this book, the author combines personal memoir with investigative journalism to tell his story of the U.S.-Russian space alliance. With unparalleled access to of-

ficial Russian archives, facilities, and key individuals associated with the Russian space
program, he describes the strengths and weaknesses that each side of the alliance brings
to the table. He also reveals the full story of Russia's decaying space program and how it
ultimately was saved from collapse by Western funds.

Slayton, Donald K., with Michael Cassutt. *Deke! U.S. Manned Space: From Mercury to the Shuttle.*
New York: Forge, 1995. This is the autobiography of the last of the Mercury astronauts to
fly in space. After being grounded from flying in Project Mercury for what turned out to
be a minor heart murmur, Slayton was appointed head of the Astronaut Office. Later, he
commanded the Apollo-Soyuz flight in 1975. He stayed with NASA through the STS-2
flight. Although he was no longer an active participant in NASA's activities, he gives
some insight into the behind-the-scenes adventures of the shuttle program.

David G. Fisher

Space Shuttle-Mir: Joint Missions

Date: March 14, 1995, to June 12, 1998
Type of mission: Piloted spaceflight

When the Russian Space Agency (RSA) became involved in the International Space Station (ISS) project run by NASA, an opportunity arose to use Russia's Mir station to gain operational experience in maintaining permanent residence in space: the so-called Phase I program.

Key Figures

George W. S. Abbey (b. 1932), director, Johnson Space Center

Frank L. Culbertson, Jr. (b. 1949), Phase 1 Shuttle-Mir program manager

Valery Victorovich Ryumin (b. 1939), Phase 1 Shuttle-Mir program manager, Russia

Mark J. Albrecht, executive secretary, National Space Council

Alexander P. Alexandrov, cochair, Crew Training and Exchange Working Group, Phase I program

Yuri P. Antoshechkin, cochair, Flight Operations and Systems Integration Working Group

Boris Artemov, RSC Energia group leader, Configuration Management Control Subgroup

James D. Wetherbee (b. 1952), STS-63 commander

Eileen M. Collins (b. 1956), STS-63 pilot

Bernard A. Harris, Jr. (b. 1956), STS-63 mission specialist

C. Michael Foale (b. 1957), STS-63 mission specialist

Janice E. Voss (b. 1956), STS-63 mission specialist

Vladimir Georgievich Titov (b. 1947), STS-63 mission specialist (Russian Space Agency, RSA)

Alexander Viktorenko, Mir Permanent Crew 17 commander

Elena V. Kondakova (b. 1957), Mir Permanent Crew 17 flight engineer

Vladimir Nikolaevich Dezhurov (b. 1962), Mir Permanent Crew 18 commander

Gennady Mikhailovich Strekalov (b. 1940), Mir Permanent Crew 18 flight engineer

Norman E. Thagard (b. 1943), Mir Permanent Crew 18 astronaut-investigator

Robert L. "Hoot" Gibson (b. 1946), STS-71 commander

Charles J. Precourt (b. 1955), STS-71 pilot

Ellen S. Baker (b. 1953), STS-71 mission specialist

Gregory J. Harbaugh (b. 1956), STS-71 mission specialist

Bonnie J. Dunbar (b. 1949), STS-71 mission specialist

Anatoly Yakovlevich Solovyev (b. 1948), STS-71 payload specialist, Mir Permanent Crew 19 commander

Nikolai Mikhailovich Budarin (b. 1953), STS-71 payload specialist, Mir Permanent Crew 19 flight engineer

Yuri Pavlovich Gidzenko (b. 1962), Mir Permanent Crew 20 commander

Sergei Avdeyev, Mir Permanent Crew 20 flight engineer

Yuri Onufrienko (b. 1961), Mir Permanent Crew 21 commander

Yuri Vladimirovich Usachev (b. 1957), Mir Permanent Crew 21 flight engineer

Shannon W. Lucid (b. 1943), Mir Permanent Crew 21 and 22 astronaut-investigator
Valery Grigorievich Korzun (b. 1953), Mir Permanent Crew 22 commander
Alexander Yurievich Kaleri (b. 1956), Mir Permanent Crew 22 flight engineer
John E. Blaha (b. 1942), Mir Permanent Crew 22 astronaut-investigator
Jerry M. Linenger (b. 1955), Mir Permanent Crew 22 and 23 astronaut-investigator
Vasili Tsibliyev, Mir Permanent Crew 23 commander
Alexander Lazutkin, Mir Permanent Crew 23 flight engineer
C. Michael Foale (b. 1957), Mir Permanent Crew 23 and 24 astronaut-investigator
Anatoly Yakovlevich Solovyev (b. 1948), Mir Permanent Crew 24 commander
Pavel Vinogradov, Mir Permanent Crew 24 flight engineer
David A. Wolf (b. 1956), Mir Permanent Crew 24 astronaut-investigator
Andrew S. W. Thomas (b. 1951), Mir Permanent Crew 24 and 25 astronaut-investigator
Talgat Musabayev, Mir Permanent Crew 25 commander (Kazakhstan)
Nikolai Mikhailovich Budarin (b. 1953), Mir Permanent Crew 25 flight engineer

Summary of the Missions

Russia's Soyuz TM-21 lifted off on March 14, 1995, with Commander Vladimir Nikolaevich Dezhurov, Flight Engineer Gennady Mikhailovich Strekalov, and American Norman E. Thagard. Two days later, Soyuz TM-21 docked to the Kvant Module's aft port. The cosmonauts and astronaut were greeted by the resident Russian team: Alexander Viktorenko and Elena V. Kondakova.

Although Thagard began his Phase I increment flying aboard a Russian spacecraft, he would return home aboard NASA's shuttle *Atlantis* after NASA's first Mir docking. *Atlantis* lifted off on mission STS-71 on June 27, 1995. The commander, Robert L. "Hoot" Gibson, flew the same rendezvous profile STS-63 had when *Discovery* blazed a trail to Mir the previous year. Gibson docked *Atlantis* to the Kristall ("crystal") Module's docking port on June 29. *Atlantis* carried supplies supplementing deliveries by robotic Progress freighters, as well as Cosmonauts Anatoly Yakovlevich Solovyev and Nikolai Budarin, who would replace Strekalov and Dezhurov—the latter returning home on *Atlantis* along with Thagard and STS-71's crew. Thagard's time aboard Mir ended on July 3, when hatches between *Atlantis* and Kristall closed. *Atlantis* undocked the next morning and landed on July 7 at Kennedy Space Center.

After STS-71, the cosmonauts moved Kristall to a different docking port, where it would perma-

nently reside. The Russians built a special docking module for addition to Kristall, providing a port for future shuttle dockings farther away from Mir's solar arrays. *Atlantis* delivered that feature on STS-74 when, on November 12, 1995, with the docking module mounted atop *Atlantis*'s own Orbiter Docking System, Commander Chris A. Hadfield slowly attached the docking module's upper end to Kristall. No NASA astronaut remained behind this time, but when *Atlantis* undocked, it separated from the docking module's bottom end, leaving that addition permanently affixed.

The next Phase I action delivered Shannon W. Lucid for a planned five-month habitation. *Atlantis* lifted off on March 22, 1996, on STS-76. Two days later, Commander Kevin P. Chilton docked *Atlantis* to Mir's docking module, and Lucid's increment began. *Atlantis* undocked on March 28, returning to Earth at Edwards Air Force Base on March 31. Lucid would be picked up by STS-79. Lucid had anticipated the Priroda Module's arrival before her own, but Priroda's launch became delayed until well after she boarded Mir. That limited her research, as much of her equipment would be inside Priroda. She used Spektr (Russian for "spectrum") as both a living space and a research laboratory. Priroda finally launched on April 23. Three days later Priroda ("nature") docked to Mir's forward axial port and was later moved to the final available radial port. Mir's completed construction surpassed 110 metric tons in total mass.

Shuttle program delays forced Lucid to remain aboard Mir longer than originally planned. STS-79 lifted off on September 16, 1996, carrying John E. Blaha, Lucid's replacement. *Atlantis* docked on September 18, onboard television showing Lucid anxiously waiting inside the docking module. After several days' joint activities, *Atlantis* undocked, effecting the first Phase I participant exchange, and spent several days flying independently before returning to Kennedy Space Center on September 26. Lucid completed 188 days in space, a new NASA record.

Blaha completed important biotechnology experiments during his increment. He set up a wheat crop inside Mir's greenhouse. Blaha became NASA's first astronaut since Skylab 4 to spend both the Thanksgiving and Christmas holidays in orbit. STS-81 launched on January 12, 1997, carrying Blaha's replacement, Jerry M. Linenger. *Atlantis* docked two days later.

After several days' joint operations and logistical transfers, *Atlantis* undocked on January 19, removing Blaha and leaving Linenger behind, thereby continuing an uninterrupted NASA presence. When *Atlantis* landed at Kennedy Space Center on January 22, Blaha agreed to be taken off *Atlantis* via stretcher, as doctors wished. He proved to be the only Phase I participant who agreed to that request. All others preferred to walk off under their own power. Despite his conscientious in-flight exercise regimen, Blaha had experienced more postlanding difficulty than most. (Russian cosmonauts routinely are carried from Soyuz spacecraft after landing.)

Linenger experienced the first in 1997's seemingly endless series of Mir mishaps and malfunc-

Yuri V. Usachev (right), Mir 21 Flight Engineer, bids STS-76 Mission Commander Kevin P. Chilton a warm farewell as Atlantis *is about to be separated from Russia's Mir Space Station.* (NASA)

tions when his long-term residence overlapped a short-term international visit that temporarily raised Mir's human complement to six. Housing that many, Mir's Elektron oxygen-generation system needed supplementation by burning lithium perchlorate canisters. While Alexander Lazutkin attempted to activate one on February 23, leaking chemicals ignited uncontrollably. Although fire extinguishers were employed to suppress the flame, they proved ineffective; eventually the fire went out when the chemicals were totally consumed. Dense smoke lingered, however, and Mir's occupants wore oxygen and gas masks until the interior atmosphere proved safe again.

Linenger became the first astronaut to wear the Russian space suit when he spacewalked with Commander Vasili Tsibliyev on April 28. Problems with the life-support equipment continued throughout much of Linenger's increment, so special replacement items were manifested on STS-84, and the next Phase I participant, C. Michael Foale, received special training to assist with repairs.

Atlantis launched on May 15, 1997, carrying Foale and Cosmonaut Elena V. Kondakova among the crew. It landed at Kennedy Space Center on May 24, returning Linenger, who had just spent 132 days in space. Foale encountered 1997's worst station problem when, during a cosmonaut-controlled Progress freighter docking test on June 25, Commander Tsibliyev neither received range and range-rate data nor had the proper control over Progress M-34 as it sailed past Mir's forward port and impacted Mir, damaging radiators and solar arrays and breaching Spektr's hull. The cosmonauts sprang into action as alarms alerted them to pressure loss. Lazutkin and Foale sealed off Spektr's hatch and powered up their Soyuz in the event that they would need to abandon Mir. Meanwhile, Mir lost attitude and power dropped seriously. Essential power-hungry systems had to be deactivated, and it took several days to reestablish power, achieve stable attitude, and restore life-support equipment.

Over several subsequent weeks, Mir's computer periodically malfunctioned, sending cosmonauts repeatedly into a lengthy sequence of restoring

power, attitude control, and life support. Additionally, both the carbon dioxide scrubbers and Elektron oxygen-generation units required frequent maintenance. With Spektr abandoned, much of Foale's personal property and access to research equipment were lost.

Soyuz TM-26 launched on August 5, bringing fresh cosmonauts up to Mir to replace Tsibliyev and Lazutkin. Cosmonauts Solovyev and Pavel Vinogradov launched with new tools and replacement parts. A spacewalk inside Spektr provided an opportunity to insert electrical lines through the hatch and make connection to Spektr's functional solar arrays, providing additional power. Attempts to isolate Spektr's hull breach, some done inside Spektr and some done outside the hull, all proved unsuccessful.

Despite Solovyev's and Vinogradov's efforts to restore Mir's faltering systems, providing a measure of normality, Mir's diminished condition provided cause for serious Phase I review, as some believed Mir to be unfit for further astronaut increments. Despite external criticism, NASA determined it safe to send David A. Wolf as Foale's replacement.

Atlantis lifted off on September 25 to begin STS-86 and perform yet another docking, logistical resupply effort, and crew exchange. Another joint spacewalk was performed before *Atlantis* left Mir and Wolf behind. When *Atlantis* touched down on October 6, it ended Foale's 144 days in space.

Mir continued experiencing problems, periodic computer disruptions, life-support systems failures, and attitude control loss, but Wolf never suffered life-threatening situations, as had Linenger and Foale. Throughout his increment it appeared Mir could support further occupations beyond Phase I's expected end, thanks in large part to Solovyev and Vinogradov's efforts.

On January 21, 1998, NASA announced that former cosmonaut and Energia's shuttle-Mir program director, Valery Victorovich Ryumin, would join STS-91, Phase I's final mission. During his brief visit, Ryumin would conduct a management appraisal of Mir's aging condition. Meanwhile, Wolf ardently worked to complete his research.

The astronauts and cosmonauts of STS-74, which saw the second shuttle and Mir docking. (NASA)

Endeavour lifted off on January 22, starting STS-89. Two days later, Commander Terrence W. "Terry" Wilcutt gently docked to Mir. After some initial concern over how well Andrew Thomas fit into a Russian space suit, Thomas was cleared to remain behind. Wolf joined the STS-89 crew, and *Endeavour* departed Mir on January 29, returning home just about five hours after Soyuz TM-27 docked to Mir, replacing Solovyev and Vinogradov and providing a three-week research opportunity for French Astronaut Léopold Eyharts. Wolf had spent 119 days aboard Mir.

Soyuz TM-26 landed on February 19, returning Solovyev and Vinogradov, the two men who had worked diligently to restore Mir. This left Mir under the direction of Cosmonauts Talgat Musabayev and Budarin.

Thomas made his home inside Priroda. His research proceeded without major difficulties, finishing late in May. *Discovery* launched at 6:06 P.M. on June 2. Two days later, STS-91 Commander Charles J. Precourt performed the final shuttle-Mir docking. *Discovery* undocked on June 8 following final ceremonies marking Phase I's completion. Thomas and the rest of the crew landed at Kennedy Space Center on June 12. Thomas had spent 130 days aboard Mir.

Contributions

The International Space Station represents a collaboration of NASA, RSA, the European Space Agency, the Canadian Space Agency, and the Japanese Space Agency. Perhaps the biggest lesson learned during Phase I was that technological problems were not the only ones such a program would encounter. Learning to coordinate scientific, engineering, and management teams from different countries and cultures was key to Phase I success, and it would be so during the ISS era as well.

There were problems, to be sure. NASA's and RSA's responses to 1997's serious difficulties, specifically the onboard fire during Linenger's increment and the Progress freighter collision during Foale's increment, were quite dissimilar. With regard to the former, the Russians considered the fire less serious than did NASA and did not feel obliged to notify their NASA counterparts immediately. Russian space stations had previously suffered onboard fires, although not quite as bad as this one. With regard to the collision, the lesson was clear. Multinational crews needed to be well prepared to diagnose and recover from potentially life-threatening situations. Tsibliyev, Lazutkin, and Foale worked fastidiously, stopping the pressure leak and keeping Mir from drifting uncontrollably so that a reasoned damage assessment could be made. Their swift action saved Mir from being abandoned.

Other Phase I lessons essentially related to human factors, such as the need to maintain a positive living and working environment so far from home. Each NASA participant suffered some degree of isolation, both physical and cultural, although to varying degrees. John E. Blaha recognized symptoms of depression developing. He had trained with one team of cosmonauts but found himself living aboard Mir with an unfamiliar group. Thagard frequently felt isolated from news and contact with his ground team and loved ones for lengthy periods. Linenger found himself working independently of his cosmonaut comrades. Lucid, and even more so Foale, Wolf, and Thomas, integrated better, but the latter three had the advantage of learning from experiences and difficulties undergone by their predecessors. One key to success was a good working understanding of the Russian language. Another was development of good working relationships with ground controllers and the establishment of a satisfying work schedule. Space

shuttle missions are governed by flight plans detailing every waking hour. Long-term station operations, it was found, would benefit from greater flexibility and freedom for adjustment, relaxation, and entertainment.

Seven NASA astronauts lived and worked aboard Mir. Some experienced few problems, whereas others faced life-threatening ordeals. Together approximately one thousand days of experience aboard the space station were logged, the majority accumulated continuously—experience highly valuable for the coming International Space Station, which would certainly experience its own share of minor and major difficulties over its expected twenty-year lifetime.

Context

Although numerous cooperative space projects had involved Russians and Americans, the first time that astronauts and cosmonauts met in space came in 1975 with the last flight of an Apollo spacecraft. During the Apollo-Soyuz Test Project, three astronauts and two cosmonauts linked independently launched spacecraft together to perform symbolic ceremonies and joint experiments for several days before landing separately. It would take two decades before American and Russian space programs again cooperated in piloted flight. With Russian inclusion in the development, construction, and operation of ISS, it became imperative to gain experience working together. The combination of Russia's Mir and NASA's space shuttle became known as Phase I, the first stage of the ISS program, wherein American astronauts would spend time aboard Mir gaining long-duration experience not possible on brief shuttle missions. Phase I also offered experience at station logistical resupply, as well as rendezvous and docking exercises. Originally the Phase I program included ten shuttle-Mir dockings, but it was briefly cut back, then again expanded so that nine were eventually accomplished.

With his Phase I experience, Thagard easily surpassed NASA's spaceflight duration record, which had been unchallenged for twenty-one years since

Skylab's final mission. He temporarily became NASA's most experienced astronaut, but it must be noted that at this point thirty-seven cosmonauts had accumulated greater total time in space than had Thagard. Lucid and the other Phase I participants quickly surpassed Thagard's experience.

Prior to Phase I, NASA's only previous space station flight experience had been Skylab 2, decades earlier. Skylab had been occupied for only 171 days, and, although it certainly merited the label of space station, NASA did not learn how to maintain a space station during permanent occupation through logistical transfer. In that regard, Skylab had more in common with the first four Russian Salyut stations (Salyuts 1, 3, 4, and 5). To be sure, piloted Skylab missions did bring along supplies, but those supplies were meant to last for the duration of individual, non-overlapping habitations. Between missions, Skylab transitioned into what might be characterized as hibernation; during Mir's uninhabited periods, it was referred to as being mothballed. During Phase I, a total of 58,000 pounds of logistical resupplies—including water, air, and food—were delivered to Mir by shuttles. This was a very necessary resupply, but Mir consumables relied primarily upon resupply by reliable Progress robotic freighters, a technology that would be incorporated into ISS alongside visits by NASA shuttles to change out crews and transport new hardware. Without Phase I, the transition to ISS operations would have been even more difficult.

Mir was constructed in orbit by connecting different modules, each launched separately from 1986 to 1996. During the shuttle-Mir program, Russia's Mir combined its capabilities with America's space shuttles. The orbiting Mir provided a large and livable scientific laboratory in space.

Magnificent to behold through the windows of a space shuttle, the 100-ton Mir was as big as six school buses. Inside, it looked more like a cramped labyrinth, crowded with hoses, cables and scientific instruments, as well as articles of everyday life, such as photographs, children's drawings, books, and a guitar. It commonly housed three crew members, but it sometimes supported as many as six, for up to a month. Except for two short periods, Mir was continuously occupied until August, 1999. Its journey ended March 23, 2001, as Mir reentered the Earth's atmosphere near Nadi, Fiji, and fell into the South Pacific.

See also: International Space Station: Development; International Space Station: Modules and Nodes; International Space Station: U.S. Contributions; Space Shuttle Mission STS-63; Space Shuttle Flights, 1996; Space Shuttle Flights, 1997.

Further Reading

Burrough, Bryan. *Dragonfly: NASA and the Crises Aboard Mir.* New York: HarperCollins, 1998. A nonchronological account of Mir operations during NASA's Phase I involvement with the Russian space program, largely focusing on behind-the-scenes interactions of principal managers, astronauts, and cosmonauts. Devotes much space to discussing Mir's lengthy list of major problems encountered in 1997.

Fisher, David G. "Shannon W. Lucid's 188 Days in Space." *QUEST: The Magazine of Spaceflight* 5, no. 3 (1996): 22-25. This article details the extended visit to Mir of Dr. Shannon W. Lucid, including aspects of STS-76 and STS-79. The article is preceded by a pair of articles by other authors—an STS-79 overview and STS-79 postflight data—and followed by a biographical article on Lucid, also by a different author.

Hall, Rex, and David J. Shayler. *Soyuz: A Universal Spacecraft.* Chichester, England: Springer-Praxis, 2003. The authors review the development and operations of the reliable Soyuz family of spacecraft, including lesser-known military and unmanned versions. While most works on Soviet/Russian space operations focus on space station activities, the story of the Soyuz spacecraft has been largely neglected.

Harland, David M. *The Space Shuttle: Roles, Missions, and Accomplishments.* Hoboken, N.J.: John Wiley, 1998. Topics include shuttle operations and payloads, weightlessness, materials processing, exploration, Spacelabs and free-flyers, and the shuttle's role in the International Space Station.

_____. *The Story of Space Station Mir.* Chichester, England: Springer-Praxis, 2005. The book tells how the Soviet Union's experience with a succession of Salyut space stations led to the development of Mir, which became an international research laboratory whose technology went on to form the "core modules" of the International Space Station. The book runs through to Mir's deorbiting in March, 2001, providing the definitive account of the Mir Space Station. The book reviews the origins of the Soviet space station program, in particular the highly successful Salyuts 6 and 7, describes Mir's structure, environment, power supply and maneuvering systems, and provides a comprehensive account of how it was assembled and how it operated in orbit.

Harrington, Philip S. *The Space Shuttle: A Photographic History.* San Francisco, Calif.: Brown Trout, 2003. With one hundred full-color photographs by Roger Ressmeyer and others and with text by popular astronomy writer Phil Harrington, this book tells the story of the space shuttle program from 1972 to 2003. Its beautiful photographs allow the general reader to survey the history of the space shuttle program and be uplifted by the pioneering spirit of one of humanity's grandest enterprises.

Jenkins, Dennis R. *Space Shuttle: The History of the National Space Transportation System: The First 100 Missions.* Stillwater, Minn.: Voyageur Press, 2001. This is a concisely written technical reference account of the space shuttle and its ancestors, the aerodynamic lifting bodies. It details some of the advantages and inherent disadvantages of using a reusable space vehicle. Each of the vehicles is illustrated by line drawings. The book follows the space shuttle from its original concepts and briefly chronicles its first one hundred flights.

Linenger, Jerry M. *Off the Planet.* New York: McGraw-Hill, 2000. A personal account of Astronaut Linenger's Phase I increment. It must be noted that his account paints a more serious situation with regard to the fire than either the official Russian or the official NASA account. Provides insight into interpersonal relationships between cosmonauts and astronauts, particularly with those with whom Linenger served.

National Aeronautics and Space Administration. *Space Shuttle Mission Press Kits.* http://www-pao.ksc.nasa.gov/kscpao/presskit/presskit.htm. Provides detailed preflight information about each of the space shuttle missions. Accessed March, 2005.

Oberg, James. *Star-Crossed Orbits: Inside the U.S.-Russian Space Alliance.* New York: McGraw-Hill, 2001. In this book, the author combines personal memoir with investigative journalism to tell his story of the U.S.-Russian space alliance. With unparalleled access to official Russian archives, facilities, and key individuals associated with the Russian space program, he describes the strengths and weaknesses that each side of the alliance brings to the table. He also reveals the full story of Russia's decaying space program and how it ultimately was saved from collapse by Western funds.

David G. Fisher, updated by Russell R. Tobias

Space Shuttle Mission STS-71/Mir Primary Expedition Eighteen

Date: June 27 to July 7, 1995
Type of mission: Piloted spaceflight

STS-71, the one hundredth human spaceflight conducted by NASA, accomplished the United States' first docking with Russia's space station Mir, bringing it supplies and a new crew from Earth, returning other materials and the old crew, and gaining valuable experience for both countries in complex, joint operations in space. Already on board the Mir Complex were two Russians and one American participating in the Mir Primary Expedition Eighteen mission. The Russian abbreviation for Primary Expedition (Ekspeditsya Osnovnaya) is EO. This flight was the eighteenth long-duration mission aboard the Mir Complex.

Key Figures

Robert L. "Hoot" Gibson (b. 1946), STS-71 commander
Charles J. Precourt (b. 1955), STS-71 pilot
Ellen S. Baker (b. 1953), STS-71 mission specialist
Gregory J. Harbaugh (b. 1956), STS-71 mission specialist
Bonnie J. Dunbar (b. 1949), STS-71 mission specialist
Anatoly Yakovlevich Solovyev (b. 1948), STS-71 payload specialist
Nikolai Mikhailovich Budarin (b. 1953), STS-71 payload specialist
Vladimir Nikolaevich Dezhurov (b. 1962), Soyuz TM-21/Mir-18 commander
Gennady Mikhailovich Strekalov (b. 1940), Soyuz TM-21/Mir-18 flight engineer
Norman E. Thagard (b. 1943), Soyuz/Mir-18 cosmonaut researcher

Summary of the Mission

After decades of rivalry in the exploration and utilization of space, interrupted by brief collaborations, the United States and Russia agreed in the early 1990's to conduct joint operations in human spaceflight. Following STS-60, with the first flight of a cosmonaut on a space shuttle, and STS-63, which demonstrated the ability to rendezvous and orbit as close as 11 meters from Russia's space station Mir ("peace"), STS-71 would be the first of a series of space shuttle missions to dock with the station.

When the space shuttle lifted off from the Kennedy Space Center (KSC) in Florida at 19:32:18.988 Coordinated Universal Time (UTC, or 3:32 P.M. eastern daylight time) on June 27, 1995, Mir was in orbit over Iraq. When *Atlantis* reached orbit eight minutes later, it was 13,000 kilometers behind Mir, but in an orbit that carried it around Earth faster than the space station. Initially, with each orbit the distance closed by 1,630 kilometers. Over the course of the next two days, the astronauts would modify their orbit to match that of Mir.

Robert L. "Hoot" Gibson commanded STS-71 and his pilot was Charles J. Precourt. Bonnie J. Dunbar, Ellen Baker, and Gregory J. Harbaugh were the mission specialists. The two other crew members, Cosmonauts Anatoly Yakovlevich Solovyev and Nikolai Budarin, would take up residence aboard Mir after the two ships had docked.

Already in space were Vladimir Nikolaevich Dezhurov, Gennady Mikhailovich Strekalov, and NASA Astronaut Norman E. Thagard, conducting the eighteenth long-duration mission on Mir, denoted EO-18 by Russia. They had been living there since two days after they left Earth on March 14 and would exchange places with Solovyev and Budarin during STS-71. In addition to performing experiments and maintaining the station, they had finalized the reconfiguration of its movable modules and solar arrays in preparation for *Atlantis*'s arrival.

On the second day of STS-71, as *Atlantis* continued to close the distance to Mir, Gibson, Precourt, and Dunbar activated the Spacelab module carried in the orbiter's payload bay. This laboratory's extensive complement of scientific instruments would be used to measure the effects of the Mir crew's long exposure to the microgravity of orbit. Meanwhile, Baker verified that the Russian-built docking system, also in the payload bay, was in good condition. By the end of the day, *Atlantis* was within 2,500 kilometers of Mir, and the crew was ready to focus on performing the complicated rendezvous and docking the following day.

On June 29, as Gibson flew the orbiter to Mir, Precourt monitored *Atlantis*'s position and orientation on a computer, Harbaugh measured the range and speed between the two spacecraft, and Dunbar transmitted these data to the crew on Mir. When they were about 75 meters from Mir, Gibson halted the approach to allow time for Russian controllers to verify that Mir was in the correct orientation for docking. In addition, the network of solar arrays on the station had to be adjusted so that they would be edge-on to *Atlantis*. This would ensure that exhaust from the orbiter's maneuvering thrusters would cause minimal disturbance to these sensitive appendages.

Continuing to close slowly, Gibson used a camera mounted in the docking apparatus to provide a view of the docking target attached to the Kristall ("crystal") Module of Mir. When he had maneuvered *Atlantis* to about 9 meters from the station, he halted his approach again for five minutes. This

would allow the actual link-up to occur within range of a Russian tracking station and permitted a final check on all critical systems before joining the two massive spacecraft. Finally, Gibson nudged *Atlantis* up to Mir at 3.3 centimeters per second. Precourt immediately fired *Atlantis*'s thrusters to engage the docking mechanism and make a firm connection.

Besides providing a strong coupling, the docking system included a tunnel that allowed the astronauts and cosmonauts to move freely between the linked spacecraft. After pressurizing the passageway and confirming that both spacecraft were in good condition, the two commanders, Gibson and Dezhurov, opened the hatches and greeted each other.

Following a short welcoming ceremony, the resident crew on Mir gave a safety briefing, including emergency evacuation procedures, to their guests. *Atlantis* carried custom-made contoured seat liners that Solovyev and Budarin would use when they returned to Earth the following October aboard the Soyuz ("union") spacecraft the EO-18 crew had used to reach the station. Before the EO-19 crew could take up residence on Mir, these liners were installed in the return craft, and those used by the EO-18 crew on their ascent were transferred to *Atlantis*. With the completion of the safety briefing and the seat-liner exchange, Dezhurov, Strekalov, and Thagard officially became members of the *Atlantis* crew, and Solovyev and Budarin took over as the newest Mir crew. That night, after 105 days on board Mir, the EO-18 crew slept in *Atlantis*.

Apart from some ceremonies commemorating the first American-Russian docking in twenty years, most of the next four days was devoted to transferring supplies between the two craft and conducting experiments designed to help understand the effects of the long stay in space on the physiology of the EO-18 crew members. The crews also took measurements to assess the hygienic and radiation conditions on Mir and conducted joint engineering experiments on the most massive and complex spacecraft ever assembled.

A substantial effort went into the time-consuming transfer of equipment and supplies. The shuttle orbiter would return the bounty from the EO-18 crew's nearly four months in space. They had accumulated many samples of blood, saliva, and urine from their human physiology investigations, and these were carefully stowed in *Atlantis* for the ride back to Earth. Cassettes and disks with scientific and engineering data were removed from Mir, as was some equipment no longer needed aboard the station.

Atlantis resupplied Mir with many of the items it needed to support humans in space for extended periods. Water, produced in excess as a by-product of the orbiter's generation of electricity, was a precious resource on Mir. Transferring it was slow, as it required the astronauts to fill containers in *Atlantis*'s galley and carry them to Mir, but they managed to deliver nearly 500 kilograms. By raising the

air pressure in *Atlantis* (and allowing the air to mix through the docking tunnel), the astronauts increased the oxygen and nitrogen supply on board the station.

During their stay on Mir, Solovyev and Budarin planned to try to free a jammed solar array on Mir's newest module, Spektr ("spectrum"). Special tools for this task had been constructed in the United States and Russia and were ferried to Mir by *Atlantis*.

Some items delivered during STS-71 were specifically intended to make life on board an orbiting station more pleasant. Fresh fruits, chocolates, flowers, and even replacement strings for the guitar on board Mir would boost the spirits of people spending long times away from Earth.

Engineering tests of the operation of the large structure were conducted to aid in the design of future space stations. By firing the thrusters on *Atlantis*, the crew could measure how firmly the docking system gripped the two spacecraft, and the stability of the complex could be inferred by measurements of vibrations induced in the solar arrays.

On July 3, after a farewell ceremony in Mir, the final transfers were completed and the crews left each other's spacecraft for the last time. The next day, at 10:55 UTC, Solovyev and Budarin separated their Soyuz craft from Mir so they could film *Atlantis*'s undocking. Fifteen minutes later, springs gently pushed *Atlantis* away from the complex, and Gibson and Precourt began a flight around the station to make detailed assessments of its condition after more than nine years in space. A computer problem on Mir forced the cosmonauts to redock sooner than planned, but they managed to capture some spectacular views first.

As *Atlantis* receded from Mir, the crew conducted more medical examinations on the EO-18 crew members, who also exercised to prepare their bodies for the return to gravity. By the next day, Mir appeared to them as only a distant point of light.

The day before landing, the astronauts performed a routine prelanding evaluation of *Atlantis*, deactivated Spacelab, and installed special seats

Orbiting over Lake Baikal in southern Siberia, Atlantis *and* Mir *are docked to Mir's Kristall Module, which joins the orbiter to the space station.* (NASA)

that would allow the EO-18 crew to be recumbent during the landing. This position minimizes the physiological stress of returning to gravity, a necessity after a long absence from the effects of that constant force. A communications test from NASA's new flight control room showed that it operated as intended, so subsequent flights could be controlled from this modernized facility.

On July 7, Gibson guided *Atlantis* to a smooth landing at 14:54:36 UTC (10:55 A.M. eastern daylight time) at KSC, completing two missions at once. The five STS-71 astronauts added more than nine days and 19 hours to their spaceflight experience. At the same time, the EO-18 crew completed nearly 115 days, 9 hours of spaceflight. Yet their work was not complete. Several weeks of medical tests to monitor their bodies' readaptations to gravity awaited them. More immediate, however, were some rewards, including Thagard's being reunited with his family and hot-fudge sundaes for the humans who had been deprived of many Earthly pleasures during their extended stay in space.

Contributions

More than two years of engineering analysis and planning preceded the flight of STS-71. The success of the mission not only was a testament to the skill and dedication of the Americans and Russians who made the preparations but also served as a valuable validation of the techniques they developed and used to prepare for such a complex operation. This new knowledge would serve well for future, even more ambitious flights with the space shuttle and Mir and, it was hoped, with the International Space Station, which was in advanced design and construction during STS-71.

Activities that would be important for the International Space Station and future Mir docking missions were evaluated for the first time on STS-71. For example, the transfer of several hundred items from each spacecraft to the other proved to be more time-consuming than expected. This experience led to methods for improving efficiency, thus helping to ensure that subsequent flights would be as productive as possible.

Over many years, NASA had honed its skills in the complex tasks of planning before and replanning during space missions to make optimal use of the crews and the systems. STS-71 illuminated an important difference between what NASA had become accustomed to, with space shuttle missions generally lasting one to two weeks, and the longer missions that would be typical for people on the International Space Station. When circumstances necessitated a change in plans on a short space shuttle flight, ground controllers frequently developed a new plan overnight so the crew could incorporate it into their activities as soon as possible. Such a hectic pace could not be sustained during extended stays on Mir, and a delay of a few days was not that important, so cosmonauts were more accustomed to receiving new plans several days in advance of implementing them. This difference in style contributed to some difficulty in working with the EO-18 crew during STS-71 and provided a valuable lesson for NASA.

After their long stay in the station, the EO-18 crew members were occasionally irritable and not as cooperative as ground personnel expected them to be. The men were not always willing to participate, complaining that some experiments were too strenuous or uncomfortable. To accommodate this reticence, ground controllers were forced to make quick changes in the plans and schedules. Nevertheless, Baker and Dunbar were able to perform extensive cardiovascular, pulmonary, metabolic, neurological, and behavioral experiments on the EO-18 crew, providing scientists in the United States their first opportunity since Skylab, more than twenty-one years earlier, to investigate the effects of prolonged stays in space on the human body. Experiments performed during and after the flight added to a small but growing database on humans' adaptation to space and readaptation to Earth. The EO-18 crew's extensive program of exercising served the dual purpose of aiding in the evaluation of their physiological changes during spaceflight and acting as a countermeasure to those changes to ease the return to Earth.

Context

Although NASA had gained extensive experience in orbital docking during the Gemini and Apollo Programs, STS-71 was its first docking since the Apollo-Soyuz Test Project, twenty years earlier. Because of the tremendous mass of the space shuttle orbiter and Mir, and their distinct lack of symmetry, their docking was much more complicated than that of the spacecraft in earlier programs. Nevertheless, Gibson and his crew executed a flawless docking, and their structural experiments showed that the docking system formed a very secure attachment.

The second shuttle-Mir docking mission, STS-74, incorporated several changes based on the results of its predecessor. For example, after the orbiter's propellant usage, when controlling the docked assembly, was discovered to be higher than predicted on STS-71, computer control systems were improved, and this led to reduced expenditure of precious propellant during STS-74. Mir had experienced power shortages after it assumed the orientation required for docking with *Atlantis*, still about 75 meters away. To reduce the time the station would spend without pointing its solar arrays at the Sun, on STS-74 *Atlantis* closed to about 50 meters before halting to allow Mir to orient itself. In addition, the geometry of the approach of the two spacecraft was modified to improve the lighting on Mir's solar arrays rather than to guarantee coverage by Russian ground stations.

The excellent cooperation demonstrated during STS-71 was a stark contrast to most of the history of space exploration. During the Cold War, the space programs of the United States and the Soviet Union were strong political weapons, helping each country demonstrate to the world its technological strengths. By the beginning of the 1990's, however, priorities had shifted, and economic aspects of the space programs grew in importance compared with the political rivalries. The United States Congress became increasingly reluctant to fund NASA's planned International Space Station and other expensive programs, and a collapsing economy in Russia made it harder and harder for the new government to sustain the former Soviet Union's costly space program.

As both sides recognized that to accomplish their goals they could benefit more from cooperation than competition, they signed a set of agreements in 1992 and 1993 to conduct three phases of joint operations in human spaceflight. The principal objectives of Phase I, of which STS-71 was a key element, were to gain experience in working together on complex space missions; to resupply Mir with water, air, food, and other essentials; and to test hardware, software, and techniques to be used on the International Space Station, which already included the United States, Canada, the European Space Agency, and Japan. Phases II and III, extending from 1997 to the early 2000's, would cover the joint construction and initial operation of the vast orbiting complex.

Phase I would provide the opportunity for each country to gain experience with the complex rules, procedures, and organizations that the others had developed during more than thirty years of human spaceflight. This would help greatly as they entered into the challenging task of building and operating the International Space Station. Russia had a well-developed capability to resupply Mir and return equipment to Earth, but the payload capacity was limited. The space shuttle, as demonstrated on STS-71 and subsequent missions, could provide a very large capacity. At the same time, both countries were interested in developing countermeasures to the physiological and psychological effects of extended spaceflight. Russia had greater experience in long-duration flights on the Soviet Salyut ("salute") stations, operated between 1971 and 1986, and Mir, inhabited nearly continuously from 1986 to 1999. By combining this experience with the comprehensive space-borne scientific facilities, the United States could apply to this research in Spacelab, and both countries expected to garner better data. This was expected to serve the future well, as people from those countries and others worked together on Earth and in space.

See also: Space Shuttle-Mir: Joint Missions.

Further Reading

Hall, Rex, and David J. Shayler. *Soyuz: A Universal Spacecraft.* Chichester, England: Springer-Praxis, 2003. The authors review the development and operations of the reliable Soyuz family of spacecraft, including lesser-known military and unmanned versions. While most works on Soviet/Russian space operations focus on space station activities, the story of the Soyuz spacecraft has been largely neglected.

Harland, David M. *The Space Shuttle: Roles, Missions, and Accomplishments.* Hoboken, N.J.: John Wiley, 1998. This book is written thematically, rather than purely chronologically. Topics include shuttle operations and payloads, weightlessness, materials processing, exploration, Spacelabs and free-flyers, and the shuttle's role in the International Space Station.

Harrington, Philip S. *The Space Shuttle: A Photographic History.* San Francisco, Calif.: Brown Trout, 2003. With one hundred full-color photographs by Roger Ressmeyer and others and with text by popular astronomy writer Phil Harrington, this book tells the story of the space shuttle program from 1972 to 2003. Its beautiful photographs allow the general reader to survey the history of the space shuttle program and be uplifted by the pioneering spirit of one of humanity's grandest enterprises.

Jenkins, Dennis R. *Space Shuttle: The History of the National Space Transportation System: The First 100 Missions.* Stillwater, Minn.: Voyageur Press, 2001. This is a concisely written technical reference account of the space shuttle and its ancestors, the aerodynamic lifting bodies. It details some of the advantages and inherent disadvantages of using a reusable space vehicle. Each of the vehicles is illustrated by line-drawings with important features pointed out with lines and text. The book follows the space shuttle from its original concepts and briefly chronicles its first one hundred flights.

Oberg, James. *Star-Crossed Orbits: Inside the U.S.-Russian Space Alliance.* New York: McGraw-Hill, 2001. In this book, the author combines personal memoir with investigative journalism to tell his story of the U.S.-Russian space alliance. With unparalleled access to official Russian archives, facilities, and key individuals associated with the Russian space program, he describes the strengths and weaknesses that each side of the alliance brings to the table. He also reveals the full story of Russia's decaying space program and how it ultimately was saved from collapse by Western funds.

Slayton, Donald K., with Michael Cassutt. *Deke! U.S. Manned Space: From Mercury to the Shuttle.* New York: Forge, 1995. This is the autobiography of the last of the Mercury astronauts to fly in space. After being grounded from flying in Project Mercury for what turned out to be a minor heart murmur, Slayton was appointed head of the Astronaut Office. Later, he commanded the Apollo-Soyuz flight in 1975. He stayed with NASA through the STS-2 flight. Although he was no longer an active participant in NASA's activities, he gives some insight into the behind-the-scenes adventures of the shuttle program.

Marc D. Rayman

Space Shuttle Flights, 1996

Date: January 11 to December 7, 1996
Type of mission: Piloted spaceflight

During seven shuttle voyages in 1996, one shuttle docked twice with the Russian space station Mir, deposited an American astronaut on it, and later retrieved her. Another launched into space an antenna the size of a tennis court. Yet another accomplished the longest shuttle flight in history by staying aloft for eighteen days.

Key Figures

Brian Duffy (b. 1953), STS-72 commander
Brent W. Jett (b. 1958), STS-72 pilot
Leroy Chiao (b. 1960), STS-72 mission specialist
Daniel T. Barry (b. 1953), STS-72 mission specialist
Winston E. Scott (b. 1950), STS-72 mission specialist
Koichi Wakata (b. 1963), STS-72 mission specialist
Andrew M. Allen (b. 1955), STS-75 commander
Scott J. Horowitz (b. 1957), STS-75 pilot
Franklin R. Chang-Díaz (b. 1950), STS-75 mission specialist
Maurizio Cheli (b. 1959), STS-75 mission specialist (European Space Agency, ESA)
Jeffrey A. Hoffman (b. 1944), STS-75 mission specialist
Claude Nicollier (b. 1944), STS-75 mission specialist (European Space Agency, ESA)
Umberto Guidoni (b. 1954), STS-75 payload specialist (Italian Space Agency, ASI)
Kevin P. Chilton (b. 1954), STS-76 commander
Richard A. Searfoss (b. 1956), STS-76 pilot
Linda M. Godwin (b. 1952), STS-76 mission specialist
Michael R. "Rich" Clifford (b. 1952), STS-76 mission specialist
Ronald M. Sega (b. 1952), STS-76 mission specialist
Shannon W. Lucid (b. 1943), STS-76 mission specialist, STS-79 mission specialist, Mir Permanent Crew 21 and 22 astronaut-investigator
John H. Casper (b. 1943), STS-77 commander
Curtis L. Brown, Jr. (b. 1956), STS-77 pilot
Daniel W. Bursch (b. 1957), STS-77 mission specialist
Mario Runco, Jr. (b. 1952), STS-77 mission specialist
Marc Garneau (b. 1949), STS-77 mission specialist (Canadian Space Agency, CSA)
Andrew S. W. Thomas (b. 1951), STS-77 mission specialist
Terence T. "Tom" Henricks (b. 1952), STS-78 commander
Kevin R. Kregel (b. 1956), STS-78 pilot
Susan J. Helms (b. 1958), STS-78 mission specialist
Richard M. Linnehan (b. 1957), DVM, STS-78 mission specialist
Charles E. Brady, Jr. (b. 1951), MD, STS-78 mission specialist

Jean-Jacques Favier (b. 1949), STS-78 payload specialist (French Space Agency, CNES)
Robert Brent Thirsk (b. 1953), STS-78 payload specialist (Canadian Space Agency, CSA)
William F. Readdy (b. 1952), STS-79 commander
Terrence W. "Terry" Wilcutt (b. 1949), STS-79 pilot
Thomas D. "Tom" Akers (b. 1951), STS-79 mission specialist
John E. Blaha (b. 1942), STS-79 mission specialist
Jerome "Jay" Apt (b. 1949), STS-79 mission specialist
Carl E. Walz (b. 1955), STS-79 mission specialist
Kenneth D. Cockrell (b. 1950), STS-80 commander
Kent V. Rominger (b. 1956), STS-80 pilot
Tamara E. Jernigan (b. 1959), STS-80 mission specialist
Thomas D. "Tom" Jones (b. 1955), STS-80 mission specialist
F. Story Musgrave (b. 1935), STS-80 mission specialist

Summary of the Missions

Seven shuttle flights were completed in 1996—three by the shuttle *Columbia* and two each by *Endeavour* and *Atlantis*. There was just one major launching delay: The departure of *Atlantis*, scheduled for July, was pushed back six weeks to September in order to replace its rocket boosters.

The first shuttle flight of 1996 was on January 11, when, despite unusually cold weather, *Endeavour* blasted off from the Kennedy Space Center (KSC), Florida, at 09:41:00.015 Coordinated Universal Time (UTC, or 4:41 A.M. eastern standard time). The nine-day STS-72 mission had several objectives, among them was to capture the 3,600-kilogram Japanese Space Flyer Unit (SFU), launched aboard a Japanese H-2 booster on March 18, 1995, and return it to Earth. Japanese Astronaut Koichi Wakata deftly snagged the satellite. The mission objectives of the SFU were to validate the recoverable free-flyer system by demonstrating launch, in-orbit experiments, and retrieval, and by implementing science and engineering experiments and astronomical observation in orbit.

The crew deployed the $10 million Office of Aeronautics and Space Technology Flyer (OAST-Flyer) for the National Aeronautics and Space Administration (NASA). It orbited for two days before being retrieved, performing experiments involving lasers, space contamination, and amateur radios.

This mission involved two spacewalks. One tested the insulation of thermal gear to protect astronauts from frigid space temperatures. The other tested construction tools to be used in space and experimented with building techniques necessary to assemble a space station. The mission ended on January 20 at 07:41:45 UTC with a landing at KSC.

Columbia's voyage began on February 22 at 20:18:00.004 UTC (3:18 P.M. eastern standard time) and ended on March 9 at 13:58:21 UTC (8:58 A.M. eastern standard time). The sixteen-day STS-75 mission included the reflight of the $450 million Italian Tethered Satellite System. The satellite was reeled out from the payload bay on a tether 20.7 kilometers long. *Columbia* was to retrieve it and return it to Earth. This experiment would study the characteristics of the ionospheric plasma and how an electrical field can be extracted. The tether snapped about five hours after deployment, at a distance of 19.7 kilometers, just as a similar tether deployed by *Atlantis* in 1992 had snapped. It was hoped that the Italian experiment would demonstrate the capability of such a tethered satellite to provide power for a space station. Indeed, before the tether snapped, the system generated as much as 3,500 volts of electricity while moving through Earth's magnetic field. NASA extended *Columbia*'s flight for a fifteenth day for additional experiments. The flight ended normally, as *Columbia*'s commander, Andrew Allen, guided the orbiter to a smooth landing at KSC.

The next shuttle, *Atlantis*, was launched on March 22 at 08:13:03.999 UTC (3:13 A.M. eastern standard time). The main mission of STS-76 was to dock with the Russian space station Mir and to carry water and other supplies to it. It also transported Shannon W. Lucid, an American biochemist and experienced astronaut, to the space station, where she was scheduled to spend five months.

While *Atlantis* was docked with Mir, its crew members made a spacewalk, hanging panels on Mir to collect space dust. The landing, scheduled for March 29, was twice waved off due to bad weather at the Kennedy Space Center. After the wave off, a problem developed while reopening the doors of the spacecraft's cargo bay, which was necessary to cool crucial equipment. The crew manually opened the doors and, later, before the deorbit burn, they closed them without incident. The shuttle landed successfully at 13:28:56.8 UTC (5:29 A.M. Pacific standard time) at Edwards Air Force Base in California after its nine-day flight.

Endeavour was launched on a ten-day scientific mission on May 19 at 10:30:00.009 UTC (6:30 A.M. eastern daylight time), during which its crew deployed two satellites and retrieved one of them, the SPARTAN satellite. This satellite carried with it an antenna the size of a kitchen tabletop that was inflated (by nitrogen) to the size of a tennis court. The structure, made of thin Mylar with a reflective surface, tested the feasibility of using in space similar structures that are lighter, less expensive, and more flexible than solid structures and have fewer moving parts.

Another of the STS-77 mission's experiments involved a four-satellite rendezvous. One small satellite stabilized itself without the help of jets, using only Earth's magnetic field and the thin atmosphere of space. Twelve major industrial experiments were also conducted in a laboratory contained in a module owned by SPACEHAB, Inc. A space aquarium that contained one thousand mussel larvae, thirty-two thousand eggs from sea urchins, and six thousand starfish embryos was closely observed. This experiment studied marine feeding behavior and the development of calcified

tissue. *Endeavour* landed at KSC on May 29 11:09:24 UTC (7:09 A.M. eastern daylight time), following its journey of more than 6.4 million kilometers.

Columbia was launched from KSC on June 20 at 14:49:00.019 UTC (10:49 A.M. eastern daylight time), on what became the longest shuttle mission to date. STS-78 was designed to test how weightlessness and space travel affected humans. Four astronauts aboard the shuttle slept in wired nightcaps that provided information about insomnia in space as well as on Earth. The motion sickness that affected two-thirds of all previous astronauts was also studied. The mission ended on July 7 12:36:36 UTC (8:37 A.M. eastern daylight time), when the shuttle landed safely at KSC after seventeen days in space.

During *Columbia*'s mission a gas leak was discovered in *Atlantis*'s rocket boosters that caused a six-week delay in its launch, originally scheduled for mid-July. Until its boosters were replaced, it could not return to Mir to retrieve Shannon W. Lucid, who had been there since March 24, or to bring her replacement, John E. Blaha. Finally, on September 16 at 08:54:48.991 UTC (4:55 A.M. eastern daylight time), *Atlantis* was launched on mission STS-79. Two days later, *Atlantis* docked with Mir, where it spent five days transferring supplies and equipment to the space station. By the time *Atlantis* returned to Earth on September 26, Shannon W. Lucid had spent 188 days in space, the longest such flight for any woman. *Atlantis* safely touched down on Runway 15 at KSC at 12:13:13 UTC (8:13 A.M. eastern daylight time).

The mid-November launch of *Columbia* was delayed for several days by strong winds. On November 19 at 19:55:46.990 UTC (2:56 P.M. eastern standard time), STS-80 blasted off, carrying among its crew F. Story Musgrave, who, at sixty-one, was the oldest human at that time to undertake a space mission. *Columbia*'s crew deployed two scientific satellites. The first of the two free-flying payloads carried aboard *Columbia* was the Orbiting Retrievable Far and Extreme Ultraviolet Spectrometer (ORFEUS) satellite. The ORFEUS instruments were mounted on the reusable Shuttle Pallet Satel-

Members of the STS-72 crew participated in the Crew Equipment Interface Test (CEIT), which gave them the opportunity to look at the payloads they would be working with on orbit. The primary payloads are the Space Flyer Unit-Retrieval and the Office of Aeronautics and Space Technology-Flyer. (NASA)

lite (SPAS) and studied the origin and makeup of stars. The second—the Wake Shield Facility—is a free-flying stainless steel disk designed to generate an "ultra-vacuum" environment in space in which to grow semiconductor thin films for use in advanced electronic devices.

The mission also included a spacewalk to test tools to be used in assembling the International Space Station. A second spacewalk had to be scrubbed because of a stuck hatch. The shuttle's return was delayed for two successive days because of heavy clouds. It finally touched down in Florida on December 7 at 11:49:06 UTC (6:49 A.M. eastern standard time) after eighteen days in space, the longest such flight to date.

Contributions

Ten years after the *Challenger* mishap, NASA was keenly aware of the hazards involved in launching space shuttles in cold weather. When *Endeavour* was scheduled to lift off in early January, 1996, the weather was unusually and threateningly cold. Over the years, ground crews had learned a great deal about making safer launches under these conditions.

As *Endeavour* stood poised to lift off from KSC, the support team on the ground blew hot air into it and pumped warm nitrogen over the steering devices of the shuttle's boosters. An automatic system had been installed that consisted of warming bands wrapped around the O-rings that circle and seal the joints of the solid rocket boosters. Each joint also had three O-rings. All these precautions were designed to provide optimal safety during the launch.

During *Endeavour*'s January flight, astronauts tested thermal gear during spacewalks, determining that the gloves, boots, and space suits they wore could sustain them in temperatures below 100° Fahrenheit. They also tested some of the tools required for assembling the proposed International Space Station, proving categorically that the work needed for this assembly job could be successfully accomplished in space's low gravity under conditions of weightlessness.

Although *Columbia*'s February flight resulted in the loss of an Italian satellite when the 20.7-kilometer tether to which it was attached snapped, NASA gained considerable information about why the tether snapped. Before the satellite was lost, it had begun to generate electrical power of up to 3,500 volts, proving that its basic idea was workable.

Each of *Atlantis*'s flights, one in March, the other in September, carried an astronaut to the Russian space station Mir and left the crew member there for extended periods. *Atlantis* also carried considerable food, water, and other supplies to the space station and accomplished most of what it set out to do and was returned successfully to Earth.

The six-week delay of *Atlantis*'s second mission clearly demonstrated that Mir could function without an immediate replenishing of supplies. It also demonstrated that a human who spent more than six months in a weightless condition, as Shannon W. Lucid did, could quite quickly resume normal activity upon returning to Earth.

Endeavour's May mission showed how a relatively compact Mylar structure, the SPARTAN-207/ IAE (Inflatable Antenna Experiment) could be launched in space and then inflated with nitrogen to the size of a tennis court. The large antenna deployed by *Endeavour*'s crew functioned satisfactorily and led NASA officials to conclude that other such structures, much lighter than their solid counterparts, could serve useful functions in the projected International Space Station. The space aquarium aboard *Endeavour* provided scientists with information about calcification and feeding habits in weightless environments. The module carried for SPACEHAB, Inc., pointed the way to future commercial ventures connected with space.

Columbia's June flight was designed to test the effects of weightlessness on humans over an extended period of time using the Life and Microgravity Spacelab. This longest shuttle flight to date also conducted productive studies concerned with space sickness among space travelers and the insomnia that often results on space missions during which astronauts pass through the day-night cycle every hour.

Columbia made history again when its November launch began the longest shuttle voyage in history, eighteen days, and carried the oldest person who had ventured into space up to that time. The crew deployed two scientific satellites and learned a great deal about the possible hazards involved in doing this.

When a telescope launched by *Columbia* had alignment problems, NASA sent computer commands to correct the problem, demonstrating that land-based computers can function effectively in dealing with problems that occur in space. Finally, a crucial spacewalk designed to enable two astronauts to test how effectively a construction crane could be used in space had to be canceled because a hatch through which they had to pass to leave the space shuttle was jammed and could not safely be freed. This disappointment led to the redesign of such hatches with backup systems for their release.

Context

Memories of the *Challenger* accident a decade earlier cast a long shadow upon all subsequent space ventures, and particularly upon those undertaken in 1996, which marked the tenth anniversary of the loss of that space shuttle. Memories of *Challenger* were particularly pervasive as *Endeavour* was prepared for launch at about the same time in January that *Challenger* had originally been scheduled for liftoff. A sense of déjà vu accompanied the preparations for *Endeavour*'s departure. Florida was experiencing the same sort of unusually cold weather that had plagued the region at the time of *Challenger*'s ill-fated departure. The public, especially those who were present for the launch of *Endeavour*, drew frightening analogies between the two flights.

NASA officials, however, exuded confidence and optimism. They knew that the agency, despite budgetary cutbacks, was functioning more harmoniously and apolitically than it had been in 1986. They were also keenly aware of the strategic changes that had been made to assure the safety of future shuttle flights.

Not only did onboard and external systems keep vital parts of the shuttles at optimum temperatures, but the shuttles had also been redesigned to make their controls easier to read and to use. Escape systems that did not exist in 1986 were now in place. Although these escape mechanisms would not have saved the *Challenger* crew, they could save shuttle crews in the first seconds after launch if the launch had to be aborted.

All in all, NASA was now more streamlined than in the past. Its members had learned from the mistakes of the past and had gained a confidence that increasingly made space travel, despite its obvious complexities, seem more or less routine.

See also: Space Shuttle; Space Shuttle Flights, 1995; SPACEHAB.

Further Reading

Ahrens, C. Donald. *Essentials of Meteorology: An Invitation to the Atmosphere.* 4th ed. Pacific Grove, Calif.: Thomson Brooks/Cole, 2005. This is a text suitable for an introductory course in meteorology. Comes complete with a CD-ROM to help explain concepts and demonstrate the atmosphere's dynamic nature.

Andrade, Alessandra A. L. *The Global Navigation Satellite System: Navigating into the New Millennium.* Montreal: Ashgate, 2001. Provides an international view of issues of availability, cooperation, and reliability of air navigation services. Attention is specifically paid to the American GPS (Global Positioning System) and Russian GLONASS systems, although the development of the Galileo civilian system in Europe is also presented.

Brandt, John C., and Robert D. Chapman. *Introduction to Comets.* New York: Cambridge University Press, 2004. Provides a detailed exposé about virtually every cometary phenomenon.

Burrows, William E. *This New Ocean: The Story of the First Space Age.* New York: Random House, 1998. Burrows offers one of the most comprehensive presentations in print of the beginnings and progress of space exploration. His lively and provocative account is strongly recommended.

Fifty Years of Rockets and Spacecraft in the Rocket City: NASA-Marshall Space Flight Center. Atlanta: Turner Publishing Company, 2003. This concise history of rocketry and the development of space travel over the past half-century centers on the contributions made to that effort by the Marshall Space Flight Center. Thoroughly laced with photographs, many of which are not typically included in other such works.

Hardensen, Paul S. *The Case for Space: Who Benefits from Explorations of the Last Frontier?* Shrewsbury, Mass.: ATL Press, 1997. In this book, those who question the wisdom of the expenditures of public funds for space exploration will find many justifications. Hardensen enumerates the many practical outcomes that such exploration can yield in the day-to-day lives of human beings.

Harland, David M. *The Space Shuttle: Roles, Missions, and Accomplishments.* Hoboken, N.J.: John Wiley, 1998. *The Space Shuttle* is written thematically, rather than purely chronologically. Topics include shuttle operations and payloads, weightlessness, materials processing, exploration, Spacelabs and free-flyers, and the shuttle's role in the International Space Station.

Harrington, Philip S. *The Space Shuttle: A Photographic History.* San Francisco, Calif.: Brown Trout, 2003. With one hundred full-color photographs by Roger Ressmeyer and others and with text by popular astronomy writer Phil Harrington, this book tells the story of the space shuttle program from 1972 to 2003. Its beautiful photographs allow the general reader to survey the history of the space shuttle program and be uplifted by the pioneering spirit of one of humanity's grandest enterprises.

Horta, Lucas G. *A Historical Perspective on Dynamics Testing at the Langley Research Center.* Washington, D.C.: National Aeronautics and Space Administration, 2000. This brief report comes from an article in the NASA Technical Memorandum series. It provides insights into the type of aeronautical engineering performed at the Langley Research Center.

Jenkins, Dennis R. *Space Shuttle: The History of the National Space Transportation System: The First 100 Missions.* Stillwater, Minn.: Voyageur Press, 2001. This is a concisely written technical reference account of the space shuttle and its ancestors, the aerodynamic lifting bodies. It details some of the advantages and inherent disadvantages of using a reusable space vehicle. Each of the vehicles is illustrated by line-drawings with important features pointed out with lines and text. The book follows the space shuttle from its original concepts and briefly chronicles its first one hundred flights.

Klerkx, Greg. *Lost in Space: The Fall of NASA and the Dream of a New Space Age.* New York: Pantheon Books, 2004. The premise of this work is that NASA has been stuck in Earth's orbit since the Apollo era, and that space exploration has suffered as a result.

Lambright, W. Henry, ed. *Space Policy in the Twenty-First Century.* Baltimore: Johns Hopkins University Press, 2003. This book addresses a number of important questions: What will replace the space shuttle? Can the International Space Station justify its cost? Will Earth be threatened by asteroid impact? When and how will humans explore Mars?

Launius, Roger D. *Frontiers of Space Exploration.* Westport, Conn.: Greenwood Press, 1998. Directed primarily toward adolescent readers, this clear explanation of many aspects of space exploration may prove a fruitful beginning for those unfamiliar with the field, regardless of age. Launius writes with clarity and vigor about a topic that obviously intrigues him.

Mari, Christopher, ed. *Space Exploration.* New York: H. W. Wilson, 1999. Of the twenty-five essays in this book, the five contained in Section III, "The International Space Station," are most relevant to the 1996 spaceflights, because some of them were specifically designed to test the feasibility of assembling such a station. The five essays range from Andrew Lawler's upbeat "Onward into Space" to Gregg Esterbrook's dyspeptic "Cosmic Clunker," which questions the validity of the expenditures involved in building a space station whose annual bill for bottled water alone he estimates at $817 million.

Martin, Donald H. *Communication Satellites.* 4th ed. New York: American Institute of Aeronautics and Astronautics, 2000. This work chronicles the development of communications satellites and worldwide networks over the past four decades, from Project Score to modern satellite communication systems.

National Aeronautics and Space Administration. *Space Shuttle Mission Press Kits.* http://www-pao.ksc.nasa.gov/kscpao/presskit/presskit.htm. Provides detailed preflight information about each of the space shuttle missions. Accessed March, 2005.

Ordway, Frederick I., III, and Mitchell Sharpe. *The Rocket Team.* Burlington, Ont.: Apogee Books, 2003. A revised edition of the acclaimed thorough history of rocketry from early amateurs to present-day rocket technology. Includes a disc containing videos and images of rocket programs.

Parks, George K. *Physics of Space Plasmas: An Introduction.* 2d ed. Boulder, Colo.: Westview Press, 2004. Provides a scientific examination of the data returned during what might be called the "golden age" of space physics (1990-2002) when over two dozen satellites were dispatched to investigate space plasma phenomena. Written at the undergraduate level for an introductory course in space plasma, there is also detailed presentation of NASA and ESA spacecraft missions.

Tassoul, Jean-Louis, and Monique Tassoul. *A Concise History of Solar and Stellar Physics.* Princeton, N.J.: Princeton University Press, 2004. A comprehensive study of the historical development of humanity's understanding of the Sun and the cosmos, written in easy-to-understand language by a pair of theoretical astrophysicists. The perspective of the astronomer and physicist are presented.

Tompkins, Phillip K., and Emily V. Tompkins. *Apollo, Challenger, and Columbia: The Decline of the Space Program.* New York: Roxbury Publishing Company, 2004. A subheading for this book's title says "A Study in Organizational Communications." As such the work examines changes in NASA's internal communications since its inception through the space shuttle accidents. Focuses more on sociology than engineering.

Van Biema, David. "High-Tech Pie in the Sky." *Time,* July 15, 1996, 58. This brief article discusses the spacecraft of the future and how it will differ from the kinds of space shuttles launched between 1981 and 1996. It provides excellent details about the future of space shuttles.

Vaughan, Diane. *The Challenger Launch Decision: Risky Technology, Culture, and Deviance at NASA.* Chicago: University of Chicago Press, 1996. This book, better than any other, provides extensive, well-documented information about the *Challenger* accident, which has had broad implications for the future of the space program, particularly during 1996, which marked the tenth anniversary of that catastrophe.

Yenne, Bill. *Secret Weapons of the Cold War: From the H-Bomb to SDI.* New York: Berkley Books, 2005. A contemporary examination of Cold War superweapons and their influence on American-Soviet geopolitics.

Zimmerman, Robert. *The Chronological Encyclopedia of Discoveries in Space.* Westport, Conn.: Oryx Press, 2000. Provides a complete chronological history of all crewed and robotic spacecraft and explains flight events and scientific results. Suitable for all levels of research.

R. Baird Shuman

Space Shuttle Flights, 1997

Date: January 12 to December 5, 1997
Type of mission: Piloted spaceflight

In 1997, NASA conducted eight piloted Earth-orbiting spaceflights. Each mission obtained important information relating to possible advancements in the quality of human life on Earth, including the efficacy of modern medicine and technology and the development of an international space station.

Key Figures

Michael A. Baker (b. 1953), STS-81 commander

Brent W. Jett (b. 1958), STS-81 pilot

John M. Grunsfeld (b. 1958), STS-81 mission specialist

Marsha S. Ivins (b. 1951), STS-81 mission specialist

Peter J. K. "Jeff" Wisoff (b. 1958), STS-81 mission specialist

Jerry M. Linenger (b. 1955), STS-81 mission specialist, STS-84 mission specialist, Mir Permanent Crew 22 and 23 astronaut-investigator

John E. Blaha (b. 1942), STS-81 mission specialist, Mir Permanent Crew 22 astronaut-investigator

Kenneth D. Bowersox (b. 1956), STS-82 commander

Scott J. Horowitz (b. 1957), STS-82 pilot

Mark C. Lee (b. 1952), STS-82 mission specialist

Steven A. Hawley (b. 1951), STS-82 mission specialist

Gregory J. Harbaugh (b. 1956), STS-82 mission specialist

Steven L. Smith (b. 1958), STS-82 mission specialist

Joseph R. Tanner (b. 1950), STS-82 mission specialist

James D. Halsell, Jr. (b. 1956), STS-83 and STS-94 commander

Susan L. Still, STS-83 and STS-94 pilot

Janice E. Voss (b. 1956), STS-83 and STS-94 mission specialist

Donald A. Thomas (b. 1955), STS-83 and STS-94 mission specialist

Michael L. Gernhardt (b. 1956), STS-83 and STS-94 mission specialist

Roger K. Crouch, STS-83 and STS-94 payload specialist

Greg Linteris (b. 1957), STS-83 and STS-94 payload specialist

Charles J. Precourt (b. 1955), STS-84 commander

Eileen M. Collins (b. 1956), STS-84 pilot

Jean-François Clervoy (b. 1958), STS-84 mission specialist

Carlos I. Noriega (b. 1959), STS-84 mission specialist

Edward T. Lu (b. 1963), STS-84 mission specialist

Elena V. Kondakova (b. 1957), STS-84 mission specialist (Russian Space Agency)

C. Michael Foale (b. 1957), STS-84 mission specialist, STS-86 mission specialist, Mir Permanent Crew 23 and 24 astronaut-investigator

Curtis L. Brown, Jr. (b. 1956), STS-85 commander

Kent V. Rominger (b. 1956), STS-85 pilot

N. Jan Davis (b. 1953), STS-85 mission specialist

Robert L. Curbeam, Jr. (b. 1962), STS-85 mission specialist

Stephen K. Robinson (b. 1955), STS-85 mission specialist

Bjarni V. Tryggvason (b. 1945), STS-85 payload specialist (Canadian Space Agency)

James D. Wetherbee (b. 1952), STS-86 commander

Michael J. Bloomfield (b. 1959), STS-86 pilot

Vladimir Georgievich Titov (b. 1947), STS-86 mission specialist (Russian Space Agency)

Scott E. Parazynski (b. 1961), STS-86 mission specialist

Jean-Loup Chrétien (b. 1938), STS-86 mission specialist (French Space Agency, CNES)

Wendy B. Lawrence (b. 1959), STS-86 mission specialist

David A. Wolf (b. 1956), STS-86 mission specialist, STS-89 mission specialist, Mir Permanent
    Crew 24 astronaut-investigator

Kevin R. Kregel (b. 1956), STS-87 commander

Steven W. Lindsey (b. 1960), STS-87 pilot

Winston E. Scott (b. 1950), STS-87 mission specialist

Kalpana Chawla (1961-2003), STS-87 mission specialist

Takao Doi (b. 1954), STS-87 mission specialist (Japanese Space Agency, NASDA)

Leonid K. Kadenyuk (b. 1951), STS-87 mission specialist (Ukraine Space Agency, NSAU)

Summary of the Missions

A total of eight spaceflights occurred in 1997 relating to the future of human explorations in space and to medical advancements affecting human life on Earth. Space Transportation System mission STS-81 launched from the Kennedy Space Center on January 12, 1997, at 09:27:22.984 Coordinated Universal Time (UTC, or 4:27 A.M. eastern standard time). Its duration was ten days and it was the fifth mission between the U.S. space shuttle and the Russian space station Mir. STS-81 used the space shuttle *Atlantis,* transporting Jerry M. Linenger to replace John E. Blaha, a mission specialist from STS-79 (which launched on September 16, 1996). Linenger became the first American to conduct a spacewalk from a foreign space station and in a non-American-made space suit. He and his Russian colleague tested the Orlan-M Russian-built space suit as they installed the Optical Properties Monitor (OPM) and Benton dosimeter on the outer surface of Mir. This series of missions expanded U.S. research on Mir by providing resupply materials for experiments to be performed aboard the station and returning samples and data to Earth. In five days, more than three tons of food, water, equipment, and samples were moved between the two spacecraft. *Atlantis* also carried the SPACEHAB module in its payload bay. The module configuration housed experiments and studies performed by *Atlantis*'s crew along with logistics equipment that was transferred to Mir. STS-81 carried KidSat, a three-year pilot education program designed to bring the frontiers of space exploration to fifteen U.S. middle school classrooms via the Internet. STS-81 returned to Earth at Kennedy Space Center in Florida on January 22, 1997, at 14:22:44 UTC (9:23 A.M. eastern standard time).

On February 11, 1997, at 08:55:17.017 UTC (3:55 A.M. eastern standard time), *Discovery* was launched, marking the beginning of the STS-82 mission. The objective of this seven-member piloted mission was to service the Hubble Space Telescope (HST). This was the second servicing mission to the HST—the first was performed on STS-61. Members of STS-82 performed numerous spacewalks (also called extravehicular activities or EVAs) to replace the Goddard High Resolution Spectrometer and the Faint Object Spectrograph with two new instruments, the Space Telescope Imaging Spectrograph and the Near Infrared Camera and Multi-Object Spectrometer. Various maintenance repairs and services needed to keep the telescope functioning were performed. The ten-day mission ended on February 21, 1997, at 07:32:27.3 UTC (3:32 A.M. eastern standard time) when *Discovery* landed at the Kennedy Space Center.

STS-83 was a sixteen-day mission that began on April 4, 1997, from the Kennedy Space Center at 19:20:32.019 UTC (2:21 P.M. eastern standard time). Shuttle *Columbia* and seven astronauts spent more than two weeks in orbit doing experiments within the primary payload, the Microgravity Science Laboratory 1 (MSL-1). They examined how

Astronaut Takao Doi works with a 156-pound crane during this flight of Columbia *(STS-87) in preparation for work on the International Space Station.* (NASA)

various materials and liquids behave in the weightless environment of space and tested hardware, procedures, and facilities for the future International Space Station. Scientists from the European Space Agency (ESA), German Space Agency (DARA), National Space Development Agency of Japan (NASDA), and NASA developed thirty-three experiments for this mission, including studies of combustion and growth of protein crystals in space. The mission included the Shuttle Amateur Radio Experiment (SAREX), in which radio operators and students made radio contacts with the astronauts. On April 6, the STS-83 mission was shortened due to problems with one of the three fuel-cell power-generation units. Two days later, the crew landed at 18:33:11 UTC (2:33 P.M. eastern daylight time), shortening STS-83 to a four-day mission.

Atlantis returned to Mir with the crew of STS-84. Liftoff occurred at 07:07:48.003 UTC (4:08 A.M. eastern daylight time) on Thursday, May 15, from NASA's Kennedy Space Center. STS-84 transferred 3,318 kilograms of water and logistics equipment to and from Mir. STS-84 Mission Specialist C. Michael Foale replaced Astronaut Jerry M. Linenger on Mir. *Atlantis* also carried the SPACEHAB module, housing experiments performed by the crew. *Atlantis* returned on May 24, at 13:27:44 UTC (9:28 A.M. eastern daylight time).

On Tuesday, July 1, the shuttle *Columbia* was launched again from the Kennedy Space Center to complete the microgravity science mission. The returning seven crew members of this STS-94 mission blasted off at 18:10:59.993 UTC (2:02 P.M. eastern daylight time) to complete more than thirty experiments in the Microgravity Science Laboratory. *Columbia* landed fifteen days later, on July 17 at 10:46:36 UTC (6:47 A.M. eastern daylight time).

The six astronauts of the shuttle *Discovery* were launched on August 7, at 14:41:00.013 UTC (10:41 A.M. eastern daylight time) from the Kennedy Space Center. The payload for the STS-85 mission included Cryogenic Infrared Spectrometers and Telescopes for the Atmosphere-Shuttle Pallet Satellite-2 (CRISTA-SPAS-2), which comprises three telescopes and four spectrometers to measure trace gases and dynamics of the Earth's middle atmosphere. STS-85 marked the fourth mission in a cooperative venture between the German Aerospace Center (Deutsches Zentrum für Luft- und Raumfahrt, or DLR) and NASA. The flight ended August 19, 1997, at 11:08:00 UTC (7:08 A.M. eastern daylight time), with the crew gliding to a safe landing at the Kennedy Space Center.

On September 25, 1997, the seventh Mir docking mission, STS-86, launched at 14:34:19.000 UTC (10:34 P.M. eastern daylight time). The ten-day mission included five days of docked operations between *Atlantis* and Mir and the exchange of crew members Foale (returning) and David A. Wolf (boarding). *Atlantis* again carried the SPACEHAB double module to support the transfer of supplies and equipment and the return of hardware and specimens to Earth. The crew landed on October 6, at 21:55:12 UTC (5:55 P.M. eastern daylight time) at the Kennedy Space Center.

The final mission of 1997, STS-87, was made up of a multinational crew of six astronauts prepared to conduct investigations in microgravity science, satellite-based studies of the Sun, and a spacewalk to prepare for the assembly of the International Space Station. Shuttle *Columbia*, carrying the United States Microgravity Payload-4 (USMP-4), launched on November 19, at 19:45:59.993 UTC (2:46 P.M. eastern daylight time). The sixteen-day flight included numerous and extensive experiments that studied how the weightless environment of space affected various materials and observed the Sun's outer atmospheric layers (the solar corona). Astronauts Winston E. Scott and Takao Doi evaluated equipment and procedures to be used during the construction and maintenance of the International Space Station. STS-87 also included an end-to-end demonstration of a maintenance task simulating the changing out of Orbital Replacement Units (ORUs) on the International Space Station. STS-87 mission ended on December 5, when the shuttle *Columbia* landed on Runway 33 at the Florida spaceport at 12:20:02 UTC (7:20 A.M. eastern daylight time).

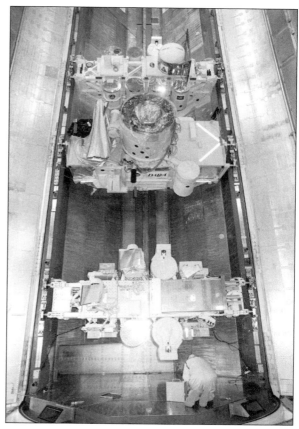

A loaded payload canister in the Payload Changeout Room at Launch Pad 39A at Kennedy Space Center, to be transferred into the payload bay of the space shuttle orbiter Discovery *for mission STS-85.* (NASA)

Contributions

On all three NASA missions (STS-81, STS-84, and STS-86) to the Russian space station Mir, the SPACEHAB double module was used to return experiment hardware and specimens to Earth. Numerous specimens of diffusion-controlled crystallization were transferred to study the effects of the low-gravity environment on the growth rate and physical structure of the protein crystal samples. The absence of gravity allowed for the growth of larger, purer crystals of greater structural integrity. Protein crystal samples were also gathered in experiments performed by the members of STS-94 on the MSL-1.

STS-84 Pilot Eileen Collins after flying in the Shuttle Training Aircraft at Kennedy Space Center. Collins would be the first woman to command an STS mission, STS-93. (NASA)

Effects of the low-gravity environment on the human body and immune system were a main objective of the STS-81, STS-84, and STS-86 missions. The human immune system involves both blood-borne (humoral) and cell-mediated responses to invading substances known as antigens. Endurance and strength tests performed while in orbit and after the mission provided information on sensory and muscle-motor performance and the effects that extended exposure to microgravity has on the human body.

By studying the effects of microgravity on plants, researchers gained knowledge of the effects of the space station environment on certain processes of plants such as photosynthesis, respiration, transpiration, stomatal conductance, and water use.

STS-83 and STS-94 carried the MSL-1, which was involved in thirty-three experiments. Two combustion experiments were the Laminar Soot Processes (LSP) Experiment and the Structure of Flame Balls at Low Lewis Number (SOFBALL) experiment. LSP collected data on flame shape, the type and amount of soot produced under various conditions, and the temperature of the soot components, which aided in the research of fire safety considerations such as containing unwanted fires and decreasing the number of fatalities from carbon monoxide emissions. SOFBALL experiments were aimed at determining if stable balls of flame actually exist, which could lead to improvements in lean-burn internal combustion engines, increased efficiency, and reduced emissions. The Droplet Combustion Experiment collected information on the burning rates of flames, flame structures, and conditions when extinguishing a flame, leading to cleaner and safer ways to burn fossil fuels and more efficient methods of generating heat power on Earth.

On STS-85, the CRISTA-SPAS-2 used three telescopes and four spectrometers to measure infrared radiation emitted by the Earth's middle atmosphere. The Technology Applications and Science-01 (TAS-01) held a total of seven separate experiments that provided data on the Earth's topography and atmosphere, studied the Sun's energy, and tested new thermal control devices that were to be included on the International Space Station. The four experiments utilizing the International Extreme Ultraviolet Hitchhiker-02 (IEH-02) studied ultraviolet radiation from the stars, the Sun, and various other sources in the solar system. Building from past discoveries from its previous flight on STS-66, the Middle Atmosphere High Resolution Spectrograph Investigation (MAHRSI) gathered data on the distribution of hydroxyl (OH) and nitric oxide (NO) in the mesosphere and upper stratosphere by sensing UV radiation emitted and scattered by the atmosphere to see if there were any noticeable patterns of change when conditions (such as the change in season) changed. Hydroxyl and nitric oxide are chemicals called free radicals; they contain unpaired electrons and are implicated in many atmospheric processes, including the depletion of the stratospheric ozone layer.

The Shuttle Ozone Limb Sounding Experiment (SOLSE) and the Limb Ozone Retrieval Experiment (LORE) used on STS-87 also gathered vital information on the Earth's ozone layer. The USMP-4 focused on increasing our understanding of the basic properties and behavior of various materials and liquids in outer space. The near-weightless environment of space provides an environment free of gravitational effects, allowing researchers to obtain a clearer understanding of the laws of nature. On STS-87 the SPARTAN 201-04 free-flyer determined physical conditions and processes of the hot outer layers of the Sun's atmosphere (solar corona). While deployed from the shuttle, it gathered various data relating to the solar corona and solar winds which directly influence orbiting satellites and weather conditions on Earth and which also have an impact on communications, including consumer-related television and phone communications.

Context

The various experiments and research that occurred on the space shuttle missions of 1997 played a vital role in the advancements of future explorations in space. Numerous studies were performed and observations were made of the crew members of Mir missions to gain information on the durability of the human body in space. The effects of space travel were also observed in plant life. This study was vital due to the fact that plants could eventually be a major contributor to life-support systems for spaceflight. Plants produce oxygen and food while eliminating carbon dioxide and excess humidity from the environment. These functions are essential for sustaining life in a closed environment such as that of Mir or the International Space Station.

A greater understanding of the Earth and the surrounding universe was attained from numerous space explorations. The HST has made many important scientific discoveries since it became operational in 1990, such as providing enough visual detail to construct the first surface map of the planet Pluto, discovering oxygen atmospheres on the moons Europa and Ganymede, finding conclusive evidence for the existence of black holes, and making the deepest visual looks into the universe, revealing thousands of galaxies. The continuous servicing of such an instrument is vital in order to continue the advancement to new discoveries. STS-85 utilized the CRISTA-SPAS-2 to measure trace gases, such as hydroxyl and nitric oxide, and dynamics of the Earth's middle atmosphere. The data gathered have helped scientists to understand how small-scale tracer "filaments" in the stratosphere contribute to the transport of ozone and chemical compounds that affect the distribution of ozone. The SPARTAN 201-04 free-flyer of STS-87 retrieved data on the physical conditions, processes, and measurements of the hot outer layers of the solar corona and solar winds, which directly influence orbiting satellites and weather conditions on Earth.

The protein crystals that were grown on STS-81, STS-84, STS-86, and the MSL missions on STS-83 and STS-94 may help scientists better understand the processes of various diseases and promote the development of cures. Data gathered from the combustion experiments aboard STS-94 could lead to ways of increasing the safety, efficiency, and cleanliness of the burning of fossil fuels.

See also: Cooperation in Space: U.S. and Russian; Hubble Space Telescope: Servicing Missions; International Space Station: Development; Russia's Mir Space Station; Shuttle Amateur Radio Experiment; Space Shuttle: Microgravity Laboratories and Payloads; Space Shuttle-Mir: Joint Missions; Space Shuttle Flights, 1996; Space Shuttle Flights, 1998; Space Stations: Origins and Development; Spacelab Program.

Further Reading
Golden, Frederic. "A Close Shave in Orbit." *Time*, July 21, 1997. An article that briefly touches on the shape of the Mir Space Station during the missions of 1997, and U.S.

involvment with the Space Station. C. Michael Foale was the American on board during this time.

Harland, David M. *The Space Shuttle: Roles, Missions, and Accomplishments.* Hoboken, N.J.: John Wiley, 1998. *The Space Shuttle* is written thematically, rather than purely chronologically. Topics include shuttle operations and payloads, weightlessness, materials processing, exploration, Spacelabs and free-flyers, and the shuttle's role in the International Space Station.

Harrington, Philip S. *The Space Shuttle: A Photographic History.* San Francisco, Calif.: Brown Trout, 2003. With one hundred full-color photographs by Roger Ressmeyer and others and with text by popular astronomy writer Phil Harrington, this book tells the story of the space shuttle program from 1972 to 2003. Its beautiful photographs allow the general reader to survey the history of the space shuttle program and be uplifted by the pioneering spirit of one of humanity's grandest enterprises.

Jenkins, Dennis R. *Space Shuttle: The History of the National Space Transportation System: The First 100 Missions.* Stillwater, Minn.: Voyageur Press, 2001. This is a concisely written technical reference account of the space shuttle and its ancestors, the aerodynamic lifting bodies. It details some of the advantages and inherent disadvantages of using a reusable space vehicle. Each of the vehicles is illustrated by line-drawings with important features pointed out with lines and text. The book follows the space shuttle from its original concepts and briefly chronicles its first one hundred flights.

Kerrod, Robin. *Hubble: The Mirror on the Universe.* Richmond Hill, Ont.: Firefly Books Ltd., 2003. The book covers the observable universe in six sections: "Stars in the Firmament," "Stellar Death and Destruction," "Gregarious Galaxies," "The Expansive Universe," "Solar Systems," and "The Heavenly Wanderers." Clear and concise text explains the fascinating history of astronomy and the development of the HST. Hubble transports readers to the planets of our solar system and on to galaxies millions—even billions—of light years away. These dramatic, unforgettable images will bring into sharp focus how the universe is unfolding in new and astonishing ways.

Kluger, Jeffrey. "Time to Jump Ship?" *Time,* May 12, 1997. Details the conditions of Mir as well as various bad occurrences, such as a fire, that Astronaut Jerry M. Linenger experienced while aboard the Space Station.

National Aeronautics and Space Administration. *Space Shuttle Mission Press Kits.* http://www-pao.ksc.nasa.gov/kscpao/presskit/presskit.htm. Provides detailed preflight information about each of the space shuttle missions. Accessed March, 2005.

Thompson, Dick. "Eyes on the Storm-Tossed Sun." *Time,* September 8, 1997. Details the data gathered from mission STS-87 concerning the physical conditions and process of the hot outer layers of the Sun's atmosphere, or solar corona, as well as information concerning solar winds.

Adam Lee and Massimo D. Bezoari

Space Shuttle Flights, 1998

Date: January 22 to December 15, 1998
Type of mission: Piloted spaceflight

Space shuttle flights in 1998 consisted of Space Transportation System missions STS-89, STS-90, STS-91, STS-95, and STS-88. Missions to Mir were concluded, and studies relating to solar winds, as well as experiments on the effects of microgravity environments on biological systems, were continued. Astronaut John H. Glenn, Jr., returned to space. The first Space Station Assembly Flight was conducted.

Key Figures

Terrence W. "Terry" Wilcutt (b. 1949), STS-89 commander
Joe F. Edwards, Jr. (b. 1958), STS-89 pilot
Bonnie J. Dunbar (b. 1949), STS-89 mission specialist
Michael P. Anderson (1959-2003), STS-89 mission specialist
James F. Reilly (b. 1954), STS-89 mission specialist
Salizhan Shakirovich Sharipov (b. 1964), STS-89 mission specialist
Andrew S. W. Thomas (b. 1951), STS-89 mission specialist, STS-91 mission specialist, Mir
    Permanent Crew 24 and 25 astronaut-investigator
David A. Wolf (b. 1956), STS-86 mission specialist, STS-89 mission specialist, Mir Permanent
    Crew 24 astronaut-investigator
Richard A. Searfoss (b. 1956), STS-90 commander
Scott D. Altman (b. 1959), STS-90 pilot
Richard M. Linnehan (b. 1957), veterinarian and STS-90 mission specialist
Dafydd Rhys Williams (b. 1954), medical doctor and STS-90 mission specialist, Canadian
    Space Agency
Kathryn P. Hire (b. 1959), STS-90 mission specialist
Jay Clark Buckey (b. 1956), STS-90 payload specialist
James A. Pawelczyk (b. 1960), STS-90 payload specialist
Charles J. Precourt (b. 1955), STS-91 commander
Dominic L. Gorie (b. 1957), STS-91 pilot
Wendy B. Lawrence (b. 1959), STS-91 mission spcialist
Franklin R. Chang-Díaz (b. 1950), STS-91 mission specialist
Janet L. Kavandi (b. 1959), STS-91 mission specialist
Valery Victorovich Ryumin (b. 1939), STS-91 mission specialist (Russian Space Agency, RSA)
Curtis L. Brown, Jr. (b. 1956), STS-95 commander
Steven W. Lindsey (b. 1960), STS-95 pilot
Scott E. Parazynski (b. 1961), STS-95 mission specialist
Pedro Duque (b. 1963), STS-95 mission specialist
Stephen K. Robinson (b. 1955), STS-95 mission specialist
Chiaki Mukai (b. 1952), STS-95 payload specialist (Japanese Space Agency, NASDA)
John H. Glenn, Jr. (b. 1921), STS-95 payload specialist

Robert D. Cabana (b. 1949), STS-88 commander

Frederick W. Sturckow (b. 1961), STS-88 pilot

Jerry L. Ross (b. 1948), STS-88 mission specialist

Nancy J. Currie (b. 1958), STS-88 mission specialist

James H. Newman (b. 1956), STS-88 mission specialist

Sergei Konstantinovich Krikalev (b. 1958), STS-88 mission specialist (Russian Space Agency, RSA)

Summary of the Missions

Space Transportation System mission STS-89 began with the launch of the shuttle *Endeavour* on January 22, 1998, at 02:48:15.017 UTC (9:48:15 P.M. eastern standard time). The mission completed the eighth of nine scheduled missions to the Russian space station Mir. It was *Endeavour*'s first docking with the Russian station and *Endeavour*'s first flight since STS-77 in May, 1996. Astronaut David A. Wolf, who had been aboard Mir since late September, 1997, was replaced by *Endeavour* Mission Specialist Andrew Thomas, who became the seventh U.S. astronaut to fly aboard Mir. His exchange marked STS-89 as the fifth mission of the nine to involve the exchange of U.S. astronauts. Thomas was to remain aboard Mir until his return to Earth aboard *Discovery* during STS-91 in late May, 1998. The mission ended when *Endeavour* landed at Kennedy Space Center on January 31 at 22:35:10 UTC (5:36 P.M. eastern standard time). The duration of mission STS-89 was eight days, 19 hours, 46 minutes, and 55 seconds; it covered a distance of 5.8 million kilometers.

STS-90 completed the fifteen-year-old U.S.-European Spacelab program. Space shuttle *Columbia* transported Neurolab, a Spacelab module mission focusing on the effects of microgravity on the nervous system, in its payload bay. The goals of the scheduled experiments were to study basic research questions and to increase the current understanding of the mechanisms responsible for neurological and behavioral changes in space. The mission was a joint venture of six space agencies from various countries and seven U.S. research agencies. Shuttle *Columbia* was launched on April 17, 1998, at 18:18:59.988 UTC (2:19 P.M. eastern daylight time). It returned to Earth, landing at

Kennedy Space Center on May 3, at 16:08:58 UTC (12:09 P.M. eastern daylight time). The STS-90 flight lasted fifteen days, 21 hours, 49 minutes, and 58 seconds.

STS-91 marked the final shuttle-Mir docking mission. Astronaut Thomas joined the crew of *Discovery* after having been a member of the Mir crew since his exchange for Astronaut Wolf during STS-89 in January, 1989. STS-91 carried as part of its cargo (in the payload bay) the Alpha Magnetic Spectrometer investigation (AMS). The major objective of this investigation was to search for antimatter and dark matter in space and to study astrophysics. The mission was the first to make use of the Super Lightweight External Tank (SLWT). Improvements in the tank would provide additional payload capacity to the International Space Station while decreasing the amount of fuel required for launches. Shuttle *Discovery* was used for this mission, which began with the launch on June 2, 1998, at 21:06:24.008 UTC (6:06:24 P.M. eastern daylight time). The mission ended with the landing at Kennedy Space Center on June 12, at 18:00:24 UTC (2:00:17 P.M. eastern daylight time). STS-91 lasted nine days, 19 hours, and 54 minutes.

The primary objectives of STS-95 included conducting a variety of science experiments in the pressurized SPACEHAB module. The mission comprised eighty-eight experiments, more than conducted on any previous shuttle mission. A series of experiments sponsored by NASA and the National Institute on Aging gathered information that may provide a model system to help understand bone and muscle loss as well as balance and sleep disorders that are associated with aging. Astronaut John H. Glenn, Jr. served as mission specialist and geriatric test subject, making him the oldest spaceman at age seventy-seven. Glenn's historic contributions to U.S. exploration efforts began on February 20, 1962, when he piloted the Mercury-Atlas 6 *Friendship 7* spacecraft on the first American piloted orbital mission in space. STS-95 began thirty-six years later, when shuttle *Discovery*, launched on October 29, 1998, at 19:19:33.984 UTC (2:19 P.M. eastern standard time). The mission ended with the landing at Kennedy Space Center on November 7, at 17:03:30 UTC (12:04 P.M. eastern standard time). STS-95 flight lasted eight days, 21 hours, 43 minutes, and 56 seconds, and covered a distance of 5.8 million kilometers.

STS-88 began the construction of the International Space Station (ISS). Five U.S. astronauts and one Russian cosmonaut conducted the first stage of the project. The first American-built component of the ISS, an 11,300-kilogram, six-sided connecting unit named Unity (technically referred to as Node 1), would be connected to the 20,000-kilogram Zarya Control Module. Zarya ("sunrise," a word whose symbol is common in Russian and Soviet spaceflight), also referred to as the Functional

The connected Zarya and Unity modules after they were released from the cargo bay of Endeavour *(STS-88) a bit earlier.* (NASA)

Cargo Block (Russian *funktsionalya-gruzovod blokor*, FGB), is a U.S.-funded and Russian-built ISS component. FGB was launched on November 20, 1997, by a Russian Proton rocket from the Baikonur Cosmodrome in Kazakhstan. *Endeavour* launched on December 3, 1998, at 08:35:34.019 UTC (3:36 A.M. eastern standard time) and returned to the Kennedy Space Center on December 14, at 03:53:32 UTC (10:53 P.M. eastern standard time). The STS-88 mission flight lasted 11 days, 19 hours, and 18 minutes and covered a distance of 7.6 million kilometers. The landing was the tenth night landing in the history of the space shuttle program, and the fifth Kennedy Space Center night landing.

Contributions

The STS-89 mission was NASA's final chance to bring back large items from the Mir station, because the *Discovery* STS-91 flight to Mir (to retrieve Thomas) in May was to have limited cargo-return capability. The STS-89 mission and the work performed by Thomas during his time on the Mir station included investigations in the fields of advanced technology, Earth sciences, fundamental biology, human life sciences, International Space Station risk mitigation, microgravity sciences, and space sciences. One of the most important studies involved the growth of three-dimensional human tissues. Thomas expanded Wolf's work in this area by growing human breast-cell cancer tissues and observing the formation of blood vessels in the tissue, which, in the regular environment on Earth, can cause the spreading of cancer cells. STS-89 also carried China's first space shuttle payload, a collection of six experiments carried in a Get-Away Special canister. The investigations developed by the Chinese Academy of Sciences include crystal growth and high-temperature semiconductor research.

During the STS-90 mission, the crew used themselves as test subjects in a variety of experiments associated with studying functions such as blood pressure regulation, balance, coordination, and sleep patterns. Neurolab spent more than two weeks in orbit hosting experiments with a menagerie of test subjects ranging from rats to crickets. A variety of animals was studied in order to gain additional insight into the effects the weightless environment of outer space may have on the development and performance of the nervous system. Buckey and Williams dissected eight young rats that were nine days old at the launch in an experiment to look at changes occurring in the vestibular receptors of rats at different stages of growth. Without the presence of gravitational forces during development, receptors and neuronal circuits that process neural information on balance, movement, and sense of position may develop differently from those of animals developing on Earth. It was noted that an unexpected number of animals, especially rats, died during the mission.

STS-91 carried the Alpha Magnetic Spectrometer, which collected a total of about 1,000 minutes of data on tape that scientists hope will help them find and understand the nature of the dark matter

Astronauts Jerry L. Ross (left) and James H. Newman on the final of three spacewalks of the STS-88 mission. One of the solar panels of the Russian-built Zarya Module is in the background. (NASA)

and antimatter that are currently believed to make up a large portion of the universe's overall mass. The particle detector continued collecting data in the cargo bay until only a few hours before *Discovery* began its descent to Earth.

The mission was also the first to use the Super Lightweight External Tank (SLWT). Although this tank is the same size (47 meters long and 8.2 meters in diameter) as the External Tank used on previous launches, it is 3,400 kilograms lighter, reducing the thrust required to launch and allowing the transportation of larger and heavier cargoes. The tank is made of an aluminum-lithium alloy and its structural design was improved, making it 30 percent stronger and 5 percent less dense than previous tanks. The walls of the redesigned hydrogen tank are machined in an orthogonal waffle-like pattern, providing more strength and stability than the previous design.

Astronaut Thomas joined the crew of STS-91 after *Discovery* made its final docking of the nine-mission series with the Mir Space Station. Thomas had been aboard Mir since STS-89 in early January, 1998.

During the STS-95 mission, the SPARTAN 201 free-flyer was deployed and retrieved using the shuttle's mechanical arm. SPARTAN 201 was designed to investigate the physical conditions and processes occurring in the hot outer layers of the Sun's atmosphere, which is called the solar corona. Information collected will lead to a better understanding of the solar winds that directly influence orbiting satellites, communication systems, and weather conditions on Earth. The International Extreme Ultraviolet Hitchhiker (IEH) payload involved a half dozen various experiments mounted on a support structure being carried in *Discovery*'s payload bay.

Astronaut John H. Glenn, Jr. participated in ten experiments studying space and age-related problems. He spent four

Processing activities for STS-91 continue as installation of two Get-Away Special (GAS) canisters begins. This mission saw the ninth and final docking with the Russian space station Mir. (NASA)

nights in a wired-up sleep suit, provided seventeen blood samples, and wore a mini data recorder for twenty-four hours to monitor his heart rate. He also swallowed a capsule holding a tiny radio transmitter and temperature sensor.

The seven-day STS-88 mission was highlighted by the joining of the U.S.-built Node 1 station element to the FGB, which was already in orbit. This marked an initial step in the construction of the International Space Station, the largest international aerospace project ever undertaken. Sixteen nations and nearly three hundred prime and subcontractors from around the world are involved in the project. Node 1 was the first International Space Station hardware delivered by the space shuttle. The crew had to conduct a series of rendezvous maneuvers to reach the orbiting FGB.

Context

The five shuttle missions of 1998 represented an end for some projects and the beginning for others. STS-89 and STS-91 continued to open international space relations with Russia, France, and China. STS-90 added insight into the chances for survival with regard to the normal healthy development of infants, a major concern of future space colonization efforts. STS-95 expanded the knowledge of the limits that an adult human body can endure in a microgravity environment. STS-88 began the first real steps in the assembly of the International Space Station.

NASA continues to move forward and pioneer new steps toward human exploration of space, but concerns have begun to arise concerning just how long NASA can maintain its momentum. The concerns are a result of federal budget cuts that began in 1993. Since that time, the Kennedy Space Center has reduced its workforce by about 50 percent, with additional cuts occurring at the Johnson Space Center.

According to a special assessment performed by NASA's Office of Safety and Mission Assurance at NASA Headquarters in Washington, D.C., the reduced workforce is able to maintain the current flight rate of about five missions per year without jeopardizing shuttle safety. However, the same review panel also found there are concerns about whether the reduced contractor workforce could maintain that level of performance to process an average of eight or more missions per year that could be needed to maintain a strong assembly flow. The review stated that "at some point in the future, the International Space Station launch and build demand will drive the Kennedy Ground Operations organization to a point of saturation."

See also: Astronauts and the U.S. Astronaut Program; Cooperation in Space: U.S. and Russian; Get-Away Special Experiments; International Space Station: Crew Return Vehicles; International Space Station: Design and Uses; Russia's Mir Space Station; Space Shuttle: Life Science Laboratories; Space Shuttle-Mir: Joint Missions; Space Shuttle Mission STS-95; Space Shuttle Flights, 1999.

Further Reading

Buckey, Jay C., and Jerry L. Homick, eds. *The Neurolab Spacelab Mission: Neuroscience Research in Space: Results from the STS-90, Neurolab Spacelab Mission.* Washington, D.C.: U.S. Government Printing Office, 2003. The book reveals the results of Neurolab, a sixteen-day space shuttle mission dedicated to studying how weightlessness affects the brain and nervous system. This book shows the complex and sometimes surprising changes in the brain and nervous system that allow astronauts to adapt to weightlessness. The results suggest that the developing nervous system may need gravity to develop normally and that some concept of how gravity works may be "built in" to the brain.

Couvalt, Craig. *Aviation Week and Space Technology,* January 19, 1998.

_____. *Aviation Week and Space Technology,* May 4, 1998.

_____. *Aviation Week and Space Technology,* December 7, 1998. In articles by Craig Couvault, these issues cover the main events in the Space Transportation System schedule during 1998.

Harland, David M. *The Space Shuttle: Roles, Missions, and Accomplishments.* Hoboken, N.J.: John Wiley, 1998. *The Space Shuttle* is written thematically, rather than purely chronologically. Topics include shuttle operations and payloads, weightlessness, materials processing, exploration, Spacelabs and free-flyers, and the shuttle's role in the International Space Station.

Harrington, Philip S. *The Space Shuttle: A Photographic History.* San Francisco, Calif.: Brown Trout, 2003. With one hundred full-color photographs by Roger Ressmeyer and others and with text by popular astronomy writer Phil Harrington, this book tells the story of the space shuttle program from 1972 to 2003. Its beautiful photographs allow the general reader to survey the history of the space shuttle program and be uplifted by the pioneering spirit of one of humanity's grandest enterprises.

Holden, Constance. *Science,* May 15, 1998. Details various new technologies. Semitechnical descriptions used in the articles.

Jane's Intelligence Review. London: Jane's Information Group, March 1, 1999. A technical review of shuttle mission STS-88. Lists personnel and cargo specifications.

Jenkins, Dennis R. *Space Shuttle: The History of the National Space Transportation System: The First 100 Missions.* Stillwater, Minn.: Voyageur Press, 2001. This is a concisely written technical reference account of the space shuttle and its ancestors, the aerodynamic lifting bodies. It details some of the advantages and inherent disadvantages of using a reusable space vehicle. Each of the vehicles is illustrated by line drawings. The book follows the space shuttle from its original concepts and briefly chronicles its first one hundred flights.

National Aeronautics and Space Administration. *Space Shuttle Mission Press Kits.* http://www-pao.ksc.nasa.gov/kscpao/presskit/presskit.htm. Provides detailed preflight information about each of the space shuttle missions. Accessed March, 2005.

Scholastic News, May 11, 1998. Aimed at a younger generation, articles are about hot topics and larger items of interest in the news, including the 1998 shuttle flights.

James Thornton

Space Shuttle Mission STS-95

Date: October 29 to November 7, 1998
Type of mission: Piloted spaceflight

John H. Glenn, Jr., the first American to orbit the Earth, left NASA in the mid-1960's to enter the worlds of business and politics, eventually becoming a U.S. senator from his native Ohio. He always desired to return to orbit. At age seventy-seven, Glenn fulfilled that dream on shuttle mission STS-95.

Summary of the Mission

Media representatives assembled at Kennedy Space Center (KSC) on October 29, 1998, in greater abundance than typical for shuttle coverage, anticipating John H. Glenn, Jr.'s triumphant return to space. President Bill Clinton and First Lady Hillary Rodham Clinton joined enthusiastic masses of invited VIPs and interested private citizens gathered at every available viewing location along Florida's Space Coast to witness this historic spaceflight. Many were returning to recapture a feeling of national pride they experienced when witnessing Glenn's 1962 *Friendship 7* flight launched from Cape Canaveral. This was the first time a sitting president had viewed a piloted launch since President Richard M. Nixon attended Apollo 12's liftoff in November, 1969.

Activities on the morning of the launch proceeded smoothly until the countdown's final moments, when an unplanned hold was called in order to clear aircraft that had encroached upon restricted airspace surrounding KSC. *Discovery* lifted off at 19:19:33.984 Coordinated Universal Time (UTC, or 2:19 P.M. eastern standard time). Seconds before, as *Discovery*'s main engines built up proper thrust levels, a small door covering the compartment in which the drag chute was stored fell off *Discovery*'s aft section and impacted one nozzle of the main engine before falling through the vehicle's engine exhaust.

Eight and one-half minutes later, *Discovery* and John H. Glenn, Jr., were back in orbit. It had been

13,400 days since John H. Glenn, Jr., had been in space. Glenn had experienced ascent strapped down into a jump seat on *Discovery*'s middeck, unable to see outside. Commander Curtis L. Brown called Glenn up to the flight deck to view the mission's first orbital sunset, a sight Glenn had not enjoyed in thirty-six years. Despite intense public interest in Glenn's condition, he stayed in the background as nominal postinsertion activities were carried out by the rest of the crew. Glenn checked in shortly after three hours of flight had transpired, restating his famous line from *Friendship 7*, "Zero G and I feel fine!" Early in the flight, Glenn exceeded his previous experience in orbit. Later, ten hours after launch, the crew entered their first sleep period.

The crew awoke the next morning, October 29, to a rendition of Louis Armstrong's "What a Wonderful World." Back on *Friendship 7* Glenn had only applesauce to eat from squeeze tubes—mostly as a test of eating in weightlessness rather than for nutritional value. However, this morning Glenn dined on a breakfast of oatmeal that had more in common with earthly breakfast fare. He managed to get some stray oatmeal and raisins on his glasses, something he kiddingly joked about during public affairs events throughout the flight. He teased that on Earth elderly gentlemen often dribble food on their ties, but in space things do not fall down.

In 1962, citizens of Perth, Australia, turned on just about every available light to signal *Friendship 7* when John H. Glenn, Jr., passed overhead. Around noon, Perth repeated that action, signaling to Glenn once more.

October 30 was largely devoted to setting up and activating the myriad experiments carried on board *Discovery*'s middeck or in the SPACEHAB module back in the payload bay connected to the cabin by a pressurized tunnel. Also, *Discovery*'s Remote Manipulator System (RMS) arm was powered up, and preparations were made for the deployment of an independently flying astrophysics package that was to come several days later. A Get-Away Special (GAS) canister in the payload bay housed the Naval Postgraduate School's Petite Amateur Naval Satellite (PANSAT). The lid covering the canister was raised, and at the appropriate time a switch thrown inside the cabin resulted in PANSAT's spring ejection. This nonrecoverable small satellite provided an educational experience for military students. For the next year selected students would monitor PANSAT's position, and they would later prepare an amateur radio operator handbook on how to use this satellite as a space-based radio bulletin board.

Life sciences experiments began in earnest on October 31. Blood was drawn, and other bodily fluids were collected. Glenn also injested an amino acid pill so that alanine and histodine, absorbed in muscles, could be traced throughout his body. The RMS arm was powered up in a test of the system's capability to support the Shuttle-Pointed Autonomous Research Tool for Astronomy (SPARTAN) and to perform a technology test of the Orbiter Space Vision System (OSVS), a positional aid to be used by astronauts on the next shuttle flight when joining the first two modules of the International Space Station (ISS). Research was initiated both on the middeck and inside the SPACEHAB module

housed in *Discovery*'s payload bay and connected to the shuttle's air lock by a pressurized tunnel.

The primary high-profile activity of November 1 involved preparations of, deployment of, and separation from the SPARTAN free-flying solar physics platform. This particular scientific payload had also flown on STS-87, but on that mission improper deployment left it incapable of collecting data. Fortunately, it was retrieved for refurbishment, and this time on STS-95 the solar physics community had another opportunity to investigate key questions about the nature of the Sun's high-temperature corona. SPARTAN incorporated a white-light coronagraph and an ultraviolet coronal spectrometer. Coronal observations from SPARTAN would be compared to similar measurements taken by the SOHO and Ulysses spacecraft.

Early in the afternoon Mission Specialist Robinson grappled SPARTAN and released latches securing it to a payload bay berth. SPARTAN was used as a target in support of an OSVS technology demon-

The International Extreme Ultraviolet Hitchhiker-3 (IEH-3) is prepared for launch aboard the STS-95 mission. IEH-3 comprised seven experiments, mounted on a hitchhiker bridge in Discovery's *payload bay.* (NASA)

stration test for video guidance sensing before it was placed in proper attitude for release from the RMS arm. Precisely on time at 15:59 UTC, Robinson released RMS end-effector snare wires that had been holding SPARTAN securely, and SPARTAN gently separated, automatically initiating a programmed pirouette maneuver, which indicated that the payload could support its forty-eight-hour independent operation. With that visual confirmation, Commander Brown executed an engine firing designed to increase *Discovery*'s separation from SPARTAN slowly up to about 64 kilometers (40 miles). Several adjustments were performed during the rest of the day, as were additional public affairs events.

November 2 represented the halfway point of the STS-95 flight, and although research continued, each crew member received a half day's off-duty time for relaxing and enjoying views of the Earth below. Mission Specialist Scott E. Parazynski performed technology tests in support of upcoming Hubble Space Telescope servicing missions, and he and Pilot Steven W. Lindsey began setting up rendezvous tools in anticipation of retrieving SPARTAN the next day. Glenn had an opportunity to speak privately with his wife, son, daughter, and grandchildren.

The primary activity of November 3 at 19:48 UTC involved rendezvousing back to the independent free-flyer. Commander Brown flew *Discovery* into position within reach of the RMS arm so that Mission Specialist Robinson could reach up and grapple SPARTAN. Within half an hour, SPARTAN was latched securely in its restraint in the payload bay. Further use of this payload would support other STS-95 research efforts, but this was the final shuttle mission on which SPARTAN would perform autonomously while acquiring astrophysical data. For the evening's sleep period, Glenn donned special sensing gear to monitor his physiological responses.

Medical work was continued on November 4, and Robinson and Parazynski used SPARTAN grappled to the end of the RMS arm in support of the OSVS study. At the conclusion of that work,

SPARTAN was returned to its berthing location. That evening Commander Brown, Pilot Lindsey, and Glenn made a remote-taped appearance on NBC's *Tonight Show*, trading jokes with Jay Leno.

Although some experiments were winding down, medical research continued on November 5. Early in the afternoon the crew held the traditional in-flight press conference. As had been the case in numerous public affairs events thus far during this flight, the majority of reporters addressed their questions to John H. Glenn, Jr.

A thorough check of *Discovery*'s flight control system and Reaction Control System (RCS) thrusters on November 6 revealed one failed RCS thruster, but that did not pose a problem for the coming reentry. Mission Control decided that it would be prudent not to deploy the suspect drag chute after touchdown. Other than that, reentry was expected to be quite normal. Glenn conversed with President Clinton via e-mail, thanking the First Couple for personally witnessing his launch.

Discovery touched down at KSC's Shuttle Landing Facility runway on November 7 at 17:03:30 UTC (12:03 P.M.) without using the suspect drag chute. *Discovery* had completed 135 Earth orbits, traveling 5.8 million kilometers (3.6 million statute miles) over the course of eight days, 21 hours, and 44 minutes. This was the sixteenth consecutive landing at KSC, the forty-fifth overall in the shuttle era.

There was much anticipation surrounding the crew's exit from the orbiter, specifically to ascertain how well Glenn had fared upon return to Earth-normal gravity. Concerns over his condition increased when the crew's egress from a portable transporter device was significantly delayed. When Glenn did come down the stairs, he walked on his own but clearly displayed an uncertain gait. However, the effects of microgravity routinely cause many astronauts to experience uneasiness with regard to balance and a perceived heaviness in the legs upon returning to Earth.

Contributions

STS-95 included one of the most intense and diverse scientific regimens of any shuttle flight. It included some of the science-dedicated Spacelab missions and manifested a total of eighty-three separate investigations. Categories included life sciences research, basic biological investigations, materials science experiments, crystal growth studies, pharmaceutical research, solar astrophysics, and ultraviolet astronomy. STS-95's crew represented individuals from several national space programs. That multicultural aspect was shared by the mission's manifested research, involving the National Aeronautics and Space Administration (NASA), the Canadian Space Agency, the European Space Agency, and the Japanese Space Agency, and included some jointly sponsored experiments. Many of these astronauts, Glenn in particular, served as test subjects for life sciences research conducted both on board and as postflight follow-ups.

NASA also manifested three technology demonstrations of key hardware that was eventually installed on the Hubble Space Telescope (HST): specifically, a radiation-hardened 486 computer system, a solid-state data recorder (SSR), and a newly designed electromechanical-cryocooler which, it was hoped, would restore HST's Near Infrared Camera and Multi-Object Spectrometer (NICMOS) capability to collect data. The latter device had prematurely run out of cryogenic coolant needed for infrared astronomy. These devices received a thorough space-based test during STS-95, clearing them to be manifested on subsequent shuttle HST servicing missions.

A large percentage of STS-95's experiments became part of continuing research programs, especially the basic biological ones and protein crystal growth studies. Tests involving technology upgrades for HST proved that the radiation-hardened 486 microprocessor and SSR would function in the Observatory's radiation environment and that the electromechanical cryocooler should be able to keep NICMOS sufficiently cooled for restoration of infrared observations.

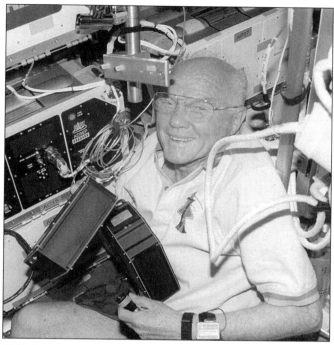

U.S. Senator John H. Glenn inside the SPACEHAB facility on board Discovery (STS-95). At age seventy-seven, Glenn was the oldest astronaut to date. (NASA)

Six months after *Discovery* returned to Earth, preliminary SPARTAN solar physics data had been reduced, indicating mechanisms that provided answers to questions about the nature of the solar wind. These questions were first raised by Mariner 2 on its way to pass near Venus in 1962, the same year that Glenn first flew into orbit. That spacecraft, the first successful interplanetary probe, indicated that free electrons and ions stream away from the Sun at speeds greater than one million miles per hour, much faster than solar physicists had previously predicted. SPARTAN provided data explaining why heavier ions such as oxygen and hydrogen could stream from the solar corona faster than much lighter free electrons. STS-95 data revealed that the ions in question spiraled around vibrating solar magnetic fields. A resonant situation appeared to exist in which the frequency of spiraling motion around field lines nearly matched the frequency of magnetic field temporal variations.

Also, Glenn's intense biomedical regimen proved fruitful, a circumstance critics of his inclusion on STS-95 had previously doubted would be possible. On May 25, 1999, a number of leading academicians involved in aging research gathered with Glenn at the National Press Club in Washington, D.C. Scientists reported a strong link between microgravity adaptation in younger, career astronauts and aging-related health concerns of elderly Earth-bound individuals. Surprisingly, Glenn's response to weightlessness proved no worse than that of younger astronauts, and his readaptation to Earth-normal gravity, although more severe than many astronauts, initially proved to be remarkably rapid. The STS-95 research hinted at the desirability of sending additional elderly individuals into space to provide useful data with applications toward easing the effects of aging and osteoporosis.

Context

On February 20, 1962, John H. Glenn, Jr., became the first American to orbit the Earth. Glenn had served as backup to both Alan Shepard and Virgil I. "Gus" Grissom, who, in that order, had flown brief suborbital missions down the Atlantic Missile Range in 1961. On Glenn's landmark spaceflight he circled the Earth three times, splashed down safely in the Atlantic Ocean, and jumped into the history books, earning the adulation of a majority of Americans. During that Mercury flight, Glenn encountered a faulty signal, suggesting that his spacecraft might not be properly configured for safe passage through the heat of the upper atmosphere that would be created by friction. As a precaution, he was ordered not to jettison the retropack prior to reentry, and, as a direct result, probably calmly endured the most stressful return to Earth of any American astronaut, with the possible exception of the Apollo 13 crew.

Glenn became such a national treasure that under unofficial orders from the Kennedy White House, he was removed from consideration for future flights. Glenn left NASA in 1964, entering the business world and eventually politics, becoming the Democratic senator from his native Ohio, a seat he held for four consecutive terms. Glenn attempted an abortive run for the 1984 Democratic Party's presidential nomination. Throughout his post-NASA period, Glenn always maintained that he would like to return to space. That opportunity became possible when in the late 1990's he devised a proposal to use him as an elderly test subject for some space-based medical tests. A vigorous campaign involving peer review at the National Institutes of Aging and acquiring other scientific backing ultimately led to Glenn's selection as a payload specialist for STS-95, an announcement that gathered much public attention.

Glenn's two spaceflights spanned the beginning and end of an era. On *Friendship 7* he had flown for only 4 hours, 55 minutes, and 24 seconds as an American Cold Warrior traveling into space in his country's race for technological supremacy over Russia. In 1998 he flew as a statesman advocating international cooperation in scientific research. Glenn's STS-95 experience represented the final American shuttle flight before an expected period of intense cooperation between those two competitor space-faring nations. The next Russian launch (November, 1998) included the Zarya Control Module, and the next shuttle flight (December, 1998) took NASA's Unity Node up to attach to Zarya, thereby initiating orbital construction of the International Space Station, a project involving contributions from seventeen member nations.

See also: Space Shuttle; Space Shuttle Flights, 1998; SPACEHAB.

Further Reading

Carpenter, M. Scott, L. Gordon Cooper, Jr., et al. *We Seven, by the Astronauts Themselves.* New York: Simon and Schuster, 1962. This account of Project Mercury includes highlights of the historic flights of Shepard, Grissom, Glenn, and Carpenter. Individual chapters are written by different members of the original seven Mercury astronauts.

Glenn, John H., with Nick Taylor. *John Glenn: A Memoir.* New York: Bantam Books, 1999. This autobiography spans Glenn's life, including his childhood in New Concord, Ohio, college days at Muskingum College, experiences during World War II and the Korean War, test pilot days, participation in Project Mercury, and political career, as well as a firsthand report of his history-making flight on space shuttle *Discovery.*

Harland, David M. *The Space Shuttle: Roles, Missions, and Accomplishments.* Hoboken, N.J.: John Wiley, 1998. *The Space Shuttle* is written thematically, rather than purely chronologically. Topics include shuttle operations and payloads, weightlessness, materials processing, exploration, Spacelabs and free-flyers, and the shuttle's role in the International Space Station.

Harrington, Philip S. *The Space Shuttle: A Photographic History.* San Francisco, Calif.: Brown Trout, 2003. With one hundred full-color photographs by Roger Ressmeyer and others and with text by popular astronomy writer Phil Harrington, this book tells the story of the space shuttle program from 1972 to 2003. Its beautiful photographs allow the general reader to survey the history of the space shuttle program and be uplifted by the pioneering spirit of one of humanity's grandest enterprises.

Jenkins, Dennis R. *Space Shuttle: The History of the National Space Transportation System: The First 100 Missions.* Stillwater, Minn.: Voyageur Press, 2001. This is a concisely written technical reference account of the space shuttle and its ancestors, the aerodynamic lifting bodies. It details some of the advantages and inherent disadvantages of using a reusable space vehicle. Each of the vehicles is illustrated by line drawings. The book follows the space shuttle from its original concepts and briefly chronicles its first one hundred flights.

Montgomery, Scott, and Timothy R. Gaffney. *Back in Orbit: John Glenn's Return to Space.* Marietta, Ga.: Longstreet Press, 1998. Provides biographical background on Glenn, detailing primarily his Mercury days and the STS-95 flight of *Discovery.* Contains numerous large photographs.

National Aeronautics and Space Administration. *Space Shuttle Mission STS-95 Press Kit.* http://www.shuttlepresskit.com/sts-95/index.htm. Provides detailed preflight information about the STS-95 space shuttle mission. Accessed March, 2005.

Riper, Frank Van. *Glenn: The Astronaut Who Would Be President.* New York: Empire Books, 1983. The author interviewed Glenn extensively. Although little of the book involves Glenn's spaceflight career, it provides a thorough exposition of Glenn's political life, particularly detailing Glenn's 1984 presidential bid.

David G. Fisher

Space Shuttle Flights, 1999

Date: May 27 to December 27, 1999
Type of mission: Piloted spaceflight

The year 1999 witnessed many advances in space exploration along with several setbacks, the most notable being a decrease in the number of missions undertaken. Advances included learning about working rapidly in a weightless environment, successfully launching the Chandra X-Ray Observatory, and deploying a base for the Russian construction crane. During this time NASA's focus shifted from saving money to making safety its highest priority.

Key Figures

Kent V. Rominger (b. 1956), STS-96 commander
Rick D. Husband (1957-2003), STS-96 pilot
Ellen Ochoa (b. 1958), STS-96 mission specialist
Tamara E. Jernigan (b. 1959), STS-96 mission specialist
Daniel T. Barry (b. 1953), STS-96 mission specialist
Julie Payette (b. 1963), STS-96 mission specialist (Canadian Space Agency, CSA)
Valery Ivanovich Tokarev (b. 1952), STS-96 mission specialist (Russian Space Agency, RSA)
Eileen M. Collins (b. 1956), STS-93 commander
Jeffrey S. Ashby (b. 1954), STS-93 pilot
Steven A. Hawley (b. 1951), STS-93 mission specialist
Catherine G. "Cady" Coleman (b. 1960), STS-93 mission specialist
Michel Tognini (b. 1949), STS-93 mission specialist (French Space Agency, CNES)
Curtis L. Brown, Jr. (b. 1956), STS-103 commander
Scott J. Kelly (b. 1964), STS-103 pilot
Steven L. Smith (b. 1958), STS-103 mission specialist
C. Michael Foale (b. 1957), STS-103 mission specialist
John M. Grunsfeld (b. 1958), STS-103 mission specialist
Claude Nicollier (b. 1944), STS-103 mission specialist (European Space Agency, ESA)
Jean-François Clervoy (b. 1958), STS-103 mission specialist (European Space Agency, ESA)

Summary of the Missions

The year 1999 was among the most troubled in NASA's history. In February, the agency had to re-arrange its schedule of shuttle launches, cutting it to five flights, one of which, a map-making mission planned for *Endeavour,* was not launched until early 2000. The *Columbia* shuttle was to have carried the $1.5 billion Chandra X-Ray Observatory into space in April. The telescope, however, had faulty circuit boards, necessitating a delay to July.

This change resulted in NASA's scheduling the next shuttle launch for May. *Discovery*'s mission in May was to deliver supplies to the International Space Station. This launch was to be followed in August by *Atlantis*'s flight to take more supplies and materials to the International Space Station's service module, which the Soviets had promised to launch in July. It had to be canceled when the service module was not launched.

Added to these rescheduling difficulties was a suspension of all shuttle flights between August and late October because of potentially dangerous wiring problems that surfaced during *Columbia*'s July flight. All four shuttles were grounded when Ronald D. Dittemore, the shuttle program's manager, ordered a stand-down until some two hundred miles of wiring in each shuttle were thoroughly checked.

Discovery was the first shuttle launched in 1999. It was scheduled for May 25, but delayed until May 27, when it streaked into space at 10:49:42.021 Coordinated Universal Time (UTC, or 6:50 A.M. eastern daylight time). Early on May 28, it docked with the fledgling International Space Station, the first two components of which had been joined by astronauts in December, 1998.

STS-96 was to prepare the 250-meter space station for the delivery of its first living quarters, the Zvezda Service Module being built by the Russians. *Zvezda* is the Russian word for "star." The crew of seven, through a corridor between the shuttle and the Space Station, delivered 1,800 kilograms of tools, laptop computers, drinking water, spare parts, clothing, and trash bags to the Space Station, scheduled to be occupied by a permanent crew in late 2000. Astronauts repaired a broken radio, replaced faulty battery packs, and installed foam mufflers over fans so noisy that they might, over time, damage the hearing of permanent crew members.

Astronauts Tamara E. Jernigan and Daniel T. Barry spent more than six hours in space installing an American-built construction crane (Orbital Replacement Unit Transfer Device, or OTD) and the mount for the Strela ("Arrow") crane that the Russians would provide later. NASA's deputy program manager for space station operations, Frank L. Culbertson, Jr., heralded the successful docking of the shuttle with the Space Station, calling it "a very significant event—one that we're going to repeat many, many times in the future." It will take 160 spacewalks and eighty-six American and Russian spaceflights, including two dozen more shuttle flights, to complete the $60 billion International Space Station. STS-96 landed on KSC's Runway 15 at 06:02:45 UTC (2:03 A.M. eastern daylight time) on June 6.

In July, *Columbia*, commanded by Eileen M. Collins, the first woman commander of a shuttle flight, was launched. On July 20, computers aborted the launch 6.6 seconds before liftoff when sensors detected a hydrogen leak. Liftoff of STS-93 was further delayed by threatening weather on July 22 but was finally accomplished on July 23 at 04:30:59.984 UTC (12:31 A.M. eastern daylight time).

Columbia was to launch the Chandra X-Ray Observatory, the most sophisticated instrument of its kind to date. Weighing more than 20 tons, the telescope and its booster were ejected from the shuttle seven hours into the flight,

Astronauts Kent V. Rominger and Julie Payette help prepare the International Space Station for occupancy during the STS-96 mission in 1999. The two are in the U.S.-built Unity node near the hatch leading to the Russian-built Zarya Control Module. (NASA)

The International Space Station moves away from the space shuttle Discovery *(STS-96). The Unity node (left) and the Zarya Control Module (with the solar array panels) were joined during a 1998 mission.* (NASA)

when Astronaut Catherine G. "Cady" Coleman activated a spring that catapulted Chandra from its cargo bay. The shuttle, moving 56 kilometers away, ignited Chandra's boosters, sending it into an elliptical orbit 140,000 kilometers above Earth. The five-day mission concluded with a smooth touchdown at KSC at 03:20:36 UTC, July 28 (11:21 P.M. eastern daylight time, July 27).

NASA was under pressure to advance *Discovery*'s mission to repair the Hubble Space Telescope. Shuttle launches are generally not scheduled close to major holidays, but the urgency to launch *Discovery* was substantial, resulting in a December 19 launch and a December 27 return to Earth. Edward Weiler, NASA's associate administrator for science, noted that with three of Hubble's gyroscopes not working, "We are one failure away from losing all science from the Hubble Space Telescope."

Fears that the "Year 2000" (Y2K) problem could disrupt some of NASA's computer programs forced NASA to launch *Discovery* so that its mission would end before the last day of 1999. It was feared that some NASA computers might read "00" as 1900 rather than 2000 (a fear, like most related to the Y2K problem, that did not materialize). The ninety-sixth shuttle mission blasted off into the night sky above KSC at 00:49:59.986 UTC (7:50 P.M. eastern standard time) on December 20. The astronauts aboard the shuttle were aloft over Christmas, which added considerable expense to the mission because NASA had to pay the ground-based support team extra holiday pay and overtime.

On December 22, *Discovery*, 595 kilometers (370 miles) above Earth, rendezvoused with Hubble and grappled it during the final mission of 1999, STS-103. Astronauts Steven L. Smith and John M. Grunsfeld, during a spacewalk that lasted 8 hours and 15 minutes, replaced Hubble's six gyroscopes and equipped their batteries with voltage regulators to prevent overheating. Astronauts C. Michael Foale and Claude Nicollier replaced Hubble's computer and installed a new fine guidance sensor. This mission rescued the Hubble from what was termed a "science emergency."

Contributions

Both *Discovery*'s mission to take supplies to the International Space Station and its mission to repair the Hubble Space Telescope demonstrated that astronauts can work rapidly and efficiently for extended periods in microgravity. The extravehicular activity (EVA) involved in repairing the Hubble Space Telescope was the second longest in history. It yielded gratifying results. The EVA involved in erecting a construction crane on the Space Station demonstrated that heavy equipment can realistically be transported into space and assembled or installed there.

The delayed launch of *Columbia* in July, 1999, as a result of a computer shutdown less than seven seconds from liftoff, was based on inaccurate data. The launch could have taken place safely, but, according to NASA's launching commentator, Bruce

Buckingham, the shuttle's sensors might have provided faulty readings.

During *Columbia*'s liftoff, only five seconds into the flight, Commander Collins reported to Mission Control in Houston a short circuit that disabled computer controllers for two of the shuttle's three main engines. Backup computers took over the work of the short-circuited machines, so the problem did not compromise the safety of the flight. It did, however, raise concerns that NASA pursued vigorously on *Columbia*'s return. An initial examination of *Columbia*'s electrical circuitry, confined to the cockpit and engine compartment, revealed twenty-six spots in which wire was exposed or damaged. *Endeavour* had at least thirty-eight such spots.

NASA's shuttle program manager, Ronald D. Dittemore, concluded that a more extensive inspection, including that of wiring beneath the decks of the cargo bays, was urgent. He grounded all the shuttles until a thorough inspection could be made. *Columbia*, already scheduled for an $80 billion overhaul by Boeing, was sent off to Palmdale, California, for that overhaul. The three remaining shuttles were grounded until inspections could be completed and essential repairs made. These inspections revealed that much of the flawed wiring had apparently been rubbed or stepped on or had had heavy objects placed upon it. Work rules were altered to prevent future damage, and considerable wiring was relocated to places where it would be less vulnerable to such damage.

During this period, quality control became NASA's major concern. The agency had already upgraded the four existing shuttles, replacing their original screens and gauges with liquid crystal displays that were much easier to read. Extra insulation was added to protect the shuttles' wings and radiator lines from damage by orbital junk or tiny meteors. New cooling systems were being investigated. New super-lightweight external fuel tanks permitted carrying heavier payloads into space. Copper wire has been replaced by fiber optics, permitting an increase in the amount of data they can

transmit. New digital radios block out interference from radio stations on Earth during orbit.

NASA was also planning to replace its alkaline fuel cells with more efficient proton exchange membrane fuel cells. It is investigating the possibility of using nontoxic fuels rather than the present hydrazine fuel, which is toxic and flammable. Each flight of the three shuttles currently in operation has yielded enormous quantities of information about how to make space travel safer and more efficient. Each problem that has occurred has provided new knowledge about space travel.

Despite the agonizing launch delays and the reduction in the number of launches in 1999, the year can be viewed as quite successful because during it the Chandra X-Ray Observatory, one in NASA's Great Observatories Program, was sent into orbit. This most sophisticated telescope ever made was scheduled to orbit for five years, providing extremely important information about areas of the cosmos that have not been visible previously and possibly about the black holes that have for so long puzzled physicists.

Such threatening lapses as the short circuit on *Columbia* have demonstrated the effectiveness of backup systems while pinpointing potential dangers that need to be eliminated. Between the first piloted shuttle launch on April 12, 1981, and the end of 1999, ninety-six shuttle flights carried 242 humans into space. Each shuttle has an estimated potential of one hundred flights, so about one-quarter of their projected, collective lives had been expended by year's end. The major question NASA now had to ask was whether it wished to continue using shuttles as we know them or would instead create a new generation of shuttles that could safely carry much greater payloads.

Context

The exploration of space has always been a politically controversial matter. President John F. Kennedy made it a matter of national pride to put an astronaut on the Moon. His successor, Lyndon B. Johnson, followed through on the Kennedy mandate, although many Americans thought the money

Astronaut Tamara E. Jernigan on a spacewalk carries part of a Russian-built crane, called Strela. Her feet are anchored on a mobile foot restraint connected to the shuttle's Remote Manipulator System (RMS). (NASA)

involved in space exploration might better be spent on domestic programs. With the Cold War still raging, however, even some skeptical Americans felt that the United States had to outdo the Soviets in space.

President Ronald W. Reagan gave grudging support to the space program. He had a minimal understanding of its importance and failed to seek the kind of detailed information that he might have from experts. His appointments to high administrative positions in the space program tended to be more political than scientific. He supported sending a schoolteacher into space, but after the 1986 loss of *Challenger* and its crew, it came to light

that NASA had become a demoralized, chaotic agency.

Following the discovery of the wiring problems in the shuttles after *Columbia*'s flight in May, 1999, NASA made safety its highest priority. Citing the need for much greater quality control, the agency sought, but did not receive, substantial budget increases for fiscal year 2001. On February 1, 2003, *Columbia* broke up approximately 16 minutes before landing, during reentry over Texas, en route to the Kennedy Space Center. The crew was returning home after a successful sixteen-day scientific research mission. There were no survivors.

Between 1995 and 1999, there was a 50 percent decrease in the number of shuttle engineers and managers employed by NASA. Federal budget cuts and draconian economic reductions by the United Space Alliance had resulted in the loss of about two thousand contractor jobs since 1991. From 1992 to 1997, NASA carried out seven or eight shuttle launches every year. Early in that period about 7,850 people were employed by the space program. This workforce gradually shrank to about 3,725 toward the end of this period. In 1999, about seventy-five new hires were approved, but that hardly brought the agency up to the level it had reached early in the decade. Howard DeCastro, United Space Alliance's vice president and program manager, publicly vowed to hire as many additional people as it would take to ensure safety.

During NASA's period of retrenchment, despite a reduction in the number of launches, each shuttle trip was yielding spectacular results. Putting Chandra into orbit was a virtually flawless operation carried out precisely as scheduled. Erecting a construction crane on the International Space Station received less publicity than it might have, but it was a remarkable accomplishment, as was the deployment of a base for the Russian construction crane.

More publicized was *Discovery*'s mission to repair the Hubble Space Telescope, which, had it not been fixed within a small window of time, would have become useless. In March, 2002, the STS-109 crew successfully serviced Hubble. The upgrades and servicing by the crew left Hubble with a new power unit, a new camera, and new solar arrays.

Endeavour's delayed flight to make more intricate maps of Earth than ever existed took place early in 2000 on STS-99. This flight would provide humankind with the most detailed representations of Earth's surface that ever existed. The information provided by these maps would have significant implications for geologists and physicists as well as extremely practical uses for airlines and other transportation systems.

See also: Chandra X-Ray Observatory; Cooperation in Space: U.S. and Russian; Far Ultraviolet Spectroscopic Explorer; Funding Procedures of Space Programs; International Space Station: 1999; Space Shuttle: Radar Imaging Laboratories; Space Shuttle-Mir: Joint Missions; Space Shuttle Flights, 1998; Space Shuttle Mission STS-93.

Further Reading

Chien, Philip. "Shuttle Gets a Boost." *Popular Science* 255 (August, 1999): 70-73. Chien emphasizes the internal changes that have taken place in the shuttle since the first one was launched in 1981. Especially valuable is a sidebar that contains statistical information about the space shuttle. The article also indicates future changes to improve the program and make space travel safer.

Couvalt, Craig. "Shuttle Quality Control Now a Major Concern." *Aviation Week and Space Technology* 50 (December 20, 1999): 10-12. This brief article offers information about improvements planned for the space program. It also provides some shocking details about personnel cuts in NASA and related agencies at a time when NASA came under pressure to increase substantially the number of shuttle launches it undertakes every year.

Harrington, Philip S. *The Space Shuttle: A Photographic History.* San Francisco, Calif.: Brown Trout, 2003. With one hundred full-color photographs by Roger Ressmeyer and others and with text by popular astronomy writer Phil Harrington, this book tells the story of the space shuttle program from 1972 to 2003. Its beautiful photographs allow the general reader to survey the history of the space shuttle program and be uplifted by the pioneering spirit of one of humanity's grandest enterprises.

Ivereigh, Djuna. "Lift Off! Watching a Space Shuttle Launch Is Always a Thrill." *Astronomy* 27 (August, 1999): 74-79. Ivereigh not only recounts her experience of witnessing a launch of space shuttle *Endeavour* but also provides useful information about how to arrange to witness a launch from the Kennedy Space Center at Cape Canaveral, Florida. She presents the projected list of shuttle launches scheduled for 1999 and 2000.

Jenkins, Dennis R. *Space Shuttle: The History of the National Space Transportation System: The First 100 Missions.* Stillwater, Minn.: Voyageur Press, 2001. This is a concisely written technical reference account of the space shuttle and its ancestors, the aerodynamic lifting bodies. It details some of the advantages and inherent disadvantages of using a reusable space vehicle. Each of the vehicles is illustrated by line drawings. The book follows the space shuttle from its original concepts and briefly chronicles its first one hundred flights.

Kerrod, Robin. *Hubble: The Mirror on the Universe.* Richmond Hill, Ont.: Firefly Books Ltd., 2003. The book covers the observable universe in six sections: "Stars in the Firmament," "Stellar Death and Destruction," "Gregarious Galaxies," "The Expansive Universe," "Solar Systems," and "The Heavenly Wanderers." Clear and concise text explains the fascinating history of astronomy and the development of the HST. Hubble transports readers to the planets of our solar system and on to galaxies millions—even billions—of light years away. These dramatic, unforgettable images will bring into sharp focus how the Universe is unfolding in new and astonishing ways.

McDowell, Jonathan. "Mission Update." *Sky and Telescope* 98 (October, 1999): 30-32. McDowell focuses on the *Discovery* mission to repair the Hubble Space Telescope. He points out the "Year 2000" (Y2K) fears that made necessary a launch that would permit *Discovery* to return to Earth by the end of December, 1999.

Mari, Christopher, ed. *Space Exploration.* New York: H. W. Wilson, 1999. Although the twenty-five essays in this well-selected collection do not deal specifically with any of the 1999 shuttle launches, such essays as William E. Burrows' "Why Build a Space Station?" and John M. Logsdon's "Building a Space Station Still Makes Sense" provide interesting background material.

National Aeronautics and Space Administration. *Space Shuttle Mission Press Kits.* http://www-pao.ksc.nasa.gov/kscpao/presskit/presskit.htm. Provides detailed preflight information about each of the space shuttle missions. Accessed March, 2005.

R. Baird Shuman

Space Shuttle Mission STS-93

Date: July 23 to July 27, 1999
Type of mission: Piloted spaceflight

STS-93 achieved two landmark distinctions: the first flight of a female space shuttle commander and orbital deployment of the Advanced X-Ray Astrophysics Facility (AXAF), the third space-based telescope in NASA's Great Observatories Program.

Key Figures

Eileen M. Collins (b. 1956), STS-93 commander

Jeffrey S. Ashby (b. 1954), STS-93 pilot

Steven A. Hawley (b. 1951), STS-93 mission specialist

Catherine G. "Cady" Coleman (b. 1960), STS-93 mission specialist

Michel Tognini (b. 1949), STS-93 mission specialist (French Space Agency, CNES)

Subrahmanyan Chandrasekhar (1910-1995), Indian American Nobel laureate, known to the world as Chandra ("moon" or "luminous" in Sanskrit), widely regarded as one of the foremost astrophysicists of the twentieth century

Fred S. Wojtalik, Chandra program manager

Keith Hefner, Chandra program manager

Jean Olivier, Chandra deputy program manager

Martin C. Weisskopf, Chandra project scientist

Summary of the Mission

NASA announced on March 4, 1998, that Eileen M. Collins would be assigned command of STS-93, a flight of space shuttle *Columbia*. Her remaining crew members included Pilot Jeffrey S. Ashby and Mission Specialists Dr. Steven A. Hawley, Dr. Catherine G. "Cady" Coleman, and Michel Tognini, a French Space Agency (Centre National d'Études Spatiales, CNES) astronaut. Collins had something in common with Tognini. Both had spent time briefly aboard Russia's Mir Space Station, although on quite different missions.

The next day Collins appeared at a White House ceremony where special notice of the first American woman named to command a spaceflight was made by both President Bill Clinton and First Lady Hillary Rodham Clinton. Collins paid tribute to American female astronauts who paved the way for her present assignment and hoped that she too might inspire young girls to follow lofty, challenging goals through perseverance, hard work, and dedicated study, both in general and in mathematics and science in particular.

STS-93 was planned as one of the shortest shuttle missions since flight resumption back in 1988, but it included a prized mission objective. *Columbia*'s principal payload was the Advanced X-Ray Astrophysics Facility (AXAF), which had been renamed the Chandra X-Ray Observatory (CXO), or Chandra for short, at the conclusion of an international competition administered by the Smithsonian Astrophysical Observatory's AXAF Science Center in Cambridge, Massachusetts, which would assume primary scientific oversight of the telescope once it was operationally deployed in space. Contest participants were required to submit appropriate names pertinent to AXAF's mission. Two individuals submitted winning entries proposing the same name. They were Tyrel Honson, a student from Priest River Lamanna High School in Priest River, Idaho, and Jatila van der Veen, an instructor of physics and astronomy at Adolfo Camarillo High School in Camarillo, California. Both individuals wrote brief essays proposing that AXAF be renamed in honor of the late Nobel laureate Indian American astrophysicist Dr. Subrahmanyan Chandrasekhar, someone whom Dr. Martin Rees, Great Britain's Astronomer Royal, once commented thought longer and deeper about the universe than anyone since Albert Einstein.

Originally set to launch in December, 1998, STS-93 experienced numerous delays due to problems encountered with final testing and preflight preparations of the Chandra Observatory. Chandra finally arrived at Kennedy Space Center (KSC) on February 4, 1999. Then, more systems problems surfaced, and a thorough investigation of a booster stage needed by Chandra followed after a similar stage failed to perform properly on an expendable Titan IV launch. Together, these delays pushed *Columbia*'s scheduled launch to coincide with the agency's thirtieth anniversary of the first moonwalk, July 20.

STS-93 suffered a pair of consecutive launch scrubs, the first due to a computer-ordered cutoff just seconds prior to scheduled main engine ignition and the second due to potential for lightning near the launch pad. First Lady Hillary Clinton attended the first two attempts, bringing members of the triumphant National Women's Soccer Team along with her. Unfortunately, the First Lady's schedule did not permit a return for the next launch attempt.

Columbia lifted off Pad 39-B at 04:30:59.984 Coordinated Universal Time (UTC, or 12:31 A.M. eastern daylight time) on July 23, 1999, lighting up the area surrounding KSC briefly before heading out over the Atlantic Ocean as it accelerated toward orbit. The initial stage of ascent provoked some anxious moments. Just seconds after liftoff, Commander Collins reported observing a fuel cell warning light. Simultaneously, an electrical tran-

sient caused two main engine controllers to go down. Automatically, control of the center and right main engines switched over to backup controller units. Fortunately none of these anomalies affected main engine performance for the remainder of the ascent. When the main engines cut off, *Columbia* entered an initial 75-by-262-kilometer orbit with an underspeed of 4.6 meters per second, which was a curious (though minor) discrepancy. Later it was determined that a small hydrogen leak in one of *Columbia*'s main engine nozzles had been responsible for the underspeed. Forty-five minutes after launch an orbital maneuvering system (OMS) engine burn circularized *Columbia*'s orbit at the desired altitude for deployment of Chandra.

Unlike the Hubble Space Telescope (HST) and Compton Gamma Ray Observatory (GRO), which were stationed in low-Earth orbits, Chandra was designed to operate in an elliptical orbit well outside most of Earth's radiation belts, thereby permitting uninterrupted observations as long as fifty-five hours. In its planned orbit, Chandra would range from about 10,000 kilometers in altitude to a point nearly one-third of the way toward the Moon, an orbit having a period of around sixty-four hours. To get there the telescope was attached to an Air Force two-stage Inertial Upper Stage (IUS). The combined length of Chandra and its solid-fueled upper stage nearly spanned *Columbia*'s payload bay.

STS-93 was scheduled to be one of the shortest space shuttle missions since the return to flight after the 1986 *Challenger* accident. It must be recognized that the short duration resulted from the size and weight of the mission's principal payload. Chandra left only a little room near *Columbia*'s forward bulkhead and some space behind the telescope's support ring near the aft bulkhead. Thus *Columbia* could carry little else. Although STS-93 manifested several lower-priority experiments, there was little point in keeping the vehicle aloft longer than necessary for astronaut adaptation to their change in environment before returning to Earth.

The mission's first workday schedule was heavily loaded, including Chandra's deployment at the earliest opportunity after *Columbia*'s computers and

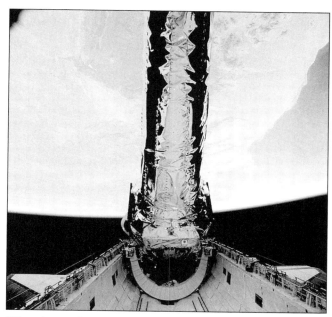

The primary duty of the STS-93 crew was to deploy the world's most powerful x-ray telescope, the Chandra X-Ray Observatory, shown here just before its release from Columbia*'s payload bay.* (NASA)

systems transitioned to the needed on-orbit configuration. Backup deployment opportunities existed in the flight plan, but the first proved to be the charm. After Chandra was raised upward to a 58° inclination relative to *Columbia*'s payload bay, Mission Specialist Catherine G. "Cady" Coleman, using spring ejection, gently pushed the telescope out of its support ring and safely up and over *Columbia*'s flight deck as the vehicle and telescope jointly passed over the Indian Ocean near Sri Lanka at 11:47 UTC. The astronauts' workday was not yet over. Collins had to perform a separation maneuver to move *Columbia* off to a position safe from Chandra's IUS motor exhaust.

Chandra's IUS first-stage motor ignited at 12:47 UTC, providing 44,000 pounds of thrust for 125 seconds, setting up the first part of Chandra's journey away from *Columbia*'s orbit. The IUS second-stage motor ignited at 12:51 UTC, providing 18,200 pounds of thrust for 117 seconds and completing the upper stage's job. Over the next few

hours, controllers began the task of initial setup of many of Chandra's systems, such as deployment of its twin solar arrays, and monitored the telescope's slow climb out to apogee. Five additional small burns by the telescope's own propulsion system were required to finish Chandra's transfer from low-Earth orbit to its operational orbit.

Over the next few days, the astronauts performed several secondary experiments, participated in a number of public affairs and educational events, and began preparations for their return to Earth. One experiment that Mission Specialist Hawley performed involved a small ultraviolet telescope called the Southwest Ultraviolet Imaging System (SWUIS) operated from windows inside *Columbia*'s crew compartment. Meanwhile, three days after Chandra's deployment, ground-based controllers began activating the telescope's science instruments. The hinged door covering Chandra's imaging camera opened on July 26. That was a major step, but several weeks were needed before the telescope could begin collecting valuable scientific data.

Columbia touched down at 03:20 UTC on July 28 (11:20 P.M. on July 27 local time) at KSC, accomplishing the twelfth night landing of the shuttle program at the Shuttle Landing Facility just a few miles from the Launch Complex and Orbiter Processing Facilities. *Columbia* had completed seventy-nine orbits over the course of four days, 22 hours, 49 minutes, and 35 seconds.

Contributions

Some of the most intriguing objects in the universe are also the most energetic, producing enormous high-energy x-ray fluxes. NASA's Great Observatories Program was designed to make a coordinated investigation of the universe detecting a variety of segments of the electromagnetic spectrum. The HST investigated visible emissions and portions of the infrared and ultraviolet spectrum. The Compton Gamma Ray Observatory was designed to investigate high-energy gamma-ray emissions. The job of x-ray astronomy fell to the Advanced X-Ray Astrophysics Facility (AXAF), renamed the Chandra X-Ray Observatory (CXO), or Chandra for short. One more telescope was planned for this series, the Space Infrared Telescope Facility (SIRTF). Ironically, shortly after Chandra's deployment, Congress initially seriously cut NASA funding to the point where SIRTF would have had to be canceled. Outrage from the scientific community coupled with White House support led to restoration of those targeted funds.

Like its sister Great Observatories before it, Chandra held the promise of rewriting present knowledge in its subfield of astronomy. Chandra was equipped with a high-resolution camera and an imaging spectrometer. The former would provide x-ray images of selected objects, whereas the latter separates x-ray emissions into individual wavelengths, thereby providing insight into chemical and thermal aspects of studied objects.

Although Chandra's x-ray detection instruments pushed the state of the art in that technology, an even bigger challenge in designing and building Chandra involved the telescope's cylindrical mirrors. X rays are much more difficult to bring into focus than visible, ultraviolet, infrared, or radio emissions, and require a grazing incidence into x-ray detection devices. Building and verifying Chandra's special mirrors, manufactured smooth to within the thickness of just three atomic layers, took more than six years.

On August 26, NASA held its first Chandra science briefing, thereby revealing initial images taken by the new x-ray observatory. Unlike HST, Chandra was able to start science work immediately upon conclusion of its certification cycle. Initial images revealed that Chandra indeed would provide x-ray images far superior to any previously obtained. One of the first images taken was of the much-studied Cassiopeia A supernova. Chandra's image clearly identified for the first time the central neutron star remnant of that giant explosion. Throughout Chandra's first year, virtually every image taken provided additional understanding that had previously eluded astrophysicists.

Astronomers have used an x-ray image from Chandra to make the first detailed study of the behavior of high-energy particles around a fast-moving pulsar. The image shows the shock wave created as a pulsar plows supersonically through interstellar space. These results will provide insight into theories about the production of powerful winds of matter and antimatter by pulsars.

Data from Chandra, along with infrared observations, have uncovered evidence that a gamma-ray burst—one of nature's most cataclysmic events—occurred in our galaxy, the Milky Way, a few thousand years ago.

Some of Chandra's most intriguing discoveries involve black holes. The observatory has revealed new details about x-ray jets produced by black holes, and discovered two black holes flourishing in a single galaxy 400 million light years from Earth. For the first time, humans have been able to track the life cycles of large-scale x-ray jets produced by a black hole. Chandra has revealed that, as the jets evolved, the material in them traveled near the speed of light for several years before slowing and fading.

By revealing two active black holes in the nucleus of the extraordinarily bright Galaxy NGC 6240, another Chandra image showed that two supermassive black holes could coexist in the same galaxy. The observatory offered new insights into pulsars—small and extremely dense stars. Whipping about like an untended firehose at about half the speed of light, the jet of high-energy particles has offered new insights into the nature of jets from pulsars and black holes.

Chandra has revealed the most distant x-ray cluster of galaxies, identified a pulsating hot spot of x rays in Jupiter's upper atmosphere, uncovered a "cool" black hole at the heart of the Andromeda galaxy, and found an x-ray ring around the Crab Nebula.

Context

X-ray astronomy is impossible from Earth's surface, and therefore is a relatively young scientific discipline. NASA launched the first x-ray observatory, called Uhuru, in late 1970. Later that decade there followed the more capable Einstein Observatory. Both telescopes identified a host of fascinating high-energy x-ray emissions, some coming from the far reaches of the universe. Russian and European x-ray telescopes also provided groundbreaking x-ray observations, continuing the preliminary work of Uhuru and Einstein.

When *Columbia* departed KSC, celebrations were already under way commemorating the thirtieth anniversary of Apollo 11, NASA's first lunar landing. Many Apollo-era astronauts participated in some of the festivities, providing STS-93 a nostalgic air, as had STS-95 nearly nine months earlier when John H. Glenn, Jr., returned to space. Adding to the tremendous volume of space news around KSC was the successful retrieval of Virgil I. "Gus" Grissom's *Liberty Bell 7* Mercury spacecraft, which sank shortly after Grissom safely completed the nation's second suborbital spaceflight in July, 1961. On a sad note, the day before Collins entered the history books, ex-Astronaut Charles "Pete" Conrad, Jr., a veteran of two Gemini flights, the Apollo lunar landing, and one Skylab mission, was buried at Arlington National Cemetery following a fatal motorcycle accident.

In the three decades since Apollo, a large number of American women had gone into space aboard space shuttle orbiters in the role of mission specialist. Sally K. Ride was the first, flying in 1983 on STS-7, the second flight of *Challenger.* That flight came twenty years after the very first woman to be sent into space, the Russian cosmonaut Valentina Tereshkova, flew alone for three days aboard Vostok 6. However, very few women had joined the Astronaut Corps under the category of shuttle pilot. Eileen M. Collins was among the first three women to earn that status. Indeed, she became the first female astronaut assigned the pilot's responsibilities, which she carried out on STS-63 from *Discovery*'s right-hand flight deck seat. On STS-63 she served as pilot on the mission that blazed the trail to the Russian space station Mir. Collins returned to Mir again on STS-84, this time docking, and she

was given the responsibility for a fly-around maneuver about Mir before *Atlantis* departed from the station's vicinity for eventual return to Earth. While flying on that mission, Collins received word that she was in line for a command of her own, an unprecedented means of notification.

Columbia's ascent anomalies were traced to electrical wiring deficiencies in the vehicle's midbody that probably resulted from incidental contact during preflight processing. In STS-93's aftermath, the entire shuttle fleet was grounded for extensive safety inspections. Three shuttle flights had been scheduled for the remainder of 1999, but only one flew, STS-103, during the second half of December. That was the third HST servicing mission.

Columbia's next mission, STS-109, was its last completely successful flight. In March, 2002, the STS-109 crew successfully completed the mission's objectives of servicing the Hubble Space Telescope. The upgrades and servicing by the crew left Hubble with a new power unit, a new camera, and new solar arrays. This was the fourth shuttle mission dedicated to servicing Hubble.

On February 1, 2003, *Columbia* broke up approximately 16 minutes before landing, during reentry over Texas, en route to the Kennedy Space Center. The STS-107 crew members—Commander Rick D. Husband, Pilot William C. McCool, mission specialists Michael P. Anderson, David M. Brown, Laurel B. Clark and Kalpana Chawla, and payload specialist Ilan Ramon of Israel—were returning home after a successful sixteen-day scientific research mission. *Columbia*, the first reusable piloted space vehicle, was lost to history.

By sheer coincidence or by a stroke of fate, Eileen M. Collins was slated to command the flight of *Discovery* on the STS-114 mission, scheduled to be launched shortly after STS-107's landing. Collins was designated to command the STS-114 "Return to Flight" mission, scheduled to launch more than two years after the accident.

See also: Chandra X-Ray Observatory; Compton Gamma Ray Observatory; Ethnic and Gender Diversity in the Space Program; Space Shuttle; Space Shuttle Flights, 1999.

Further Reading

Anselmo, Joseph C. "Chandra's Successor in the Works." *Aviation Week and Space Technology,* August 2, 1999, 35-36. This article describes follow-up on x-ray observatories and how they could coordinate investigations with Chandra.

_____. "Shuttle Scrubs Delay Observations." *Aviation Week and Space Technology,* July 26, 1999. This article details the problems that forced two launch scrubs for STS-93. Surrounding articles provide descriptions of *Liberty Bell*'s retrieval and Apollo 11 celebrations that occurred in and around Kennedy Space Center before, during, and after STS-93.

Anselmo, Joseph C., and Craig Covault. "Chandra Deployed After Liftoff Anomalies." *Aviation Week and Space Technology,* August 2, 1999, 34-35. This article details *Columbia*'s ascent anomalies, providing a high-resolution photograph of one engine's hydrogen leak. Also provides details concerning Chandra's capabilities and postdeployment operations.

Covault, Craig. "Chandra Liftoff Set on Orbiter *Columbia.*" *Aviation Week and Space Technology,* July 12, 1999, 31-33. This article details Chandra's capabilities and previews *Columbia*'s anticipated flight operations.

Elvis, Martin. "NASA's Chandra X-Ray Observatory: A Revolution Through Resolution." *Sky and Telescope,* August, 1999, 44-53. This article details Chandra's development, testing, and resolution capabilities. Detailed photographs and diagrams.

Harland, David M. *The Space Shuttle: Roles, Missions, and Accomplishments.* Hoboken, N.J.: John Wiley, 1998. *The Space Shuttle* is written thematically, rather than purely chronologically. Topics include shuttle operations and payloads, weightlessness, materials processing, exploration, Spacelabs and free-flyers, and the shuttle's role in the International Space Station.

Harrington, Philip S. *The Space Shuttle: A Photographic History.* San Francisco, Calif.: Brown Trout, 2003. With one hundred full-color photographs by Roger Ressmeyer and others and with text by popular astronomy writer Phil Harrington, this book tells the story of the space shuttle program from 1972 to 2003. Its beautiful photographs allow the general reader to survey the history of the space shuttle program and be uplifted by the pioneering spirit of one of humanity's grandest enterprises.

Hawley, Steven A. "How We'll Deliver Chandra to Orbit." *Sky and Telescope,* August, 1999. This article, written by the astronaut who deployed the Hubble Space Telescope and flew as part of the second servicing mission, describes how Chandra was launched aboard and deployed from the space shuttle *Columbia.*

Jenkins, Dennis R. *Space Shuttle: The History of the National Space Transportation System: The First 100 Missions.* Stillwater, Minn.: Voyageur Press, 2001. This is a concisely written technical reference account of the space shuttle and its ancestors, the aerodynamic lifting bodies. It details some of the advantages and inherent disadvantages of using a reusable space vehicle. Each of the vehicles is illustrated by line-drawings with important features pointed out with lines and text. The book follows the space shuttle from its original concepts and briefly chronicles its first one hundred flights.

Schuiling, Roelof L. "STS-93: Launch Delays: Problems on Ascent." *Spaceflight: The Magazine of Astronautics and Outer Space* 12, no. 41, 500-505. This article provides detailed descriptions of Chandra, day-by-day STS-93 flight activities, secondary payloads, crew biographies, and postflight analysis of *Columbia*'s ascent anomalies.

Still, Martin. "X-Ray Astronomy's Golden Age." *Sky and Telescope*, August, 1999, 56-57. This article previews what insight Chandra might provide in x-ray astronomy in concert with the European X-Ray Multi-Mirror Mission and the Japanese Astro-E Observatory. Note that the latter was lost in a launch accident six months after Chandra was placed in space.

Tucker, Wallace H., and Karen Tucker. *Revealing the Universe: The Making of the Chandra X-Ray Observatory.* Cambridge, Mass.: Harvard University Press, 2001. *Revealing the Universe* takes you easily through some basic physics and the history of x-ray astronomy. The authors delve into the Chandra project and take the reader through the often-dramatic process of getting such a complicated and costly project through the cogs of bureaucracy and politics.

David G. Fisher, updated by Russell R. Tobias

Space Shuttle Flights, 2000

Date: January to December, 2000
Type of mission: Piloted spaceflight

Five space shuttle flights were made in the year 2000. These flights included only one research mission to study the Earth through radar imaging. The other four missions were construction missions to the International Space Station (ISS).

Key Figures

Kevin R. Kregel (b. 1956), STS-99 commander
Dominic L. Gorie (b. 1957), STS-99 pilot
James D. Halsell, Jr. (b. 1956), STS-101 commander
Scott J. Horowitz (b. 1957), STS-101 pilot
Terrence W. "Terry" Wilcutt (b. 1949), STS-106 commander
Scott D. Altman (b. 1959), STS-106 pilot
Brian Duffy (b. 1953), STS-92 commander
Pamela A. Melroy (b. 1961), STS-92 pilot
Brent W. Jett (b. 1958), STS-97 commander
Michael J. Bloomfield (b. 1959), STS-97 pilot

Summary of the Mission

The first space shuttle mission of the year 2000 was STS-99. Originally scheduled for September 16, 1999, the mission had been postponed as engineers inspected wiring on the orbiter *Endeavour.* The inspection was spurred by concerns regarding wiring throughout the four-spacecraft shuttle fleet. The rescheduled launch date was set for January 31, 2000. However, a combination of poor weather conditions at launch and hardware problems resulted in a scrub of the launch, and the launch was rescheduled for February 11, 2000. On that day, after a nearly fourteen-minute delay to study an anomalous pressure sensor reading, *Endeavour* lifted off at 17:43:39.997 UTC (Coordinated Universal Time, or 12:44 P.M. eastern standard time, or EST) from Pad 39A at the Kennedy Space Center (KSC) to begin its scientific research mission. Within twelve hours of launch, the Shuttle Radar Topography Mission (SRTM) equipment was deployed and mapping had begun. The radar map-

ping continued until February 21. Astronauts also took 2,715 digital photos of Earth through the overhead flight-deck window in an experiment called EarthKAM. On the 181st orbit, *Endeavour* landed at the Shuttle Landing Facility at Kennedy Space Center on Runway 33 at 22:50:08 UTC (6:22 P.M. EST) on February 22, 2000. This was one orbit later than originally intended, because crosswind conditions at the landing site were too strong at the time of the earlier orbit.

The second space shuttle mission of the year was STS-101, which was also the third shuttle mission to assist in construction of the International Space Station (ISS). The mission began with the launch of the shuttle *Atlantis* from Pad 39A at 10:11:09.994 UTC (6:11 A.M. eastern daylight time, or EDT) on May 19, 2000. Launch had been scrubbed on three earlier dates during the third week of April due to unsuitable weather conditions at either the launch site or the emergency landing sites. One 6-hour, 44-

minute extravehicular activity (EVA), or spacewalk, was used to install parts on two cranes previously attached to the ISS and to install handrails on the exterior of the Space Station to be used in later EVAs. *Atlantis* touched down at 06:20:19 UTC (2:21 A.M. EDT) on Runway 15 at the Kennedy Space Center. (Runway 15 is actually the same runway as Runway 33, on which *Endeavour* had landed at the end of STS-99. The runway is designated Runway 33 for landings headed toward the northwest, azimuth 330°, and it is designated Runway 15 for landings headed toward the southeast, azimuth 150°.)

STS-106 was the fourth shuttle mission to the ISS and the third shuttle mission of 2000. STS-106 also marked the second trip into space for the orbiter *Atlantis* in less than four months. After a perfect countdown, *Atlantis* lifted off from Pad 39B of the Kennedy Space Center at 12:45:47.008 UTC (8:46 A.M. EDT) on September 8, 2000. The main purpose of the STS-106 mission was to deliver several tons of equipment and supplies to the ISS and to boost the Space Station's orbit to a higher altitude. The astronauts entered the ISS to install equipment needed for station operation. The STS-106 crew also unloaded supplies from a Russian Progress supply craft previously launched to the ISS. After completing its mission, *Atlantis* landed at the Kennedy Space Center on Runway 15 at 07:56:43.9 UTC (3:57 A.M. EDT) on September 20, 2000.

The fifth ISS construction mission was STS-92, which began with the launch of the space shuttle *Discovery* from Pad 39A at 23:17:00.011 UTC (7:17 P.M. EDT) on October 11, 2000. The mission had been postponed from October 5 because of concerns about the external fuel tank and again from October 9 because of high winds. An additional twenty-four-hour delay on October 10 resulted from observations that a ground-support pin had not been removed prior to launch. Four EVAs were performed to prepare the ISS for its first occupants, to be launched to the ISS soon thereafter. Most of the work was electrical in nature. Supplies were delivered to the Space Station, and astronauts made final preparations inside the station for occupation by the first resident crew. Landing was de-

layed approximately three days because of poor weather conditions at the Shuttle Landing Facility at Kennedy Space Center. Initially winds were too high at the backup landing site at Edwards Air Force Base, in California, but eventually the winds subsided and *Discovery* landed at Edwards on Runway 22 at 20:59:42 UTC (2:00 P.M. Pacific daylight time, or PDT) on October 24, 2000.

The final shuttle mission of 2000 was STS-97. After a flawless countdown, the space shuttle *Endeavour* lifted off from Pad 39B to begin the STS-97 mission at 03:06:00.986 UTC on December 1, 2000 (10:06 P.M. EDT on November 30). STS-97 was the sixth shuttle mission to the ISS and the first mission to the station after it had been occupied by its first resident crew. Three EVAs were performed to install solar power arrays to the ISS. Though *Discovery* had been linked to the ISS for five days, the hatches between the two craft were not opened together until December 8. Prior to that time supplies had been left in the docking port for the ISS crew to retrieve. However, on December 8 both hatches were opened and the crews were able to visit face-to-face for the first time. More supplies were transferred to the ISS and refuse from the ISS was loaded onto the shuttle. The *Endeavour* undocked from the ISS on December 9 at 15:51 UTC. Landing occurred December 11, 2000, at 23:03:23 UTC (6:04 P.M. EST) on Runway 15 at the Kennedy Space Center.

Contributions

Though a few ride-along experiments went with each shuttle mission to study the properties of materials in space, and every mission studied the effects of weightlessness on the bodies of the astronauts, the only major science experiment of the year was the Shuttle Radar Topography Mission (SRTM) of STS-99. This radar-mapping mission produced the best radar studies of Earth's surface that have ever been made. Some portions of Earth's surface had never been mapped so thoroughly because of frequent poor weather, obscuration by vegetation, and difficult access. Additionally, many small islands were mapped that had never before undergone detailed topographical studies. *Endeavour*'s

Part of the Shuttle Radar Topography Mission (SRTM) hardware as seen from space shuttle Endeavour *'s flight deck windows during the eleven-day SRTM mission (STS-99) in 2000.* (NASA)

orbit was inclined in such a way that the spacecraft passed over nearly 80 percent of the Earth's landmass. SRTM mapped virtually all of Earth's surface between 56° south latitude to 60° north latitude. Though the radar array deployed by *Endeavour* had better resolution, the public data released showed surface details slightly smaller than about 90 meters across (295 feet) over most of this area.

These data, especially if combined with future studies of this sort, can allow scientists to study changes in land use and land topography. Furthermore, the data collected could be particularly important to national and international policy planners if global warming were to result in a change in sea level, since the effect of such a change in sea level on coastal areas depends very heavily on local topography. Following STS-99, that topography is now known with higher precision than ever before.

The data can also be used to compute flooding risks near rivers, as well as to study changes in topography resulting from natural and human-made causes.

The remaining space shuttle missions for the year were engineering missions rather than scientific missions. These missions were largely involved in fitting assemblies onto the International Space Station to provide living capabilities and power for the future inhabitants of the station. Additionally, the shuttle missions brought instruments to install in the ISS and installed equipment, such as a robotic arm attached to the exterior of the ISS, that would allow the occupants of the station themselves to assist in its construction and maintenance in later missions.

In addition to construction activities, the space shuttle missions of 2000 brought tons of supplies to the ISS for use by astronauts and cosmonauts who were to partake in a continuous operation of the station beginning later in the year.

Context

Several milestones were reached for the space shuttle program during 2000. STS-92 marked the 100th space shuttle mission. This mission also marked the end of a long streak of landings at the primary landing site at Kennedy Space Center, as it was the first Edwards Air Force Base landing since 1996. The previous mission, STS-106, had made the 23d consecutive landing at KSC.

The year 2000 marked a major shift in space shuttle operations for the National Aeronautics and Space Administration (NASA). For the first time since the early days of the space shuttle program, NASA had only one planned mission that year dedicated to science operations: STS-99. The rest of the missions were dedicated to construction and outfitting of the International Space Station. When the space shuttle was first proposed, it was designed to facilitate the construction of a future, yet-to-be designed space station. Finally, after two decades, the shuttle fleet was being used for precisely that purpose. The fact that only one shuttle mission during the year was not an ISS mission shows NASA's shift in priorities from other missions to the construction of the Space Station. While some ISS construction activities are possible using the robotic arms on the shuttle and the ISS, many construction activities still require astronauts to perform spacewalks. All four of the ISS missions of 2000 required EVAs, resulting in a total of 59 hours and 37 minutes' worth of EVAs. That was a new NASA yearly record for time that astronauts spent outside of an orbiting spacecraft.

See also: International Space Station: Design and Uses; International Space Station: Living and Working Accommodations; International Space Station: Modules and Nodes; International Space Station: 2000; Russia's Mir Space Station; Space Shuttle: Radar Imaging Laboratories; Space Shuttle-Mir: Joint Missions; Space Shuttle Flights, 1999; Space Shuttle Flights, 2001.

Further Reading

Burrows, William E. *This New Ocean: The Story of the First Space Age.* New York: Random House, 1998. This book traces the development of space exploration from the first rockets through the decision to build the International Space Station. The role of the space shuttle and the ISS is given in the context of the whole of space exploration.

Caprara, Giovanni. *Living in Space.* Buffalo, N.Y.: Firefly Books, 2000. Discusses the development of ideas of building space stations from the first concepts proposed more than a century ago through the plans for the International Space Station.

Jenkins, Dennis R. *Space Shuttle: The History of the National Space Transportation System—The First 100 Missions.* Stillwater, Minn.: Voyageur Press, 2001. To celebrate the twentieth anniversary of the first space shuttle launch, this edition adds many previously uncovered early designs, details the latest modifications to the operational vehicles, and provides expanded coverage of the first hundred space shuttle missions.

Oberg, James. *Star-Crossed Orbits: Inside the U.S.-Russian Space Alliance.* New York: McGraw-Hill, 2002. Taking a more controversial dissenting opinion than that of the other books listed here, the author presents his highly critical assessment of the International Space Station and of the entire NASA cooperative effort with the Russian space agency, suggesting that NASA could have used the money allocated to the ISS in a better fashion.

Reichhardt, Tony, ed. *Space Shuttle: The First Twenty Years.* New York: DK, 2002. A collection of short essays by the astronauts themselves about living and working in space.

Raymond D. Benge, Jr.

Space Shuttle Flights, 2001

Date: January to December, 2001
Type of mission: Piloted spaceflight

Six space shuttle flights were made in the year 2001, the most flights in one year since 1997. All six missions were to the International Space Station (ISS). The year 2001 was the first since the beginning of the space shuttle program in which NASA had no shuttle missions planned primarily as science missions.

Key Figures

Kenneth D. Cockrell (b. 1950), STS-98 commander
Mark L. Polansky (b. 1956), STS-98 pilot
James D. Wetherbee (b. 1952), STS-102 commander
James M. Kelly (b. 1964), STS-102 pilot
Kent V. Rominger (b. 1956), STS-100 commander
Jeffrey S. Ashby (b. 1954), STS-100 pilot
Steven Linksey, STS-104 commander
Charles O. Hobaugh (b. 1961), STS-104 pilot
Scott J. Horowitz (b. 1957), STS-105 commander
Frederick W. Sturckow (b. 1961), STS-105 pilot
Dominic L. Gorie (b. 1957), STS-108 commander
Mark E. Kelly (b. 1964), STS-108 pilot

Summary of the Mission

On February 7, 2001, at 23:13:01.990 UTC (6:13 P.M. eastern standard time, or EST) the space shuttle *Atlantis* lifted off from Pad 39A at the Kennedy Space Center (KSC) to begin the first space shuttle mission of the year. This mission, STS-98, like all of the others that year, was a mission to support the International Space Station (ISS), which had been orbiting the Earth since the previous year.

The primary objective of the mission was to deliver the U.S. Destiny Laboratory Module to the ISS. This module was to be the living and working space for National Aeronautics and Space Administration (NASA) astronauts working on the Space Station. Two days after launch, *Atlantis* rendezvoused with ISS. After docking with the station, astronauts began unloading the nearly 1,400 kilograms of supplies, equipment, and water that the

space shuttle had brought for the Space Station, a task that was to last for several days. On February 11, with the aid of the shuttle's Remote Manipulator System (RMS) and an extravehicular activity (EVA), astronauts attached the Destiny Laboratory Module to the ISS. Two further EVAs were needed to finish connecting the Destiny and installing communication antennae to the exterior of the ISS. In total, astronauts spent 19 hours and 49 minutes on EVAs during the mission. Before they left the ISS, they loaded 386 kilograms of trash onto the space shuttle. For three consecutive days, the shuttle was unable to land at the Shuttle Landing Facility (SLF) at Kennedy Space Center because of clouds and unacceptably high crosswinds at the SLF runway. Finally, *Atlantis* landed at Edwards Air Force Base (EAFB) on Runway 22, touching down at 20:33:06 UTC (12:34 P.M. Pacific standard time, or PST) on February 20 to end the STS-98 mission.

The second space shuttle mission of the year was STS-102, which began when the shuttle *Discovery* lifted off from Pad 39B at KSC on March 8, 2001, at 11:42:09.004 UTC (6:42 A.M. EST). The focus of this mission was to carry supplies to the ISS and to bring a replacement crew to the Space Station. Though some supplies were carried in the crew compartment of the shuttle, most of the cargo was carried aboard the Leonardo Multi-Purpose Logistics Module (MPLM) in the shuttle's cargo bay. Weighing 4 metric tons, the MPLM functions as both a cargo carrier and a temporary ISS module. After riding into space in the shuttle's cargo bay, the MPLM is attached to the ISS. Crew aboard the Space Station can then unload the supplies from

the module. In this mission, approximately 4.5 metric tons of supplies and equipment were unloaded from Leonardo into the ISS. After unloading the MPLM, the ISS crew loaded a ton of garbage and experiments that had been aboard the ISS. The Leonardo module was then detached from the ISS and loaded back aboard *Discovery* for return to Earth. While the MPLM was being unloaded and loaded, astronauts performed two EVAs to attach a platform to the ISS that was to house a Canadian-built robotic arm to be delivered on a later mission.

Riding aboard *Discovery* to the ISS was the Expedition Two crew, which replaced the three-man Expedition One crew that had been living on the ISS for the past few months. Upon completion of the mission, *Discovery* undocked from the ISS with the

Atlantis lands on February 20, 2001, at Edwards Air Force Base after delivering the Destiny Laboratory Module to the International Space Station. *(NASA)*

Expedition One crew aboard and flew partially around the Space Station performing a visual inspection. *Discovery* landed on Runway 15 of the SLF at KSC on March 21, 2001. The shuttle touched down at 07:31:43 UTC (2:33 A.M. EST), making this the twelfth night landing of the space shuttle program.

STS-100 was the third space shuttle mission of the year. It began with the liftoff of the shuttle *Endeavour* from Pad 39A at KSC on April 19, 2001, at 18:40:41.988 UTC (2:41 P.M. eastern daylight time, or EDT) to deliver supplies and equipment to the ISS. Some of the supplies were delivered to the ISS with the aid of the Raffaello MPLM, a counterpart of the Leonardo module that had flown on STS-102. After attaching the MPLM to the ISS, astronauts unloaded nearly 2,700 kilograms of supplies and loaded more than 700 kilograms of trash and science experiments that had been aboard the

ISS to return to Earth. In the meantime, astronauts from the shuttle performed two EVAs to attach the Canadian-built Canadarm2 robotic arm to the ISS's Destiny Laboratory Module and to install another UHF communications antenna on the ISS. Upon completion of the mission, Raffaello was loaded back aboard *Discovery* and the shuttle undocked from the ISS to return to Earth. Weather conditions were unfavorable at KSC, so *Discovery* landed at EAFB on Runway 22 at 16:10:43 UTC (9:11 A.M. Pacific daylight time, or PDT) on May 1, 2001.

The fourth shuttle flight of the year was STS-104, which began with the launch of the *Atlantis* from Pad 39B at KSC on July 12, 2001, at 09:03:58.991 UTC (5:04 A.M. EDT). This mission was to deliver the Quest Joint Airlock Module (JAM) to the ISS. The 6,045-kilogram air lock was to allow ISS crew to conduct their own EVAs without using the air

The Leonardo Multi-Purpose Logistics Module resting in the payload bay of Discovery. *(NASA)*

lock on a docked space shuttle. Three EVAs were needed to install the JAM and storage tanks for nitrogen and oxygen. Upon completion of the mission, *Atlantis* returned to Earth, landing on Runway 15 of the SLF at KSC at 03:38:55 UTC on July 25, 2001 (11:39 P.M. EDT on July 24, 2001).

August 10, 2001, the fifth space shuttle mission of the year, began with the launch of *Discovery* from Pad 39A at KSC at 21:10:14.019 UTC (5:10 P.M. EDT) to begin the STS-105 mission. STS-105 was primarily a crew rotation mission, carrying Expedition Three crew members to the ISS and returning the Expedition Two crew to Earth. Additionally, *Discovery* carried the MPLM Leonardo back to the ISS loaded with 3,073 kilograms of supplies. Performing two EVAs, astronauts attached an ammonia tank to the ISS to support the thermal subsystems of the station. They also attached the Materials International Space Station Experiment (MISSE), a passive materials science experiment, to the exterior of the ISS. After loading the Leonardo module with trash and equipment to return to Earth and loading Leonardo back into *Discovery*, the astronauts undocked the shuttle and returned to Earth, landing on Runway 15 of the SLF at KSC on August 22, 2001, at 18:22:59 UTC (2:23 P.M. EDT).

The final space shuttle mission of the year was STS-108, beginning December 5, 2001, at 22:19:27.987 UTC (5:19 P.M. EST) with the liftoff of *Endeavour* from Pad 39B at KSC. *Endeavour* carried more than 1,600 kilograms of supplies to the ISS aboard the MPLM Raffaello. Additional supplies and equipment for the ISS were stowed in the shuttle's mid-deck. Furthermore, the Expedition Four crew arrived aboard the *Endeavour* to replace the Expedition Three crew, who had been on the ISS since August. A single EVA was performed to install insulation on the ISS solar power arrays. On December 11, the astronauts held a brief memorial for the victims of the September 11 terrorist attacks three months before. Prior to undocking from the ISS, *Endeavour* used her thrusters to boost the orbit of the ISS away from a piece of space debris that flight controllers had calculated would otherwise have passed uncomfortably close to the station in a

future orbit. *Endeavour* landed on Runway 15 of the SLF at KSC December 17, 2001, at 17:55:12 UTC (12:55 EST).

Contributions

The emphasis of this year's space shuttle missions was on engineering rather than science. Shuttle missions were used to attach major components of the International Space Station, including the Destiny Laboratory Module and the Quest Joint Airlock Module. Shuttle missions also carried a significant amount of supplies and equipment to the ISS and returned tons of equipment and refuse.

However, some science missions rode along on the shuttle missions to the ISS. These experiments studied such things as cell growth, combustion, heat exchange, and material fabrication in the microgravity environment of the space shuttle. Additionally, an experiment aboard STS-108 studied radiation produced by dust collisions with the space shuttle in near-Earth orbit. In addition, a mirrored passive microsatellite called Starshine 2 was deployed from STS-108. Observations of this satellite for the next few weeks as its orbit decayed provided data on the density fluctuations in the Earth's outermost atmospheric layers.

STS-105 carried the MISSE container to the ISS. Attached to the exterior of the ISS, this experiment was designed to expose materials to the harsh environment of space for more than one year before being returned to Earth. Analysis of the materials on MISSE allows for a greater understanding of the effects of the radiation, micrometeorite damage, and thermal cycling experienced by materials in space. The MISSE samples have had a much longer exposure than anticipated, since the grounding of the space shuttle fleet following the *Columbia* accident in early 2003. Routinely photographed by the ISS crew, the samples seem to be holding up well.

Context

NASA administrators and engineers knew that construction of the ISS would require a large number of dedicated space shuttle flights. With the de-

On STS-105, Expedition Two flight engineer Susan Helms prepares to return home after five months aboard the International Space Station. (NASA)

cision to go forward with the International Space Station program, NASA had to change its emphasis for the space shuttle program from science missions to engineering missions. For the second year in a row, the space shuttle program was fulfilling one of its original design goals to shuttle crew and material between Earth and a space station. Even though the ISS was designed to be supplied periodically by the robotic Russian Progress modules, these remotely operated spacecraft do not have the cargo capability of a space shuttle. Therefore, space shuttles must fly several supply missions per year to support the ISS at fully crewed levels. The year 2001 was also notable in that a record eighteen EVAs were performed, twelve of them from the space shuttle in support of the ISS.

The six shuttle missions of 2001 represented the most piloted American spaceflights in a single year since 1997. However, for the second year in a row, science took a back seat to ISS operations in the space shuttle program. With all missions of the year heading to the ISS, the year 2001 marked the first time in nearly twenty years, other than the stoppage of flights after the *Challenger* accident in 1986, that no shuttle missions during a year were scheduled solely for scientific purposes. This did not mean that no science was done, however. Although no science-dedicated missions were flown, small experiments rode along with the shuttle missions, and STS-105 delivered the MISSE unit to the ISS. Ironically, an experiment similar to MISSE called the Long Duration Exposure Facility, which was also designed to expose materials to the harsh space environment for about a year, was left stranded in orbit for far longer than originally planned after *Challenger* accident.

See also: Cooperation in Space: U.S. and Russian; International Space Station: Crew Return Vehicles; International Space Station: Development; International Space Station: Living and Working Accommodations; International Space Station: Modules and Nodes; International Space Station: 2001; Russia's Mir Space Station; Space Stations: Origins and Development.

Further Reading

Burrows, William E. *This New Ocean: The Story of the First Space Age.* New York: Random House, 1998. Traces the development of space exploration from the first rockets through the decision to build the International Space Station. The role of the space shuttle and ISS is given in the context of the whole of space exploration.

Caprara, Giovanni. *Living in Space.* Buffalo, N.Y.: Firefly Books, 2000. Discusses the development of ideas of building space stations from the first concepts proposed more than a century ago through the plans for the International Space Station.

Oberg, James. *Star-Crossed Orbits: Inside the U.S.-Russian Space Alliance.* New York: McGraw-Hill, 2002. Taking a more controversial dissenting opinion than that of the other books listed here, the author presents his highly critical assessment of the International Space Station and of the entire NASA cooperative effort with the Russian space agency, suggesting that NASA could have used the money allocated the ISS in a better fashion.

Reichhardt, Tony, ed. *Space Shuttle: The First Twenty Years.* New York: DK, 2002. A collection of short essays by the astronauts themselves about living and working in space.

Raymond D. Benge, Jr.

Space Shuttle Flights, 2002

Date: January to December, 2002
Type of mission: Piloted spaceflight

Five space shuttle flights were made in the year 2002. For the second year in a row, no shuttle flight had scientific research as a primary mission. The first mission was the last Hubble Space Telescope servicing mission, and the other missions were support missions for the International Space Station (ISS).

Key Figures

Scott D. Altman (b. 1959), STS-109 commander
Duane G. Carey (b. 1957), STS-109 pilot
John M. Grunsfeld (b. 1958), STS-109 payload commander
Michael J. Bloomfield (b. 1959), STS-110 commander
Stephen N. Frick (b. 1964), STS-110 pilot
Kenneth D. Cockrell (b. 1950), STS-111 commander
Paul S. Lockhart (b. 1956), STS-111 and STS-113 pilot
Jeffrey S. Ashby (b. 1954), STS-112 commander
Pamela A. Melroy (b. 1961), STS-112 pilot
James D. Wetherbee (b. 1952), STS-113 commander

Summary of the Mission

The first space shuttle mission of 2002 was STS-109, a planned-service mission to the Hubble Space Telescope (HST). The mission began on March 1, 2002, at 11:22:02.021 UTC (6:22 A.M. eastern standard time, or EST), with the launch of the shuttle *Columbia* from Pad 39A at the Kennedy Space Center. The launch had been delayed one day because the forecast temperature on February 28 was too low for an adequate measure of safety. Servicing HST required five extravehicular activities (EVAs) totaling nearly thirty-six hours of spacewalks. The first three spacewalks were largely dedicated to replacing the power systems on the HST. Both solar arrays were replaced with new, higher-capacity arrays, and the Power Control Unit (PCU) was replaced with a new one capable of handling the additional power generated by the new solar panels. The final two EVAs focused on upgrading the HST science instruments. HST was re-leased from *Columbia*'s robotic arm at 10:04 UTC on March 9. *Columbia* landed on Runway 33 of the Shuttle Landing Facility (SLF) at the Kennedy Space Center (KSC) on March 12, 2002, at 09:31:53 UTC, rolling to a stop at 09:33 UTC (4:33 A.M. EST).

Originally scheduled for April 4, 2002, the launch of STS-110, the second space shuttle mission of the year, was scrubbed because of a leak in a liquid-hydrogen vent line on the Mobile Launcher Platform at Pad 39B at KSC. After engineers repaired the leak, the launch was rescheduled for April 8. Software problems, however, caused a hold in the launch countdown until the problems could be corrected. The launch finally occurred at 20:44:18.983 UTC (4:44 P.M. eastern daylight time, or EDT) on April 8, as *Atlantis* lifted off only seconds before the close of the available launch window. The primary objective of the mission was to install onto the International Space Station a truss that contained equipment permitting additional laboratories and solar panels to be attached to the station later. Installation of the truss required four EVAs over several days, with astronauts spending more than twenty-eight hours outside the spacecraft. On the fourth EVA, astronauts also installed electrical equipment as well as lights, handholds, and a platform to assist later EVAs. Astronauts transferred equipment and supplies between *Atlantis* and the ISS, including nearly 50 kilograms of oxygen. After completing its mission, *Atlantis* touched down on SLF Runway 33 at KSC at 16:26:58 UTC and rolled to a stop at 16:28 UTC (12:28 P.M. EDT) on April 20.

STS-111 was the third shuttle mission of the year. Originally scheduled for May 30, it was delayed until June 5 because of weather problems and the need to replace a piece of equipment on the orbiter. The shuttle *Endeavour* finally lifted off from Pad 39A at KSC on June 5, 2002, at 21:22:49.008 UTC (5:23 P.M. EDT). STS-111 was primarily a crew transfer mission, transporting the Expedition Five crew to the ISS and bringing the Expedition Four crew back to Earth. Carried in *Endeavour*'s cargo bay was the Multi-Purpose Logistics Module (MPLM) Leonardo. The MPLM acts as both a cargo carrier and a temporary ISS module. After arriving at the ISS, the space shuttle's crew used the robot arm of the shuttle to attach the MPLM to the ISS. Astronauts unloaded about 3,700 kilograms of supplies and equipment from Leonardo. After unloading the MPLM, the Space Station crew loaded more than 2,100 kilograms of garbage and science experiments aboard Leonardo to be returned to Earth. While the onboard crew were working, other astronauts performed three EVAs to install equipment on the exterior of the ISS and to repair the station's robotic arm. Micrometeoroid shields to be installed later on the Russian-built Zvezda module were unloaded and temporarily stored. *Endeavour* undocked from the ISS on June 15 and prepared to land at KSC; however, weather conditions were unsuitable for landing. Weather did not

The external fuel tank falls toward Earth after it was jettisoned from Atlantis *during STS-110.* (NASA)

After STS-111, Endeavour *takes off from Edwards Air Force Base on a Boeing 747, heading for Kennedy Space Center.* (NASA)

improve over the next several days, and so *Endeavour* was forced to land at Edwards Air Force Base (EAFB) in California. The spacecraft touched down at 17:57:41 UTC and rolled to a stop on Runway 22 at EAFB on June 19, 2002, at 17:59 UTC (10:59 A.M. Pacific daylight time, or PDT).

The fourth space shuttle mission for the year was STS-112, another ISS construction mission. Launch was delayed from the planned October 2 date because of uncertainty of the movement of Hurricane Lili, which was in the Gulf of Mexico at the time. While the hurricane did not threaten the launch from KSC directly, it was feared that it might move in the direction of Houston, Texas, and interfere with Mission Control Center operations at the Johnson Space Center. The launch was rescheduled for October 7, 2002, when the space shuttle *Atlantis* lifted off exactly on schedule from Pad 39B

at 19:45:51.018 UTC (3:45 P.M. EDT). Though the liftoff looked perfect to onlookers, two incidents occurred during the launch that could have been disastrous for the mission. First, one set of explosive bolts that release connections between the shuttle and the launch platform did not fire. Backup explosives fired and released the bolts. The second incident happened 33 seconds into the flight, when a large piece of insulating foam ripped loose from the external fuel tank to strike and damage the Solid Rocket/External Tank Attachment ring. Although the impact did damage to flight hardware, it affected neither the remainder of the launch nor any part of the flight. The primary mission for STS-112 was to install the Crew and Equipment Translation Aid (CETA), a cart used to provide a mobile work platform external to the ISS, and another truss segment. These tasks were completed with three EVAs. Before undocking, *Atlantis* used her thrusters to raise the ISS to a higher orbit. Upon completion of its mission, *Atlantis* landed at KSC on October 18, 2002, at 15:43:40 UTC, rolling to a stop at 15:45 UTC (11:45 A.M. EDT) on Runway 33 of the SLF.

Endeavour lifted off from Launch Pad 39A at KSC to begin STS-113 on November 24, 2002, at 00:49:47.021 UTC (7:50 P.M. EST, November 23). The launch had been delayed because of a leaking oxygen hose in the payload bay of the orbiter, as well as poor weather at the Trans-oceanic Abort Landing sites. One of the primary mission goals of STS-113 was to deliver the P1 Truss to the ISS, which was installed during two EVAs. A third EVA was used to perform other station maintenance. In addition to accomplishing station construction, the mission delivered supplies and equipment to the ISS. The ISS Expedition Six crew rode the shuttle to the station to replace the Expedition Five crew that had been aboard the ISS the past six months. Landing was delayed because of poor weather conditions at both primary and backup landing facilities for four days. Finally, *Endeavour* was able to land on Runway 33 of the SLF at KSC on December 7, 2002, at 19:37:13 UTC, rolling to a stop at 19:38 UTC (2:38 P.M. EST).

Contributions

Though some experiments rode along on the space shuttle missions of 2002, for the second year in a row none of the missions was specifically designed as a research mission. STS-109 was the second mission of the third scheduled servicing cycle for the Hubble Space Telescope. Part of the original third mission had been moved forward to STS-103 in 1999 due to gyroscope failures on HST.

Most of the shuttle missions of the year were to the ISS to further construction of the Space Station. Truss segments were delivered to the ISS on which to attach further components carried on later missions, as well as work platforms to assist in later construction efforts. Much of the emphasis was on adding to the ISS or preparing the ISS to re-

ceive further components in order to complete the initial construction phase by February, 2004.

STS-112 carried a camera on the external fuel tank that recorded images during the launch of *Atlantis*. This camera recorded the large foam piece coming loose and striking flight hardware during launch.

A few experiments did ride along on the year's shuttle missions. As on nearly every piloted spaceflight, the reaction of the human body to microgravity was studied. Other experiments studied some material properties in space. Most of these experiments did not require astronaut activity but were passive payloads along for the ride.

Context

For the third year in a row, science objectives were not a main emphasis for the National Aeronautics and Space Administration (NASA) space shuttle fleet, and for the second year no mission had science as a primary objective. The last dedicated science mission had been STS-99 in 2000. In keeping with visions for the Space Transportation System (STS) in its planning days, the missions of this year were primarily shuttle missions to and from a space station, with an additional mission to repair a satellite. For the first time, every shuttle mission for the year included EVAs.

With the decision several years earlier to go forward with the ISS, NASA knew that a greater proportion of its spaceflight resources would be shifted to station activities. However, ISS was far behind schedule and over budget. NASA administrators were applying pressure to avoid further delays.

It was in this climate that a decision was made not to classify the event of foam shedding and striking hardware on STS-112 as an in-flight anomaly (IFA). Through the history of the space shuttle program, every other major foam-shedding event, and certainly every significant foam impact event, had been designated an IFA. Such a designation would have required that this hazard be addressed before additional missions were flown. However, such delays would have made it virtually impossible to complete ISS by the target date. This decision would prove fatal for the crew of the next shuttle mission, STS-107.

See also: Hubble Space Telescope: Servicing Missions; International Space Station: Crew Return Vehicles; International Space Station: Design and Uses; International Space Station: Living and Working Accommodations; International Space Station: Modules and Nodes; International Space Station: 2002; Russia's Mir Space Station; Space Shuttle Flights, 2001.

Further Reading

Burrows, William E. *This New Ocean: The Story of the First Space Age.* New York: Random House, 1998. Traces the development of space exploration from the first rockets through the decision to build the International Space Station. The role of the space shuttle and the ISS is given in the context of the whole of space exploration.

Caprara, Giovanni. *Living in Space.* Buffalo, N.Y.: Firefly Books, 2000. Discusses the development of ideas of building space stations from the first concepts proposed more than a century ago through the plans for the International Space Station.

Kerrod, Robin. *Hubble: The Mirror on the Universe.* Buffalo, N.Y.: Firefly Books, 2003. Though concentrating on the magnificent pictures sent back by the Hubble Space Telescope, this book also explains the astronomical findings associated with the pictures, and it provides a good description of the Hubble telescope itself.

Oberg, James. *Star-Crossed Orbits: Inside the U.S.-Russian Space Alliance.* New York: McGraw-Hill, 2002. Taking a more controversial dissenting opinion than that of the other books listed here, the author presents his highly critical assessment of the International Space Station and of the entire NASA cooperative effort with the Russian space agency, suggesting that NASA could have used the money allocated to the ISS in a better fashion.

Reichhardt, Tony, ed. *Space Shuttle: The First Twenty Years.* New York: DK, 2002. A collection of short essays by the astronauts themselves about living and working in space.

Raymond D. Benge, Jr.

Space Shuttle Mission STS-107

Date: January 16 to February 1, 2003
Type of mission: Piloted spaceflight

STS-107 was the last flight of the orbiter Columbia. *The spacecraft broke apart during atmospheric reentry over Texas, killing all seven crew members. This was the second major loss to the space shuttle fleet and the third accident for NASA in which crew members were lost.*

Key Figures

Rick D. Husband (1957-2003), commander
William C. "Willie" McCool (1961-2003), pilot
Michael P. Anderson (1959-2003), payload commander
Kalpana Chawla (1961-2003), mission specialist
David M. Brown (1956-2003), mission specialist
Laurel B. Clark (1961-2003), mission specialist
Ilan Ramon (1954-2003), payload specialist

Summary of the Mission

The first of several planned space shuttle missions for 2003, the STS-107 mission began January 16, 2003, with the liftoff of *Columbia* (OV-102) from Pad 39A of the Kennedy Space Center (KSC). STS-107 was a dedicated scientific research mission, the first such mission in three years. As with all space missions, preparations began long before liftoff. In 1997, plans were made for an extended microgravity science mission to be modeled after STS-90, flown in 1998. This mission, designated STS-107, was planned for late 2000. STS-107 was rescheduled for launch in 2001 and was delayed more than a dozen times before finally being scheduled for launch in early 2003.

Preparations began on the orbiter in March, 2001, and *Columbia* was moved to the Vehicle Assembly Building (VAB) at KSC on November 20, 2002. After mating the orbiter with its external fuel tank and the solid rocket boosters, the shuttle was moved on its mobile launch platform to Pad 39A on December 9, 2002, for final launch preparations. The launch countdown proceeded without incident, and *Columbia* lifted off January 16 at 15:38:59.994 UTC (10:39 A.M. eastern standard time, or EST). The liftoff appeared to be routine to observers watching the launch.

All later shuttle missions for the year were to support the International Space Station (ISS). In fact, STS-107 was the last mission the National Aeronautics and Space Administration (NASA) currently had planned as a science mission. Thus, there was great pressure to include various experiments on the flight. In all, there were approximately ninety-two different experiments to be performed, including sixty biomedical experiments, ten physics experiments, ten materials science experiments, seven engineering experiments, and one experiment to study the Sun. Some of the more notable experiments included several studies involving combustion and the Mediterranean Israeli Dust Experiment (MEIDEX), for which Israeli astronaut Ilan Ramon was payload specialist. Many of the biomedical experiments studied the astronauts themselves, but several involved samples grown in space, many of which were to be returned to Earth for a complete study.

To accommodate the large number of experiments, *Columbia* carried the SPACEHAB-Research Double Module (RDM) and Fast Reaction Experiments Enabling Science, Technology, Applications and Research (FREESTAR). Located in the cargo bay, the RDM was an expanded version of space laboratory payloads flown on earlier missions and was accessible from the shuttle's crew module via a tunnel. The RDM provided accommodations for a large number of experiments that required astronaut involvement. FREESTAR was located in the cargo bay and was not accessible to the astronauts. Most of the FREESTAR experiments were self-contained and did not require the astronauts' attention.

Shortly after launch, *Columbia* opened her cargo bay doors and was positioned to fly tail-first around Earth. Meanwhile, a NASA team called the Intercenter Photo Working Group (IPWG) began studies of the STS-107 launch photos. This is a standard procedure used to help evaluate the performance of the launch vehicle. Initial images showed nothing unusual. However, the day after launch, higher-resolution images showed that a large piece of

The crew of STS-107 (from left): David M. Brown, Rick D. Husband, Laurel B. Clark, Kalpana Chawla, Michael P. Anderson, William C. McCool, and Ilan Ramon. All perished when the orbiter Columbia *broke apart during reentry into Earth's atmosphere in February, 2003.* (NASA)

foam insulation had separated from the external fuel tank left bipod ramp 81.9 seconds after liftoff. The bipod ramp shields the upper connector that attaches the External Tank to the orbiter. The IPWG team members were quite alarmed to see video of the foam appearing to strike *Columbia*'s left wing. This was one of the largest pieces of foam ever observed to strike a shuttle during launch. The team was concerned that the impact may have damaged the critical heat-resistant tiles on the wing, which protected the shuttle during reentry into Earth's atmosphere at the end of its mission. The IPWG declared this to be a major event and unanimously agreed that NASA should request the Department of Defense to use its surveillance satellites to take high-resolution images of the *Columbia* while in orbit to assess possible critical damage, as had been done on some earlier flights. Such information would permit flight directors and engineers to develop a plan to deal with the damage if needed. No action was taken at that time on their request for images.

Unknown to any NASA personnel at that time, the Air Force Space Command radar had detected a small object drifting away from *Columbia* the day after liftoff. The unknown object did not pose a hazard to any satellites or spacecraft, and was in a decaying orbit that caused it to reenter the atmosphere two and a half days later, so it was simply assigned a catalog number and forgotten. Meanwhile, astronauts performed routine duties and conducted the scheduled experiments, encountering only minor problems involving environmental systems on the RDM.

A Mission Management Team (MMT), which is supposed to meet every day to review mission operations and issues, monitors every space mission. The MMT for STS-107, chaired by Linda Ham, did not meet over the Martin Luther King, Jr., federal holiday weekend, so the IPWG did not present its findings to the MMT until Tuesday, January 21. However, over the weekend, engineers from the United Space Alliance, a partnership of shuttle contractors Boeing and Lockheed, had become concerned about possible damage to the heat-

resistant tiles. Bob White, the shuttle manager for the alliance, called Mission Control at the Johnson Space Center (JSC) asking if it were possible to get on-orbit images of the potentially damaged area. After the IPWG presented their findings to the MMT, a separate Debris Assessment Team (DAT) was formed to study the possible damage from the foam impact. Almost immediately, they too requested on-orbit images from Mission Control.

Both requests for images were passed to the Air Force, which agreed to take the images. However, when Ham was informed of the requests, she noted that the requests had not gone through the proper channels and that the persons contacting the Air Force did not have the authority to do so. Thus, Ham notified her superiors at NASA to cancel the request for the Defense Department images. When the DAT reported their findings to the MMT, Ham and the MMT members had already decided that the foam strike did not pose a safety risk since previous, though smaller, foam strikes had not resulted in major damage. Mission Control did e-mail commander Husband aboard *Columbia* telling him about the foam incident in case reporters asked about it in a news conference, but the e-mail also told Husband that the MMT had deemed that the foam strike posed no safety-of-flight issues.

On February 1, at the conclusion of its mission, *Columbia*'s crew closed the payload bay doors and fired rockets to bring the shuttle back to Earth. As *Columbia* began to enter the upper atmosphere over the Pacific Ocean, ground crew stood ready at KSC for *Columbia* to land on Runway 33 of the Shuttle Landing Facility. As *Columbia* passed over Nevada, engineers in Mission Control began to see temperature sensors fail in the orbiter's left wing. Autopilot computers on board *Columbia* were simultaneously adjusting for unusual drag on the orbiter's left side. As *Columbia* crossed New Mexico, sensors indicated failures associated with the left landing gear, and observers on the ground reported seeing debris shed from the orbiter. As *Columbia* began to cross over Texas, controllers radioed the orbiter that they were looking at anomalous tire sensor readings. Mission Control received

Mission specialist Kalpana Chawla looks through the SPACEHAB module on January 18, 2002, during STS-107. (NASA)

a partial transmission from Husband, and then all contact was lost. At about that time, *Columbia*'s computers had adjusted flight control systems to their maximum extent and had commanded orbital thrusters to fire in a desperate attempt to hold the orbiter's attitude. Observers near Fort Worth and Dallas watching the shuttle pass overhead heard an unusually extended sonic boom and watched as the shuttle disintegrated overhead at about 14:00 UTC (8:00 A.M. CST). *Columbia* was destroyed and all aboard perished. Pieces of debris were found from Fort Worth, Texas, to near the Texas-Louisiana border. Radar tracked smaller pieces of debris floating downward for much of the rest of the day.

Contributions

Much, but not all, of the scientific data was lost when *Columbia* was destroyed upon reentry. Some experiments, such as MEIDEX, relayed much of the data to the ground during the mission. Other experiments, however, required study of samples grown in space, and these samples were lost. In addition to studies of dust in the Mediterranean, MEIDEX detected dust clouds over the Atlantic for the first time and even found smoke plumes over the Amazon. *Columbia*'s astronauts also captured some of the first images of certain atmospheric electrical disturbances associated with lightning discharges. Much physiological data on the crew members was also relayed to Earth before reentry, allowing for the continued study of how the human body responds to weightlessness.

Investigation of telemetry and debris allowed the *Columbia* Accident Investigation Board (CAIB) to reconstruct the events leading to the demise of the orbiter. The CAIB concluded that there was a breach in the thermal protection system on the leading edge of the left wing, caused by a piece of insulating foam which separated from the left bipod ramp section of the External Tank at 81.7 seconds after launch. The foam is sprayed on the skin of the External Tank to prevent the formation of ice on the tank and to reduce heat flow into the tank. During launch, a large chunk of ice would pose a serious threat to the orbiter. Although this foam is very light in mass, it struck the wing traveling at about 855 kilometers per hour near the vicinity of the lower half of reinforced carbon-carbon (RCC) panel number 8. Just prior to separating from the External Tank, the foam was traveling with the shuttle stack at about 2,500 kilometers per hour. Visual evidence shows that the foam debris impacted the wing approximately 0.161 seconds after separating from the tank. In that time, the velocity of the foam debris slowed from 2,500 kilometers per hour to about 1,645 kilometers per hour. Therefore, the orbiter hit the foam with a relative velocity of about 855 kilometers per hour. In essence, the foam debris slowed down and *Columbia* did not, so *Columbia* ran into the foam. The foam slowed down rapidly because such low-density objects have low ballistic coefficients, which means their speed rapidly decreases when they lose their means of propulsion.

The breach in the RCC panel, close to the orbiter body, permitted hot plasma to enter the wing during reentry. The plasma then caused more tiles to peel away, shedding the initial debris observed from the ground in California and Nevada. The damage to the left wing produced drag that the *Columbia*'s autopilot computer tried to compensate for with increased movement of elevons on the shuttle's wings. The hot gases cut through wiring, leading to the initial sensor dropouts that Mission Control observed. The hot gases continued to burn through the lightweight aluminum wing structures until they burned into the left wheel well, resulting in loss of the landing gear tire pressure sensors. The destruction became so extensive that large sections of the left wing collapsed and the wing separated from the orbiter. Unstable, *Columbia* began an uncontrollable death roll. Somewhere over Texas, at nearly 25 times the speed of sound (Mach 25), aerodynamic forces caused the orbiter to disintegrate. The pressurized crew compartment, however, likely remained intact for an-

other half-minute or so before breaking up from structural and thermal stresses. Investigations of the astronauts' remains indicate that they died of decompression and blunt force trauma when the crew module disintegrated, with heat damage to their bodies coming post mortem. The crew likely did not know that the *Columbia* was even in trouble until the left wing failed.

Context

The decision to build the ISS prompted a change in NASA's thinking about the space shuttle program. For three years, nearly every shuttle mission had been to the ISS, with STS-107 being the first science mission since January, 2000, and the last science mission that NASA had scheduled. *Columbia*, heavier than the other orbiters, was not desirable for ISS missions and thus was used for non-ISS missions. Such missions were not a high priority as they conflicted with ISS construction timetables.

The schedule was important, and throughout MMT meetings, greater concern was placed on how the foam impact might affect scheduling for later missions than on how it might affect *Columbia*. The CAIB faulted NASA culture for such decisions as the canceling of on-orbit imaging requests because the requests had not gone through proper channels, also noting that only five required daily MMT meetings were scheduled for the sixteen-day mission. Furthermore, CAIB faulted a cavalier attitude toward foam-shedding events that had developed among shuttle managers, citing a very similar event that had occurred on STS-112 only two missions earlier.

During the ensuing hiatus in shuttle flights of more than two years, NASA upgraded flight hardware, as well as visual tracking and inspection equipment, to ensure that the STS-114 "Return to Flight" mission would be successful. NASA engineers made dozens of changes to the External Tank design. The "bipod fitting"—which joins it with the orbiter—now excludes using foam and instead relies on electric heaters to keep the area clear.

The orbiters' new imaging equipment makes the most of current consumer photography equipment. The External Tank camera, which is in the underbelly of the orbiter, was replaced with a digital, not a film, camera. With the simplicity and increased speed of a digital system, the image files are transmitted to Earth shortly after the orbiter reaches space.

Once in orbit, the orbiter's visual inspection will continue with the help of a new piece of robotic technology. The Remote Manipulator System (RMS) will include a Canadian-built Orbiter Boom Sensor System. The boom extension houses a camera and laser-powered measuring device that astronauts will use to scan the orbiter's exterior. The boom attaches to the end of the existing robotic arm and doubles its length to 30 meters. The extra length will allow the arm to reach around the spacecraft for the best possible views. With the new boom, astronauts will take a good look at features like the orbiter's leading wing edges.

The orbiters' leading wing edges are outfitted with twenty-two temperature sensors to measure how heat is distributed across their spans. Both wings also have sixty-six accelerometers apiece to detect impacts and gauge their strength and location. The sensors are highly sensitive and take twenty thousand readings per second. This new network of sensors running along the wings provides an electronic nervous system that gives engineers a valuable way to monitor their condition.

See also: Escape from Piloted Spacecraft; Ethnic and Gender Diversity in the Space Program; Space Shuttle; Space Shuttle Flights, 2002; Space Shuttle Activity, 2003-2004; Space Shuttle Mission STS-114; SPACEHAB.

Further Reading

Burrows, William E. *This New Ocean: The Story of the First Space Age.* New York: Random House, 1998. This book traces the development of space exploration from the first rockets through the decision to build the International Space Station. The role of the space shuttle is given in the context of the whole of space exploration.

Cabbage, Michael, and William Harwood. *Comm Check: The Final Flight of Shuttle Columbia.* New York: Free Press, 2004. A well-researched and informative look at the decisions and events of STS-107 written by two veteran space reporters in an easy-to-read style.

Columbia Accident Investigation Board. *Columbia Accident Investigation Report.* Vol. 1. Washington, D.C.: Government Printing Office, 2003. The official government report on the accident. Very thorough, it explains not only the accident but also the context of the mission and the events surrounding STS-107. A reprint was produced in 2004 by Apogee Books of Burlington, Ontario, that includes additional documents and images related to STS-107.

Joels, Kerry Mark, Gregory P. Kennedy, and David Larkin. *The Space Shuttle Operator's Manual.* New York: Ballantine Books, 1982. Though written in the early days of the space shuttle program, this work, aimed at the general public, still gives the reader an excellent feeling for the events involved in launch, operation, and landing of a space shuttle, including an overview of the spacecraft itself.

Raymond D. Benge, Jr., with Russell R. Tobias

Space Shuttle Activity, 2003-2004

Date: January, 2003, to December, 2004
Type of mission: Piloted spaceflight

The only space shuttle flight in 2003 ended in tragedy when the shuttle Columbia *disintegrated upon reentry into Earth's atmosphere on February 1, 2003. All shuttle flights were postponed until the* Columbia *Accident Investigation Board could determine the cause of the accident and NASA could fix the design flaws leading to the accident. Space shuttle activity during 2003-2004 consisted primarily of working to ensure that the* Columbia *accident would never be repeated.*

Key Figures

Rick D. Husband (1957-2003), STS-107 commander
William C. "Willie" McCool (1961-2003), STS-107 pilot
Michael P. Anderson (1959-2003), STS-107 payload commander
Kalpana Chawla (1961-2003), STS-107 mission specialist
David M. Brown (1956-2003), STS-107 mission specialist
Laurel B. Clark (1961-2003), STS-107 mission specialist
Ilan Ramon (1954-2003), STS-107 payload specialist

Summary of the Activity

On January 16, 2003, the *Columbia* space shuttle launched on flight STS-107, seemingly without incident. The crew consisted of Rick D. Husband, William C. (Willie) McCool, Michael Anderson, Kalpana Chawla, David M. Brown, Laurel B. Clark, and Ilan Ramon. During liftoff, a piece of foam insulating material about the size of a briefcase broke off from the left bipod ramp section of the external fuel tank and struck the leading edge of the left shuttle wing. This small piece of debris was considered insignificant at the time. The mission proceeded normally until February 1, 2003. Then, while flying 19,000 kilometers per hour at an altitude of nearly 75 kilometers, *Columbia* disintegrated on reentry. Pieces of the shuttle were scattered over a 5,000-square-kilometer area of Texas and Louisiana. The entire crew perished. Fortunately, none of the 40 metric tons of falling debris killed anyone on the ground; there were, however, a few close calls. This tragedy reshaped the course of the American piloted spaceflight program.

Within hours, a massive search effort was mounted in Texas and Louisiana to find the pieces of the ill-fated *Columbia*. The search effort lasted more than three months and involved as many as 25,000 people from 450 organizations. Many but not all pieces were found. The pieces found ranged from the size of postage stamps to as large as 400 kilograms. Pieces struck the Earth as fast as 1 kilometer per second and in some cases finally came to a stop only after penetrating several meters into the ground. Accident investigators then began to sift through the pieces, searching for clues to the cause of the accident.

All shuttle flights were postponed until the *Columbia* Accident Investigation Board (CAIB) could determine the cause of the accident and recommend steps to prevent a recurrence. In the meantime, all resupply and crew rotation missions for the

International Space Station were completed by Russian Progress and Soyuz flights. The board's report, filed in August, 2003, identified both the immediate physical cause of the accident and a complex chain of political, budgetary, and management causes. The report also prescribed specific actions to fix both the immediate and root causes of the accident.

That seemingly insignificant 750-gram piece of foam insulation, seen by video cameras that were monitoring the launch, turned out to be the culprit. The external fuel tank, filled with liquid oxygen and hydrogen propellants, is very cold and, therefore, covered with foam insulation to prevent ice from forming. Just 81.7 seconds after liftoff, a piece of this foam insulation broke off, slammed against the left wing, and damaged the wing's reinforced carbon-carbon (RCC) thermal insulating panels. Not considering this event serious, officials at the National Aeronautics and Space Administration (NASA) did not ask military surveillance satellites to examine the shuttle for possible damage. In any case, they did not have the capability of fixing any damage while in space. The damage caused no problem during the orbital portion of the mission. On reentry, however, the damaged panels were unable to do their job of protecting the interior of the shuttle from the extreme heat of reentry. The heat entered the wing, melting the wiring and aluminum structural spars in the wing. The weakened wing then failed, causing the shuttle to lose control and break apart.

The prescription: Redesign the external fuel tank to reduce the possibility of foam and other debris breaking off; develop a way to inspect the exterior of the shuttle for damage and to repair any damage found while in orbit; photograph the shuttle in orbit during all flights; increase the photographic monitoring of shuttle launches; and for the long term, redesign the shuttle's exterior to make the thermal insulating system more resistant to impacts. The International Space Station was earmarked to function as a safe haven when necessitated by shuttle damage.

NASA implemented these recommendations for its Return-to-Flight mission, STS-114, launched on July 26, 2005. Some of the foam insulating the fuel tank was replaced by heating elements to prevent ice buildup. Additional cameras with better resolution were added on the ground, aircraft, and the shuttle itself to monitor the debris flow during launch. NASA designed a 15-meter-long Orbiter Boom Sensor System to inspect the shuttle's exterior from orbit. NASA engineers also designed a procedure to repair the external thermal insulating tiles while in orbit. During the two-week mission, crew members Steve Robinson and Soichi Noguchi ventured outside the shuttle three times on spacewalks. The two demonstrated repair techniques on the shuttle's protective tiles.

In addition to the physical cause of the accident, the *Columbia* Accident Investigation Board's report cited a long and complex chain of political, budgetary, and management causes of the accident. This chain dates back to the shuttle's inception during the presidencies of Richard M. Nixon and Lyndon B. Johnson. Neither of these presidents was interested in grand plans for a permanent human presence in space and further exploration of the Moon and Mars. Hence political pressures constrained NASA's budget, reducing the capabilities originally planned for the shuttle. The board stated that the shuttle never reached its full operational status, despite the myth that it had. The report also criticized NASA management. For example, NASA's upper management was unaware of engineering concerns about the foam hitting the wing and was unconcerned about foam strikes that had occurred on previous missions, despite a safety requirement that no debris hit the shuttle.

The board therefore recommended changes to NASA management, including the Mission Management Team (MMT) and the safety organization. For example, the MMT will be required to meet daily to ensure that engineering and safety concerns are heard and to receive annual training for dealing with problems. The MMT met only five times during the sixteen-day *Columbia* mission. NASA also began to examine many safety procedures, such as waivers for shuttle requirements that often affect safety.

Debris from the STS-107 accident: an 800-pound powerhead from the main engine buried in mud. Parts of the shuttle were found from Fort Worth, Texas, to the Texas-Louisiana border. (NASA)

Dominated by these concerns, space shuttle activity for 2003 and 2004 was limited to one flight that ended in tragedy, followed by an investigation into the causes of the accident and the considerable work required to fix the identified problems before returning to flight.

Contributions

Most of the knowledge gained during this time period concerned procedures to increase safety during shuttle flights. NASA flight management thought a relatively small piece of debris hitting the shuttle would not cause significant damage. The lesson that the debris from launch could damage the thermal protection panels unfortunately came at a tragically high price.

To prevent a repetition of the same accident, NASA developed techniques to examine and repair the shuttle's external thermal protection system during flight. Developing these new safety procedures required considerable computer analysis of the debris flow striking the shuttle during liftoff. NASA learned that its computing power was inadequate to do the task properly. As a result, NASA started a program to upgrade its computing ability for solving complex scientific and engineering problems. Still, procedures that theoretically should work often do not work in practice. During the Return-to-Flight (RTF) STS-114 mission in July, 2005, these procedures were put into practice. Visual and electronic examination of *Discovery*'s critical heat-protection areas and inflight video of the external tank during launch revealed minor foam loss and minimal damage to the underbody tile area. A second RTF mission, STS-121, was scheduled for 2006. Part of its mission would be to repeat the tests and observations conducted by STS-114.

The *Columbia* Accident Investigation Board was harshly critical of NASA management, but with the criticism came lessons on improving management policies. A primary lesson here is that the Mission Management Team must listen to safety concerns raised by the engineering teams. In addition, the board's report pointed out that the shuttle had never reached the fully operational status envisioned in the design stage. Factors involved in this status include number of flights, flight turnaround time, and cost-effectiveness as well as safety issues. The accident and subsequent investigation drove home the lesson that the shuttle fleet is getting old and was built with technology available in the 1970's and suggested that it is time to retire and replace the space shuttles with a more modern fleet.

The large amount of possibly dangerous debris from *Columbia*, falling over a wide area of Texas and Louisiana, presented an unprecedented problem in emergency management. More than 450 organizations had to learn very quickly to cooperate with each other in order to deal with the problem of finding all the debris and ensuring the public safety. Such agencies can rehearse for common natural disasters, but this was neither anticipated nor rehearsed. Emergency management experts have gained insight into interagency cooperation, from the national to the local level, by examining the *Columbia* emergency.

Context

For NASA's space shuttle program, 2003 and 2004 were rebuilding years. Whenever any human endeavor suffers a tragic accident, it is necessary to reexamine goals and priorities as well as to learn from the tragedy to avoid a repeat. Like the Apollo accident in 1967 and the *Challenger* accident in 1986, the *Columbia* accident prompted a close examination of both immediate and systemic causes of the tragedy and, it is hoped, means to correct them. All three of these accidents reinforce the fact that spaceflight is a very risky business. As stated in the *Columbia* Accident Investigation Board's report:

> *Columbia*'s failure to return home is a harsh reminder that the space shuttle is a developmental vehicle that operates not in routine flight but in the realm of dangerous exploration.

The *Columbia* accident reignited the debate on the value of piloted spaceflight. Is piloted spaceflight worth the cost and risk? Throughout history, humans have always explored the unknown. Exploration is always risky and many have lost their lives willingly while taking these risks.

See also: Escape from Piloted Spacecraft; Ethnic and Gender Diversity in the Space Program; Funding Procedures of Space Programs; National Aeronautics and Space Administration; Rocketry: Modern History; Space Shuttle; Space Shuttle Flights, 2002; Space Shuttle Mission STS-114.

Further Reading

Cabbage, Michael, and William Harwood. *Comm Check: The Final Flight of Shuttle Columbia.* New York: Free Press, 2004. This work is devoted to the only shuttle flight during 2003-2004. Chapter 11 is about returning to flight after the *Columbia* accident and the work required for this return.

Donahue, A. "The Day the Sky Fell: The Space Shuttle Columbia Disaster." *Public Management* 86, no. 8 (2004): 8. This article examines the emergency management aspects of the agencies on the ground in Texas and Louisiana that worked to recover the pieces of the shuttle and to protect the public safety.

Godwin, R. *Columbia Accident Investigation Report.* Burlington, Ont.: Apogee Books, 2003. This Apogee edition of the accident investigation report not only reproduces the official government report but also contains a CD-ROM not found in the government edition. Reading the original report gives insight into the cause of the *Columbia* accident and the extra safety components developed during 2003 and 2004. This reprint includes additional documents and images related to STS-107.

Jenkins, Dennis R. *Space Shuttle: The History of the National Space Transportation System—The First 100 Missions.* 3d ed. Cape Canaveral, Fla.: Author, 2001. This book is a well-written history of the space shuttle. The third edition eliminates some history of the development found in the second edition but nonetheless provides comprehensive background on the space shuttle, placing the activities of 2003-2004 in context.

World Spaceflight News. *2003 Space Shuttle Columbia Tragedy: NASA Plan for Space Shuttle Return to Flight, Response to the Columbia Accident Investigation Board (CAIB). Report by the Gehman Board.* Progressive Management, 2003. The rather lengthy title says it all. Most of the shuttle activity in 2003 and 2004 was in response to the *Columbia* tragedy.

Paul A. Heckert

Space Shuttle Mission STS-114

Date: Summer, 2005 (projected)
Type of mission: Piloted spaceflight

STS-114, originally scheduled to fly in March, 2003, on Atlantis, *was delayed by the February, 2003, Columbia reentry accident. The mission, rescheduled for 2005 on board* Discovery, *represented NASA's return to piloted spaceflight. The payload module contained supplies and equipment for the International
Space Station (ISS), and the mission made repairs to the ISS and test new safety features.*

Key Figures

Eileen M. Collins (b. 1956), mission commander
James M. Kelly (b. 1964), pilot
Stephen K. Robinson (b. 1955), flight engineer
Charles J. Camarda (b. 1952), mission specialist
Wendy B. Lawrence (b. 1959), mission specialist
Soichi Noguchi (b. 1965), mission specialist
Andrew S. W. Thomas (b. 1951), payload specialist

Summary of the Mission

The STS-114 mission was originally scheduled
to fly during March, 2003, on the space shuttle
Atlantis. It carried an Italian-made Raffaello Multi-
Purpose Logistics Module to the International
Space Station (ISS). In addition, STS-114 rotated
the ISS crew by delivering Astronauts Aleksandr
Kaleri, Edward T. Lu, and Yuri Malenchenko (Expedition Seven) to the Space Station and returning
Kenneth Bowersox, Nikolai Budarin, and Donald
R. Pettit (Expedition Six) to Earth. Eileen M. Collins, James M. Kelly, Soichi Noguchi, and Stephen
K. Robinson made up the originally scheduled
crew. From October 18, 2002, when *Atlantis* landed
after the STS-112 mission, until January 31, 2003,
preparations of *Atlantis* for STS-114 proceeded
normally.

On February 1, the *Columbia* orbiter on STS-107
disintegrated during reentry, killing the entire
crew. In the wake of the accident, all shuttle flights,
including STS-114, were postponed until the cause
of the accident could be determined and fixed.

With the shuttle fleet grounded, Russian Soyuz
spacecraft were used for crew transfers and Progress spacecraft for resupplying the ISS.

Shuttle flight STS-114 was both postponed and
modified. For greater safety, launch windows are
constrained by the additional requirement that
both the launch and fuel tank separation occur in
daylight. In October, 2004, officials at the National
Aeronautics and Space Administration (NASA)
postponed STS-114 from the projected launch-
planning window in March, 2005, to a launch window extending from May 12 to June 3, 2005. Four
hurricanes struck Florida in rapid succession during the 2004 hurricane season and hindered
launch preparations.

Discovery was designated to replace *Atlantis* for
flight STS-114, with some of the mission tasks transferred to a newly added flight, STS-121, on *Atlantis.*
The transfer allowed STS-114 to perform tasks related to testing new safety systems. In addition,
STS-114 did not carry a replacement crew to the

ISS. Hence, there was room for three additional crew members to help perform these new tasks: Mission Specialists Charles J. Camarda, Wendy B. Lawrence, and Andrew S. W. Thomas.

Many of the modifications to flight STS-114 were related to new safety requirements and testing new safety equipment. The *Columbia* Accident Investigation Board had concluded that there was a breach in *Columbia*'s Thermal Protection System on the leading edge of the left wing, caused by a piece of insulating foam that separated from the left bipod ramp section of the External Tank at 81.7 seconds after launch. Although this foam is very light in mass, it struck the wing traveling at

about 855 kilometers per hour near the vicinity of the lower half of Reinforced Carbon-Carbon (RCC) panel number 8. The breach in the RCC panel, close to the orbiter body, permitted hot plasma to enter the wing during reentry. The hot gases continued to burn through the lightweight aluminum wing structures, large sections of the left wing collapsed, and the wing separated from the orbiter.

The daylight launch requirement allowed better monitoring of similar debris during launch. To reduce the amount of debris hitting the shuttle during launch, some of the foam insulating the fuel tank, in the region where it attaches to the

The "return to flight" crew of STS-114 (from left): Stephen K. Robinson, James M. Kelly, Andrew S. W. Thomas, Wendy B. Lawrence, Charles J. Camarda, Eileen M. Collins, and Soichi Noguchi. (NASA)

shuttle, was replaced with heating elements to prevent ice from forming on the cold liquid-fuel tank. The modified shuttle also included sensors on the wings and several new cameras to monitor the debris striking the shuttle during launch. For the first two Return-to-Flight missions, STS-114 and STS-121, NASA decided to use, on an experimental basis, ground-based radar and two jets to monitor the debris flow. The jets also monitored the reentry. To ensure that there would be no external damage, the crew was assigned to spend the second day of the mission inspecting the external wing and nose structure via the newly designed 30-meter-long Orbiter Boom Sensor System. During the docking with the ISS, space station crew members photographed the underside of the shuttle to look for possible damage.

In the event of external damage to the shuttle on this or future flights, procedures for repairing the thermal protection tiles were developed by NASA. Crew members Stephen K. Robinson and Soichi Noguchi were assigned to test these tile repair procedures during the first of three planned extravehicular activities (EVAs, or spacewalks) on day five of the mission.

The first Return-to-Flight mission began with a near-perfect launch on July 26, 2005, at 14:39:00.07 Coordinated Universal Time (UTC) or 10:39 A.M. eastern daylight time) from the Kennedy Space Center's Launch Complex 39B. Launch video taken from a digital camera mounted on the External Tank revealed two "debris events." One image showed what appeared to be a small fragment of tile coming from *Discovery*'s underside, on or near the nose gear doors. A later image, taken about the time of solid rocket booster separation, showed an unidentified fragment separating from the tank without striking the orbiter. The crew was notified of these observations and told that imaging experts would be analyzing the pictures.

On day two of the mission, *Discovery* crew members completed a camera survey of the heatshields (on the leading edges of the orbiter's wings) and the orbiter's nose cone using the Orbiter Boom Sensor System (OBSS) laser scanner. Preparations for

docking included an investigation of rendezvous tools and the extension of the Orbiter Docking System ring, which would make first contact with the ISS.

Discovery, which was the first shuttle to visit the International Space Station since late 2002, linked with the ISS's orbiting laboratory over the southern end of the Pacific Ocean, west of the South American coast, on July 28. The approach included the first rendezvous pitch maneuver, a slow backflip performed about 200 meters below the station immediately before docking. After the initial hugs and handshakes, Station Commander Sergei Krikalev gave a safety briefing for the new arrivals. Among early tasks for the joint crews was preparation for additional robotic arm surveys of the orbiter. Kelly and Lawrence, with help from Station Science Officer John Phillips, used the station's Canadarm2 to lift the Orbiter Boom Sensor System from *Discovery* and hand it to the shuttle arm. Camarda and Thomas steered the shuttle arm, which cannot grasp the boom directly while the station lies in its way.

Shuttle and station crew members installed the Raffaello Multi-Purpose Logistics Module and began unloading the pressurized cargo carrier on the fourth day of the mission. They also carried out a survey of selected areas of *Discovery*'s Thermal Protection System (TPS) and continued preparations for the next day's spacewalk.

Astronauts Steve Robinson and Soichi Noguchi of the Japan Aerospace Exploration Agency (JAXA) made a successful 6-hour, 50-minute spacewalk on Saturday, July 30, completing a demonstration of TPS repair techniques and enhancements to the station's attitude control system. For the repair demonstration, they worked with tiles and RCC intentionally damaged on Earth and brought into space in *Discovery*'s payload bay. They tested an Emittance Wash Applicator for tile repair and Non-Oxide Adhesive Experimental (NOAX) for the RCC samples. They also installed a base and cabling for a stowage platform and rerouted power to the ISS's Control Moment Gyroscope-2 (CMG-2), one of four 600-pound gyroscopes that control the orientation of the station in orbit. *Discovery*'s heat-protective tiles and thermal blankets were pronounced fit for entry after engineers reviewed the imagery and other data to judge their health. About 25 dings have been seen on *Discovery*, compared to a mission average of 145 in missions before the loss of *Columbia*.

Mission managers also decided to extend the STS-114 mission by one day to spend more time docked with the station. Astronauts transferred additional water and supplies to the ISS in case the next mission (STS-121) was delayed. During the sixth and seventh days of the mission, approximately six metric tons of hardware and equipment were moved from *Discovery* to the station. Just over three and a half tons of material, including experiment packages and trash, would return to Earth aboard *Discovery*.

On Sunday, July 31, Robinson and Noguchi replaced a 275-kilogram gyroscope on the ISS, giving the orbiting laboratory a complete, functional set of four gyroscopes. These Control Moment Gyros, or CMGs, maintain the station's orientation in space. The 7-hour and 14-minute spacewalk began at 8:42 UTC. After leaving the *Discovery* airlock, Noguchi and Robinson made their way, hand over hand, to the station's Z1 Truss (atop the Unity Node), where the four CMGs are housed. There the spacewalkers removed CMG-1, which had failed in June, 2002, and installed its replacement. Later, controllers on the ground began the process of spinning the CMG to normal operating speed of 6,600 revolutions per minute.

On Wednesday, August 3, astronaut Steve Robinson made the first-ever inflight repair of a shuttle orbiter. Despite—or perhaps because of—days of anticipation and intense planning, Robinson made it look easy as he gently pulled two protruding gap fillers from between thermal protection tiles on *Discovery*'s underside. Gap fillers are used in areas to restrict the flow of hot gas into the gaps between Thermal Protection System components. They consist of a layer of coated Nextel fabric and normally have a thickness of about 0.5 millimeter. Pilot Jim Kelly worked with Mission Specialist

Charlie Camarda on an inspection of the repair demonstration tiles in *Discovery*'s cargo bay. Using the Orbiter Boom Sensor System, they looked at tiles brought up for an experimental repair by Robinson and Noguchi on the mission's first spacewalk.

On mission day ten, *Discovery* and International Space Station crew members delivered a moving tribute to members of the *Columbia* crew and other astronauts and cosmonauts who lost their lives exploring space. Each crew member wore a red shirt with *Columbia*'s STS-107 mission patch and spoke as the docked spacecraft flew over the southern end of the Indian Ocean, approaching a sunset. Phillips said, "To the crew of *Columbia*, as well as the crews of *Challenger*, Apollo 1, Soyuz 1 and 11, and to those who have courageously given so much, we now offer our enduring thanks." Mission Specialist Soichi Noguchi repeated Phillips's words in Japanese, and Station Commander Sergei Krikalev translated them into Russian.

Using the station arm, the *Discovery* astronauts undocked the Raffaello Multi-Purpose Logistics Module from the station's Unity Node on August 5 and reberthed it in the shuttle's cargo bay. After Raffaello was secured in *Discovery*'s cargo bay, Camarda and Thomas used the shuttle arm to hand off the Orbiter Boom Sensor System to the station arm. Lawrence and Kelly reberthed the OBSS in its position on the starboard sill of the cargo bay.

After more than a week of working together in space, the *Discovery* and International Space Station crews bid each other farewell during a ceremony on August 7, at 04:36 UTC. Hatches between the spacecraft were closed thirty minutes later. Pilot Jim Kelly was at the controls as latches between the two vehicles were released, and *Discovery* began to back gently away from the station. Undocking occurred at 07:24 UTC as the two spacecraft flew high over the Pacific Ocean, west of Chile. As *Discovery* moved away to a distance of about 125 meters, Kelly began a slow fly-around of the station. Cameras on each spacecraft captured video and still images of the other. After the fly-around, Kelly

Rollout of the shuttle Discovery *toward Pad 39B at Kennedy Space Center.* (NASA)

executed the first of two separation burns to move *Discovery* away from the station and begin its trip home. Onboard the station, Expedition Eleven crew members Krikalev and Phillips returned to their normal schedule.

Discovery's first two landing opportunities to Florida were waved off on Monday, August 8, due to unpredictable cloud cover at the landing site. Persistent thunderstorms the following day resulted in a second wave-off of two opportunities to return to Kennedy Space Center. Flight controllers decided to land *Discovery* in California to avoid any further delay in completing the mission.

Commander Collins and Pilot Kelly, assisted by Mission Specialist Robinson, began *Discovery*'s return to Earth by firing the spacecraft's orbital maneuvering system engines to slow its speed and begin its descent. *Discovery*'s ground track took it over the western Indian Ocean, around Australia,

northeast across the Pacific, across the California coast north of Los Angeles, and then to Edwards Air Force Base. *Discovery* glided to a landing at Edwards just before dawn on Tuesday morning, August 9, concluding a journey of 9.3 million kilometers. It touched down at 12:11:22 UTC. The landing marked the sixth night landing at Edwards Air Force Base and the fiftieth time that a shuttle concluded its mission in the California desert. The flight, which lasted thirteen days, 21 hours, 32 minutes, and 22 seconds, was considered a complete success, despite the loss of insulating foam on the External Tank and minor damage to *Discovery*'s heatshield. The orbiter returned to the Kennedy Space Center atop a modified Boeing 747, called the Shuttle Carrier Aircraft, on August 21. It was taken to the Orbiter Processing Facility to be readied for mission STS-121, which was scheduled for 2006.

Contributions

STS-114 carried important equipment and supplies to continue the construction of the International Space Station. The external storage platform will store equipment needed future assembly and wiring of solar panels to provide electrical power to the station. The Multi-Purpose Logistics Module also carried badly needed supplies and equipment to the Space Station. The supplies are especially important because the Space Station crew began to run low on supplies during late 2004 and early 2005.

The mission also retrieved experimental packages containing materials that had been exposed to space. Upon their return to Earth, analysis of these materials could reveal how various materials react to the harsh environment of space. Such knowledge is important for engineers designing equipment to be used in space.

The most important result of STS-114, however, came from the tests of the new safety features. Analysis of the performance of new camera and sensor systems, tested for the first time on this mission, allowed NASA engineers to improve their designs to increase the safety of future missions. The additional monitoring by these systems of debris flow during the launch phase of the mission told NASA that additional design modifications were needed to reduce the flow of debris during launch.

The mission also determined that a tile repair system could be used, in the event that tile damage was detected. The highest-priority EVA during STS-114 tested the tile-repair procedures. The results will guide NASA engineers seeking to improve the tile repair process.

Context

The STS-114 mission was originally scheduled to fly about a month after *Columbia* completed the STS-107 mission in 2003. After the accident occurred, all shuttle flights were postponed until the *Columbia* Accident Investigation Board could determine the cause of the accident and recommend the changes needed to prevent a recurrence. NASA also needed time to implement the recommenda-

tions, and an unusually heavy hurricane season in 2004 slowed the work. Hence, STS-114 became the first shuttle flight in more than two years.

This special status placed an extra burden on the mission and crew of STS-114. While the shuttles were grounded, Russian Soyuz craft were the only flights available to rotate station crew and ferry supplies. By late 2004, station supplies had started to run low. The resupply mission therefore became especially important.

In addition to the original mission of ferrying equipment and supplies to the International Space Station, STS-114 assumed the mission of testing and implementing new safety features, and considerable crew time was spent inspecting the exterior tiles using equipment previously untested in real orbital conditions. Finally, STS-114 was the first shuttle mission to have last-resort contingency plans of using the Space Station as a safe haven in the event of an emergency.

After the *Challenger* accident in 1986, NASA also suspended shuttle flights until the accident cause could be determined and fixed. As the first flight after a major accident, STS-114 shared many characteristics with *Discovery*'s September, 1988, flight, STS-26, which was the first after the *Challenger* accident. Once again, *Discovery* was designated to lead NASA's return to space after a major setback.

The completion of all mission objectives and an in-flight repair made STS-114 as successful a mission as NASA could hope to attain. A pair of NASA engineering "Tiger Teams" investigated the External Tank's foam loss during *Discovery*'s launch. The teams identified the major areas of concern and made good progress on dealing with the problems.

The STS-121 crew would continue to test new equipment and procedures that increase the safety of space shuttles. They would also deliver more supplies and cargo for future Station expansion. This, the second Return-to-Flight test mission, would continue the analysis of STS-114's findings on safety improvements and would build upon those tests. STS-121 would bring a third crew member to the Station, European Space Agency Astronaut Thomas Reiter. His arrival would create the first three-person crew since the Expedition 6 crew returned to Earth on May 4, 2003. After the *Columbia* accident, the absence of vehicles needed to ferry equipment to the Station meant that only two people could be supported on board.

See also: International Space Station: 2004; National Aeronautics and Space Administration; Space Shuttle Flights, 2002; Space Shuttle Mission STS-107; Space Shuttle Activity, 2003-2004.

Further Reading

Cabbage, Michael, and William Harwood. *Comm Check: The Final Flight of Shuttle Columbia.* New York: Free Press, 2004. Chapter 11 is about returning to flight after the *Columbia* accident. This book is based in part on interviews with people involved, including interviews with the crew of flight STS-114.

Godwin, R. *Columbia Accident Investigation Report.* Burlington, Ont.: Apogee Books, 2003. This Apogee edition of the accident investigation report not only reproduces the official government report but also contains a CD-ROM not found in the government edition. Reading the original report gives insight into the cause of the *Columbia* accident and the extra safety components of the STS-114 mission.

Jenkins, Dennis R. *Space Shuttle: The History of the National Space Transportation System—The First 100 Missions.* 3d ed. Cape Canaveral, Fla.: Author, 2001. This book is a well-written history of the space shuttle. The third edition eliminates some history of the development found in the second edition but nonetheless provides comprehensive background on the space shuttle.

World Spaceflight News. *2003 Space Shuttle Columbia Tragedy: NASA Plan for Space Shuttle Return to Flight, Response to the Columbia Accident Investigation Board (CAIB). Report by the Gehman Board.* Progressive Management, 2003. The rather lengthy title says it all. Covers NASA's plans for the return to space, including the STS-114 mission.

Paul A. Heckert, updated by Russell R. Tobias

Space Stations: Origins and Development

Date: Beginning March, 1969
Type of program: Space station, piloted spaceflight

A space station is a structure in space large enough to allow long-term human habitation. The term "space station" also implies a mother ship, able to receive and send smaller craft.

Summary of the Technology

Early space station concepts appeared in the work of Edward Everett Hale's "The Brick Moon" (1869) as a thirty-seven-crew, 60-meter-diameter navigational satellite and in Konstantin Tsiolkovsky's descriptions of space stations in *The Exploration of Cosmic Space by Means of Reaction Motors* (1903). Hermann Oberth described a launch station to the Moon and Mars in his book *Rocket into Interplanetary Space* (1923), and Herman Noordung gave a detailed description of an orbiting space observatory in *The Problem of Space Travel: The Rocket Motor* (1929). Serious studies started with Sputnik developer Sergei Korolev, who proposed an orbiting military station made of rocket parts (1963), which became the Almaz station, and Wernher von Braun, who published his concept of a space station in a *Collier's Weekly* magazine series of articles entitled "Man Will Conquer Space Soon."

The problems of building and living on Earth-orbiting stations are similar to those of ships for deep space travel, except that the latter need large engines and fuel. Early thinkers chose spinning wheel-and-spoke designs to simulate gravity. Building or launching massive radiation-shielded stations remained impractical.

Designers focused on low-Earth-orbit (LEO) stations to answer the important questions for humans in space: the effects and uses of microgravity, human adaptation, and developing reliable space hardware. LEO stations enable emergency escape and real-time control from Earth, unlike craft in deep space. To avoid in-space assembly, early stations were preassembled modules that fit on large launch vehicles.

The USAF Manned Orbiting Laboratory (MOL, also known as Dorian and KH-10) was to be a 21.92-meter-long, 3.05-meter-diameter, 11.3-cubic-meter module with a mass of 14,476 kilograms that could carry a 2,760-kilogram payload and two astronauts in shirtsleeve comfort for stays of forty days in LEO. One mock-up flight (November 3, 1966) of a modified Titan II propellant tank released three satellites and a Gemini 2 module for a 33-minute suborbital test flight. Budgetary cuts and developmental cost overruns caused the MOL to be canceled in 1969.

Orbit rendezvous and docking, prerequisites for stations, were demonstrated by the American Gemini VIII-Agena rendezvous (with Neil Armstrong, David Scott on March 16, 1966); the Soviet rendezvous of Soyuz 4 and Soyuz 5 (with Vladimir Shatalov on Soyuz 4 and Boris Volynov, Alexei Yeliseyev, and Yevgeny Khrunov on Soyuz 5) in January, 1969; and the Apollo-Soyuz mission in 1975. Vladimir Komarov died on April 24, 1967, at the end of the first Soviet docking mission, Soyuz 1-2, when Soyuz 1's parachutes failed. The first mothership function in Earth orbit was performed by Apollo 9 (James McDivitt, Russell Schweickart, and David Scott) in March of 1969, and in lunar orbit by Apollo 10 (Thomas Stafford, John Young, and Eugene Cernan) in May, 1969, the latter sending two astronauts in Lunar Modules on missions away from the Command and Service Module and back.

The first orbiting station was the 15-meter-long, three-compartment Soviet Almaz ("diamond" in Russian) orbiting piloted station. It was designated Salyut 1. (*Salyut* is Russian for "salute"; it was scheduled to be placed into orbit on the tenth anniversary of Yuri A. Gagarin's flight as a "salute" to his accomplishment as the first man in space.) Salyut 1 was launched as a unit on a Proton on April 19, 1971. The Almaz, planned by Vladimir Chelomey's design bureau to compete with the U.S. MOL, was to include the station, return capsules, a supply ship, and a rapid-fire cannon. Almaz was the only piloted military space station ever actually flown. With the MOL canceled, the Almaz was merged with Korolev's Soyuz program. The first three-person crew returned six hours after docking without occupying Salyut 1. The second crew stayed for twenty-four days, conducting experiments. Georgi

Dobrovolsky, Vladislav Volkov, and Viktor Patsayev died on June 24, 1971, when their reentry module lost pressure. Salyut 2 deorbited in two weeks after an engine explosion (May, 1973), as did Kosmos 557, a military station (May, 1973).

The first American space station, Skylab, was also launched as a single unit atop a Saturn V to a 435-kilometer orbit on May 14, 1973. It was a modified Saturn S-IVB third stage, 17 meters long and 6.5 meters in diameter. Its 85,000-kilogram mass included a 10,000-kilogram solar observatory. Skylab's thermal protection was damaged during launch, and a solar panel failed to deploy, reducing power and raising inside temperature. The first crew—Charles Conrad, Joseph Kerwin, and Paul Weitz—launched on May 25. They opened an improvised parasol to shade the craft and deployed the solar panel during a spacewalk. Skylab was a

Alan Chinchar's rendition of Space Station Freedom in orbit (1991). (NASA)

successful laboratory. The second and third crews stayed for fifty-eight and eighty-four days respectively, through February, 1974. Lack of independent reboost or refueling, delays in the shuttle program, and air drag forced Skylab to reenter Earth's atmosphere and burn up on July 11, 1979.

The Soviets accumulated knowledge on human adaptation to space travel through a series of Salyut-Soyuz missions. Salyut 6's two access ports left a Soyuz emergency escape ship attached while docking Soyuz craft or Progress cargo ships (1978). These transitioned to the Mir ("peace") Space Station, the first to be assembled from multiple modules and to incorporate a robot construction and docking arm. The Mir Core Assembly had living quarters, a transfer compartment, an intermediate working module, and an assembly bay. Six modules were added by 1996. These were Kvant 1 (astrophysics); Kvant 2 (science) for biology, Earth observations, and spacewalks; Kristall (technological) for biological and materials experiments; a shuttle-compatible docking port; Priroda (remote sensing); and a docking module. Mir orbited for fifteen years beginning in February, 1987, and cosmonaut endurance was now limited by lifetime radiation exposure rather than microgravity adaptation. Valery Polyakov set the record with 438 days in 1994-1995.

With Russian participation in the International Space Station, Mir became an equipment test bed. Faced with age, failures, and hazards—a fire in February of 1997, collision damage during docking the following June, and tight Russian finances—Mir was deorbited on March 23, 2001.

The American space station effort started when Congress approved a 1959 National Aeronautics

Herman Noordung's design for a space station. (NASA)

and Space Administration (NASA) recommendation that a space station follow Project Mercury. This was superseded in 1961 by the Apollo lunar landing goals. In 1969, following the Apollo 11 lunar landing, NASA proposed a 100-person Space Base to be completed by 1975, to serve nuclear-powered craft flying between LEO and the Moon. The unpopularity of nuclear reactors forced NASA to pursue the Space Transportation System (STS)—the space shuttle—with a projected cost of $220 per kilogram to LEO. The STS, however, did not achieve high flight frequency, eventually costing more than $29,000 per kilogram to LEO. This dictated revisions of the space station program. John D. Hodge's NASA Space Station Task Force, started in May, 1982, proposed international partnership, and President Ronald W. Reagan mandated the station in January, 1984.

The Europeans had agreed in 1973 to supply Spacelab modules for shuttle experiments. Japan and Canada had joined the program by May, 1985. In 1987, the design was downsized from dual keel to single keel. President Reagan named the space station Freedom in 1988. The initial plan for an Assured Crew Return Vehicle (ACRV) was modified when the Rosaviakosmos Mir 2 design was merged with Freedom. The Soyuz became the station's crew escape vehicle, though it could carry only three of the station's full complement of seven. In 1993, the "Alpha" redesign was selected, with Johnson Space Center as the host center and Boeing the prime contractor. The new International Space Station (ISS) included the United States, Russia, Canada, France, Japan, Belgium, Brazil, Denmark, Germany, Italy, the Netherlands, Norway, Spain, Sweden, Switzerland, and the United Kingdom.

The name "International Space Station" (abbreviated MKS in Russian) represents a neutral compromise that ended a disagreement about a proper name for the station. The initially proposed name, Space Station Alpha, was rejected by Russia, because it would have implied that the station was something fundamentally new, whereas the Soviet Union already had operated eight orbital stations long before the ISS launch. The Russian proposal to name the space station *Atlant* was in turn rejected by the United States, which was worried about that name's similarity to *Atlantis*, the name of a shuttle orbiter.

From 1992 to 1998, American and Russian crews accumulated long-duration spaceflight experience using the STS and the Mir/Soyuz. ISS assembly began with the Russian Control Module Zarya (meaning "sunrise"), launched on November 20, 1998, by a Proton rocket from Baikonur. The Unity Node went up in STS-88 on December 4. The SPACEHAB Logistics Modules—delivered by STS-96 and STS-101 in May, 1999, and May, 2000, respectively—helped perform assembly. The Zvezda Service Module went up on a Proton on July 12, 2000. STS-92 installed the Integrated Truss Struc-

ture Z1 for solar panels in October. On October 31, a Soyuz took up the first ISS crew (William M. Shepherd, Sergei Krikalev, and Yuri Gidzenko), returning on March 21, 2001. Integrated Truss Structure P9, the Photovoltaic Module, and radiators went up in STS-97 on November 30, 2000.

The Destiny Laboratory Module arrived on STS-98 on February 7, 2001, and the Leonardo Multi-Purpose Logistics Module (MPLM) on March 8 via STS-102 along with Expedition Two: Yury Usachev, James Voss, and Susan Helms. On April 19, the Raffaello MPLM arrived on STS-100, along with the Canadian SSRMS mechanical assembly arm. The Joint Airlock and High Pressure Gas Assembly arrived on STS-104 on July 12. On September 14, a Soyuz brought the Docking Compartment 1 (DC-1) and Strela Boom. On April 8, 2002, STS-110 brought the Central Truss Segment (ITS S0) and Mobile Transporter (MT). STS 112, on October 7, brought the First Right-Side Truss Segment (ITS S1) with radiators and the Crew and Equipment Translation Aid (CETA) Cart A. STS-113, on November 23, 2002, brought the First Left-Side Truss Segment (ITS P1) and the Crew and Equipment Translation Aid (CETA) Cart B.

In February, 2003, the *Columbia* accident grounded STS flights. The ISS crew has been reduced to two, because of the crew-size limitations of Soyuz, which is used to transport crews to and from the station.

As of 2005, the ISS had a mass of 187,016 kilograms (of the June, 1999, planned 473,000 kilograms) and 425 cubic meters of living space. It is 73 meters wide, 52 meters long and 27.5 meters high. Sixteen STS and twenty-two Russian flights had occurred. Construction had taken 51 of the estimated 160 spacewalks, totaling 318 of 1,920 hours.

The first space "hotel stays" by paying customers were by American businessman Dennis Tito, who flew on Soyuz TM-32, and South African Mark Shuttleworth, who arrived on Soyuz TM-34, for eight days each. Each is reported to have paid $20 million to Russia for training and the launch. Each trained for a year and participated in ISS experiments. The first "space wedding" occurred on August 10, 2003, between Cosmonaut Yuri Malenchenko on the ISS and Ekaterina Dmitriev in Texas.

In 2004, the presidential administration of George W. Bush announced the Moon-Mars Initiative, with ISS construction to be completed by 2010 with the fulfillment of U.S. obligations. NASA funding for ISS was to stop by 2016. Unfortunately, the project is trapped in a dilemma. Its present orbit is too low to permit clean long-duration microgravity experiments for three reasons: perturbations in Earth's gravitational field are felt; air drag varies; and construction and thrust corrections disturb the environment. However, ISS cannot be boosted to its intended high orbit because the cost of subsequent construction and resupply would rise drastically and there are no suitable human-carrying vehicles at present. Commercial interest has been damped by extreme costs and schedule slippage. The project cost is now to be capped at roughly $100 billion.

Contributions

Apart from the notable achievements in space-based assembly and medical and psychological data, space stations have served many experiments in tissue culture, protein crystals and plant growth, materials science, combustion and fluid mechanics, Earth observation, astronomy, and the fundamental physics of weak forces. The main present mission of the ISS is to help develop the space-validated components and techniques that will be used in missions to the Moon, Mars, and beyond.

Context

Perhaps the most lasting lesson from these first efforts to move into space will be the experience of more than 100,000 people across so many nations spanning numerous cultures and politics around the world, working together toward a grand dream of humanity. Future plans include a large station at the Earth-Moon L-1 or L-5 Lagrangian points, and larger stations at Earth-Sun L-4, L-1, or in Mars orbit.

See also: Astronauts and the U.S. Astronaut Program; Cooperation in Space: U.S. and Russian; International Space Station: Design and Uses; International Space Station: Development; International Space Station: U.S. Contributions; Manned Orbiting Laboratory; National Aeronautics and Space Administration; Skylab Space Station.

Further Reading

Dick, Steven J. *Aeronautics and Space Report of the President.* Available at the NASA Headquarters Library, http://www.hq.nasa.gov. Accessed February, 2005. Mandated by law, these reports summarize government aerospace activities each year, and have extensive appendices with data on launchers, spacecraft and other vehicles, and systems. This one is by NASA's chief historian.

Dyson, Marianne J. "Building on What We Know: The History of Space Stations." By the author of *Space Station Science* (Windward Publishing), this unpublished chapter is posted on the Web by the author at http://www.mariannedyson.com/stationhistory.html. Accessed March, 2005.

Harland, David, and John E. Catchpole. *Creating the International Space Station.* London: Springer-Verlag London, 2002. A comprehensive review of the historical background, rationale behind, and events leading to the construction and commissioning of the International Space Station. The authors describe the orbital assembly of the ISS on a flight-by-flight basis, listing all the experiments planned in the various laboratory modules, and explain their objectives. They also provide an account of the long-term stresses and strains of building the ISS on the U.S.-Russia relationship, especially after 1997.

Smith, Marcia. *Space Stations.* CRS Issue Brief IB93017, updated November 3, 2003. Washington, D.C.: Congressional Research Service, Library of Congress. Covers the history of space stations, starting with Skylab, and contains programmatic information on the International Space Station and international agreements.

Zack, Anatoly. Russian SpaceWeb.com. http://www.russianspaceweb.com. This Web site offers news and history of astronautics in the former Soviet Union. Contains extensive documentation, images, and lists of references to original documents on the Soviet space program. Accessed February, 2005.

Zimmerman, Robert. *Leaving Earth: Space Stations, Rival Superpowers, and the Quest for Interplanetary Travel.* Washington, D.C.: Joseph Henry Press, 2003. This 544-page document, available online at http://books.nap.edu, starts with a history of the Soviet and U.S. space programs and a comprehensive study of the space station development, including many vignettes about the human side of such programs.

Narayanan M. Komerath with Russell R. Tobias

Space Suit Development

Date: Beginning May 5, 1961
Type of program: Piloted spaceflight

Space suits initially provided astronauts with an emergency intravehicular backup system in case of the loss of spacecraft cabin pressure. Early space suit designs were based on the technology developed for high-altitude aircraft pressure suits. More complex space suit mobility systems now allow astronauts to venture beyond the protective limits of the spacecraft.

Key Figures

John Scott Haldane, a British physiologist
Mark Edward Ridge, an American balloonist
Wiley Post (1898-1935), an American aviator
Russell Colley, a pressure suit designer
Yuri A. Gagarin (1934-1968), a Soviet cosmonaut
Alan B. Shepard, Jr. (1923-1998), an American astronaut
Alexei Leonov (b. 1934), a Soviet cosmonaut
Edward H. White II (1930-1967), an American astronaut
Neil A. Armstrong (b. 1930), an American astronaut
Joseph Kosmo, subsystem manager, Space Suit Development, NASA
Hubert Vykukal, senior research scientist, NASA

Summary of the Technology

It has long been recognized that humans cannot survive the conditions of space without special protection. The development of space suits contributed significantly to making space exploration possible.

In 1920, John Scott Haldane, a British physiologist, first proposed the use of a pressure suit to provide protection for flight crew members against the lack of oxygen at altitudes above 12,200 meters. This idea, however, was not tested until November 30, 1933, when American balloonist Mark Edward Ridge donned a modified deep-sea diver's suit and was exposed to 25,600-meter-altitude conditions in a low-pressure chamber.

In the same year that Ridge was testing the British-built "high-altitude" suit, Wiley Post, an American aviator, initiated the development of a full pressure suit. Post needed the pressure suit to help him break the existing world aircraft altitude record. Wearing a suit designed by Russell Colley of the B. F. Goodrich Company, Post made a number of high-altitude flights during 1934 and 1935. Unfortunately, because of problems with the recording barographs, no official altitude record could be verified. Unofficially, however, Post had reached altitudes in excess of 12,200 meters with the help of the full pressure suit. More important, the efforts made and risks taken by Ridge and Post proved that pressure suits were practical systems that would enable humans to fly safely at high altitudes and, in time, in space.

Before and during World War II, full pressure suits were developed by various European countries and the United States. All the early suits were very cumbersome to wear and, when inflated, caused serious mobility restrictions. These suits were primarily developed to protect high-altitude flight crew members from lack of oxygen, and, as such, they were regarded as precursors of aircraft cabin pressurization systems. With the onset of World War II, the pressure cabin became standard in most aircraft, and from then on, pressure suit research focused on emergency situations in military and experimental high-altitude aircraft.

In 1961, the first men to wear full pressure suits in space were Russian cosmonaut Yuri A. Gagarin and American astronaut Alan Shepard. The pressure suits they used were worn uninflated in a pressurized cabin and would have been inflated only in the event of a failure of the Vostok or Mercury Capsule pressurization system. The space suit configurations developed by the National Aeronautics and Space Administration (NASA) for Project Mercury and the Gemini Program originated from the earlier high-altitude aircraft full pressure suit.

The Project Mercury and Gemini Program space suits were essentially modified versions of existing military full pressure suits. Project Mercury utilized the Navy's Mark 4 full pressure suit, built by B. F. Goodrich Company, and the Gemini Program space suit was derived from the Air Force's AP/22 full pressure suit, manufactured by David Clark Company. The early Soviet space suits also originated from military aircraft pressure suit technology. These early space suits lacked sophisticated mobility systems; because the suits served primarily as backup systems against the loss of cabin pressure, only limited pressurized intravehicular mobility was required.

The development of the mobile space suit was spurred by the requirement for astronauts to perform tasks outside the spacecraft. In 1965, Cosmonaut Alexei Leonov, of Voskhod 2, and Astronaut Edward White, of Gemini IV, performed the world's first "spacewalks." During this phase of the Gemini Program, U.S. scientists recognized that astronauts needed improved mobility systems and protection from extravehicular environmental hazards.

The Apollo space suit was designed to function as an emergency intravehicular suit and as an extravehicular suit that would enable lunar surface exploration. A variety of prototype Apollo space suit configurations evolved between 1965 and 1970. The International Latex Corporation was responsible for the design, development, and fabrication of the A71 and A71B Apollo space suits, which were selected for use. On July 20, 1969, Neil A. Armstrong, wearing an A71 space suit, was the first human to set foot on an extraterrestrial surface and collect data while being sustained in and protected from a hostile environment. Later Apollo astronauts wore the improved A71B space suit when they explored the lunar surface. The Skylab program adopted the basic Apollo A71 space suit with minor modifications for use in various planned orbital extravehicular activities.

During the early 1960's, as space suits were being developed for the Gemini Program and the planned Apollo and Skylab programs, Joseph Kosmo of the Johnson Space Center (JSC) and Hubert Vykukal of the Ames Research Center embarked on the development of advanced space suit mobility systems for potential future application. NASA long-range program planners had been studying the feasibility of establishing lunar surface bases and conducting piloted Mars planetary exploration. In support of these envisioned post-Apollo operations, NASA initiated a series of space suit technology development programs.

The first of the JSC-sponsored advanced technology suit concepts was the rigid experimental suit assembly, or RX-1, developed by Litton Industries' Space Sciences Laboratory. The suit was a radical departure from the basic, soft-fabric space suits of early 1962. The RX-1 was developed to demonstrate the feasibility of low-force mobility joint systems in the arms and legs. Additional suit features included hard torso structure and an easily fastened single-plane body seal closure. (The Mercury, Gemini, and Apollo space suits used pressure-sealing zippers.) Between 1962 and 1968, numerous RX models were developed. Each version incorporated mobility joint and structural improvements made possible by evaluation and testing of the previous model. The RX-5A was the final configuration of the RX series.

In conjunction with the development of JSC's RX, Ames Research Center initiated investigations into a hard space suit that would use a combination of bearings and metal bellows for mobility joint systems. Between 1964 and 1968, two "Ames experimental" hard suit assemblies, identified as AX-1 and AX-2, were developed by Ames.

As the Apollo Program matured, it became apparent that spacecraft payload weight and stowage volume limitations were constraints on the various hard suit concepts. This realization resulted in the initiation of new approaches to space suit design, and JSC produced a family of advanced space suit configurations representing a hybridization of hard suit and soft suit technologies. With the completion of the Apollo Program, much of the advanced mobility system technology that had been developed earlier was shelved. In the 1970's and 1980's, various elements of the advanced space suit technology base were incorporated in the space suits designed for the U.S. space shuttle and space station programs.

Unlike previous flight program space suits, the shuttle space suit was designed for extravehicular use only. Emphasis was therefore placed on providing astronauts with a high degree of extravehicular operational capability, uncompromised by other requirements. The shuttle space suit incorporated a hard upper-torso shell of fiberglass, a horizontal single-plane body seal closure ring, pressure-sealed bearings in the shoulder, upper arm, and lower torso areas, and flat, all-fabric arm, waist, and leg joints. All these elements had evolved from earlier advanced space suit technology.

The space station program focused on expanding extravehicular activity (EVA) capabilities beyond those of previous space programs. The space station suit was to have a higher operating pressure—8.3 pounds per square inch—so that astronauts would not need to spend costly time prebreathing pure oxygen before performing an EVA. Previously, prebreathing operations served to wash

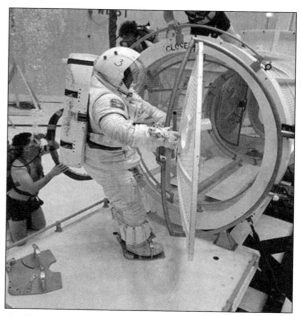

A Johnson Space Center MK-III space suit is tested in the center's Weightless Environment Test Facility (WETF). (NASA)

nitrogen gas from an astronaut's bloodstream so that nitrogen bubbles would not form when the astronaut moved from the shuttle cabin, pressurized at 14.7 pounds per square inch, to the space suit, pressurized at 4.3 pounds per square inch. If, however, the space suit could operate at higher pressure, prebreathing could be eliminated and EVA operations would be able to be conducted more routinely. The core technology for advanced mobility systems established over the previous years enabled the development of higher operating pressure space suits. JSC's Mark 3 and the Ames Research Center's AX-5, designs that eliminated the need for prebreathing, were developed in response to needs of the space station program.

Minimizing energy expenditure by the astronaut, made difficult by the tendency of the volume of gas in joint elements to change during pressurized operations, continues to be the primary impetus behind research in improved mobility systems. The systems developed between 1960 and the 1980's have demonstrated the technical viability of certain design features and have served as a base for the development of future lunar and planetary exploration space suits.

As assembly of the International Space Station proceeds, NASA's astronauts are spending more time outside the safe environment of their spacecraft. The Extravehicular Mobility Unit of the shuttle program is the garment of choice in the harsh environment of space. When astronauts take a stroll on the Martian landscape, a new suit will be available. The space suit is constructed primarily of fabric, with ball bearings that allow the wearer to move more easily when the suit is inflated to 0.26 kilogram-force per square centimeter above the local pressure, as it would be on the Moon or Mars. A self-contained liquid air backpack provides life support, cooling, communications, and power. The suit and backpack have a weight of about 68 kilograms on Earth.

Contributions

As space missions became more complex, so did space suits. The experience gained from a variety of space missions has influenced space suit development.

From the piloted EVAs of the Gemini Program, it was learned that improved cooling techniques to remove astronaut-produced metabolic heat would be required if longer and more involved EVAs were to be conducted. As a result, NASA scientists developed an undergarment, to be worn inside the space suit, that contained small tubes through which water could be pumped. The liquid-cooled garment became a standard design feature of the Apollo space suit and all subsequent suits.

In the Apollo and Skylab programs, the space suit fulfilled two roles. It was worn at launch and during critical spacecraft docking operations, and its function during these phases was to provide a backup pressurized environment in case of cabin pressure failure. It also acted as a pressurized

mobility system and a portable life-support system during orbital and lunar surface EVAs. The environmental hazards and hostile conditions of extended orbital operations and lunar surface exploration meant that scientists had to develop improved materials and designs to protect the space suited astronaut. The requirements of intravehicular versus extravehicular operations posed a number of design problems, including limitations on bulk, operational complexity, and mobility.

For the shuttle program, the space suit was designed for the single purpose of supporting piloted extravehicular operations, and features were optimized for enhanced EVA performance and reduced cost. The shuttle space suit, with its corresponding life-support system, represented the first completely integrated extravehicular assembly. No cumbersome external life-support hoses, connec-

Gemini VI astronauts Wally Schirra (seated) and Thomas P. Stafford suited up for the flight. Tubing for the life support system is shown connected to the upper half of Stafford's space suit. (NASA)

tions, or harness straps were required to allow an astronaut to leave the spacecraft. Previously, each space suit was custom built to fit the astronaut and was used on only one mission. The shuttle space suit design featured modular components that could be combined in various ways, enabling both male and female astronauts to be fitted with a minimum number of space suits and reducing overall program cost.

The elimination of prebreathing operations through the development of space suit mobility systems that operate at higher pressure levels make EVAs more routine and easier to perform. The modular design of the candidate space station pressure suits, being similar to that of the shuttle space suit, enables the suits to be reused numerous times. In addition, in-orbit maintenance and replacement of various fabric components and the reuse of hardware components reduce overall space suit and flight program costs.

Continued research conducted outside the space shuttle and the Russian space station Mir have shown that a rigid space suit is not necessary for work on the International Space Station. The flexibility and thermal control of the shuttle extravehicular suit makes it appropriate for the tasks required to assemble the station. Additional development costs were saved by the continued use of existing hardware.

The next generation of Manned Maneuvering Unit (MMU), the Simplified Aid for Extravehicular Activity Rescue (SAFER), is being used to aid astronauts in the construction of the International Space Station. SAFER was first tested on STS-64 in September, 1994, ten years after the last MMU mission. Astronauts Mark C. Lee and Carl J. Meade performed an engineering evaluation, an EVA self-rescue demonstration, and an overall flight quality evaluation, which included a demonstration of precision flying by tracking the Remote Manipulator System arm.

While docked to the Mir Space Station in March, 1996, astronauts Linda M. Godwin and Michael R. "Rich" Clifford attached four experiments, known collectively as the Mir Environmen-

tal Effects Payload (MEEP), on the outside of the Mir Docking Module. As a precaution, they each wore a SAFER pack. Astronaut Scott E. Parazynski and Russian cosmonaut Vladimir Georgievich Titov tested the first flight production model of SAFER on STS-86 during the September, 1997, mission.

During the third STS-88 spacewalk to assemble the International Space Station, Astronaut Jerry Ross achieved only 50 percent of the evaluation objectives for SAFER. Still, the tests were considered successful, and SAFER will be worn by astronauts during the station's construction.

Context

Throughout Project Mercury and until the feasibility of EVAs was established on the Gemini IV mission, space suits were simply backups for cabin pressure systems. The development of the true space suit occurred almost simultaneously in two separate parts of the world. In 1965, Soviet cosmonaut Leonov and American Astronaut White performed independent spacewalks outside the confines of their respective spacecraft. The Gemini Program provided the first EVA in the U.S. piloted space effort.

The Gemini Program's accomplishments were significant for space suit development. For the first time, space suits were used to allow humans to work in space. It was recognized that piloted EVAs would increase spacecraft's capabilities and enable the development of new operational techniques. EVA technology from the Gemini Program was incorporated wherever possible in the Apollo space suit. Improved body-cooling and mobility systems were direct results of the Gemini experiences.

Space suits make EVAs possible in three ways. First, in combination with a portable life-support system, the space suit maintains the physiological well-being of the astronaut, which includes supplying oxygen for breathing and ventilation and removing carbon dioxide and metabolic heat. Second, the space suit incorporates various mobility joint system features that enable the astronaut to perform tasks in the extravehicular environment.

Finally, the space suit provides protection against the hazards of space, which include thermal extremes, meteoroid and debris particles, radiation, and, on the lunar surface, sand and dust. The pressure retention layer of the space suit is both a protective barrier and a structural foundation for various mobility systems. A separate outer garment comprising layers of various materials protects the astronaut from hostile environments.

None of the materials used in the early space suits were originally developed with space exploration in mind. As more complex space suit systems evolved, special needs were identified that required the development of new materials or combinations of materials to provide structural integrity and increased protection.

See also: Apollo Program; Apollo Program: Developmental Flights; Cooperation in Space: U.S. and Russian; Escape from Piloted Spacecraft; Extravehicular Activity; Gemini Program; Lunar Exploration; Manned Maneuvering Unit; Skylab Program; Space Shuttle.

Further Reading

Abramov, Isaac P., and A. Ingemar Skoog. *Russian Spacesuits.* London: Springer-Verlag London Limited, 2003. The authors, part of the original Zvezda team that manufactured space suits for the first Russian spaceflights, still play an integral role in space suit research and development. The book covers the technical innovations of the past forty years, which enabled Gagarin's first flight in 1961, the first spacewalk in 1965, and the Mir missions of the 1980's and 1990's, culminating in today's International Space Station.

Burrows, William E. *This New Ocean: The Story of the First Space Age.* New York: Random House, 1998. This is a comprehensive history of the human conquest of space, covering everything from the earliest attempts at spaceflight through the voyages near the end of the twentieth century. Burrows is an experienced journalist who has reported for *The New York Times, The Washington Post,* and *The Wall Street Journal.* There are many photographs and an extensive source list. Interviewees in the book include Isaac Asimov, Alexei Leonov, Sally K. Ride, and James A. Van Allen.

Cortright, Edgar M., ed. *Apollo Expeditions to the Moon.* NASA SP-350. Washington, D.C.: Government Printing Office, 1975. The personal accounts of eighteen men, including NASA managers, scientists, engineers, and astronauts, who directed, developed, and conducted the Apollo missions. Suitable for general audiences, it describes the various political events and engineering projects that influenced the Apollo Program. Includes numerous illustrations and photographs covering the historical period of the Apollo Program, along with pictures transmitted from Apollo spacecraft showing the use of space suits.

Kozloski, Lillian D. *U.S. Space Gear: Outfitting the Astronauts.* Washington, D.C.: Smithsonian Institution Press, 1999. Covers the space suits developed for each of the U.S. piloted spaceflight projects, from the Mercury pressure suit to the current suit in the space shuttle program.

National Aeronautics and Space Administration. *Simplified Aid for Extravehicular Activity Rescue (SAFER) Operations Manual.* Washington, D.C.: Government Printing Office, 1994. Written as the training manual for SAFER, this is a very good technical reference on the inner workings of the backpack. It is filled with detailed drawings and specifications, as well as operating procedures.

Oberg, James E. *Mission to Mars: Plans and Concepts for the First Manned Landing.* Harrisburg, Pa.: Stackpole Books, 1982. Describes the feasibility of a crewed Mars mission and discusses topics such as spaceship design, propulsion, life-support systems, space suits, Martian surface exploration, cost factors, and political and social issues relating to future plans for colonization. Contains photographs and illustrations. Suitable for general audiences.

Shayler, David J. *Walking in Space: Development of Space Walking Techniques.* Chichester, England: Springer-Praxis, 2003. Shayler provides a comprehensive overview and analysis of EVA techniques, drawing on original documentation, personal interviews with astronauts with experience in EVAs, and accounts by those involved in suit design and EVA planning and operations.

Wendt, Guenter, and Russell Still. *The Unbroken Chain.* Burlington, Ont.: Apogee Books, 2001. Wendt is the only person who worked with every astronaut bound for space. This story is filled with important accounts, many of which are published here for the first time.

Joseph J. Kosmo

Space Task Group

Date: November 5, 1958, to November 1, 1961
Type of organization: Aerospace agency

The Space Task Group was the United States' first civilian agency for piloted spaceflight. It was the core team responsible for the Mercury, Gemini, and Apollo Programs, and it was the seed from which grew the Manned Spacecraft Center, now the Johnson Space Center, near Houston, Texas.

Key Figures

Robert R. Gilruth (1913-2000), manager and director, Space Task Group (STG)
Charles J. Donlan, assistant manager, STG
Maxime A. Faget (1921-2004), chief designer, STG

Summary of the Organization

Unofficially established by the brand-new National Aeronautics and Space Administration (NASA) on October 8, 1958, the Space Task Group (STG), created by a memorandum bearing thirty-five names and dated November 5, 1958, was destined to place the first humans on the Moon.

The thirty-five scientists from the old aeronautical laboratory at Langley Field, Virginia, and the ten additional professionals soon to join them from Lewis Field in Cleveland, Ohio, foresaw the possibilities that might arise from combining aviation with rocket and missile technologies. Their initial charge was simply to create a team of humans and machines for piloted (or in those days, "manned") space exploration, and their early projects involved one-person ballistic and orbital spaceflight. Soon, however, they were responsible for a two-person maneuverable spacecraft and for three-person circumlunar and lunar-landing vehicles.

In the mid-1950's, well before Soviet Sputniks 1 and 2 spurred the creation of NASA from the National Advisory Committee for Aeronautics (NACA), certain farsighted engineers within and outside the government were studying rockets' potential to send humans beyond Earth's atmosphere. The U.S. Air Force had long been interested in expanding its flight regime into space, and NACA's X-15 rocket research airplane was another example of future-oriented work. Perhaps the most vigorous group of aerospace visionaries, however, was gathered around Robert R. Gilruth and his Pilotless Aircraft Research Division (PARD) at Langley Field and Wallops Island in Virginia. Another group at the NACA Lewis Propulsion Laboratory, led by Abe Silverstein, also vied for attention, but when NACA became NASA in October, 1958, President Dwight D. Eisenhower appointed T. Keith Glennan as the first NASA administrator, and Glennan needed Silverstein's assistance. They delegated authority to Gilruth's group to proceed with a piloted satellite program. Officially designated Project Mercury on November 26, 1958, the piloted satellite program began the first American series of flights into space.

During the spring and summer of 1958, a series of competitive planning conferences around the country gradually led NACA engineers to a consensus that the best proposal for a method of piloted spaceflight was the one championed by Maxime A. Faget and his colleagues at Langley, near Norfolk. After years of experience with Gilruth's PARD testing drones and guided missiles, Faget and his associates advocated a wingless, nonlifting, nose cone configuration for the first piloted satellite. Rather than follow the pattern of the X-15 rocket research airplane, they wanted to adapt a small, inhabitable cockpit to the first operationally tested intercontinental ballistic missile (ICBM). This idea was at first received without enthusiasm by the Air Force and by General Dynamics Astronautics, whose Atlas ICBM was the only viable candidate at that time for the job of launching a person into orbit. While the orbital flight plan rapidly took shape at the field centers, NASA Headquarters expanded and helped STG to complete preliminary designs, to issue specifications, to choose the prime contractors, and to manage the entire project. Criticism abated as creative engineering activities moved ahead rapidly.

At the beginning of 1959, McDonnell Aircraft Corporation in St. Louis was chosen, out of a competitive bidding group of a dozen companies, to manufacture a dozen piloted satellite spacecraft according to the Faget concept. John F. Yardley of McDonnell quickly assumed leadership in the development of the Mercury hardware. He and his corporation, together with Faget and his STG colleagues, became the core of the Mercury team. By midyear, when seven military test pilots, to be called astronauts, joined the project, most of the basic decisions as to how NASA would try to put a human in space were firm.

Three central principles guided Project Mercury: Use the simplest and most reliable approach, attempt a minimum of new developments, and conduct a progressive series of tests. In the hope of saving time and money and ensuring safety, NASA's policymakers tried to minimize trial and error. Five approaches to major aspects of the project were determined as soon as the government-industry team began to cooperate: The piloted satellite spacecraft would be launched into orbit by the Atlas ICBM; it would be equipped with a tractor escape system, in case the booster malfunctioned; it would be a frustum-shaped vehicle with an attitude control system; it would be braked in orbit by retrorockets; and it would be slowed on descent by

parachutes. Although these plans and the mission profiles were remarkably well laid, nearly all the details of their implementation were yet to be incorporated and verified.

Patents for inventions made in the course of work on Project Mercury were conferred only after the designs were proved in practice, so that official awards tended to obscure the actual process of innovation. Seven men were credited by NASA with designing the Mercury spacecraft: Faget, Andre J. Meyer, Jr., Robert G. Chilton, Willard S. Blanchard, Jr., Alan B. Kehlet, Jerome B. Hammack, and Caldwell C. Johnson. For their conceptual designs and preliminary tests of components, these members of Faget's team were recognized some eight years later in the issuance of U.S. Patent 3,270,908. In addition, Faget and Meyer were credited with the tractor-pylon emergency rocket escape system, and Meyer was credited with the parachute and jettison system design; along with Faget, William M. Bland, Jr., and Jack C. Heberlig were recognized for the pilot's contour couch. Later still, R. Bryan Erb and Kenneth C. Weston shared honors with Meyer for the ablative heatshield, and Matthew I. Radnofsky and Glenn A. Shewmake were recognized for their inflatable life rafts and radar reflectors.

McDonnell employees, led by Raymond A. Pepping, Edward M. Flesh, Logan T. MacMillan, John F. Yardley, and, later, Walter F. Burke, took an active and at times initiating role in the creation of the Mercury spacecraft. NASA's policy of retaining ownership of inventions was highly controversial at first, but it did not stanch industrial initiative; STG grew from 35 to more than 350 members within its first nine months of existence, but the industrial team grew even faster.

Many subcontractors and third-tier vendors, as well as the prime contractor working with STG, suggested and completed systems engineering studies and components for Project Mercury. Especially noteworthy examples were the McDonnell "pig-drop" impact studies of the aluminum honeycomb shock absorber, the research work of Brush Beryllium Company and of Cincinnati Testing

Laboratories on the heatsink and ablative heatshield, and the extraordinarily careful design and development of the environmental control system by AiResearch Manufacturing. The contractors were not limited to hardware development; new techniques and procedures, notably human factors engineering led by Edward R. Jones of McDonnell, originated as often from contractors as from NASA workers.

Because of concerns over the microgravity environment (or, loosely, "weightlessness") and its effects on humans and mechanical parts in orbit, the automation experts held sway over the development of Mercury during 1959. By the end of 1960, however, the automatons had failed so often and the astronauts had been trained so well that Mercury's managers were beginning to place more reliance on people than on machines for mission success. At all critical points, redundant, automatic safety features were built in, but the pilots were given manual control over their vehicle wherever feasible. Missile and aircraft technology were rapidly converging.

Meanwhile, STG was continuously testing each part and the whole Mercury configuration in the laboratory and in flight. Three levels of testing had originally been specified: development, qualification, and performance. To these were added, in mid-1960, reliability tests of many varieties to ascertain the life and limits of all the systems. Most dramatic was the extensive flight-testing program, which used the unique Little Joe boosters for several tests and the Atlas booster for a single Big Joe shot that demonstrated reentry capability. The Big Joe mission was accomplished successfully on September 9, 1959, and so paved the way for a series of seven more Little Joe missions during the next two years.

By the beginning of 1960, a presidential election year, Gilruth's STG was in high gear and accelerating. Military liaisons had been established, a worldwide tracking and data network was being arranged, an industrial priority rating for Mercury was obtained, a class of seven military test pilots had been chosen and were undergoing astronaut training, and intensive studies and renovations were under way to "man-rate" the booster rockets (that is, to make them safe enough for humans). Politically, however, 1960 was to be a rough year. STG's personnel roster contained about five hundred names, and its prime contract with

A Little Joe booster on a rocket launcher at Wallops Island in 1959. (NASA)

McDonnell, already modified in more than 120 particulars, was nearing $70 million and rising. Gilruth's group, still housed and hosted by Langley, was supposed to be moving to the new Goddard Space Flight Center being built at Beltsville, Maryland, between the capital and Baltimore. It was unclear, however, whether construction would go forward on the Marshall Space Flight Center at Huntsville, Alabama, which would be occupied by Wernher von Braun's team of rocket experts, at work on the Saturn series of engines and boosters. Political rhetoric about the so-called missile gap, the U-2 incident in May, continuing Soviet launches of dogs and robots in orbital spacecraft, and several widely publicized failures of NASA flight tests in October and November helped make 1960 a most suspenseful year.

The appointments of a number of senior engineers, who distinguished themselves further as Gilruth's group evolved, added to STG's strength during this critical year. Walter C. Williams, Kenneth S. Kleinknecht, Robert O. Piland, James A. Chamberlin, and G. Merritt Preston were a few of the managers. George M. Low and others at NASA Headquarters decided between administrations, in January, 1961, to make STG separate from the Goddard center. By then, STG employed 680 persons.

After monkeys had survived flights in boiler-plate spacecraft propelled by Little Joe solid rockets, the McDonnell-built spacecraft were mated to Atlas and Redstone liquid-fueled rockets for their combination qualification flight tests. The first two attempts at mated flight failed, because of the boosters more than the spacecraft. By February, 1961, however, successful flights of both the Mercury-Redstone (MR) and the Mercury-Atlas (MA) combinations had gone far toward man-rating the machines. The performance and recovery of the chimpanzee Ham in MR-2 seemed to indicate that a human could make a similar suborbital hop. On April 12, 1961, however, Yuri A. Gagarin orbited Earth in 108 minutes aboard the Soviet Union's Vostok 1. Thus the parabolic test flights in May of Alan B. Shepard, Jr., in *Freedom 7* (MR-3), and

in July of Virgil I. "Gus" Grissom, in *Liberty Bell 7* (MR-4), set no world records, but merely tested the ability of the Mercury people and machines to work in space for a few minutes. Shepard and Grissom did prove, however, that STG had designed and developed a primitive spacecraft and not merely a piloted bullet.

By mid-1961, the tiny Mercury spacecraft, encasing forty thousand components and 11 kilometers of wiring, was widely publicized around the world. Designed for a reference mission of three orbits, the basic systems in Mercury were advertised openly and often described as falling into ten categories: heat protection, mechanical, pyrotechnical, control, communication, instrumentation, life-support, electrical, sequential, and network. Some sixteen major subsystems were novel and critical enough to worry reliability experts and STG managers. STG was upstaged again when, on August 6 and 7, 1961, the Soviet cosmonaut Gherman S. Titov made a seventeen-orbit, day-long circumnavigation of Earth in Vostok 2. In contrast to Mercury, the details of the Vostok spacecraft were shrouded in secrecy.

Difficulties and delays in manufacturing the Mercury Capsule and in man-rating its boosters had afforded the seven American astronauts more than an extra year of training. Because they were active as consulting engineers as well as test pilots, the Mercury astronauts contributed to quality control, mission planning, and operational procedures before they ventured into space. Their specialty assignments indicated another way of categorizing the most critical features of Project Mercury. Shepard became the expert on tracking and recovery operations, Grissom studied the complicated electromechanical spacecraft control systems, and John H. Glenn, Jr., worked on the cockpit layout. M. Scott Carpenter specialized in the communications and navigation systems, Walter M. Schirra, Jr., handled the life-support systems and space suits, and L. Gordon Cooper, Jr., and Donald K. "Deke" Slayton analyzed the Redstone and Atlas boosters.

For all the exotic training and trips undergone by the astronauts, only three activities proved to have been indispensable: weightlessness conditioning, accomplished through flights of Keplerian parabolas; acceleration endurance tests in human centrifuges; and, most important, the overlearning of mission tasks in McDonnell-built spacecraft procedures trainers. Many other training aids were helpful in bolstering the astronauts' confidence that they could endure and overcome any eventuality, but they were confident men; learning to live and work within the pressure suits and within the sealed pressure vessels was an exceedingly difficult job in itself.

At times in 1960 and 1961, all members of the Mercury teams were stymied by some recalcitrant system, process, or device. The recurrent balkiness of the smaller thrusters in the Reaction Control System, the overassigning of pilot tasks in flight planning, and difficulties with the Department of Defense in scheduling support operations typified tendencies that threatened to become permanent. Both STG and McDonnell underwent several reorganizations of personnel and divisions of labor to meet changing program situations. Moving to the Cape Canaveral launch site, establishing an operations team, responding to new hardware integration needs, and riding the tide of a new political administration all caused confusion and elicited new organization.

Nevertheless, the flight test series began to experience success. By late 1961, it was obvious that STG was to become institutionalized as a permanent, separate NASA installation devoted to the long-term development of piloted spaceflight and space exploration. John H. Glenn, Jr., was ready to fulfill the Mercury mission, the Capsule had evolved from a container into a spacecraft, and the boosters had been refined to the point of deserving to be called human-launching vehicles.

On May 25, 1961, President John F. Kennedy called upon Congress to approve a decade-long lunar-landing-and-return program. Already funds had been approved for a site selection process. STG itself proposed a piloted spacecraft development center. On September 19, NASA announced that a site near Houston, Texas, had been selected,

and by October 13, NASA Headquarters had approved construction plans for at least eighteen buildings. More important, STG's responsibility for Project Mercury had escalated into responsibility for the Apollo spacecraft Mercury Mark II, soon to be renamed Gemini. Thus it was no surprise when on November 1, 1961, STG personnel, now numbering about one thousand, learned that "the Space Task Group is officially redesignated the Manned Spacecraft Center."

Context

The Space Task Group, headquartered in Virginia from 1958 to 1961, fulfilled its initial mission with Glenn's three-orbit flight in *Freedom 7* (MA-6) on February 20, 1962. By that time, the STG that had become virtually synonymous with Project Mercury was anticipating relocating under its new name, the Manned Spacecraft Center (MSC), to southeast Texas around Houston and Galveston Bay. There, its members would design, develop, manage, and control the missions for several new generations of spacecraft. The influence of Faget's flight systems design team and Gilruth's directorship pervaded the next decade of U.S. spacecraft developments, as attested by the similarities in the Mercury, Gemini, and Apollo Command Modules. In addition to twelve people brought back to Earth safely after six lunar landings, fifteen astronauts had circumnavigated the Moon and returned in Apollo Command Modules by the end of 1973. On February 17 of that year, the MSC was officially renamed the Lyndon B. Johnson Space Center (JSC).

A space task group of a different sort passed quickly into obscurity during this period. In January, 1969, President Richard M. Nixon appointed his vice president, Spiro Agnew, to chair a special advisory committee on future directions for piloted spaceflight. This commission was formed in the wake of celebrations of humankind's first circumnavigation of the Moon, in Apollo 8, and met amid the excitement of the Apollo 9, 10, and 11 achievements. In September of 1969, it published a report titled *Post-Apollo Space Program: Directions for the Future.* The group advocated piloted missions to Mars, but it was so marred by the political scandals that soon enveloped its chairperson that its recommendations were quickly forgotten.

See also: Apollo Program; Funding Procedures of Space Programs; Gemini Program; Mercury Project; National Commission on Space; Space Suit Development; Vandenberg Air Force Base.

Further Reading

Brooks, Courtney G., James M. Grimwood, and Loyd S. Swenson, Jr. *Chariots for Apollo: A History of Manned Lunar Spacecraft.* NASA SP-4205. Washington, D.C.: Government Printing Office, 1979. This is the semiofficial history of the initial achievements of the Apollo spacecraft as seen from Houston. Part of the NASA History series, the work is a sequel to two earlier books that cover the Mercury and Gemini Projects. It stops short of considering the Apollo 12 through 17 missions. Several more volumes in the series deal with other aspects of the Moon-landing program.

Burrows, William E. *This New Ocean: The Story of the First Space Age.* New York: Random House, 1998. This is a comprehensive history of the human conquest of space, covering everything from the earliest attempts at spaceflight through the voyages near the end of the twentieth century. Burrows is an experienced journalist who has reported for *The New York Times, The Washington Post,* and *The Wall Street Journal.* There are many photographs and an extensive source list. Interviewees in the book include Isaac Asimov, Alexei Leonov, Sally K. Ride, and James A. Van Allen.

Catchpole, John. *Project Mercury: NASA's First Manned Space Programme.* London: Springer-Verlag London Limited, 2001. This work offers a developmental resume of Project Mercury and its associated infrastructure, including accounts of space launch vehicles. The book highlights the differences in Redstone/Atlas technology, drawing similar comparisons between ballistic capsules and alternative types of spacecraft. The book also covers astronaut selection and training, as well as tracking systems, flight control, basic principles of spaceflight, and detailed accounts of individual flights.

Ertel, Ivan D., and Mary Louise Morse. *Through November 7, 1962.* Vol. 1 in *The Apollo Spacecraft: A Chronology.* NASA SP-4009. Washington, D.C.: Government Printing Office, 1969. This first of four volumes covering the Apollo Program chronicles key events from the 1920's through the lunar-orbital rendezvous decision of November 7, 1962. Includes a foreword by Robert O. Piland as well as forty-four illustrations, abstracts of key events, seven appendices, and an index.

Ezell, Linda Neuman. *Programs and Projects.* Vol. 2 in *NASA Historical Data Book, 1958-1968.* NASA SP-4012. Washington, D.C.: Government Printing Office, 1988. This reference work complements the volume by Van Nimmen, with five chapters documenting launch vehicles, piloted spaceflight, space science and applications, advanced research and technology, and tracking and data acquisition. Charts, tables, maps, diagrams, and drawings abound, but there are no photographs.

Grimwood, James M. *Project Mercury: A Chronology.* NASA SP-4001. Washington, D.C.: Government Printing Office, 1963. This is the first of a series of historical chronologies and programmatic accounts of U.S. piloted spaceflight projects. Features a preface by Kenneth S. Kleinknecht and a foreword by Hugh L. Dryden. Includes sixty-eight illustrations, ten appendices, and a good index.

Grimwood, James M., Barton C. Hacker, and Peter J. Vorzimmer. *Project Gemini, Technology and Operations: A Chronology.* NASA SP-4002. Washington, D.C.: Government Printing Office, 1969. Focuses on the technology and operations of Gemini, from its concept and design in April, 1959, to its abolition and the summary conference in February, 1967. This book would serve as a good introduction to *On the Shoulders of Titans* (see below). The foreword is by Charles W. Mathews. Includes 131 illustrations, eight appendices, and a thorough index.

Hacker, Barton C., and James M. Grimwood. *On the Shoulders of Titans: A History of Project Gemini.* NASA SP-4203. Washington, D.C.: Government Printing Office, 1977. A volume in the NASA History series, this work is a history of the Gemini Program. It describes how the Mercury Mark II became a first-class maneuverable spacecraft, suitable for rendezvous and docking in orbit.

Harland, David. *How NASA Learned to Fly in Space—An Exciting Account of the Gemini Missions.* Burlington, Ont.: Apogee Books, 2004. The nuts and bolts of the Gemini Program are explained in this well-written book. The launch vehicles and spacecraft are detailed, as are the astronauts who flew them and the missions they flew.

Heppenheimer, T. A. *Countdown: A History of Space Flight.* New York: John Wiley, 1997. A detailed historical narrative of the human conquest of space. Heppenheimer traces the development of piloted flight through the military rocketry programs of the era preceding World War II. Covers both the American and the Soviet attempts to place vehicles, spacecraft, and humans into the hostile environment of space. More than a dozen pages are devoted to bibliographic references.

Lee, Wayne. *To Rise from Earth: An Easy to Understand Guide to Spaceflight.* New York: Checkmark Books, 1996. This is a good introduction to the science of spaceflight. Although written by an engineer with the NASA Jet Propulsion Laboratory, it is presented in easy-to-understand language. In addition to the theory of spaceflight, it gives some of the history of the human endeavor to explore space.

Mari, Christopher, ed. *Space Exploration.* New York: H. W. Wilson, 1999. Twenty-five articles (reprinted from magazines), covering the state of the space program at the time of publication, are divided into five sections: John H. Glenn, Jr.'s return to space, the exploration of Mars, the International Space Station, recent mining efforts by commercial industries, and new types of space vehicles and propulsion systems.

Rosholt, Robert L. *An Administrative History of NASA, 1958-1963.* NASA SP-4101. Washington, D.C.: Government Printing Office, 1966. With an interesting foreword by James E. Webb, this book presents a political scientist's analysis of the first five years of NASA administration. It is heavily documented but poorly illustrated, and it focuses almost exclusively on NASA Headquarters.

Siddiqi, Asif A. *Sputnik and the Soviet Space Challenge.* Gainesville: University Press of Florida, 2003. This is the first comprehensive history of the Soviet piloted space programs, covering a period of thirty years, from the end of World War II, when the Soviets captured German rocket technology, to the collapse of their Moon program in the mid-1970's.

Swenson, Loyd S., Jr., James M. Grimwood, and Charles C. Alexander. *This New Ocean: A History of Project Mercury.* NASA SP-4201. Washington, D.C.: Government Printing Office, 1966. This 681-page narrative is the first program history to be published in the NASA History series. Organized in three parts—"Research," "Development," and "Operations"—this book is the semiofficial account of Project Mercury. It emphasizes the history of Mercury's technology and field management. Profusely illustrated and fully documented, the work was designed as a model for NASA spaceflight histories and is aimed at the intelligent layperson.

Van Nimmen, Jane, Leonard C. Bruno, and Robert L. Rosholt. *NASA Resources.* Vol. 1 in *NASA Historical Data Book, 1958-1968.* NASA SP-4012. Washington, D.C.: Government Printing Office, 1976. This reference work traces the growth of NASA over its first decade, with six topical chapters and two appendices. With a brief foreword by George M. Low, the book presents tabular and graphical data on NASA's facilities, personnel, finances, procurement, installations, awards, and organization. The largest section is chapter 6, which details basic facts about NASA's fourteen-largest field installations.

Loyd S. Swenson, Jr.

Spaceflight Tracking and Data Network

Date: Beginning 1972
Type of facility: Spacecraft tracking network

The Spaceflight Tracking and Data Network (STDN)—now the Space Network (SN)—was fifteen ground communications and tracking stations, located in countries around the world, which provided data relay, data processing, communications, and command support to the U.S. space shuttle program and to other orbital and suborbital spaceflights. The SN was established in the early 1980's to replace the STDN. It consists of a constellation of geosynchronous satellites and associated ground systems.

Key Figures

Daniel A. Spintman, division chief, Ground and Space Networks, Goddard Space Flight Center (GSFC)

Vaughn E. Turner, chief, NASA Communications Division, GSFC

Robert T. Groves, chief, Flight Dynamics Facility, GSFC

John T. Dalton, chief, Data Systems Technology Division, GSFC

Summary of the System

The Spaceflight Tracking and Data Network (STDN) was part of a complex and rapidly changing group of programs designed to provide a two-way communications and command link between flight control centers on the ground and piloted and robotic space missions. In the 1980's, STDN also provided primary support for U.S. space shuttle missions.

The Space Network (SN) is a data-communication system composed of a constellation of Tracking and Data-Relay Satellites (TDRS's) in geostationary orbit and a ground terminal complex employing high-gain microwave antennae. The ground stations send and receive commands and data to and from the TDRS's, which in turn receive and relay data from a multitude of low-Earth-orbit (LEO) satellites. The SN provides global telecommunication services for telemetry, tracking, and command between LEO spacecraft and customer-control and data-processing facilities.

The space segment of the SN consists of up to four operational TDRS's in geostationary orbit at allocated longitudes. The TDRS's are deployed in geostationary orbit to provide the broadest possible coverage. Each TDRS provides two-way data communications relay between spacecraft and the White Sands Complex (WSC) for data transfer and tracking. Each of the satellites is extremely sophisticated: They are able to track and communicate with up to twenty-one separate low-Earth-orbit satellites from positions 35,900 kilometers above Earth.

In 2000, the first of the second generation of TDRS's, TDRS-8, was launched. Two years later, it was joined by TDRS-9 and TDRS-10. The still operational TDRS-1 and TDRS's 3, 4, 5, 6, and 7 serve as backups. The newer satellites offer the same S- and Ku-band single access (SA) services and tracking capabilities as the TDRS 1-7 fleet. Additionally, they offer Ka-band SA services, which feature extremely high data rates and enhanced multiple-access capability, allowing for a lower cost for some customers who require lower-rate SA sevices. The availability of Ka-band and SA services also frees the more heavily utilized SA system for customers requiring higher data rates.

The ground portion of the SN consists of the WSC, the Merritt Island Launch Area (MILA) TDRSS Relay, and the Bilateration Ranging Transponder System (BRTS). WSC is a major element of the ground segment and is located near Las Cruces, New Mexico. It consists of two functionally identical ground terminals, the White Sands Ground Terminal (WSGT) and the Second TDRSS Ground Terminal (STGT). Each ground station employs three 18-meter-diameter high-gain microwave antennae. Another component of the WSC is the Guam Remote Ground Terminal (GRGT), which provides coverage in the area of the globe not "seen" by the TDRS system.

The BRTS are totally automated transponders that provide metric tracking data for the accurate determination of the ephemerides for each in-orbit TDRS.

The ground segment ensures uninterrupted communications between the spacecraft and the control center. Data are transmitted between ground locations and the WSC using domestic communications satellites (DOMSATs), the ground segment of the SN, and other supporting elements.

SN operates in cooperation with the National Aeronautics and Space Administration (NASA)—specifically, with NASA's Communications Division (NASCOM) and the Flight Dynamics Facility at Goddard Space Flight Center (GSFC). NASCOM is the communications link for launch and landing sites, for mission and network control centers, and for all U.S. spacecraft. It provides voice, low- and high-speed telemetry, and television transmissions to more than one hundred NASA facilities. The Flight Dynamics Facility receives the tracking data relayed by SN and calculates the information necessary to orient the spacecraft being tracked.

By 1988, SN was providing tracking, communications, and command services to a total of nineteen scientific, weather, communications, and environmental U.S. satellites in Earth orbit. SN also had the capability to support European, Soviet, and Chinese satellites with similar services.

As part of Space Network, the NASA Ground Terminal at White Sands, New Mexico, served as a backup space shuttle Mission Control facility if Edwin P. Hubble Space Center, in Houston, was rendered inoperative for any length of time. GSFC served as an interim Mission Control center while the flight control personnel transferred from Houston to White Sands.

Each STDN station was able to track and communicate with a spacecraft only during the period when the spacecraft's orbit brought it into the station's "line of sight," or when Earth's curvature did not block direct radio and radar contact. Each station could track or remain in contact with a space-

A Tracking and Data Relay Satellite (TDRS) right before it was released from the cargo bay of shuttle orbiter Endeavour *during STS-54.* (NASA)

craft for a maximum of approximately 15 percent of its orbit. When one station lost contact, responsibility for tracking and communications passed to the next ground station in the network.

Through the use of Tracking and Data-Relay Satellites, only one major ground station is required. Whether it is the International Space Station or a Jovian probe, the Space Network keeps space travelers in constant contact with their Earthbound support.

Contributions

On both piloted and robotic missions, the data received by STDN, and later SN, gave mission managers and technicians a complete picture of the health and reliability of the spacecraft in orbit, something that the astronauts on piloted missions often did not have the time or opportunity to do. The information lets the mission managers on Earth serve as "extra crew members" who could help prevent or overcome problems with the spacecraft.

The space shuttle *Challenger*'s launch of the first Tracking and Data-Relay Satellite (TDRS) in April, 1983, demonstrated the interaction between ground control and spacecraft made possible by STDN. TDRS-A was successfully released from the space shuttle on April 5, 1983, but the booster rocket attached to the satellite failed to fire, leaving it in a uselessly low-Earth orbit. SN allowed TDRS mission managers to assess TDRS's situation and devise an alternative way for it to reach a geosynchronous orbit (an orbit wherein a satellite travels once around Earth every twenty-four hours) 35,900 kilometers above Earth. Sending commands via SN, ground control workers used the satellite's tiny reaction control thrusters to move it slowly to the proper altitude.

STDN, SN, and other NASA-operated tracking networks have allowed the United States to participate in the growth of the international space community by providing launch and data-tracking support for the French Ariane rocket program. SN is also capable of providing support to other foreign satellites.

Context

Space Network works in conjunction with the Deep Space Network, which is controlled by the Jet Propulsion Laboratory in Pasadena, California. SN is part of an effort to develop U.S. communication and tracking capabilities that began in the earliest days of the nation's space program.

In 1958, as part of the country's plan to launch an artificial satellite into orbit as the United States' contribution to the eighteen-month International Geophysical Year (July 1, 1957, to December 31, 1958), NASA took over the U.S. Naval Research Laboratory's Minitrack network of ground stations. These facilities were designed only to track satellites and receive data and did not have the capacity to transmit commands to spacecraft from the ground.

There were only ten stations in the Minitrack system when NASA first began using the network; by 1963, however, eighteen ground facilities were in use. Their locations were San Diego; Goldstone, California; Blossom Point, Maryland; Fort Meyers, Florida; East Grand Forks, Minnesota; Fairbanks, Alaska; Rosman, North Carolina; Antigua, West Indies; Quito, Ecuador; Lima, Peru; Antofagasta and Santiago, Chile; Canberra and Woomera, Australia; Saint John's, Newfoundland; Winkfield, England; and Eselen Park and Johannesburg, South Africa.

During the years that the Minitrack network was in operation, NASA began expanding the technological capabilities of its ground stations, adding new and more powerful antennae and better data retrieval and processing systems. With additions in 1963 of 12- and 26-meter antennae to several Minitrack stations, the system was renamed the Satellite Network. By 1964, NASA had brought into use the Satellite Telemetry Automatic Reduction (STAR) system, which not only provided better tracking and data processing but also enabled ground stations to issue commands to robotic satellites. The improved network, which operated from 1964 to 1972, was known as the Space Tracking and Data Acquisition Network (STADAN). STADAN operated ten ground stations at former Minitrack locations, with an additional station at Tananarive, Madagascar.

In 1962, NASA had separated tracking and communications functions into a satellite division and a piloted division, creating the Manned Space Flight Network (MSFN), which operated concurrently with the STADAN satellite-tracking system. In addition to land-based stations, MSFN used eight aircraft and five ships to provide a comprehensive network of facilities that could communicate with Mercury, Gemini, and Apollo astronauts, receive telemetry signals, and command both piloted spacecraft and robotic target vehicles such as those used during Gemini flights VIII through XII.

A total of twenty-two MSFN ground stations were located in White Sands, New Mexico; Corpus Christi, Texas; Eglin Air Force Base and Merritt Island, in Florida; Point Arguello and Goldstone, in California; Kauai, Hawaii; Antigua; Ascension Island; the Canary Islands; Bermuda; Canton Island; Grand Bahama Island; Grand Turk Island; Guam; Canberra, Carnarvon, and Muchea, in Australia; Guaymas, Mexico; Kano, Nigeria; Madrid; and Tananarive, Madagascar. In 1972, the STADAN and MSFN systems were unified to create the STDN system.

Because of the complexities of receiving data from piloted and robotic spacecraft and relaying data among the several STDN facilities, NASA inaugurated the TDRSS system with the 1983 launch of TDRS-A, which became TDRS-1 when it was successfully placed in orbit. TDRS-B was on board the space shuttle *Challenger* when it exploded shortly after launch on January 28, 1986. TDRS-C was the payload on the space shuttle *Discovery*, launched in September, 1988. TDRS-D was deployed in March, 1989, and TDRS-E in August, 1991. In January, 1993, the fifth TDRS (TDRS-F) was deployed from the cargo bay of *Endeavour*. TDRS-G (TDRS-7) was deployed from *Discovery* in July, 1995.

TDRS-8 lifted off at 12:56 UTC (8:56 A.M. eastern daylight time) on June 30, 2000 from Cape Canaveral Air Force Station, Florida, aboard an Atlas-2A rocket. It was the first of three new satellites featuring improved multiple-access and S-band single-access performance, along with a new high-frequency Ka-band service. The Advanced TDRS satellites are 21 meters long with solar arrays deployed and 13 meters wide with antennae deployed.

TDRS-9 was placed into orbit on March 8, 2002, by Atlas-2A AC-143. The third and final Advanced TDRS, TDRS-10, separated from the Centaur upper stage 30 minutes after launch on December 4, 2002. This completed the $800 million, three-satellite system.

The TDRSS network was designed to replace STDN as NASA's primary tracking system, using satellites in geostationary orbits above the equator to receive data from other spacecraft and relay them to the White Sands Ground Terminal.

See also: Deep Space Network; Goddard Space Flight Center; Orbiting Solar Observatories; Space Shuttle Flights, 1983; Space Task Group; Tracking and Data-Relay Communications Satellites; Vandenberg Air Force Base.

Further Reading

Burrows, William E. *This New Ocean: The Story of the First Space Age*. New York: Random House, 1998. This is a comprehensive history of the human conquest of space, covering everything from the earliest attempts at spaceflight through the voyages near the end of the twentieth century. Burrows is an experienced journalist who has reported for *The New York Times*, *The Washington Post*, and *The Wall Street Journal*. There are many photographs and an extensive source list. Interviewees in the book include Isaac Asimov, Alexei Leonov, Sally K. Ride, and James A. Van Allen.

Butrica, Andrew J. *Beyond the Ionosphere: Fifty Years of Satellite Communications*. NASA SP-4217. Washington, D.C.: Government Printing Office, 1997. Part of the NASA History series, this book looks into the realm of satellite communications. It also delves into the technology that enabled the growth of satellite communications. The book includes many tables, charts, photographs, and illustrations, a detailed bibliography, and reference notes.

Elbert, Bruce R. *Introduction to Satellite Communication.* Cambridge, Mass.: Artech House, 1999. This is a comprehensive overview of the satellite communication industry. It discusses the satellites and the ground equipment necessary to both the originating source and the end-user.

National Aeronautics and Space Administration. *Entering the Era of the Tracking and Data-Relay Satellite System: NASA Facts/Goddard Space Flight Center.* Washington, D.C.: Government Printing Office, 1987. This brochure introduces STDN and TDRSS to the layperson. It also discusses the importance of ground stations to the success of both piloted and robotic space missions.

_____. *Space Network Users' Guide (SNUG).* Washington, D.C.: Government Printing Office, 2002. The *Space Network User's Guide* is intended as a guide to the user community for obtaining communication support from the Space Network. The emphasis in the *User's Guide* is on the interfaces between the user ground facilities and the Space Network. Other topics include the radio frequency interface between the user spacecraft and the Tracking and Data-Relay Satellite; procedures for working with the Goddard Space Flight Center (GSFC) Space Communications program; space network capabilities and service characteristics; and general information pertaining to the operational aspects of the Space Network.

Rosenthal, Alfred. *The Early Years, Goddard Space Flight Center: Historical Origins and Activities Through December, 1962.* Washington, D.C.: Government Printing Office, 1964. This commemorative manual provides a precise and comprehensive look at the founding of Goddard Space Flight Center, the Minitrack network, and the beginnings of the Satellite Network.

Eric Christensen, updated by Russell R. Tobias

SPACEHAB

Date: Beginning 1983
Type of program: Scientific platform

SPACEHAB is a commercial mini-laboratory that fits into the cargo bay of the space shuttle. Designed to be leased to private corporate interests, it represents the expansion of outer space from a governmental research arena to a locale for business and commerce.

Key Figures

Robert Citron (b. 1932), founder of SPACEHAB, Inc.
James M. Beggs (b. 1926), SPACEHAB, Inc., chairperson
Richard K. Jacobson (d. 2001), president and chief executive of SPACEHAB from 1987 to 1991
Janice E. Voss (b. 1956), mission specialist aboard first SPACEHAB flight
G. David Low (b. 1956), payload commander aboard first SPACEHAB flight

Summary of the Facility

SPACEHAB (Space Habitat Module) is a privately produced, pressurized, cylindrical research module designed to fit in the space shuttle's cargo bay. It reflects the National Aeronautics and Space Administration's (NASA) increased emphasis on cost savings through use of private money.

The intent of SPACEHAB is to promote the use of the microgravity environment of space to produce products such as drugs, crystals, and fine machine parts more efficiently, perfectly, and economically than Earth manufacturing. For example, some materials can be produced at high temperatures in space without the use of containers, which on Earth can add contaminants to the product.

Robert Citron, a former Smithsonian Institution scientist, who at first intended to build a module that would carry tourists into orbit, founded SPACEHAB in 1983. This idea was soon abandoned in favor of a module for scientific experiments. SPACEHAB, Inc., contracted the actual building of the modules out to McDonnell Douglas Space Systems Company, under the leadership of Richard K. Jacobson, SPACEHAB president and chief execu-

tive during SPACEHAB's design period. Under an agreement with NASA, SPACEHAB buys launch services from the space agency and in turn leases its capacity to users, such as corporations and universities.

A major hurdle in initiating for-profit laboratory space on the shuttle was the reluctance of private industry to invest in such a new and untested venture. To address this reality, the U.S. government, in 1985, promoted the establishment of the Centers for Commercial Development of Space (CCDS), a nonprofit consortium for conducting space-based, high technology research and development. The intent was to encourage U.S. industry leadership in commercial space-related activities. The seventeen CCDS centers operate with government grants to provide American companies and universities with opportunities to carry out low-cost, space-based commercial research and development. Often, a company interested in SPACEHAB works with one of the CCDS centers.

A SPACEHAB Middeck Augmentation module consists of up to sixty-one separate compartments or "lockers," each capable of housing a separate experiment. An individual locker has a volume of 0.06 cubic meters (2.2 cubic feet). There is also space for accommodating larger experiments in one or more single or double "racks" (volume limit 0.63 cubic meters or 22.5 cubic feet, and 1.3 cubic meters or 45 cubic feet, respectively). The module is 2.7 meters long and 4.1 meters in diameter and weighs 4,220 kilograms. It has a truncated top and flat end caps and sits in front of the shuttle's cargo bay, occupying about a fourth of the bay's space. Astronauts are able to enter and leave the module through a special hatch. A single SPACEHAB module doubles the available living and working space on the otherwise cramped shuttle and quadruples the experimentation space. It can be flown on any of the four shuttle orbiters, which are modified with special attachments to accommodate the unit. Each SPACEHAB mission has one full-time mission specialist to tend the module's experiments, as well as one mission specialist required part time for this task. Before a launch, the experiments are processed, integrated into lockers, and installed in the modules at SPACEHAB, Inc.'s payload processing facility in Port Canaveral, Florida.

In 1988, NASA signed an agreement with SPACEHAB, Inc., allowing the company to load SPACEHAB on six shuttle flights. NASA itself purchased back two hundred of the three hundred experiment lockers that would be available. SPACEHAB 1 was launched on June 21, 1993, aboard the shuttle *Endeavour* mission STS-57.

SPACEHAB 1 was the maiden flight of the program and module Flight Unit One. A six-person crew operated SPACEHAB experiments, with the majority being carried out by Mission Specialist Janice E. Voss and Payload Commander G. David Low. Twenty-two experiments were flown in the module and the shuttle middeck. These experiments were designed by both American and European scientists and involved investigations in the biomedical and materials sciences. Most were of a scale small enough to fit into SPACEHAB's bulkhead-mounted lockers.

As it has long been known that gravity causes the growth of imperfect crystals on Earth, several SPACEHAB experiments took advantage of the shuttle's microgravity environment to study the growth of crystals in space. Five other experiments measured the environment within SPACEHAB, including the light, sound, high-energy particle, and acceleration levels. Four payloads investigated the growth and separation of living cells. One experiment used rodents to test the effects of drugs on adaptation to space. Two others examined the effects of weightlessness on the astronauts themselves, including a study of the so-called "zero-g crouch," a postural change that affects humans in space. To document this phenomenon over the duration of a space mission, still and video images of crew members in a relaxed position were recorded at both early and late stages in the mission. Three hardware experiments were directly related to space sta-

The SPACEHAB-4 payload that flew on space shuttle mission STS-77 was installed in the payload canister transporter before it headed to Launch Pad 39B for liftoff. (NASA)

tion concepts: One dealt with water filtration; another involved the in-orbit repair of electronics, which included the first soldering ever conducted on a spaceflight; and a third examined the lighting and nutrient needs of plants.

In addition, experiments were carried out dealing with the transfer of fluids in weightlessness without creating bubbles. This experiment, called the Fluid Acquisition and Resupply Experiment (FARE), was designed to study filters and processes connected with the refueling of orbiting spacecraft.

The majority of the first SPACEHAB experiments achieved their scientific objectives, although nine required resources and support from the crew that had not been foreseen at the time of initial request. All of the support systems cooling, AC and DC power, computer monitoring, and video functioned without problem.

SPACEHAB 2 was the maiden flight of module Flight Unit Two, carried aboard the space shuttle *Discovery* launched on February 3, 1994, for an eight-day mission of STS-60. Twelve experiments were housed in the SPACEHAB module and the middeck of the shuttle itself.

SPACEHAB 2 contained several experiments that were reflown from SPACEHAB 1: two acceleration measurement setups, an upgraded test of a plant growth system, protein crystal growth hardware, and an investigation of the organic separation of cells. New experiments examined space adaptation in rodents as well as cell and crystal growth. A hallmark for SPACEHAB 2 was the first externally mounted experiment, which collected cosmic dust from the top of the module. As with SPACEHAB 1, the support systems of SPACEHAB 2 functioned flawlessly, supplying electricity, cooling, computer data, and video.

The third SPACEHAB flight, involving the module Flight Unit Three, was also aboard the space shuttle *Discovery*, on STS-63, launched February 3, 1995. This mission was unique in that its primary objective was to perform a rendezvous and fly around of the Russian space station Mir. NASA had already signed a contract leasing SPACEHAB to resupply Mir in 1996 and 1997.

The experiences with SPACEHAB 1 and 2, although by and large successful, reemphasized the precious nature of the astronauts' time. Due to the demands of operating the shuttle itself and tending to mission priorities, a plan for reducing human interface time with the SPACEHAB module was developed prior to the launch of SPACEHAB 3. SPACEHAB, Inc.'s response was to develop improved equipment to automate a number of manual tasks. The first of these new features was a video switch to reduce the time the crew had to dedicate to video operations. Another was the installation of a system to relieve the astronauts of some of the responsibilities for the downlink of data. By far the most creative, interesting, and efficient of these labor-saving innovations was the development of a robot named Charlotte to carry out many of the experiment-tending tasks previously done by the astronauts themselves. The robot moves along cables and can perform many routine procedures, such as changing experiment samples. Charlotte was controlled by a mission specialist using a laptop computer while scientists on Earth observed their experiments on television transmitted through cameras on the front of the robot. Charlotte was also able to digitize, compress, and downlink still images taken from the video system. The experience with Charlotte marked the first time that a robot has worked together with astronauts in the same area in a space vehicle.

During SPACEHAB 3's flight, the *Discovery* crew carried out some twenty experiments, mostly associated with the research and development of commercial products, including experiments for new pharmaceuticals and advanced materials for improved contact lenses. The *Discovery* crew also conducted an experiment that examined how materials burn in weightlessness, using Plexiglas in this instance.

As result of SPACEHAB's superior performance on three shuttle flights through 1999, its near-term future looked promising. SPACEHAB, Inc., developed a Double Module and a connecting tunnel to the shuttle. The first Double Module flew on STS-79 in September, 1996, carrying supplies to the

Russian Mir Station and retrieving experiments from it. Five additional resupply flights were undertaken in 1997 and 1998.

The SPACEHAB module flown on STS-95 provided additional pressurized workspace for experiments, cargo and crew activities. The module flew in the forward portion of *Discovery*'s payload bay with the crew gaining access to the module through the air-lock tunnel system. A variety of experiments sponsored by NASA, the Japanese Space Agency (NASDA), and the European Space Agency (ESA) focused on life sciences, microgravity sciences, and advanced technology during the flight.

The seven-person crew of STS-96, an International Space Station (ISS) logistics and resupply mission, carried internal and resupply cargo for station outfitting. The SPACEHAB Logistics Double Module's (LDM) standard experiment accommodations include up to 61 bulkhead-mounted middeck locker locations as well as floor storage for large unique items and soft stowage. The LDM has a capacity of up to 4,500 kilograms with the ability to accommodate powered payloads. It also has four external rooftop stowage locations and four double-rack locations (two powered). SPACEHAB's Integrated Cargo Carrier (ICC) carried the Russian cargo crane, Strela, which was mounted to the exterior of the Russian station segment; the SPACEHAB Oceaneering Space System Box (SHOSS) and a U.S.-built crane (ORU Transfer Device, or OTD). The ICC is an externally mounted, unpressurized, aluminum flatbed pallet, coupled with a keel-yoke assembly that expands the shuttle's capability to transport cargo. The ICC can carry up to 2,700 kilograms of unpressurized payload. DaimlerChrysler Aerospace of Bremen, Germany, and RSC Energia of Korolev, Russia, built it for SPACEHAB.

In July, 1999, STS-93 flew the first mission of the Space Technology and Research Students (S*T*A*R*S) Program. Two schools designed life science experiments and received live shuttle downlink video that they used to compare space experiment results with those of classroom experiments. Five additional schools conducted ground-based experiments in conjunction with those flown by the lead schools.

U.S. and Russian hardware for the ISS is carried in the SPACEHAB logistics double module. The seven-member STS-101 crew transferred more than 1,250 kilograms of U.S. supplies and more than 1,000 kilograms of Russian supplies from the module to the Unity and Zarya Modules of the ISS. The logistics included clothing and personal hygiene articles, health care supplies, exercise equipment, food, television and film equipment, a fire detection and suppression system, computers, and sensors. In addition to the logistics and maintenance cargo, SPACEHAB carried a commercial payload, the Self-Standing Drawer—Morphological Transition and Model Substances. STS-101 was the fourteenth flight of SPACEHAB.

One of the primary objectives of the STS-106 mission included outfitting tasks in the Zvezda Service Module of the ISS. The crew transferred approximately 3,674 kilograms of hardware, equipment, and logistical supplies to outfit the Space Station. On STS-106, the ICC carried 1,300 kilograms of cargo to orbit.

On STS-102 the first 9.1-metric ton, Italian-built Multi-Purpose Logistics Module (MPLM) was delivered to the ISS. Named Leonardo, it was lifted out of *Discovery*'s payload bay and attached directly to the Destiny Laboratory Module. Its cargo included six systems racks and the Human Research Facility experiment rack. The MPLM essentially replaced the SPACEHAB double module with a cargo capacity of 9,000 kilograms of cargo. It can hold sixteen International Standard Payload Racks (ISPRs). *Discovery*'s payload bay also held the Integrated Cargo Carrier, which included the Lab Cradle Assembly with Module Truss Structure Attach System installed, Pump Flow Control Subassembly with attached Flight Support Equipment External Stowage Platform, and the Rigid Umbilical. The Lab Cradle Assembly was installed on Destiny's aft zenith trunnion during the mission's first spacewalk.

STS-105 brought the third MPLM to the ISS. The Early Ammonia Servicer (EAS) and the Materials International Space Station Experiment (MISSE) were taken into space in two Passive Experiment Carriers on the Integrated Cargo Carrier in *Discovery*'s cargo bay. The power generation system on the ISS uses ammonia for cooling and refrigeration and the EAS is the backup system if the primary system fails. MISSE (pronounced "missy") tested the durability of hundreds of material samples ranging from lubricants to solar cell technologies. The samples, engineered to better withstand the punishing effects of the Sun, extreme temperatures, and other elements, were flown outside the ISS and, where they were unprotected by Earth's atmosphere. By examining how the coatings fare in the harsh environment of space, researchers hope to gain information that will help them develop materials for future spacecraft as well as materials that last longer on Earth.

Columbia, on STS-107, carried the SPACEHAB Research Double Module (RDM) in its payload bay. The RDM is a pressurized environment that is accessible to the crew while in orbit via a tunnel from the shuttle's middeck. Together, the RDM and the middeck housed the majority of the mission's payloads and experiments. STS-107 marked the first flight of the RDM. Experiments in the SPACEHAB RDM included nine commercial payloads involving twenty-one separate investigations: four payloads for the European Space Agency with fourteen investigations, one payload/investigation for ISS Risk Mitigation, and eighteen payloads supporting twenty-three investigations for NASA's Office of Biological and Physical Research. When *Columbia* broke up during reentry on February 1, 2003, the RDM was destroyed.

STS-114, the first of two "Return to Flight" missions, was launched on July 26, 2005. *Discovery*'s payload bay carried the Raffaello Multi-Purpose Logistics Module, an Italian-built module that can deliver supplies and cargo to the International

The first flight of the commercially developed SPACEHAB module aboard the space shuttle Endeavour *(STS-57).* (NASA)

Space Station. *Discovery* also carried SPACEHAB's External Stowage Platform 2 (ESP2), a modified version of the company's Integrated Cargo Carrier system. ESP2 was designed to carry replacement parts, known as Orbital Replacement Units (ORU), to the orbiting ISS. This resupply platform was attached to the Space Station's air lock and marked SPACEHAB's first permanent hardware residence on the ISS. The ESP2 consists of two major components: the keel yoke assembly and the deployable pallet.

Plans call for SPACEHAB to have payloads on at least two more shuttle flights to the ISS: STS-116 and STS-118. STS-116 will be a crew rotation and logistics mission that utilizes the SPACEHAB Single Module. STS-118 will be the first mission for Astronaut Barbara Morgan. The mission will also use the SPACEHAB Single Module and ICC as the shuttle resupplies the ISS.

Contributions

The intent of the SPACEHAB program is to determine whether outer space can serve as a part of the national economy, a place where resources are produced rather than consumed.

SPACEHAB performed flawlessly on its first three flights, the research experiments it carried generating revolutionary advances in biotechnology, materials sciences, and other technologies. This type of product-oriented research is essential to preparing U.S. industry for the era of the Space Station. The successes achieved aboard SPACEHAB have demonstrated the viability of a partnership in space between government and the private sector.

Although many of SPACEHAB's experimental results, such as those from the external cosmic dust collector, are still undergoing evaluation, many findings became clear almost immediately. In the area of advanced materials science, Paragon Vision Sciences Corporation of Mesa, Arizona, a leader in the production of oxygen-permeable contact lenses, placed a polymerization experiment on board the first SPACEHAB flight. The experiment mixed raw materials currently in use on Earth with a new type of material that demonstrated a high permeability. The resulting product proved to be more permeable than what was produced on Earth, allowing almost four times as many lenses to be sliced from the same amount of material.

SPACEHAB 2 carried an experiment dealing with the growing of insulin crystals in microgravity. The results revealed new information about insulin's molecular structure, prompting the development of a time-release substance that can be combined with insulin, reducing the frequency of injections for diabetics.

SPACEHAB 3 carried an experimental unit called Astroculturex, sponsored by the Wisconsin Center for Space Automation and Robotics, a NASA Center for the Commercial Development of Space. This setup demonstrated the successful flowering of plants in space, a powerful indication of the degree of environmental control attainable in a compact and reliable flight package.

The small robot Charlotte, which flew aboard the third SPACEHAB mission, was unique in that it was an experiment that interacted with the operations of the shuttle itself. Charlotte was suspended from eight cables emanating from the corners of SPACEHAB, creating a "web" along which the robot moved. This type of suspension configuration eliminated heavy gantries, reducing the robot's weight and increasing its flexibility. This represented a revolutionary advance in robotics technology. The flawless performance of Charlotte demonstrated the ability of robots to assist astronauts in space with routine tasks. The robot's camera equipment also allowed scientists on Earth to monitor their experiments during the astronauts' sleep period.

The performance of SPACEHAB provided guidance for NASA as it developed plans for the age of the Space Station. SPACEHAB's agreement with NASA, allowing it to ferry supplies to the Russian space station Mir in 1996 through 1998, gave U.S. industry opportunities to participate in the build-up phase of the space-station era, furthering the development of space as an area of private and commercial interest.

Context

When SPACEHAB 1 blasted off aboard the space shuttle *Endeavour* on June 21, 1993, hopes were high that the microgravity found only in space would allow experiments to be conducted that would lead to improved manufacturing processes, better electronic components, and life-saving drugs.

There were also tremendous doubts about the time being right for the successful commercialization of space. Cost was the main concern that caused hesitation on the part of corporations to invest in SPACEHAB's promise: The price of the transportation is still exorbitant. As long as products can be manufactured more cheaply on Earth, space will not be an option. At present it costs about $100,000 per pound to ship materials into space for processing and return a commercial product. Therefore, this product would have to be worth much more than $100,000 per pound.

Another complication is that no one at present has a clear idea of which space-made products will be commercially feasible. SPACEHAB, then, cannot be considered at this time to be a manufacturing plant in the commercial sense. Rather, it is a host for corporate experiments to determine whether space can, in fact, manufacture products more efficiently, economically, and perfectly than on Earth. Assuming positive answers to these questions, the next phase would be full-scale manufacturing in space.

Because of these doubts, SPACEHAB, Inc., initially had great difficulty securing financing from banks. It had no customers and therefore no immediate way to repay its loans. The breakthrough came when NASA bought most of the locker space on the first three SPACEHAB flights, which inspired confidence in a number of private entities to invest in space experimentation. This enabled SPACEHAB, Inc., to secure the funding it needed to proceed with the program.

There is still one fundamental disadvantage, however, that SPACEHAB has not overcome: It is dependent on buying government launch services to get its modules into space. NASA, in turn, is unable to guarantee SPACEHAB's launch agreements, dependent as it is on fluctuating government support. This means that SPACEHAB cannot promise its corporate customers that its modules will be launched on schedule. For this reason many potential clients remain leery, and SPACEHAB's continued success cannot be realized without ongoing support from NASA.

In spite of these realities, SPACEHAB's first three flights proved the feasibility of space-based manufacturing and increased enthusiasm for such ventures. Research in SPACEHAB's laboratories has generated revolutionary advances in biotechnology, advanced materials, and other technologies. NASA's purchase of locker space on board at least twelve subsequent flights is further guarantee that continued opportunities for such experimentation will be there. SPACEHAB's gamut of successes has created a sense of urgency about creating space facilities for manufacturing on a larger scale, specifically the International Space Station, in which SPACEHAB is playing a significant role. The ESA, the NASDA, and corporations around the world are developing payloads for future SPACEHAB missions.

Although supplanted by the larger Multi-Purpose Logistics Modules, SPACEHAB continues its contributions to the ISS by providing flexible-configuration pressurized and open cargo carriers. With the cancellation of all shuttle flights not related to the ISS, no plans have been made to replace the Research Double Module lost on STS-107.

See also: Space Shuttle; Space Shuttle Flights, 1994; Space Shuttle Mission STS-63; Space Shuttle Flights, 1996; Space Shuttle Flights, 1997; Space Shuttle Flights, 1998; Space Shuttle Mission STS-95; Space Shuttle Mission STS-107; Space Stations: Origins and Development; Spaceflight Tracking and Data Network.

Further Reading

Banham, Russ. "Insuring the Next Frontier." *Risk Management* 40 (November 11, 1993): 29 (6). The only comprehensive article available on SPACEHAB. It offers detailed information about some of the difficulties involved in getting SPACEHAB off the ground, as well as a discussion of the feasibility of space-based manufacturing. Also examines the private sector's view of SPACEHAB's potential. Nontechnical, accessible reading.

Harland, David M. *The Space Shuttle: Roles, Missions, and Accomplishments.* Hoboken, N.J.: John Wiley, 1998. Topics include shuttle operations and payloads, weightlessness, materials processing, exploration, Spacelabs and free-flyers, and the shuttle's role in the International Space Station.

Haskell, G., and Michael Rycroft. *International Space Station: The Next Space Marketplace.* Boston: Kluwer Academic, 2000. Addresses issues of ISS utilization and operations from all perspectives, especially the commercial viewpoint, as well as scientific research, technological development, and education in the widest sense of the word. Of interest to those working in industry, academia, government, and particularly public-private partnerships.

Jenkins, Dennis R. *Rockwell International Space Shuttle.* Osceola, Wis.: Motorbooks International, 1989. Includes a brief yet concise history of the orbiter and its predecessors. Contains dozens of close-up color and black-and-white photographs, detailing the orbiter's exterior and interior features, including the major subsystems. Some of the text, especially the data tables, is printed in extremely small type.

McCurdy, Howard E. *The Space Station Decision.* Baltimore: Johns Hopkins University Press, 1990. A historical account of the growth of thought from the space program's early days regarding the construction of a space station. This is a thoroughgoing study of the politics, people, and plans for an orbiting space station, written almost like a novel, full of strongly defined personalities and moments of high drama.

National Aeronautics and Space Administration, Space Station Task Force. *Space Station Program Description, Applications, and Opportunities.* Park Ridge, N.J.: Noyes Publications, 1985. This formidable volume, written in a matter-of-fact style, is for the informed layperson with a serious interest in the Space Station program. Pages 160-190 are relevant to the SPACEHAB program, as they detail the commercial possibilities for space-based manufacturing. The book suffers, however, from lack of an alphabetized index.

SPACEHAB. http://www.spacehab.com. The Web site of SPACEHAB. It provides information about the company and the various modules. It also provides historical data on the SPACEHAB missions and plans for the future. Accessed April, 2005.

Summers, Carolyn. *Toys in Space: Exploring Science with the Astronauts.* Blue Ridge Summit, Pa.: TAB Books, 1994. Directed at upper elementary grades through high school, this book combines concise, well-written doses of theory with experiments young people can carry out to emulate the experiments conducted aboard the space shuttle as part of the Toys in Space project, which explored how common toys behave in the zero-gravity conditions of space.

Robert Klose, updated by Russell R. Tobias

Spacelab Program

Date: August, 1973, to December, 1997
Type of program: Scientific platform

Spacelab was a major space shuttle payload designed to provide scientists with facilities approximating those of a terrestrial laboratory.

Key Figures

James C. Fletcher (1919-1991), NASA administrator
Alexander Hocker, director general of ESRO
Roy Gibson, director general of ESA
Douglas R. Lord, Spacelab program manager
Thomas J. Lee, Spacelab project manager
John Thomas, Spacelab project manager
James Downey, Spacelab missions manager
Jesse W. Moore, Spacelab missions manager

Summary of the Program

Because Spacelab was designed to operate within the payload bay of the space shuttle orbiter, the configuration interface between it and the concurrently designed shuttle was sometimes problematic. Components of Spacelab often had to be redesigned in order to meet changing shuttle requirements. In particular, a major redesign of Spacelab's instrument pointing system was required. Starting in 1974, Spacelab passed through many tests and design reviews as hardware was planned and built. These led to final acceptance reviews in 1981 and 1982, when the elements of what was termed Flight Unit I were delivered to the National Aeronautics and Space Administration (NASA).

Meanwhile, NASA had organized the management of Spacelab within its network of facilities. Marshall Space Flight Center (MSFC) in Huntsville, Alabama, was to oversee the work of the European Space Agency (ESA) on Spacelab and assure that agency's compliance with shuttle standards. (Later, MSFC was given responsibility for develop-ing additional missions and for providing the hardware to other NASA centers that also prepare and conduct Spacelab missions.) NASA issued an "announcement of opportunity" to space scientists, asking them to propose experiments that might be performed aboard the first two Spacelab missions. Because these were verification flights, NASA tried to accommodate as many scientific disciplines as possible. The payload mass was allocated equally between NASA and the ESA for Spacelab 1, while Spacelab 2 was primarily an American mission (but European scientists were invited to propose experiments). Researching a path for the complete Spacelab, NASA flew engineering models of Spacelab pallets on the STS-2 and STS-3 shuttle missions in 1981 and 1982. As part of the exercise, the pallets carried science instruments that gathered useful data.

The final configuration for Spacelab comprised pressure modules and open pallets in addition to equipment designed to join these components and provide supports for the experimental gear they would carry. Spacelab is controlled by crew members operating a computer either in the module or in the aft flight deck of the shuttle.

The module was designed with core and experiment components. Each segment is 2.70 meters wide and 2.88 meters long. With end cones, a short module measures 4.27 meters in length and a long module 6.96 meters in length. The interior arrangement includes a floor to cover the support systems, equipment racks placed on each side of the module, overhead storage areas, and a small access science port. Designed as "singles" and "doubles," the racks were 1.48 centimeters wide and capable of holding up to 290 and 580 kilograms of experimental gear, respectively.

In the Spacelab core module, the two forwardmost double racks were dedicated as the control station (starboard) and the workbench (port). That left two double racks and two single racks (one each, port and starboard) for use by experimenters. The experiment segment added another four double racks and two single racks.

A space shuttle mission STS-9 onboard view shows the European Space Agency's Spacelab 1 (SL-1) module in orbiter Columbia's *payload bay.* (NASA)

Additional experiments could be accommodated by an optical quality viewport mounted over the core segment and a small science air lock in a similar position in the experiment segment. The viewport had a removable exterior cover so that it was protected from the space environment and shuttle contamination, except when used by medium-sized cameras mounted by the crew. The air lock allowed the payload crew to expose equipment up to 1 meter long and 0.98 meter wide into space.

Linking the module to the shuttle cabin was the transfer tunnel, 1.02 meters in diameter. It was assembled from a set of cylindrical sections to match different module lengths and locations. It also had flexible sections to allow for slight bending in the airframe during ascent and entry.

The other major element of the Spacelab system was the pallet, a U-shaped platform that provides an interface between experiment hardware and the shuttle itself. Each pallet was 4 meters wide and 2.9 meters long and, like the modules, could be joined with similar components. Each pallet was made up of five angular U-shaped frames joined by longitudinal members and covered with metal plates. The inner plates had a pattern of bolt holes in a 14-by-14-centimeter grid for mounting lightweight hardware; twenty-four hard points were provided for heavy equipment. The pallets also provided routing for cooling equipment, electrical cables, and other support services.

Both the pallets and the modules were held in the payload bay by sill and keel trunnions, 8.25-centimeter pins locked down by special clamps bolted to the orbiter structure. The modules and pallets could be grouped in almost a dozen configurations depending on mission needs. The module could be flown "long" or "short," with or without one to three pallets. Up to three pallets could be joined in a train, and up to five flown at once.

Spacelab was totally dependent upon the space shuttle for electrical power, environmental control, and life-support. Power and environmental gear provided in Spacelab's subfloor area were designed to assist the shuttle in that respect. Spacelab did have its own Command and Data Management System (CDMS), through which the crew could control experiments. The CDMS commanded experiment apparatus and collected data from them by way of Remote Acquisition Units (RAUs), which functioned somewhat like sophisticated telephone exchanges. The rapid advance of micro-electronics in the late 1970's, however, relegated the CDMS to the role of traffic controller for the various experiments, which often had their own microprocessors. The Spacelab CDMS actually comprised three central processing units: one to operate Spacelab proper, one to operate the experiments, and a third held as a manually selected backup. In missions using the module, the CDMS was housed in the starboard forward double rack. For pallet-only missions, it was housed in a pressurized container, the "igloo," mounted on the forwardmost pallet. The igloo provided a sea-level environment for the CDMS, thus eliminating the need to prepare the computer for the environment of raw space.

For pointing large telescopes or telescope clusters at targets, an Instrument Pointing System (IPS) was provided. The IPS had three electrically driven gimbals that could point the IPS payload within extremely close range of a target. The IPS was mounted on a support framework on a pallet, and, in turn, provided a large, circular equipment platform for the payload. Payloads could weigh up to 7,000 kilograms and could be several meters long. Not all payloads requiring pointing could justify use of the IPS, so experimenters developed smaller pointers tailored to their investigations.

Assembly of a Spacelab mission was a long, complex process involving several levels of effort. After the science community identified important investigations, NASA performed a preliminary study of the kinds of instruments that might satisfy their needs. Instruments generally fell into two classes: the principal investigator and the facility. In the first, an individual scientist or science team developed an instrument for a narrow investigation. In the second, NASA and a contractor developed an instrument that could serve a number of scientists on many missions. Experiments on Spacelabs 1, 2, and 3 were developed from announcements asking specifically for them. In 1978, NASA issued a broader announcement soliciting instruments in physics and astronomy. Forty instruments were selected, some of which were grouped for Spacelab or other missions and some of which were later canceled. Other announcements were issued for life science missions and facility-class instruments.

After the science investigations were selected, an Investigators' Working Group (IWG) was formed from the lead scientists. NASA appointed a mission manager and mission scientist from its own ranks. The IWG and NASA engineers worked closely together to develop the flight plan and details of how and when each investigation was to be conducted during the mission. It was not unusual to discover that some experiments would not fit in or would be late. This normally resulted in an instrument's being moved to a later mission rather than its being canceled.

As it became ready, experiment hardware would be delivered to Kennedy Space Center, Florida, for integration into the complete Spacelab. The first step in the process was to install the experiment elements in racks or on pallets. The racks and floor were then fitted into the module, the module was closed, and the module and pallets were physically and electrically joined. The complete assembly was placed inside the Cargo Integrated Test Equipment (CITE) stand, where all the components were exercised as they would be in flight. Finally, the complete Spacelab was installed in the space shuttle, and an "end-to-end" test was conducted to validate all links from the experiment to the control center.

Typically, a Spacelab mission included three types of crew members: pilot astronauts, the mission commander and pilot, flying the shuttle itself; mission specialists, career NASA scientist astronauts with overall responsibility for the payload; and payload specialists, members of the IWG se-

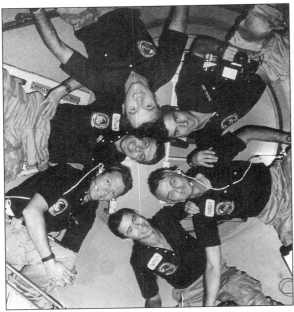

The STS-9 crew members inside Spacelab. (NASA)

lected to fly on the mission and to conduct the experiments. Two payload specialists, prime and alternate, were selected for each flight opening.

The inclusion of payload specialists on the Spacelab missions was a major point used by NASA in selling Spacelab and the shuttle to the science community. Previously, scientists could only listen or watch from the ground while their experiments were conducted by career NASA astronauts. With the routine operations to be provided by the shuttle, scientists could fly, almost passenger-like, with the experiments that they had developed. The process turned out to be slightly more complex, but the basic philosophy held.

Spacelab missions would start a few hours after the shuttle achieved orbit and would last until about four hours before reentry. When the shuttle's in-orbit time was shorter than originally planned—for example, ten days instead of thirty—mission activities were intense and would go around the clock. Typically, a six- or seven-person crew would operate in three-person, twelve-hour shifts.

Spacelab missions were directed from two control centers. The first, Mission Control at Johnson Space Center, retains overall control of and responsibility for the completion of the flight. For the most part, Mission Control defers to the Payload Operations Control Center (POCC), where the science phase of the mission is directed. Thus, Spacelab was heavily dependent on the Tracking and Data-Relay Satellite System (TDRSS) to relay telemetry from the experiments to the POCC and commands back from the POCC.

Contributions

Spacelab proved a versatile and useful facility for conducting space science research. Spacelabs 1, 2, 3, and D-1 (this last set of experiments were sponsored by West Germany)—and the various single-pallet payloads flown on STS-2, STS-3, and STS-41G—were all successful. Unfortunately, in the view of many scientists, NASA made use of the facility too difficult. In fact, in the era preceding the 1986 *Challenger* accident, the agency replaced its own science payloads with commercial and military equipment. Thus, scientists soon found themselves in a sort of inflationary spiral where the cost of a mission required extreme efforts to guarantee success, which, in turn, raised the cost of the mission.

To combat this problem, NASA conducted a Spacelab mission integration cost analysis and developed a concept known as the Dedicated Discipline Laboratories (DDLs). Each DDL would comprise a group of experiments with similar or complementary mission requirements. For example, it would be logical to carry astronomical instruments and solar instruments on one mission, because they would have similar pointing requirements during a mission. One would not, perhaps, think to carry materials and life science experiments together until one compared their needs: heavy electrical power demands and intermittent tuning for materials experiments, and intense people-power and low-power demands for life sciences. Yet procedures often required of biomedical experiments can be disruptive to crystal-growth and other fluid experiments. Thus, carrying them together would require innovative scheduling to avoid conflicts.

At the very least, the DDL concept would reduce the integration cost of Spacelab missions by reducing the analysis and paperwork required for each mission. At best, much work could be avoided by allowing clusters of instruments to remain intact until their next flight. Even requests to upgrade instruments were disregarded in order to cut costs.

The Spacelab and shuttle experiences also contributed to a better understanding of what is required to support a vigorous experiment program. This knowledge led to the development of intermediate payload carriers between Spacelab and the Get-Away Specials, an innovative payload system. The effort required to replace even a single rack inside the module affected the design of the U.S. space station, so its racks were better designed for easy replacement in orbit.

Space shuttle Columbia *(STS-65) astronauts Richard J. Hieb (right) and James D. Halsell work on Spacelab experiments in the International Microgravity Laboratory (IML-2). (NASA)*

Context

Spacelab proved difficult for scientists from some disciplines to use. Materials scientists need as smooth a ride as possible so that samples are not jostled (excessive motion disrupts the formation of crystals and the study of fluid flows). These required conditions are at odds with necessary crew exercise periods and even with pumps and fans that cool Spacelab. Early shuttle missions discovered a phenomenon known as "shuttle glow," an eerie luminescence that peaks in the infrared spectrum. The cause remains under debate but appears to be some chemical reaction between the shuttle itself and rare molecular species in the upper atmosphere. The shuttle glow hampers observations in the infrared and low-light levels under certain conditions.

Spacelab grew out of a 1969 invitation by NASA for the European Space Research Organization (ESRO) to become involved in the post-Apollo space program. European involvement in U.S. space activities had been commonplace since the origins of NASA but rarely had been larger than limited partnerships on small satellite projects. The European Space Conference in 1970 authorized studies with the United States in the post-Apollo area. In 1972, NASA selected the space shuttle program as its major effort for the 1970's.

As conceived, the space shuttle was to have a reusable third stage, called the Space Tug, to carry satellites to and from geostationary orbit and other destinations. ESRO was very interested in developing this vehicle, which it saw as having potential uses aboard European launch vehicles then under study and possibly providing more jobs for the European aerospace industry. Yet because the shuttle also was to serve a number of U.S. military payloads, the Department of Defense opposed any foreign role in the Space Tug, especially since ESRO might try to veto launches of defense satellites it found objectionable. In 1972, both the Department of Defense and the State Department formally denied ESRO a role in the Space Tug, and an alternate was sought by NASA and ESRO. The two possibilities were structural elements of the shuttle orbiter and a science lab that would fit in the payload bay. Of the two, ESRO found the latter more attractive because it would provide research opportunities for European scientists and provide the community with direct experience in piloted spaceflight.

NASA had for some time been studying a Research and Applications Module (RAM), which would function as a lab facility and turn the shuttle into a temporary space station. Because a permanent space station was on indefinite hold, that was seen as necessary to continue piloted space research.

Between December, 1972, at the ministerial meeting of the European Space Conference, and August, 1973, NASA and ESRO officials conducted concept and definition studies of the laboratory facility, soon called Spacelab. An intergovernmental agreement was reached in August, 1973, and a memorandum of understanding was signed by NASA Administrator James C. Fletcher that month and by the ESRO director general, Alexander Hocker, in September.

Under the terms of the memorandum, ESRO would design and build a complete Spacelab flight unit for use by NASA and ESRO aboard the space shuttle "for peaceful purposes," and NASA agreed to buy a second flight unit at a price to be negotiated later. Although the term "peaceful purposes" is subject to debate, it has been interpreted by NASA and ESRO (later ESA) as permitting Department of Defense research missions but not weap-

ons missions. In 1974, a West German consortium was selected as the prime contractor for Spacelab. In keeping with ESRO's international nature, contracts were awarded to ESRO member nations in proportion to their contributions to Spacelab. In this manner, each nation recouped most of the money that it had invested in Spacelab. Finally, in 1975, ESRO merged with the European Launcher Development Organization (ELDO) to become the European Space Agency (ESA).

ESA's experience in developing Spacelab, and in flying it less often than expected, led that agency to assume a tougher negotiating stance on participation in the Space Station missions and to demand treatment as an equal partner in the space community. It also provided the basis for ESA's own Columbus program to develop a human-tended station.

The Spacelab Program ended in 1997, when NASA redirected its efforts toward the International Space Station. Between 1983 and 1997, twenty-four Spacelab missions were flown successfully.

See also: Atmospheric Laboratory for Applications and Science; Get-Away Special Experiments; Manned Maneuvering Unit; Space Shuttle; Space Shuttle: Life Science Laboratories; Space Shuttle: Living Accommodations; Space Shuttle: Microgravity Laboratories and Payloads; Space Shuttle: Radar Imaging Laboratories; SPACEHAB.

Further Reading

Buckey, Jay C., and Jerry L. Homick, eds. *The Neurolab Spacelab Mission: Neuroscience Research in Space: Results from the STS-90, Neurolab Spacelab Mission.* Washington, D.C.: U.S. Government Printing Office, 2003. The book reveals the results of Neurolab, a sixteen-day space shuttle mission dedicated to studying how weightlessness affects the brain and nervous system. It shows the complex and sometimes surprising changes in the brain and nervous system that allow astronauts to adapt to weightlessness. The results suggest that the developing nervous system may need gravity to develop normally and that some concept of how gravity works may be "built in" to the brain.

Burrows, William E. *This New Ocean: The Story of the First Space Age.* New York: Random House, 1998. This is a comprehensive history of the human conquest of space, covering everything from the earliest attempts at spaceflight through the voyages near the end of the twentieth century. Burrows is an experienced journalist who has reported for *The New York Times*, *The Washington Post*, and *The Wall Street Journal*. There are many photographs and an extensive source list. Interviewees in the book include Isaac Asimov, Alexei Leonov, Sally K. Ride, and James A. Van Allen.

Dooling, Dave. "Future Spacelab Missions." *Space World* T-10-238 (October, 1983): 33-37. Describes efforts by NASA to reduce the cost of future Spacelab missions and plans for dedicated discipline laboratories. Written for the general reader. Illustrated.

_____. "Spacelab 1." *Space World* T-8-9-236/237 (August/September, 1983): 8-14. This article provides an overview of how a Spacelab mission is developed and traces the plans for Spacelab 1. Describes preliminary results from the Spacelab pallets carried on STS-2 and STS-3.

Froelich, Walter. *Spacelab: An International Short-Stay Orbiting Laboratory.* NASA EP-165. Washington, D.C.: Government Printing Office, 1983. A booklet designed for teachers and students. Describes the development of Spacelab and the work required to assemble a mission. Includes color illustrations.

Harland, David M. *The Space Shuttle: Roles, Missions, and Accomplishments.* New York: John Wiley, 1998. The book details the origins, missions, payloads, and passengers of the Space Transportation System (STS), covering the flights from STS-1 through STS-89 in great detail. This large volume is divided into five sections: "Operations," "Weightlessness," "Exploration," "Outpost," and "Conclusions." "Operations" discusses the origins of the shuttle, test flights, and some of its missions and payloads. "Weightlessness" describes many of the experiments performed aboard the orbiter, including materials processing, electrophoresis, phase partitioning, and combustion. "Exploration" includes the Hubble Space Telescope, Spacelab, Galileo, Magellan, and Ulysses, as well as Earth observation projects. "Outpost" covers the shuttle's role in the joint Russian Mir program and the International Space Station. Contains numerous illustrations, an index, and bibliographical references.

Jenkins, Dennis R. *Space Shuttle: The History of the National Space Transportation System: The First 100 Missions.* Stillwater, Minn.: Voyageur Press, 2001. This is a concisely written technical reference account of the space shuttle and its ancestors, the aerodynamic lifting bodies. It details some of the advantages and inherent disadvantages of using a reusable space vehicle. Each of the vehicles is illustrated by line drawings. The book follows the space shuttle from its original concepts and briefly chronicles its first one hundred flights.

National Aeronautics and Space Administration. *Spacelab 1.* NASA MR-009. Washington, D.C.: National Aeronautics and Space Administration, 1984. A NASA publication written for teachers and reporters, with color illustrations and spacecraft summaries of experiments. It includes a discussion of how an IWG functions and how the mission was conducted.

National Research Council. *Future Materials Science Research on the International Space Station.* Washington, D.C.: National Academy Press, 1997. This composite work describes the International Space Station and materials science research in the following areas: microgravity research and the Space Station Furnace Facility Core, NASA's microgravity research solicitation and selection process, and the ability of the Space Station Furnace Facility Core to support materials science experiments that require a microgravity environment. There is a table of references, acronyms, and biographical sketches of committee members.

Shapland, David, and Michael Rycroft. *Spacelab: Research in Earth Orbit*. New York: Cambridge University Press, 1984. A broad description of the development of Spacelab through the first mission, with descriptions of various scientific disciplines it can serve. Written for a general audience.

Dave Dooling

SpaceShipOne

Date: November 15, 2004
Type of program: Piloted spaceflight

SpaceShipOne and the TierOne system of which it was a part are significant for three major reasons. First, SpaceShipOne made the first successful entry into space by a private company solely financed by private investment. Second, the spacecraft used a unique feathering system designed to allow a low-speed, controlled reentry into Earth's atmosphere. Third, the total cost of the program could be measured in millions of dollars, which is significantly lower than the billions of dollars spent on government-sponsored space programs.

Key Figures

Burt Rutan (b. 1943), manager
Brian Binnie (b. 1953), flight crew and astronaut
Mike Melvill (b. 1941), flight crew and astronaut
Pete Siebold, flight crew
Doug Shane, flight crew
Paul G. Allen, financier

Summary of the Mission

SpaceShipOne is a spacecraft that was designed by Burt Rutan at his company, Scaled Composites, as part of the TierOne system that competed for the $10 million Ansari X Prize, which it won on October 4, 2004. The X Prize was created through private donations for the purpose of boosting the space tourism industry through competition. The prize was given to the first team that privately financed, built, and launched a spaceship capable of carrying three people (or a pilot and ballast simulating two passengers) to 100 kilometers in altitude, return to Earth safely, and then repeat the launch with the same spacecraft within two weeks. An altitude of 100 kilometers is the boundary of space as defined by the Fédération Aéronautique Internationale. Scaled Composites, owned by Rutan, started development work on SpaceShipOne in 2001; however, the company did not reveal its plans to the public until April of 2003.

SpaceShipOne is part of a system known as TierOne. The TierOne system includes an airborne launcher known as the *White Knight* and the spacecraft. The launcher is an aircraft with a wingspan that can be varied from 25 meters to 28.3 meters and is propelled with afterburning J-85-GE-5 engines. It can carry loads of 3,629 kilograms, along with three people, above 16,154 meters in altitude. The engines and the spacecraft are attached to the center pod that carries the crew within. The wings are of a modified gull-wing shape and have twin booms attached that contain the landing gear. Attached to each boom is a T-shaped tail section containing the rudders and elevators. *White Knight* takes off and lands like a conventional airplane and is designed to carry SpaceShipOne underneath the fuselage to 15,240 meters, where it will release the spacecraft. The *White Knight* has controls and instruments identical to SpaceShipOne. This way, pilots can use the launcher to simulate flying the spacecraft.

SpaceShipOne is a spacecraft propelled with a rocket motor using a solid fuel and a liquid oxidizer and is designed to carry three people to a suborbital altitude of 100 kilometers. After being carried aloft to 15,240 meters, it releases from its launcher and fires its rocket motor. With the rocket burning, the spacecraft rotates to a vertical attitude and climbs straight up to 100 kilometers. Once at peak altitude, and out of fuel, the ship begins reentry into the Earth's atmosphere. SpaceShipOne pneumatically reconfigures itself into a configuration known as "feather" during the first phase of reentry to create stability and high drag, which keeps the speed and temperature low. The maximum indicated airspeed upon reentry is about 240 kilometers per hour. Expensive heatshielding is not required on SpaceShipOne as it is on the National Aeronautics and Space Administration (NASA) space shuttle. Once into the atmosphere, the spacecraft glides to a runway to land like most other airplanes.

The spacecraft is approximately 9 meters long, has a wingspan of 5 meters, and has a 1.524-meter-diameter cabin designed to hold the astronaut and two passengers. The wings are very short in span and long from front to back (this is known as low aspect ratio). Control is maintained with the use of elevons and rudders while the spacecraft is in the atmosphere. In space, control is maintained with the use of a ring of gas thruster nozzles built into the nose section. For reentry, the wings and tail section pivot into the vertical position on a hinge that runs from one side of the wing to the other just past the midpoint. With the wing in this feathered position, SpaceShipOne will automatically move a high-drag, nose-high orientation as air density increases. The design concept, according to Rutan, is based on a shuttlecock. This method of reentry requires very little work on the part of the pilot. Once back into the atmosphere, the astronaut pivots the wings and tail back to horizontal.

Both the *White Knight* and SpaceShipOne are pressurized craft that do not require occupants to wear pressure suits or space suits. Those riding within the craft wear masks that contain special filters to scrub the air of carbon dioxide and reduce moisture.

The rocket motor is considered a hybrid design because it is a cross between a liquid-fueled and solid-fueled rocket motor, manufactured by California-based SpaceDev. The motor uses hydroxyl-terminated polybutadiene (HTPB), a solid propellant that has been used in the upper stages of the Delta II and Titan IV launch vehicles. HTPB is a stable and easily stored synthetic rubber, often used in tire manufacturing. HTPB propellant was first used and test-flown in 1970 in Aerojet's Astrobee D meteorological sounding rocket. Small igniters in the front of the motor ignite the propellant. Nitrous oxide is added at high pressure. Under normal circumstances, HTPB burns slowly; however, the high-pressure nitrous oxide drastically increases the burn rate of the propellant, which is ejected through the rocket nozzle as it expands. The motor is approximately 50.8 centimeters in diameter and is wrapped with electrical wire, connected to a shutoff valve. If the rocket fuel burns through the side of the rocket, it will burn the wire, which in turn will close the valve and shut down the motor.

Both SpaceShipOne and its rocket motor underwent several ground tests. The rocket had twelve ground tests beginning in November of 2002, and the spacecraft was tested on the ground several times from the summer of 2002 to May of 2003.

Starting on May 20, 2003, Scaled Composites began a series of test flights. These included piloted and unpiloted captive carry flights, during which the spacecraft remained attached to the launcher, and release and glide flights, during which handling characteristics in both feather and glide conditions were tested.

On December 17, 2003, the one hundredth anniversary of powered flight, pilot Brian Binnie flew SpaceShipOne through the sound barrier during a fifteen-second burn of the rocket motor. For this test flight, *White Knight* carried the spacecraft to 14,630 meters, where it was released. Upon release, Binnie steered the spacecraft into a dive until his speed reached 600 kilometers per hour. Then he raised the nose and fired the rocket motor. Nine

SpaceShipOne in flight. (AP/Wide World Photos)

seconds after Binnie started the motor, while climbing at a 60° angle, the spacecraft broke the sound barrier. The pilot continued the climb into a vertical angle until the airspeed dropped to zero. At that point, he tested the feathering mechanism before gliding back for a landing. After the landing, the left landing gear accidentally retracted, and the spacecraft departed off the left side of the runway; however, no one was hurt in the incident, and the repairs required were minor. That same day, Microsoft cofounder Paul Allen announced that he was sponsoring the project, a fact previously kept secret. Both powered and glide test flights continued until June of 2004.

On June 21, 2004, Mike Melvill flew SpaceShip-One into space. Binnie flew the *White Knight* to 14,325 meters and released the spacecraft. Melvill fired the rocket immediately. The rocket motor ran for 76 seconds, pushing SpaceShipOne to 3,500 kilometers per hour. The rocket burned out at 54,864 meters; however, the spacecraft had enough speed to coast to an apogee of 100.124 kilometers. The new astronaut experienced 3.5 minutes of weightlessness as SpaceShipOne decelerated to apogee and slowly began to fall back to Earth. Melvill kept the spacecraft in the feathered condition until reaching 17,373 meters. During reentry,

SpaceShipOne was subjected to forces greater than 5g as it decelerated.

This flight was not without incident. While the spacecraft was under rocket power, the primary pitch trim control system malfunctioned. Melvill was able to switch to a backup system to continue the mission. As a result, SpaceShipOne did not climb as high as planned, and the reentry point was farther south than planned. Fortunately, the spacecraft was still able to glide back to base for a normal landing.

Scaled Composites made no more flights until September 29, 2004, when Melvill attempted the first X Prize flight. For this flight, the crew placed enough ballast on board SpaceShipOne to simulate two passengers. With Binnie at the controls, the *White Knight* took off at 15:12 Coordinated Universal Time (UTC) carrying the spacecraft underneath. SpaceShipOne was released 58 minutes later, and the rocket motor started. The burn lasted for 77 seconds and lifted the vehicle to 54,864 meters and 3,485 kilometers per hour. From there, the spacecraft coasted into space to an apogee of 103 kilometers. During the ascent, the space vehicle began rolling. Melvill used both aerodynamic controls while still in the atmosphere and the thrusters as SpaceShipOne entered space to stop the rolling motion. Melvill was weightless for 3.5 minutes before accelerating to 3,580 kilometers per hour during reentry. The peak deceleration was 5.1g. The ship was reconfigured out of feather at 18,592 meters. The return glide lasted 18 minutes.

On October 4, 2004, Binnie attempted the second flight required to win the X Prize. Again, the spacecraft was loaded with ballast to simulate two passengers. With Melvill at the controls of the *White Knight*, the launcher and spacecraft took off at 14:49 UTC. One hour later, at 14,356 meters, SpaceShipOne was released and Binnie started the rocket motor immediately. On this flight, to prevent the roll problems experienced in the first X Prize attempt, Binnie pulled up to vertical flight in a less aggressive manner. The rocket burned for 83 seconds, sending the vehicle to 64,922 meters and 3,687 kilometers per hour. The spacecraft continued coasting upward to a maximum altitude of 112 kilometers. Maximum speed during reentry was 3,878 kilometers per hour, and the maximum deceleration was measured at 5.4g. Binnie was weightless for 3.2 minutes. The ship was reconfigured out of feather at 15,545 meters. The return glide lasted 18 minutes. Not only did this flight win the X Prize for Scaled Composites, but it also broke the August, 1963, altitude record of 107.8992 kilometers for an air-launched winged spacecraft, set by NASA pilot Joseph A. Walker in an X-15.

Contributions

These missions sought to prove that government subsidies are not necessary for spaceflight. A secondary goal was to prove that spaceflight could be achieved at a relatively low cost. The success of the flights proved both. Furthermore, the flight proved that with low-speed reentry, expensive heatshields are not necessary.

Context

Since the dawn of the Space Age, the process of sending a vehicle outside Earth's atmosphere has been very expensive—so expensive that only governments, or government-subsidized programs, have been able to reach space. These programs have grown so costly that governments have difficulty justifying the expense. As technologies have advanced, cost-effective ideas are coming forward that may motivate private industry to create a commercially viable space program. This idea led to the Ansari X Prize, which Burt Rutan and Paul G. Allen won.

See also: Ansari X Prize; Launch Vehicles: Reusable; Private Industry and Space Exploration; Space Centers, Spaceports, and Launch Sites.

Further Reading

Adams, Eric. "The New Right Stuff." *Popular Science* 265, no. 5 (November, 2004): 60ff. This article is quite comprehensive and makes several comparisons between the SpaceShipOne program and NASA's Mercury program. Includes several photos.

Davisson, Bud. "Inside SpaceShipOne." *EAA Sport Aviation* 53, no. 11 (November, 2004): 34ff. This article describes the launcher, spacecraft, and simulator. It includes some technical information and a description of what it is like to fly both the spacecraft and the aircraft.

Dornheim, Michael A. "Affordable Spaceship." *Aviation Week and Space Technology* 158, no. 16 (April 21, 2003): 64ff. This article was written early in the program; however, it includes many technical specifications and provides a good overview of the project.

_____. "Trials of SpaceShipOne." *Aviation Week and Space Technology* 161, no. 15 (October 18, 2004): 36ff. This article focuses on the problems that faced the SpaceShipOne team as it tested the spacecraft, rocket motor, and control system.

Thomas D. Inman

Spitzer Space Telescope

Date: Beginning August 25, 2003
Type of spacecraft: Space telescope

The Spitzer Space Telescope, originally designated the Space Infrared Facility (SIRTF), completed NASA's Great Observatories Program. Designed to detect celestial infrared emissions, it provided a different window on the universe for collecting data complementary to observations made by the other space-based Great Observatories.

Key Figures

Giovanni Fazio, manager of the development of Spitzer's infrared array camera
David Gallagher, SIRTF project manager at JPL
Anne Kinney, division director of astronomy and physics at NASA Headquarters
Lia La Piana, program manager at NASA Headquarters
Michael Werner, Spitzer project scientist at JPL

Summary of the Mission

Responding to budget cuts, the Space Infrared Telescope program suffered a large launch delay. Originally it was supposed to fly aboard a shuttle in 1990 as a payload bay-mounted instrument capable of multiple flights each of two weeks duration. The Space Infrared Telescope Facility (SIRTF) then developed into a free-flyer and Great Observatories telescope, and was redesigned to launch on an expendable booster. By the time of launch in August, 2003, SIRTF had cost taxpayers approximately $1.2 billion.

Perhaps the biggest advance in Spitzer's design involved its infrared detection technology, which reduced the telescope's launch weight by using a lightweight 85-centimeter primary and a secondary mirror, both constructed out of beryllium, a lightweight metal with a very high stiffness-to-density ratio, high thermal conductivity, and a quite low specific heat at cryogenic temperatures. Altogether, Spitzer telescope optics amounted to 50 kilograms. The whole facility had a mass of 950 kilograms. This diffraction-limited telescope was designed to cover wavelengths between 3 and 180 microns with

imaging and photometry. Spectroscopy could be performed in wavelengths from 5 to 40 microns, and spectrophotometry between 50 and 100 micron wavelengths.

Spitzer's optics were enclosed in a cryostat chilling critical components to 1.5° above absolute zero at launch and maintaining 5.5° above absolute zero in its operational orbit. Spitzer launched with 360 liters of liquid helium, but because liquid helium boils off at 4.2° above absolute zero, the observatory's lifetime was limited to about five years. Previous infrared observatories had not lasted as long.

Spitzer's high-gain antenna was fixed to the telescope's aft end. It would be necessary to interrupt science once or twice daily for an hour to point Spitzer toward Earth in order to download data. Despite the need to record and download data, it was expected that 100,000 observations could be made during a minimum five-year lifetime.

Orbital placement of Spitzer was quite innovative and maximized the time the telescope could be used for observations. Approximately 35 percent of the sky would be available to Spitzer at any time

In the visible light spectrum, these stars in the constellation Cygnus would be hidden by interstellar dust. Spitzer, taking images in the infrared spectrum, has revealed a star "factory" including some of the brightest stars of the Milky Way. (NASA-JPL)

during the year. Some portions would be unavailable for varying periods during the year. Also, the telescope could not be pointed within 80° of the Sun, and it could not be pointed more than 120° away from the Sun, which would cut off solar panel illumination needed for power generation. Trailing Earth in heliocentric orbit, Spitzer would drift away annually one-tenth of the distance from Earth to Sun (0.1 astronomical unit). This would keep Spitzer cold enough to last five years.

Once scheduled, Spitzer's launch suffered delays amounting to four and a half months as a result of problems with its booster and the requirement to launch two Mars Exploration Rovers during high-priority Mars windows of opportunity during the summer of 2003. Spitzer used a launch vehicle similar to those used to send the Spirit and Opportunity rovers to Mars, except Spitzer's Delta II booster was equipped with higher-performance Delta III-class solid rocket strap-on boosters because of its greater weight as well as the orbital altitude that was required by Spitzer.

Spitzer was launched atop the three hundredth Delta rocket at 05:35:39 Coordinated Universal Time (UTC) on August 25, 2003. After an ascent lasting slightly less than eight minutes and with the booster's second stage still attached, Spitzer achieved a nearly circular initial orbit at an altitude of 144 kilometers inclined 31.5° to the equator. The second-stage engine fired again 41 minutes after liftoff to boost Spitzer out of this parking orbit. That burn lasted about four minutes. Fifty minutes after liftoff, Spitzer separated from its spent booster. Sixty-seven minutes after liftoff, the Deep Space Network picked up the telescope's data, and the Spitzer team could relax knowing that their spacecraft was healthy and moving along its transfer orbit toward the operational location.

In early September after initial checks of all systems, Spitzer achieved "first light," the term astronomers use to celebrate the initial observations made by a new telescope. A series of images were released on September 4, but the National Aeronautics and Space Administration (NASA) qualified the images by noting that Spitzer was still in the process of being finely focused and was still being cooled to its operational temperature. Thus the quality of these initial photographs was well beneath the full capability of the telescope. Still, first light indicated that the onboard pointing calibration and reference sensor was properly able to detect light from a star cluster. A two-month in-orbit checkout was then followed by a one-month science verification period before Spitzer embarked on its mission to study stars, dust clouds, planetary disks, and galaxies.

Contributions

Everything in the universe radiates electromagnetic energy. Radiation from stellar and galactic objects is rich in information about the nature and physical attributes of the emitting objects. For example, the wavelength of an object's most intense emission is indicative of the object's temperature. Also spectral lines can reveal information about line-of-sight speed and pressure of stars. Objects radiating the shortest wavelengths have the highest temperatures; the Compton Gamma Ray Observatory (GRO) and Chandra X-Ray Observatory (CXO) detected emissions from high-temperature, hence high-energy, objects. Whereas such objects can involve temperatures in excess of millions of kelvins, visible light such as that seen in optical telescopes and the Hubble Space Telescope (HST) originates from radiating objects with temperatures of only several thousands of degrees. The Sun appears yellow to the naked eye in large part because of the photosphere's temperature of 5,000 kelvins. Cooler objects radiate in the long-wavelength infrared region of the electromagnetic spectrum or even in the longer-wavelength radio portion; these involve temperatures of only hundreds of degrees and further down toward absolute zero, respectively.

Whereas a great deal of radio astronomy can be conducted from Earth's surface, atmospheric absorption precludes much of incoming celestial infrared emissions from reaching the ground. Therefore, infrared astronomy must be done from extreme altitudes above the densest portion of the atmosphere or, even better, in space. Hence it was necessary to place the Spitzer Space Telescope in orbit far from contaminating background infrared radiation from Earth. With data from Compton, Chandra, Hubble, and Spitzer, astrophysicists would be able to coordinate activity of celestial objects across a large portion of the electromagnetic spectrum. Compton had been deliberately deorbited before Spitzer's launch, but it had amassed a tremendous data archive of objects upon which astrophysicists would also train Spitzer's gaze. What types of objects would Spitzer investigate over its five-year lifetime?

Spitzer was designed to observe objects the science team referred to as the old, the cold, and the dirty. The "old" refers to the earliest stars and galaxies; the finding that galaxies appear to have formed far earlier in the universe than previously expected provided astronomers with another enigma that defied explanation. In looking at early galaxies—ones so far away that their light has taken billions of years to reach the solar system—Spitzer is providing clues about the character of the universe from the time it became optically transparent through the first billion or two years after the Big Bang. Radiation originally in the visible and ultraviolet emitted from these galaxies has been shifted into the infrared by the Doppler effect because of these objects' tremendous speed of recession from us as the universe expands.

The "cold" refers to brown dwarfs, objects with insufficient mass to ignite thermonuclear fires in their interiors, and circumstellar disks. The "dirty" refers to processes such as those that form stellar and planetary systems that are obscured by dust. Visible light is absorbed by such dust, but these processes glow in the infrared, and Spitzer can provide important clues about them.

Spitzer was also designed to investigate ultraluminous infrared galaxies. Such objects emit more infrared radiation than at all other wavelengths combined. It is believed that these galaxies are powered by intense star formation or central black holes, or could be stimulated into emission by collision with other galaxies.

Scientific observations carried out using the Spitzer Space Telescope are conducted at the Spitzer Science Center, located in Pasadena, California, at the California Institute of Technology.

Context

Infrared radiation was discovered in 1800 by astronomer William Herschel when he used a thermometer to measure the heating power of solar radiation split into a spectrum. Herschel noted increase in temperature from the portion of the spectrum beyond the red, something invisible to his eyes. The so-called invisible heat rays involved were termed "infrared." When Charles Piazzi Smythe climbed Mount Teide on Tenerife in 1856, he took a crude infrared detection system with him and in so doing invented infrared astronomy. The first significant infrared survey of the night sky was done by Gerry Neugebauer and Robert Leighton in 1965 using declassified military infrared technologies to discover ten infrared sources. Within four years, the catalog of infrared sources numbered thousands of entries.

Even before the dawn of the Space Age, astronomer Lyman Spitzer, Jr., strongly advocated the notion of placing a telescope in orbit about Earth, where it would be free from the obscuring effects of the life-sustaining atmosphere. His proposal involved a telescope with a large-diameter reflecting mirror collecting visible light. In the early 1970's, NASA vigorously pursued Spitzer's original idea under the concept of the Large Space Telescope. After a downsizing the mirror, this eventually led to the Hubble Space Telescope (HST) and also NASA's Great Observatories Program, the latter being a coordinated investigation using cutting-edge space-based instruments to examine celestial emissions across the electromagnetic spectrum. To supplement Hubble, NASA developed GRO and

SIRTF. HST was launched on April 24, 1990, aboard the space shuttle *Discovery* on the STS-31 mission. GRO was launched on April 7, 1991, aboard the space shuttle *Atlantis* on the STS-37 mission. The Chandra X-Ray Observatory (CXO) was launched on July 23, 1999, aboard the space shuttle *Columbia* on the STS-93 mission. Soon after SIRTF was launched, it was renamed the Spitzer Space Telescope in honor of one of the twentieth century's greatest astronomers and the strongest early proponent of space-based astronomical observations. Alas, Spitzer did not live to see his name honored in this way, having died in 1997 at the age of eighty-three, but had been present to witness the launch of Hubble Space Telescope, which more closely matched his original Large Space Telescope proposal.

Spitzer followed several earlier space-based infrared astronomical observatories. One of the most prodigious in terms of the number of infrared-emitting objects it discovered was the Infrared Astronomical Satellite (IRAS). Many IRAS sources were studied with Hubble prior to the launch of Spitzer. Hubble was designed to be serviced by space shuttle astronauts, and on one servicing mission astronauts removed an earlier instrument and replaced it with an advanced infrared camera designated the Near Infrared Camera and Multi-Object Spectrometer (NICMOS). NICMOS continued to collect data after Spitzer was launched, and the two observatories coordinated investigations in infrared astronomy.

See also: Chandra X-Ray Observatory; Compton Gamma Ray Observatory; Delta Launch Vehicles; Funding Procedures of Space Programs; Galaxy Evolution Explorer; Hubble Space Telescope; Hubble Space Telescope: Science; Mars Reconnaissance Orbiter; Telescopes: Air and Space.

Further Reading

Chaikin, Andrew. *Space.* New York: Carlton Books, 2002. A photographic history by a noted science journalist investigates the human understanding of the universe through crewed spaceflight and observations through space-based facilities such as Spitzer.

Leverington, David. *New Cosmic Horizons: Space Astronomy from the V2 to the Hubble Space Telescope*. New York: Cambridge University Press, 2001. Lavishly illustrated, this volume provides a history of astrophysical investigations since the early days of the Space Age.

Livio, Mario, and Michael Fall. *The Dark Universe: Matter, Energy, and Gravity*. New York: Cambridge University Press, 2003. Explains how portions of the electromagnetic spectrum other than the visible can provide insights into the nature of the universe.

Voit, Mark. *Hubble Space Telescope: New Views of the Universe*. New York: Harry N. Abrams, 2000. Although this book focuses on HST and its findings, it includes descriptions of infrared astronomy and discusses the role Spitzer would play in the Great Observatories Program.

David G. Fisher

Spy Satellites

Date: Beginning February 28, 1959
Type of spacecraft: Military satellites

Spy satellites provide countries with an accurate and fast means of gathering sophisticated information where other means of reconnaissance are less effective and more dangerous. The United States and the Soviet Union created the most highly developed spy satellite programs.

Summary of the Satellites

Although reconnaissance airplanes had been in use since the outbreak of World War I, the development of antiaircraft weapons made the airplane increasingly vulnerable. It was not until the late 1950's that technology was advanced enough to permit an alternative method by which intelligence could be gathered.

In 1946, the Research and Development (RAND) Corporation published a report in which the feasibility of launching a reconnaissance satellite into orbit was discussed. Additional reports were published in 1956 and 1957. These reports played a large part in the eventual development and launching of spy satellites.

In 1958, President Dwight D. Eisenhower approved a reconnaissance program that was to be operational by 1959. Under this program, the Missile Defense Alarm System (MiDAS), Discoverer, and the Satellite and Missile Observation System (SAMOS) were developed.

MiDAS, later renamed Program 239A, was first launched February 26, 1960. Relying on an infrared scanner that was sensitive to the heat emitted by a rocket, it was to provide warnings of any intercontinental ballistic missile (ICBM) attack. The advantage of MiDAS over earlier warning systems, such as the Canadian Distant Early Warning System, was that it was capable of detecting ICBMs more quickly.

Discoverer 1 was launched on February 28, 1959, and Discoverer 38 on February 27, 1962. Of the thirty-eight launches attempted, twenty-six were successful. During the time that this program was in effect, emphasis was placed on the biomedical experiments conducted, such as the one involving the orbiting of the chimpanzee Pale Face. The first U.S. photo reconnaissance satellite to be launched was Discoverer 14. The first twelve Discoverer enterprises had all ended in failure; finally, the successful launch of Discoverer 13 on August 10, 1960, convinced scientists that it was feasible to include a camera in Discoverer 14.

The average perigee of the Discoverer satellites was 220.3 kilometers. The perigee is that point at which a satellite makes its closest approach to Earth and, owing to increased gravitation, the point at which the speed is greatest. Photographs are therefore usually taken before or after the perigee. The apogee, or farthest point from Earth, was 706.3 kilometers. At a later stage, the resolution of the film was, according to the director of the program, on the order of 30.48 centimeters; in other words, the satellite was able to detect objects that were 30.48 centimeters or larger. This capability enabled the detection and identification of Soviet ICBMs. The lifetime of these satellites averaged 108 days.

Until the Kennedy administration, the Discoverer program had had two sides, the public and the official. The public name for the program was Discoverer, and the official CORONA. President John F. Kennedy phased out the Discoverer program by simply removing the public name, after which essentially the same program continued for six more launches, but now under the name CORONA. CORONA was designated Keyhole 4 (KH-4), but commonly called Close Look. Six KH-4's were launched between March 7 and November 11, 1962. The perigee was reduced slightly, while the apogee now was only 337.63 kilometers. The lifetime was shortened to a mere 2.8 days.

SAMOS began as Weapon System-1171 (WS-1171), with the code name Pied Piper, which was changed to Sentry and finally, during the Kennedy Administration, to KH-1. KH-1, which was operated by the United States Air Force, carried a conventional camera to photograph the target. The film was then developed and scanned by a fine beam of light, after which the signal was transmitted to a station on the ground, where it was used to construct a picture. Although this system was intended to pioneer spy activities, owing to unforeseen delays it was the last of the three to be launched. SAMOS was in operation from October 11, 1960, until November 27, 1963.

A second generation of U.S. spy satellites had been initiated with the launch of Keyhole 5 (KH-5) on February 28, 1963, to replace SAMOS. On July 12, 1963, KH-6 replaced CORONA. The success rate of these new satellites was considerably greater than that of their predecessors, with forty-six of fifty KH-5's succeeding between February 28, 1963, and March 30, 1967, and thirty-six of thirty-eight KH-6's between July 12, 1963, and June 4, 1967.

In the summer of 1966, a new, third generation of satellites was introduced. On July 29, KH-8 was launched, followed almost immediately on August 9 by KH-7. The third-generation satellites supplemented their ordinary cameras with infrared scanners and with a new antenna that allowed a faster transmission rate. KH-7 satellites were phased out in 1972, but KH-8, with its excellent resolution, continued functioning until the early 1980's. It eventually became known that KH-8 was 7.3 meters long and 1.73 meters in diameter, with a weight of approximately 2,990 kilograms.

The fourth generation of satellites consisted of KH-9, KH-10, and KH-11. Of these, KH-9 is com-

monly dubbed Big Bird. KH-10 was to be a piloted orbiting laboratory, but the success of KH-11, coupled with the growing cost of KH-10, resulted in the cancellation of the latter.

Big Bird was first launched on June 15, 1971. It weighed about 11,340 kilograms, was 15.25 meters in length, and had a diameter of 3.048 meters. The satellite was equipped with an ordinary camera and an infrared scanner. It was also believed to contain a multispectral camera, for use in detecting camouflage, and sensitive listening devices that would allow the Pentagon to intercept radio and microwave telephone signals as well as transmissions from Soviet satellites. On board there were four returnable film canisters. The average lifetime of KH-9 was 130 days. Its mean perigee was 166 kilometers and its apogee 269 kilometers.

KH-11 was the first satellite to report events in real time (as they occurred). This feat was achieved by using a charge-coupled device (CCD), first developed in the New Jersey Bell Telephone Laboratory in 1970. A CCD is activated when light strikes a silicon sheet divided into millions of small pixels. The silicon sheet converts photons into electrons; the electrons in each pixel are counted, and the information is then transmitted to the ground. The greater the number of electrons, the greater the light intensity. Because the electrons are captured for only milliseconds before they drain away, the thousands of pictures transmitted each second can be put together in much the same way as films are.

The resolution of KH-11 was inferior to that of KH-8 but exceeded that of KH-9, in the range of 6.6 to 8.8 centimeters. KH-11 transmitted its information directly to a communications satellite and from there either to Fort Meade, Florida, or to Fort Belvoir, Virginia.

The KH-11 program received a severe setback when a technical manual was lost to Soviet intelligence in 1977. According to some, the Soviets had been completely unaware that KH-11 was a reconnaissance spacecraft (because of its indirect transmission methods), imagining instead that it was a "ferret" satellite (a term used to describe a spacecraft used to probe foreign radar and to detect microwave signals). Others disagree, arguing instead that the Soviets were aware of its nature but had simply underestimated its capabilities.

To replace KH-11, researchers designed KH-12. It was projected that this fifth-generation reconnaissance satellite would carry advanced versions of the KH-11 sensors, as well as extra fuel that would allow it to move from a low to a high orbit when not in use. Operating in fours, the new satellites would provide instant coverage of any locale within twenty minutes. When the fuel supply was exhausted, the satellite was to be refueled by the space shuttle.

Apart from the KH series, the United States placed three other types of reconnaissance satellite in orbit: electronic, ocean surveillance, and early warning. In the first, electronic devices pick up radio and microwave signals, which are used to determine what type of radar waves are sent out, so that correct methods can be used to penetrate them in the event of war. Ocean surveillance satellites play the same role as those that gather information over land. These satellites are used in detecting ships and submarines. Early-warning satellites depend on infrared sensors to detect missile and rocket launches. Sophisticated radar then tracks each object until it becomes clear that it does not pose a threat to the United States.

Of the more than two thousand military satellites that had been launched, more than half were reconnaissance spacecraft. These reconnaissance satellites were usually launched by modified Titan rockets. It was projected that satellites would also be sent into orbit by the space shuttle. The loss of the Titan 34D rockets in August, 1985, and April, 1986, left the United States with only one KH-11 in orbit at that time.

Since then, three Advanced Keyhole satellites, sometimes referred to as the KH-12 or Improved Crystal, have been launched. Although the KH-12 was originally designed to be placed into orbit (and perhaps serviced and refueled in orbit) by the shuttle, the Titan IV is now the primary launch vehicle

for them. It is believed that KH-12 is in fact not the designation that is used by the National Reconnaissance Office for this satellite, which may be designated the KH-11B or the KH-11/Improved. Like other intelligence satellites, this satellite has a Byeman code name (such as Kennan or Lacrosse), but this Byeman code name has not been publicly compromised. To avoid confusion over nomenclature, however, this satellite will be referred to here as the Improved Crystal. Three of the older KH-11's were in orbit in the year 2000, as well as three of the more capable Advanced Keyholes, flying in orbits nearly twice as high as those of their predecessors.

Despite its many advances, these optical imaging satellites suffer a common shortcoming: the inability to see through clouds. With many areas of interest frequently covered with clouds, this has always posed a problem for intelligence collection. In the past, this problem was primarily one of directing the satellite's coverage toward cloud-free areas or awaiting improved visibility in cloudy regions. While this procedure may have been adequate for peacetime operations, it is clearly inadequate for wartime target acquisition. A space-based imaging radar can see through clouds, and utilization of synthetic aperture radar (SAR) techniques can potentially provide images with a resolution that approaches that of photographic reconnaissance satellites. A project to develop such a satellite, named Lacrosse, was initiated in the early 1980's, with the first satellite launched in 1988. Three of these satellites were in orbit in the year 2000. The United States continued operations of three KH-11 photographic intelligence satellites through the year. The sixth KH-11, launched in December of 1984, remained in orbit at the end of 1990, surpassing by three years the demonstrated orbital life for these satellites. On February 28, 1990, the space shuttle *Discovery* on flight STS-36 deployed what appeared to be the second new generation of photographic reconnaissance satellites, popularly referred to as the Advanced Keyhole or the KH-12. This spacecraft was placed in an orbit with an initial inclination of 62°. On March 3, 1990, this spacecraft executed a small maneuver to raise its altitude, and on March 7, it executed a much larger maneuver, which raised its inclination to 65° and its altitude to a roughly circular orbit at 811 kilometers.

Imaging intelligence satellites were widely used in Desert Shield operations in 1990. In addition, civilian Landsat and Spot images were used to develop up-to-date maps of the theater of operations. Intelligence reports provided warning of the Iraqi invasion nearly a week before it occurred, including both the timing and magnitude of the assault. A few days after the invasion, satellite photography showing the Iraqi military buildup on the Kuwaiti-Saudi border was instrumental in convincing Saudi King Fahd to permit the introduction of American troops. These systems were not infallible, since the United States lost track of four Iraqi divisions for a twenty-four-hour period on August 7-8. By late August satellite photographs showed new troops being deployed in lines parallel to the Saudi border and along the Persian Gulf. In early September, satellite imagery detected the deployment of a new type of missile mobile launcher mounted atop flatbed trucks. In late September, U.S. intelligence identified chemical decontamination sites in Iraq along the Kuwaiti border with Iraq. In early October, several vans photographed by a U.S. spy satellite were identified as Soviet jamming gear, codenamed Paint Can, at several locations inside Kuwait and along Iraq's border. By late October, Iraqi forces were shifting position frequently to evade satellite intelligence. Imaging intelligence systems were also used to monitor the effectiveness of the embargo, and by early December satellite showed a steady stream of trucks entering Iraq from Iran. Imaging intelligence satellites have played similar roles in the conflict in Iraq.

The United States operates four constellations of signal intelligence satellites in geostationary, elliptical and low-Earth orbits. The geostationary SigInt constellation consists of three or four satellites. The National Reconnaissance Office and National Security Agency launched a Magnum signal intelligence satellite on the space shuttle on No-

vember 15, 1990, on STS-38. This spacecraft joined the Chalet (also known as Vortex) launched on May 10, 1989, and the second Magnum launched on November 23, 1989. In addition to these 1989 launches the constellation also includes the first Magnum, launched on the shuttle in 1985. The older Chalet launched in 1981 has probably left service, having more than surpassed its five-year design life.

Signal intelligence provided one of the first warnings that the Iraqi invasion of Kuwait was likely, when a Soviet-built Tall King radar resumed operation July 29, 1990. The 560-kilometer range radar had been out of service for a number of months prior to the invasion. By early October, U.S. electronic intelligence had some success in monitoring Iraqi military communications, but the Iraqi army was also using underground cables to communicate, making it difficult to determine Iraqi military intentions.

The White Cloud Naval Ocean Surveillance System (NOSS) performed wide-area ocean surveillance, primarily for the Navy. White Cloud was used to determine the location of radio and radar transmissions, using triangulation. The identity of naval units could be deduced by analysis of the operating frequencies and transmission patterns of the emitters. Each NOSS launch placed a cluster of one primary satellite and three smaller subsatellites (that trail along at distances of several hundred kilometers) into low polar orbit. This satellite array could determine the location of radio and radar transmitters and the identity of naval units, by analysis of the operating frequencies and transmission patterns.

NOSS used the ELINT technique called "time difference of arrival" (TDOA) rather than true interferometry. Conceptually, TDOA and interferometry are very similar, though distinct, techniques. They may also use the frequency-domain version of TDOA, FDOA, which exploits Doppler shifts somewhat in the way the COSPAS/SARSATs do. Although there did not appear to be a definitely fixed constellation size for White Cloud, at the beginning of 1990 the constellation apparently consisted of two clusters of primary and secondary satellites, launched on February 9, 1986, and May 15, 1987. No launches under this program were conducted after 1987.

A Space Based Wide Area Surveillance System (SB-WASS) replaced White Cloud. This program, which was initiated in the early 1980's, was used to track ships and aircraft on a global basis. Originally, SB-WASS was supposed to be a joint Air Force-Navy program, but the cost as well as disagreement over the type of sensors resulted in postponement of full-scale development. In the interim, the Navy developed a White Cloud successor, referred to as the SB-WASS (Navy) for convenience, but its real designation is not known. Originally it was to have been launched by the shuttle, but after the *Challenger* accident in 1986, it was decided to use the Titan IV booster instead.

The satellites are larger and heavier than their predecessors. The White Cloud system massed about 1.5 metric tons in orbit, with the main satellite about 600 kilograms and each subsatellite about 200 kilograms. In the new system, the whole lot masses up to 8 tons in orbit. This extra mass is consistent with the presence of an advanced scanning infrared sensor on the subsatellites. The apparent size of the primary (3 or more tonnes) suggests that it may have a secondary SigInt mission; this idea is reinforced because the primary in each launch (USA-59, USA-72, and USA-119) has disappeared. It appears that the program was to have consisted of four clusters, one in each orbital plane. However, the third launch attempt, in 1993, went spectacularly wrong when the Titan IV booster blew up shortly after launch. At the time this was described as the most expensive U.S. space failure since *Challenger.* Operation of SB-WASS is similar to White Cloud. Orbits of 1,000 by 1,100 kilometers at about 63° are employed, but the subsatellites are much closer together (30 to 110 kilometers). It is possible that the subsatellies communicate with one another by laser, rather than microwave link as White Cloud did. A laser used for this purpose would be virtually impossible to intercept or jam. It would also have a much higher band-

width, which could be needed to handle the data generated by the infrared sensors. Martin Marietta and Lockheed were the prime contractors for this program.

Refurbished Titan II's launched three satellites in 1988, 1989, and 1992. These seem to have been the Air Force version of SB-WASS. They were launched into 780-by-800-kilometer, 85° orbits and exhibited regular flashing reflection patterns, as would be expected from the rotating reflector radar antenna planned for the Air Force SB-WASS. The project appears to have been terminated. TRW was the prime contractor.

There appears to be the intention within the United States defense establishment to go ahead with the development of a single SB-WASS system to be used by the Air Force and Navy. The apparent intention is to consolidate operations at one site in an attempt to save money. Martin Marietta (now Lockheed Martin) was awarded the development contract in 1994. The SB-WASS spacecraft may be a 7.5-metric-ton platform that could be launched into a nominal 180-kilometer orbit at 63°, from either Vandenberg or Cape Canaveral.

The U.S. Air Force, the Defense Department's Advanced Research Projects Agency (ARPA), and the National Reconnaissance Office (NRO) jointly selected three contractor teams on February 22, 1999, for Phase One of the Discoverer II space-based radar technology demonstrator program. Lockheed Martin Astronautics, Spectrum Astro, and TRW Defense Systems Division lead these teams. The goal of the demonstration program is to develop, build, and launch two research and development satellites capable of detecting and tracking moving targets on Earth's surface, producing high-resolution imagery and collecting high-resolution digital terrain-mapping data.

Contributions

Reconnaissance satellites regularly provide information on new airplanes, ships, and submarines. Such spying often encounters difficulties. For example, in order to stop satellites from monitoring the construction of their submarines, the

Soviets placed netting over them. The Americans temporarily overcame this obstacle—until they inadvertently divulged their tactics during the Strategic Arms Limitation Treaty (SALT) II negotiations—by using slant photography to view the submarines from the side.

Reconnaissance satellites continually monitor troubled areas of the world. Photo reconnaissance is believed to have been used in 1967 during the Arab-Israeli conflict, for the attempted rescue in 1979 of the American hostages in Iran, and, in 1986, in advance of the Libyan bombing and at the time of the nuclear accident at Chernobyl. The Persian Gulf War of the 1980's was presumably scrutinized by American satellites. When ships and submarines are at sea, infrared pictures allow their paths to be plotted.

Technology produced for spy satellites has had important ramifications for other space programs. For example, it is believed that military photographic technology was made available to the National Aeronautics and Space Administration (NASA) for its missions to Mars and the Moon.

Advanced radar has made it possible to plot the ground of the former Soviet Union and other countries accurately. Shots of the same area taken at different times are subsequently fed into a computer, which removes similar objects and highlights the discrepancies by using a process known as electro-optical subtraction. The remaining objects either are identified and noted or, if they are blurred, are enhanced by digital restoration.

The CCD technology developed in Bell Laboratories for use in spy spacecraft is now used in many other fields, including astronomy, medical imaging, and plasma physics—even in video cameras. The Galileo spacecraft and the space telescope also rely on the CCD.

An obvious use of reconnaissance photographs is to study cloud formations and weather around the world. It is possible that pictures of clouds are used by meteorologists at the Central Intelligence Agency (CIA) to determine the nature of the weather, and this information is then passed on to the armed forces.

Context

In the years preceding the launch of the first U.S. reconnaissance satellite, it was widely believed that the Soviet Union had as many as two hundred ICBMs. As a result, millions of dollars were spent by the United States in an effort both to develop ICBMs and to modernize the airplanes that carried nuclear bombs. After the launch of the Discoverer spacecraft, the number of Soviet first-generation ICBMs was found to have been only around twenty. By exposing the true number of first-generation Soviet ICBMs, Discoverer saved millions of dollars on further U.S. military expenditures. It is possible that if Premier Nikita Khrushchev had not pushed the United States into the development of ICBMs with his constant hints that the Soviets would without hesitation use them on unfriendly countries, the United States might not have manufactured its own weapons for ten years, and the launching of reconnaissance spacecraft could well have been delayed by at least five years.

The significance of spy satellites to the United States during the Cold War can be gauged by the amount of money that was spent on the program. In March, 1967, President Lyndon B. Johnson told a small group of people that the total space-program expenditure up to that time was between $35 billion and $40 billion. At that time, the space program was less than twenty years old. He concluded that if nothing else had been achieved, the photographs from the spy satellites would have made the program worthwhile.

Since the beginning of the U.S. reconnaissance program, great advances have been made in photography. This photographic technology has been used in other programs, most notably in the space telescope, the Galileo spacecraft, and the probes sent to view the Moon and Mars. The excellent resolution of KH spacecraft cameras permits the identification of missiles and hence provides a means of verifying compliance with arms control treaties. In a political climate of mutual mistrust between the United States and the Soviet Union, spy satellites were relied on heavily to ensure that treaties were honored.

Today, spy satellites are utilized in scientific research when they are not spying on our neighbors. The pinpoint accuracy they possess allows scientists to study plate tectonics in greater detail. They also can be used to study the effects of natural disasters, such as floods and wildfires. Scientists at the Los Alamos National Laboratory in New Mexico are currently studying lightning from space. Flashes and thunder are but a fraction of the energy blasts from lightning: The bolts are fountains of radio waves as well—so powerful, in fact, that they are the closest thing on the radio dial to a nuclear explosion. Their minisatellite called FORTE (Fast On-Orbit Recording of Transient Events) is capturing millions of flashes and radio blasts from lightning all over the tropics. What FORTE is hearing by radio challenges a long-held notion that lightning occurs up to ten times more often over continents than oceans. That is why Los Alamos scientists launched FORTE in August, 1997. Early nuclear-detection satellites were so overwhelmed by radio noise from Earth that their radio detectors got a bad reputation.

See also: Attack Satellites; Early-Warning Satellites; Electronic Intelligence Satellites; Meteorological Satellites: Military; Ocean Surveillance Satellites; Telecommunications Satellites: Military; Titan Launch Vehicles; United States Space Command; Vandenberg Air Force Base.

Further Reading

Andrade, Alessandra A. L. *The Global Navigation Satellite System: Navigating into the New Millennium.* Montreal: Ashgate, 2001. Provides an international view of issues of availability, cooperation, and reliability of air navigation services. Attention is specifically paid to the American GPS (Global Positioning System) and Russian GLONASS systems, although the development of the Galileo civilian system in Europe is also presented.

Baker, David, ed. *Jane's Space Directory, 2005-2006.* Alexandria, Va.: Jane's Information Group, 2005. Extensive bibliographic presentation of all space programs, broken down into programatic categories.

Burrows, William E. *Deep Black: Space Espionage and National Security.* New York: Random House, 1986. This carefully researched and well-written book documents the progression of the American reconnaissance satellite from the early years until the mid-1980's, presenting its strengths and weaknesses. Contains photographs and a list of references.

Butrica, Andrew J. *Beyond the Ionosphere: Fifty Years of Satellite Communications.* NASA SP-4217. Washington, D.C.: Government Printing Office, 1997. Part of the NASA History series, this book looks into the realm of satellite communications. It also delves into the technology that enabled the growth of satellite communications. The book includes many tables, charts, photographs, and illustrations, a detailed bibliography, and reference notes.

Cox, Christopher. *The Cox Report: U.S. National Security and Military/Commercial Concerns with the People's Republic of China.* Washington, D.C.: Regnery Publishing, 1999. Investigates U.S.-Chinese security interaction and reports that China successfully engaged in harmful espionage and obtained sensitive military technology from the United States. Some of the technology obtained includes information on American reconnaissance and attack satellites.

Gavaghan, Helen. *Something New Under the Sun: Satellites and the Beginning of the Space Age.* New York: Copernicus Books, 1998. This book focuses on the history and development of artificial satellites. It centers on three major areas of development—navigational satellites, communications satellites, and weather observation and forecasting satellites.

Jasani, Bhupendra. *Space Weapons: The Arms Control Dilemma.* London: Taylor and Francis, 1984. Features a chapter on reconnaissance satellites and antisatellite (ASAT) weapons. A bibliography is included.

Klass, Philip J. *Secret Sentries in Space.* New York: Random House, 1971. Klass has written one of the best books available on early spy satellites. He discusses their significance and use, covering reconnaissance spacecraft from their inception up to the ill-fated Manned Orbiting Laboratory. Klass also includes a chapter on the Soviet effort. Contains photographs.

Richelson, Jeffrey. *American Espionage and the Soviet Target.* New York: William Morrow, 1987. The book gives a detailed view of how American espionage functions. There is a chapter on the Keyhole program, showing how the different methods of gathering information fit together. A large bibliography, a section on acronyms, and photographs are included.

_____. "The Keyhole Satellite Program." *Journal of Strategic Studies* 7 (June, 1984): 121-153. This essay is probably the most complete analysis of the Keyhole program. Includes, among other things, sections on sensors and resolution. All satellites from CORONA to KH-11 are mentioned, and their contributions to the reconnaissance program are listed. Bibliography included.

Taylor, John W. R., ed. *Jane's All the World's Aircraft, 1978-1979.* London: Jane's Publishing Company, 1978. Contains a small section on the Big Bird satellite, giving dimensions and orbit specification. Directly below are notes on the Titan rocket, which was used to launch KH-9.

Zimmerman, Robert. *The Chronological Encyclopedia of Discoveries in Space.* Westport, Conn.: Oryx Press, 2000. Provides a complete chronological history of all piloted and robotic spacecraft and explains flight events and scientific results. Suitable for all levels of research.

John Newman

Stardust Project

Date: February 7, 1999, to January 15, 2006 (projected)

Type of program: Planetary exploration

Stardust is the first U.S. space mission designed from the beginning to study a comet and only the second sample return mission since the Apollo missions to the Moon. Stardust will return samples of Comet Wild 2 and samples of an interplanetary dust stream, making these the first samples returned from beyond the region of the inner solar system near Earth as well as possibly the first samples originating from beyond the solar system.

Key Figures

Donald E. Brownlee, principal investigator

Peter Tsou, deputy principal investigator

Thomas Duxbury, Stardust project manager

Robert Ryan, Stardust mission manager

Barry Geldzahler, Stardust program executive

Mark Saunders, Stardust program manager

Joseph Vellinga, Stardust program manager at Lockheed Martin Space Systems

Summary of the Mission

Stardust was launched February 7, 1999, at 21:04:15 Coordinated Universal Time (UTC) from Pad 17A of the Cape Canaveral Air Force Station, adjacent to the Kennedy Space Center, in Florida. The launch vehicle was a Delta II 7426 rocket, a variant of the Delta rocket with four strap-on solid rocket boosters. The upper stage of the rocket put Stardust into an elliptical orbit that carried it farther from the Sun than the planet Mars. However, this orbit was not far enough out to intercept Comet Wild 2. To achieve a rendezvous with the comet, Stardust's orbit was designed to pass by Earth again on January 15, 2001. This flyby used Earth's gravity to add energy to the spacecraft, making its orbit larger so that it could intercept the comet in 2004. This gravity assist to adjust an orbit is often called the "slingshot effect." The gravity-assist method of changing orbits uses far less fuel and requires a smaller and less expensive rocket than would be needed to go directly into the final orbit desired. On November 2, 2002, on the second orbit, Stardust passed within 3,100 kilometers of the asteroid named Annefrank. The asteroid flyby was used as a test of guidance and control operations prior to the comet rendezvous.

Prior to the Stardust mission, astronomers had discovered a stream of interstellar dust particles passing through the solar system. Since Stardust was already being planned to collect samples of dust particles from a comet, it was a natural extension of the mission to design the collector with two collecting surfaces. One side of the collector, side A, was reserved for comet samples. However, side B was to be used to sample these interstellar particles. This side of the dust collector was deployed two times, at the outer portions of the two orbits of the Stardust spacecraft. The collector was deployed from March to May, 2000, and again August to December, 2004. The collector was deployed to capture interstellar dust particles for a total of 195 days.

By the end of 2003, Stardust's orbit had carried it into position in front of Comet Wild 2. The comet was moving faster than the spacecraft, and it passed within 236 kilometers of Stardust on January 2, 2004, at 19:41 UTC, with a relative speed of 6.1 kilometers per second. Stardust deployed side A of its dust collector to sample comet material from nine days until about five hours after the closest encounter, after which the collector was stowed and the sample return capsule was sealed. As Wild 2 passed Stardust, cameras on board the spacecraft took photographs of the comet's nucleus, while other instruments determined the composition and density of dust particles that Stardust was passing through.

Following the encounter with Wild 2, Stardust continued on its orbit and will return to Earth on January 15, 2006. At an altitude of 110,728 kilometers, Stardust's sample return capsule will be separated from the main spacecraft. The spacecraft will continue past Earth, and the capsule will reenter the Earth's atmosphere at a speed of 12.8 kilometers per second. Friction with the atmosphere will slow the capsule, and it will deploy a drogue parachute at an altitude of nearly 30 kilometers and a larger parachute at an altitude of about 3 kilometers. The capsule will float down on the parachute to land at the Utah Test and Training Range near Salt Lake City, Utah. The capsule will then be recovered and sent to the Johnson Space Center in Houston, where it will be opened under the strictest of quarantine conditions for analysis. The quarantine is necessary to avoid cross-contamination of the sample with Earth. The comet dust is unlikely to be hazardous to Earth, but certainly Earth can contaminate the sample that was collected with such great difficulty.

The Stardust spacecraft is based on a deep-space frame developed by Lockheed Martin Astronomics. The main body of the spacecraft is essentially a rectangular box measuring 0.66 meter in height and width and 1.7 meters long. On either side of the box, two large solar panels are positioned parallel to the long axis of the spacecraft, giving the spacecraft the shape of an "H"

when viewed from above. A 0.6-meter-diameter low-profile high-gain communication dish is set on top of the spacecraft. The sample return capsule, a 0.81-meter-diameter clamshell-shaped device, is fitted to the back end of the spacecraft.

The space around a comet is filled with small particles shed by the comet. The goal of the Stardust mission is to sample these particles; however, they also pose a major threat to the spacecraft. Impacting the spacecraft at high velocity, these dust particles strike with the force of bullets. To protect against these impacts, Stardust is fitted with armored shields called Whipple shields, named after Fred Whipple, the astronomer who first proposed the model of comet nuclei that is, with only small revisions, almost universally accepted. The Whipple shields consist of rectangular bumpers fitted to the front ends of the two solar panels and a larger square plate fitted to the front of the spacecraft's main body.

To record the size and frequency of dust particle impacts, two vibration sensors are fitted to the

Comet Wild 2 as seen from the Stardust spacecraft on January 2, 2004. (NASA/JPL-Caltech)

Whipple shields to detect particles large enough to penetrate the outer layers of the shields. To detect smaller particles, two film sensors are used. The film sensors generate an electrical signal when struck by a dust particle. The larger the particle, the larger the signal generated. In addition to recording dust impacts, Stardust has a mass spectrometer aboard to record the composition of dust particles intercepted by the spacecraft.

Stardust carries a navigation camera designed to allow the spacecraft to find Comet Wild 2 and adjust its trajectory and orientation for the best flyby encounter. However, the camera is also capable of making scientifically useful images. The camera has a filter wheel with three gas filters, three dust filters, a high-resolution filter, and a clear filter.

The main scientific instrument on board Stardust is its aerogel dust collector. Fitted to the sample return capsule, the dust collector looks much like a long-handled tennis racket that can be extended away from the spacecraft, beyond the shadow of the Whipple shields. Aerogel is a glass-like material that is very porous. In fact, aerogel is more than 99 percent empty space. Aerogel looks like pale blue smoke, and it is the least-dense human-made material yet manufactured. The dust particles striking the aerogel, rather than being destroyed on impact, are slowed by passing through the aerogel and are thus captured intact and undamaged. The dust collector has two sides, each with 1,000 square centimeters of collecting surface. One side is used to collect samples from the comet, and the other side is used to collect interstellar dust grains. These samples are then returned to Earth for extensive study in the laboratory.

Contributions

The comet samples returned by Stardust should help astronomers understand comets better. Comets are believed to be relics from the formation of the solar system, having changed very little since that time. Also, comets are known to contain water and organic compounds, and some scientists believe that they may have been important in

bringing to Earth water and the basic prebiological organic material needed for life to develop.

Since Stardust visited Comet Wild 2 on only its fifth orbit after an encounter with Jupiter in 1974 perturbed the comet into an orbit nearer the Sun, astronomers expected the nucleus to be rather pristine. However, the seventy-two photographs of the nucleus taken by Stardust show a ragged and tortured landscape filled with giant spires 100 meters tall and craters more than 150 meters deep, with one crater having a diameter nearly 20 percent that of the entire nucleus. Only two other comet nuclei have been photographed before, and neither showed any features like these. Astronomers are not yet sure what to make of these features.

As the ice and gases that make up the comet are heated by the Sun, they sublimate and often shoot away from the comet's nucleus in jets. Stardust recorded about twenty-five jets coming from Comet Wild 2. The high-resolution images showed where the jets originated on the surface of the nucleus, something never before seen. These jets also did not dissipate as close to the nucleus as astronomers had expected, and Stardust actually flew through three streams of jet material.

Comet jets contain not only gases but also dust particles from the comet. As Stardust passed through the jets, it was pummeled with far more dust particles than expected, with nearly a dozen breaking through the outer layers of the Whipple shields. However, the region between the jets had only about 1 percent of the dust that had been expected.

As exciting as was the data radioed back to Earth by Stardust during and after its encounter with the comet, astronomers are eagerly awaiting the samples being brought back to Earth. In the laboratory, these samples can be subjected to tests much more exhaustive than can be performed by any spacecraft.

Context

The Stardust mission was selected to be part of the Discovery Program, funded by the National

Aeronautics and Space Administration (NASA). The Discovery Program is a series of low-cost, unpiloted space missions to explore the solar system. These missions were selected from proposals submitted to NASA. Two other comet missions were selected as part of the Discovery Program: the Comet Nucleus Tour (CONTOUR) mission, which lost contact with the CONTOUR spacecraft six weeks after launch, and the Deep Impact mission, which launched spacecraft on January 8, 2005, to blast a crater into comet Tempel 1.

Stardust was only the third American space probe to visit a comet and was the first, other than CONTOUR, designed from the beginning as a comet mission. One of the previous spacecraft, the International Comet Explorer, had originally been named the International Sun-Earth Explorer until it had been diverted from its original mission to study Comet Giacobini-Zinner. The other spacecraft was Deep Space 1, which was designed primarily as an engineering test of an ion propulsion system. Although the United States had not sent a spacecraft to visit a comet as its prime mission before, the European Space Agency (ESA) had sent Giotto to Halley's comet, along with two spacecraft from the Soviet Union and two from Japan. Unlike any of these other spacecraft, however, Stardust was designed as the very first spacecraft to attempt to return a sample of comet material to Earth for detailed study. Stardust is also only the second sample return mission attempted by NASA since Apollo 17. Before its mission could be concluded, the Genesis spacecraft returned solar wind particles to Earth when its sample return canister crashed into the Utah desert on September 8, 2004. The Deep Impact spacecraft launched after Stardust, but, on July 4, 2005, after a six-month voyage, it slammed an impactor probe into Comet Tempel 1 and generated both a large crater and debris plume that the flyby spacecraft observed and studied. Deep Impact data were expected to be quickly combined with Stardust samples to provide a better understanding of the nature of comets.

See also: Asteroid and Comet Exploration; Dawn Mission; Deep Impact; Delta Launch Vehicles; Johnson Space Center; National Aeronautics and Space Administration; Planetary Exploration; Search and Rescue Satellites.

Further Reading

Brandt, John C., and Robert D. Chapman. *Introduction to Comets.* Boston: Cambridge University Press, 2004. Provides a detailed examination of virtually every cometary phenomenon.

Crovisier, Jacques, and Thérèse Encrenaz. *Comet Science.* Translated by Stephen Lyle. New York: Cambridge University Press, 2000. An excellent treatise on comet structure and the study of comets, though somewhat technical in places.

Davies, John. *Beyond Pluto.* New York: Cambridge University Press, 2001. Covers the Kuiper Belt, a region beyond Pluto from which many comets are believed to originate.

Godwin, Robert. *Deep Space: The NASA Mission Reports.* Burlington, Ont.: Apogee, 2005. A collection of NASA documents relating to various deep space probes, including Stardust.

Levy, David. *Comets: Creators and Destroyers.* New York: Touchstone, 1998. This book discusses some of the theories of comet impacts bringing life-giving materials to Earth and perhaps resulting in some of the past mass extinctions.

Sagan, Carl. *Comet.* Rev. ed. New York: Ballantine Books, 1997. Written by the noted popularizer of science, this work describes the pristine remnants left over from the origin of the solar system called comets. Liberally illustrated with photographs, Sagan's book humanizes the study of science and teaches a history of astronomy along the way.

Verschur, Gerrift L. *Impact: The Threat of Comets and Asteroids.* London: Oxford University Press, 1997. Describes the nature of Near Earth asteroids and comets, and their threat of devastating impacts on Earth.

Whipple, Fred L. *The Mystery of Comets.* Washington, D.C.: Smithsonian Institution Press, 1985. An older book, but it gives an excellent history of comet studies, as well as an easy-to-understand basic presentation of the standard model for comets as explained by Fred Whipple himself.

Raymond D. Benge, Jr.

Strategic Defense Initiative

Date: 1983 to 1993
Type of spacecraft: Military satellites

The United States' Strategic Defense Initiative (SDI) and similar efforts in the former Soviet Union were multiphased programs designed to counter nuclear missile attacks. Both nations' projects involved the use of high-technology, space- and Earth-based, directed energy weapons to defeat intercontinental ballistic missiles.

Key Figures

Ronald W. Reagan (1911-2004), fortieth president of the United States, 1981-1989
Edward Teller (1908-2003), associate director emeritus, Lawrence Livermore National Laboratory
William R. Graham, Jr. (b. 1937), director, Office of Space and Technology Policy
James A. Abrahamson, Jr. (b. 1933), director, SDI organization
George A. Keyworth, former science adviser to President Reagan
Leonid Ilyich Brezhnev (1906-1982), president of the Soviet Union, 1977-1992
Mikhail Sergeyevich Gorbachev (b. 1931), general secretary of the Communist Party of the Soviet Union

Summary of the Program

The Strategic Defense Initiative (SDI), as first defined by U.S. President Ronald W. Reagan in 1983, was a research program designed to create an effective defense against nuclear missile attacks. SDI, or "Star Wars," as it was popularly termed, has been the subject of immense controversy in scientific and political circles.

On March 23, 1983, President Reagan used a nationally televised speech to announce a major research effort to discover ways to protect the United States from a strategic nuclear missile attack by the Soviet Union. The president stated his hope that technology developed through SDI could be used to make missile-delivered nuclear weapons obsolete in the twenty-first century. Achieving this end would require tremendous advancements in existing technology and numerous breakthroughs in hardware applications.

At the time of the president's announcement, both the superpowers were depending on the prin-ciple of mutual assured destruction (MAD) to prevent nuclear war. MAD was predicated on the belief that, because both the United States and Soviet Union possessed enough nuclear weapons (six thousand for the United States, twelve thousand for the Soviet Union) to destroy each other several times over, and because neither nation could reasonably expect to survive a nuclear exchange, neither would be willing to risk its own destruction to defeat the other.

Strategic defense is based on the premise that high technology will allow the building of a defensive system to bear the brunt of a first-strike nuclear attack so that the defender nation is then able to use its own weapons to counterattack. The threat of such a counterattack would, according to the theory, be an adequate deterrent to a first strike.

For more than thirty years prior to the strategic defense program, both the United States and the Soviet Union had worked to find ways to defend against a nuclear missile attack. These efforts, known as BMD, for ballistic missile defense, provided impetus for the Anti-Ballistic Missile Treaty signed by President Richard M. Nixon and Soviet premier Leonid Ilyich Brezhnev in 1972. The ABM Treaty allowed the two nations to build two antiballistic missile sites of no more than one hundred missiles each and to continue research into BMD. That research led to some of the major developments in SDI.

The Strategic Defense Initiative program was to be deployed in three stages beginning in the 1990's and continuing through about 2115. It consisted largely of two types of technology: kinetic energy devices and lasers. When fully deployed, SDI was designed to use these weapons systems and a complex array of sensors, relay satellites, and battle management computers to construct ground-based and Earth-orbiting "screens" to "filter out" intercontinental ballistic missiles (ICBMs) and their multiple nuclear warheads, known as reentry vehicles, before detonation.

Kinetic energy devices are nonexplosive projectiles that, when launched at an object, rely on their momentum, or kinetic energy, to destroy the target. Two types of kinetic energy weapon, space-based interceptors (SBIs) and the ground-launched Exoatmospheric Reentry Interceptor System (ERIS), were to be used in SDI. SBIs are weapons platforms to be placed in Earth orbit that would fire kinetic energy projectiles at just-launched ICBMs, destroying the missiles before the release of their multiple reentry vehicle payloads. ERIS interceptors launched from the ground would destroy reentry vehicles outside Earth's atmosphere after they have been released from an ICBM. High endoatmospheric interceptors would be employed at later stages to destroy warheads after they entered the atmosphere but before they reached their targets and could inflict significant damage.

These kinetic energy weapons would be used in SDI's first phase of deployment, with refinements being put in place as development of the technology permitted. In 1984, the United States successfully conducted the first test of a ground-launched interceptor designed to destroy an ICBM, and 1995 was a target date for the first phases of a ground-based kinetic energy defense system.

The second weapon to be used in SDI was the ground-based or space-based laser. Any or all of several types of laser, possibly including free electron lasers, nuclear powered x-ray lasers, chemical lasers, and laser guided particle beams, would be fired from bases on Earth at reflecting mirrors in Earth orbit that would reflect and enhance the beams, directing them toward incoming ICBMs or reentry vehicles. An experiment in which a harmless laser beam was aimed at the space shuttle *Discovery* was successfully completed in 1985.

Research was also being conducted into the possible use of particle beams as defensive weapons against nuclear missiles. Although this plan had numerous technological drawbacks, some scientists believed it would be possible to use beams of highly charged hydrogen protons to destroy an ICBM in flight.

It was also thought that orbiting platforms equipped with particle beam or laser weaponry for an advanced BMD program would be feasible at some point, perhaps as late as 2115.

In addition to the weapons systems, SDI required a highly sophisticated command and communications system that would use massive satellites equipped with Earth-orbiting heat and radar sensors, ground-based sensors, aircraft, and specially launched sensor rockets. This complex network would detect, analyze, and relay information on incoming ICBMs to battle command centers and then relay and execute the commands of military and civilian leaders. So that those leaders could effectively recognize and react to a potential nuclear threat, command of the weapons systems and the sensor network would rely on a blend of computerized analysis and human judgment.

As originally conceived by the Reagan administration, SDI served at least two major military and diplomatic purposes. First, SDI was a defensive system designed to make the cost of a nuclear attack by either nation against the other prohibitive and its success uncertain, and second, from the United

States' viewpoint, SDI gave the Soviet Union an added impetus to negotiate further arms control and disarmament agreements on every class of nuclear weapon. How successful SDI would be in bringing about either objective was a matter of great controversy in the American and international political arenas.

Critics of SDI, who called the program Star Wars, believe the concept to be too heavily dependent on uncertain technology to be fully or even reasonably effective. Because a full-scale missile attack might involve hundreds of missiles, thousands of reentry vehicles, and tens of thousands of decoy reentry vehicles launched simultaneously at hundreds of targets, there is considerable doubt whether a tremendously complex system such as SDI would work in the first minutes of an attack.

Some experts were concerned that, if the technology were deployed and was even partially successful, it might have either provoked the Soviet Union into launching a preemptive nuclear strike or created the dangerous and false assumption among American leaders that a nuclear war could be survived and even won. Many European nations would have raised the question of how SDI affected the balance of power between the United States and the Soviet Union, because the Soviet Union had a large numerical superiority in conventional military forces.

Opponents of the program also pointed out that space-based weapons or weapons-support systems are highly vulnerable to attack by a variety of different methods. This weakness, the critics argued, could render SDI useless and leave the United States more vulnerable to nuclear attack. Satellites in space can be destroyed by the detonation of a nuclear bomb above the atmosphere. They can be attacked by specially designed hunter-killer satellites that can be placed in orbit near the original satellites and used at the first sign of hostilities. Moreover, the same laser or particle beam technology that can be used to destroy an ICBM would be equally effective against satellites in Earth orbit.

The vulnerability of space-based weaponry led SDI scientists and designers to include satellite de-

fense and antisatellite weapons systems in the SDI blueprints. These systems, in turn, made the whole SDI system more complex and, its critics suggested, more unreliable, because of the greater potential for technical problems and hardware failure.

Proponents of SDI, however, believed that SDI technology would enable the United States and the Soviet Union to control massive defense spending and that it would reduce the threat of nuclear war. They believed SDI to be more rational and moral than mutual assured destruction as a nuclear strategy for the twenty-first century.

With the 1992 election of Bill Clinton to the presidency, SDI was essentially canceled following the collapse of the Soviet Union. The Ballistic Missile Defense Organization, a scaled-back successor to SDI, is prohibited by treaty to practice its methods on real missiles and has rechanneled its resources into such research projects as Clementine.

Contributions

The knowledge gained from SDI falls into two categories: scientific and political. As a scientific endeavor, SDI significantly increased the funding and resources available to researchers working on directed energy (lasers and neutral and charged particle beams) and on satellite defense, communications, and support systems. Ways were found to create free electron lasers and nuclear-powered x-ray lasers, to guide a charged particle beam with a laser, and to accelerate neutrons into neutral particle beams, all of which have potential commercial and civilian applications.

Advances in kinetic weaponry can help reduce the threat of nuclear war by making the effect and success of a first strike more uncertain. That, in itself, made SDI an important part of the defensive structure of both the United States and the Soviet Union.

Star Wars also changed the scope of political debate about the arms race. The Reagan Administration used it as a bargaining chip; some observers claim that it helped bring about the 1988 Intermediate Nuclear Forces (INF) Treaty, an agreement between the two superpowers to destroy a whole class of intermediate range nuclear missiles. SDI is also credited with helping advance talks on the reduction of other classes of nuclear missiles.

SDI, by virtue of its high visibility and President Reagan's open and vocal support of the program, also helped to increase public awareness of the arms race. Americans gained a greater understanding of how, and how well, the nation is prepared to defend itself in the event of a nuclear war. This understanding, and the sentiment it generated, may help to prevent a nuclear attack as effectively as the technology behind the SDI program.

Context

The Strategic Defense Initiative was spurred by the desire to reduce the threat of nuclear war. When President Reagan proposed SDI, ballistic missile defense technology, particularly surface-to-air antiballistic missiles, had been under development for many years. BMD was the impetus for the ABM Treaty of 1972 between the Soviet Union and the United States. This treaty, in turn, created the avenue for BMD research that led to President Reagan's announcement of the SDI program.

In a broader sense, SDI was an attempt to move the United States away from its reliance on mutual assured destruction—a primarily offensive strategy—and toward the defensive posture of SDI, which would ultimately eliminate the need for a first-strike capability.

From a political perspective, SDI was initiated in the context of a massive increase in defense spending on the part of both the United States and the Soviet Union. Some experts believed that SDI, as an effort to diminish the effect of nuclear weapons, would allow the two nations with the most nuclear weapons in their arsenals to reduce their stockpiles and divert the funds for those weapons systems to other, productive uses. The SDI came to an inauspicious conclusion in 1993 with the inauguration of Bill Clinton as President of the United States. This was made possible by the collapse of the Soviet Union and a variety of strategic arms limitations treaties.

Only history will determine what the true significance of SDI has been. Some believe that SDI was an honest but misguided attempt to make nuclear war obsolete. Others believe it was an attempt to force the Soviet Union to the negotiating table in a position of weakness. Some maintain that it was a foolish and dangerous scheme that, if it had succeeded, might have resulted in an arms race in outer space. Others believe it was a serious diversion of funds and resources that left the United States militarily unprepared to meet external threats to the country in the early twenty-first century.

See also: Attack Satellites; Early-Warning Satellites; Electronic Intelligence Satellites; Manned Orbiting Laboratory; Meteorological Satellites: Military; Nuclear Detection Satellites; Ocean Surveillance Satellites; Spy Satellites; Telecommunications Satellites: Military; United States Space Command; Vandenberg Air Force Base.

Further Reading

Adragna, Steven P. *On Guard for Victory: Military Doctrine and Ballistic Missile Defense in the U.S.S.R.* Foreign Policy Report. Cambridge, Mass.: Institute for Foreign Policy Analysis, 1987. An overview of Soviet military policies relating to the Soviet ballistic missile defense program. Written from a conservative viewpoint, but very informative.

Broad, William J., et al. *Claiming the Heavens: The New York Times Complete Guide to the Star Wars Debate.* New York: Times Books, 1988. A compilation of an exhaustive series of articles on SDI that appeared in *The New York Times*, this resource is essential reading for the beginner interested in learning the basics before going on to more complex aspects of the SDI controversy.

Codevilla, Angelo. *While Others Build: The Common Sense Approach to the Strategic Defense Initiative.* New York: Free Press, 1987. Codevilla, a noted expert on SDI, provides a strong, easy-to-read overview of the program and a critical look at its advantages and disadvantages from both a political and a technological viewpoint. The subject is presented mostly from a proponent's perspective, but the text is filled with interesting information.

Davis, Jacquelyn K., and Robert L. Pfaltzgraff. *Strategic Defense and Extended Deterrence: A New Transatlantic Debate.* Cambridge, Mass.: Institute for Foreign Policy Analysis, 1986. This booklet examines the debate among the United States' allies over the benefits and drawbacks of SDI from an international political perspective. It also examines the Soviet strategic defense program and the European defense structure. A scholarly work, yet understandable to the nonspecialist.

Godson, Dean. *SDI: Has America Told Her Story to the World?* Cambridge, Mass.: Institute for Foreign Policy Analysis, 1987. This brief treatise looks at how the United States presented the arguments for SDI to the world community, particularly its European allies. Somewhat scholarly and technical, the work nevertheless provides a good geopolitical overview of a complex question.

Lambright, W. Henry, ed. *Space Policy in the Twenty-First Century.* Baltimore: Johns Hopkins University Press, 2003. This book addresses a number of important questions: What will replace the space shuttle? Can the International Space Station justify its cost? Will Earth be threatened by asteroid impact? When and how will humans explore Mars?

Mikheyev, Dmitry. *The Soviet Perspective on the Strategic Defense Initiative: Foreign Policy Report.* Elmsford, N.Y.: Pergamon Press, 1987. This book, written by a former Soviet physicist, looks at the Soviet government's views on the SDI program. Provides interesting insights into Soviet military and political policies regarding strategic defense. Written from a conservative perspective.

Reiss, Edward. *The Strategic Defense Initiative.* New York: Cambridge University Press, 1992. This history of the Strategic Defense Initiative shows how political, economic, strategic, and cultural factors have interacted to shape SDI. It examines new research into the SDI interest groups, the distribution of contracts, the Hardback politics of influence, and SDI's alliance management, popular culture, and military spin-offs. Throughout the book, the author tests the theoretical literature on the dynamics of the arms race against the reality of "Star Wars" and draws important conclusions about the motives behind SDI and its prospects for the 1990's.

United States Congress. Office of Technology Assessment. *SDI: Technology, Survivability and Software.* Princeton, N.J.: Princeton University Press, 1988. This is an unclassified version of a classified report prepared by the Office of Technology Assessment on SDI progress and feasibility as of 1985. Although somewhat technical in nature, the report gives a clear picture of "Star Wars" technology and reviews some of the arguments surrounding SDI.

Yenne, Bill. *Secret Weapons of the Cold War: From the H-Bomb to SDI.* New York: Berkley Books, 2005. A contemporary examination of Cold War superweapons and their influence on American-Soviet geopolitics.

Eric Christensen

Stratospheric Aerosol and Gas Experiment

Date: Beginning February 18, 1979
Type of program: Scientific platform

The Stratospheric Aerosol and Gas Experiment (SAGE) instrument was designed to measure the concentration of some of the constituents of Earth's atmosphere. The SAGE satellite was one of the first remote-sensing satellites to provide estimates of the global distribution of stratospheric aerosols, ozone, nitrogen dioxide, and water vapor.

Summary of the Technology

The Stratospheric Aerosol and Gas Experiment (SAGE) class of instruments was developed by the National Aeronautics and Space Administration (NASA) primarily through the efforts of M. Patrick McCormick of the Atmospheric Sciences Division at NASA's Langley Research Center in Virginia. The prototype to the SAGE experiments was conceived and built by Theodore J. Pepin of the University of Wyoming. This prototype was a small, manually operated instrument flown on the Apollo-Soyuz Test Project in 1975; it consisted of a handheld package containing a telescope, a Sun sensor, and an external electronics package. Its purpose was to measure the extinction of sunlight caused by aerosols in the atmosphere. The instrument was named the Stratospheric Aerosol Measurement (SAM) instrument and was operated by an astronaut as he pointed it directly at the Sun during a sunrise or sunset aboard a spacecraft.

It was shown that aerosols, which are small solid or liquid particles on the order of a micrometer in diameter, could be accurately measured in the atmosphere. After the successful results of SAM were revealed, SAM 2 was designed by NASA engineers and built at the University of Wyoming. SAM 2 was a much more sophisticated instrument than SAM and was built to fly aboard Nimbus 7. SAM 2 was to measure atmospheric aerosols in the polar regions of Earth between the latitudes of 64° to 84° south and 64° to 84° north. SAM 2 began operation in November, 1978, and would operate far longer than its initial design lifetime.

The SAGE I through III instruments were even more sophisticated than the SAM 2 instrument. Their primary objectives were to determine the spatial distribution of not only stratospheric aerosols but also ozone, nitrogen dioxide, and water vapor on a global scale. The principle of operation of the SAGE instruments is identical to that of the SAM instruments. On a spacecraft that orbits approximately 600 kilometers above Earth, SAGE instruments receive sunlight that is focused onto a set of detectors. Depending on the point in the orbit, this sunlight enters the instrument unobstructed (full-Sun condition), enters the instrument partially obstructed by Earth's atmosphere (occultation), or does not enter the instruments because of Earth's being in the path between the instrument and the Sun (total occultation). Scientists call the sequence just described an "event."

The SAGE instrument makes measurements of solar irradiance during the full-Sun condition in order to obtain a reference value for the intensity of the Sun as it is viewed unobstructed by the atmosphere. The actual atmospheric measurement occurs during the time that the Sun's disk appears through the atmosphere. During this portion of the event, the instrument focuses on a very small area of the solar surface with a telescope. Simultaneously, a mirror sweeps the image of the Sun across the field of the telescope. The actual spatial extent of the solar disk as viewed from the instruments varies in size depending upon how high the Sun appears above Earth. As the Sun is viewed in the higher layers of the atmosphere, the vertical extent of the Sun is approximately 32 kilometers. As the solar disk appears to move lower into Earth's atmosphere, the disk becomes flattened because of the refraction by Earth's atmosphere. The entire event lasts approximately one minute.

Events are classified as either sunrises or sunsets. Sunrise events are observed as the instrument orbits out from behind Earth and into the full sunlight, while sunset events are observed as the instrument orbits out of the full-Sun condition and into Earth's shadow. The small region on the surface of the Sun observed by the SAGE instrument during an event corresponds to a region in Earth's atmosphere that is approximately 1 kilometer high, 1 kilometer wide, and 200 kilometers long. This 200-cubic-kilometer volume of atmosphere lies along the line of sight joining the instrument and the Sun and, at any time during an event, is within 1 kilometer of a fixed altitude above the surface of Earth.

Three flights of the SAGE III instruments are planned, including one aboard the Russian Meteor 3M satellite. It will be launched in a Sun-synchronous orbit that provides high-latitude coverage. A second flight is scheduled aboard the International Space Station, and a third is in development.

At the heart of the SAGE instruments is a grating, a flat plate on which are ruled closely spaced lines. This grating splits the incident sunlight into a rainbow of colors. Each color corresponds to a particular wavelength of light emitted from the Sun. The SAGE instruments measure the intensity of several of these wavelengths of light. As the sunlight is occulted by the atmosphere, each individual wavelength is diminished in intensity more or less independently of the other wavelengths being measured. The measure of intensity that a particular wavelength loses because of the intervening atmosphere is directly related to the type of aerosol or gas present in the atmosphere.

The presence of aerosols and gases in the atmosphere causes incident light to scatter or be absorbed. When a light ray is scattered, it is simply redirected from its incident direction into another direction. When light is absorbed, it is taken up by the molecules in a gas or aerosol and converted into energy. The conversion process may change the energy into molecular rotation or vibration, kinetic energy, or new photons of light may be emitted at wavelengths different from that of the incident light. The processes of scattering and absorption are also called extinction. Very little absorption occurs for aerosols at the wavelengths where SAGE operates. SAGE III was designed to monitor global distribution of aerosols and gaseous constituents using the solar occultation approach. It uses instruments similar to its predecessors and incorporates an advanced imaging system, similar to that in a digital camera. It uses a 16-bit analog to digital converter for higher resolving capability. SAGE III can make multiple measurements of the absorption features of each target.

Basic principles of optics, atmospheric physics, and orbital mechanics are used to convert the relative instrumental quantities measured by the SAGE instruments into the absolute quantities of extinction and gas concentration. The information provided for experimenters by the SAGE I and SAGE II instruments comes in the form of vertical profiles of aerosol extinction and gas concentration. Each profile contains approximately sixty measurements, one measurement per kilometer, over an altitude range from cloud top to the top of the stratosphere. During any given day, fifteen sunrise and fifteen sunset measurements are made at equally spaced intervals around the world. Because the instruments take measurements at a local sunrise or sunset, the set of measurement profiles represent atmospheric conditions at the terminator, or the transition region on Earth that separates night and day.

Richard Rawls stands next to a Stratospheric Aerosol and Gas Experiment (SAGE III) instrument during operational checks. (NASA)

On February 18, 1979, SAGE I was launched by NASA on the Applications Explorer Mission 2 (AEM 2). It operated successfully until November 18, 1981, when a spacecraft power problem caused the instrument to stop functioning. The mission provided more than thirteen thousand events from which atmospheric profiles of aerosol extinction, ozone, and nitrogen dioxide were retrieved. The latitudes over which measurements were made ranged from about 72° south to 72° north.

The SAGE II instrument was launched on October 5, 1984, on the Earth Radiation Budget Satellite (ERBS). Along with SAGE II, the ERBS platform contains the Earth Radiation Budget Experiment. The ERBS is a key part of NASA's climate program. The SAGE II instrument makes atmospheric measurements from approximately 80° south to 80° north, depending on the season. It has recorded more than thirty-four thousand events from which scientists can retrieve profiles of extinction and gas concentration.

SAGE III was launched on December 10, 2001, from the Baikonur Cosmodrome on a Zenit 2 rocket. The instrument was aboard the Russian Meteor-3M spacecraft and represented a joint effort by the United States, Ukraine, and Russia.

Contributions

The long-term record of SAM 2 and SAGE II aerosol extinction data reveals periodic phenomena that occur in the Southern Hemisphere. The summertime aerosol extinction is relatively constant. During fall and winter, however, the temperatures decrease, and a large polar low-pressure system with swirling winds at its boundaries, called the polar vortex, establishes itself over the Antarctic. When the stratospheric temperatures reach their lowest values in the year, polar stratospheric clouds form inside the vortex and persist throughout the winter and early spring. Such clouds are always detected by SAM 2 and SAGE II aerosol extinction sensors during the winter and early spring. A high degree of variability in the set of daily aerosol extinction values is recorded as SAM 2 measures inside and outside the vortex. This high variability is also observed in the SAGE II ozone, nitrogen dioxide, and water vapor measurements taken at this time. With the onset of springtime and the consequent evaporation of polar stratospheric clouds, the polar stratospheric aerosol extinction measurements reach very low values relative to the rest of the year. Furthermore, as the polar vortex is replaced by warmer springtime air, the values of aerosol extinction increase to levels more typical of the summertime.

During the periods immediately after the eruptions of volcanoes, the aerosol extinction data of SAGE I, SAGE II, and SAM 2 show high values of aerosol. Volcanic eruptions typically place large volumes of gas and dust high into the atmosphere. The newly injected material perturbs the background aerosol distribution, which is normally hemispherically symmetric with a maximum above the equatorial latitudes. For some months after the volcanic injections, the aerosol extinction values continue to rise and eventually reach a maximum. This phenomenon is attributed to the conversion of gas to particles. That is, the large gas quantity emitted by volcanic eruption is converted to aerosols.

The SAGE I and II data reveal that the distribution of global ozone concentration is hemispherically symmetric with a maximum above the equator and minima near the polar latitudes. The temporal variation of the ozone in the midlatitudes is characterized by a strong annual cycle. Below 25 kilometers at these latitudes the ozone attains its maximum value of the year in summer, while between 25 and 37 kilometers the ozone attains its maximum value in winter. The temporal variation of the ozone at the equator is dominated by a semiannual cycle. Below approximately 60 kilometers there is no significant difference between sunrise ozone and sunset ozone. Above this altitude, the sunrise ozone is greater than the sunset ozone because of photochemical reactions. These types of reactions are controlled by the presence of sunlight. The chemical bonds of certain gases in the atmosphere break apart when light is absorbed.

The SAGE data reveal that the distribution of global nitrogen dioxide is essentially hemispherically symmetric. The temporal variation of the nitrogen dioxide in the midlatitudes is characterized by a strong annual cycle. Over the altitudes from 20 to 45 kilometers at these latitudes, the maximum value of the year occurs in summer. Furthermore, the sunset nitrogen dioxide concentration is always greater than that for sunrise because of photochemical reactions.

Because the occultation method of measuring atmospheric constituents is dependent on sensing the solar disk as it rises or sets in the atmosphere, any cloud that might be along the line of sight from the satellite to the Sun affects the measurement. Thick clouds, such as cumulonimbus, tend to block the sunlight completely, whereas thin clouds, such as cirrus, may only partially obscure the sunlight. Thus, the global distribution of the frequency of occurrence of clouds and cloud altitudes may also be estimated from the SAM 2, SAGE I, SAGE II, and SAGE III data.

Context

A phenomenon called the ozone hole has been observed during the Antarctic springtime since the 1980's. The ozone hole is a decrease in the abundance of Antarctic ozone as sunlight returns to the pole in early springtime. Minimum monthly values of ozone concentration are observed in October. The ozone hole occurs inside the polar vortex and is believed to be related to chemical reactions that take place on the surface of polar stratospheric clouds. Human industrial activities have caused chlorine compounds to be released into the atmosphere. The unique atmospheric conditions that exist inside the polar vortex during the Antarctic springtime cause the chlorine from these compounds to be released in a reactive form that destroys ozone. The SAGE II measurements have been shown to agree with ozone measurements taken by balloon-borne instruments in the same region. The large gradients in ozone concentration as one moves from outside the vortex to inside the vortex are recorded in the SAGE II data.

The SAGE instruments measured ozone in the midlatitudes more frequently than near the polar regions. In 1985, a report, based on satellite data, was presented to the United States Congress stating that large global ozone decreases had occurred since 1978. The satellite dataset discussed was that of the Solar Backscattered Ultraviolet (SBUV) instrument aboard Nimbus 7. It indicated that ozone had decreased by approximately 20 percent at 50 kilometers since 1978. An independent investigation of the SAGE ozone datasets over the same period revealed no trend at 50 kilometers and a small decrease (about 3 percent) at 40 kilometers. The trend calculated from the SAGE datasets agreed more closely with theoretical predictions of ozone and temperature change than the trend computed from the SBUV data.

It is important to know the magnitude of any long-term decrease in the amount of ozone in the stratosphere because of the potential damage to life on Earth. The stratosphere contains most of the ozone in the atmosphere, with approximately 10 percent of the atmospheric ozone contained in the troposphere (even though human activities generate ozone). Loss of ozone means that more solar ultraviolet light would penetrate the atmosphere to Earth's surface. Ultraviolet light is absorbed by stratospheric ozone and prevented from penetrating to the surface. The damage to humans, animals, and vegetation by unshielded solar ultraviolet light is potentially large.

The global distribution of aerosols mapped from the SAGE data shows that during volcanically active periods, gas and particles from eruptions are transported to different parts of the atmosphere by the general circulation. The data from the four aerosol extinction wavelengths measured by SAGE II have shown that it is possible to discriminate between the particle sizes injected into the atmosphere by a volcano. The aerosols emitted by volcanoes also act as tracers for the atmospheric circulation. In turn, researchers can further understand the dynamic motion of Earth's atmosphere.

The SAGE I, SAGE II, SAGE III, and SAM 2 satellite instruments have served as important tools in the understanding of Earth's atmosphere. Using their data, scientists have been able to map aerosol and ozone concentrations on a time scale shorter than major stratospheric changes; to locate stratospheric aerosol, ozone, nitrogen dioxide, and water vapor sources and sinks; to estimate long-term global ozone trends; to study the Antarctic ozone hole; to monitor circulation and transport phenomena; to observe hemispheric differences; and to investigate the optical properties of aerosols and assess their effects on global climate.

See also: Earth Observing System Satellites; Heat Capacity Mapping Mission; Private Industry and Space Exploration; Space Shuttle Flights, 1984.

Further Reading

Anthes, Richard A., Hans A. Panofsky, John J. Cahir, and Albert Rango. *The Atmosphere.* 2d ed. Columbus, Ohio: Charles E. Merrill, 1978. A nonmathematical introductory book on meteorology. Many illustrations and photographs show examples of tropospheric phenomena. Suitable for a general audience.

Craig, Richard A. *The Upper Atmosphere: Meteorology and Physics.* New York: Academic Press, 1965. Most of chapters 2 and 3 of this work are nonmathematical in nature and deal exclusively with the stratosphere. The emphasis is on historical discoveries and observed phenomena. Many charts and graphs illustrate the subject matter.

Finlayson-Pitts, Barbara J., and James N. Pitts, Jr. *Chemistry of the Upper and Lower Atmosphere: Theory, Experiments and Applications.* San Diego, Calif.: Academic Press, 2000. Subjects include atmospheric and environmnental chemistry. Tables on front lining paper, bibliographical references, and an index are included.

National Research Council of the National Academies, Committee on Microgravity Research. *Assessment of Directions in Microgravity and Physical Sciences Research at NASA.* Washington, D.C.: National Academies Press, 2003. Provides detailed reports on microgravity research conducted on robotic missions, the space shuttle, and the International Space Station.

Pitts, M. C., L. W. Thomason, and W. P. Chu. *Satellite Remote Sensing of Temperature and Pressure by the Stratospheric Aerosol and Gas Experiment 3, Presented at the 9th Conference on Satellite*

Meteorology and Oceanography, Paris, France, May, 1998. Washington, D.C.: Government Printing Office, 1998. This is a brief discussion of one of the experiments to be conducted on board the SAGE III platform. It discusses the methods the experiment will use to test the Earth's atmosphere in relationship to its temperature and atmospheric pressure. There is a brief bibliography of other works relating to SAGE.

Scorer, Richard S. *Cloud Investigation by Satellite.* New York: Halstead Press, 1986. Provides six hundred satellite images of cloud structures of the Northern Hemisphere. Written for a general audience.

Wallace, John M., and Peter V. Hobbs. *Atmospheric Sciences: An Introductory Survey.* New York: Academic Press, 1977. An excellent introduction to physical and dynamical meteorology, this college-level book assumes a first-year calculus and chemistry background. The authors alternate the subject matter in the chapters between dynamical meteorology and physical meteorology.

Watson, R. T., et al. *Present State of Knowledge of the Upper Atmosphere, 1988: An Assessment Report.* Reference publication 1208. Washington, D.C.: National Aeronautics and Space Administration, 1988. A publication geared toward those interested in the efforts of scientists to assess the ozone trend in 1987. All aspects of ozone measurement and trend estimation are discussed. Topics include satellite measurements, ground-based measurements, instrument calibration, the Antarctic ozone hole, stratospheric temperature, trace gases, chemical modeling, and aerosol distributions and abundances.

Young, Louise B. *Earth's Aura.* New York: Avon Books, 1979. Chapter six gives an interesting description of the ozone layer and the controversy concerning its destruction. A historical account leading up to regulations on the use of certain chlorine compounds is given.

Zimmerman, Robert. *The Chronological Encyclopedia of Discoveries in Space.* Westport, Conn.: Oryx Press, 2000. Provides a complete chronological history of all crewed and robotic spacecraft and explains flight events and scientific results. Suitable for all levels of research.

Robert Veiga

Surveyor Program

Date: May 30, 1966, to February 21, 1968
Type of program: Lunar exploration

Surveyor 1 achieved the first lunar soft landing by a fully automated spacecraft. The Surveyor Program developed the technology for soft-landing on the Moon, verified the compatibility of the design of the Apollo crewed lunar-landing spacecraft with actual lunar surface conditions, and added to the scientific knowledge of the Moon.

Key Figures

Benjamin Milwitzky, Surveyor program manager, NASA
Walker E. "Gene" Giberson, Surveyor project manager until 1965, Jet Propulsion Laboratory (JPL)
Robert Parks, JPL Surveyor program manager, 1965-1966
Howard H. Haglund, JPL Surveyor program manager, 1966-1968
R. G. Forney, JPL Spacecraft Systems manager
Leonard Jaffe, Survey Scientific Evaluation Advisory Team chairperson
Eugene M. Shoemaker (1928-1997), principal investigator, imaging science experiment
A. L. Turkevich, principal investigator, alpha scattering experiment
Ronald F. Scott, principal investigator, soil mechanics surface sampler experiment

Summary of the Program

In May, 1960, the National Aeronautics and Space Administration (NASA) approved a Surveyor Program consisting of two parts: a Lunar Orbiter for photographic coverage of the Moon's surface and a lunar lander to obtain scientific information on the Moon's environment and structure. Before human beings could safely be sent to the Moon, the Surveyors were to provide spacecraft designers with information on the load-bearing limits of the lunar surface, its magnetic properties, and its radar and thermal reflectivity.

The Jet Propulsion Laboratory (JPL) was assigned project responsibility, and four Surveyor study contracts were awarded in July, 1960, to Hughes Aircraft, North American, Space Technology Laboratories, and McDonnell Aircraft. On January 19, 1961, NASA chose Hughes Aircraft's proposal for the Surveyor and began planning at JPL

for seven lunar-landing flights, the first of which was planned for launch on an Atlas-Centaur booster in 1963.

Because of development problems with the Centaur, early failures of the Ranger lunar impactor, and increasing demands for information on the lunar surface to support the Apollo Program, the orbiter portion of Surveyor was dropped in 1962 and replaced by the Lunar Orbiter project, managed by NASA's Langley Research Center. Problems with the Centaur upper stage forced postponement of the first Surveyor launch and required a reduction in the spacecraft's weight—from an original 1,134 kilograms with a 156-kilogram payload to 953 kilograms carrying only 52 kilograms of instruments.

The Atlas-Centaur became operational in 1966, and Surveyor 1 was launched from Launch Pad Complex 36A at Cape Kennedy, Florida, on May

30, 1966. Surveyor 1, the first test model of the series, carried more than one hundred engineering sensors to monitor spacecraft performance. No instrumentation was carried specifically for scientific experiments, but the spacecraft was outfitted with a survey television system and with instrumentation to measure the bearing strength, temperature, and radar reflectivity of the lunar surface.

After injection on a trajectory intersecting the Moon, the Surveyor spacecraft separated from the Centaur upper stage. Midcourse maneuvers, using vernier engines on the spacecraft, were performed to bring it within the desired target area. For the terminal descent, the main retroengine was ignited, by command of an onboard radar altimeter, to provide most of the braking. After this retroengine burned out, at about 10 kilometers above the lunar surface, it was jettisoned. A second radar altimeter, providing measurements of velocity and altitude, was used with the smaller vernier engines, in a closed loop under control of an onboard analog computer, for the final descent phase. To reduce the disturbance to the lunar surface, the vernier engines were extinguished by the computer when the spacecraft was about 4 meters above the surface and the descent rate was about 1.5 meters per second. At 06:17 Coordinated Universal Time (UTC) on June 2, 1966, Surveyor 1 soft-landed in the southwest portion of the Ocean of Storms, becoming the first U.S. spacecraft to soft-land on another celestial body.

During the next two lunar days (28 Earth days each) the Surveyor returned some eleven thousand photographs of the surrounding terrain to Earth. Surveyor 1 completed its primary mission on July 14, 1966, but engineering interrogations were conducted at irregular intervals through January, 1967.

Surveyor 2, launched on September 20, 1966, had essentially the same configuration as Surveyor 1 and was intended to land in the Central Bay, another potential Apollo landing zone. When the midcourse maneuver was attempted on September 21, one of the three vernier engines failed to ignite, and the unbalanced thrust caused the spacecraft to tumble. Although repeated commands were sent in an attempt to salvage the mission, Surveyor 2 crashed on the Moon on September 22, 1966.

Surveyor 3, though similar to the two earlier spacecraft, was equipped with two fixed mirrors to extend the view of its television camera underneath the spacecraft. A remotely controlled surface sampler arm, capable of digging trenches and manipulating the surface material in view of the television camera, was also added. The Surveyor 3 spacecraft was launched on April 17, 1967, from Launch Pad Complex 36B at Cape Kennedy— the first time the two-burn capability of the Centaur was used on an operational

Surveyor 1 lifts off with a boost from an Altas-Centaur launch vehicle. (NASA CORE/Lorain County JVS)

mission. After separation from the Atlas, the Centaur engine ignited and burned for approximately five minutes to place the vehicle into a 167-kilometer circular parking orbit. The Centaur coasted for 22 minutes, then reignited to place the spacecraft on a lunar-intercept trajectory. The use of a parking orbit greatly increased the fraction of the lunar surface to which the Surveyor could be targeted.

The midcourse correction maneuver and the firing and jettisoning of the retroengine proceeded as planned. A few seconds before touchdown, however, the onboard radar lost its lock on the surface, apparently because of unexpected reflections from large rocks near the landing site. As a result, the spacecraft guidance system switched to an inertial mode, which prevented the vernier engines from extinguishing about 4 meters above the surface as planned. The spacecraft touched down with its vernier engines still firing, lifted off, touched down a second time, lifted off again, then touched down for a third time after receiving a command from Earth 34 seconds after the initial touchdown. The spacecraft had a lateral velocity of about 1 meter per second at the first touchdown, and the distance between the first and second touchdowns was about 20 meters, while there was a distance of about 11 meters between the second and third touchdowns.

The landing occurred on April 19, 1967, in the southeast part of the Ocean of Storms, a potential Apollo landing region. Surveyor 3 landed in a medium-sized crater and came to rest tilted at an angle of about 14°. This location allowed the crater to be viewed from the inside, and the unplanned tilt permitted the camera to aim high enough to photograph an eclipse of the Sun by Earth, which would not have been possible if the landing had been on a level surface. The spacecraft returned 6,315 television pictures and operated its surface sampler for more than eighteen hours before transmissions ceased shortly after local sunset on May 3, 1967.

Surveyor 4, carrying the same payload as Surveyor 3 had, was launched on July 14, 1967. After a flawless flight to the Moon, radio signals from the spacecraft ceased abruptly during the final descent, approximately 2.5 minutes before touchdown and only two seconds before retroengine burnout. Radio contact with the spacecraft was never reestablished, and Surveyor 4 crashed into the lunar surface, possibly after an explosion.

Surveyor 5 was launched from Cape Kennedy on September 8, 1967. Because of a helium regulator leak that developed during flight, a radically new descent technique was engineered, and the Surveyor 5 performed a flawless descent and soft landing in the Sea of Tranquility on September 11, 1967. The spacecraft was similar to its two immediate predecessors, except that the surface sampler was replaced by an "alpha backscatter instrument," a device to determine the relative abundances of the chemical elements in the lunar surface material. In addition, a bar magnet was attached to one of the footpads to determine if magnetic material was present in the lunar soil.

During its first lunar day, which ended at sunset on September 24, 1967, Surveyor 5 took 18,006 television pictures, performed chemical analyses of the lunar soil, and fired its vernier engines for 0.55 second to determine the effects of high-velocity exhaust gases impinging on the lunar surface. On October 15, 1967, after exposure to the two-week deep freeze of the lunar night, Surveyor 5 responded to a command from Earth, reactivating, and transmitted an additional 1,043 pictures and data from the lunar surface.

The Surveyor 6 spacecraft, essentially identical to Surveyor 5, was launched on November 7, 1967, and landed on the Moon on November 10, 1967. The landing site, near the center of the Moon's visible hemisphere in the Central Bay, was the last of four potential Apollo landing sites designated for investigation by the Surveyor Program. From landing until a few hours after lunar sunset on November 24, 1967, the spacecraft transmitted more than 29,000 television pictures, and the alpha backscattering experiment acquired thirty hours of data on the chemical composition of the lunar soil. On November 17, 1967, Surveyor 6's vernier engines were fired for 2.5 seconds, causing the spacecraft to lift off from the lunar surface and move laterally about 3 meters to a new location. Television pictures showed the effect of rocket firings close to the lunar surface. When combined with images taken from the earlier landing site, the new photographs provided stereoscopic data of the surrounding terrain and surface features.

Because the previous Surveyors had completed the Apollo landing site survey, Surveyor 7 was targeted at the scientifically interesting rock-strewn ejecta blanket of the crater Tycho. Launched on January 7, 1968, the Surveyor 7 spacecraft landed less than 2 kilometers from its target on January 10, 1968. During the first lunar day, 21,046 pictures of the lunar surface were acquired, but the alpha backscatter package failed to deploy. The surface sampler arm was programmed from Earth to force the package into position and subsequently to move it to two additional sites. Laser beams from Earth were also detected by the television camera during a test. Forty-five more photographs and additional surface chemical data were obtained during the second lunar day of operation, before the spacecraft was deactivated on February 21, 1968.

Having met all the Apollo survey objectives, follow-on Block 2 missions were canceled because of budget constraints, and the Surveyor Project Office at JPL was closed on June 28, 1968.

Contributions

The surface sampler arms on Surveyors 3 and 7 made measurements of how much weight the lunar surface soil could bear before being penetrated as well as how depth affected the bearing capacity and shear strength in trenches up to 20 centimeters deep. The strength and density of individual lunar rocks were also determined. Strain gauges on the shock absorbers determined the loads on the legs of the spacecraft during the touchdown on the lunar surface. The radar reflectivity and dielectric constant of the lunar surface material were determined by the landing radar system. Thermal sensors determined the surface temperatures, thermal inertia, and directional infrared emission at all five landing sites.

At the mare landing sites, firings of the vernier rocket engines on Surveyors 3, 5, and 6 and the attitude control rockets on Surveyors 1 and 6 against the lunar surface provided information on the permeability of the surface to gases, the cohesion of the soil, its response to gas erosion, and its adhesion to spacecraft surfaces.

The alpha backscattering experiments on Surveyors 5, 6, and 7 determined the abundances of the major chemical elements from carbon to iron on six surface samples and one subsurface sample at two mare sites and one highland site. These analyses indicated that the most abundant element of the lunar surface is oxygen (57 atomic percent), followed by silicon (20 atomic percent), and aluminum (7 atomic percent). These are the same elements, and in the same order, that are most common in Earth's crust. The major chemical elements at the mare sites are generally similar to those found in terrestrial basaltic rocks. The similarity to basalts, as well as the morphology of the surface features determined from the Lunar Orbiter images, provides strong circumstantial evidence that some melting and chemical separation of the lunar material had occurred in the past. This surface composition is significantly different from primordial solar system material. It is also different from most known meteorites, indicating that the Moon is not a major source of the meteorites that hit Earth.

The magnets carried on Surveyors 5, 6, and 7 provided information on the magnetic particles at two mare sites and one highland site. The single highland site showed a lower abundance of iron and other chemically similar elements, demonstrating that the lunar surface is not homogeneous. These chemical differences were suggested as an explanation for the difference in albedo, or surface reflectivity, between the highland and mare regions.

Five successful Surveyor spacecraft returned 87,700 television pictures of the lunar surface, Earth, the solar corona, Mercury, Venus, Jupiter, and stars to the sixth magnitude. Lunar surface images provided information on the size-frequency distribution of lunar craters ranging from a few centimeters to tens of meters in diameter. Observations were consistent with the distribution predicted from prolonged bombardment by meteorites, and therefore allowed for the size distribution of the incoming meteoroids to be inferred.

Two and a half years after its landing, Surveyor 3 was visited by the Apollo 12 astronauts, who cut off the video camera, some camera cable, aluminum tubing, and a glass filter and returned these items to Earth for analysis to determine the extent of lunar weathering, micrometeorite impacts, and ion bombardment from the Sun.

Context

Although the Soviet Union's Luna 9 spacecraft preceded Surveyor in soft-landing on the Moon, the Surveyors provided the first quantitative data on the bearing strength of the lunar soil, its radar reflectivity, and the variations in the lunar surface temperature. The Surveyors also demonstrated the existence of loose dust on the Moon.

Prior to Surveyor 1, the photographs of the lunar surface having the best resolution were taken by the Luna 9 spacecraft. Surveyor supplemented these with photographs from five additional sites, including one in the lunar highlands. A comparison of the highland surface at the Surveyor 7 landing site with that at the mare sites showed fewer craters larger than 8 meters in diameter at the highland site, indicating that the Tycho rim material on which Surveyor 7 landed was much younger than the mare material. This provided a relative chronology for the Moon's different regions.

The computer-controlled soft landings of the Surveyor spacecraft on the Moon verified the soft-landing techniques planned for the Apollo lunar-landing missions. Four of the Surveyors successfully landed in potential Apollo landing regions and determined that the lunar surface was sufficiently strong to support the footpad loads planned for the Apollo spacecraft. They also provided information on the surface roughness and topography necessary for the Apollo flights. Another successful Surveyor landing expanded the scientific investigations of the lunar surface begun by the previous soft landers.

See also: Apollo Program; Apollo Program: Geological Results; Apollo Program: Lunar Module; Apollo Program: Orbital and Lunar Surface Experiments; Clementine Mission to the Moon; Lunar Exploration; Lunar Orbiters; Lunar Prospector.

Further Reading

Burrows, William E. *This New Ocean: The Story of the First Space Age.* New York: Random House, 1998. This is a comprehensive history of the human conquest of space, covering everything from the earliest attempts at spaceflight through the voyages near the end of the twentieth century. Burrows is an experienced journalist who has reported for *The New York Times, The Washington Post,* and *The Wall Street Journal.* There are many photographs and an extensive source list. Interviewees in the book include Isaac Asimov, Alexei Leonov, Sally K. Ride, and James A. Van Allen.

French, Bevan M. *The Moon Book.* New York: Penguin Books, 1977. Describes the explanations for the origin of the Earth-Moon system and places the contribution of the Surveyor Program into the context of the entire lunar exploration program. Includes a thorough description and illustration of the alpha backscattering experiment and its results. Suitable for general audiences.

Hartmann, William K. *Moons and Planets.* 5th ed. Belmont, Calif.: Thomson Brooks/Cole Publishing, 2005. Provides detailed information about all objects in the solar system. Suitable on three separate levels: high school student, general reader, and the college undergraduate studying planetary geology.

Hibbs, Albert R. "The Surface of the Moon." *Scientific American* 216 (March, 1967): 60-74. A well-illustrated article describing the results from the first nine spacecraft, including Surveyor 1, which provided close-up photographs of the lunar surface. Describes the various theories for the origin of the Moon in the context of scientific knowledge as it existed immediately after the landing of Surveyor 1.

McBride, Neil, and Iain Gilmour. *An Introduction to the Solar System.* New York: Cambridge University Press, 2004. This work provides a comprehensive tour of the solar system. Suitable for a high school or college course on planetary astronomy.

Mari, Christopher, ed. *Space Exploration.* New York: H. W. Wilson, 1999. Twenty-five articles (reprinted from magazines), covering the state of the space program at the time of publication, are divided into five sections: John H. Glenn, Jr.'s return to space, the exploration of Mars, the International Space Station, recent mining efforts by commercial industries, and new types of space vehicles and propulsion systems.

Morrison, David, and Tobias Owen. *The Planetary System.* 3d ed. San Francisco: Addison-Wesley, 2003. Organized by planetary object, this work provides contemporary data on all planetary bodies visited by spacecraft since the early days of the Space Age. Suitable for high school and college students and for the general reader.

Newell, Homer E. "Surveyor: Candid Camera on the Moon." *National Geographic* 130 (October, 1966): 578-592. A comprehensive collection of early Surveyor photographs and a description of the preliminary scientific interpretations of the video data.

Nicks, Oran W. *Far Travelers: The Exploring Machines.* NASA SP-480. Washington, D.C.: Scientific and Technical Information Branch, 1985. Discusses all major NASA planetary spacecraft during NASA's first quarter-century. Written by a senior NASA official involved with lunar and planetary programs during that era.

Scott, Ronald F. "The Feel of the Moon." *Scientific American* 217 (November, 1967): 34-43. The principal investigator for the soil mechanics surface sampler describes the Surveyor spacecraft, emphasizing the sampler arm and the experiments to determine the strength of the lunar surface. Includes Surveyor images of the lunar surface and diagrams of the spacecraft and the sampler arm.

_____. "Report on the Surveyor Project." *Journal of Geophysical Research* 72 (January 15, 1967): 771-856. In this series of articles, the scientists who participated in the Surveyor Program report their original data and interpretations. Contains descriptions of the Surveyor instruments and a detailed account of the scientific results.

_____. *Surveyor Program Results.* NASA SP-184. Washington, D.C.: Government Printing Office, 1969. This official program overview includes introductory articles summarizing each Surveyor mission and describing the principal scientific results. These are followed by individual articles on the imaging, surface mechanical and electrical properties, temperature experiments, and the chemical analysis. The document also contains a list of the program participants.

Wilhelms, Don E. *To a Rocky Moon: A Geologist's History of Lunar Exploration.* Tucson: University of Arizona Press, 1993. A detailed scientific history of lunar exploration from the early Pioneer probes through the final crewed Apollo lunar landing.

George J. Flynn

Swift Gamma Ray Burst Mission

Date: November 20, 2004
Type of program: Scientific platform

Swift is the first telescope of its kind designed to detect and study gamma-ray bursts in real time. Gamma-ray bursts are high-energy explosions that occur daily throughout the sky. Swift will detect them and automatically move to observe them, giving astronomers their first chance to view a gamma-ray burst in the critical first few minutes. As a result, astronomers will be able to test theories on the origin and aftermath of gamma-ray bursts, identify classifications of gamma-ray bursts, devise studies on how gamma-ray bursts affect their environments, and learn more about the history of the early universe.

Key Figures

Neil Gehrels, principal investigator, Goddard Space Flight Center (GSFC)

Joe Dezio, project manager, GSFC

Nicholas White, Science Working Group chair, GSFC

Lynn Cominsky, education and public outreach manager and press officer

John Nousek, mission operations lead, Pennsylvania State University (PSU)

Padi Boyd, science support center lead, GSFC

Lorella Angelini, High-Energy Astrophysics Science Archive Research Center lead, GSFC

Scott Barthelmy, Burst Alert Telescope lead, GSFC

Dave Burrows, X-Ray Telescope lead, PSU

Pete Roming, Ultraviolet/Optical Telescope lead, PSU

Frank Marshall, Ground Segment, GSFC

Mark Edison, with Spectrum Astro, a division of General Dynamics

Summary of the Mission

The physics of the formation of black holes is one of the most intense areas of theoretical and observational study in science. When a star collapses to form a black hole, it releases gamma-ray bursts that have been detected on Earth since the 1960's. These gamma-ray bursts contain information on the evolution of the universe as well as the high-energy processes in black holes. The Swift Gamma Ray Burst Mission, launched in 2004, was aimed at studying these gamma-ray bursts as soon as they were detected.

Theories about gamma-ray energy existing in the universe were well developed by the 1960's, but there was no technology capable of measuring it. In the early 1960's, however, the U.S. military was deeply concerned about Soviet nuclear weapons testing. Such weapons would release small amounts of gamma-ray energy into the atmosphere, so the United States launched the Vela satellites, which were equipped to detect this energy. These satellites did detect gamma rays but found that, instead of coming from the Soviet Union, the rays were originating in outer space. By the 1970's, this information had been released to the civilian scientific community, and there was much debate as to the source and strength of the gamma-ray bursts.

Gamma rays have the shortest wavelengths in the electromagnetic spectrum and energies of up to 10^{46} joules. They are produced by nuclear explosions or nuclear reactions. The radiation produced by these processes is deadly to humans but can be used in small quantities to fight diseases, especially cancer. Gamma-ray bursts from space are hard to detect on Earth because most of them are absorbed by the Earth's atmosphere, protecting the biosphere from this radiation.

The two leading hypotheses on the origins of gamma rays are that they result from neutron star mergers or hypernovae. As two neutron stars orbit each other they gain gravitational energy, losing rotational energy. Eventually their orbits decay and they collide to form a black hole. The energy released in this process is one possible method of generating gamma rays. The other leading theory is that at the end of a massive star's lifetime it explodes as a supernova to become either a neutron star or a black hole. Stars with masses greater than about forty solar masses may release about one hundred times more energy when they explode. These hypernovae may also be a source of gamma-ray bursts.

Gamma-ray bursts started to be detected by satellites in the early 1960's, but image resolution was poor. The exact location of the bursts could not be determined, and there was no way of looking at the burst immediately after it was detected. It took time to move the satellites into position, so all that was visible was the afterglow of the burst. (The afterglow is radiant energy usually emitted in the x-ray, radio, or visible parts of the electromagnetic spectrum.) The result was only limited information as to where the gamma-ray bursts originated. The scientific community was divided into two schools of thought. One group believed that gamma-ray bursts were sufficiently powerful to have been generated outside our Milky Way galaxy; the other believed that they must come from within the Milky Way.

To resolve this question, the National Aeronautics and Space Administration (NASA) launched the Compton Gamma Ray Observatory (GRO) in the early 1990's. One of the instruments on board GRO was the Burst and Transient Source Experiment (BATSE). BATSE monitored the sky for gamma-ray bursts and recorded the numbers of bursts detected. After several years of monitoring, BATSE data showed scientists that not all gamma-

At Kennedy Space Center, workers inside the mobile service tower on Pad 17A prepare the Swift spacecraft for launch. (NASA)

ray bursts were on the plane of the Milky Way, but that did not necessarily mean that they could not be coming from elsewhere in our galaxy.

In 1997 an Italian telescope, Beppo-Sax, detected an afterglow at a recent gamma-ray site. The discovery of the afterglow made redshift measurements possible, giving the distance to the gamma-ray bursts. This information made it clear that not all gamma-ray bursts came from within the Milky Way galaxy.

The Hubble Space Telescope (HST) gave scientists a more detailed view of deep-space objects than ever before. With HST, sites where gamma-ray bursts had been detected could be observed to see their effects on the environment. While Hubble had no instruments to measure gamma rays directly, the views of past gamma-ray sites helped scientists better understand the origins and the effects of gamma-ray bursts.

The pictures from HST led to a much greater understanding of gamma-ray bursts but, because the Hubble had no gamma-ray equipment and because of the amount of time it took to turn Hubble, there were still no pictures from the first few critical minutes of a gamma-ray burst site. In November of 2004, NASA launched the Swift Gamma Ray Observatory to fill this gap.

The Swift observatory is named after a small, nimble bird. The observatory is built to be able to spot a gamma-ray burst and quickly move to observe the area of its origin. It does this much the same way the bird would spot its prey and quickly move to attack. The bird is pictured in the Swift observatory logo.

There are three main telescopes aboard the observatory: the Burst Alert Telescope (BAT), the X-Ray Telescope (XRT), and the Ultra-Violet/Optical Telescope (UVOT). These three telescopes work together to gain as much information as possible about each gamma-ray burst. The BAT first detects the gamma-ray burst and slews the observatory to the area of the sky from which it came. Next, the XRT narrows in on the exact area of the bursts and finishes the alignment of the telescope. The UVOT finalizes the positioning of the telescope by producing the most accurate coordinate set.

The BAT is Swift's first line of detection. It scans the sky for any gamma-ray counts that are significantly above the background gamma-ray radiation. When the radiation reaches the detector, it passes through a filter, which produces a shadow. From the image of the shadow, BAT can calculate the position of the gamma-ray bursts to within five arc minutes in about ten seconds. This position is then immediately sent to the ground-based operations while Swift automatically starts slewing to it. BAT also has a hard x-ray detector designed to produce an all-sky survey twenty times more sensitive than the last one, done in the 1970's by High-Energy Astronomical Observatories 1 A4 (HEAO-1 A4). BAT also watches for transient hard x-ray sources and transmits their data to the ground.

Once BAT has allowed Swift to slew so that the gamma ray source is within the XRT's line of sight, the XRT takes over. X rays, unlike gamma rays, can be focused so that a more exact location of the gamma-ray burst site can be determined to within a five arc second accuracy. The XRT measures the fluxes, spectra, and light glows of both the gamma-ray bursts and their afterglows. It also calculates redshifts from these afterglows, which is useful because no distances to short gamma-ray bursts have yet been calculated. Finally, the XRT looks at absorption and emission spectra to determine the material surrounding the gamma-ray burst sources.

Once the XRT has narrowed the field of vision down to 5 arc seconds, the UVOT detects the source and starts observing. It can narrow the location down to a 0.3 arc second accuracy. The UVOT studies the gamma-ray burst and afterglow with a series of filters, tracking the changes over time in different colored filters. Also, for redshifts greater than one, the UVOT can provide redshift and distance information.

Contributions

Swift launched on schedule on November 20, 2004. Activation of the observatory finished on January 19, 2005. In the first twelve days, nine gamma-ray bursts were detected, which was more than was expected, but that was followed by two weeks of nothing being detected. On December 19, 2004, a long gamma-ray burst was detected as well as an infrared flash and its radio counterpart. On December 23, an optical afterglow was also detected. The only problem with activation was that the thermoelectric cooler did not work as expected. This means that the temperature of the observatory cannot be regulated as desired, so the spectral resolution will be shifted a few percentage points per year. Also, more on-orbit calibration time will be required and the flight parameters will have to be modified.

The primary objectives of the Swift mission had been met as of mid-2005, and most of the secondary objectives have remained unaffected by the small problems that Swift has encountered. Swift is now observing and coming fully online to develop a more complete understanding of gamma-ray bursts, their origins, and their affects on the environment in which they occur.

Classes of gamma-ray bursts will be established based on their characteristics, and bursts will be grouped into them. Swift will allow the study of what fraction and kinds of gamma-ray bursts have underlying supernovae. The universe will be looked at back to a redshift value of about 15, which will allow the study of the first stars.

Context

Gamma-ray bursts were first detected coming from all areas of the universe by the Explorer 11 satellite in 1961. More gamma-ray-burst satellites were launched by the U.S. government to monitor the Soviet Union's nuclear testing. The first satellite dedicated to gamma-ray detection was the second Small Astronomy Satellite (SAS-2). Its mission lasted only seven months because of electrical failure. The European Space Agency launched COS-B in 1972, which observed gamma rays until 1985. Both SAS-2 and COS-B confirmed the gamma-ray background radiation, and both detected point sources of gamma rays. The resolutions on these telescopes were not good enough to pinpoint exact locations of the point sources. From 1991 until 2000, the Compton Gamma Ray Observatory received information at better resolutions, helping to establish theories as to the origin of the gamma-ray bursts. Swift is the most recent in this line of gamma-ray telescopes, with the ability to detect and move to observe gamma-ray bursts at a time closer to their origin than ever before. This will allow new research on high-energy physics of black holes and the evolution of the universe.

See also: Compton Gamma Ray Observatory; Ulysses: Solar Polar Mission.

Further Reading

Cheng, K. S., and G. V. Romero. *Cosmic Gamma-Ray Sources.* New York: Springer-Verlag, 2004. Provides an introduction to gamma-ray astrophysics and in-depth explanations of known sources.

Katz, Jonathan. *The Biggest Bangs: The Mystery of Gamma-Ray Bursts.* New York: Oxford University Press, 2002. Reveals the advances in science and instrumentation that have led to the current understanding of gamma-ray bursts.

Kerrod, Robin. *Hubble: The Mirror on the Universe.* Richmond Hill, Ont.: Firefly Books, 2003. This volume explains the history of astronomy and the advances Hubble made in the observable universe. It also contains many of the images from Hubble.

Peterson, Carolyn C., and John C. Brant. *Hubble Vision: Astronomy with the Hubble Space Telescope.* New York: Cambridge University Press, 1995. A nontechnical guide to Hubble operations and discoveries, including chapters explaining the science behind the Hubble advances.

Schlegel, Eric M. *The Restless Universe: Understanding X-Ray Astronomy in the Age of Chandra and Newton.* New York: Oxford University Press, 2002. An analysis of the history, methods, and future of x-ray astronomy and the role telescopes such as Chandra play in it.

Schonfelder, V. *The Universe in Gamma Rays.* New York: Springer Verlag, 2001. Presents a technical explanation of gamma rays and how they can answer questions concerning the formation of stars and black holes.

Wheeler, J. Craig. *Cosmic Catastrophes: Supernovae, Gamma-Ray Bursts, and Adventures in Hyperspace.* New York: Cambridge University Press, 2000. Explains the science behind supernovae, black holes, gamma-ray bursts, and hyperspace in a manner suited for the general audience.

Rebecca B. Jervey

Telecommunications Satellites: Maritime

Date: Beginning February 19, 1976
Type of spacecraft: Communications satellites

A network of satellites designed to upgrade the communications capabilities of commercial and military maritime vessels was first proposed in 1972. In 1979, the International Maritime Satellite Organization was formed to create such a network.

Summary of the Satellites

In 1972 a subgroup of the United Nations, the Intergovernmental Maritime Consultative Organization (IMCO), expressed an interest in using satellites in space for maritime purposes. IMCO represented the concerns of seafaring nations in areas such as distress systems, navigation and position determination, and operation of maritime mobile services. The tremendous growth of the international maritime industry after World War II indicated a serious need for improved communication methods. The American Institute of Shipping echoed this concern when it estimated that by 1980, the number of vessels on the high seas weighing more than 1.5 million kilograms could exceed fourteen thousand at any one time. Vessels were still using the inefficient "brass key" radiotelegraph method for their ship-to-shore communications.

In a report to the U.N. Secretariat, IMCO proposed a satellite system that would allow the exchange of telegraph, telephone, and facsimile messages and improve navigation for maritime interests. A panel was formed to study the legal, financial, technical, and operational problems involved in creating an entity that would be responsible for such a system. An international convention was held in April and May of 1975, and delegates and observers from forty-five nations and fifteen international agencies attended. The conference concluded that an international organization was needed to administer a worldwide maritime satellite system, and by the end of the first session, the United States and thirteen other countries had agreed on the major elements of a system that would eventually be known as the International Maritime Satellite Organization (Inmarsat). It would be patterned after the International Telecommunications Satellite Organization (Intelsat), with some modifications for its unique maritime interests. Inmarsat would provide global telecommunications services for maritime commercial activities and safety.

After four years of study and debate, Inmarsat was formally chartered on July 16, 1979, in London. The organization comprised forty-three nation members, including the United States, Great Britain, Norway, Japan, the Soviet Union, and Canada. The system was scheduled to become operational by the early 1980's.

Between 1976 and 1979, maritime communications were enhanced through the U.S. maritime satellite (Marisat) system. The Marisat system consisted of three satellites built by Hughes Aircraft Company for the Communications Satellite Corporation (COMSAT). Launched in 1976, the satellites formed the first maritime communications system in the world. The Marisats provided rapid, high-quality communications between ships at sea and home offices; they greatly improved communications of distress, safety, search-and-rescue, and weather reports.

All three Marisats were structurally identical. At liftoff they weighed 655 kilograms and were approximately 231 centimeters long with their antennae. A panel containing seven thousand solar cells supplied the craft with primary power. A "Straight-Eight" Delta rocket launched the satellites from Cape Canaveral, Florida, into geostationary orbits, wherein they would travel around Earth once every twenty-four hours. The payloads consisted of three ultrahigh-frequency (UHF) bands, reserved for U.S. government use, and L-band and C-band channels. The latter were kept entirely separate from the UHF channels and were used to translate ship-to-shore signals. Each Marisat had a life expectancy of five years.

Marisat-A was launched on February 19, 1976, and was renamed Marisat 1 when it became operational over the equator at longitude 5° west. It served maritime traffic in the Atlantic. Marisat-B, later Marisat 2, was launched on June 9, 1976, and became operational in July. It served the major shipping lanes of the Pacific Ocean and covered a 311-million-square-kilometer (120-million-square-mile) area. It was stationed slightly west of Hawaii at longitude 176.5° west. On October 14, 1976, Marisat-C, the final satellite in the Marisat system, was launched. It was stationed over the Indian Ocean at longitude 73° east and was renamed Marisat 3 when it became operational. Initially, it served only the U.S. Navy, but it also acted as an in-orbit backup for Marisats 1 and 2.

Although the U.S. Maritime Administration and the U.S. Navy were early users of the Marisat system, the Marisats' main function was to provide communications services to ships of all nations. The satellites were owned by COMSAT, and the company reimbursed NASA for all administrative and launch costs.

Commercial satellite service to shipping continued to develop and expand during this period. Eight thousand ships were expected eventually to be equipped with satellite communications terminals. On February 15, 1979, the United States signed the convention on Inmarsat, in accordance with the International Maritime Communications Act of 1978. That summer, the existence of Inmarsat became official.

The Inmarsat system consists of nine satellites in geostationary orbit. Four of these satellites, the latest Inmarsat-3 generation, provide overlapping operational coverage of the globe (apart from the extreme polar areas). The others are used as in-orbit spares or for leased capacity. Each Inmarsat-3 satellite also operates a number of spot beam "cells," enabling the satellites to concentrate extra power in areas of high demand and to provide services to smaller, simpler terminals. Roaming users communicate direct via Inmarsat's satellites. There are about forty land Earth stations (LESs) in thirty-one countries. These stations receive and transmit communications through the Inmarsat satellites and provide the connection between the satellite system and the world's fixed communications networks.

Development of the Marecs portion of the Inmarsat system began in 1973 under the auspices of European Space Agency. Initial funding came from Belgium, France, Italy, the United Kingdom, Spain, and West Germany. Later, the Netherlands, Norway, and Sweden joined in the effort. The Marecs satellite consisted of a service module, a derivative of the European Communications Satellite design, and a payload module. The payload contained a C-band to L-band forward transponder and an L-band to C-band return transponder incorporating a Search And Rescue (SAR) channel. The system was capable of operating without continuous ground control.

Marecs 1 was launched December 20, 1981, from the French Guiana complex by an Ariane rocket; it was sent into a geostationary equatorial orbit at longitude 26° west. Marecs 2 was lost at launch because of the failure of the Ariane rocket. Its replacement, Marecs B2, was launched from the French Guiana complex on November 10, 1985; it orbited at longitude 177° east. The Marecs satellites had a life expectancy of seven years. The estimated cost of the Marecs program was $359.8 million.

Four Intelsat V satellites made up the remainder of the Inmarsat configuration. Each had a box-shaped housing and payload support system with a pair of winglike solar arrays. The three-axis stabilized satellites weighed about 1,950 kilograms at launch and about 860 kilograms in orbit. They were capable of being launched by the Atlas-Centaur rocket, the NASA space shuttle, and ESA's Ariane rocket. Ford Aerospace and Communications was the prime contractor for the Intelsat V series. The satellites were launched from 1980 through 1984 and had an expected life span of seven years.

The current Inmarsat satellites are located in geostationary orbit 35,786 kilometers out in space. Each satellite covers up to one-third of the Earth's surface and is strategically positioned above one of the four ocean regions to form a continuous "worldwide web in the sky." Every time a call is made from an Inmarsat mobile satphone, it is beamed up to one of the satellites. On the ground, distributed all around the world, giant communications antennae are listening for the return signal, which they then route into the ordinary telephone network. When someone calls an Inmarsat customer, it happens the same way, but in reverse.

The four Inmarsat-2 satellites were built by an international consortium headed by the Space & Communications division of British Aerospace (now part of the Anglo-French company Matra Marconi Space). Subcontractors included Hughes Aircraft Company (United States), Fokker (The Netherlands), Matra (France), MBB (Germany), NEC (Japan), and Spar (Canada). Satellite ground control operations contractors included CLTC (China), CNES (France), SED (Canada), Telespazio (Italy), and Intelsat. Inmarsat-2 is a three-axis-stabilized satellite design based on the Eurostar satellite platform, developed by British Aerospace and Matra Espace (both now part of Matra Marconi Space). The satellites were designed for a ten-year life. At launch, each weighed 1,300 kilograms, had an initial in-orbit mass of 800 kilograms, and had 1,200 watts of available power. They currently serve as backups to the Inmarsat-3 satellites.

U.S.-based Lockheed Martin Astro Space built the Inmarsat-3 spacecraft bus, the structure and systems that serve as the satellites' basic utilities, based on the Astro Space Series 4000. Matra Marconi Space, based in the United Kingdom, built the

communications payload, which included antennae, repeaters, and other communications electronics. The communications payload operates in the C- and L-band portions of the radio spectrum. Each satellite has its antennae and electronics tuned for optimum coverage of a particular area on Earth, known as its footprint.

The tremendous advantage of the Inmarsat-3 satellites is their ability to concentrate power on particular areas of high traffic within the footprint. Each satellite utilizes a maximum of seven spot beams and one global beam. The number of spot beams will be chosen according to traffic demands. In addition, these satellites can reuse portions of the L-band frequency for nonadjacent spot beams, effectively doubling the capacity of the satellite.

Each satellite weighs about 2,066 kilograms at launch, compared to 1,300 kilograms for an Inmarsat-2 satellite. Spacecraft 3F1, launched April 3, 1996, on an Atlas Centaur IIA, is on-station at 64.0° east. Spacecraft 3F2 was launched September 6, 1996, on a Russian Proton D-1-E; it is on-station at 15.5°. Spacecraft 3F3 was launched December 18, 1996, on an Atlas Centaur IIA and is on-station at 178° east. Spacecraft 3F4 was launched June 3, 1997, on an Ariane 4 and is on-station at 54° west. Spacecraft 3F5 was launched February 3, 1998, on an Ariane 4 and is on-station at 25° east. It carries leased services and also serves as a backup spacecraft for Inmarsat 3F2.

The Inmarsat system has greatly improved communications between ships at sea and their land-based offices. There are thousands of users of the system, and they are increasing. Tens of thousands of vessels of all types from many nations receive navigation and communications services from the Inmarsat system. Among the largest users are the United States, Great Britain, Norway, Japan, and Canada. Each Inmarsat user organization purchases its own terminal for installation on a ship or at a remote site.

Contributions

Inmarsat has proved that the use of space-based satellites to enhance marine communications is both economically and technologically sound. Shipboard antennae link ships' telephone systems via Inmarsat satellites to coastal stations. Users can route their calls through any of the coastal stations within range of their transmissions. These Earth terminals are provided by the individual member countries; Inmarsat purchases or leases only the space-based portion of the system.

Although the agency does not provide the land-based stations for its members, it does control and monitor the specifications for the equipment at these stations. It therefore pays for technical consultancy services for its members. The need for these services is determined by the organization's monitoring commitments, which fluctuate. Purchasing the services allows Inmarsat to maintain its standards without the need for a large staff of employees.

The international agency is run along strictly commercial lines. There are no requirements as to the work share of each member nation, and countries other than members can sell their goods or services to Inmarsat if they meet the specifications and have the best price.

Inmarsat also spends about $1 million per year on research and development programs. One program was designed to install a distress beacon, called an emergency pointing/indicating radio beacon (EPIRB), on every ship. The beacon is activated automatically or by a crew member. It reports the ship's location and details to rescue agencies on shore rather than at sea. It has been shown that shore-based rescue operations are better equipped to respond quickly and effectively to a maritime emergency than are other vessels at sea.

Inmarsat is also exploring the possibility of entering the television transmission and medical assistance fields. It is investigating a narrow-band video transmission system that would transmit a high-quality picture. Inmarsat has had some successes in a procedure whereby medical diagnosis is conducted from a remote location via television cameras and the Inmarsat system.

Inmarsat also conducts market research on future technical needs. One area of focus is the expansion of its technological services to include telex, computer links, remote control, automated reporting, and high-speed data transmissions.

The agency has also considered entering the aeronautical field, initially by providing air-to-ground communications. A change in the original charter that would permit Inmarsat to provide aeronautical and other nonmaritime services is being considered, because much of the technology developed by the agency can be applied to other fields.

Context

When Arthur C. Clarke, British scientist and science fiction writer, suggested in a technical paper published in 1945 that communications satellites were feasible, the Soviet launching of Sputnik 1 was still twelve years in the future. It would be fifteen years before the National Aeronautics and Space Administration (NASA) launched Echo, the silvery balloon that orbited Earth every 118 minutes and reflected radio signals back to the surface. Echo was a passive satellite, but two years later NASA launched Relay, an active satellite that received signals, amplified them, and returned them to Earth.

With the launches of the early satellites in the late 1950's, many scientists and industry leaders recognized the potential for practical applications of the space program, especially in the area of communications. Even before any formal effort was made to utilize space technology, maritime interests were beginning to reap the benefits of satellite technology. The orbiting spacecraft were providing weather forecasts, storm tracking data, and information on iceberg locations with a degree of accuracy never before possible.

By the early 1960's, private companies were producing their own communications satellites, and in 1962 Congress authorized the Communications Satellite Corporation (COMSAT). COMSAT later became the U.S. representative in and manager of the International Telecommunications Satellite Consortium (Intelsat). As the satellites' ability to transmit voice, picture, and computer data grew, maritime interests began to investigate the possibility of using these technologies to improve and expand the outdated radiotelegraph technology that still handled most ship-to-shore communications on the high seas.

Inmarsat, the maritime equivalent of Intelsat, was formally chartered in 1979 after four years of study by a large group of nations and international maritime agencies. The space-based maritime communications system was dedicated to developing and implementing the latest in satellite and communications technology to enhance the worldwide marine communications network and to ensure the safety and efficiency of the international maritime industry. Throughout the world, there are over 150,000 terminals commissioned for use with services offered via the Inmarsat system. Depending on the type of terminal used, services available include direct-dial telephone, data, facsimile, telex, electronic mail, high quality audio, compressed video and still video pictures, telephoto, slow-scan television, videoconferencing, and telemedicine.

See also: Amateur Radio Satellites; Applications Technology Satellites; Intelsat Communications Satellites; Mobile Satellite System; Telecommunications Satellites: Military; Telecommunications Satellites: Passive Relay; Telecommunications Satellites: Private and Commercial; Tracking and Data-Relay Communications Satellites.

Further Reading

Butrica, Andrew J. *Beyond the Ionosphere: Fifty Years of Satellite Communications.* NASA SP-4217. Washington, D.C.: Government Printing Office, 1997. Part of the NASA History series, this book looks into the realm of satellite communications. It also delves into the technology that enabled the growth of satellite communications. The book includes many tables, charts, photographs, and illustrations, a detailed bibliography, and reference notes.

Caprara, Giovanni. *The Complete Encyclopedia of Space Satellites.* New York: Portland House, 1986. This volume presents short entries on both civil and military satellites of all nations. The book includes line drawings and an index. Suitable for general audiences.

Gavaghan, Helen. *Something New Under the Sun: Satellites and the Beginning of the Space Age.* New York: Copernicus Books, 1998. This book focuses on the history and development of artificial satellites. It centers on three major areas of development—navigational satellites, communications satellites, and weather observation and forecasting satellites.

Gregory, William H., ed. "Inmarsat." *Commercial Space* 1 (Summer, 1985): 55-57. This is an article in a quarterly publication that focuses on the commercial applications of space technology. The magazine contains about a dozen articles per issue and includes high-quality color photographs. Suitable for general audiences.

Lee, Wayne. *To Rise from Earth: An Easy to Understand Guide to Spaceflight.* New York: Checkmark Books, 1996. This is a good introduction to the science of spaceflight. Although written by an engineer with the NASA Jet Propulsion Laboratory, it is presented in easy-to-understand language. In addition to the theory of spaceflight, it gives some of the history of the human endeavor to explore space.

Martin, Donald H. *Communication Satellites.* 4th ed. New York: American Institute of Aeronautics and Astronautics, 2000. This work chronicles the development of communications satellites and worldwide networks over the past four decades, from Project Score to modern satellite communication systems.

National Aeronautics and Space Administration. *NASA: The First Twenty-Five Years, 1958-1983.* NASA EP-182. Washington, D.C.: Government Printing Office, 1983. A brief chronological and topical history of NASA and the U.S. space program during its first twenty-five years, this book is designed for classroom teachers and features charts, graphs, drawings, color photographs, and suggested classroom activities. Topics include tracking and data-relay systems, space applications, aeronautics, and piloted and robotic missions. Suitable for general audiences.

_____. *Space Network Users' Guide (SNUG)*. Washington, D.C.: Government Printing Office, 2002. This users' guide emphasizes the interface between the user ground facilities and the Space Network, providing the radio frequency interface between user spacecraft and NASA's Tracking and Data-Relay Satellite System, and the procedures for working with Goddard Space Flight Center's Space Communication program.

National Aeronautics and Space Council. Executive Office of the President. *Aeronautics and Space Report of the President: 1979 Activities*. Washington, D.C.: Government Printing Office, 1979. A compilation in textbook format of all space-related activities carried on during 1979. Some technical data presented in general terms.

_____. *Aeronautics and Space Report of the President: 1983 Activities*. Washington, D.C.: Government Printing Office, 1983. This book has the same format as the one above, but the information is for 1983. Suitable for general audiences.

Ritchie, Eleanor H. *Astronautics and Aeronautics, 1976: A Chronology*. NASA SP-4021. Washington, D.C.: National Aeronautics and Space Administration, Scientific and Technical Information Branch, 1984. A compilation of U.S. space activities during the year 1976. This book presents technical information clearly for a general audience.

Rosenthal, Alfred, ed. *Satellite Handbook: A Record of NASA Space Missions 1958-1980*. Washington, D.C.: Government Printing Office, 1982. Brief summaries and documentation of all NASA piloted and robotic space missions undertaken between 1958 and 1980. Technical data on each mission includes descriptions of the spacecraft and launch vehicle, the payload, the mission's purpose, project results, and major participants. Suitable for a general audience.

Sherman, Madeline, ed. *TRW Spacelog: Twenty-Fifth Anniversary of Space Exploration, 1957-1982*. Redondo Beach, Calif.: TRW, 1983. This is a booklet in magazine format that focuses on international space-related activities and programs. Contains brief summaries of various topics. Suitable for a general audience.

United States Congress. Office of Technology Assessment. *Civilian Space Policy and Applications*. OTA-STI-177. Washington, D.C.: Government Printing Office, 1982. An official publication that discusses the different policies and agencies that make use of space technology and applications. Contains technical material.

Zimmerman, Robert. *The Chronological Encyclopedia of Discoveries in Space*. Westport, Conn.: Oryx Press, 2000. Provides a complete chronological history of all crewed and robotic spacecraft and explains flight events and scientific results. Suitable for all levels of research.

Lulynne Streeter

Telecommunications Satellites: Military

Date: Beginning December 18, 1958
Type of spacecraft: Communications satellites

Since the mid-1960's, communications satellites for the U.S. military have provided reliable, worldwide communications between troops and decision makers. They have also made possible a new form of conflict: advanced nuclear war.

Summary of the Satellites

Although the first artificial satellites were developed for nonmilitary, scientific purposes, it did not take long for the United States military to understand and appreciate their value. Communications satellites in particular, it was realized, could aid the Army, Navy, Marines, and Air Force strategically and tactically. Communications satellites would enable a country to wage global war by maintaining links across the planet.

The Score (Signal Communication by Orbiting Relay Equipment) program of the United States Army was the first project of the world's first military communications satellite. The Score spacecraft was launched from Cape Canaveral around Christmastime, 1958; from its elliptical Earth orbit, it transmitted messages for two weeks, including a holiday greeting from President Dwight D. Eisenhower. The satellite weighed 70 kilograms and was launched atop an Atlas B rocket into a 185-by-1,470-kilometer orbit. Next came the Courier, a military communications satellite that made few advances. In fact, the experiment was a failure, and because of it, other military communications satellites were delayed for four years.

The first operational system was named the Interim (or Initial) Defense Communications Satellite Project (IDCSP), later referred to as phase 1 of the Defense Satellite Communications System (DSCS 1). This system made use of a total of twenty-six satellites, launched between June, 1966, and June, 1968. These were simple spin-stabilized spacecraft with no complicated moving parts;

they were designed for the highest possible reliability, weighing 45 kilograms and laced with solar cells. Dispensed at six-hour intervals into slightly different orbital velocities to give them global coverage, the IDCSP satellites were placed in subsynchronous orbits at 33,915 kilometers, so for about four and one-half days they remained within view of an Earth ground station located at the equator. They were deployed in slightly subsynchronous orbits with an eastward drift in groups of seven at a time, using a massive Titan military rocket.

Because each satellite remained in view of an equatorial station for four or five days, several were accessible at any one time. Some thirty-five Earth terminals (including seven located on ships) linked with stations up to 16,000 kilometers apart using an FM band. Links between South Vietnam, Hawaii, and Washington, D.C., were used for high-speed digital data during the last seven years of the United States' involvement in the Vietnam War. Indeed, many of the sophisticated bombing raids would not have been possible without satellite communications to and from headquarters to manage the strategy. Moreover, this experience proved how valuable the satellite system could be during wartime. Communications involved single-channel voice relay, imagery (including photographs), teletype (written communications), and computerized digital data transmission. Although they were designed for a life of only eighteen months, many of the satellites lasted years.

For NATO, two satellites were built in the United States and then successfully launched during 1970 and 1971. The original NATO operational system included twelve ground stations near the capitals of the twelve NATO countries. The NATO spacecraft were of the IDCSP design, only slightly larger, with an improved antenna system. They were placed in geostationary orbits so that they were above the same spot on Earth at all times. Later, the NATO system expanded the number of ground stations so the system could cover the Northern Hemisphere from Ankara, Turkey, to the suburbs of northern Virginia, home of the Pentagon.

The DSCS 2 followed the IDCSP. It consisted of spacecraft designed to provide the Department of Defense with a reliable network of strategic communications satellites for global coverage. Managed by the United States Air Force, the DSCS 2 satellites were developed by TRW as the primary contractor. The DSCS 2 program had its first launch in 1971 and featured much larger, cylindrical spacecraft, more than 3 meters in height and 3 meters in diameter. These were designed to function for five years and were placed in geosynchronous orbits. The system provided up to 1,300 duplex voice communication channels. The DSCS 2 was designed with a multiple-channel, wideband antenna system so it could link more easily to smaller ground stations. In this way, the design resembled that of Intelsat IV. During the next eleven years, five of seven DSCS 2 missions were successfully launched.

The DSCS 3 program operated on the same geostationary equatorial orbits as DSCS 2 but offered 50 percent greater communications capability and a promised ten-year life span. Like the DSCS 2, the third phase would be controlled by the United States Air Force Satellite Control Facility near San Francisco, with communications control built into selected Defense Communications Agency terminals. Through its larger antennae, DSCS 3 would provide Earth coverage as well as spot-beam transmissions to smaller, portable ground stations.

The first DSCS 3 was launched in October, 1982, and offered greater flexibility and increased channel capacity. It weighed 1,042 kilograms. The DSCS

3 satellites were to be used by the National Command Authority, the White House Communications Agency, and the Diplomatic Telecommunications Service. By 1984, their price had risen to $150 million each. Between 1982 and 1997, nine DSCS 3 satellites were successfully deployed in orbit, two from the payload bay of *Atlantis* during its maiden flight in 1985.

The United States Navy Fleet Satellite Communications, or FLTSATCOM, system and the Satellite Data System (SDS) are part of the total Air Force Satellite Communications system, or AFSATCOM. Both were designed to establish reliable networks for communication between decision makers and military forces in the field, at sea, or in the air. FLTSATCOM was designed to provide worldwide high priority ultrahigh-frequency (UHF) communications between aircraft, ships, submarines, and ground stations—as well as between the United States military and presidential command networks. The communications provided more than thirty voice and twelve teletype channels designed to serve mobile as well as stationary centers.

The 1,860-kilogram FLTSATCOM spacecraft have twenty-three communications channels. Ten are allocated to the Navy for command of its air, ground, and ocean forces. Twelve are allocated to the Air Force as part of the AFSATCOM system of worldwide command and control, especially of nuclear weapons. One is singularly reserved for command authorities. The AFSATCOM system enables aircraft in flight to communicate with ships, Earth stations thousands of miles away, and even properly equipped submarines. Information from these satellites is used to control and monitor intercontinental ballistic missile and nuclear weapons sites. The SDS provides additional coverage in these areas.

The FLTSATCOM spacecraft is a hexagonal structure, nearly 2 meters wide and more than 1 meter high, weighing 1,005 kilograms. It is solar-powered and contains three antenna systems. Its launch vehicle was an Atlas-Centaur. Fully operational by January, 1981, the FLTSATCOM was built

A concept drawing for Syncom, the first geosynchronous satellite. (NASA)

by TRW. It was set up to establish lines of communication among the president, troops, and nuclear weapons anywhere around the world. There are relay links among nine hundred Navy ships, submarines, and aircraft of the Navy and selected ground stations; among more than one thousand United States Air Force aircraft and ground stations; and for the Strategic Air Command. Only the polar regions are not covered.

The SDS system denotes satellites in highly elliptical orbits similar to those of the Soviet Molniya satellites. The SDS spacecraft were designed to cover the polar cap areas not covered by FLTSATCOM. They also are set to link with spy satellites directly. Weighing approximately 700 kilograms, they were launched by a Titan III-B/Agena D rocket from Vandenberg Air Force Base, with a

planned life span of ten to one hundred years. Six of the eight FLTSATCOMs launched aboard Atlas-Centaur boosters from Cape Canaveral were placed into their proper orbits. The last of the fleet, FLTSATCOM F-8, was launched on September 25, 1989.

The Navy's Leased Satellite (Leasat) system consisted of three Syncom IV spacecraft leased from Hughes, which was also the satellite's manufacturer. The final launch of the Leasat program occurred in 1990. The last operational spacecraft among the constellation of Leasat communications satellites used for more than a decade by the U.S. military was retired in February, 1998. Leasat was developed to augment the Navy's Fleet Satellite Communications System. The Leasat program was a pioneering effort to provide dedicated communications services through a long-term lease arranged by the Navy for the Department of Defense. The lease provided that the U.S. military would pay for the use of communications channels aboard each spacecraft, but not until the system was built and placed in service.

The contract also specified that the Leasat spacecraft be launched by the space shuttle. Built by Hughes Communications, Inc., the "wide body" satellites were designed with a 4.27-meter diameter to take full advantage of the room available in the space shuttle orbiters' cargo bay. Leasats were the first geosynchronous communications satellites to incorporate integral propulsion. This innovation, when coupled with the satellite's folding antenna, made it possible for the spacecraft to fit compactly in the cargo bay of the space shuttle, reducing launch costs. The first two Leasats were launched into geosynchronous orbit aboard the space shuttle *Discovery* in August and November, 1984. Leasat 3 was deployed in April, 1985, but failed to boost itself into geosynchronous orbit. The spacecraft remained dormant until it was retrieved and repaired in orbit by another *Discovery* crew four months later. That same mission launched Leasat 4, but that spacecraft malfunctioned and became unusable. Leasat 5, deployed in 1990 from space shuttle *Columbia*, completed the constellation by providing four geosynchronous communications satellites approximately 90° apart.

Each Leasat spacecraft provided thirteen UHF communications channels, including a 500 kilohertz wideband channel. The satellites provided high-priority global communications for the Fleet as well as the Air Force's Strategic Airborne Command and various Army combat units. Naval Space Command served as the operational manager for the system. UHF satellite communications for U.S. military forces are now being provided by a new constellation of UHF Follow-On (UFO) spacecraft, also built by Hughes.

Contributions

Generally, military communications satellites are not designed to gain new knowledge; they are

expected to provide reliable service. Yet some appear to have had experimental components. For example, the Tactical Communications Satellite, or Tacsat, was first launched in 1967 as part of an experimental system. Tacsat went into orbit as the largest communications satellite in the West. Its size, 0.5 meter in diameter and 1 meter in height, enabled it to work with a network of very small ground stations. A cylindrical satellite, it was designed and built by Hughes Aircraft Company for a February, 1969, launch at the direction of the United States Air Force and Missile Systems Organization. It was placed in geosynchronous orbit and was the first used for the experiment between tactical communications among the Air Force, Navy, and Army ground stations and mobile field units, aircraft, and ships. Submarines, it was found, could be connected to command stations almost anywhere on the planet.

An examination of the Military Satellite Communications (MILSATCOM) system, designed to serve both strategic and tactical forces, best summarizes the knowledge gained by U.S. military communications satellites. Labeled Milstar (Military Strategic and Tactical Relay System), it includes many improvements. It is designed to cover the entire globe and have the ability to detect both the deepest submarine and activity on the highest mountain. In order to achieve worldwide communications capability, Milstar must use the extremely high-frequency radio band. Milstar is also designed to be jam-resistant, with the ability to confuse an enemy's alternative signals and prevent interference. No military tool can survive all battle conditions, especially in a nuclear age, but planners would like satellites to be as hardy as possible. Milstar has built-in components to enable it to fend off missiles or enemy satellites. Researchers have designed the Milstar to stay in orbit for a decade or more to allow time to develop and perfect a replacement system.

The operational Milstar satellite constellation will be composed of four satellites, positioned around the Earth in geosynchronous orbits, and a polar adjunct system. Each mid-latitude satellite will weigh approximately 4,500 kilograms and have a design life of ten years. The first Milstar satellite was launched February 7, 1994, aboard a Titan IV launch vehicle. The second was launched November 5, 1995. These first two Block 1 spacecraft began to be replaced by the Block 2 Milstars 3 through 6 in 1999; the full Block 2 constellation or four would allow as many as ten thousand users to access the system at any given time. The first Milstar-2 satellite, launched April 30, 1999, carried a medium-data-rate (MDR) payload. A complete constellation of LDR (low-data-rate) and MDR satellites would be achieved with the launch of the fourth Milstar-2 satellite. Replenishment of the four-satellite Milstar-2 constellation would occur between 2006 and 2009, with the exact launch dates to be determined by actual satellite longevity.

Context

The military employs satellites for many non-communications functions, including navigational assistance, weather forecasting, surveillance, nuclear testing and radiation detection, and research and technological development. Yet certainly one of the most valued benefits of these satellites has been improved communications.

In the nineteenth century, ships carried messages from Europe to the United States, Asia, and Africa. A singular revolutionary change came in the early part of the twentieth century, with radio communications from land to land, ship to ship, ship to shore, and later aircraft to shore and ship and other aircraft. Radio communications permitted the rapid and reliable exchange of information and facilitated the logistics of managing a global war on the scale of World War II.

Another generation of revolutionary change came with radio communications by satellite. Such an advanced network permitted the increase of re-liable communications worldwide. Within massive grids, planes, missiles, submarines, ships, ground troops, and bases could be linked. Fast, reliable communication has always been a military necessity, and the military communications satellites provide this. Because of their high costs, all spacecraft are designed to be as efficient as possible. Only military communications satellites are specifically required to perform uniformly well twenty-four hours a day throughout the year for a period of years.

The effect of satellites on the United States' military was evident by the late 1960's. The IDCSP, for example, provided a crucial link for the United States in its war in Vietnam. With IDCSP satellites, information and reports could flow from South Vietnam through Hawaii to headquarters in Washington, D.C. Thus it was possible to run the war from the United States, indeed from the White House itself. Still, the United States seemed not to have pushed as hard in this direction as the Soviet Union had. The Soviet Union's military launched almost twice the number of satellites as the United States' military. Virtually all uncrewed Soviet spacecraft were launched into Earth orbit by the military, many for communications purposes.

See also: Applications Technology Satellites; Attack Satellites; Early-Warning Satellites; Electronic Intelligence Satellites; Intelsat Communications Satellites; Meteorological Satellites: Military; Mobile Satellite System; Nuclear Detection Satellites; Ocean Surveillance Satellites; Private Industry and Space Exploration; Spy Satellites; Strategic Defense Initiative; Telecommunications Satellites: Maritime; Telecommunications Satellites: Passive Relay; Telecommunications Satellites: Private and Commercial; Tracking and Data-Relay Communications Satellites.

Further Reading

Baker, David, ed. *Jane's Space Directory, 2005-2006.* Alexandria, Va.: Jane's Information Group, 2005. A comprehensive guide to spacecraft. Somewhat technical but still appropriate for the layperson.

Brown, Martin P., Jr. *Compendium of Communication and Broadcast Satellites: 1958-1980.* New York: IEEE Press, 1981. This reference guide provides a listing and accessible description and analysis of the various military satellite communications systems at the end of the 1970's. In diagrammatic fashion, it examines the inner workings of all satellites that were not classified. A wonderful guide to a secretive industry.

Butrica, Andrew J. *Beyond the Ionosphere: Fifty Years of Satellite Communications.* NASA SP-4217. Washington, D.C.: Government Printing Office, 1997. Part of the NASA History series, this book looks into the realm of satellite communications. It also delves into the technology that enabled the growth of satellite communications. The book includes many tables, charts, photographs, and illustrations, a detailed bibliography, and reference notes.

Chetty, P. R. *Satellite Technology and Its Applications.* Blue Ridge Summit, Pa.: TAB Books, 1987. This book provides a useful overview of the functions of communications satellites in space and includes sections on the workings of the military satellites.

Gavaghan, Helen. *Something New Under the Sun: Satellites and the Beginning of the Space Age.* New York: Copernicus Books, 1998. This book focuses on the history and development of artificial satellites. It centers on three major areas of development—navigational satellites, communications satellites, and weather observation and forecasting satellites.

Jarett, David, ed. "Satellite Communications: Future Systems." Vol. 54 in *Progress in Astronautics and Aeronautics.* New York: American Institute of Aeronautics and Astronautics, 1977. Details the architecture of a military communications system. Offers the reader an interesting discussion of how military communications satellites differ from civilian satellites.

Lee, Wayne. *To Rise from Earth: An Easy to Understand Guide to Spaceflight.* New York: Checkmark Books, 1996. This is a good introduction to the science of spaceflight. Although written by an engineer with the NASA Jet Propulsion Laboratory, it is presented in easy-to-understand language. In addition to the theory of spaceflight, it gives some of the history of the human endeavor to explore space.

Martin, Donald H. *Communication Satellites.* 4th ed. New York: American Institute of Aeronautics and Astronautics, 2000. This work chronicles the development of communications satellites and worldwide networks over the past four decades, from Project Score to modern satellite communication systems.

Porter, Richard W. *The Versatile Satellite.* New York: Oxford University Press, 1977. This short book provides a fine introduction to the various uses of satellites. Chapter 4 deals with communications satellites and covers their military uses.

Raggett, R. J. *Jane's Military Communications.* 7th ed. London: Jane's Publishing Company, 1986. This catalog examines all sorts of military communications hardware, including satellites. Updated annually, this fascinating guide is somewhat technical.

Zimmerman, Robert. *The Chronological Encyclopedia of Discoveries in Space.* Westport, Conn.: Oryx Press, 2000. Provides a complete chronological history of all crewed and robotic spacecraft and explains flight events and scientific results. Suitable for all levels of research.

Douglas Gomery

Telecommunications Satellites: Passive Relay

Date: August 12, 1960, to June 7, 1969
Type of spacecraft: Communications satellites

The United States' Echo satellite project included the first passive relay communications satellite launched into space and the first cooperative space venture between the United States and the Soviet Union. The satellites were inflated Mylar balloons that bounced radio signals between ground stations on Earth.

Summary of the Satellites

The Echo passive relay communications satellite project was the first practical application of space technology in the field of telecommunications. The satellites were aluminum-coated Mylar spheres that were inflated in space. The concept was simple. The spheres passively reflected electromagnetic radio waves directed to them from a ground-based station at one location on Earth to a ground-based station at another location. They also provided information about the density of the upper atmosphere.

Echo was the first communications satellite project of the National Aeronautics and Space Administration (NASA). It was rooted in an earlier project conceived by William J. O'Sullivan, Jr., an aeronautical engineer at the Langley Aeronautics Laboratory, which was part of NASA's predecessor, the National Advisory Committee for Aeronautics (NACA). In 1956, O'Sullivan proposed that studies to measure Earth's atmosphere during the International Geophysical Year (IGY)—a period of intense scientific research set for July, 1957, to December, 1958—could be conducted more efficiently by using a low-density inflatable sphere that could be tracked optically. John R. Pierce of Bell Telephone Laboratories had proposed a similarly designed balloon in a 1955 article entitled "Orbital Radio Relays," but Pierce had wanted to use the orbiting inflatable spheres as reflectors for radio signals.

Several attempts to launch O'Sullivan's balloon with IGY payloads failed when the launch vehicles malfunctioned. Pierce proposed a cooperative communications experiment using O'Sullivan's inflatable spheres. According to O'Sullivan, it was a logical next step to consider using the balloons for communication purposes. In 1958, NACA's director, Hugh L. Dryden, told the U.S. Congress that the technology to orbit such a passive communications satellite existed—but certain design changes had to be made. It would be necessary to increase the size of the air-density balloon to provide a larger surface from which to bounce signals, and the surface would also have to be treated to increase its reflective capacities.

By 1959, the IGY project had become the newly formed NASA's passive communications satellite project. It was designated Project Echo. Technicians at NASA's Langley Research Center had three major requirements as they began designing a passive communications satellite that would inflate in orbit into a perfectly smooth-surfaced sphere: They needed a suitable material for the sphere, an inflation system, and a canister in which to launch the collapsed balloon.

It was decided that the balloon would be made from an aluminized polyester film manufactured by E. I. Dupont. Known as Mylar, the material was 0.5 millimeter thick. Another company cemented the Mylar into eighty-two flat gores that formed the Echo sphere. Benzoic acid was chosen as the inflating agent because it could change from a solid state to a gaseous state without going through the liquid stage. A spherical metal canister impregnated with plastic would carry the deflated balloon into space.

In October, 1959, the first test model was assembled and readied for testing. On the fourth attempt, in April, 1960, the balloon satellite was successfully inflated at an altitude of 375 kilometers. On May 13, 1960, a three-stage Thor-Delta rocket launched Echo A-10. During the vehicle's coast period, the attitude control jets on the second stage failed. The vehicle reentered the atmosphere and decomposed.

Three months later, on August 12, 1960, the world's first passive communications satellite was successfully launched from Cape Canaveral, Flor-

ida. Echo 1 was placed into space by a Thor-Delta three-stage rocket at an inclination of 47.28° to the equator. The satellite measured 30.5 meters in diameter and weighed 76 kilograms. The inflatable portion was packed in a magnesium sphere and was released about two minutes after injection into orbit. In addition to the balloon, which was designed to transmit images, music, and voice signals from one side of the United States to the other, Echo carried two small radio beacons that assisted in locating and tracking the satellite. The beacons were mounted on small disks and attached to the balloon.

A pretaped message by President Dwight D. Eisenhower was the first voice signal to be bounced off Echo. It traveled from NASA's facility at Goldstone, California, to Bell Telephone's station at Holmdel, New Jersey. The first known two-way voice communication was bounced off Echo 1 on

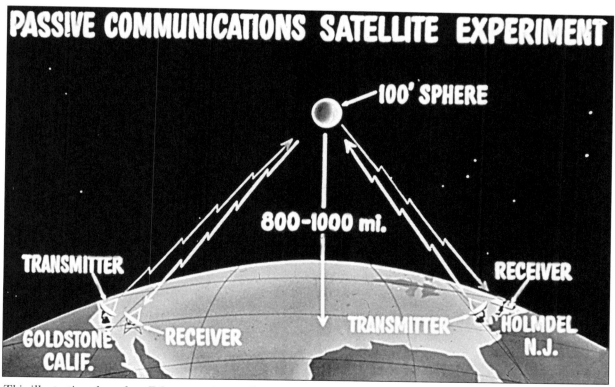

This illustration shows how Echo, a passive satellite with no instrumentation, worked simply by reflecting a signal from one site to another one on Earth. (NASA CORE/Lorain County JVS)

August 13, 1960, between Cedar Rapids, Iowa, and Richardson, Texas. The first reported image transmission via Echo 1 occurred on August 19, 1960, again between Cedar Rapids and Richardson.

For the next four months, Echo 1 was utilized by Bell Telephone Laboratories in New Jersey and the Jet Propulsion Laboratory in California. Eventually, micrometeoroids damaged the balloon's sensitive "skin," and its orbit was affected by solar winds. Yet it continued to reflect a variety of communications signals to and from Earth at ground stations all over the world. Echo 1 reentered Earth's atmosphere on May 24, 1968.

Echo 1 performed successfully, but it was apparent that some modifications of the design were necessary. In January of 1962, several suborbital tests of a modified Echo inflation system were conducted. Later that same year, plans were announced for the launch of two Echos to determine how smooth the surface area of an advanced Echo had to be. In December of 1962, the United States and the Soviet Union agreed to cooperate in the upcoming experiments planned for Echo.

Early in 1963, however, NASA officials announced that because of the formation of the Communications Satellite Corporation (COMSAT) and the Department of Defense's decision to cancel its Advent project, NASA would cancel its own plans for advanced passive and intermediate altitude communications satellite projects. In August, a private contractor was selected to build three second-generation Echos. This project was also eventually canceled.

On January 25, 1964, Echo 2 was launched from Vandenberg Air Force Base in California. Launched by a Thor-Agena B, the balloon was successfully

The inflated Echo satellite in a hangar in North Carolina. (NASA)

placed in orbit at an inclination of 81.5°. With a weight of 256 kilograms and a diameter of 41 meters, Echo 2 was somewhat larger than its predecessor. It was also more durable. The sphere was made up of 106 gores of Mylar, three layers thick, and bonded between two layers of a soft aluminum foil alloy. Its outer surface was coated with alodine, and its inner surface was coated with India ink. The improved satellite could maintain its rigidity for a longer period of time. Pyrazole crystals were used as the inflating agent. The crystals were positioned such that upon ejection from the canister into the sunlight they would gradually expand. It was estimated that with this method it would take about ninety minutes for the balloon to become fully inflated. That would permit higher pressures and produce a stronger structure and better reflecting power. Echo 2 also carried two beacon transmitters powered by solar cells and nickel-cadmium batteries.

Echo 2 had several objectives. It was to perform passive communications experiments with radio, telex, and facsimile signals, collect data concerning the spacecraft's orbital environment, and test the new inflation method. In addition, the Echo 2 spacecraft was used to conduct the first joint U.S.-Soviet space experiment, under the auspices of a 1962 cooperative space exploration experiment consisting of a communications link between stations at Jodrell Bank in Great Britain and the Zimenski Observatory at Gorki University, near Moscow. Echo 2 reentered Earth's atmosphere on June 7, 1969.

Contributions

The Echo project proved that an inflatable sphere coated with aluminum to increase reflecting abilities could be successfully launched into space and placed into orbit. The balloon could be inflated in space and remain orbiting as it transmitted radio communications and images between distant points on Earth. It also provided a means of measuring such things as the density of the atmosphere. Echo 2 also carried out experiments on the pressure created by solar radiation. During both Echo missions the orbital parameters of the satellites underwent continual variation because of the pressure effects of solar radiation.

Passive relay communications satellites were soon replaced by more efficient and technologically advanced "active" communications satellites. The Echo satellites, however, did offer two advantages over the more sophisticated communications satellites that followed. They were extremely reliable because of the simplicity of their design and the lack of electronic equipment, and they had multiple access capabilities.

Context

Echo satellites constituted the first civilian telecommunications system set up in space. They were the beginning of a complex space-based communications network that could handle telephone, tele-

vision, telex, facsimile, and radio signals. They represented the advent of communications systems that would eventually reach every portion of Earth, no matter how remote.

Passive communications satellites were not the most promising method of establishing space-based communications systems, and they were soon abandoned in favor of active systems. These active satellites were capable of receiving signals from Earth and then retransmitting them to another part of the planet. These satellites did not require the expensive ground-based stations necessary for the passive satellites.

Because it seemed clear that these so-called active repeater satellites in synchronous orbits were more viable than the passive satellites for building a commercial communications system, NASA decided to direct its research to that area, and the agency abandoned further plans to upgrade the Echo project. In fact, Relay 1, an active repeater satellite, was launched in 1962, two years before the flight of Echo 2.

At first, the use of space communications systems was an international effort through such organizations as the International Telecommunications Satellite Consortium (Intelsat), but soon more and more nations began launching their own satellites to reach remote areas. Such systems offer education, news, entertainment, and business and financial information to every citizen in possession of a receiver.

It is certain that as communications satellites become more advanced and efficient, their numbers will increase. It has even been suggested that widespread use will eventually result in overcrowding of frequencies and orbital positions. Yet whatever the future holds for communications satellites, the Echos, once visible to the naked eye as they traveled around Earth, are remembered as popular symbols of the peaceful and practical application of space research.

See also: Telecommunications Satellites: Maritime; Telecommunications Satellites: Military; Telecommunications Satellites: Private and Commercial.

Further Reading

Andrade, Alessandra A. L. *The Global Navigation Satellite System: Navigating into the New Millennium*. Montreal: Ashgate, 2001. Provides an international view of issues of availability, cooperation, and reliability of air navigation services. Attention is specifically paid to the American GPS (Global Positioning System) and Russian GLONASS systems, although the development of the Galileo civilian system in Europe is also presented.

Branigan, Thomas, ed. "Echo." *TRW Spacelog* 4, no. 2 (1964): 8-9. The magazine in which this article appears reports on and describes current international space research, development, and missions on a quarterly basis. This article features technical information, photographs, and charts. Written for a general audience.

Butrica, Andrew J. *Beyond the Ionosphere: Fifty Years of Satellite Communications*. NASA SP-4217. Washington, D.C.: Government Printing Office, 1997. Part of the NASA History series, this book looks into the realm of satellite communications. It also delves into the technology that enabled the growth of satellite communications. The book includes many tables, charts, photographs, and illustrations, a detailed bibliography, and reference notes.

Caprara, Giovanni. *The Complete Encyclopedia of Space Satellites*. New York: Portland House, 1986. This volume contains a complete listing of every civilian and military satellite launched from 1957 to 1986. It contains color and black-and-white photographs, a bibliography, an index organized by country, a general index, and a table of contents.

Divine, Robert A. *The Sputnik Challenge: Eisenhower's Response to the Soviet Satellite*. New York: Oxford University Press, 1993. This is a dramatic account of the national hysteria surrounding the Soviet Union's launching of the early Sputniks. It details America's attempts to put its own satellites into orbit and discusses Eisenhower's role in the early exploration of space.

Elbert, Bruce R. *Introduction to Satellite Communication*. Cambridge, Mass.: Artech House, 1999. This is a comprehensive overview of the satellite communication industry. It discusses the satellites and the ground equipment necessary to both the originating source and the end-user.

Gavaghan, Helen. *Something New Under the Sun: Satellites and the Beginning of the Space Age*. New York: Copernicus Books, 1998. This book focuses on the history and development of artificial satellites. It centers on three major areas of development—navigational satellites, communications satellites, and weather observation and forecasting satellites.

National Aeronautics and Space Administration. *NASA: The First Twenty-Five Years, 1958-1983*. NASA EP-182. Washington, D.C.: Government Printing Office, 1983. Designed for use by classroom teachers, this book features color photographs, charts, graphs, tables, and suggested activities. Topics include NASA history, programs, and missions.

_____. *Space Network Users' Guide (SNUG)*. Washington, D.C.: Government Printing Office, 2002. This users' guide emphasizes the interface between the user ground facilities and the Space Network, providing the radio frequency interface between user spacecraft and NASA's Tracking and Data-Relay Satellite System, and the procedures for working with Goddard Space Flight Center's Space Communication program.

Rosenthal, Alfred. *Satellite Handbook: A Record of NASA Space Missions, 1958-1980*. Washington, D.C.: Government Printing Office, 1982. This handbook contains detailed descriptions of major NASA satellite missions from 1958 to 1980. It includes a NASA launch record for the relevant dates, a list of abbreviations and acronyms, an index, and photographs.

Van Nimmen, Jane, Leonard C. Bruno, and Richard L. Rosholt. *NASA Historical Date Book: 1958-1968*. NASA SP-4012. Springfield, Va.: National Technical Information Service, 1976. A complete history of NASA's programs and projects from 1958 to 1968. Contains much detailed information. Includes illustrations, charts, and tables.

Lulynne Streeter

Telecommunications Satellites: Private and Commercial

Date: Beginning August 12, 1960
Type of spacecraft: Communications satellites

Since the early 1960's, communications satellites have been designed and built by private corporations to serve the needs of their customers. These commercial ventures in space have contributed to U.S. technological developments.

Summary of the Satellites

The satellite communications industry was born shortly after the first satellite was placed in orbit in the late 1960's. The usefulness of communications via satellite was obvious: An orbiting satellite's altitude allows it to transmit signals over very long distances; ground-based transmissions are limited by Earth's curvature. Because the microwaves that make up radio, television, and telephone signals travel in a straight line, ground relay stations must be placed about every 55 kilometers to compensate. Before satellites came into use, therefore, worldwide communications costs were prohibitive.

Scientists experimented with two types of communications satellites: the passive reflector and the active repeater. The passive reflector is typified by Echo 1, a "balloon" satellite, whose large surface was used to reflect radio signals from the ground. This sort of satellite had two advantages: Any transmitter could bounce a signal of any frequency off the reflective surface, and there were no parts to malfunction. The disadvantage was that the received signal was extremely weak. The first Echo was launched on August 12, 1960, and was useful for a little more than four months.

In 1962 the Telstar and Relay satellites were launched. These were active repeaters that received transmissions from ground stations, amplified them, and relayed them back to Earth. Telstar was designed and built by Bell Telephone Labora-

tories, and Relay was built by RCA. The early satellites were placed in relatively low orbits and moved rapidly across the sky, requiring elaborate tracking.

Telstar could handle one television channel, the equivalent of six hundred one-way voice channels. Because the satellite's signal was weak even after amplification, the ground station was located in the shelter of a ring of low mountains to reduce radio interference. A horn-shaped antenna with an opening 1,100 meters wide was used to focus the signal, which even then reached only a billionth of a watt.

Besides their usefulness as experimental communications satellites, Telstar and Relay also were important as probes of the space environment. They carried instruments that collected data on the Van Allen radiation belts, regions that encircle Earth and contain radioactive particles. The belts' existence was inferred from data returned by the first American orbiting satellite, Explorer 1.

Synchronous Communication Satellites, or Syncom satellites, built by Hughes Aircraft for NASA, represented the next generation of communications satellites. Launched in 1963, Syncom 1 was placed in an orbit with a twenty-four-hour period and therefore matched Earth's rotation. Syncom was the first "geosynchronous" satellite. The main advantage of a geosynchronous satellite is that ground control is greatly simplified. Because the satellite remains over one area of Earth's surface, expensive and complex tracking equipment is unnecessary. The satellite is accessible to any ground station within its line of sight. A geosynchronous satellite must be placed directly over the equator to maintain its position relative to the surface; complex firings of gas jets are required to maneuver it into position. From its high altitude of about 35,900 kilometers, a geosynchronous satellite can transmit signals to nearly a third of Earth's surface. With a network of three such satellites, communications service can be provided to all populated areas of Earth. Although Syncom 1 failed and the network was not completed at that time, Syncom was the first step toward a global communications system.

Three Syncoms were launched. The first was lost when a high-pressure nitrogen bottle aboard the satellite burst. Syncom 2 was successful, and it set new records in long-distance communication. Syncom 3 transmitted the first television program (a relay of the Olympic Games in Tokyo) ever to span the Pacific via a geosynchronous satellite. Syncoms 2 and 3 remained in service until 1969.

The successes of Relay, Telstar, and particularly Syncom clearly demonstrated that communications satellites were moving away from experimental projects and toward commercial ventures. In 1963, the Communications Satellite Corporation (COMSAT) was created as a private company that would develop satellite systems. Half the corporation's stock was to go to the public; the other half, to large communications companies. COMSAT is part of Intelsat, the International Telecommunications Satellite Consortium, which was set up to provide a global communications network. COMSAT's first commercial satellite was Early Bird, launched in 1965. Early Bird relied heavily on technology from the Syncom satellites. This and other early commercial satellites demonstrated the feasibility of using geosynchronous satellites for commercial communications.

Private users of the satellites were communications common carriers, broadcasters, news wire services, newspapers, airlines, computer services companies, and television companies. At first, the satellites were used simply to receive television signals or telephone communications and relay them to one or several ground stations. It was foreseen that multipurpose satellites would eventually be built to serve the needs of various customers.

Early users of communications satellites would lease the services from one of a very few satellite owners. As the industry expanded, some large businesses began to own and operate their own satellites and lease excess capabilities to smaller businesses. Satellite capacity grew rapidly with new satellite technology. Satellites with multiple transponders (devices that, when triggered by a signal, transmit another signal at the same frequency) set at different frequencies began to support many customers simultaneously. In the mid-1960's, companies began building basic "production line" satellites; customers could choose one and then refine it to meet specific needs.

Western Union developed the first U.S. domestic satellite communications system. Westars 1 and 2 were launched in April and October of 1974, and Westar 3 was launched in August, 1979. Companies wishing to operate domestic satellite systems must obtain permission from the U.S. Federal Communications Commission (FCC), and Western Union was one of the first such companies to be licensed. The Westars were designed to relay voice, video, and data communications to the continental United States, Alaska, Hawaii, and Puerto Rico. The satellite could handle twelve color television channels or seven thousand two-way voice circuits simultaneously. The new generation of Westars—Westars 4, 5, and 6—were deployed by 1982 and had twice the capacity of the earlier satellites. They could also relay signals to the U.S. Virgin Islands.

Another domestic communications system, Satcom, was placed in orbit by RCA beginning in 1975. Satcom was the first satellite system devoted to relaying signals to cable television installations in the United States.

A third domestic communications satellite system was launched in 1976. Comstar is a telephone satellite system for long-distance calling through-

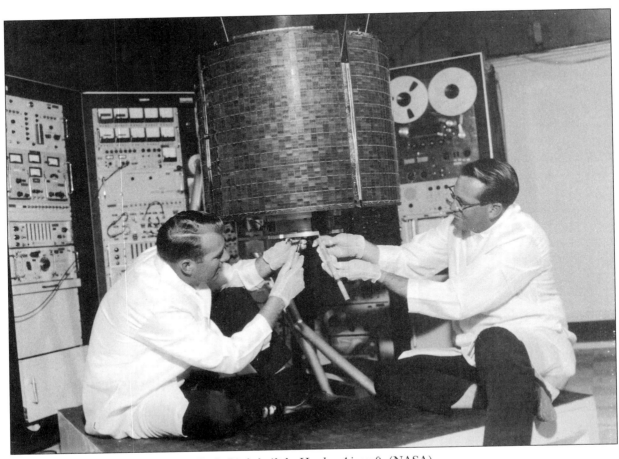

Early Bird, built by Hughes Aircraft. (NASA)

out the United States; it was designed to handle increasing domestic telephone usage. Each Comstar satellite can relay more than eighteen thousand calls simultaneously. They are leased jointly by the American Telephone and Telegraph Company and GTE Satellite Corporation, a subsidiary of General Telephone and Electronics. In the early 1980's, AT&T launched a new series of Telstar satellites: Telstars 301, 302, and 303.

In March, 1981, Satellite Business Systems (SBS) began to offer satellite services for private business communications to large U.S. companies. The third SBS satellite to be deployed was the first commercial satellite launched by the U.S. space shuttle.

Although Hughes Aircraft had been designing and building satellites for other customers since the early 1960's, it was not until 1983 that it began launching the Galaxy series, owned by Hughes Communications. The Galaxy series was dedicated to the distribution of cable television programming. Hughes offered cable programmers the opportunity to buy, rather than lease, transponders on the satellite.

Hughes Space and Communications Company has produced nearly 40 percent of the satellites now in commercial service worldwide. These spacecraft routinely relay digital communications, telephone calls, videoconferences, television news reports, facsimiles, television programming, mobile communications, and direct-to-home entertainment—truly global communications. Currently, their fleet includes a variety of communications satellites.

The smallest is the spin-stabilized HS 376. Available in several configurations, the HS 376 ranges from 800 to 2,000 watts and features twenty-four transponders in C band or Ku band, or a combination of both. Next is the body-stabilized HS 601 series. The basic configuration features as many as forty-eight transponders and offers up to 4,800 watts. A higher-power version, the HS 601HP, features as many as sixty transponders and uses dual-junction gallium arsenide solar cells and other new technologies to provide up to 8,700 watts. In 1995, Hughes introduced a more powerful body-stabilized design, the HS 702, with capacity for as many as ninety-four operating transponders and offering 15,000 watts at end of life. Nine HS 702 spacecraft have been ordered, and the first was launched in 1999.

In 1997, Hughes received its first order for an HS GEM spacecraft model. Designed for geomobile communications, the HS GEM draws upon the technical heritage of the HS 601 and the technical advances inherent in the larger, more powerful HS 702. The HS 376 model boasts customers in Asia, Australia, the Americas, and Europe. Fifty-four had been ordered by the end of 1998, making it one of the most often purchased satellite models in the world, second only to the HS 601.

PAS-9, a high-power Hughes 601HP satellite built for PanAmSat Corporation by Hughes, was successfully launched in July, 2000, on a Sea Launch rocket from the middle of the Pacific Ocean. PAS-9, once on-station at 58° west longitude, will provide at least fifteen years of video, data, and Internet services in C band and in Ku band for the Americas, the Caribbean, and Western Europe, as well as direct-to-home services for Mexico in Ku band. The satellite features forty-eight transponders, twenty-four in C band and twenty-four in Ku band. PAS-9 is PanAmSat's twenty-first satellite.

High-powered satellites built by Hughes began bringing true direct broadcast satellite (DBS) service to homes throughout North America in 1994. The spacecraft are HS 601 body-stabilized models ordered by DIRECTV, a unit of Hughes Electronics

Corporation. DIRECTV 2 and DIRECTV 3 are collocated with DIRECTV 1 at 101° west longitude and are used exclusively by DIRECTV. Collectively the three satellites have the capacity to deliver more than five hundred channels of entertainment programming to subscribers that are equipped with DSSTM digital home receiving units, which feature a 45.72-centimeter dish. To provide the high satellite power necessary for such small antennae, the DIRECTV 1 spacecraft has sixteen transponders powered by 120-watt traveling-wave tube amplifiers (TWTAs). The TWTAs were reconfigured to provide eight channels with 240 watts of power on DIRECTV 2 and DIRECTV 3. The amplifiers are suitable for analog or digital signals and are capable of transmitting high-definition television (HDTV) signals and compact disc-quality audio as well.

Hughes Network Systems (HNS), a unit of Hughes Electronics Corporation and Pegasus Communications Corporation, announced an agreement in July, 2000, that will enable Pegasus to offer high-speed Internet access by satellite to rural and underserved households served by Pegasus and its Pegasus Retail Network. The system will enable users of personal computers to obtain high-speed Internet connections virtually anywhere in the continental United States, no matter how remote. The service will provide downstream access speeds of up to 400 kilobits per second (Kb/s), far surpassing the 56 Kb/s currently available through top-of-the-line dial-up modems. Early subscribers will send data upstream using a modem and standard telephone line. However, the two companies plan to enhance the initial service with full two-way satellite Internet access.

Contributions

The major gains in knowledge made possible by private communications satellites have stemmed from advances in satellite technology and from the increased ease and speed of global and domestic communications.

The first satellites provided valuable information about the environment of Earth orbit. Technology improved almost immediately as the first satellites reported on the new conditions. Solar cells on Telstar 1 were damaged by radiation in the Van Allen belts, and satellite manufacturers learned new ways of compensating for radiation. Telstar 1 also returned valuable scientific data on the Van Allen belts as it passed through them. The densities and energies of free protons and electrons were measured, and the temperature and pressure within the satellite was recorded.

In 1959, an engineer working for Hughes Aircraft developed a satellite design that revolutionized the communications industry. Tiny rocket boosters were incorporated into the satellite to position it in a geosynchronous orbit. The rockets' periodic thrusts stabilize the satellite and orient it so that its communications antenna continually aims at Earth's surface. Hughes also developed spin-stabilized satellites. Spin stabilization previously had been used to improve rockets' accuracy.

New technology was devised for the Comstars' antenna arrangement. A vertically polarized and a horizontally polarized antenna were used. Polarization changes the form in which signals are received and sent without changing the frequency, which allows the capacity of the system to be doubled.

Another Hughes development is the shaped beam. The microwave beam transmitted by the satellite is shaped to the contours of the receiving area, allowing a more concentrated and powerful transmission. Along with the new technology, the improvement of existing technology has allowed for increasingly longer satellite lifetimes and increased power output.

In 1981, the space shuttle began to be used for launching commercial satellites. Cargo space in the payload bay is limited, so satellite designs were modified with telescoping solar panels and folding antennae. Starting from a stowed size of about 3 meters high by 2 meters wide, a satellite can expand to the height of a two-story building when deployed in orbit.

Communications satellites function as channels for information. Knowledge must be distributed to be effective. It would be difficult to gauge the spread of knowledge that has resulted from improved communications. Satellites have helped bring a variety of educational programs to persons who were previously beyond the reach of educational systems. Engineering and technological programs are broadcast instantaneously to universities across the country, and businesses use satellites to provide training seminars for their employees.

Context

The development of a country is related to its ability to communicate quickly and efficiently. Satellite communications have brought about a huge increase in the ability to transmit information economically and quickly. The transition from ground-based telecommunications to satellite telecommunications took place in less than twenty years. This growth is still taking place, not only in technologically advanced countries such as the United States, but all over the world. Many underdeveloped countries are using satellite systems to unite remote regions of the country. They are finding that the cost of employing satellite systems is far less than the cost of installing land-based relay stations. Educational systems can be improved and can compensate for teacher shortages.

Satellite usage primarily involves telephone and television transmission, but as the industry expands and costs become less prohibitive, new satellite uses evolve. Satellites regularly provide such services as computer-to-computer digital communications, video conferencing, monetary fund transfers, and air traffic control. Technology is being developed for mobile communications, and businesses are using satellite systems to handle electronic mail and data transfer.

The U.S. commercial satellite industry has been affected by unreliable launch systems. Because the shuttle program was temporarily suspended after the 1986 *Challenger* accident and the revitalized program was slow to accelerate, satellite users were forced to use other, conventional launch systems. Some of those have proved unreliable, and several multimillion-dollar satellites have been lost. Insurance costs increase dramatically as the risk associated with a launch increases, and orders for new satellites slow as launch capability lessens. Many companies have looked to European launch facilities for reliable and timely deployment of satellites.

Some companies have argued that the federal government has not offered enough support to commercial space ventures. In the late 1980's, government support began to grow along with the perception that the United States was falling behind the rest of the world in the utilization of space. One way to increase a nation's space use is through the mobilization of the private sector.

Debris in space is also a growing problem. Hundreds of satellites are in geosynchronous orbit and satellite overcrowding is possible. For many applications, there are only a few suitable locations for signal transmission. Furthermore, satellites that are less than 2° apart in orbit risk radio frequency interference.

Despite problems, the communications satellite industry has grown. Further technological developments may lead to communications satellites that can transmit power, rather than information, to Earth. It is conceivable that by 2025, one hundred power satellites could meet 30 percent of the United States' electrical needs.

See also: Amateur Radio Satellites; Applications Technology Satellites; Intelsat Communications Satellites; Mobile Satellite System; Private Industry and Space Exploration; Telecommunications Satellites: Maritime; Telecommunications Satellites: Military; Telecommunications Satellites: Passive Relay; Tracking and Data-Relay Communications Satellites.

Further Reading

Aviation Week and Space Technology 120 (June 25, 1984). This entire volume is devoted to prospects for the commercial use of space. It contains a series of authoritative articles that are readily understandable by laypersons.

Butrica, Andrew J. *Beyond the Ionosphere: Fifty Years of Satellite Communications.* NASA SP-4217. Washington, D.C.: Government Printing Office, 1997. Part of the NASA History series, this book looks into the realm of satellite communications. It also delves into the technology that enabled the growth of satellite communications. The book includes many tables, charts, photographs, and illustrations, a detailed bibliography, and reference notes.

Gavaghan, Helen. *Something New Under the Sun: Satellites and the Beginning of the Space Age.* New York: Copernicus Books, 1998. This book focuses on the history and development of artificial satellites. It centers on three major areas of development—navigational satellites, communications satellites, and weather observation and forecasting satellites.

Goldman, Nathan C. *Space Commerce: Free Enterprise on the High Frontier.* Cambridge, Mass.: Ballinger Publishing, 1985. Written for nonspecialists, this book provides a useful review of American commercial space enterprises. It is both synthetic and specific. Charts, tables, and a few photographs augment the text. Contains six appendices and an index.

Lee, Wayne. *To Rise from Earth: An Easy to Understand Guide to Spaceflight.* New York: Checkmark Books, 1996. This is a good introduction to the science of spaceflight. Although written by an engineer with the NASA Jet Propulsion Laboratory, it is presented in easy-to-understand language. In addition to the theory of spaceflight, it gives some of the history of the human endeavor to explore space.

Martin, Donald H. *Communication Satellites.* 4th ed. New York: American Institute of Aeronautics and Astronautics, 2000. This work chronicles the development of communications satellites and worldwide networks over the past four decades, from Project Score to modern satellite communication systems.

National Aeronautics and Space Administration. *Space Network Users' Guide (SNUG)*. Washington, D.C.: Government Printing Office, 2002. This users' guide emphasizes the interface between the user ground facilities and the Space Network, providing the radio frequency interface between user spacecraft and NASA's Tracking and Data-Relay Satellite System, and the procedures for working with Goddard Space Flight Center's Space Communication program.

Ohmori, Shingo, Hiromitsu Wakana, and Seiichiro Kawase. *Mobile Satellite Communications*. Cambridge, Mass.: Artech House, 1997. This relatively technical reference gives an insight to the making of a mobile satellite communications network. The authors take an easy-to-understand approach to the satellite position determination principle, and provide a compact, practical procedure for making orbit calculations. They explore vehicle antennae, their requirements, gain, beam width, phased arrays, and satellite tracking, and discuss signal propagation problems and counter-methods in marine, aeronautical and land environments.

Ordway, Frederick I., III, Carsbie C. Adams, and Mitchell R. Sharpe. *Dividends from Space*. New York: Thomas Y. Crowell, 1971. Explores the benefits to humankind that have directly resulted from the space program. Discusses products developed for mass consumption and advances in medicine, industry, research, and the study of Earth. Sparsely illustrated. Written for general audiences with an interest in space technology.

Paul, Günter. *The Satellite Spin-Off: The Achievements of Space Flight*. Translated by Alan Lacy and Barbara Lacy. Washington, D.C.: Robert B. Luce, 1975. A survey of the commercial scientific and communications applications that developed from the space research of the 1960's and early 1970's. Contains a comprehensive account of the early, politically charged days of Intelsat. This book is written from the perspective of the European community, and it is necessary reading for those desiring a broader understanding of Intelsat than might be available from the U.S. point of view.

Porter, Richard W. *The Versatile Satellite*. New York: Oxford University Press, 1977. This short book provides a fine introduction to the various uses of satellites. Chapter 4 deals with communications satellites.

Schwarz, Michiel, and Paul Stares, eds. *The Exploitation of Space: Policy Trends in the Military and Commercial Uses of Outer Space*. London: Butterworth, 1986. Consists of highly readable and informed articles written by scholars but intended for laypeople. Well illustrated. Notes and references conclude each article.

Zimmerman, Robert. *The Chronological Encyclopedia of Discoveries in Space*. Westport, Conn.: Oryx Press, 2000. Provides a complete chronological history of all crewed and robotic spacecraft and explains flight events and scientific results. Suitable for all levels of research.

Divonna Ogier

Telescopes: Air and Space

Date: Beginning 1608
Type of spacecraft: Space telescope

Telescopes, first invented in 1608, underwent enormous diversification in the twentieth century, particularly as it became possible to elevate them into and beyond Earth's atmosphere. They have played a vital role in the discovery and investigation of a wide variety of celestial bodies.

Key Figures

Hans Lippershey (1570-1619), recognized as the inventor of the telescope
Galileo Galilei (1564-1642), first to use the telescope for astronomical observations
Sir Isaac Newton (1643-1727), improved understanding of telescope optics
Edwin P. Hubble (1889-1953), astronomer who proved the universe is expanding; Hubble Space Telescope is named after him
Arthur Holly Compton (1892-1962), Nobel Prize-winning high-energy physicist; the Compton Gamma Ray Observatory was named after him
Lyman Spitzer, Jr. (1914-1997), astronomer who argued strongly for large space-based telescopes; the Spitzer Space Telescope is named after him
Subrahmanyan Chandrasekhar (1910-1995), Nobel Prize-winning astrophysicist who described the characteristics of black holes; the Chandra X-Ray Observatory is named after him

Summary of the Technology

Astronomy, one of the most ancient of sciences, was altered for all time by the invention of the telescope. The first such instrument was developed by Hans Lippershey of Middelburg, the Netherlands, in 1608; he discovered that a certain combination of concave and convex lenses worked to make distant objects seem nearer. Experimentation and refinement of technologies over the subsequent centuries made possible the proliferation of ground and space-based telescopes of the twentieth century. This combination of telescopes has permitted investigations of emissions across the electromagnetic spectrum—from gamma rays, the most energetic, to radio waves, the least energetic.

Telescopes can be divided into two broad categories: optical telescopes and radio telescopes. Optical instruments collect light energy from a distant source and focus it into an image that can be studied by a number of different techniques. Those satellite-borne telescopes that are aimed at detecting celestial x rays or ultraviolet radiation are included in this category, because they operate according to the principles of geometrical optics. Radio telescopes are used in astronomical research to detect and measure radio waves coming from various parts of the galaxy. These instruments consist of three complementary parts: a large reflecting surface that collects and focuses incident radiation, an electronic receiver that detects and amplifies cosmic radio signals, and a device that displays the information.

The diversity of modern telescope technology has its origins in the pioneering work of scientists such as Galileo, who in 1609 heard of Lippershey's "magic glass" and became fascinated with the idea of using it to study celestial bodies. The instrument that he developed consisted simply of a paper tube and a pair of appropriate lenses; it had the advantage of an erect image, but afforded only a small field of view. Johannes Kepler, Francesco Generini, and others experimented subsequently with variations on this design. Naturally, a principal goal was to increase the instrument's power, or magnification, which varies with the ratio of the focal length of the front (objective) lens to the focal length of the rear lens, or eyepiece. These early optical instruments, with two lenses centered on the same axis, are known as refracting telescopes.

In 1668, Sir Isaac Newton experimented with the use of a small, concave mirror in what became known as the reflecting telescope. Though his instrument was not particularly successful, its basic concept proved durable. The Frenchman N. Cassegrain produced a different type of reflecting instrument in 1672, with a convex mirror. These two types of reflectors, termed Newtonian and Cassegrain, remain in use in many large reflecting telescopes.

The simple, versatile Newtonian reflecting telescope is the most universally used type of telescope. It bends light rays by reflecting them from the surface of a concave mirror. Light from the object being viewed travels down the tube assembly and strikes the primary mirror; the light is then reflected toward a focal point that lies just outside the tube assembly. The focused rays are intercepted by a secondary diagonal mirror, which bends the rays at a 90° angle to the incoming light. The light rays are then brought into contact with the eyepiece, which lies outside the tube plane. As a result of the bending of the light rays, the image is seen upside down, with some loss of light.

In the Cassegrain configuration, light is reflected by a hyperbolic mirror inserted in front of the prime focus, passes through a hole in a large parabolic mirror, and comes to a new focus on the back of the mirror. The Cassegrain telescope improved upon the Newtonian design through its use of a shorter focal length in its system. Its more com-

pact formation requires a less massive mount for the same degree of stability. In addition, the eyepiece is placed more conveniently. Unfortunately, parabolizing the primary mirror to the required degree of accuracy is an expensive process.

Variations on these reflecting telescopes have been produced, each with its strengths and weaknesses. The Dall-Kirkham telescope uses an ellipsoidal primary mirror and a spherical secondary mirror. Its field of vision, however, is quite limited, and like the Cassegrain and Newtonian telescopes, it produces coma effects (a distortion in which points of light appear comet- or fan-shaped). The Ritchey-Chrétien telescope uses a hyperboloidal primary mirror and an elliptical, convex secondary mirror. It reduces coma almost completely but suffers from astigmatism and a severe curvature of its field of vision. In general, because reflecting telescopes are not sealed, dust can accumulate on the mirrors; in addition, air turbulence within the tube can disrupt viewing.

Chromatic aberrations (false colors arising from the refraction of different wavelengths of light) are not a problem with reflecting telescopes, although low-expansion glass, which can resist the tendency of large mirror-surfaces to expand and contract differently in different areas, is a necessity. This type of telescope possesses considerable advantages in spectroscopic work and is particularly useful in the observation of faint or very distant objects, such as nebulae and galaxies.

With the success of reflecting-type telescopes, their refracting counterparts became temporarily obsolete, but eventual improvements in glass technology made possible a return to refracting instruments. By 1799, a Swiss craftsman had mastered the art of making flint glass that had the optical qualities required for refraction. Innumerable further refinements have led to the development of refracting telescopes that are mechanically simple, durable, and readily available.

The refracting telescope's closed-tube design eliminates the air currents that can degrade images. The objective lens is mounted at the far end of the apparatus; its diameter ranges from about

0.05 to 1 meter. This lens is focused on the image of a distant object, capturing light, which it then focuses onto a microscopic point. The eyepiece lens then concentrates that point of light onto the human eye. Refracting telescopes are especially useful for visual observations and astrometric measurements of stellar parallax and binary stars and for the visual and photographic determination of stellar positions.

The glass of the reflecting telescope's lens must be optically homogeneous and free of bubbles. The lenses require expensive material, and unless the correct combination of lenses and the proper focal length are used, chromatic aberrations arise. The flint lens of a refractor becomes increasingly opaque to wavelengths shorter than 40 nanometers (nm); it becomes useless in this spectral region.

Catadioptric telescopes, developed in the twentieth century, combine the best features of reflecting and refracting telescopes. These instruments have closed tubes, thus eliminating image-degrading currents and making the optics almost maintenance-free. These telescopes are portable and free of chromatic aberrations, and their additional optical elements allow users greater facility in correcting foci and surmounting the faults of the mirrors.

Various other telescopes have been developed for specialized functions. A heliostat, for example, is a telescope used to study the Sun; its flat moving mirror captures sunlight and feeds it to a fixed telescope with a stationary focal plane. A similar instrument used for observing stars is known as a siderostat. The zenith telescope has a fixed lens that points toward the zenith—that is, directly overhead. Incoming light is reflected from a mercury pool onto a photographic plate placed at the focus of the lens. Such telescopes are used for the precise determination of time. Transit telescopes are refractors capable of moving in altitude but not in azimuth; they are particularly useful in making accurate observations of the stars as they cross the meridian. A chronograph is used to obtain photographs and film of the solar corona, even in daylight.

Despite all the refinements that increased the quality of the images produced by ground-based telescopes, astronomers realized that they suffered from an inherent limitation: the distortions produced by Earth's atmosphere. Earth's air absorbs essentially all ultraviolet radiation below a wavelength of 300 nm, as well as much of the infrared spectrum between wavelengths of 1 micron and 1 millimeter (mm). Moreover, the thermal currents in Earth's atmosphere deflect light and thus limit the sharpness of most Earth-based astronomical photographs.

It was a desire to observe the Sun that motivated early efforts to send telescopes into air and space. A balloon-borne refractor sent aloft in 1956 and 1957 reached an altitude of between 6,096 and 7,620 meters and enabled astronomers from the University of Cambridge in England and the Meudon Observatory in France to obtain photographs of the granulation of the Sun.

Stratoscope 1, an American project, followed on the heels of the European refractor's launch. An robotic balloon capable of lifting a payload of 635 kilograms to an altitude of 25,603.2 meters was designed. It carried a reflecting telescope that yielded an optical enlargement corresponding to an effective focal length of 60.96 meters. Several pairs of photodiodes acted as remote eyes to point the telescope toward the Sun's disk. A total of five flights of Stratoscope 1 were undertaken. The last flight furnished excellent photographs of solar granulation and sunspots, taken from an altitude of about 24,400 meters.

The success of these balloon experiments encouraged investigators to devise ways of observing other solar phenomena. Astronomers from Boulder, Colorado, organized the launch of the Coronascopes, special balloon-borne telescopes whose purpose was to photograph the solar corona in full daylight. Coronascope 1 carried a small coronagraph with an aperture of 0.033 meter. It recorded images of the Sun automatically on a red-sensitive 0.035-meter spectroscopic film. Coronal streams between two and five times the solar radius from the Sun's center were detected.

Stratoscope 2 carried elaborate instruments to an altitude ranging from 21,641 to 25,603 meters. The first flight was on March 1, 1963; it was successful in obtaining measurements of a part of the infrared spectrum of Mars. Stratoscope 2's second flight took place on November 26, 1963; it returned images of the infrared spectra of about nine red stars and new data on Jupiter and the Moon.

Balloons could carry telescopes, however, only to the threshold of space. Further penetration into space required vehicles with greater lifting power. In response to this need, the first Orbiting Astronomical Observatory was launched into space on April 8, 1966. It carried eleven telescopes, four of which were coupled with ultraviolet vidicons to obtain maps of the sky. The other seven telescopes, in conjunction with two spectrometers and five photometers, were used to obtain spectral and energy distributions of numerous celestial objects.

A series of Orbiting Solar Observatories were launched beginning in 1962. These satellites were designed to return measurements of ultraviolet, x-ray, and gamma ray emissions from the Sun. The Orbiting Geophysical Observatories, whose launches began in 1964, obtained measurements of radio noise and bursts of cosmic-ray protons from the Sun.

Beginning in 1969, various probes—Mariners, Lunas, Voyagers, Rangers, Orbiters, Surveyors—were being launched in space with telescopic imaging systems as standard equipment. Each craft carried a different type of telescope related to its function. By the late 1970's, these telescopes were being controlled by microcomputer.

With the May 14, 1973, launch of Skylab, a new era of space research began. Skylab carried eight separate solar telescopes on its Apollo Telescope Mount: two x-ray telescopes, an extreme ultraviolet spectroheliograph, an ultraviolet spectroheliometer, an ultraviolet spectrograph, a visible light coronagraph, and two hydrogen-alpha telescopes. These highly complex instruments gathered data on the Sun across the electromagnetic spectrum.

The first High-Energy Astronomical Observatory was launched on August 12, 1977; a series of

such orbiting observatories were aimed at returning data on high-energy astrophysical processes. High-Energy Astronomical Observatory 2 carried the first x-ray telescope capable of providing focused images of x-ray objects in the sky.

In the meantime, scientists continued to find that telescopes mounted on aircraft could also return data of considerable value. The Kuiper Airborne Observatory C-141, for example, was mounted with a 0.90-meter telescope that was used to study the nature of extragalactic, x-ray sources in Galaxy M31 and x-ray emissions from other clusters of galaxies.

Certain Explorer satellites were designed to provide data on sources emitting ultraviolet radiation. The International Ultraviolet Explorer, an international venture undertaken by the National Aeronautics and Space Administration (NASA), Great Britain's Science Research Council, and the European Space Agency, was launched on January 26, 1978, into an eccentric geosynchronous orbit. It was mainly used to gather data on the transmission and absorption of radiation in the atmosphere of subluminous stars, in the interplanetary medium, and around other objects within the solar system.

The Infrared Astronomical Satellite, a cooperative project of NASA, the Netherlands, and Great Britain, carried a cryogenically cooled telescope system. It surveyed various celestial sources of infrared radiation.

The Solar Maximum Mission satellite, another telescope-bearing craft, was put into space to gather information on solar flares and the globular solar corona.

The first generation of orbiting telescopes—the Orbiting Solar Observatories, the Orbiting Geophysical Observatories, and the High-Energy Astronomical Observatory, for example—were automated telescopes revolving around Earth. Such telescopes function simply as search cameras or survey instruments. Eventually, plans were made for the design of piloted orbiting telescopes, so that astronauts could go into space to do maintenance work on them; instruments could also be picked up by the space shuttle and returned to Earth for refurbishment and relaunch.

After several years of study and refinement of the design principles, the Hubble Space Telescope (HST) became an approved NASA project. This high-resolution, 2.4-meter telescope was placed in orbit in 1990 as a joint venture of NASA and the European Space Agency. A high-resolution camera, a faint-object spectrograph, an infrared photometer, and guidance and protective equipment accompany the Cassegrain telescope. The space telescope is designed to operate from wavelengths of 91.2 nanometer to about 1 millimeter—across the ultraviolet, optical, and infrared regions. Hubble was the first of four orbiting observatories known as the Great Observatories. The Compton Gamma Ray Observatory was launched in 1991 and the Chandra X-Ray Observatory was launched on

board *Columbia* during the STS-93 mission in July, 1999. The fourth, the Space Infrared Telescope Facility (SIRTF), was launched on August 25, 2003.

After Hubble was launched, scientists discovered that its primary mirror had been ground to the wrong prescription, which had resulted in spherical aberration: the blurring of starlight because the telescope could not bring all the light to a single focal point. Using image-processing techniques, scientists were able to do significant research with Hubble until an optical repair could be developed. In December, 1993, the first Hubble servicing mission carried replacement instruments and supplemental optics aboard the space shuttle *Endeavour* to restore the telescope to full optical performance. A corrective optical device, called the Corrective Optics Space Telescope Axial Replacement (COSTAR), was installed (requiring re-

Of the many spaceborne telescopes, the Hubble Space Telescope is perhaps most famous for its many awe-inspiring images of galaxies, star clusters, and other heavenly bodies. (NASA)

moval of the High Speed Photometer) so that it could improve the sharpness of the first-generation instruments. The Wide Field/Planetary Camera 2 replaced the original camera; it has a built-in correction for the aberration in the primary mirror.

In February, 1997, the space shuttle *Discovery* returned to Hubble for a second servicing mission. Two advanced instruments, the Near Infrared Camera and Multi-Object Spectrometer (NICMOS) and the Space Telescope Imaging Spectrograph (STIS), were swapped out with the two first-generation spectrographs. The astronauts also replaced or enhanced several electronic subsystems and patched unexpected tears in the telescope's shiny, aluminized thermal insulation blankets, which give the telescope its distinctive foil-wrapped appearance.

In December, 1999, the space shuttle *Discovery* rendezvoused with Hubble for a third servicing mission referred to as Servicing Mission 3A or SM3A; the original third servicing mission was too ambitious and ultimately was split into SM3A and SM3B missions. Astronauts replaced faulty gyroscopes, which had suspended science observations for nearly a month. The telescope also got a new high-tech computer and a data recorder. The astronauts left the telescope in "better than new" condition.

The space shuttle *Columbia* rendezvoused with Hubble for the SM3B mission in March, 2002. Astronauts replaced a faulty reaction wheel assembly, installed upgraded solar arrays, installed a cooling system to return NICMOS to operational capability, and swapped out an old instrument in order to incorporate the Advanced Camera for Surveys into the science payload of the telescope.

A fourth Hubble servicing mission was planned for 2003 or 2004, depending upon funding and International Space Station construction scheduling. However, in the aftermath of the *Columbia* accident on February 1, 2003, shuttle flights to orbital inclinations other than that of the International Space Station were declared too risky. Although the scientific community objected strongly, plans for astronauts to repair the Hubble were scrapped. NASA attempted to put together a major development program to service Hubble robotically be-

fore its solar batteries were expected to fail in 2007 or 2008, but when program costs rose, those plans were scrapped as well. The fate of Hubble was uncertain, the scientific community would make the ultimate use of Hubble's unique capabilities, and expressed interest in taking those advanced instruments that had been destined for installation on Hubble during the fourth servicing mission and placing them on new, smaller space-based observatories. However, as of 2005, there were no definitive plans to make use of those instruments in such a fashion.

During its first ten years of observation, more than 260,000 exposures came down from Hubble, revealing much about the age and size of the universe, proving there are black holes, showing the surface of Pluto, demonstrating how stars are born, and telling us what happens when a comet strikes a planet.

An HST Second Decade study was prompted by NASA's decision in January, 2000, to extend the HST mission until 2010, if possible, with low-cost operations. The goal was to operate Hubble in parallel with its successor, the Next Generation Space Telescope (NGST), initially planned for launch in 2007. This second decade would also be the era of the Chandra X-Ray Observatory, SIRTF, and the Space Interferometry Mission (SIM), as well as other new space-based and ground-based observing facilities. The anticipated launch of the Next-Generation Space Telescope, renamed the James C. Webb Space Telescope (JWST), slipped from 2007 to 2009, and then to 2011. As of 2005, it was not expected that Hubble would be able to operate in concert with the James C. Webb Space Telescope.

The James C. Webb Space Telescope is to observe the first stars and galaxies in the universe. This grand effort is embedded in fundamental questions that have been posed to NASA's space science program: What is the shape of the universe? How do galaxies evolve? How do stars and planetary systems form and interact? How did the universe build up its present elemental/chemical composition? and What is dark matter? Answers to most of these questions involve objects formed extremely early in the history of the universe. The radiation of such objects, by the time it is visible on Earth and to the relatively near-Earth space-based telescopes, has greatly redshifted when observed in the current epoch. This redshifting (elongating of the electromagentic waves as they spread outward) means that observations from our vantage point are best performed in the infrared (longer-wave) portion of the electromagnetic spectrum. The JWST will be capable of detecting radiation with wavelengths in the range 0.6 to 20 microns (and be optimized for the 1- to 5-micron region). Furthermore, the JWST must be able to see objects four hundred times fainter than those currently studied with large ground-based infrared telescopes (such as the one at the Keck Observatory in Hawaii) or the current generation of space-based infrared telescopes—and it must do so with a spatial resolution (image sharpness) comparable to that of the Hubble Space Telescope.

The Compton Gamma Ray Observatory was the second of NASA's Great Observatories. Compton, at 15,455 kilograms, was the heaviest astrophysical payload ever flown at the time of its launch on April 5, 1991, aboard the space shuttle *Atlantis*. Compton was safely deorbited and reentered the Earth's atmosphere on June 4, 2000. It had four instruments that covered an unprecedented six decades of the electromagnetic spectrum, from 30 kiloelectron volts to 30 gigaelectron volts. In order of increasing spectral energy coverage, these instruments were the Burst and Transient Source Experiment (BATSE), the Oriented Scintillation Spectroscopy Experiment (OSSE), the Imaging Compton Telescope (COMPTEL), and the Energetic Gamma Ray Experiment Telescope (EGRET). For each of the instruments, an improvement in sensitivity of better than a factor of ten was realized over previous missions. The observatory was named in honor of Arthur Holly Compton, who won the Nobel Prize in Physics for work on scattering of high-energy photons by electrons—a process that is central to the gamma-ray detection techniques of all four instruments.

NASA's Chandra X-Ray Observatory, which was launched and deployed by space shuttle *Columbia* in July of 1999, is the most sophisticated x-ray observatory built to date. Chandra is designed to observe x rays from high-energy regions of the universe, such as hot gas in the remnants of exploded stars. The observatory has three major parts: the x-ray telescope, whose mirrors focus x rays from celestial objects; the science instruments, which record the x rays so that x-ray images can be produced and analyzed; and the spacecraft, which provides the environment necessary for the telescope and the instruments to work. Chandra's unusual orbit was achieved after deployment by a built-in propulsion system, which boosted the observatory to a high Earth orbit. This orbit, which has the shape of an ellipse, takes the spacecraft more than a third of the way to the Moon before returning to its closest approach to the Earth of 10,000 kilometers. The spacecraft spends 85 percent of its orbit above the belts of charged particles that surround the Earth. Uninterrupted observations as long as fifty-five hours are possible and the overall percentage of useful observing time is much greater than for the low-Earth orbit of a few hundred kilometers used by most satellites.

The Chandra observatory was first proposed to NASA in 1976, and funding began in 1977 when NASA's Marshall Space Flight Center started the definition studies of the telescope. At that time, it was known as the Advanced X-Ray Astrophysics Facility (AXAF). The x-ray observatory was named the Chandra X-Ray Observatory in honor of the late Indian American Nobel laureate Subrahmanyan Chandrasekhar (pronounced soo-brah-MON-yon chahn-drah-SAY-kar). Known to the world as Chandra (which means "moon" or "luminous" in Sanskrit), he was widely regarded as one of the foremost astrophysicists of the twentieth century.

The Space Infrared Telescope Facility (SIRTF) is a spaceborne, cryogenically cooled infrared observatory capable of studying objects ranging from the solar system to the distant reaches of the universe. SIRTF consists of a 0.85-meter telescope and three cryogenically cooled science instruments capable of performing imaging and spectroscopy in the 3- to 180-micron wavelength range. Incorporating the latest in large-format infrared detector arrays, SIRTF offers orders-of-magnitude improvements in capability over existing programs. While SIRTF's mission lifetime requirement remains 2.5 years, programmatic and engineering developments have brought a five-year cryogenic mission within reach. Once in orbit in August, 2003, SIRTF was renamed the Spitzer Space Telescope.

Spitzer is the final element in NASA's Great Observatories Program, an important scientific and technical cornerstone of the new Astronomical Search for Origins Program. NASA's Origins Program seeks to answer two enduring human questions that we once considered around ancient campfires yet still keep alive in today's classrooms: Where do we come from? Are we alone? Over the course of the next two decades, the Origins Program will develop the sophisticated telescopes and technologies that will bring us the information we seek. While the questions are challenging, our generation is privileged to have the technological ability to reveal the possibilities for the first time. Just as the Greeks were known for democracy, the Egyptians for pyramids, and the Romans for roads, our civilization may well be remembered for discovering life beyond our own planet, forever changing our perception of the universe and our place within it.

Contributions

Almost all that is known regarding the Sun, the Moon, planets, stars, galaxies, and other celestial bodies has been gained with the aid of ground-, air-, or space-based telescopes.

Galileo and other early telescope builders produced drawings of the sunspots that they observed, showing the presence of umbras (dark central regions) rimmed by penumbras (lighter areas). Scientists were able to observe and measure the sunspots' drifting motion across the Sun's face and thereby to show that the Sun rotates. Eventually, it was understood that the Sun goes through peaks (maxima) and valleys (minima) of sunspot activity

approximately every eleven years as part of an overall twenty-two-year-long solar cycle during which the Sun's magnetic field reverses and then returns to the initial configuration. Telescopes carried by the Orbiting Solar Observatories, the Solar Maximum Mission satellite, and Skylab returned vital information regarding the sunspot cycle, coronal holes and loops, the solar wind, and much more.

With the help of his simple refractor, Galileo was able to observe the mountains and craters of the Moon's surface. He also discovered Jupiter's four large moons: Io, Europa, Ganymede, and Callisto. A dozen smaller Jovian satellites have been discovered. Jupiter's cloud belts have been found to be discontinuous and composed of a multitude of streamers and festoons. The 1955 discovery of radio waves being emitted from Jupiter led to the discovery of that planet's radiation belts. Mars's surface features, such as Syrtis Major and Mare Erythraeum, have been found to be discontinuous. That planet is known to be covered with volcanoes, canyons, and craters; Mercury, too, is highly cratered. Mercury and Venus were discovered to have unusual patterns of rotation. Telescopic examination has shown that Saturn has twelve concentric rings, as well as a red spot like that of Jupiter. Saturn also has some seventeen satellites of widely varying natures. Uranus, which was discovered by Sir William Herschel, has been found to possess only five moons. Rings have also been discovered around Uranus. Unexpected aspects of Uranus's orbital path led to the discovery of the planet Neptune, and distortions in Neptune's path in turn helped astronomers locate tiny Pluto.

Halley's comet, whose periodicity was determined by Edmond Halley to be seventy-six years, has been studied telescopically during its three recurrences since 1758. Much has been learned about stars as well through the use of telescopes—for example, that there are many types of stars, including double, variable, and exploding stars. Quasi-stellar objects discovered in the 1950's with the aid of radio telescopes were dubbed quasars. Pulsars, celestial objects emitting rapid radio pulses, have also been discovered. With the aid of telescopic instruments, almost all the known galaxies, including the Milky Way, have been grouped into clusters. Large numbers of asteroids have also been discovered; most of them were found to move in solar orbits between the orbits of Mars and Jupiter.

Infrared emission from the planets has provided important information about the structure and evolution of planetary interiors. Various nebulae, such as the planetary nebulae, extragalactic objects, and some quasars, have been discovered to emit unexpectedly large amounts of infrared energy. Numerous quasars have been found to have a large redshift; thus, astronomers have discovered the rapid movement of those objects away from the galaxy.

Radio observations of the interstellar medium have led to the discovery of complex molecules such as ammonia and formaldehyde, along with hydroxyl ions and water molecules. Added to these discoveries are the amazing images and data that have been returned by the Great Observatories: Hubble, Chandra, Compton, and Spitzer. This new information is still being analyzed but clearly will change our perception of the universe—its origins, expansion, and future—and already has done so.

Context

The invention of the telescope provided humankind with a window on the universe. Subsequent refinements opened that window ever wider. Yet at the beginning of the twentieth century, much remained to be discovered. For example, knowledge of solar radiation extended only slightly beyond the visible-light spectrum, a range of 400 to 700 nm. Earth's atmosphere blocks out most of the shorter and longer wavelengths. Exotic, nonthermal radiations as well as the radiation emitted by very hot and very cold objects were beyond understanding.

Thus, the ability to raise telescopes into air and space led to a remarkable amplification of astronomers' knowledge of celestial bodies and their radiations. Observations of the planetary spectra by means of telescopes carried to Earth's stratosphere in jets, for example, are far superior to those obtained from ground-based facilities. In general, measurements obtained with Earth-orbiting telescopes are free from the interference from Earth's atmosphere. Deep space probes are able to carry telescopes even farther—and further open the window to knowledge of the universe.

See also: Asteroid and Comet Exploration; Chandra X-Ray Observatory; Compton Gamma Ray Observatory; Gamma-ray Large Area Space Telescope; High-Energy Astronomical Observatories; Hubble Space Telescope; Planetary Exploration; Solar and Heliospheric Observatory; Spitzer Space Telescope.

Further Reading

Brown, Sam. *All About Telescopes.* 5th ed. Barrington, N.J.: Edmund, 1981. This volume describes in detail the qualities of a good telescope and explains how to use a telescope and mount it for best viewing of the finer details of the sky. Suitable for high school and college-level students.

Chaikin, Andrew. *Space.* London: Carlton Books, 2002. A large image picture book spanning piloted and robotic exploration of space. Provides pictures of Earth and special resources as well.

Davies, John K. *Astronomy from Space: The Design and Operation of Orbiting Observatories.* New York: John Wiley, 1997. This is a comprehensive reference on the satellites that have revolutionized twentieth century astrophysics. It contains in-depth coverage of all space astronomy missions. It includes tables of launch data and orbits for quick reference as well as photographs of many of the lesser-known satellites. The main body of the book is subdivided according to type of astronomy carried out by each satellite (x-ray, gamma ray, ultraviolet, infrared and millimeter, and radio). It discusses the future of satellite astronomy as well.

Fimmel, Richard O., James A. Van Allen, and Eric Burgess. *Pioneer: First to Jupiter, Saturn, and Beyond.* NASA SP-446. Washington, D.C.: Government Printing Office, 1980. This overview of the Pioneers' missions and data includes a brief introduction to telescopic study of the planets.

Fischer, Daniel, and Hilmar W. Duerbeck. *Hubble Revisited: New Images from the Discovery Machine.* New York: Copernicus Books, 1998. This book concentrates on the discoveries of the Hubble Telescope from 1992 through 1997. Containing over 140 spectacular images, the text explores a wide range of astronomical topics—from the births and deaths of stars to quasars and black holes—and includes self-contained portraits of astronomers, as well as explanations of astronomical topics and instruments.

Kopal, Zdenek. *Telescopes in Space.* London: Faber and Faber, 1968. This readable book covers the literature of the telescope. Illustrated.

Kuiper, Gerard P., and Barbara M. Middlehurst, eds. *Stars and Stellar Systems.* Vol. 1 in *Telescopes.* Chicago: University of Chicago Press, 1960. This well-illustrated volume describes optical and radio telescopes and their accessories. There is a particular focus on the Hale telescope and the Lick Observatory.

Leverington, David. *New Cosmic Horizons: Space Astronomy from the V2 to the Hubble Space Telescope.* New York: Cambridge University Press, 2001. This is a broad treatise exploring the development of space-based astronomical observations from the end of World War II to the Hubble Space Telescope and other major NASA space-based observatories.

Moore, Patrick, ed. *Astronomical Telescopes and Observatories for Amateurs.* New York: W. W. Norton, 1973. This practical book for amateur astronomers contains detailed descriptions of many telescopes, along with illustrations.

Page, Thornton, and Lou Williams Page. *Sky and Telescope.* New York: Macmillan, 1965-1966. Volume 1 describes various celestial bodies and defines key astronomical concepts. Volume 4, illustrated with more than 120 photographs, drawings, and diagrams, describes in detail the principles of telescope design and methods of fabrication. It also discusses some of the world's famous observatories and telescopes.

Petersen, Carolyn Collins, and John C. Brandt. *Hubble Vision: Further Adventures with the Hubble Space Telescope.* 2d ed. New York: Cambridge University Press, 1998. This picture book of astronomy is both a classroom textbook and an excellent reference. It includes illustrations of exploding stars and colliding galaxies, gravitational lenses, the impact of Comet Shoemaker-Levy on Jupiter, and pictures of other solar systems. It describes celestial objects and the instruments Hubble uses to capture images of them.

Wilkie, Tom, and Mark Rosselli. *Visions of Heaven: The Mysteries of the Universe Revealed by the Hubble Space Telescope.* London: Hodder & Stoughton, 1999. This is an excellent, well-written look at some of the most incredible images taken by the Hubble Space Telescope. The images, most of which are in color, are sharp and clear and cover a wide range of celestial objects. This book presents these remarkable pictures with a concise narrative that reveals the thrilling and moving stories behind them.

Raj Rani, updated by David G. Fisher

Tethered Satellite System

Date: Beginning 1966
Type of program: Scientific platform

During the early 1990's, NASA conducted a series of experiments involving tethered satellite systems. Thin cables ranging in length from 20 to 100 kilometers were deployed between a main satellite and a subsatellite to determine if it would be feasible to use tethered satellite systems as generators to provide electrical power to the space shuttle or to recharge failing satellite batteries. Scientists also suggested using tethered satellites to conduct upper atmosphere research.

Summary of the System

Tethered satellite systems involve using a retractable cable, or tether, to connect satellites orbiting the Earth. Tethered satellite systems have been proposed for use in generating electrical power, for deploying instruments for upper atmosphere research, and even to serve as elevators for transferring materials from a planet's surface to a space station. For most intended applications, the simplest way to visualize a tethered satellite system is to picture someone walking a pet. In the case of a tethered satellite system intended for upper atmosphere research, for example, the space shuttle or space station would be analogous to a pet owner circling the neighborhood with a dog on a leash. Walking steadily down the middle of the sidewalk the human maintains a constant distance from neighbors' homes while the dog on the leash explores their yards, with the owner shortening or lengthening the leash, or tether, depending on circumstances. Similarly, during a space mission devoted to upper atmosphere research, the space shuttle would maintain an orbit that was a constant distance from Earth while using a tether to deploy a small subsatellite. This probe's instruments would go down to regions of the upper atmosphere that the shuttle could not safely descend.

Proposals for various types of space elevators and tether systems date back to the late nineteenth century. Scientists from many nations suggested ideas that could exploit the gravity gradient, making use of the fact that the farther an object is from the Earth the less it apparently weighs, but they were frustrated by limitations in available materials and technology. For example, one of the pioneers of Russian rocket science, Konstantin Tsiolkovsky, proposed a space railroad in the 1890's. Tsiolkovsky theorized that immensely tall towers could be built that would serve as elevators to lift freight into space. A height of 36,000 kilometers would be required for the towers to reach the point where objects attain weightlessness. Tsiolkovsky envisioned eventually circling the Earth with a cosmic railway, but his ideas remained unrealized as even much shorter towers were beyond the capabilities of nineteenth century engineering. Almost one hundred years later, engineers revived the concept of a space railroad, although in a different form from that proposed by Tsiolkovsky.

As the space program progressed during the 1960's and 1970's, scientists and engineers began to discuss approaches to space elevators and space railways that would take advantage of natural forces such as gravity and the Earth's magnetic field. Launching spacecraft into orbit or deep space using rockets requires tremendous amounts of energy to lift the spacecraft beyond the attraction of the Earth's gravity. Using a tethered satellite system to serve as a skyhook to raise freight or passengers from the Earth's surface could be both more energy efficient and dependable than conventional rocket boosters could. Ideas for possible space elevators ranged from relatively simple schemes to elaborate, globe-circling systems. Rather than building a tower from the ground, engineers suggested lowering a cable from an orbiting satellite or space station. While the simplest tethered satellite systems proposed, such as that advocated by Soviet scientist Yuri N. Artsutanov in the 1950's, generally involved placing a single satellite in a geostationary orbit above a fixed point on the Earth's surface, other plans were much more elaborate. In the 1980's a British scientist, Paul Burch, and an American, Jerome Pearson, both proposed using electromagnetic forces rather than rigid towers to hold a cosmic railway in place. Burch suggested launching a series of electromagnetic coils into orbit and then connecting them to form a tube circling the globe. By using a connecting wire that conducted electricity, a magnetic field would be created that would accelerate the wire to speeds above orbital velocity. This in turn would create a force pushing the tube away from the Earth's surface. Such a railway could be placed in a relatively low orbit for use as an intermediate platform to launch research satellites into higher orbits or spacecraft into outer space. Materials would be lifted from the Earth's surface to the railway using tethers.

The advantage to the ring suggestion was that it could be placed in a fairly low orbit, perhaps only 100 kilometers or so above the Earth's surface. The system employing a geostationary satellite would require that satellite be placed in a much higher orbit, 36,000 kilometers, in order for the satellite to remain in position above one spot on the Earth's surface. Many of the proposals for a geostationary tethered satellite system also involved a series of electromagnetic rings. Rather than circling the globe, the rings would be placed to form a tube leading straight out from the Earth, creating an elevator to the satellite. By the mid-twentieth century, advances in knowledge and materials since Tsiolkovsky's time had made such elevators to the

stars seem more technologically feasible, but the incredibly high costs, the length of time needed for completion, and numerous safety and environmental considerations made it unlikely that such a project would ever be completed. Pearson estimated it would take thirty to forty years to build the globe-circling ring Burch had proposed.

In addition, the possibility for catastrophic damage on a global scale caused even the most fervent advocates of such tethered systems to admit that such systems would be best developed for use on the Moon or other planets, such as Mars. Several hundred kilometers of even micro-thin polymer cable falling from the sky could devastate a wide area around the ground station for a skyhook system. The 1974 "Skyhook report," coauthored by two Italian scientists, Giuseppe Columbo and Mario D. Grossi, suggested instead that an excellent application for a tethered satellite system would be to deliver and pick up cargo to and from the Moon's surface. Such a Skyhook system would eliminate the need for spacecraft to actually land on the Moon, theoretically reducing both the costs and the risks involved in lunar exploration and development.

Intriguing though these ideas for space elevators were, research in space with tethered satellites proved more feasible than exploring the elaborate schemes for Earth-anchored systems. Although early experiments involving spacecraft and short tethers revealed numerous technical obstacles to be overcome, engineers and scientists still believed tethered satellite systems could serve many different purposes in space. For example, engineers believed that applications might include using tethers to connect two spacecraft that would then revolve around each other to create artificial gravity.

Beginning in the 1960's both Soviet and American spaceflights involved experimentation with tethers. In 1966, during two piloted missions, Gemini XI and Gemini XII, NASA used a 30-meter-long Dacron strap to connect the Gemini spacecraft with the upper stage of an Agena rocket as part of research into docking methods in space. Astronauts reportedly could not get the tether to remain

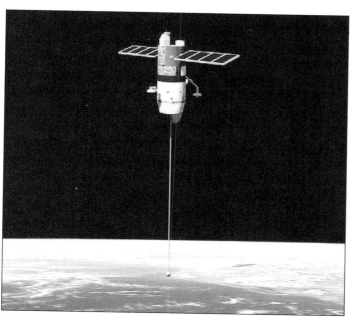

An artist's concept of an orbiting vehicle using an electrodynamic tether propulsion system. (NASA)

Small, expendable subsatellites could be connected to larger satellites or the space shuttle and lowered hundreds of kilometers into the upper atmosphere to record scientific measurements unattainable by other methods. These small subsatellites could either be retrieved by reeling them in or cut loose to burn up harmlessly as their orbit decayed on reentry. Such a tethered satellite system could provide data for a variety of scientists. Meteorologists might devise instrumentation for use in weather forecasting, while aerospace engineers have suggested using tethered satellites to obtain data for use in developing a space plane, that is, a spacecraft that could leave from the Earth's surface and fly into space rather than being lifted by external rockets as in the case of the space shuttle.

Another proposed use for tethered satellite systems was that of generating electricity. Electricity is created when a conductor, such as a copper wire, is passed through a magnetic field. Dragging a satellite connected to the shuttle with a conducting wire many kilometers long through the Earth's magnetic field theoretically would have the same results that passing a wire through a magnetic field on the Earth's surface would. Engineers suggested that if a subsatellite were connected to a main satellite via a tether composed of a conducting material, the system would generate electricity that could be used to power a satellite or to recharge existing batteries. NASA conducted a number of experiments to test this theory in 1992 and 1993.

The first experiments involving NASA's Tethered Satellite System (TSS) took place aboard the space shuttle *Atlantis* during STS-46 space shuttle mission as part of a joint research effort with the Italian Space Agency. Using a deployer system and satellite developed by Martin Marietta for the Italians, astronauts attempted to deploy the satellite upward away from the *Atlantis*. The protocol for the experiment called for deploying the satellite on a

taut as both the Gemini and the rocket upper stage tumbled erratically.

Other possible applications of tethered satellite systems include using tethers to connect storage tanks for fuel or other volatile substances a safe distance from a space station in orbit. Dangerous materials could be stored in canisters deployed well away from the space station with the canisters being reeled in to the station when needed. Fuel could be stored to be used by spacecraft that had launched from Earth and then needed to take on additional fuel to leave Earth orbit as part of research missions to other planets or to the Moon. The amount of fuel needed to boost a satellite or spacecraft out of orbit at 36,000 kilometers is extremely small compared to the amount of fuel required to lift it from the Earth's surface to that orbit to begin with. By using tethered satellites as storage, a space station could serve as a staging area for final assembly for both piloted and robotic explorations of the solar system.

Tethered satellite systems could also be utilized for conducting research in the upper atmosphere.

tether extending 20 kilometers from the shuttle and conducting 5,000 volts to the orbiter. When the astronauts attempted to deploy the satellite, however, they were unable to do so. The tether snagged shortly after the initial deployment and never extended farther than 256 meters from the shuttle and conducted only 50 volts. The *Atlantis* crew succeeded in unsnagging the tether, but ground controllers canceled any further attempts at deployment. The satellite was successfully retrieved for re-use.

Following the unsuccessful attempts of the *Atlantis* crew to deploy a tethered satellite in 1993, NASA launched two robotic tether experiments. The first employed a Small Expendable Deployer System (SEDS) and was designed to test deployer technology. The second, the plasma motor generator experiment, was meant to investigate the behavior of electrodynamic tethers and hollow-cathode plasma contactors in space. These missions, as well as the second SEDS mission the following year, were flown as secondary payloads on Air Force Global Positioning System missions.

The Air Force launched the Delta II ELV carrying SEDS-1 on March 30, 1993, as part of the Delta II GPS-31 mission. SEDS-1 was intended primarily to test the components of the deployment system such as the reel and the brake. SEDS-1 consisted of a deployer housing a lightweight spinning reel, a brake, tether-turns counter, tether cutter, and electronics and an end mass payload connected to the deployer with a 20-kilometer-long tether. The tether, which was 0.75 millimeter in diameter, was made from an eight-strand braid of a high strength polyethylene polymer, Spectra 1000. The end mass payload carried three primary science sensors: a three axis accelerometer, a three axis tensiometer, and a three axis magnetometer. The end mass payload was designed to function independently of

A space shuttle Atlantis *(STS-46) photo of the Tethered Satellite System (TSS-1) deployment. These satellites are used as observation platforms outside the orbiter.* (NASA)

the deployer and so carried its own battery and electronics.

After achieving orbital velocity, the SEDS-1 deployer successfully completed the downward deployment of the end mass payload. A brake under the control of an onboard timer was applied to the free reeling deployment of the tether as it neared its end and brought the subsatellite to a smooth stop. Following the successful deployment, the deployer's tether cutter severed the tether. The end mass payload transmitted data for approximately twenty-two hours before burning upon re-entering the atmosphere. NASA considers SEDS-1 to have been the first successful 20-kilometer space tether experiment.

In June, 1993, NASA launched the plasma motor generator as a payload on the Air Force Navstar II-21 Delta II second stage. A subsatellite containing a hollow-cathode plasma contactor was connected to a hollow-cathode plasma contactor on the Delta with a 0.5-kilometer conductive tether. The two contactors provided a low-resistance path with the ionosphere for the tether current. The experiment was designed to test a possible system for use in grounding high power systems in space. Research has shown that interactions between high-voltage power systems, such as those on the space shuttle, and the ambient Earth orbit plasma, can result in arcing and sputtering, leading to severe structural damage. A tethered system using hollow-cathode plasma contactors could be used to prevent such damage as well as serving as a source of power, or to alter an orbit. When current on the tether flowed in one direction, the tether could actually serve as a propulsive mechanism to increase the orbital velocity. If the current were reversed, it would have the opposite effect. The tether would then serve as a brake, slowing the spacecraft and eventually causing it to move closer to the Earth's surface.

SEDS-2, launched almost a year later on March 10, 1994, as part of the Delta II GPS-6 mission, also tested a downward deployment of a tethered satellite system. This mission was designed to test long-term tether dynamics. After a successful deployment, the end mass payload remained connected to the deployer with a 20-kilometer tether consisting, like SEDS-1, of a 0.75-millimeter eight-strand braid of Spectra 1000 polymer. The SEDS-2 end mass payload transmitted data for approximately ten days before exhausting its batteries, although space debris or a micrometeorite severed the tether after only five days had elapsed.

NASA again flew the Tethered Satellite System on STS-75. The mission was launched on February 22, 1996, and lasted approximately sixteen days. The 20.6-kilometer tether was deployed from *Columbia*'s cargo bay on the third day of the flight. Approximately five hours into the deployment, at a length of 19.7 kilometers, the tether unexplainably snapped and separated rapidly from the shuttle. A rendezvous and recovery operation was planned, but insufficient quantities of propellant on board *Columbia* prevented it. The satellite reentered the

atmosphere on March 19, over the Atlantic Ocean near Africa.

Contributions

By the end of the 1990's attempts at implementing tethered satellite systems had met with mixed results. The upward deployment of a tethered satellite system from the space shuttle *Atlantis* had proved unsuccessful. A NASA investigation concluded that the deployment mechanism had been insufficiently tested prior to launch. Last-minute changes to a crucial bolt caused the tether to snag, a problem that by itself need not have been fatal to the mission. However, when thrusters that investigators concluded were too small failed to provide adequate motive power to the satellite, slack in the tether contributed to binding in the reel and deployment halted at 256 meters. The *Atlantis* mission had been designed to test the theory that it was possible to generate electricity to recharge satellite and space shuttle batteries by dragging a conducting cable through the Earth's magnetic field. Because the tether stopped far short of its intended length of 20 kilometers, the theory remained untested. Crew members were able to conduct tests of various parts of the system, including seeing the effect normal shuttle operations, such as releasing cooling water overboard, might have on the conducting nature of the shuttle's surroundings, but many of the big questions regarding conductivity in space remained unanswered.

The successful deployment of SEDS-1 and SEDS-2, however, indicated that many of the problems scientists and engineers had feared would occur with tethered satellite systems were solvable. SEDS demonstrated the feasibility of postdeployment stabilization of tether dynamics by passive means, deployment control, and the long-term dynamics of satellites in space when deployed on a tether of 20 kilometers in length. While some observers of the Tethered Satellite System experiments claimed the failure of the TSS to successfully deploy from the *Atlantis* raised questions about the validity of the theories underlying the experiment, most scientists and engineers agreed that the problems lay in the hardware employed. NASA researchers remained confident that a tethered satellite system could be developed for future use in the space program.

Context

The Small Expendable Deployer System and Tethered Satellite System experiments were important to the overall development of the United States space program for several reasons. As NASA moved forward with plans for the construction of a permanent piloted space station, numerous technical hurdles remained. Tethered satellites held the potential to serve as research tools, as storage units for hazardous materials and for fuel, for generating electricity, and for use as microgravity laboratories. Connecting a subsatellite to the space station with a retractable tether could allow scientists to take advantage of the gravity gradient in conducting research. A subsatellite lowered even a few kilometers closer to the Earth than the space station itself would possess a different gravity. By moving along the tether, scientists could actually measure the differences in gravity at various points and perhaps answer some fundamental questions about the nature of gravity and its effects on both living and inanimate objects.

The possibility of using tethered satellite systems as an energy source also held great potential for the space program. In the 1990's, all satellites and spacecraft that were in long-term orbits around the Earth or engaged in deep-space exploration relied on batteries, photovoltaic cells, or a combination of the two for power. Despite advances in battery technology, the power source for a satellite still contributed considerable mass to any payload. The successful development of tethered satellite systems for recharging satellite or spacecraft batteries could reduce the weight of a satellite's power plant, thus allowing engineers to incorporate more scientific instrumentation into a satellite or to use a smaller rocket for launching the satellite into space. Further, photovoltaic panels were vulnerable to damage in space from space debris and micrometeors, which reduced their effi-

ciency. A tethered satellite system with a thin conducting cable might withstand damage that photovoltaic panels could not.

Before tethered satellite systems could be incorporated into planning for a space station, however, engineers and scientists needed to know if those systems would behave as predicted by scientific theories. If it proved impossible to maintain tautness between a main satellite and a subsatellite, for example, scientists would know that all components of a research facility in space would have to be connected using only rigid materials. Thus, experiments like SEDS-1 and SEDS-2 that tested basic components of a tethered satellite system, such as the deployer mechanism, were a vital step in advancing the goals of the United States space program as a whole. The knowledge gained regarding postdeployment stabilization meant that further research into tethered satellite systems as part of the space station development was worthwhile.

In 2003, NASA announced plans to study Momentum-eXchange/Electrodynamic Reboost (MXER) tether technology. If implemented, MXER would station kilometers of cart-wheeling cable in orbit around the Earth. Rotating like a giant sling, the cable would swoop down and pick up spacecraft in low orbits and then hurl them to higher orbits or even lob them onward to other planets. The hope is to harness momentum while dramatically lowering the cost of launching space missions. The study is managed in the Office of Space Sciences at NASA Headquarters. The MXER tether system would serve as a fully reusable transportation hub in orbit.

Tethers Unlimited, Inc. (TUI), was awarded funds to look into a MXER tether system based on deployment of a 100-kilometer-long cable in orbit around the Earth. TUI is collaborating with the Air Force Research Laboratory's Materials Laboratory to create a failsafe, multiline tether structure.

A NASA-funded study couples Tethers Unlimited with students and faculty at Stanford University. The project entails deployment of three tiny spacecraft along a lengthy tether. The Multi-

Application Survivable Tether (MAST) Experiment would show off numerous key technologies, such as tether designs that can survive in space and momentum-exchange propulsion.

See also: Gemini XI and XII; Marshall Space Flight Center; Space Shuttle; Space Shuttle Flights, 1992; Space Shuttle Flights, 1996; Telescopes: Air and Space.

Further Reading

Anderson, Loren A., and Michael H. Haddock. "Tethered Elevator Design for Space Station." *Journal of Spacecraft and Rockets* 29 (March/April, 1992): 233-238. Good discussion of the theories behind tethered systems as well as the problems engineers must consider in designing tether systems. Contains some technical jargon, but is accessible to the general reader.

Asker, James R. "*Atlantis* to Evaluate Characteristics of Tethered Satellite." *Aviation Week and Space Technology,* July 20, 1992, 40-44. A specific description of a proposed tethered satellite experiment that includes a good, easily understandable explanation of both the history and the potential of tethered satellite systems.

Baracat, William A. *Tethers in Space Handbook.* Washington, D.C.: National Aeronautics and Space Administration, Office of Space Flight, 1986. Description of tethered satellite proposals and experiments.

Beletskii, Vladimir V. *Dynamics of Space Tether Systems.* San Diego: Univelt, 1993. Discussion of theoretical considerations of tethered satellite systems. The advanced mathematical concepts may prove difficult for some readers.

Glaese, John R. *Tethered Satellite System (TSS) Dynamics Assessments and Analyses, TSS-1R Post Flight Data Evaluation Final Report.* Washington, D.C.: National Aeronautics and Space Administration, 1996. This is a comprehensive, technical report of the Tethered Satellite System and the tests aboard STS-46 and STS-75. It describes the experiments and the difficulties each mission had with the deployment of the satellite. It also gives the cause of the satellite's loss on STS-75.

Pearson, Jerome. "Ride an Elevator into Space." *New Scientist* 14 (January, 1989): 58-61. Excellent summary of the evolution of the concept of tethered satellite systems as well as the author's suggestions for future research. Includes illustrations.

Tethered Satellite System (TSS) Research on Orbital Plasma Electrodynamics (ROPE) Web page. http://science.nasa.gov/ssl/pad/ sppb/Tether. NASA Historical information about the two tethered satellite missions flown on the shuttle. Accessed April, 2005.

Nancy Farm Mannikko, updated by Russell R. Tobias

TIROS Meteorological Satellites

Date: April 6, 1960, to September 17, 1986
Type of satellite: Meteorological satellites

A TIROS satellite provided the first television picture from space on April 1, 1960. Thereafter, the satellites in this series provided high-altitude views that have increased meteorologists' capability to forecast weather.

Key Figures

William G. Stroud, TIROS project manager
Morris Tepper (b. 1916), TIROS program manager

Summary of the Satellites

The first Television Infrared Observations Satellite, known as TIROS 1, was the earliest of a series of weather satellites designed to provide environmental data for the 80 percent of the globe that was not covered by conventional means. The purpose of these satellites is to measure temperature and humidity in Earth's atmosphere, the planet's surface temperature, cloud cover, water-ice boundaries, and changes in the flow of protons and electrons near Earth. TIROS satellites have the capability of receiving, processing, and transmitting data from balloons, buoys, and remote automatic stations.

The TIROS series was the first of four generations of polar-orbiting weather satellites launched and operated by the United States. The series included ten successful launches from Cape Canaveral between April, 1960, and July, 1965. The first eight spacecraft—eighteen-sided polygons 106 centimeters in diameter, 56 centimeters high, covered by solar cells, and approximately 12 kilograms in weight—were in orbits inclined 48° and 58° to the equator. With TIROS 9, engineers made the first attempt to place a satellite in polar orbit from Cape Canaveral, using a series of dogleg maneuvers over twenty orbits to reach that orbit. Similar maneuvers were used to achieve polar orbit with TIROS 10.

Initially, the TIROS spacecraft were equipped with two miniature television cameras, a tape re-corder for each camera, two timer systems for programming future camera operations as set by a command from ground stations, and sensing devices for determining spacecraft attitude, environment, and equipment operations. As the program moved ahead, the satellites had progressively longer operational times and carried infrared measuring instruments (radiometers) to study the amount of radiation received and reradiated from Earth.

TIROS 8 had the first automatic picture transmission (APT) equipment, allowing pictures to be sent back to Earth immediately after they had been taken instead of being stored in tape recorders for later transmission. TIROS 9 and 10 were improved configurations; they led to the second generation of TIROS satellites, called ESSA after the government agency that financed and operated them, the Environmental Science Services Administration, later known as the National Oceanic and Atmospheric Administration (NOOA).

TIROS 1 through 10 enjoyed 6,630 useful days and provided 649,077 pictures. TIROS 1 proved television operation in space to be feasible. It had eighty-nine days of useful life and transmitted 22,952 pictures. TIROS 2 studied ice floes and had 376 useful days, providing 36,156 pictures. TIROS 3 made the first hurricane observation and

1734

provided the first advance storm-warning. It had 230 days of useful life and sent back 35,033 pictures. TIROS 4 permitted the first international use of its weather data. It operated for 161 days and transmitted 32,593 pictures. TIROS 5 extended the coverage of weather satellites, operated for 321 days, and sent 58,226 pictures. TIROS 6 provided support for the Mercury 8 and Mercury 9 crewed space missions, operated for 389 days, and provided 68,557 pictures.

TIROS 7 provided weather data to the International Indian Ocean Expedition, operated for 1,809 days, and sent back 125,331 pictures. TIROS 8 inaugurated the APT direct readout system, had a useful life of 1,287 days, and sent 102,463 pictures.

TIROS 9 and 10, with their improved configurations and near-polar orbits, improved global coverage. They had useful lives of 1,238 and 730 days, respectively, and sent back 88,892 and 78,874 pictures.

The ESSA satellites for the first time provided daily, worldwide observations without interruption in data. There were nine ESSA satellites. ESSA 1 was launched in February, 1966, operated for 861 days, and sent back 111,144 pictures. The average useful life of each of the nine ESSA satellites was nearly three years. Transmitting more than a quarter of a million television pictures, ESSA 8 lasted the longest, from December, 1968, until March, 1976. Altogether, ESSA 1 through 9 had 10,494 useful days and returned 1,006,140 television pictures.

In the 1970's, the third generation of meteorological satellites, known as the Improved TIROS Operational Satellite (ITOS), was developed. ITOS-1, launched in January, 1970, was the first of the TIROS system satellites equipped with television cameras for daytime coverage of the sunlit portion of Earth and infrared sensors sensitive to temperatures of the land, sea, and cloud tops for both daytime and nighttime coverage. A single ITOS spacecraft fur-

nished global observation of Earth's cloud cover every twelve hours, as compared to every twenty-four hours with two of the ESSA satellites.

A second ITOS satellite was launched successfully in December, 1970. Following a launch failure in October, 1971, four ITOS satellites were launched successfully, the last one being ITOS-H in July, 1976. When the Environmental Science Services Administration became the National Oceanic and Atmospheric Administration, the satellites were renamed: ITOS-A became NOAA-1; ITOS-D, NOAA-2; ITOS-F, NOAA-3; ITOS-G, NOAA-4; and ITOS-H, NOAA-5. The ITOS (NOAA) satellites were launched from the Western Test Range at Vandenberg Air Force Base in California.

The TIROS-N/NOAA series is the fourth generation of the meteorological satellites built by RCA. These carry a wide variety of new and considerably more advanced sensors. Besides providing weather imagery data, these satellites transmit atmospheric

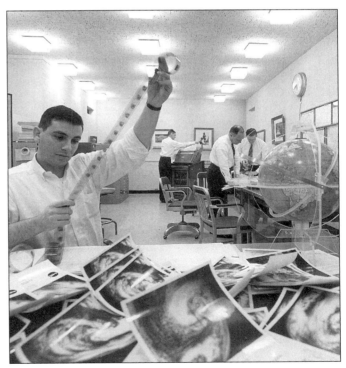

Engineers at NASA's Wallops Station in Virginia review weather photos received from the TIROS satellite. (NASA)

and sea surface temperatures, water vapor soundings, measurements of particle activity surrounding Earth, and data gathered from balloon-borne, ocean-based, and land-based weather-sensing platforms.

Sensor systems on the TIROS-N satellites include the TIROS Operational Vertical Sounder (TOVS), which combines data from three complementary sounding units to provide improved temperature and moisture data from Earth's surface up through the stratosphere; the Advanced Very High-Resolution Radiometer (AVHRR), which gathers and stores visible and infrared measurements and images, permitting more precise evaluation of land, surface water, and ice as well as information on cloud conditions and sea surface temperatures; the Space Environment Monitor (SEM), which allows the measurement of energetic particles emitted by the Sun over essentially the full range of energies and magnetic field variations in Earth's near-space environment (these measurements are tremendously helpful in determining solar radiation activities); and ARGOS, which permits the collection and transmission of environmental data from platforms on land, at sea, and in the air (ARGOS also determines the geographic location of platforms in motion on the sea surface or in the air).

The first satellite in the TIROS-N series was launched in October, 1978. It operated for twenty-eight months, twice its designed mission life. The second in the series, NOAA-6, was launched in June, 1979, followed by NOAA-7 in June, 1981. NOAA-8 was launched in March, 1983, NOAA-9 in December, 1984, and NOAA-10 in September, 1986. All these launches were from the Western Test Range in California, using an Air Force Atlas launch vehicle.

These NOAA satellites operate in a near-polar, circular orbit at an altitude of 870 kilometers. In their operational configuration, two satellites are positioned with a normal separation of 90°. One of the satellites operates in an afternoon ascending orbit, crossing the equator at 3:00 P.M. local solar time. The second operates in a morning descending or-

bit with an equator crossing at 7:30 A.M. local solar time. The satellites take 102 minutes to circle Earth.

These TIROS series satellites include instruments from the United Kingdom and France. The United Kingdom, through its Ministry of Defense Meteorological Offices, provides the stratospheric sounding unit, one of the three atmospheric sounding instruments on each satellite. The soundings provide three-dimensional observations of the atmosphere from the surface to approximately 65.5 kilometers. The two TIROS-N satellites provide roughly 16,000 timely soundings a day. The Centre National d'Études Spatiales of France supplies the data-collection and location system for the satellite (ARGOS) as well as the ground facilities to process data and make the information available to users. The French operate a receiving station for weather data at Lannion, on the English Channel, about 80 kilometers northeast of Brest.

The TIROS-N/NOAA satellites have been designed with sufficient space, power, and data-handling capability to allow additional instruments to be installed—including a search-and-rescue system made up of repeater and processor units, a solar backscatter ultraviolet radiometer for ozone mapping, and two Earth Radiation Budget Experiment instruments. NOAA-8, NOAA-9, and NOAA-10 have these instruments and are known as Advanced TIROS-N (ATN) satellites.

The search-and-rescue equipment installed on ATN enables detection of emergency transmissions from downed aircraft or ships in distress. The devices support an international search-and-rescue effort in which the principal partners are Canada, France, the former Soviet Union, and the United States. From its beginning in September, 1982, until July, 1988, more than eleven hundred lives had been saved as a result of the program. NOAA, the National Aeronautics and Space Administration (NASA), the U.S. Air Force, and the U.S. Coast Guard are the principal U.S. agencies supporting this international program.

For more than a quarter-century, these meteorological satellites have shown improvement in quality, quantity, and reliability. Since 1966, the entire Earth has been photographed at least once daily on a continuous basis. These data are considered almost indispensable for short-range weather forecasts by meteorologists and environmental scientists.

Contributions

TIROS 1 demonstrated the practicality of meteorological observations from space and the ability to make observations with television. Observations from TIROS 1 confirmed and, in some cases, shed new light on the large-scale organization of clouds, showing that they were arranged in a highly organized pattern. The cloud vortex stood out dramatically. Meteorologists learned that these cloud vortices could be used to pinpoint accurately atmospheric storm regions. Of the nearly twenty-three thousand cloud-cover pictures transmitted by TIROS 1, more than nineteen thousand were usable for weather analysis. This opened a new era in weather observation by providing data that covered vast areas of Earth that previously had not been available to meteorologists.

Some storms were tracked for as long as four days, underscoring the value of television satellite pictures for weather forecasting. In one case, TIROS 1 observations of a storm near Bermuda in May, 1960, provided a series of pictures recording the degeneration of the storm. The weather maps drawn from conventional data were correlated with the pictures, affording a unique set of data on which to judge future pictures.

TIROS satellite data were used to establish, modify, and improve cloud analysis and to brief pilots on weather. Pictures of cloud cover, transmitted by teletype, were used to refine weather analysis in the region from Australia to Antarctica, and the Australian Meteorological Service used the data in conjunction with special research projects on storms over Australia.

Pictures of ice-pack conditions in the Gulf of Saint Lawrence proved that weather satellites could locate ice boundaries in relation to open seas. TIROS 2 observations were used in selecting proper weather conditions for the suborbital flight of Alan B. Shepard in May, 1961—America's first crewed spaceflight.

The TIROS 3 satellite provided observations of all six of the major hurricanes in 1961. It detected Hurricane Esther two days before it was observed by conventional methods.

In 1963, three TIROS satellites (TIROS 5, 6, and 7) were in operation simultaneously and provided for the first time pictures of the cloud cover of various parts of the world on an almost continuous basis. These

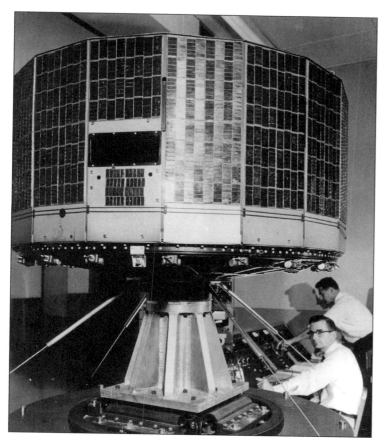

A TIROS satellite undergoing a vibration stress test in 1960. (NASA)

pictures formed the basis for adapting satellite data to forecasting methods and the use of numerical prediction models. Special projects surveying the Atlantic and Indian Oceans brought clues to the origin and development of tropical cyclones. TIROS satellites have detected sandstorms in Saudi Arabia and have provided storm advisories in many areas of the world.

As a result of their analyses of the satellite pictures, meteorologists have been able to categorize cloud vortex patterns and to determine from those patterns the structure of moisture fields, the cloud structure of frontal zones, and the relationship of cloud patterns associated with jet streams. A series of atlases has been developed and are being used with the satellite data by research meteorologists to investigate a wide variety of theoretical conclusions.

Context

Until the 1960's, when meteorological satellites made their debut, the science of weather prediction left much to be desired. Meteorologists could not forecast with any great degree of accuracy the weather four or five days ahead. Forecasts can now be made for a number of days ahead, most of the time with great accuracy. A number of technological factors have contributed to this improvement. One was the development of the computer, which permits the analysis of millions of bits of information at high speeds so that the physical laws governing atmospheric motion can be calculated more accurately. The development of new communications links, enabling the rapid dissemination of meteorological data over great distances, is another.

Another advancement, equally important in the eyes of many, was the development of meteorological satellites. The development and operation of TIROS and other meteorological satellites have opened the inaccessible areas of the world to observation.

Improved weather forecasting extends to many daily activities of society. For example, improved forecasts are of inestimable value to agriculture.

Better forecasting means better crop management and protection. A reliable forecast during the fruit-growing season in California or Florida could save fruit growers millions of dollars.

Improved forecasting offers benefits to transportation, to construction, to manufacturing, and to recreation. Better forecasts allow the routing of aircraft so that safety and comfort are enhanced. Thus, delays are avoided and money is saved as a result of improved fuel consumption. In construction, improved forecasting permits contractors to plan and schedule construction work as environmental conditions dictate. Improved forecasting also permits manufacturers to plan production, particularly when sales are sensitive to weather conditions.

Satellite pictures provide unique assistance to many. Images of the polar ice have helped vessels that otherwise would have been trapped find a way through the ice and have been used to assist ships that were caught in ocean storms. Weather services for Antarctica and for offshore oil-drilling platforms make it possible for activities in these locations to be undertaken in comparative safety. The infrared sensors can scan the ocean surface and determine the temperature, providing extremely useful information to the fishing industry, for example—temperatures can help locate schools of certain kinds of fish.

Meteorological support for humanity's daily activities on land, on sea, and in the air underscores the importance of continued measurements from space. Such measurements are being carried out on a global scale by all seafaring nations. Canada, France, Japan, India, Russia, the European Space Agency, the United States, and others all have active weather satellite programs.

See also: Environmental Science Services Administration Satellites; Global Atmospheric Research Program; Interplanetary Monitoring Platform Satellites; ITOS and NOAA Meteorological Satellites; Landsat 1, 2, and 3; Landsat 4 and 5; MESSENGER; Meteorological Satellites; Nimbus Meteorological Satellites; Seasat; SMS and GOES Meteorological Satellites.

Further Reading

Ahrens, C. Donald. *Essentials of Meteorology: An Invitation to the Atmosphere.* 4th ed. Pacific Grove, Calif.: Thomson Brooks/Cole, 2005. This is a text suitable for an introductory course in meteorology. Comes complete with a CD-ROM to help explain concepts and demonstrate the atmosphere's dynamic nature.

Bader, M. J., G. S. Forbes, and J. R. Grant, eds. *Images in Weather Forecasting: A Practical Guide for Interpreting Satellite and Radar Imagery.* New York: Cambridge University Press, 1997. The aim of this work is to present the meteorology student and operational forecaster with the current techniques for interpreting satellite and radar images of weather systems in mid-latitudes. The focus of the book is the large number of illustrations.

Burrows, William E. *This New Ocean: The Story of the First Space Age.* New York: Random House, 1998. This is a comprehensive history of the human conquest of space, covering everything from the earliest attempts at spaceflight through the voyages near the end of the twentieth century. Burrows is an experienced journalist who has reported for *The New York Times, The Washington Post,* and *The Wall Street Journal.* There are many photographs and an extensive source list. Interviewees in the book include Isaac Asimov, Alexei Leonov, Sally K. Ride, and James A. Van Allen.

Grieve, Tim, Finn Lied, and Erik Tandberg, eds. *The Impact of Space Science on Mankind.* New York: Plenum Press, 1976. Contains edited summaries of papers presented at the Nobel Symposium in Spatind, Norway, in September, 1975. In a paper presented by Robert M. White, administrator of the National Oceanic and Atmospheric Administration, the use of environmental satellites is described. The use of satellites and their benefits to humankind are outlined.

Heppenheimer, T. A. *Countdown: A History of Space Flight.* New York: John Wiley, 1997. A detailed historical narrative of the human conquest of space. Heppenheimer traces the development of piloted flight through the military rocketry programs of the era preceding World War II. Covers both the American and the Soviet attempts to place vehicles, spacecraft, and humans into the hostile environment of space. More than a dozen pages are devoted to bibliographic references.

Jakes, John. *TIROS: Weather Eye of Space.* New York: Julian Messner, 1966. A nontechnical review of the significance of TIROS and the benefits of the meteorological satellites. Reviews history of meteorological research as well as the development of the TIROS payload. A list of accomplishments is given, underscoring the benefits of satellite observations.

Lee, Wayne. *To Rise from Earth: An Easy to Understand Guide to Spaceflight.* New York: Checkmark Books, 1996. This is a good introduction to the science of spaceflight. Although written by an engineer with the NASA Jet Propulsion Laboratory, it is presented in easy-to-understand language. In addition to the theory of spaceflight, it gives some of the history of the human endeavor to explore space.

Mari, Christopher, ed. *Space Exploration.* New York: H. W. Wilson, 1999. Twenty-five articles (reprinted from magazines), covering the state of the space program at the time of publication, are divided into five sections: John H. Glenn, Jr.'s return to space, the exploration of Mars, the International Space Station, recent mining efforts by commercial industries, and new types of space vehicles and propulsion systems.

National Aeronautics and Space Administration. *Significant Achievements in Satellite Meteorology, 1958-1964*. NASA SP-96. Washington, D.C.: Government Printing Office, 1966. This volume is one in a series that summarizes the progress made by NASA's Space Science and Applications Program from 1958 through 1964. It reviews NASA's meteorological program and its benefits. Describes TIROS, Nimbus, and the meteorological sounding-rocket program.

Parkinson, Claire L. *Earth from Above: Using Color-Coded Satellite Images to Examine the Global Environment*. Sausalito, Calif.: University Science Books, 1997. A book for nonspecialists on reading and interpreting satellite images. Explains how satellite data provide information about the atmosphere, the Antarctic ozone hole, and atmospheric temperature effects. The book includes maps, photographs, and fifty color satellite images.

Widger, W. K., Jr. *Meteorological Satellites*. New York: Holt, Rinehart and Winston, 1965. As a member of the team that developed TIROS, the author provides an interesting overview of the TIROS program itself and a definitive account of the value of satellite information, the equipment that provides the satellite weather data, and early results of the scientific investigations obtained from TIROS.

Zimmerman, Robert. *The Chronological Encyclopedia of Discoveries in Space*. Westport, Conn.: Oryx Press, 2000. Provides a complete chronological history of all crewed and robotic spacecraft and explains flight events and scientific results. Suitable for all levels of research.

James C. Elliott

Titan Launch Vehicles

Date: Beginning December 20, 1957
Type of technology: Expendable launch vehicles

The Titan series of rockets provided the United States with a means of launching satellites into orbit. During its early years, the Titan served as an intercontinental ballistic missile for the United States Air Force.

Key Figures

Albert C. Hall, director of Technical Activities for Titan Intercontinental Ballistic Missile (ICBM)
Roger Chamberlain, manager of commercial Titan program

Summary of the Technology

During the years immediately following World War II, American officials had predicted that the Soviet Union would not have enough information to create an atomic bomb for at least another three years, and probably for five. It was, therefore, a great blow when in August, 1949, the Soviet Union detonated its first bomb. The Soviets again proved that they were an emerging superpower by performing intercontinental ballistic missile (ICBM) tests in 1957.

Although ICBMs were among the new weapons the United States decided to build in the postwar years, several factors prevented their production before 1954. The major cause can be traced to the lack of conviction on the part of many influential scientists that a rocket could possess enough accuracy, let alone enough strength, to make the system feasible. In addition, such an undertaking would have required enormous financial resources and skilled personnel, neither of which was available to the military until it had become clear that the United States was lagging behind the Soviets.

As a Cold War response to Soviet ICBMs, Americans designed several rockets. In this field the United States had a great advantage, for during the closing stages of World War II, many German rocket experts had surrendered to the Americans

rather than face capture by the Soviets; these researchers became American citizens, willing to help their new country advance its weaponry.

The first American ICBM was the Atlas. It stood 24.4 meters tall and had two stages. During the early part of the program, both stages were ignited at liftoff to avoid having the rocket fail once the first stage had exhausted its fuel. The special Air Force Strategic Missiles Evaluation Committee and the Research and Development (RAND) Corporation published reports in 1954 and 1955 in which they cited two reasons that Atlas should be replaced: the necessity of igniting both boosters at the same time and the unsuitable nature of a thin pressurized skin for high-acceleration liftoff.

President Dwight D. Eisenhower ordered the development of a new ICBM, to be called Titan, in mid-1955. The contract was awarded to the Martin Company (later the Martin Marietta Corporation). It was to have a range of approximately 11,000 kilometers. The total thrust produced at liftoff was to be 1,343 kilonewtons, while the weight of the rocket (including fuel) was to be 99,790 kilograms. Its height was to be 29.87 meters, its maximum diameter 3.048 meters.

Titan was to have two stages that would ignite separately, using a propellant of kerosene and liquid oxygen. Two huge tanks were located inside the first stage. The bottom tank housed the kerosene while the top housed the liquid oxygen. Two pumps pushed the liquids to the combustion chamber, where they were ignited. Once the fuel was consumed, the separation rocket fired and the first stage was freed. The second stage was much like the first, except that it contained, in addition to the main engine, four smaller rockets to guide it. Above the second stage were the nuclear warhead and the guidance systems.

After liftoff, Titan rose vertically for twenty seconds. It then arched into a curved trajectory for one hundred seconds, to a speed of 8,530 kilometers per hour. Its fuel exhausted, the first stage separated. The second stage ignited and propelled the rocket to a speed of up to 28,970 kilometers per hour. Four vernier engines (stabilizers) corrected velocity and trajectory. Once this had been accomplished, the second stage was released, and the nose cone fell toward Earth. As a missile, the Titan reached a height of 805 kilometers.

In the early 1960's, researchers realized that the Titan system had some flaws. For example, it was necessary to fuel the rockets immediately before the launch so that the liquid oxygen would not evaporate. Furthermore, the range of the Titan was considered insufficient. Martin Marietta and the National Aeronautics and Space Administration (NASA) were contracted to build a successor to Titan.

Titan II was completed in 1964. It stood 33.22 meters tall, with a diameter of 3.048 meters. It weighed 185,034 kilograms. For its fuel it used a combination of nitrogen tetroxide and Aerozine-50. The two stages together produced a total thrust of more than 2,223 kilonewtons.

When NASA built Gemini, a successor to the Mercury Capsule, the Titan was seen as the most suitable launch vehicle, with one exception: The Titan suffered from what was known as the "pogo effect" (named for the "pogo stick"), violent, up-and-down motions as the rocket rose. These oscillations mattered little when the Titan carried warheads; for human passengers, however, the pogo

effect would create unacceptable conditions. Designers were able to eliminate the worst of the pogo effect, and on April 8, 1964, using Titan II, the first Gemini spacecraft was launched, followed on March 25, 1965, by the first piloted launch.

The Gemini Program as a whole was considered a success. The Gemini spacecraft carried two astronauts, an improvement over the crowded, one-person Mercury Capsule. During this program several records were set. The first docking with another spacecraft occurred, and the first American walked in space.

From 1965 onward, Titan III replaced Titan II. Titan III, the modern-day heavy launch vehicle, can propel into orbit loads of 15,875 kilograms and handle as much as 3,175 kilograms for interplanetary missions. The core section of the rocket, sometimes known as Titan III-A, has two stages and is commonly flanked by boosters. The third stage, or transtage, contains the payload and the final fuel. Titan III had twelve configurations during its operational period. The most often used variation was the Titan III-C. It utilized the Titan III core vehicle—an advanced Titan II two-stage vehicle—and two strap-on solid-propellant boosters. It launched a variety of military payloads and several civilian satellites. Titan III was also intended to launch the Dyna-Soar, a military predecessor of the space shuttle.

The Titan's first stage is 22.25 meters tall and provides approximately 2,356 kilonewtons of thrust. The second stage measures slightly more than 7 meters and is capable of more than 445 kilonewtons of thrust. Both stages use nitrogen tetroxide and a mixture of unsymmetrical dimethylhydrazine (UDMH) and hydrazine. The transtage is of varying heights and can be as tall as 15.24 meters when launching Air Force payloads. It uses the same type of fuel as the first two stages. The two boosters are 26 meters long and use solid propellant.

The Titan III-B core has two stages and permits a variety of upper stages to be fitted, the most common being the Agena. It is not launched with boosters. Its cargoes have included Air Force reconnaissance satellites. The Titan III-B uses the standard inertial guidance system, which enables the satellite to maintain a predetermined course.

The Titan III-C, first launched in 1965, has a takeoff thrust of 10,490 kilonewtons. Its height is 47.85 meters, and it carries two boosters that are 26 meters in length with diameters of approximately 3 meters. At takeoff, the two boosters are the only source of thrust. The core is ignited just before the boosters are released. The fuel consists of nitrogen tetroxide and Aerozine-50. Approximately 80 percent of the satellites launched by the United States into geosynchronous equatorial orbit use a Titan III-C rocket. A geosynchronous orbit is one that keeps precise pace with the rotation of Earth, thereby maintaining a stable position relative to any given site.

The Manned Orbiting Laboratory (MOL), which was to have undertaken reconnaissance projects, was slated for launch by the Titan III-C. Growing costs and the high performance of a previously launched spy satellite resulted in the cancellation of the project; yet, the Titan III-C continued to launch signal intelligence satellites (SigInts).

Although the Titan III-D is similar to the III-C in many respects, it does not have a transtage. Instead of depending on inertial guidance, it depends on radio guidance. Titan III-D is the rocket used to launch the so-called Big Bird satellite.

In February, 1974, the Titan III-E-Centaur, a modified version of the Titan III-D, was first tested. Although the test ended in failure, the Titan III-E-Centaur performed flawlessly when it launched the Viking spacecraft to Mars in 1975.

The Titan III-4D, yet another model, lacks the traditional transtage and uses the Boeing Inertial Upper Stage instead (a stage originally developed for the space shuttle). Its first stage has been lengthened to 25.22 meters. Titan III-4D was completed February 28, 1981; however, the program suffered from a series of failures in the mid-1980's. On August 18, 1985, after being launched from Vandenberg Air Force Base, the engine shut down prematurely, and the rocket fell into the Pacific Ocean. Disaster struck again on April 18, 1986. Mere seconds after launch, at 91.44 meters above ground, a Titan exploded, destroying a classified satellite.

The approximate cost of using a Titan III-4D to launch a 12,700-

On September 12, 1966, Titan lofts Gemini toward a rendezvous with an Agena target vehicle. (NASA)

kilogram satellite into a polar orbit of 185 kilometers is approximately $60 million if six are purchased at the same time. To place a satellite of 1,815 kilograms into geosynchronous orbit costs $125 million. In the event that only one launcher per year is ordered, the price would be raised by $121 million and $125 million, respectively, per flight. The average cost of the satellites that use the Titan III-4D is more than $100 million.

In March, 1985, the U.S. Air Force awarded a $5 million contract to Martin Marietta to build a new and more powerful Titan rocket. The Titan IV is the largest robotic space booster used by the United States. The vehicle is designed to carry payloads equivalent to the size and weight of those carried on the space shuttle. The Titan IV consists of two solid-propellant motors, a liquid propellant two-stage core, and a 5-meter-diameter payload fairing. The system has three upper-stage configurations that include a cryogenic wide-body Centaur Upper Stage (CUS), but also may be flown with an Inertial Upper Stage (IUS), or No Upper Stage (NUS). Overall length of the system is 62 meters when flown with a 26-meter payload fairing. The Titan IV Centaur is capable of placing 4,500-kilogram payloads into geosynchronous orbit. The Titan IV system is also capable of placing 18,000 kilograms into a low-Earth orbit at 28.6° inclination or 14,000 kilograms into a low-Earth polar orbit. The Solid Rocket Motor Upgrade, used only on Titan IV-Bs, incorporates modern technology to provide increased performance and enhanced reliability. With SRMU, the Titan IV Centaur is capable of placing 5,700-kilogram payloads into geosynchronous orbit or 21,700 kilograms into a low-Earth orbit. SRMU production started in November, 1993.

Titan launch operations came to a close at Cape Canaveral when a Titan IV booster departed Complex 40 at 00:50 Coordinated Universal Time (UTC) on April 30, 2005. This booster delivered a National Reconnaissance Office classified payload to orbit and put on a spectacular show in the skies along the East Coast of the United States and Canada as it ended more than five decades of heavy-lifting launch services from Cape Canaveral. This

was the 168th Titan to launch. In total, there were forty-seven Titan I ICBMs, twenty-three Titan II ICBMs, twelve Titan II Gemini Program support boosters, four Titan III-As, thirty-six Titan III-Cs, seven Titan III-Es, eight Titan 34Ds, four Commercial Titans, and twenty-seven Titan IVs. One final Titan remained to be launched from the Vandenberg site in 2005 or 2006. Then the venerable Titan family would be retired to a rich history that served the national interest in wartime and peacetime, having dispatched both astronauts and highly ambitious robotic probes into space.

Contributions

The early Titan rocket served as a test vehicle. Various alternatives to the Atlas's thin casing were tested on the Titan, as were different fuels. Before a rocket is launched, thousands of modifications are made, ranging from the design of a better engine to the replacement of a first stage. The Atlas was designed so that both stages were ignited at once. When the Titan was built, a means of separating the first stage was devised. A system of igniting the second stage, after separation had occurred, was also designed.

In the process of choosing new propellants it was discovered that nitric acid and turpentine ignited upon contact. This fact proved to be very useful: Instead of having to reignite the fuel every time the rocket was started, it was possible, by mixing the fuel, to produce thrust once again. The combination of a fuel and an oxidizer that ignites spontaneously is called a hypergol.

For the Titan II, a mixture of UDMH, hydrazine, and nitrogen tetroxide replaced the old mixture of kerosene and liquid oxygen. Before it could be placed in the rocket, many tests were performed. First, the pumps, which were designed to work well with liquid oxygen, had to be replaced with pumps that functioned at higher temperatures. The guidance system, containing some flaws, was replaced by an inertial system. Now that it was no longer necessary to add propellant a few minutes before launch, many of the ICBM silos could be simplified.

With the introduction of the even more powerful Titan III, and the gradual phasing out of the Titan ICBMs in favor of more modern systems, Titan became primarily a launch vehicle. As a launch vehicle, the Titan was required to do much more than propel a warhead into space and guide it back to Earth. For example, during the Gemini Program the Titan had to boost the spacecraft into orbit, a job once likened to walking a mile with one's eyes shut and arriving no more than 10 centimeters from one's destination. The job was made simpler during later flights by the use of an onboard computer that could be used to plot the flight path.

On interplanetary missions, such as the Viking mission to Mars, the Titan required a greater accuracy than on any of its previous missions. The distance from Earth to Mars is about 55,520,000 kilometers. The Titan vehicle benefited directly from the information gained from previous space missions.

When the Titan reached the correct orbit with a satellite, a mechanism in the transtage allowed it to separate from the satellite. The malfunction of this mechanism resulted in the loss of many satellites. With the cost of spacecraft running into hundreds of millions of dollars, it proved an expensive mistake.

Context

Titan was the second American ICBM to be produced. The Atlas and Jupiter rockets had been built not out of necessity but by scientists who were interested in doing something new. Engineers soon became content with traditional V-2 rocket design. It was only when the need for ICBMs arose that progress was made.

The Titan provided engineers with a chance to put into use some new practices. Previously it was considered too risky to ignite a second stage after the first had been jettisoned. The use of alcohol and oxygen, though somewhat safer, was thought to provide less thrust than other fuels. Experimentation resulted in the discarding of Titan's first stage before the second stage was ignited, the use of oxygen and kerosene, and finally the mixture of UDMH, hydrazine, and nitrogen tetroxide as fuel.

During the Gemini Program, the Titan was found to suffer from the pogo effect. Previously, the condition, though recognized, was regarded as unfortunate but unpreventable. Using the Titan in the piloted space program prompted a rocket redesign. When the solution was found, it was applied to other rockets as well.

Until the 1980's, the Titan III rockets were used often in the U.S. space program. Yet with the deployment of the space shuttle, an increasing number of companies showed an interest in the more advanced reusable orbiter as a launcher. The U.S. Air Force, however, had regarded the shuttle as too experimental. Expressing its confidence in the basic Titan design, the Air Force awarded Martin Marietta a contract for the successor to the Titan III series. With the *Challenger* space shuttle accident in January, 1986, the Air Force's fears were realized, and the Titan was once again in demand for the launching of satellite payloads. After *Challenger*, all satellites that carry volatile fuel were banned from the shuttle. This ban created a new market for the Titan, and ultimately led to the Titan IV commercial launcher.

See also: Atlas Launch Vehicles; Cape Canaveral and the Kennedy Space Center; Delta Launch Vehicles; Gemini VII and VI-A; Launch Vehicles; Manned Orbiting Laboratory; Saturn Launch Vehicles; Soyuz Launch Vehicle; Space Shuttle Flights, 1992; Space Shuttle Mission STS-49; TIROS Meteorological Satellites; Vandenberg Air Force Base.

Further Reading

Binder, Otto O. *Victory in Space*. New York: Walker, 1962. One of the best books written in its period, *Victory in Space* answers many key questions that were being asked during the early part of the space program. The author does not set out to write about the Titan rocket, but it is mentioned in the context of other subjects.

Burrows, William E. *This New Ocean: The Story of the First Space Age.* New York: Random House, 1998. This is a comprehensive history of the human conquest of space, covering everything from the earliest attempts at spaceflight through the voyages near the end of the twentieth century. Burrows is an experienced journalist who has reported for *The New York Times, The Washington Post,* and *The Wall Street Journal.* There are many photographs and an extensive source list. Interviewees in the book include Isaac Asimov, Alexei Leonov, Sally K. Ride, and James A. Van Allen.

Emme, Eugene M., ed. *The History of Rocket Technology: Essays on Research, Development, and Utility.* Detroit: Wayne State University Press, 1964. This collection of essays provides a useful survey of the U.S. space program as a whole. Essays on the Atlas, Thor, Titan, and Minuteman ICBMs are followed by a concluding essay on the Titan III-C. Includes a bibliography.

Heppenheimer, T. A. *Countdown: A History of Space Flight.* New York: John Wiley, 1997. A detailed historical narrative of the human conquest of space. Heppenheimer traces the development of piloted flight through the military rocketry programs of the era preceding World War II. Covers both the American and the Soviet attempts to place vehicles, spacecraft, and humans into the hostile environment of space. More than a dozen pages are devoted to bibliographic references.

Isakowitz, Steven J., Joseph P. Hopkins, Jr., and Joshua B. Hopkins. *International Reference Guide to Space Launch Systems.* 3d ed. Reston, Va.: American Institute of Aeronautics and Astronautics, 1999. This best-selling reference has been updated to include the latest launch vehicles and engines. It is packed with illustrations and figures and offers a quick and easy data retrieval source for policymakers, planners, engineers, launch buyers, and students. New systems included are Angara, Beal's BA-2, Delta III and IV, H-IIA, VLS, LeoLink, Minotaur, Soyuz 2, Strela, Proton M, Atlas-3 and -5, Dnepr, Kistler's K-1, and Shtil, with details on Sea Launch using the Zenit vehicle.

King, Benjamin, and Timothy Kutta. *Impact: The History of Germany's V-Weapons in World War II.* New York: Sarpedon Press, 1998. The story of the V-1 and V-2 rocket-powered, explosive delivery systems is told from their design concepts through their production and use. It talks about their effectiveness as a weapon and the methods the Allies used to destroy them.

"Launch Vehicles." *Aviation Week and Space Technology,* January 17, 2000, 144-145. This table details the specifications for each of the current (2000) launch vehicles and spacecraft, as well as the current status.

Levine, Alan J. *The Missile and the Space Race.* Westport, Conn.: Praeger, 1994. This is a well-written look at the early days of missile development and space exploration. The book discusses the Soviet-American race to develop intercontinental ballistic missiles for defense purposes and their subsequent use as satellite launchers.

Ley, Willy. *Events in Space.* New York: David McKay, 1969. This book includes sections covering the beginning of satellites and rockets. Ley covers Titan, Titan II, and Titan III-C. It should be noted, however, that the primary topic is space exploration, not the Titan series.

Newlon, Clarke. *One Thousand and One Answers to Questions About Space.* New York: Dodd, Mead, 1962. Newlon answers some of the questions most frequently asked about the Titan ICBM and Titan II. Information is also given about Titan's contemporaries. Capsule biographies of important people in the field of rocketry are included.

Ordway, Frederick I., III, and Mitchell Sharpe. *The Rocket Team.* Burlington, Ont.: Apogee Books, 2003. A revised edition of the acclaimed thorough history of rocketry from early amateurs to present day rocket technology. Includes a disc containing videos and images of rocket programs.

Peterson, Robert A. *Space: From Gemini to the Moon and Beyond.* New York: Facts on File, 1972. This volume shows the involvement of the Titan launcher in the American space program. The Gemini Project, MOL, and various other payloads are discussed. Suitable for high school and college reading.

Sobel, Lester A., ed. *Space: From Sputnik to Gemini.* New York: Facts on File, 1965. Gives additional information on the use of the Titan rocket as a launching vehicle during the beginning of the space program. The book ends with the introduction of the Titan IIIC.

Taylor, John W. R., ed. *Jane's All the World's Aircraft, 1981-1982.* London: Jane's Publishing Company, 1981. Written for a general audience, this volume includes information on the later series of Titan rockets.

Zimmerman, Robert. *The Chronological Encyclopedia of Discoveries in Space.* Westport, Conn.: Oryx Press, 2000. Provides a complete chronological history of all piloted and robotic spacecraft and explains flight events and scientific results. Suitable for all levels of research.

John Newman

Tracking and Data-Relay Communications Satellites

Date: Beginning April 4, 1983
Type of spacecraft: Communications satellites

The United States developed and implemented a communications satellite system designed to improve tracking and data-relay capabilities from low-Earth-orbiting piloted and robotic spacecraft. The system was operational for more than a decade, and NASA began upgrading its aging satellites in 2000.

Key Figures

Dale W. Harris, project manager for the Tracking and Data-Relay Satellite System (TDRSS) at Goddard Space Flight Center
Charles M. "Chuck" Hunter, TDRSS deputy project manager

Summary of the Satellites

As piloted and robotic missions in both the U.S. and the Russian space programs increased in number and complexity, it became apparent that former tracking and data-relay systems were becoming obsolete. Based on expensive, complicated networks of ground stations, communications ships, and outmoded equipment, the old systems provided only minimal contact with piloted spacecraft. They also were unable to process and relay the enormous volumes of data generated by a new era of scientific applications satellites. Upgrading and expanding the old systems was economically unfeasible and inefficient for the task at hand.

In December of 1976, the National Aeronautics and Space Administration (NASA) contracted with the Space Communications Company (Spacecom) to develop and implement a tracking and data-relay system that would be able to transmit data efficiently and continuously from low-Earth-orbiting piloted and robotic spacecraft and maintain almost uninterrupted contact with the crews. Spacecom, a joint venture between Continental Telecom and Fairchild Industries, developed the Tracking and Data-Relay Satellite System (TDRSS). Its basic de-

sign consisted of two operational satellites, one in-orbit spare, and a ground station. The satellites would function as repeaters, neither processing nor altering any communications. The uplink channels would receive transmissions from the ground, amplify them, and transmit them to the spacecraft. The downlink channels would receive, amplify, and transmit signals from the craft to the ground station. When completely operational, the system could provide almost continuous contact with piloted space vehicles such as Skylab and the space shuttle. Data-transmitting equipment could transfer the entire contents of a medium-sized library in seconds. It would be the largest and most advanced space tracking and data-relay system ever developed. TDRSS would form a vital link between Earth and near space.

Two of the three TDRSS ground terminals are located at White Sands Test Facility in southern New Mexico; the third is on the island of Guam in the South Pacific Ocean. TRW Defense and Space Systems Group designed, built, and tested the first seven satellites and integrated the ground terminal equipment. Hughes Space and Communications

Company built the three upgraded satellites designed to replace the original seven. By 1983, the first satellite was ready for launch and the first White Sands ground station, with its three 18-meter-diameter dish antennae, was completed. The New Mexico and Guam locations were chosen for their arid climate and clear line of sight to the relay satellites. The Tracking, Telemetry and Command (TT&C) system collects, processes, and transmits telemetry data to the White Sands Complex for monitoring TDRS's performance and operating modes.

On April 4, 1983, at 18:30 Coordinated Universal Time (UTC, or 1:30 P.M. eastern standard time), space shuttle mission STS-6 was launched from Cape Canaveral, Florida. After the first of the three TDRSS satellites, TDRS-A, was launched from the *Challenger* payload bay and proper orbit was achieved, it became known as TDRS-1. The satellite, weighing 2,250 kilograms, measured 17 meters across at the two winglike solar panels. The panels provided the craft with electrical power. The design of the satellite consisted of three modules. The equipment module, located in the lower part of the core hexagon, encompassed the subsystems that controlled and stabilized the satellite. It also contained the equipment that stored and managed the power supply and the machinery that operated the telemetry and tracking equipment. The communications payload module was in the upper portion of the hexagon, just below the multiple-access antenna elements. It contained electronic equipment that regulated the flow of transmissions between the various antennae and other communications functions. Finally, the antenna module consisted of a platform holding various antennae, including thirty helices of the multiple-access phased array.

An Inertial Upper Stage (IUS) booster rocket was to place TDRS-1 in a geostationary orbit (an orbit in which a satellite revolves around Earth once every twenty-four hours). A malfunction in the IUS left the satellite tumbling in a 35,317-by-20,273-kilometer orbit, forcing ground controllers to separate the satellite from the IUS, and the craft stabilized itself in an elliptical orbit. A NASA/TRW team devised a way to use the tiny onboard attitude control thrusters (designed for minor station-keeping duties) to nudge the satellite into its proper orbit through a series of controlled firings. The maneuver took extensive engineering and technical analysis and thirty-nine separate burns to get the spacecraft into its appointed orbit of 67° west longitude on June 29, 1983. Following checkout, it was moved to its operational position of 41° west. In its position 35,690 kilometers above the equator on the northeast coast of Brazil, TDRS-1 was considered the east station in the TDRSS network.

With the launch of TDRS-1, the new system became partially operational. The satellite, assisted by a few of the remaining ground stations from the Spaceflight Tracking and Data Network (STDN), began transmitting data to White Sands. Eventually, TDRS-1 would be replaced by another satellite and moved to 79° west to function as the in-orbit spare for the system.

On January 28, 1986, at 16:40 UTC (11:40 A.M. eastern standard time), TDRS-B, which would have become TDRS-2 and the west link in the TDRSS network, was destroyed when the space shuttle *Challenger* was destroyed during launch at Cape Canaveral. The *Challenger* accident caused many delays in the U.S. space program, and these included the TDRSS launch schedule. It finally moved forward when TDRS-C (after launch, TDRS-3), designated to replace the destroyed satellite, was launched from the space shuttle *Discovery* on mission STS-26 on September 29, 1988. The satellite was placed in orbit at 151° west for testing and support of shuttle mission STS-27. After STS-27 completed its mission, TDRS-C (TDRS-3) would be moved to a final position of 171° west over the Pacific Ocean. It would become the west station in its geostationary orbit southwest of the Hawaiian and Gilbert Islands.

With the successful launch of TDRS-D (TDRS-4) from STS-29, and the replacement of TDRS-1 as an in-orbit spare, TDRSS would become fully operational.

On January 13, 1993, the fifth TDRS (TDRS-F) was deployed from the cargo bay of *Endeavour.* It was the primary payload on the STS-54 mission and later successfully achieved geostationary orbit at 62° west longitude. It officially became known as TDRS-6 and assumed the role of backup to TDRS-East (TDRS-4).

During STS-70, TDRS-G (TDRS-7) was deployed from *Discovery* on July 13, 1995. By the end of 1995, the TDRS constellation was in place. TDRS-1, partially operational and known as TDRS-Z, was moved to 139° west longitude to support National Science Foundation investigations of the South Pole. TDRS-3, also partially operational, was at 85° east, used for space shuttle-Mir rendezvous missions and Compton Gamma Ray Observatory operations. TDRS-4 was located at 41° west operating as TDRS-East. TDRS-5 was at 174° west as TDRS-West. TDRS-6 was placed at 46° west to serve as backup for TDRS-East. TDRS-7 was located at 171° west, performing as backup to TDRS-West.

As of July, 2000, TDRS-1 (TDRS-A) was in storage over Brazil and could be called into service if needed. TDRS-3, deployed during STS-26 in September, 1988, was in orbit over the Indian Ocean, west of Indonesia, filling a communication gap between the two active satellites, known as TDRS-East (TDRS-4) and TDRS-West (TDRS-5). TDRS-4 (STS-29, March, 1989) was hovering over the mid-Atlantic Ridge, providing coverage from west of the United States to east of Africa. TDRS-5 (STS-43, August, 1991) was over Polynesia covering the area from west of Australia to Central America. TDRS-6 (STS-54, January, 1993), an operational spare, was in orbit near TDRS-1. TDRS-7 (STS-70, July, 1995) was functioning as the other operational spare, parked near TDRS-5.

TDRS-H, lifted off on June 30, 2000, at 12:56 UTC (8:56 A.M. eastern daylight time) from Cape Canaveral Air Force Station, Florida, aboard an Atlas-2A rocket. TDRS-H (known as TDRS-8 in orbit) was the first of three new satellites featuring improved multiple-access and S-band single-access performance, along with a new high-frequency Ka-band service. The TDRS-H, TDRS-I, and TDRS-J

spacecraft were specified to replicate TDRS-1 through TDRS-7 services. Additional services would include data return rates up to 3 megabits per second, allowing some lower-rate single-access (SA) customers to use the multiple-access system at a lower cost and free up the more heavily utilized single-access system for higher data rate customers.

The Advanced TDRS's are 21 meters long with solar arrays deployed and 13 meters wide with antennae deployed. Each weighs an estimated 1,777 kilograms at the beginning of life on orbit. Payload services include S-band single access (SSA), S-band multiple access (MA), Ku-band single access (KuSA), and Ka-band single access (KaSA). Power is provided by silicon solar cell arrays that generate 2,300 watts; nickel-hydrogen batteries supply payload power during eclipses.

TDRS-I (TDRS-9) was placed into orbit on March 8, 2002, by the Atlas-2A AC-143. Four days after launch, an anomaly occurred with one of the four fuel tanks. The pressure had dropped when a faulty valve prevented helium from pressurizing a tank. Once a neighboring tank was empty, helium could be transferred to the faulty tank and little burns were possible. The satellite finally arrived in geosynchronous orbit in early October, 2002, after months of effort. Enough fuel is available for the fifteen-year lifetime.

The third and final Advanced Tracking and Data-Relay Satellite, TDRS-J (TDRS-10), separated from the Centaur upper stage 30 minutes after launch on December 4, 2002. This completed the $800 million, three-satellite contract. It also marked the last launch of the Atlas-2A booster.

Contributions

Even before the launch of TDRS-C, with only one satellite, TDRS-1, in orbit and the ground station operational, the Tracking and Data-Relay Satellite System had significantly influenced U.S. space research and operations. Alone, the first satellite and ground station had stretched communications between Earth and the space shuttle from 15 percent to 50 percent of mission time. Augmented with fifteen STDN ground stations, TDRS-1 and its ground station allowed the astronauts on STS-9, launched in November, 1983, to be the first U.S. shuttle crew to enjoy almost continuous communications with Earth. The European-built orbital research module Spacelab, carried in STS-9's cargo bay, was able to transmit volumes of data to scientists on Earth instantaneously. Researchers were able to respond to project results while many experiments were still ongoing. On that mission, more data were retrieved through space-to-ground communications than on all other previous U.S. spaceflights combined because of the partially operational TDRSS. Fifty times more information was transmitted on Spacelab's ten-day mission than on Skylab's twenty-four-day mission a decade before. TDRS-1 also returned outstanding pictures from the Landsat 4 satellite, in what was the first satellite-to-satellite-to-Earth data relay.

The TDRS system is designed to allow almost uninterrupted voice and data exchange between Earth and U.S. piloted spacecraft. Only three ground stations are involved: two in New Mexico and one on Guam. All other STDN stations, except those used during shuttle launches and those used to track satellites incompatible with TDRSS, have been shut down. TDRSS enables almost continuous command and telemetry communications between the ground control centers and robotic, automatic research and applications spacecraft orbiting several thousand kilometers above Earth. Its ability to transmit vast amounts of data also eliminates the necessity to send additional bulky equipment aboard these craft for data storage.

A "typical" message from a low-orbiting research spacecraft would be transmitted to the proper TDRSS satellite. The TDRS would then relay those data instantly to the ground station at White Sands. From there, the data might be sent back up to a commercial communications satellite and relayed to Goddard Space Flight Center (GSFC) in Greenbelt, Maryland. TDRSS is under the aegis of GSFC, and all data are transmitted there for processing and further routing or storage.

At GSFC, the message might again be relayed to another commercial satellite and sent to an appro-priate data center such as the Payload Operations Control Center (POCC), where researchers can obtain instant results from their research equipment in space. Many POCCs are located at the home laboratories of researchers on university campuses or in research centers in the United States and abroad. The amount of information transmitted and the information gained under TDRSS are more than impressive. The system has revolutionized the United States' abilities to increase and develop the commercial, scientific, and industrial potentials of space.

Context

The Goddard Space Flight Center has played a pioneering role in both piloted and robotic spacecraft tracking and data communications since the earliest days of the U.S. space program. From the first minitrack system to the Spaceflight Tracking and Data Network, GSFC has played a prominent role in keeping track of and communicating with U.S. spacecraft. Until the mid-1970's, these systems consisted of a number of complex, sometimes overlapping networks of ground-based antennae, tracking stations, and communications ships.

In 1972, the Satellite Data Relay Network (STDN) was founded when the Space Tracking and Data Acquisition Network (STADAN) merged with the Manned Space Flight Network (MSFN). The move to update and streamline the old system resulted in a network of fifteen international stations. Twelve of the stations tracked piloted and robotic Earth orbital and suborbital missions, and three were used to support the infant space shuttle program. During the 1970's, STDN was continually upgraded to provide greater data-processing capabilities and increase piloted flight communications time. Most of the STDN equipment had been installed in the mid-1960's to support the Apollo Program. Obsolescence and maintenance difficulties continued to increase. Some of the projects that benefit from the Space Network include shuttle missions, space stations, commercial satellites, research and data-gathering satellites, and the Hubble Space Telescope. Although the Air Force makes some use of TDRSS, the system does have some disadvantages from a military point of view: The lack of security equipment and the fact that NASA only leases the satellites are two reasons for the armed forces' reluctance to make use of them.

With the advent of the Space Transportation System (shuttle) program and its schedule of frequent flights and the increasing sophistication of robotic applications and research satellites, it became apparent that either STDN would have to go through a major upgrade or a new system would have to be developed. Studies revealed that it would be less expensive and more efficient to devise a new system using the latest technological advances than to attempt to refurbish the outmoded STDN.

Plans were initiated for a new Space Network that would provide global coverage for U.S. spacecraft using a low-Earth orbit. The system would be able to support the ambitious projects scheduled for the future. The key component of the Space Network would be TDRSS. The system would be cost-effective, would provide almost uninterrupted coverage, and would be able to transmit mountains of data. It would also eliminate the need to go through the political maneuvers necessary to maintain or set up additional ground stations in foreign countries.

The International Space Station requires coordination among all of the participating nations around the globe. While direct station-to-ground communications are more abundant, the TDRS system will be utilized in those parts of the globe not covered by a ground station. TDRS will also provide a backup for normal operations.

As of 2005, there were ten TDRS satellites in the constellation. Due to increasing orbit inclination, TDRS-1 was the first satellite able to see both poles. In cooperation with the National Science Foundation (NSF), an uplink/downlink station for TDRS-1 was installed in January, 1998, at the exact South Pole. TDRS-1 is located at 48.8° west longitude. TDRS-3, deployed during STS-26 in September, 1988, is in orbit at 84.8° east longitude, TDRS-4 (STS-29, March, 1989) is at 41.5° west longitude, and TDRS-5 (STS-43, August, 1991) is at 171.4° west longitude. TDRS-6 (STS-54, January, 1993) is located at 46.8° west longitude, and TDRS-7 (STS-70, July, 1995) is at 150.8° west longitude. TDRS-8 and TDRS-10 are primary satellites, with TDRS-9 serving as a spare. TDRS-8 is at 174.4° west longitude, TDRS-9 is at 64.3° west longitude, and TDRS-10 is at 42.2° west longitude. TDRS's 8, 9, and 10 were launched by Atlas-2A launch vehicles.

See also: Intelsat Communications Satellites; Mobile Satellite System; Private Industry and Space Exploration; Telecommunications Satellites: Maritime; Telecommunications Satellites: Military; Telecommunications Satellites: Passive Relay; Telecommunications Satellites: Private and Commercial.

Further Reading

Burrows, William E. *This New Ocean: The Story of the First Space Age.* New York: Random House, 1998. This is a comprehensive history of the human conquest of space, covering everything from the earliest attempts at spaceflight through the voyages near the end of the twentieth century. Burrows is an experienced journalist who has reported for *The New York Times, The Washington Post,* and *The Wall Street Journal.* There are many photographs and an extensive source list. Interviewees in the book include Isaac Asimov, Alexei Leonov, Sally K. Ride, and James A. Van Allen.

Butrica, Andrew J. *Beyond the Ionosphere: Fifty Years of Satellite Communications.* NASA SP-4217. Washington, D.C.: Government Printing Office, 1997. Part of the NASA History series, this book looks into the realm of satellite communications. It also delves into the technology that enabled the growth of satellite communications. The book includes many tables, charts, photographs, and illustrations; a detailed bibliography; and reference notes.

Elbert, Bruce R. *Introduction to Satellite Communication*. Cambridge, Mass.: Artech House, 1999. This is a comprehensive overview of the satellite communication industry. It discusses the satellites and the ground equipment necessary to both the originating source and the end-user.

Froelich, Walter. *The New Space Network: The Tracking and Data-Relay Satellite System*. NASA EP-251. Washington, D.C.: Government Printing Office, 1986. A booklet published by NASA that describes the Tracking and Data-Relay Satellite System and the Space Network. Presents technical information in terms comprehensible to the layperson. Includes color photographs, graphs, charts, and illustrations. Suitable for general audiences.

Goddard Space Flight Center. National Aeronautics and Space Administration. http://tdrs.gsfc.nasa.gov. This is Goddard Space Flight Center's Web site for information on the TDRSS. Current and historical data are available, along with photographs and charts. Accessed March, 2005.

Karas, Thomas. *The New High Ground: Strategies and Weapons of Space-Age War*. New York: Simon & Schuster, 1983. An in-depth look at the latest in space-age military weapons and systems and their proposed deployment in space. The book briefly discusses the role of TDRSS in military applications. Some technical material but written for a general audience. Contains an index, notes, and acknowledgments.

National Aeronautics and Space Administration. *Goddard Space Flight Center*. Washington, D.C.: Government Printing Office, 1987. A pamphlet describing the history and programs of the Goddard Space Flight Center.

_____. *NASA: The First Twenty-Five Years, 1958-1983*. NASA EP-182. Washington, D.C.: Government Printing Office, 1983. A chronological history of NASA and the U.S. space program during its first twenty-five years. Designed for use by teachers in the classroom, it features color photographs, charts, graphs, tables, and suggested classroom activities. Topics include tracking and data-relay systems, applications, aeronautics, piloted and robotic spacecraft, and missions. Suitable for a general audience.

Rosenthal, Alfred, ed. *A Record of NASA Space Missions Since 1958*. Washington, D.C.: Government Printing Office, 1982. Brief summaries and documentation of all NASA piloted and robotic space missions since 1958. Technical data include descriptions of the spacecraft and launch vehicle, payload, purpose, project results, and major participants and key personnel. Suitable for a general audience.

Thomas, Shirley. *Satellite Tracking Facilities: Their History and Operation*. New York: Holt, Rinehart and Winston, 1963. A somewhat dated text, but useful for its history of early spacecraft tracking and data-relay systems and the relationship to the early U.S. space program. Contains an index and footnotes. Suitable for a general audience.

Wallace, Lane E. *Dreams, Hopes, Realities. NASA's Goddard Space Flight Center: The First Forty Years*. NASA SP-4312. Washington, D.C.: Government Printing Office, 1999. Part of the NASA History series, this is the official story of the Goddard Space Flight Center from its inception to the present day. There are numerous illustrations and photographs, detailed references, tables, charts, and a bibliography.

Lulynne Streeter

Ulysses: Solar Polar Mission

Date: Beginning October 6, 1990
Type of program: Scientific platform

Ulysses is a joint mission between NASA and the European Space Agency (ESA) and the first spacecraft to explore interplanetary space out of the plane of the ecliptic, the imaginary plane above the Sun's equator where all the planets except Pluto orbit.

Key Figures

Lennard A. Fisk, associate administrator, NASA Headquarters

Wesley T. Huntress, Jr., director, Solar System Exploration Division

Robert F. Murray, program manager, NASA Headquarters

J. David Bohlin, ESA program scientist

Derek Eaton, ESA program manager

Willis G. Meeks, project manager, Jet Propulsion Laboratory

Edward J. Smith, project scientist, Jet Propulsion Laboratory

Summary of the Mission

The Ulysses mission is unique in that it is the first spacecraft to explore the region around the Sun out of the plane of the ecliptic. The plane of the ecliptic is the imaginary plane above the Sun's equator where all of the planets, except Pluto, orbit about the Sun. Before the Ulysses mission, all human-made spacecraft were confined to this plane, and our understanding of the region around the Sun, the heliosphere, was limited to what we could observe from within the plane of the ecliptic.

In 1977, the European Space Agency (ESA) added an "Out of Ecliptic" mission to its scientific program, later renamed the International Solar-Polar Mission (ISPM). Originally, it consisted of two spacecraft, one provided by ESA and one provided by NASA, but NASA canceled its portion in 1981 due to budgetary restraints. This decision caused anti-NASA feelings in ESA and canceled a primary mission objective of simultaneously observing both solar polar regions. In 1984, the project was named Ulysses. In the *Inferno*, Dante tells the story of the legendary Greek hero Ulysses, King of Ithaca, who, restless for adventure, decided to explore beyond the known world, which at that time ended with Gibraltar. He challenged his companions "To venture the uncharted distances . . . of the uninhabited world beyond the Sun . . . to follow after knowledge and excellence."

The primary objectives of the mission are to study the properties of the heliosphere at high latitudes away from the plane of the ecliptic. Specifically, Ulysses is to study the properties of the solar corona (the hot, thin outer regions of the solar atmosphere), the solar wind (the flow of high-speed charged particles from the Sun), the heliospheric magnetic field, solar energetic particles, galactic cosmic rays (very high energy charged particles), and solar radio bursts. A Jupiter flyby also provided an opportunity to study the Jovian magnetosphere (the region surrounding Jupiter dominated by the planet's magnetic field). Additionally, radio-science investigations were conducted at specific times during the mission using the spacecraft and ground communications to conduct scientific measurements.

NASA provided the spacecraft power supply; launch vehicle; mission operation center at the Jet Propulsion Laboratory (JPL) in Pasadena, California; and tracking via the Deep Space Network (DSN). ESA provided the spacecraft and is responsible for its operation. The scientific teams responsible for analyzing data collected by instruments on Ulysses come from both Europe and the United States. Ulysses was originally scheduled to be launched in 1983 but was delayed until 1986. The *Challenger* accident further delayed the mission until it was launched from the space shuttle *Discovery* on October 6, 1990, at 17:48 Coordinated Universal Time (UTC), using a two-stage solid-fueled rocket. It was traveling at 11.3 kilometers per second when it left Earth, making it the fastest interplanetary spacecraft ever launched. However, even this high speed was not sufficient to achieve a solar-polar orbit. In fact, no human-made rocket could provide the velocity to reach the high solar latitudes because the Earth orbits the Sun at 30 kilometers per second, so any rocket would have to

cancel out this motion in addition to provide the necessary velocity to achieve the solar-polar orbit. Therefore Ulysses was sent to the planet Jupiter, where that planet's large gravitational field could accelerate Ulysses out of the ecliptic plane.

The spacecraft weighed 367 kilograms at launch, including 33.15 kilograms of hydrozine rocket fuel and 55 kilograms of payload that included nine instruments. Because the spacecraft operates at a great distance from the Sun, solar panels would not supply sufficient amounts of power. Instead, power for the spacecraft is supplied by a radioisotope thermoelectric generator (RTG) that produces electricity via the nuclear decay of radioactive isotopes. Ulysses' most visible features are a 1.65-meter, Earth-pointing high-gain antenna (HGA) for communications and a 5.6-meter boom mounted opposite of the RTG, which carries several scientific sensors. The purpose of the boom is to minimize electromagnetic and radiation contamination of the instruments by the RTG. Most of the remaining scientific instruments are mounted on the main body as far removed from the RTG as possible. There is also a 7.5-meter boom and a 72-meter dipole wire boom that serve as electrical antennae. The spacecraft rotates five times per minute (5 rpm) with the HGA centered on the axis of rotation.

The HGA communicates with Earth with 20 watt X-band and 5 watt S-band transmitters. All commands from Earth are carried on the S band, while the satellite sends telemetry data with the X band and ranging codes with the S band. The downlink bit rates can be selected at either 128 bits per second, or at 8,192 bits per second. Continuous coverage by ground stations was considered impossible for such a long-duration mission, so data are stored on two onboard redundant tapes for sixteen hours, then replayed over an eight-hour period, with real-time data interleaved in with the recorded data. There are also systems on board for the detection of system failures and for safe reconfiguration. This is required for expected periods of nontracking

A model of the satellite Ulysses at one of the Sun's poles. (NASA)

and because of the long period of time it takes for a signal to travel between the spacecraft and Earth. There is also a search mode to reacquire the Earth if no signal is received within thirty days.

The instruments on board are the Vector Helium Magnetometer/Flux Gate Magnetometer (VHM/FGM), for studying the Sun's magnetic field around Ulysses; a solar wind plasma instrument for Solar Wind Observations Over the Poles of the Sun (SWOOPS); the Solar Wind I on Composition Spectrometer (SWICS), for measuring the composition of ions in the solar wind encountered during the mission; the Unified Radio and Plasma Wave Experiment (URAP), which measures the electric field in the vicinity of the spacecraft; the Heliosphere Instrument for Spectra, Composition and Anisotropy and Low Energies (HI-SCALE), for measuring the interplanetary ions and electrons throughout the entire Ulysses mission; the Energetic Particles Composition instrument (EPAC), which was used to study energetic particles in interplanetary space; Cosmic Rays and Solar Particles (COSPIN) for studying high-energy charged particles encountered by the spacecraft; the Cosmic Dust Experiment (DUST), for direct measurement of particulate matter in interplanetary space; the Gamma Ray Burst (GRB) instrument for studying those unexplained phenomena; and the radio antennae, which will be used in the study of gravity waves and for coronal soundings. The NASA spacecraft was to carry a camera, so no cameras were included in the instrument package on the ESA spacecraft. All instruments were activated shortly after launch, and the spacecraft was declared in commission in January, 1991. All instruments were functional throughout the mission.

Closest approach with Jupiter occurred at 12:02 UTC on February 8, 1992, when Ulysses was about 5.3 astronomical units (AU) from the Sun (an AU is the average distance between Earth and the Sun

The United States and the European Space Agency launched the Ulysses spacecraft in 1990 to study the Sun's polar regions. (NASA CORE/Lorain County JVS)

and is equal to about 150 million kilometers). Measurements by instruments on board Ulysses provided a wealth of information concerning Jupiter's magnetosphere. While similar information had been provided by Pioneer 10 in 1973, Pioneer 11 in 1974, and Voyagers 1 and 2 in 1979 when they passed Jupiter, Ulysses took a path through a region of Jupiter's magnetosphere that was not observed by the four previous spacecraft. Specifically, Ulysses passed through the dusk sector of the magnetosphere, as well as the doughnut-shaped ring of highly charged ions encircling Jupiter in the orbit of the moon Io, known as the Io Plasma Torus.

The spacecraft trajectory and the powerful Jovian gravity field worked together to insert Ulysses into a six-year orbit perpendicular to the ecliptic plane that first took the spacecraft over the Sun's southern polar regions and provided the first view of the regions over the Sun's poles. The maximum southern latitude of 80.2° was achieved on September 13, 1994, at a range from the Sun of 2.3 AU. Ulysses then crossed the plane of the ecliptic at a distance from the Sun of 1.3 AU in February, 1995, and reached its maximum northern latitude of 80.2° 2.1 AU above the Sun, on July 31, 1995. These maximum latitudes were limited to about 80° due to the requirement that it should not pass Jupiter closer than 6 Jupiter radii in order to avoid that planet's high radiation fields.

The nominal mission ended on September 30, 1995, when the spacecraft left the northern solar high-latitude region to return to Jupiter. The space-craft and all instruments were functioning well, however, and Ulysses returned to the southern solar high-latitude region in September, 2000, where it would detect a very different environment the second time due to the fact that the Sun was in the quiet period of its eleven-year solar cycle for the first orbit but would be in its active period for the second orbit. Ulysses and the Solar and Heliospheric Observatory (SOHO) both observed a highly active solar maximum, with huge coronal mass ejec-tions, but from vastly different perspectives: SOHO in the ecliptic plane, and Ulysses approximately 80° below.

Between November, 2003, and April, 2004, the Ulysses spacecraft was again in the vicinity of Jupi-ter and able to make observations of the Jovian environment. Solar observations continued, and in February, 2004, ESA's Science Program Commit-tee approved an extension of the Ulysses mission through March, 2008.

Contributions

Once off the ecliptic plane, the pattern of the so-lar wind velocity followed a twenty-six-day cycle that was close to the rotation rate of the Sun. This pat-tern lasted approximately one year, during which time the solar wind speed varied between about 400 kilometers per second and 750 kilometers per second. After this pattern ended, the solar wind speed was a relatively constant 750 kilometers per second, with only small changes from that value. This pattern persisted as the probe went over the southern pole of the Sun and returned toward the ecliptic plane. The twenty-six-day cycle was not re-encountered. As the velocity of the solar wind in-

creased, the density decreased. As a result, the product of solar wind velocity times density re-mained nearly constant. Solar wind measurements in the northern hemisphere mirrored measure-ments in the southern hemisphere.

While going southward, Ulysses encountered north and south magnetic fields equally, until it reached the point where the twenty-six-day pattern was detected in the solar wind. After this point the magnetic field was mainly a southward magnetic field, which continued until the point where the so-lar wind became constant. After this, the magnetic field was only southward in nature. It was expected that Ulysses would find the magnetic field strength to increase over the poles because all magnetic field lines would be funneling in to the surface of the Sun through this region, but measurements showed that the field was actually uniform throughout the polar region. Regular fluctuations in the number of charged particles were also de-tected. These fluctuations are caused by changes in the magnetic field strength, which are in turn caused by periodic oscillations that originate deep within the Sun. This was the first detection of these oscillations, which had been predicted by theory but never detected on Earth.

It was expected that Ulysses would find larger numbers of cosmic-ray particles in the high lati-tudes, but this did not occur. An unexpected and unexplained observation was the detection of a twenty-six-day cycle in the number of cosmic rays detected by the spacecraft that matched the ob-served twenty-six-day solar wind cycle. However, the observed cycle in the number of cosmic rays continued even after the solar wind cycle ceased and the solar wind became steady. Ulysses also made the first detection of neutral particles arriv-ing from outside our solar system.

Jupiter is a very large radio source, and during Ulysses' flyby, instruments were able to measure in-dividual radio signals and identify their sources. In-struments designed to measure the composition of the solar wind were able to measure the composi-tion of the plasma within Jupiter's magnetosphere. Material was identified from three separate sources:

the solar wind, Jupiter itself, and the volcanically active moon Io. Plasma from the solar wind and Io were detected in all regions investigated by Ulysses.

Context

Ulysses, a joint project between the ESA and NASA, placed an instrument-laden spacecraft in a six-year orbit with a gravitational assist from Jupiter. It is the first spacecraft to go into a solar-polar orbit out of the ecliptic plane and investigate the heliosphere at all latitudes of the Sun.

Ulysses' measurements of the solar wind velocity off the ecliptic plane are considerably different from measurements near the ecliptic plane, where the solar wind velocity is highly and irregularly structured: The solar wind velocity averages about 450 kilometers per second and gusts as high as 1,000 kilometers per second. Southern hemisphere measurements were mirrored in the northern hemisphere, indicating that the solar wind is faster at latitudes higher than about 20° and slower and more structured at latitudes less than about 20°. Ulysses measurements also indicated that there is more than one kind of solar wind, supporting previous theories that had suggested different kinds of solar wind with the differences being in the characteristic ion composition, velocity patterns, and density.

Measurements from Ulysses during the Jupiter encounter indicate how the magnetosphere can change in response to changing solar wind conditions, as well as the importance of large-scale current systems in determining the configuration and dynamics of the magnetosphere. Data on the composition of the plasma within Jupiter's magneto-sphere provided new information concerning the origin and physical cycle of the Jovian plasma. The ability to identify the sources of particular regions of plasma shows that solar wind plasma is able to penetrate deep into the magnetosphere, while the Io plasma is able to move outward—valuable data for any model of plasma circulation within the magnetosphere.

Changes in the solar wind and related magnetic disturbances give rise to changes in the Earth's magnetosphere that propagate down to the atmosphere and cause effects that impact our society. Some of the effects are radio communication problems, interference with radar signals and satellite communications, and loss of power in large power grids. These effects are most severe in high-latitude regions but can be experienced at much lower latitudes. The study of Jupiter's magnetosphere provides us with insights into the behavior of Earth's magnetosphere. Obviously the study of the solar wind is of great importance to our increasingly technology-dependent society.

Ulysses was very successful in making measurements of Jupiter's magnetosphere and exploring the high-latitude regions of the heliosphere. It has provided valuable information concerning several theories, but it has also provided some interesting, and unexpected findings. As of 2005, Ulysses continued to observe the Sun's polar regions and the surrounding heliosphere. It instruments were still functioning despite the harsh environment the spacecraft had encountered during its journey.

See also: Planetary Exploration; Space Shuttle; Tracking and Data-Relay Communications Satellites.

Further Reading

Covault, Craig. "European Ulysses Fired to Jupiter, Sun as *Discovery* Returns to Space." *Aviation Week and Space Technology* 133 (October 15, 1990): 22. Provides interesting details for general audiences.

Davies, John K. *Astronomy from Space: The Design and Operation of Orbiting Observatories.* New York: John Wiley, 1997. This is a comprehensive reference on the satellites that have revolutionized twentieth century astrophysics. It contains in-depth coverage of all space astronomy missions. It includes tables of launch data and orbits for quick reference as well as photographs of many of the lesser-known satellites. The main body of the book is subdivided according to type of astronomy carried out by each satellite (x-ray, gamma ray, ultraviolet, infrared, and radio). It discusses the future of satellite astronomy as well.

Golub, Leon, and Jay M. Pasachoff. *Nearest Star: The Surprising Science of Our Sun.* Boston: Harvard University Press, 2002. Although written by two of the most active research astrophysicists, this book is accessible to a general audience. It describes most contemporary advances in solar physics.

Hartmann, William K. *Moons and Planets.* 5th ed. Belmont, Calif.: Thomson Brooks/Cole Publishing, 2005. Provides detailed information about all objects in the solar system. Suitable on three separate levels: high school student, general reader, and college undergraduate studying planetary geology.

Hathaway, David H. "Journey to the Heart of the Sun." *Astronomy* 23 (1995): 38. Provides an overview of what we know about how the Sun works, including solar oscillations and solar wind. Suitable for general audiences.

Hoyt, Douglas V., and Kenneth H. Shatten. *The Role of the Sun in Climate Change.* Oxford, England: Oxford University Press, 1997. This book discusses the interaction between the Sun and Earth's atmosphere and how the latter is shaped by solar activity. It describes many of the different cyclic events that affect our climate and how they can be used or abused. It contains an extensive bibliography.

Kallenrode, May-Britt. *Space Physics: An Introduction to Plasmas and Particles in the Heliosphere and Magnetospheres.* New York: Springer-Verlag, 1998. This illustrated book is an introduction to the physics of space plasmas and its applications to current research into heliospheric and magnetospheric physics. The book uses a new approach, interweaving concepts and observations to give basic explanations of the phenomena, to show limitations in these explanations, and to identify fundamental questions.

Mecham, Michael. "After Long Delay, Ulysses Mission Begins Five-Year Voyage to Expand Solar Data Base." *Aviation Week and Space Technology* 133 (October 22, 1990): 111. Discusses the mission and the instrument package in more detail. Suitable for general audiences.

Talcott, Richard. "Seeing the Unseen Sun." *Astronomy* 18 (1990): 30. Provides a good overview of the mission objectives and expected results. Suitable for general audiences.

Tassoul, Jean-Louis, and Monique Tassoul. *A Concise History of Solar and Stellar Physics.* Princeton, N.J.: Princeton University Press, 2004. A comprehensive study of the historical development of humanity's understanding of the Sun and the cosmos, written in easy-to-understand language by a pair of theoretical astrophysicists. The perspectives of the astronomer and physicist are presented.

"Ulysses Spacecraft Prepared for Long-Delayed Mission to the Sun." *Aviation Week and Space Technology* 131 (November 6, 1989): 25. A good review of the program and its problems in the years leading up to its launch. Suitable for general audiences.

Gordon A. Parker

United Space Alliance

Date: Beginnning 1996
Type of organization: Aerospace agency

In 1996, NASA ceded the responsibility for launching new space shuttle missions and training astronauts to the United Space Alliance (USA) under a six-year, $7 billion contract. This resulted in cutting costs and freed NASA to pursue more vigorously research and development relating to its shuttle program in preparation for constructing the International Space Station.

Summary of the Organization

When Daniel S. Goldin took over as the chief administrator of the National Aeronautics and Space Administration (NASA) in 1992, the agency's annual budget was declining. By 1995, it had been reduced by $3.2 billion. Part of Goldin's charge had been to bring about reductions in cost without compromising safety. The number of NASA employees was reduced by twenty-eight hundred because of budget cuts. After 1995, these reductions decreased substantially, although the retrenchment during the preceding three years left the agency streamlined in the eyes of some but seriously underfunded and understaffed in the eyes of many.

From the time Goldin arrived at NASA, one of his chief goals was to discontinue the agency's role as shuttle designer, owner, and operator, and instead to make it a shuttle customer, privatizing NASA's operations. He called for the creation of a new type of shuttle, pledging $1 billion to the winner of the design contest that, in the end, involved three major aeronautical companies: Lockheed Martin, Rockwell International, and McDonnell Douglas. Boeing Aircraft already held the prime contract for the United States' involvement in the construction of the International Space Station and would work closely with the corporation that finally won the new contract for designing and constructing new space shuttles.

Boeing, McDonnell Douglas, and a Florida management company, Bamsi Incorporated, expressed an interest in taking over the day-to-day operations of the space program. Under contracts negotiated earlier, however, Rockwell International and Lockheed Martin already were responsible for 69 percent of the contract work involving NASA's space shuttles. Rockwell, under its Space Operations Contract for Flight Support, conducted most of the processing and training work that NASA required. Lockheed Martin held the Shuttle Processing Contract for ground operations.

By late 1995, specifications were being studied by corporations interested in bidding for the NASA contract. Forty companies responded to the announcement of an open competition. In the end, however, NASA announced the creation of the United Space Alliance (USA), a joint venture involving Lockheed Martin and Rockwell International, and indicated that it hoped to complete contract talks by the end of September, 1996, which it ultimately did. In December, 1996, Boeing acquired Rockwell International, so Rockwell's share of USA was transferred to Boeing North America.

NASA officials, explaining why they offered the contract to USA without officially opening the competition, cited Lockheed Martin and Rockwell International's vast experience with space shuttles, concluding that no bidders other than the two corporate members of this alliance could meet NASA's schedule while adhering to its stringent safety standards. Goldin constantly emphasized that safety was NASA's paramount concern. USA's chief operating officer, Kent Black, listed safety as his first priority, followed by dependability in meeting launch schedules and cost reductions, in that order.

Under the terms of the new contract, two two-year extensions could be granted if NASA's experience with USA was satisfactory, meaning that the contract could run for as long as ten years. It could bring as much as $12 billion to USA, which aimed to cut NASA's cost by at least $400 million over the six years of the initial contract.

The contract called for a total of at least forty-two missions over six years. It consolidated former commitments, eliminating the need for twelve separate contracts with twelve separate entities, as was the situation at NASA in 1996. Under the terms of the contract, USA assumed responsibility for shuttle ground processing at the Kennedy Space Center and for astronaut training and Mission Control activities at the Johnson Space Center in Houston, Texas.

As an incentive for USA to perform well, NASA agreed to give it 35 percent of any savings in excess of $400 million as long as safety standards were met. If, however, NASA did not rate it very good on safety, USA would lose the 35 percent bonus. Also, a penalty clause could withhold from the company all its fees for periods of up to six months in the event of an accident resulting in the loss of human life. Other penalties would be assessed if the contractor were responsible for launch delays, last-minute cancellation of launches, or post-launch problems that result in the shortening of a mission.

USA manages more than one-third of NASA's annual budget. Safety concerns have made the transition from NASA management to USA management gradual, but every year, USA takes on more of the responsibilities for which NASA was once accountable. NASA has assigned to USA its remaining contracts for major hardware elements, such as the main shuttle engines, the solid rocket boosters and motors, and the External Tank.

The first launch of a shuttle under USA direction was that of *Columbia* in November, 1996. During its first full year of operation, USA successfully directed eight shuttle missions. Between 1996 and 1998, it succeeded in saving NASA more than $3 million. *Columbia* was lost on February 1, 2003, on mission STS-107 while returning home. Although a failure in the heat-shielding thermal protection system tiles led to the breakup, the demise was caused by the impact of debris coming free from the External Tank during launch. Reduction in the numbers of safety personnel, which had been as severe as that of the Goldin era, came under question once again.

Contributions

When Daniel S. Goldin took over as NASA's chief administrator in 1992, he was faced with the harsh reality of a vastly shrinking budget for his agency. His first three years in office involved such serious retrenchment that many of the engineers and technicians working at the Kennedy and Johnson Space Centers made dark predictions of impending disasters caused by the draconian cuts that had necessarily been made in personnel when NASA's budget underwent drastic cutting.

Goldin had to examine every aspect of NASA's operations in an attempt to work within budget limitations without seriously compromising the quality and safety of the program. One obvious problem was that NASA did business with twelve major contractors, which was time-consuming and wasteful. Eventually, Goldin realized, NASA would need to have one prime contractor that would consolidate the functions of the twelve with which the agency had been working.

He also realized that, because of the great risks involved in any NASA space operation, any move toward consolidation had to be sufficiently gradual to assure the safety of the overall operation. On August 21, 1995, NASA made public a notice that encouraged contractors interested in serving as NASA's prime contractor in the shuttle program to prepare proposals. About forty companies responded to this notice.

As time passed, however, Goldin came to acknowledge that only Lockheed Martin and Rockwell International had sufficient experience working with space shuttles to qualify as prime contractors. Rather than continue the application process, he moved toward shaping a contract that would bring the two experienced corporations together as a new corporation, USA. He justified this unilateral decision by saying that had he proceeded otherwise, precious time would have to be spent educating the participants.

Goldin and his colleagues worked assiduously to formulate a contract that would offer incentives for work well done but would provide protections against failures for which the new corporation was responsible. The monetary benefits offered for all savings more than $400 million were counterbalanced by the penalties assessed for unsatisfactory performance.

NASA was able to cut its staff of engineers and managers by 50 percent between the time USA took over the operation and late 1999. Through working out the arrangement it did with USA, NASA's administrators learned that they could enhance their operation while reducing expenditures.

During this period NASA was also learning that problems in launching and flying shuttles were often not caused by faulty hardware that was breaking, but by those working with the hardware who unwittingly broke it. NASA's shuttle program manager, Ronald D. Dittemore, called upon those who work for NASA to fix those problems, saying, "We need our design folks to get in step with our processing folks and see what we can do to make the shuttle easier to process and launch." When one considers that current space shuttles have 2.5 million parts, it is not surprising that some of them fail to work in the vacuum of space or are broken prior to or during launch. NASA and USA now sought to minimize human error.

The main lesson that NASA learned in the 1990's was that devising less expensive means of processing shuttle flights does not necessarily mean that their quality will be reduced. By reducing duplica-tion of services, the agency saved considerable money without sacrificing quality. Having a sole source provider undoubtedly streamlined the operation and made it considerably more efficient. Moving toward consolidation as gradually as NASA did ensured that safety would not be sacrificed during the transfer of responsibility.

NASA and USA both were fully aware of the wisdom of placing former astronauts in important administrative positions in USA. These administrators had firsthand knowledge of the hazards and complexities of spaceflights. They were also the best possible people to assume responsibility for the safety of spaceflights.

Context

Because the space program is a high-profile program that receives considerable and almost constant media attention, it is one fraught with political implications. Shortly after Daniel S. Goldin became NASA's chief administrator, Bill Clinton, a Democrat, became president and a largely Republican Congress was sworn in. This was just the sort of situation in which a legislative logjam could occur.

That, indeed, was the case. The resultant cuts in NASA's budget came at a time when the United States was looking ahead to the establishment of an International Space Station, which would require a well-coordinated effort by all of the participating nations. If launch schedules were not met, as had been the case with Russia's inability to launch Phase III of the Space Station on time, then other schedules would become skewed and a scheduling quagmire would begin to develop. NASA was under great pressure in the 1990's to hold to its established schedule.

Also, with the development of space shuttles by other nations, the United States risked falling behind in its ability to pursue commercial ventures into space. A major stumbling block has been an order issued by President Ronald W. Reagan shortly after the *Challenger* accident prohibiting shuttles from carrying commercial payloads into space if such payloads can be launched instead from robotic vehicles. Congress passed a commercial space bill with similar prohibitions.

The Clinton administration, in its updated statement of space policy, continued to prohibit commercial launches from piloted flights, arguing that lives are at risk every time a piloted vehicle is launched into space. Meanwhile, Daniel S. Goldin proclaimed that the shuttle could be commercialized "as fast as United Space Alliance can take us."

USA has not made public the sort of price structure that it might achieve for delivering its payloads, but it is reliably estimated that new space vehicles, which will be capable of completing many more flights each year than current shuttles do, will reduce the cost of delivering a pound of payload into space to about $1,000 as compared to a present cost of $10,000 per pound.

A great deal of the future of commercialized flights into space will depend upon the development of space vehicles that can be operated more frequently and less expensively than the existing shuttles. Six to eight launches a year are presently projected by NASA. USA expected to add commercial missions to these, allowing for as many as twelve launches a year by 2002. It had hoped to have ten launches in 1999 but fell short of the goal by five in a year that was plagued with delays and launch problems.

Former Astronaut James C. Adamson, who became USA's chief operating officer, noted that shuttle infrastructure could handle no more than ten launches per year with its present personnel. This could be increased to twelve with slight staff increases. Moving beyond twelve would have required another mobile launch platform at the Kennedy Space Center in Florida. Adamson thought that USA could finance such an expansion if revenues from commercial sources justified more than twelve launches per year. All of this had to be rethought in the aftermath of the *Columbia* accident. Shuttle missions were restricted to those dedicated to the International Space Station, which further reduced the opportunities for commercial space efforts based from the shuttle. The Bush administration was committed to retiring the shuttle fleet by 2010.

See also: Funding Procedures of Space Programs; International Space Station: Development; International Space Station: U.S. Contributions; National Aeronautics and Space Administration; Private Industry and Space Exploration; Space Shuttle; Ulysses: Solar Polar Mission.

Further Reading

Anselmo, Joseph C. "NASA to Seek Major Shift in U.S. Shuttle Policy." *Aviation Week and Space Technology* 147 (October 13, 1997): 26. Anselmo discusses the possibilities and probabilities of commercial ventures into space, noting the three areas that are most likely to be exploited commercially. These include the delivery of satellites to low-Earth orbit, the on-orbit rescue and repair of spacecraft, and the return of satellites or construction platforms from space to Earth.

Covault, Craig. "Shuttle Quality Control Now a Major Concern." *Aviation Week and Space Technology* 149 (December 20/27, 1999): 10-11. Covault enumerates recent budget cuts and staff reductions at NASA. He points to the grounding of the shuttle fleet in mid-1999 following the discovery that many of the shuttles had dangerously frayed wiring in them. He calls for more extensive inspection of each shuttle to anticipate problems that might severely compromise their safety in outer space.

_____. "United Space Alliance Leads Shuttle Operations Revolution." *Aviation Week and Space Technology* 147 (June 16, 1997): 204-207. In this report of the Paris Air Show of 1997, Covault discusses in considerable depth the transfer of many of NASA's operations to USA, spelling out the terms of the government's contract with USA. He notes employee reductions at NASA and provides valuable information about the conclusions of the fact-finding team from the independent Aerospace Safety Advisory Panel.

Mari, Christopher, ed. *Space Exploration*. New York: H. W. Wilson, 1999. Two of the twenty-five essays in this collection refer to USA. Although Jonathan Alter's "Eject Button on Cynicism" mentions it only in passing, Joseph C. Anselmo's "Kill the Shuttle? RLV Debate Heats Up" considers it at greater length, going into the possibility of using a new space vehicle, VentureStar, and alluding to the findings of the Hawthorne report concerning the future of the space program.

National Aeronautics and Space Administration. *Columbia Accident Investigation Board*. http://www.caib.us. This Web site is the official Web site of the *Columbia* Accident Investigation Board (CAIB). The board's final report as well as press releases can be obtained here. Accessed June, 2005.

President's Commission on Implementation of U.S. Space Exploration Policy. *President's Commission on the Moon, Mars, and Beyond*. http://govinfo.library.unt.edu/moontomars. This is the official Web site of the President's Commission on the Moon, Mars, and Beyond. Provides information about designs for a piloted vehicle to take human being beyond low-Earth orbit once again. Accessed June, 2005.

R. Baird Shuman

United States Space Command

Date: Beginning September 23, 1985
Type of organization: Aerospace agency

Space technology has developed to the point that systems first devised for exploration have become essential to national defense, providing critical functions. A unified command across all armed services now provides the needed operational focus, consolidating control of space assets and activities in support of nonspace missions.

Summary of the Organization

Military space systems are critical elements in national defense. The military applications of space technology were recognized very early in the Space Age. As American policy expanded to include national defense as well as scientific concerns, the need became apparent for a space command, unified across all armed services, to provide an operational focus. The need intensified as plans for the Strategic Defense Initiative (SDI) became more concrete.

The U.S. Air Force had consolidated its space-related efforts in the Air Force Space Command (AFSC), formed in September, 1982. The parallel Naval Space Command was established in October, 1983. President Ronald W. Reagan's announcement in March, 1983, that he would endorse plans for the development of SDI once again focused public attention on the military uses of space. The Joint Chiefs of Staff began to consider the formation of a unified space command. After studying a June 7, 1983, proposal from the Air Force chief of staff, the Joint Chiefs recommended the establishment of a unified space command to Secretary of Defense Caspar Weinberger on November 8, 1983. Also, Congressman Ken Kramer presented a letter, signed by fifty-three members of Congress, to President Reagan on November 18, 1983, recommending that a unified space command be a vital part of the SDI organization.

On November 26, 1983, Secretary Weinberger presented the Joint Chiefs' proposal to the National Security Council, recommending that President Reagan approve it. The Joint Chiefs created the Joint Planning Staff for Space (JPSS) in February, 1984, to plan the transition. On November 20, 1984, President Reagan announced activation of a unified space command by October, 1985. The JPSS worked from late 1984 through 1985 establishing organizational roles and relationships and assigning missions for the U.S. Space Command.

One concern of the planners was the North American Air Defense Command (NORAD). The United States and Canada had formed NORAD to provide early warning of invasions by bombers and centralized operation of air defense forces. As the Soviet danger evolved to ballistic missile and antisatellite capabilities, NORAD's atmospheric early-warning line and jet interceptors became less valuable. (Canada participates in the U.S. missile warning and space surveillance system but declines to participate in missile defense research and development.)

The Joint Chiefs addressed Canada's concerns by specifying that the U.S. Space Command would not be a component of NORAD but would provide NORAD with space surveillance and missile warning capabilities. NORAD is responsible to the National Command Authorities of both the United States and Canada for warning of any aerospace attack on the North American continent but does not have any responsibility for ballistic missile defense.

The Joint Chiefs of Staff approved the NORAD commander in chief as the commander in chief of the U.S. Space Command. NORAD headquarters is located at Peterson Air Force Base, with key warning operations inside Cheyenne Mountain, 16 kilometers to the west, near Colorado Springs, Colorado. The U.S. Space Command headquarters is also at Peterson. The commander in chief is directly responsible to the president of the United States through the Joint Chiefs and the secretary of defense. The Joint Chiefs of Staff completed assignment of responsibilities and missions in August, 1985. On August 30, President Reagan gave final approval, with the proviso for a one-year review.

The U.S. Space Command, activated September 23, 1985, has three components: the Air Force Space Command, the Naval Space Command, and the Army Space Command. The Air Force Space Command operates most military space systems for the U.S. Space Command, with headquarters at Peterson and other facilities at Cheyenne Mountain Air Force Base and the Consolidated Space Operations Center at Falcon Air Force Base, 14 kilometers east of Peterson. The Air Force Space Division, another subgroup, is not part of the U.S. Space Command but part of the Air Force Systems Command. It serves the Department of Defense, planning activities concerning the space shuttle's military use. Its staff also researches, develops, and performs in-orbit testing of Air Force satellites prior to their operation by the Air Force Space Command.

The Naval Space Command, based in Dahlgren, Virginia, has a senior liaison staff at the U.S. Space Command headquarters and two major space-oriented units: the Naval Space Surveillance System (NSSS), based in Dahlgren, and the Navy Astronautics Group (NAG), based in Point Mugu, California. The NSSS functions as a dedicated space-tracking sensor forming an electronic "fence" extending 1,609 kilometers out from both the Pacific and Atlantic coasts, and 4,827 kilometers across the Gulf Coast. It reaches 24,139 kilometers into space and disseminates information to

forces at sea. The NSSS also acts as the Alternate Space Defense Operations Center and Alternate Space Surveillance Center. The NAG operates Transit, the oldest operational satellite system.

The Army Space Command, activated on August 1, 1986, as an agency (and as a command in April, 1988), is based at Peterson. It provides military satellite communications and space surveillance support to the U.S. Space Command and support to Army field forces.

The staff of the U.S. Space Command spent most of its first year organizing and hiring new personnel, initiating its planning and operations responsibilities, and establishing relationships with its component commands. These activities included transferring to it the space mission areas of Aerospace Defense Command before the latter's deactivation (NORAD was assigned the air defense responsibilities).

The Joint Chiefs conducted the required review in October, 1986. Besides approving the continuation of one commander in chief for both the U.S. Space Command and NORAD, the Joint Chiefs authorized the U.S. Space Command's deputy commander in chief to be NORAD's vice commander in chief. This ensured continuity of responsibility in supporting NORAD.

On August 14, 1987, the commander in chief reported that the U.S. Space Command Center had reached initial operational capability. The center monitors space events globally, around the clock. It transmits to the National Command Authorities both information and warnings. In the fall of 1987, the Army Space Agency took operational control of space-tracking radar functions and the Defense Satellite Communications System, beginning its transformation into the Army Space Command. Also in 1987, the Army Space Command trained personnel in the use of Global Positioning System (GPS) terminals. These employees received data on position and navigation that allowed speedy, precise response.

During that same year, the Air Force Satellite Control Facility was transferred to the Air Force Space Command from the Air Force Space Division. The satellite control network consists of the Consolidated Space Operations Center (CSOC); the Satellite Test Center in Sunnyvale, California; seven worldwide tracking stations; and worldwide ground stations for the Navstar GPS, Defense Meteorological Satellite Program, and Defense Support Program. Automation of the Air Force Space Command tracking stations, completed in 1988, allowed CSOC and satellite test centers to control the stations remotely.

The U.S. Space Command employs fewer than twelve thousand men and women. The U.S. Space Command has responsibility in three broad areas: space operations, space surveillance and warning, and ballistic missile defense planning. Space operations covers space control, directing space support operations, and operating space systems in support of other commands.

Space control, the primary mission of space operations, includes ensuring interference-free access to and operations in space; denying an enemy the use of space-based systems supporting hostile forces; protecting space-related assets; and surveillance of space objects.

Ensuring access to space includes the U.S. Space Command's effort to devise requirements for a space launch infrastructure. Denying an enemy the use of hostile space-based systems occurs when needed. Research and development in antisatellite (ASAT) systems technology is ongoing.

Protection of the space-related assets of both the United States and its allies involves passive protection, including such measures as adding maneuvering fuel and hardening electronics, and the U.S. Space Command's Space Defense Operations Center, or Spadoc. Spadoc, located in the Cheyenne Mountain Air Force Base, monitors space activities, detects and verifies potentially hostile acts, and warns space systems owners and operators so they can take appropriate defensive measures if necessary.

The surveillance of space objects is managed by the Space Surveillance Center, also at Cheyenne. The center makes daily observations, predicts approximately where and when human-made objects will reenter the atmosphere, and warns of possible collisions between space objects. With its Space Surveillance Network, it makes thirty to fifty thousand observations daily, detecting, identifying, tracking, and cataloging more than seven thousand human-made orbiting objects. The network uses Earth-based sensors such as nonmechanical phased-array radars that track multiple satellites simultaneously, cameras with telescopes that can detect satellites thousands of kilometers away, and the ground-based electro-optical deep-space surveillance system, linked to video cameras to enable transmission to the Space Surveillance Network in minutes.

The second function under the category of space operations, directing space support operations and supporting launch and in-orbit requirements, consists of satellite support and terrestrial support. Satellite support includes both action in support of space-borne forces and operation of satellite systems. Actions consist of launch; telemetry, tracking, and commanding; in-orbit maintenance; crisis operations planning; and recovery. Operations incorporates the Transit satellite operation, the Defense Meteorological Satellite Program (DMSP), and the Satellite Early-Warning System (SEWS). Transit's successor is the Navstar GPS, a twenty-one-satellite radio navigation network designed to provide precise navigation and positioning information for both civilian and military use. The Air Force Space Command operates those GPS satellites already in place. The DMSP, a two-satellite network operated by the Air Force Space Command, provides oceanographic, meteorological, and solar-geophysical data to the Navy, the Air Force, and worldwide weather terminals. SEWS monitors the oceans and known ballistic missile sites worldwide for missile attacks.

Terrestrial support, parallel to satellite support, involves command control communications; surveillance from space; navigation; warnings and indications, including the Ballistic Missile Early-Warning System (BMEWS) of large stationary radars and phased-array radars; and environmental monitoring for all sea, air, and land forces.

Another aspect of space operations, operating space systems in support of other commands, embraces "force-enhancement" support of communications, navigation, and surveillance for both U.S. and allied ground-based forces.

Space surveillance and warning, the second major responsibility of the U.S. Space Command, involves support to NORAD through providing space surveillance and missile warning data and to commanders in chief needing warnings of attacks for areas outside North America. The third responsibility of the command is ballistic missile defense planning. The U.S. Space Command develops requirements and plans for engaging attacking missiles.

The staff of the U.S. Space Command acts as both operational support and headquarters management. For example, the systems integration, logistics, and support director has important acquisition functions in supporting the Air Force Space Command and ensures operation of command electronics and communications. The intelligence director provides intelligence support to both command headquarters and the space operational intelligence watch crews, who identify space objects and supply strategic-launch warning information. The operations director supports several operational centers in Cheyenne and the U.S. Space Command Center, also providing for the worldwide component operational plans, coordination, and guidance.

The Center for Aerospace Analysis provides the U.S. Space Command and other defense agencies with analyses of space systems, air-breathing defense, and missile warning and defense. The command enhances the support provided to operational commanders by space systems and improves support to combatant commanders by codifying operational requirements for such areas as communications, wide-area surveillance, precision navigation, and environmental information.

The U.S. Space Command has established a space annex for other commands' contingency and operational plans and developed procedures to allow theater commanders to request tactical data support. It has devised worldwide requirements for missile warning information and created procedures to distribute this information quickly. The command has also deployed a mobile command control system designed to survive an attack and provide jam-resistant tactical warning and assessment, as well as data on missile events.

Since its inception in 1985, the command has integrated its components. For example, GPS operational crews come from the Army, and Naval Transit experience shapes GPS operational techniques. Also, critical early-warning sites include Navy personnel. In short, the U.S. Space Command is well established to protect the United States in space.

Context

President and former Army general Dwight D. Eisenhower formulated the first space policy, maintaining that space should be used for peaceful, scientific activities. He favored using space satellites for "open skies" reconnaissance, instead of high-flying U-2 aircraft, which the Soviets successfully attacked. Eisenhower's emphasis on peace led to the United States' early separation of military from civilian space programs. The first proposal for a unified command of military space activities came in 1959; it had been developed by the chief of Naval Operations but was basically ignored by the other services.

Each service developed different space interests. The Army focused on launch and booster vehicles; the Navy, high-altitude rockets and early satellite technology, growing into space-based fleet-support communications and navigation systems; and the Air Force, intercontinental ballistic missiles at first, and then extensive satellite applications in communications, surveillance, meteorology, and navigation.

President John F. Kennedy did not want the deployment of reconnaissance satellites publicized, and he had classified military launches by 1962. Yet all services were allowed to conduct space research; the Air Force became responsible for research and development and for testing of all Department of Defense space projects.

The United States ratified three space treaties under President Lyndon B. Johnson: the Nuclear Test Ban Treaty, prohibiting nuclear explosions in space; the Outer Space Treaty, banning the orbiting of mass-destruction weapons; and the Astronaut Rescue and Return Agreement. To further his Great Society programs, programs aimed at expanding government's role in social welfare, Johnson emphasized the commercial and domestic benefits of space. Better capabilities in communications and meteorology resulted, aiding both civilian and military uses. The Air Force received some funding (in short supply because of the demands posed by the Vietnam War) for research on the Manned Orbiting Laboratory (MOL), a potential surveillance post that was later canceled.

Major space policy changes began with President Jimmy Carter's Presidential Directive DD-37. It stated the objective of cooperating with other nations to ensure the freedom to pursue activities in space that enhance humankind's security. It also shifted policy by pursuing the survivability of space weapons systems, creating a program to identify which civilian space resources would be incorporated into military operations during national emergencies, and researching A-sat capabilities permitted by international agreements.

President Reagan built on Carter's foundation with National Security Decision Directive (NSDD) 42, issued in July, 1982. It stated that the United States would conduct national security space activities (including communications, navigation, surveillance, warning, command and control, environmental monitoring, and space defense), develop and operate ASATs, and deny enemies the use of space-based systems in support of hostile forces. Reagan also supported the space shuttle, giving priority to national security missions.

President Reagan's NSDD 85, issued on March 25, 1983, announced his plans for the development of SDI, with a goal of defending against missile attack. Reagan promulgated a national space strategy on August 15, 1984. The shuttle would be the primary launch vehicle for national security missions. The Department of Defense was required to implement the 1982 policy and SDI, ensure access to space by supplementing the shuttle with expendable launch vehicles, and emphasize advanced technologies to provide new capabilities and improve space-based assets.

With the collapse of the Soviet Union, newly elected President Bill Clinton essentially canceled SDI. His successor, George W. Bush, attempted a revival of certain aspects of SDI without much success. In an age involving an asymmetrical war on terror, SDI had little to offer. However, a limited SDI (if deployed) still held the potential to deal with a rogue state that might be capable of launching only a few nuclear-tipped missiles toward targets in North America.

See also: National Aeronautics and Space Administration; Nuclear Energy in Space; United Space Alliance.

Further Reading

Burrows, William E. *This New Ocean: The Story of the First Space Age.* New York: Random House, 1998. This is a comprehensive history of the human conquest of space, covering everything from the earliest attempts at spaceflight through the voyages near the end of the twentieth century. Burrows is an experienced journalist who has reported for *The New York Times, The Washington Post,* and *The Wall Street Journal.* There are many photographs and an extensive source list. Interviewees in the book include Isaac Asimov, Alexei Leonov, Sally K. Ride, and James A. Van Allen.

Covault, Craig. "Ground Troops to Benefit from Army Space Command." *Aviation Week and Space Technology* 128 (April 25, 1988): 80-82. Discusses the formation of the Army Space Command and its work with the U.S. Space Command. Also discusses the Army astronauts at Johnson Space Center.

_____. "New Space Operations Center Will Improve Threat Assessment." *Aviation Week and Space Technology* 126 (May 25, 1987): 50, 52. This article offers a description of Spadoc's facilities and operations.

_____. "NORAD: Space Command Request System for Surveillance of Soviet Weapons." *Aviation Week and Space Technology* 126 (April 6, 1987): 73-76. Discussion of the request for development of space-based intelligence and surveillance systems to counter a range of new Soviet strategic weapons.

_____. "Space Command: NORAD Merging Missile, Air, and Space Warning Roles." *Aviation Week and Space Technology* 122 (February 11, 1985): 60-62. Includes explanations of how NORAD and the U.S. Space Command will cooperate, the new strategies to be implemented, and what the Command will control.

_____. "U.S. Space Command Focuses on Strategic Control in Wartime." *Aviation Week and Space Technology* 126 (March 30, 1987): 83-84. Descriptions of various operations centers in the U.S. Space Command.

_____. "USAF Initiates Broad Program to Improve Surveillance of Soviets." *Aviation Week and Space Technology* 122 (January 22, 1985): 14-17. This article describes the Air Force Space Command before the U.S. Space Command was formed and the surveillance systems the Air Force operated. Offers information on Soviet space activity as monitored by the Air Force.

Cox, Christopher. *The Cox Report: U.S. National Security and Military/Commercial Concerns with the People's Republic of China.* Washington, D.C.: Regnery Publishing, 1999. Investigates U.S.-Chinese security interaction and reports that China successfully engaged in harmful espionage and obtained sensitive military technology from the United States. Some of the technology obtained includes information on American reconnaissance and attack satellites.

Day, Dwayne A., John M. Logsdon, and Brian Latell, eds. *Eye in the Sky: The Story of the Corona Spy Satellites.* Washington, D.C.: Smithsonian Institution Press, 1998. The top-secret Corona spy satellites and their photographs were kept out of public sight until 1992. The Corona satellites are believed by many experts to be the most important modern development in intelligence gathering. The book is based upon previously classified documents, interviews, and firsthand accounts from the participants in the program.

Lambright, W. Henry, ed. *Space Policy in the Twenty-First Century.* Baltimore: Johns Hopkins University Press, 2003. This book addresses a number of important questions: What will replace the space shuttle? Can the International Space Station justify its cost? Will Earth be threatened by asteroid impact? When and how will humans explore Mars?

McDougall, Walter A. *The Heavens and the Earth: A Political History of the Space Age.* 2d ed. Baltimore: Johns Hopkins University Press, 1997. This scholarly text provides the history, especially the political history, of space exploration. The military uses of space are thoroughly covered.

Michaud, Michael A. G. *Reaching for the High Frontier: The American Pro-Space Movement, 1972-1984.* New York: Praeger, 1986. The pro-space movement (more than fifty advocacy groups, involving more than 200,000 Americans) has developed since the end of the Apollo Program. Michaud traces key groups, identifying their origins and goals and explaining the ways in which they have influenced space policy. Includes a bibliography, with many sources on military space activity.

National Aeronautics and Space Administration. *Space Network Users' Guide (SNUG)*. Washington, D.C.: Government Printing Office, 2002. This users' guide emphasizes the interface between the user ground facilities and the Space Network, providing the radio frequency interface between user spacecraft and NASA's Tracking and Data-Relay Satellite System, and the procedures for working with Goddard Space Flight Center's Space Communication program.

Reiss, Edward. *The Strategic Defense Initiative*. New York: Cambridge University Press, 1992. This history of the Strategic Defense Initiative shows how political, economic, strategic, and cultural factors have interacted to shape SDI. It examines new research into the SDI interest groups, the distribution of contracts, the Hardback politics of influence, and SDI's alliance management, popular culture, and military spin-offs. Throughout the book, the author tests the theoretical literature on the dynamics of the arms race against the reality of "Star Wars."

Richelson, Jeffrey T. *America's Space Sentinels: DSP Satellites and National Security*. Lawrence: University Press of Kansas, 1999. This is the story of America's Defense Support Program satellites and their effect on world affairs. Richelson has written a definitive history of the spy satellites and their use throughout the Cold War. He explains how DSP's infrared sensors are used to detect meteorites, monitor forest fires, and even gather industrial intelligence by "seeing" the lights of steel mills.

"Several U.S. Military Spacecraft Operating on Final Backup Systems." *Aviation Week and Space Technology* 126 (March 30, 1987): 22-23. A description of what the U.S. Space Command is doing to maintain its surveillance in the light of several military satellites operating without backups.

Smith, Bruce A. "Air Force Supports Demonstration of Surveillance Technology in the 1990's." *Aviation Week and Space Technology* 128 (March 14, 1988): 93, 95. Two examples of space-based surveillance technologies are surveillance of space objects from a space platform and a space-based radar system. The Air Force Space Division is working with U.S. Space Command to define operational requirements. In this article, details are given on these and other ongoing similar projects.

"Space Command Completes Acquisition of Pave Paws Warning Radar Installations." *Aviation Week and Space Technology* 126 (May 18, 1987): 128-129. A description of Pave Paws (Phased-Array Warning System), which tracks intercontinental and submarine-launched missiles as well as space objects for the U.S. Space Command.

Patricia Jackson

Upper Atmosphere Research Satellite

Date: Beginning September 12, 1991
Type of spacecraft: Meteorological satellites

The Upper Atmosphere Research Satellite (UARS) was deployed to measure chemical composition and changing conditions as they relate to and effect changes in the Earth's environment, especially the loss of ozone over Antarctica.

Summary of the Satellite

Atmospheric ozone depletion is recognized as a serious environmental problem. Chemical reactions occurring primarily in the stratosphere, approximately 15 to 55 kilometers above the surface of the Earth, both generate ozone by the interaction of oxygen with solar radiation and, in the presence of other chemical reagents, deplete the ozone which is formed. Chlorine-containing compounds, notably chlorofluorocarbons commonly used as refrigerants, are among the most serious of the ozone-depleting chemicals. Production of these human-made chemicals has generally been prohibited with the hope of lessening their contribution to ozone depletion. Other chemicals containing chlorine and nitrogen, some of which are naturally occurring, also contribute to ozone depletion.

In 1976 the United States Congress directed the National Aeronautics and Space Administration (NASA) to develop an Upper Atmosphere Research Satellite (UARS) to study the chemical and physical properties of the upper atmosphere, namely the stratosphere, with special emphasis on ozone and ozone-depleting chemicals. Prior studies directed toward this goal were of a limited nature. Instruments designed to measure chemical and physical properties in the atmosphere were contained aboard conventional aircraft. With their limited altitude capability, only partial observations could be made, relatively near the surface of the Earth. Instruments mounted aboard rockets were also limited, as they gave adequate height profiles of the various measurements but only over a relatively narrow land mass. The UARS was developed to obtain data over the entire circumference of the Earth and at a height suitable to map upper atmospheric chemical and physical changes at distances from 10 to 60 kilometers above the Earth's surface. The satellite was designed to provide data for three years and was built at a cost of $740 million dollars.

The Mission to Planet Earth (MTPE) program, conceived in the late 1980's, was a worldwide cooperative venture to study global problems. The first United States contribution to this program was the launch of the UARS as part of STS-48 aboard the space shuttle *Discovery* on September 12, 1991, with John O. Creighton acting as commander. The satellite itself was placed into circular orbit 600 kilometers above the Earth on September 15, 1991. It circled the Earth every ninety-seven minutes with observations covering an area from 80° north to 80° south latitude. The spacecraft had a total mass of 6,800 kilograms, which included 2,500 kilograms of scientific instruments.

The scientific instruments aboard UARS were designed to measure specific aspects of the upper atmosphere, namely chemical composition, temperature, wind speed and direction, intensity of solar radiations, and particulates. Data from these observations would then be tabulated, correlated with observations from other sources (satellites and Earth observation stations), and used to construct a model of the chemical and physical changes occurring above the Earth. From this model, predictions of future changes could be made.

Four spectrometers aboard UARS measured the chemical composition of the upper atmosphere. The first of these was the Cryogenic Limb Array Etalon Spectrometer (CLAES), an infrared-sensing instrument recording data over a wavelength range of 5.5 to 12.7 micrometers from infrared energies. It was capable of simultaneously recording twenty data points over a vertical distance of 50 kilometers. It detected the presence of nitrogen and chlorine compounds, ozone, water vapor, methane, and carbon dioxide.

The Improved Stratospheric and Mesospheric Sounder (ISAMS) was another type of infrared instrument employing telescopes that allowed it to measure in both vertical and horizontal planes over the wavelength region of 4.6 to 16.6 micrometers. It monitored the presence of nitrogen compounds, ozone, water vapor, methane, and carbon monoxide.

The Halogen Occultation Experiment (HALOE) instrument was a third type of infrared spectrometer capable of recording wavelength data from 2.4 to 10.5 micrometers. It was designed to measure hydrogen fluoride, hydrogen chloride, nitrogen compounds, ozone, water vapor, methane, carbon dioxide, and, indirectly, the gas density within the stratosphere.

The Microwave Limb Sounder (MLS) instrument recorded signals in the microwave region of the electromagnetic spectrum rather than in the infrared region. Responding to energies at 1.46-, 1.64-, and 4.8-millimeter wavelengths, it monitored the concentrations of chlorine oxide, hydrogen peroxide, water vapor, and ozone.

Knowledge of the movement of chemical substances in the stratosphere is essential in understanding the interactions of the various chemical species present and in relation to various events on Earth that alter the natural abundance of these chemicals. Indirect knowledge of the movement of

chemicals in the stratosphere is available from the observed displacement of various chemicals over time, as recorded by the spectrometers described in the preceding paragraphs. Two instruments aboard UARS, however, for the first time gave direct measurements of stratospheric winds. This was accomplished by monitoring the Doppler energy change in the electromagnetic radiation observed from selected molecules and unstable chemicals in the upper atmosphere. All else being equal, energy emitted by a stationary object is constant, but when the object sensing the energy is moving either toward or away from the detector, the frequency at which the energy strikes the detector will vary. This variation in frequency of the electromagnetic radiation emitted by particles propelled by stratospheric winds, when measured from the moving satellite, is used to calculate the wind velocity. The instrument capable of detecting slight changes in the frequency of electromagnetic radiation is the interferometer. Two interferometers were aboard the UARS. They were the High Resolution Doppler Imager (HRDI) and the Wind Imaging Interferometer (WINDII). These instruments were capable of measurement in both vertical and horizontal directions relative to the satellite. As the satellite circled the Earth, a three-dimensional profile of stratospheric wind direction and speed was generated.

Energy distribution within the stratosphere arises from solar radiation and from charged particles, the former as ultraviolet radiation and the latter as high-energy charged particles, namely electrons and protons, originating from solar flares. The Solar Ultraviolet Spectral Irradiance Monitor (SUSIM) responded to ultraviolet radiation from the Sun over the energy range from 120 to 400 nanometers. The Solar/Stellar Irradiance Comparison Experiment (SOLSTICE) also monitored changing ultraviolet energies over the wavelength range from 115 to 430 nanometer. For a fixed energy source to which changing ultraviolet energy levels can be compared, one of its detection channels focused on a distant blue star as a constant source of ultraviolet energy. The Particle Environ-

A concept drawing of the Upper Atmosphere Research Satellite in orbit. (NASA)

ment Monitor (PEM) detected and recorded data from electrons and protons present in the stratosphere; in addition, it monitored x rays produced by high-energy electrons as they passed through the upper atmosphere. Electron energy was detected over the range from 1 to 5 million electron volts (megaelectron volts), protons were detected over the energy range from 1 to 150 million electron volts, and x rays were monitored in the 2- to 50-kiloelectron volt range. An electron volt is the energy gained by one electron as it accelerates in an electric field of one volt potential difference. (One electron volt of energy is equivalent to 1.6×10^{-19} joules.)

Data from the various instruments aboard URAS were collected from onboard computers. From these they were transferred to a Central Data

Handling Facility (CDHF) at the Goddard Space Flight Center. From Goddard, the information was passed to the appropriate Remote Analysis Computer (RAC) located at the research site of each principal investigator responsible for refining and interpreting the data from a particular measurement instrument for which she or he was responsible. These sites were located at Georgia Institute of Technology, Jet Propulsion Laboratory, Lawrence Livermore National Laboratory, NASA Goddard Space Flight Center, NASA Langley Research Center, National Center for Atmospheric Research, National Oceanic and Atmospheric Administration, State University of New York at Stony Brook, United Kingdom Meteorological Office, University of Colorado, and University of Washington. This arrangement allowed for very rapid data transfer to the destination for which they were intended and where they could be addressed in the minimum amount of time.

There was a brief period in June, 1992, when eight of the ten instruments aboard UARS were inoperative because of improper adjustments in the solar array panels. This was corrected by ground control. As of 2005, a decade beyond the expected shutdown date for receiving transmission from UARS, six of the ten instruments aboard the satellite were still operational and sending data back to Earth. The observation time of several of the experiments had been reduced to conserve battery power, but scientifically useful data were still being transmitted.

Contributions

Data from the UARS provided information regarding the chemical composition of the stratosphere, measuring concentrations of both naturally occurring components and, more important, artificially produced chemicals released into the air as a result of human activity. In addition, solar wind direction, intensity, and seasonal variations provided information about the migration of these chemicals and locations where they could be expected to concentrate at various times of the year. Temperature measurements gave insight into the reactivity of the various chemical species, certain chemical changes being favored over specific temperature ranges. Data on the daily fluctuations in the solar wind, as measured by the High Resolution Doppler Imager (HRDI) from 90 to 105 kilometers above the Earth and varying both north and south from the equator, were recorded.

Recorded also were data pertaining to the Polar Stratospheric Clouds (PSCs) as monitored by the Improved Stratospheric and Mesospheric Sounder (ISAMS). Differences in the formation, composition, and duration of these clouds over the Antarctic as compared to those over the Arctic were used to explain the differences in extent of ozone depletion above the Antarctic as compared to that in the Northern Hemisphere. UARS, gathering data from both regions at considerable height variations and

for an extended period of time, provided the necessary information to distinguish this difference. The type, intensity, and seasonal variation of radio activity were followed, measurements being recorded from both particle and x-ray emissions. Particulate monitoring capability of UARS proved also to play a significant role in devising theories to explain, for example, ozone depletion over Antarctica.

It was fortunate that UARS was operational when the volcanic eruption of Mount Pinatubo on the island of Luzon in the Republic of the Philippines occurred in 1991. It was estimated that 30 million metric tons of ash were forced into the atmosphere as a result of this eruption. UARS and other satellites were able to monitor the presence of this increased particulate concentration above the Earth.

The measurements obtained by UARS correlated and extended those of earlier studies from ground-based observation and satellites launched previously. Perhaps most important, UARS data gave a rather complete three-dimensional overview of all variables recorded above the entire Earth surface through various seasonal changes.

Context

Ozone is formed in the upper atmosphere by the interaction of molecular oxygen and atomic oxygen, the latter produced by photograph decomposition from the ultraviolet radiation emitting from the Sun. Ozone depletion results naturally by several processes, including photograph decomposition, interaction with nitrogen oxide present in the atmosphere, and interaction of hydrogen oxygen radicals also present. With the widespread use of human-made chlorofluorocarbons (numerous compounds containing carbon, chlorine, and fluorine), the destruction of ozone has markedly increased. The chlorofluorocarbons eventually find their way into the atmosphere where they are decomposed by ultraviolet radiation from the Sun. The liberated chlorine atoms rapidly attack ozone molecules, forming chlorine oxide and oxygen molecules. Only through formation of a nonreacting chlorine compound can the conversion of ozone to oxygen be stopped. The presence of sulfuric acid, too, can promote chlorine oxide formation and thus ozone depletion.

By the late 1980's, ozone loss over Antarctica was a recognized and growing concern. The UARS and numerous other probes launched to study the conditions causing this loss provided data from which theories were developed and steps proposed to reverse the depletion. The data also greatly increased understanding of the composition and changes taking place in the stratosphere and at other atmospheric elevations. Polar stratospheric clouds residing over the polar regions at an altitude of about 20 kilometers and extending to a height of 10 to 100 kilometers are sometimes called nacreous, or mother-of-pearl, clouds because of their iridescent appearance from light reflecting off ice crystals.

Polar stratospheric cloud formation over Antarctica is favored during the austral (southern hemispheric) spring and winter because of the extremely low temperatures present during that time. It is during this time of extreme cold that chlorine oxide formation from chlorofluorocarbons reaches its maximum. Similar cloud formation over the Arctic region during its coldest period is less extensive because of the relatively warmer temperatures. UARS data indicate a denser cloud formation over Antarctica. The greater amount of particulate matter provides a greater area upon which ozone-depleting reactions can occur. Cloud formation in Antarctica also persists for a longer period of time than in the relatively warmer Arctic region. Particulates contained within the clouds, ice crystals and others, serve as sites for chemical reaction between ozone and chlorine-containing compounds. The unfortunate eruption of Mount Pinatubo in 1991, with its massive thrust of particles into the atmosphere, greatly accelerated this ozone depletion.

See also: Atmospheric Laboratory for Applications and Science; ITOS and NOAA Meteorological Satellites; Mission to Planet Earth; Nimbus Meteorological Satellites; Nuclear Detection Satellites; Tethered Satellite System; United States Space Command.

Further Reading

Allen, D. R., et al. *The Four-Day Wave as Observed from the Upper Atmosphere Research Satellite Microwave Limb Sounder.* NASA CR-207746. Washington, D.C.: National Aeronautics and Space Administration, 1998. Describes the Upper Atmosphere Research Satellite program and topics related to its findings.

Andrews, David G. *An Introduction to Atmospheric Physics.* Boston: Cambridge University Press, 2000. An excellent overview of atmospheric physics suitable for an introductory level college course.

Cramer, Jerome. "A Mission Close to Home." *Time,* September 16, 1992, 53. Background details leading up to and through the launch of UARS are presented, including some controversy over the significance of space probes and their importance to national well-being.

McCormick, M. Patrick, Larry W. Thomason, and Charles R. Trepte. "Atmospheric Effects of the Mt. Pinatubo Eruption." *Nature* 373 (1995): 399-404. The dramatic increase in airborne particulates and their relation to polar stratospheric cloud formation and subsequent ozone depletion are described.

"NASA Launches Mission to Planet Earth." *Sky and Telescope* 83 (1992): 9. This brief note describes the Upper Atmosphere Research Satellite, its purpose, and its capabilities for studying ozone depletion above the surface of the Earth.

National Aeronautics and Space Administration. *Upper Atmosphere Research Satellite (UARS): A Program to Study Global Ozone Change*. NASA N92-11067. Linthicum Heights, Md.: NASA Center for AeroSpace Information, 1990. This NASA publication details the purpose of the UARS project. It describes in detail each of the onboard instruments, and it details the manner in which information from these instruments was relayed back to Earth for distribution.

Rodriguez, José M. "Probing Stratospheric Ozone." *Science* 261 (1993): 1128-1129. This brief overview provides a summary of the several efforts made to monitor polar stratospheric clouds from 1987 through 1993 by various satellites.

Rowland, F. Sherwood. "Stratospheric Ozone in the Twenty-First Century: The Chlorofluorocarbon Problem." *Environmental Science and Technology* 25 (1991): 622-628. This article, written before the launch of UARS, describes the manner in which various chlorofluorocarbons interact to destroy ozone in the upper atmosphere.

Schoeberl, Mark R., et al. *Investigation of Chemical and Dynamical Changes in the Stratosphere Up to and During the EOS Observing Period*. Washington, D.C.: Government Printing Office, 1997. This report is highly technical, but full of facts that the casual researcher might find interesting. It includes the assessment of the dynamical processes of ozone changes and the ability to model the atmospheric chemical process. The EOS instrument capabilities and the production of tools to increase their capacity are also discussed. Comparisons are made between ground-based observations and the in-flight measurements made by UARS.

Taylor, F. W., et al. "Properties of Northern Hemisphere Polar Stratospheric Clouds and Volcanic Aerosol in 1991/92 from UARS/ISAMS Satellite Measurements." *Journal of the Atmospheric Sciences* 51 (1994): 3019-3026. This research paper details the data transmitted from UARS relating to polar stratospheric cloud formation and the role it plays in ozone-depleting reactions, especially over Antarctica.

Webster, C. R., et al. "Chlorine Chemistry in Polar Stratospheric Cloud Particles in the Arctic Winter." *Science* 261 (1993): 1130-1134. This article explains the theories proposed to account for the difference in ozone depletion over the Arctic pole as compared with the greater depletion over the Antarctic region.

Woodard, Stanley E. *The Upper Atmosphere Research Satellite In-Flight Dynamics*. Hampton, Va.: National Aeronautics and Space Administration, Langley Research Center, 1997. Covers the Upper Atmosphere Research Satellite (UARS) program, solar arrays, and spacecraft orbits, among other subjects.

Zimmerman, Robert. *The Chronological Encyclopedia of Discoveries in Space*. Westport, Conn.: Oryx Press, 2000. Provides a complete chronological history of all crewed and robotic spacecraft and explains flight events and scientific results. Suitable for all levels of research.

Gordon A. Parker

Vandenberg Air Force Base

Date: Beginning October 4, 1958
Type of facility: Spacecraft tracking network

Vandenberg Air Force Base has served as the site for more than five hundred orbital and one thousand nonorbital launches of American rockets and ballistic missiles. Between 1965 and 1969, a launch complex for the Manned Orbiting Laboratory was built at Vandenberg. From 1979 to 1986, these facilities were expanded to form a West Coast launch complex for the space shuttle. They were later modified for other robotic launch purposes.

Summary of the Facility

Because of its ideal position for launching satellites into polar orbits, Vandenberg Air Force Base (VAFB) has become a prime launching site for orbital payloads. As a military facility, Vandenberg has seen the launching of most American reconnaissance satellites and the firing of ballistic missiles for test purposes. In the early 1980's, the base underwent massive construction for a planned West Coast launch and landing site for the space shuttle.

In January, 1956, the U.S. Department of Defense (DoD) had decided that the United States needed a facility to train the people who were handling intercontinental and intermediate range ballistic missiles (ICBMs and IRBMs). On June 7, 1957, the DoD allocated to the U.S. Air Force the northern two-thirds of Camp Cooke, an inactive World War II Army training camp located along 40 kilometers of the Pacific Coast just 88 kilometers northwest of Santa Barbara, California. The portion of the land south of the Santa Ynez River fell to the Navy, which installed the Naval Missile Facility, Point Arguello. The Air Force made its new base the headquarters of the First Strategic Aerospace Division and integrated it into the Strategic Air Command.

On October 4, 1958, the Air Force's land was officially dedicated as Vandenberg Air Force Base, and on December 16, 1958, the first Thor missile was launched, thus making Vandenberg the first operational ICBM facility in the United States.

The Air Force, however, had even more ambitious plans than the testing of missiles, because the location of Vandenberg made it ideal for launching satellites into polar orbit. Flying from pole to pole, a satellite will pass over every part of Earth while the planet rotates beneath it. Because VAFB is situated on a promontory jutting west from the California coastline into the Pacific, a rocket that has been launched toward the South Pole from VAFB will not fly over land until it reaches Antarctica. The risks involved for such a launch are low, because fallout from failed missions will not hit inhabited areas.

In cooperation with the National Aeronautics and Space Administration (NASA) and the United States Navy, the Air Force immediately constructed control centers and pads for the launching of satellites from Vandenberg. The first of these facilities was erected on a small, round peninsula extending west into the Pacific Ocean, situated near the base's airport. Vandenberg, together with the Navy's Point Arguello, became the Western Test Range of the Space and Missile Test Organization (SAMTO), an umbrella organization for Pacific and Atlantic aerospace test ranges.

On February 28, 1959, the first successful liftoff of a launch vehicle took place at Vandenberg: A Thor-Agena A rose from launch pad 75-3-4 (now Space Launch Center 1 West, or SLC-1W) and ejected its payload, Discoverer 1, the first American satellite to reach polar orbit. Soon, the first of the powerful Atlas launch vehicles arrived at Vandenberg, where facilities grew and military and scientific personnel began to populate the village of Lompoc, east of the base.

The Discoverer series carried data capsules, which were designed to fall back to Earth to be recovered and analyzed by Air Force specialists. It took one and a half years before the capsule ejected from Discoverer 13 was recovered from the Pacific Ocean west of Vandenberg. Eight days later, the Air Force succeeded in retrieving the data capsule of the next Discoverer in midair. From that point onward, the military was able to launch payloads at Vandenberg or Point Arguello and safely recover the data packs; throughout the years, about three-quarters of all the midair recovery attempts have been successful.

In 1960 and 1961, a new class of reconnaissance satellites was launched from the Western Test Range. Enthusiastically promoted as "spy-in-the-sky-satellites" by General Bernard A. Schriever, then Head of Air Force Systems Command, this hardware forced a closer cooperation between the Air Force and the Central Intelligence Agency (CIA) at Vandenberg. As a military base, Vandenberg had a higher-level security classification than did other American space facilities. Yet the introduction of the new payloads of spy satellites was not kept a secret at first. Both the failure of the first SAMOS (Satellite and Missile Observation System) satellite, which was launched from Point Arguello on October 11, 1960, and the relative success of SAMOS 2 were publicized. Similarly, the MiDAS, or Missile Defense Alarm System, satellites were announced publicly at first.

Late in 1961, however, a shroud of secrecy descended on Vandenberg. Military reconnaissance satellites were no longer given names, but a CIA code—KH (for "keyhole") and a number—was ap-

plied to them. Discoverer 38 was the last named spy satellite to reach orbit from Vandenberg, in March, 1962. Thereafter, Air Force and Navy officials would not release any information other than the standard statement, "A classified payload went into orbit today."

In 1963 and 1964, the Air Force developed plans for a military Manned Orbiting Laboratory (MOL). Riding atop a new Titan III-C (later revised to Titan III-M) rocket, the modified Gemini spacecraft would be launched from VAFB and circle Earth for thirty days. The main goal of the MOL was to test how well piloted spacecraft could perform military space operations. The predecessors of the launch vehicle for the MOL, the Titan I and II series, had already been test-fired from Vandenberg in their military applications as ICBMs.

On February 13, 1969, President Richard M. Nixon established his Space Task Group, which recommended that MOL be canceled and the facilities that had already been erected be mothballed. This left SLC-6 with a launch pad, a mobile service tower, and a launch control center in the immediate vicinity.

Throughout the 1960's, Vandenberg had served as the prime launching site for both military and scientific satellites. ICBMs were tested and stored in silos at the northwestern edge of the base. Some of the older pads were decommissioned as advanced facilities were constructed. The Air Force cooperated closely with NASA for many successful launches of rockets delivering communications and scientific satellites such as Echo 2 and the Explorer series.

The 1970's brought new excitement after the frustration over the cancellation of MOL. On June 15, 1971, the first Big Bird (KH-9) Air Force reconnaissance satellite was successfully launched atop the first known Titan III-D-Agena rocket. Before the end of the program in 1984, Vandenberg saw nineteen launches of this type of classified satellite. The base also facilitated launching the more recent KH-11, the maiden launch of which occurred on December 19, 1976.

To equip Vandenberg with a landing strip for

The B16-10 spacecraft processing center at Vandenberg Air Force Base. (NASA-KSC)

the shuttle orbiter, the runway at Vandenberg's airstrip was expanded to 4,500 meters. To transport the orbiter on the ground, a mobile orbiter lifting frame was developed. Finally, a special maintenance facility was built. In this facility, the shuttle was placed on seismic jacks, designed to protect the orbiter from the effects of an earthquake (since VAFB lies in an earthquake-prone area). Curtains hanging from the ceiling provided a clean environment (known as the "clean room") from the orbiter's nose to the rear end of its cargo bay. A sixty-ton bridge crane was designed to lift payloads from the shuttle into a pit 16 meters below the ground. Because of cost overruns, Vandenberg was only

equipped to perform minor repairs or service the shuttle after an aborted launch. Normally, the shuttle is prepared for launching at the Kennedy Space Center.

Plans to build a facility for the refurbishment and subassembly of the solid-fueled rocket boosters, and a related installation for the refurbishment of their parachutes, never reached the construction stage. Also, the building reserved for processing of the External Tank was used only as a storage area; one empty tank was stored there in 1984.

From its checkout stand, the refurbished orbiter was placed on a special seventy-six-wheel, self-leveling transporter and sent on a three- to four-hour journey of 25 kilometers to the launch complex. There, in a procedure different from that at Kennedy Space Center, the final assembly of shuttle, payload, External Tank, and solid-fueled rocket boosters occurred directly on the launch pad.

Payloads for a shuttle mission were prepared in the payload preparation room, from which they were lifted into the mobile payload "changeout" room. This structure, at 52 meters tall, could roll the 250 meters to the shuttle assembly building. There, payloads were put into the cargo bay of the shuttle.

Once the shuttle was fueled on the launch pad, warm air was blown over the liquid hydrogen section of the External Tank to prevent icing. The air came from two jet engines housed in a concrete shack 33 meters away and was ducted around and away from the tank, warming its surface but not contaminating it. During liftoff, the exhaust gases from the boosters and the liquid fuel engines escaped through two exhaust ducts.

In 1979, the launch control center of the abandoned MOL program was transformed into a modern facility of 13,800 square meters, its outer walls consisting of a solid structure 65 centimeters thick. Electronic equipment was protected from possible seismic shocks and potential overpressure during launch.

In January, 1985, Vandenberg's shuttle launch facilities underwent a verification check with the test flight orbiter *Enterprise*. The *Enterprise* was guided through the various stations, from the orbiter lifting frame to the launch pad. While still hoping for a more active role in an actual shuttle mission after the tragic accident of the *Challenger* space shuttle in January, 1986, the base suffered from a new accumulation of financial and technical difficulties. A design flaw was discovered in the two exhaust and flame ducts at the launch pad. In those ducts enough hydrogen could be trapped to cause an explosion that could destroy the orbiter on the pad. The safest redesign required radical alteration of the ducts at a cost of millions of dollars.

During the ensuing shuttle program delays, more safety concerns were expressed. Fog and occasional temperatures below 10° Celsius at Vandenberg could stress the reliability of some critical systems of the shuttle. Some quality-control personnel believed that the distance of 350 meters between the pad and the launch control center was not enough to safeguard the delicate electronic equipment from vibrations during liftoff, or even protect the center itself in case of a shuttle explosion on the launch pad.

Finally, in July, 1986, after an intense debate about the need for a West Coast launch center for the shuttle program, the Air Force recommended mothballing Vandenberg's shuttle complex. President Ronald W. Reagan adopted the most extreme option outlined by Congress: All shuttle-related facilities at Vandenberg were to be shut down until at least the mid-1990's.

In the meantime, launches of orbital payloads as well as missile tests continued. In January, 1988, Congress cut the budget for the shuttle complex at Vandenberg to $40 million and recommended cancellation of the Air Force's shuttle plans. Instead, some of the existing facilities at Vandenberg would be used for the Strategic Defense Initiative (SDI), the Advanced Launch System (ALS, a robotic alternative to the space shuttle), the Titan program, and minor NASA missions.

By 1988, forty-eight different types of launch vehicles, sounding rockets, and missiles had been launched from fifty-one different pads at Vandenberg. The base housed more than one thousand different buildings, which were connected by more than 830 kilometers of roads. After layoffs in 1986, it was estimated that more than ten thousand workers, both military and civilian, were employed there. While most launching activity has been redirected to South Vandenberg, North Vandenberg harbors eighteen vertical silo launchers for Minuteman 3 and Titan II ICBMs.

After the *Challenger* accident in January, 1986, NASA scrubbed plans to launch the shuttle from Vandenberg, and SLC was placed in mothballs. In 1995, the Air Force awarded a twenty-five-year lease of SLC-6 to Spaceport Systems International (SSI), including the payload processing facility and more than 100 acres of land for commercial launch facility construction. In 2000, SSI launched two satellites from the facility. The Joint Air Force Academy-Weber State University Satellite (JAWSAT) was launched aboard the Orbital Suborbital Program Space Launch Vehicle (dubbed Minotaur), a combination of rocket motors from the Minuteman II and Pegasus XL launch vehicles, on January 26. The MightySat II satellite was successfully launched on July 19 atop the Minotaur.

Context

Vandenberg Air Force Base has served well in its dual function as testing site for ballistic missiles and as launching base for military and scientific satellites. Because of its geographical position, the base has a natural edge over Kennedy Space Center, where solid land to the north and south prohibits direct launches into a polar orbit. Thus, during the 1960's, more satellites were launched from Vandenberg than from Kennedy, and the American space program achieved splendid results from the California launches.

Despite their being military satellites, the Discoverers, the first satellites launched from Vandenberg, conveyed a series of important scientific findings about the dynamics and mechanics of atmospheric reentry and space radiation. These probes were crucial to the development of orbital reconnaissance for national security p urposes.

Together with their Soviet counterparts, the Kosmos series satellites, the reconnaissance satellites that were launched mostly from Vandenberg helped to make the world a safer place by providing each superpower with the means of gathering more exact knowledge of the other's military and nuclear capabilities. SAMOS 2, launched from Vandenberg on January 31, 1961, stayed in orbit for one month and proved with its photographs that the United States had vastly overestimated the so-called missile gap between itself and the Soviets and that there was far less to fear from Soviet ICBM superiority.

Also, during grave international crises such as the Arab-Israeli wars and the war between Iraq and Iran, both superpowers rapidly launched reconnaissance and surveillance satellites to gather reliable information about the areas of concern. Then, the Big Birds launched from Vandenberg helped American politicians and military officials to make informed decisions and avoid haphazard guesswork. In terms of verification of arms accord treaties, spy satellites were equally helpful in the years before on-site inspections became politically possible.

In cooperation with NASA, Vandenberg has made possible the launch of the Landsat and Seasat satellites. Both programs have delivered invaluable data about the geography of Earth and have made remote sensing a reliable tool for the geological sciences. Agricultural projects can be assessed more easily, flood warnings can be served very quickly, and the effects of natural and human-made changes on the face and structure of Earth can be studied from a sharp, bird's-eye view.

In the future, Vandenberg's geographical advantages will guarantee the base its share of launch traffic. In addition to the commercial launch facilities operated by Spaceport Systems International, Space Launch Complex 3 East pad is used for Atlas-2 space boosters. The Titan IV-B booster was launched from Space Launch Complex 4 East, while the Titan II lifted off from SLC-4 West. SLC-2 is used for the Minuteman III missiles. The Air Force continues to use Vandenberg to spacelift satellites to polar orbit, operate and maintain the western range, support ICBM test launches, provide host base support, respond to worldwide contingencies, and support commercial space launch.

See also: Earth Observing System Satellites; Electronic Intelligence Satellites; Interplanetary Monitoring Platform Satellites; Meteorological Satellites: Military; Nuclear Detection Satellites; Ocean Surveillance Satellites; Space Centers, Spaceports, and Launch Sites; Spy Satellites; Telecommunications Satellites: Military; Telecommunications Satellites: Passive Relay.

Further Reading

Burrows, William E. *This New Ocean: The Story of the First Space Age*. New York: Random House, 1998. This is a comprehensive history of the human conquest of space, covering everything from the earliest attempts at spaceflight through the voyages near the end of the twentieth century. Burrows is an experienced journalist who has reported for *The New York Times*, *The Washington Post*, and *The Wall Street Journal*. There are many photographs and an extensive source list. Interviewees in the book include Isaac Asimov, Alexei Leonov, Sally K. Ride, and James A. Van Allen.

De Ste. Croix, Philip. *Space Technology*. London: Salamander, 1981. An exhaustive look at space exploration, with many cross-references to specific points of interest such as spy satellites, launch vehicles, and ballistic missiles. Places Vandenberg in the context of the U.S. space effort. Informative and ideal for a general audience, with many color and black-and-white photographs and a detailed bibliography.

Diamond, Edwin. *The Rise and the Fall of the Space Age*. Garden City, N.Y.: Doubleday, 1964. An early critique of the military's role in space and the relationship of NASA to the military-industrial complex. Describes and criticizes the programs situated at Vandenberg and places them in a national and international context. Argumentative but informative and readable. Contains no illustrations, but a few tables.

Heppenheimer, T. A. *Countdown: A History of Space Flight*. New York: John Wiley, 1997. A detailed historical narrative of the human conquest of space. Heppenheimer traces the development of piloted flight through the military rocketry programs of the era preceding World War II. Covers both the American and the Soviet attempts to place vehicles, spacecraft, and humans into the hostile environment of space. More than a dozen pages are devoted to bibliographic references.

Isakowitz, Steven J., Joseph P. Hopkins, Jr., and Joshua B. Hopkins. *International Reference Guide to Space Launch Systems*. 3d ed. Reston, Va.: American Institute of Aeronautics and Astronautics, 1999. This best-selling reference has been updated to include the latest launch vehicles and engines. It is packed with illustrations and figures and offers a quick and easy data retrieval source for policymakers, planners, engineers, launch buyers, and students. New systems included are Angara, Beal's BA-2, Delta III and IV, H-IIA, VLS, LeoLink, Minotaur, Soyuz 2, Strela, Proton M, Atlas-3 and -5, Dnepr, Kistler's K-1, and Shtil, with details on Sea Launch using the Zenit vehicle.

Jenkins, Dennis R. *Rockwell International Space Shuttle*. Osceola, Wis.: Motorbooks International, 1989. Includes a brief yet concise history of the orbiter and its predecessors. Contains dozens of close-up color and black-and-white photographs, detailing the orbiter's exterior and interior features, including the major subsystems. Some of the text, especially the data tables, is printed in extremely small type.

Klass, Philip J. *Secret Sentries in Space*. New York: Random House, 1971. Promilitary, this text emphasizes the technology behind spy satellites. It stresses the importance of these satellites for global safety. Klass is very good at placing Vandenberg in the broader context of an international espionage race. Includes photographs of the satellites, related hardware, and what they detect. For the technically inclined reader.

"Launch Vehicles." *Aviation Week and Space Technology*, January 17, 2000, 144-145. This table details the specifications for each of the current (2000) launch vehicles and spacecraft, as well as the current status.

Levine, Alan J. *The Missile and the Space Race*. Westport, Conn.: Praeger, 1994. This is a well-written look at the early days of missile development and space exploration. The book discusses the Soviet-American race to develop intercontinental ballistic missiles for defense purposes and their subsequent use as satellite launchers.

Sharpe, Mitchell R. *Satellites and Probes*. Garden City, N.Y.: Doubleday, 1970. Close description and analysis of the international development of robotic spaceflight. Stresses the contribution of the Vandenberg launch facilities to the success of U.S. military and scientific satellites. Compares this facility with its worldwide counterparts. Includes color and black-and-white photographs.

Shelton, William Roy. *American Space Exploration: The First Decade*. Rev. ed. Boston: Little, Brown, 1967. Chronicles the history of U.S. spaceflight and provides good background information about the first decade at Vandenberg. Full of relevant anecdotes and biographies of persons important to the space effort. Written in a very readable, journalistic style, this book includes illustrations.

Sloan, Aubrey B. "Vandenberg Planning for the Space Transportation System." *Astronautics and Aeronautics* 19 (November, 1981): 44-50. A detailed description of the original, ambitious plan for the Vandenberg shuttle complex. Written by the man who was largely responsible for overseeing the development of this facility in its early stages. Good history of the decision-making process involved in bringing the shuttle complex to Vandenberg. Supplemented with illustrations, diagrams, and a useful bibliography for further, more specialized studies.

Stockton, William, and John Noble Wilford. *Spaceliner: The New York Times Report on the Columbia Voyage*. New York: Times Books, 1981. A journalistic account of the space shuttle program from its conception to the first flight of the orbiter *Columbia* in April, 1981. Delineates the planned role of VAFB for further missions and talks about the decision to create a shuttle program with two major facilities in the eastern and western regions of the United States. Anecdotal and easy to read, with some fine black-and-white illustrations.

Yenne, Bill. *Secret Weapons of the Cold War: From the H-Bomb to SDI*. New York: Berkley Books, 2005. A contemporary examination of Cold War superweapons and their influence on American-Soviet geopolitics.

Zimmerman, Robert. *The Chronological Encyclopedia of Discoveries in Space*. Westport, Conn.: Oryx Press, 2000. Provides a complete chronological history of all piloted and robotic spacecraft and explains flight events and scientific results. Suitable for all levels of research.

Reinhart Lutz

Vanguard Program

Date: September 9, 1955, to September 18, 1959
Type of program: Scientific platform

Destined to be remembered for its failed attempts to launch the first U.S. artificial satellites, the Vanguard program generated important developments in rocket propulsion, satellite design, and satellite telemetry and tracking and eventually succeeded in launching three Vanguard satellites.

Key Figures

William H. Pickering (1910-2004), director, Jet Propulsion Laboratory
John P. Hagen (1908-1990), Vanguard program manager
Milton W. Rosen, Vanguard technical director
Donald J. Markarian, Martin Company's Vanguard program engineer
James A. Van Allen (b. 1914), professor of physics, University of Iowa
Wernher von Braun (1912-1977), technical director, Army Ballistic Missile Agency
T. Keith Glennan (1905-1995), NASA administrator
Dwight D. Eisenhower (1890-1961), thirty-fourth president of the United States, 1953-1961

Summary of the Program

Project Vanguard consisted of fourteen multistage launches, including test vehicles, with the stated purpose of placing the United States' first artificial satellite into Earth orbit.

In the early part of the twentieth century, Robert H. Goddard designed, tested, and successfully launched both liquid- and solid-fueled rockets. Rocket designers in the United States, the Soviet Union, Germany, and Austria were busy throughout the 1920's, 1930's, and 1940's developing the skills and technology that would later be used by rocket scientists of the post-World War II era. After World War II, a global awareness of the effective use of rockets forced the U.S. military to alter the scope and direction of its ballistic missile research. After the May, 1945, surrender to the Allies of roughly 120 German rocket scientists at Peenemünde—led by Wernher von Braun—the academic research community, industry, and military of the United States became engaged in dissecting, modifying, and eventually using German V-2 rockets for basic research. The U.S. Army, Air Force, and Navy were independently developing sounding rockets (rockets capable of suborbital flight) and spacecraft capable of orbital flight (research satellites). Because of the outbreak of the Korean War, however, military research was largely aimed at accurate delivery of nuclear or conventional weapon payloads.

On September 9, 1955, with the backing of President Dwight D. Eisenhower, the Department of Defense authorized the Naval Research Laboratory (NRL) to administrate a far-reaching program to design, build, and launch at least one artificial satellite during the International Geophysical Year, or IGY—a period of a year and a half running from July 1, 1957, to December 31, 1958. The United States' participation in the IGY, an international peacetime research effort, was problematic, because only military agencies had the hardware, financial backing, and personnel necessary to launch a satellite. Nevertheless, with the support of the Department of Defense, the Bureau of Aeronautics, the Office of Naval Research, the National Academy of Sciences, and the National Science Foundation, the NRL began implementing Project Vanguard. The NRL was backed by the Glenn L. Martin Company (GLM), Aerojet General Corporation, Grand Central Rocket Company (GCR), Allegany Ballistics Laboratory, General Electric (GE), International Business Machines (IBM), Minneapolis-Honeywell, and myriad U.S. universities and research facilities.

The planning and construction of the Vanguard launch vehicle's main stage was facilitated with testing in the late 1940's of the Viking rocket built by GLM for the NRL. Nurturing a working association with one of the nation's largest rocket builders, it was only natural that the NRL turn to GLM for help with the development and deployment of Vanguard test vehicles and satellite launch vehicles. Unfortunately, GLM also won a contract with the Air Force to construct the Titan missile, thus diluting the human resources it could expend on the Navy's Project Vanguard.

Despite these problems and others involving questions of responsibility and decision making, the NRL and GLM effort was successful. Rocket design specifications were completed by February, 1956, but again, disagreement between the NRL and GLM resulted in amendments and delay. The early Vanguards, Viking M-15's manufactured by GLM, were modified Viking M-10 missiles with a GE first-stage motor, an Aerojet Aerobee-Hi liquid-propellant second stage, and a GCR solid-fueled upper stage. The specter of the military dissipated, as the Viking missile was a renowned research vehicle that had long been used for atmospheric sounding.

Kerosene and liquid oxygen were used for the first stage of the Vanguard rocket, which generated 120,096 newtons of thrust. The fuel and oxidant were supplied to the engines by a hydrogen peroxide decomposition technique, which produced superheated steam and oxygen to drive turbine-driven fuel and liquid oxygen pumps. Helium gas supplied pressure for the fuel tanks. The second

stage produced 33,360 newtons of thrust with a mixture of white fuming nitric acid and unsymmetrical dimethylhydrazine, an explosive rocket fuel. The third stage was powered by a solid-propellant motor and generated roughly 13,344 newtons of thrust.

With the development of the transistor and miniaturized electronic circuitry, satellite instrumentation design programs at the NRL, Jet Propulsion Laboratory (JPL), and university laboratories swung into full gear. Satellites were built under the directorship of the National Academy of Sciences. Meanwhile, NRL telemetering and tracking systems were in advanced stages of development. The deployment and success of the Minitrack system was a result of rigorous testing and research; this highly accurate tracking system formed the backbone of satellite tracking during and after the Vanguard era. GLM developed computer programs that balanced weight against anticipated flight trajectory. IBM offered the NRL free computer time at the Massachusetts Institute of Technology. Because of loans and contracts, both optical and electronic tracking and telemetry stations were built and piloted largely by civilian personnel.

As originally planned, Vanguard test vehicles—numbered TV 0 through TV 5—would precede the production of Vanguard model satellite launch vehicles that would be used for missions SLV 1 through SLV 6; there would be a total of twelve launches. Because of engineering changes in payload shape, weight, and size; a moderate degree of launch failure; and the globally transmitted 20- to 40-megahertz beep of Sputnik 1, the original Project Vanguard firing schedule was accelerated and ultimately expanded to fourteen attempts with the launches of two backup test vehicles, TV 3BU and TV 4BU. In all, three highly successful satellites were placed into Earth orbit, and abundant new geophysical, atmospheric, and near-space data were gathered for processing and analysis.

The first Vanguard launch, TV 0, occurred on December 8, 1956, and successfully tested the Viking 13 first stage and the telemetry and tracking systems, which reached an altitude of 203.5 kilome-

Vanguard explodes on the launch pad after losing thrust during a 1957 test launch. (U.S. Navy)

tors, it toppled and exploded on the launch pad. This failure was a crushing blow to American pride, but it was also a stimulant for more careful engine system tooling and rocket construction techniques by GE and GLM, respectively, in preparation for future Vanguard firings.

In the interim, the Army Ballistic Missile Agency (ABMA) at the Redstone Arsenal in Huntsville, Alabama, had been researching modified German V-2 rockets with the assistance of the California Institute of Technology's Jet Propulsion Laboratory (JPL). On January 31, 1958, the ABMA-JPL team launched the Juno 1, a four-stage version of the Jupiter-C rocket, which carried the first U.S. satellite, Explorer 1. Project Vanguard's second attempt, with TV 3BU, ended in failure on February 5, 1958, when a control system problem resulted in loss of attitude control and eventual breakup of the TV 3BU Vanguard after fifty-seven seconds of flight.

Finally, on March 17, 1958, TV 4 placed Vanguard 1 into orbit with an apogee (the point farthest from Earth) of 3,966 kilometers and a perigee (the point nearest to Earth) of 653 kilometers. Vanguard 1 was designed to measure Earth's shape and atmospheric density. It is expected to orbit, with its third-stage motor casing, for more than two centuries. The 25.8-kilogram, 16.26-centimeter spacecraft fulfilled the Project Vanguard goal of launching an artificial satellite within the IGY. The TV 5 launch attempt on April 28, 1958, failed to orbit a 50.8-centimeter, spherical, 9.8-kilogram x-ray and environmental satellite because of second-stage shutdown problems.

The first production version of a satellite launch vehicle, SLV 1, was, on May 27, 1958, to carry a satellite nearly identical to that destroyed during the TV 5 attempt. Again, during second-stage burnout,

ters and a range of 157 kilometers. TV 1, or Viking 14, followed on May 1, 1957, and after a test of third-stage propulsion reached a range of 726 kilometers with a 195-kilometer peak altitude. Sputnik 1, launched October 4, 1957, interrupted the proposed Vanguard test launch schedule; TV 2 restored confidence in the project with a better-than-expected performance, involving the three-stage Vanguard prototype with inert second and third stages, on October 23, 1957. The Soviet Union launched the dog named Laika into space on Sputnik 2 on November 3, 1957. The Vanguard TV 3, a three-stage missile complete with the United States' first artificial satellite, was fired on December 6, 1957, but to the amazement of all spectators attitude control problems arose that resulted in firing of the third stage at an angle unsatisfactory for orbit. On June 26, 1958, SLV 2 encountered second-stage propulsion system shutdown after eight seconds. Failure to achieve orbit also plagued the SLV 3 launch on September 26, 1958, when the second stage underperformed.

With passage of the National Aeronautics and Space Act of 1958, the civilian National Aeronautics and Space Administration (NASA) was created to direct the U.S. space program. Vanguard 2 was successfully placed into orbit by NASA on February 17, 1959, with the launch of SLV 4. Weighing a total of 32.4 kilograms, the 10.7-kilogram payload and 21.7-kilogram third-stage motor casing attained an initial orbital apogee of 3,319 kilometers and a perigee of 556.7 kilometers. The Vanguard 2 payload, destined to orbit for roughly two hundred years, is a 50.8-centimeter spherical satellite that measures cloud distribution and the terrestrial energy cycle budget.

SLV 5 was fired on April 13, 1959; it attempted to launch a 33-centimeter satellite magnetometer and a 76.2-centimeter expandable sphere. Problems arose during the separation of the second stage and resulted in an aborted flight. SLV 6 was fired on June 22, 1959, but again, a second-stage failure sabotaged the launch with a rapid drop in fuel tank pressure, faulty ignition, and explosion of the helium tank because of overheating. The 50.8-centimeter satellite on board, designed to measure solar radiation and its reflection from Earth, failed to orbit.

NASA decided to use a spare backup launch vehicle, TV 4BU, in a final attempt to launch a third Vanguard satellite. Using a new solid-propellant third-stage motor, built by Allegany Ballistics Laboratory, the rocket successfully launched Vanguard 3 on September 18, 1959. The final Vanguard weighed 42.9 kilograms and included a 23.7-kilogram payload and 19.2-kilogram motor casing. The initial orbital apogee was 3,743 kilometers, and the perigee was 510 kilometers. The payload included a magnetometer, an x-ray device, and environmental measuring systems. In large part, Van-

guard 3 fulfilled the agenda of previous unsuccessful Vanguard launch attempts.

Contributions

With the end of NRL control of Project Vanguard in late September, 1958, and the creation of NASA in October, many key Vanguard personnel joined the NASA staff. As such, the knowledge gained during the Vanguard era was applied toward all subsequent U.S. space ventures. Growth in the fields of vehicle engineering, construction, fueling, and launch were predictable outcomes of Project Vanguard. The three Vanguard satellites that were successfully launched investigated energy fields in the boundary between Earth's atmosphere and space and carried out Earth-directed research.

Vanguard 1 achieved a high-apogee orbit and provided a tracking signal until 1965, thanks to its pioneering use of solar cells. By analyzing changes in orbital acceleration, the satellite detected a bulge in the atmosphere caused by solar heating and recorded a bulge in the Southern Hemisphere of Earth itself, thus confirming that Earth is nonspherical and Earth's interior is inhomogeneous.

Vanguard 2 performed an experiment that measured variations in cloud-top reflectivity. The results were not conclusive, but they contributed some data and perfected meteorological techniques used in later missions.

Vanguard 3 was a 50.8-centimeter sphere with sensors for solar x-ray and Lyman-alpha radiation measurements, environmental sensors, and a 66-centimeter projection supporting a magnetometer. Because of the Van Allen radiation belts, the radiation detectors were overloaded and failed to provide accurate data, but accurate temperature monitoring was accomplished over seventy days. Measurements of interplanetary cosmic dust showed a variable but significant influx of particulate matter estimated at 9,072,000 kilograms per day. The magnetometer provided accurate measurements of Earth's magnetic field, plus data on magnetic disturbance events and upper atmosphere lightning ionization.

Context

The technological impact of Project Vanguard has permeated every aspect of piloted and robotic space exploration and discovery. First and foremost, Project Vanguard developed budgeting, command, and scheduling techniques for effective launching of missiles. Advances in missile guidance, tracking, telemetry, and antennae systems; developments in electronic miniaturization; and solar cell and mercury cell use in satellites were made possible by Project Vanguard. The use of fiberglass casings and the eventual design of the Air Force Thor-Ablestar booster and NASA's highly successful Delta and Atlas launch vehicles are all direct descendants of Viking and Vanguard technology.

In addition, the Vanguards provided new views of Earth's geologic cycles, thus promoting environmental awareness. The subsequent research in electronics, computers, communication, and optics has changed the quality of human life. Finally, Project Vanguard demonstrated that the American military-industrial-academic complex was capable of far-reaching outer space missions.

See also: Delta Launch Vehicles; Explorers: Ionosphere; Explorers 1-7; Geodetic Satellites; Jet Propulsion Laboratory; Launch Vehicles; Marshall Space Flight Center; National Aeronautics and Space Administration; Pioneer Missions 1-5; Rocketry: Modern History; Vandenberg Air Force Base.

Further Reading

Bergaust, Erik, and W. Beller. *Satellite!* Garden City, N.Y.: Hanover House, 1956. Details for the layperson the planning for satellite launches during the International Geophysical Year (IGY) and includes detailed drawings of the Vanguard missile. Dated by post-IGY satellite development, this volume discusses the visionary goals of project scientists for the Vanguard and early Explorer missions.

Bille, Matt, and Erika Lishock. *The First Space Race: Launching the World's First Satellites.* Austin: University of Texas A&M Lightning Source Titles, 2004. A thorough historical perspective of the Army's efforts to launch a satellite (Explorer 1), the Navy's efforts to launch a satellite (the not-so-well-known NOTS), and the Vanguard program.

Braun, Wernher von, and Frederick I. Ordway III. *Space Travel: A History.* Rev. ed. New York: Harper & Row, 1985. A comprehensive, superbly illustrated history of post-World War II rocket research and abundant tables of data on missiles, missions, satellites, and piloted spacecraft. It contains a detailed bibliography and is recommended to rocketry enthusiasts.

Burrows, William E. *This New Ocean: The Story of the First Space Age.* New York: Random House, 1998. This is a comprehensive history of the human conquest of space, covering everything from the earliest attempts at spaceflight through the voyages near the end of the twentieth century. Burrows is an experienced journalist who has reported for *The New York Times, The Washington Post,* and *The Wall Street Journal.* There are many photographs and an extensive source list. Interviewees in the book include Isaac Asimov, Alexei Leonov, Sally K. Ride, and James A. Van Allen.

Caidin, Martin. *Vanguard! The Story of the First Man-Made Satellite.* New York: E. P. Dutton, 1957. A layperson's account of the developmental history of Project Vanguard up to, but not including, the launch of Vanguard 1. This well-illustrated but somewhat dated volume details missile development during Project Vanguard and traces the development of payloads, tracking, and telemetry.

Divine, Robert A. *The Sputnik Challenge: Eisenhower's Response to the Soviet Satellite.* New York: Oxford University Press, 1993. A thorough history of the political and scientific decisions made in regard to the International Geophysical Year and the efforts to launch an American Earth-orbiting satellite.

Green, Constance M., and Milton Lomask. *Vanguard: A History.* NASA SP-4202. Washington, D.C.: Government Printing Office, 1970. This detailed history of Project Vanguard describes the people, agencies, and administrative programs that led to the launchings of Vanguard missiles and satellites. Contains numerous photographs and diagrams. Mission goals and successes are described in detail. The appendices contain flight summaries for the Vanguard and Explorer programs and IGY satellite launches.

Hall, R. Cargill, ed. *Essays on the History of Rocketry and Astronautics.* NASA CP-2014. 2 vols. Washington, D.C.: Government Printing Office, 1977. A compilation of papers and memoirs written by active participants, this work traces international efforts in rocketry. Volume 2 concentrates on liquid- and solid-propellant rocket research before and after World War II. Accounts of the early phases of the Vanguard and Explorer projects are noteworthy.

Heppenheimer, T. A. *Countdown: A History of Space Flight.* New York: John Wiley, 1997. A detailed historical narrative of the human conquest of space. Heppenheimer traces the development of piloted flight through the military rocketry programs of the era preceding World War II. Covers both the American and the Soviet attempts to place vehicles, spacecraft, and humans into the hostile environment of space. More than a dozen pages are devoted to bibliographic references.

Lee, Wayne. *To Rise from Earth: An Easy to Understand Guide to Spaceflight.* New York: Checkmark Books, 1996. This is a good introduction to the science of spaceflight. Although written by an engineer with the NASA Jet Propulsion Laboratory, it is presented in easy-to-understand language. In addition to the theory of spaceflight, it gives some of the history of the human endeavor to explore space.

Mari, Christopher, ed. *Space Exploration.* New York: H. W. Wilson, 1999. Twenty-five articles (reprinted from magazines), covering the state of the space program at the time of publication, are divided into five sections: John H. Glenn, Jr.'s return to space, the exploration of Mars, the International Space Station, recent mining efforts by commercial industries, and new types of space vehicles and propulsion systems.

Parkinson, Claire L. *Earth from Above: Using Color-Coded Satellite Images to Examine the Global Environment.* Sausalito, Calif.: University Science Books, 1997. A book for nonspecialists on reading and interpreting satellite images. Explains how satellite data provide information about the atmosphere, the Antarctic ozone hole, and atmospheric temperature effects. The book includes maps, photographs, and fifty color satellite images.

Siddiqi, Asif A. *Sputnik and the Soviet Space Challenge.* Gainesville: University Press of Florida, 2003. This two-volume set provides a comprehensive history of Soviet space efforts at the dawn of the Space Age.

Charles Merguerian

Viking Program

Date: August 20, 1975, to November 13, 1982
Type of program: Planetary exploration

The Viking program, using a pair of heat-sterilized landers, acquired the first data from the surface of Mars. Relay communications equipment on the landers and orbiters enhanced the ability to send high-rate data to Earth and improved the scientific value of these first landings.

Key Figures

James S. Martin, Jr. (1920-2002), project manager
A. Thomas Young, mission director
Gerald A. Soffen (1926-2000), project scientist
Wright Howard, project manager
Israel Taback, project manager
Harper E. Van Ness, project manager
B. Gentry Lee, director of Science Analysis and Mission Planning
Conway W. Snyder, orbiter scientist, Primary Mission, and project scientist, Extended Mission
Harold Masursky (1923-1990), team leader, Landing Site Certification
Thomas A. Mutch (1931-1980), team leader, Lander Imaging
Michael H. Carr, team leader, Orbiter Imaging

Summary of the Program

Plans to conduct robotic missions to Mars were initiated shortly after the National Aeronautics and Space Administration (NASA) was established in the fall of 1958. At that time, the Jet Propulsion Laboratory (JPL) was a U.S. Army laboratory operated by personnel from the California Institute of Technology (CalTech). JPL was transferred by executive order from the Army to NASA on December 3, 1958. JPL and NASA had reached agreements that the Laboratory would be principally involved in robotic exploration of the Moon and the planets.

The JPL staff started to plan a series of missions; the early ones involved small spacecraft, with larger, more complicated craft intended for subsequent missions, when more powerful launch vehicles were scheduled to be available. The initial goals were to demonstrate to the country that NASA was an aggressive organization and that it was not necessary to use the military to conduct the civil space program. The lunar and planetary missions included Ranger missions to the Moon and a series of Mariner spacecraft for exploration of Venus and Mars. The Mariners were to be followed by Voyager spacecraft that would require the development of Saturn launch vehicles before they could be flown.

During the early 1960's, developmental problems in the improved launch vehicles and launch failures resulted in the postponement of the earliest missions and deferred the more complicated ones.

The Mariner 2 spacecraft in 1962 and the Mariner 5 spacecraft in 1967 were sent to Venus. The

The Viking Lander discovers Martian Dunes. (NASA CORE/ Lorain County JVS)

Mariner 4 probe was successfully sent to Mars in 1964. Ambitious plans were supported by NASA for large Mars landers and orbiters to be launched by a single Saturn V in 1973. Yet on August 30, 1967, the Voyager project was canceled—because of the lack of adequate support from Congress, which at that time was concerned about the war in Southeast Asia. At the time, a dual flyby mission to Mars for 1969 was under development at JPL. A subsequent deep-space mission to Jupiter, Saturn, and Uranus, launched in 1977, was designated by NASA as Voyagers 1 and 2.

For several years prior to the cancellation of the Mars Voyager program, both the Langley Research Center and JPL had been planning robotic missions to Mars. The research center was interested in the technical challenge of the landing vehicles' passing through the thin Martian atmosphere and in the opportunity to develop and manage an important flight project following Langley's successful role on the Lunar Orbiter project. The design studies that had been performed by the Voyager staff (NASA, JPL, and contractors) were of considerable value in preparing plans for a Mars landing

mission at a cost considerably lower than estimates had been for the canceled Voyager project.

In December, 1968, NASA supported a soft-lander mission with a ninety-day surface lifetime goal along with a Mariner 1971-class orbiter to be launched in 1973 to continue the exploration of Mars. The project's responsibility was assigned to Langley. NASA assigned the responsibility for the development of the soft-lander system to Langley Research Center, the development of the orbiter system and the tracking and data acquisition system to JPL, and the development of the Titan-Centaur launch vehicle systems to the Lewis Research Center. Further, the control of the flights, following launch from Cape Kennedy, Florida, was to be conducted from JPL facilities in California.

In the spring of 1969, the Viking project office at Langley issued a request for proposals for the development of the Viking lander. The Martin Marietta Corporation's Aerospace Division in Denver, Colorado, won the contract and immediately set about developing the state-of-the-art lander system.

Following the successful Apollo landings on the Moon in 1969, there was an antitechnology sentiment and a subsequent shortage of federal funds, causing the Bureau of the Budget to make substantial cuts in the NASA budget for fiscal year 1971. To operate with these fiscal restraints, NASA slipped the Viking project from a launch in 1973 to 1975, when the next launch window opened. At Langley, James Martin made the decision to slow the work on the Viking orbiter substantially, because the risks associated with its development were considered lower than those of the new lander system. The JPL project staff was reduced by almost two hundred people in less than a month to be able to apply the available funds to the new technology. (It was fortunate that JPL was able to reassign these people to the Mariner 8, 9, and 10 projects.)

The Viking lander was considerably more complex than the Surveyor spacecraft that had successfully landed on the Moon. The Viking lander had to use the thin Martian atmosphere to slow it initially so that it could successfully open a large parachute that carried the vehicle to within 1,400 meters of the surface, where the terminal descent engines took over to soft-land the craft. Additionally, all the equipment contained on the lander had to be sterilized to ensure that the life-detection instruments on the lander did not detect life-forms that it had carried from Earth to Mars. The lander had to survive a sterilization cycle of forty hours in which the minimum temperature in the lander was 112° Celsius. This included all the scientific and electronic equipment, parachutes, liquid propellants, and a radioactive power source.

During the 1975 launch window, the flight time to Mars was greater than it would have been in 1973, and the spacecraft would be farther from the Sun and Earth. Thus, larger solar panels were necessary, along with a larger high-gain antenna to communicate with Earth. Some additional redundancy of engineering subsystems was included to increase the chance of success during the longer flight.

The Viking project was the most complicated robotic mission in space conducted by NASA. Activities at four NASA centers—the Langley and Lewis research centers, JPL, and Kennedy Space Center (KSC)—had to be coordinated. The Titan III-E-Centaur launcher was used on the two Viking flights. Four separate spacecraft were flown (two orbiters and two landers). Communication links were established between Earth and each of the four spacecraft; in addition, radio links were provided between the orbiters and landers to recover high-rate data from the landers when they were on the surface of Mars.

To ensure that all elements of the project were efficiently coordinated, the project's manager, James Martin, established a management council that met every month at Langley. The managers of each of the systems developed for the project spent roughly two days a month describing the progress and problems of their systems along with plans for activities currently under way. Martin also instituted a "Top Ten Problem List" to highlight the most significant problems that might put the program's success at risk. These difficulties were reviewed in detail at each management council meeting, and Martin had to be satisfied that each problem had been resolved before it could be removed from the list. A total of forty were identified and eventually resolved during the developmental phase of the project.

In a few cases, Martin supported the parallel development of competing designs to ensure that suitable equipment would be available for the mission. The opportunity to fly to Mars occurs about every twenty-five months, and if some part of the flight systems were not available, the project would miss its launch opportunity during August-September, 1975. Ensuring that costs were not overrun was a challenge. Viking had been NASA's most expensive robotic project to date; even so, it was necessary to reduce the number of available subsystems to cut costs.

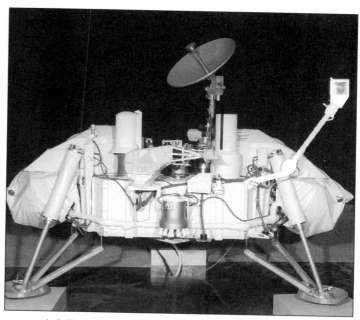

A full view of the Viking Lander spacecraft. (NASA)

As a further step to reduce costs, NASA conducted extensive joint tests with structural models of the lander and orbiter and several separate interface tests of the prototype equipment that provided either data or electrical power to the lander or orbiter. The first full physical and operational tests utilizing complete landers and orbiters did not occur until the flight systems had been delivered to the KSC for the final launch preparations. Fortunately, the project manager had included schedule reserves of roughly one month for planned activities at the Cape to ensure that the first Viking would be launched on time. This reserve became vital when a severe thunderstorm activated the electrical power of the initial orbiter on the launch pad, and the flight orbiter and lander had to be switched with those intended for the second launch to replace the discharged batteries.

Contributions

The Viking mission provided the first and second soft landings on the surface of Mars. Observations and measurements of the surface and atmospheric conditions were obtained for far longer than the three months for which the landers had been designed. The two orbiting spacecraft also operated significantly beyond their design goals; as a consequence, substantially more scientific and visual data were obtained from Mars than had been originally expected.

The flight operations activities were complicated. More than five hundred people were required to conduct these operations during the prelanding and initial postlanding operations. A large team of scientists was present at the JPL operations center, along with personnel from the Viking project office, lander technical specialists from Martin Marietta, and orbiter and tracking and data acquisition personnel from JPL.

The primary mission ended on November 15, 1976, during the solar conjunction (when Mars was on the opposite side of the Sun from Earth and no communications were possible with any of the spacecraft). When communications were again possible in mid-December, 1976, an extended mission was originated. On April 1, 1978, the extended mission was concluded and project management responsibility was transferred from Langley to JPL for a continuation mission that was designed to acquire scientific information as long as the spacecraft continued to operate.

Viking Orbiter 2 developed leaks in its attitude control gas equipment, and its operations were terminated on July 25, 1978. Viking Lander 2 was shut down on April 12, 1980. Orbiter 1 was silenced on August 7, 1980. Communications with Lander 1 became limited to a once-a-week transmission to Earth at a low data rate, and limited science data were produced. A total of 51,539 Orbiter images and more than 4,500 Lander images had been returned to Earth.

Context

The Viking missions were the fifth and sixth to visit Mars and the first two soft landings on the surface that provided data. The Soviets had tried earlier, but with minimal success. In 1971 the Soviets launched two identical missions to Mars with orbiters and landers that weighed eight times more than the U.S. orbiter Mariner 9, which was also flown in 1971. The Soviet Mars 2 crashed on Mars on November 27, 1971. The Mars 3 lander landed on December 3, 1971, but communicated with its orbiter for only twenty seconds. The large dust storm on Mars at that time outlasted the Soviet orbiters' lifetimes, and little useful data resulted. The Soviets also received some preliminary data from their Mars 6 probe in 1974.

Mariner 4, launched in 1964, had provided the first flyby views of Mars. In 1969, Mariners 6 and 7 conducted flyby missions which supplied roughly two hundred times more imaging data than had been acquired by Mariner 4. In 1971, Mariner 9—the first American Mars orbiter—was successfully flown. When Mariner 9 arrived at Mars on November 13, 1971, a severe dust storm was in progress, which obscured all surface features. After several days, the peaks of the four large volcanic craters on Mars appeared above the pall of dust. In February,

1972, the dust cleared at the lower altitudes, which were of particular interest for Viking landing sites, and useful photographs were obtained. In all, Mariner 9 provided 7,329 pictures covering 85 percent of the Martian surface. As a result, an initial mapping of Mars became possible.

The Mariner missions had showed that Mars has a small magnetic field. Its atmospheric density, altitudinal variations, and surface and atmospheric temperatures had also been determined by the Mariner probes. Before the Viking program, Mars was known to have volcanoes, and water was observed in its polar caps. It was known that the atmosphere consisted principally of carbon dioxide, but the Viking lander determined that a small amount of nitrogen was also present. The 4,000-kilometer-long Valles Marineris had been photographed by Mariner 9.

The scientific instruments on the lander were intended to determine whether life existed on Mars. As a result of the Viking mission, scientists concluded that life does not exist at the two locations where the spacecraft landed. The severe ultraviolet radiation on the planet's surface would have destroyed any organic compounds. It is probable that the severe radiation has highly oxidized chemicals on the surface of Mars. Nevertheless, several scientists still believe that the possibility for life on Mars exists, either farther below the surface of the planet or nearer the polar regions, where it is

clear that more water is present than exists at the Viking landers' equatorial locations.

A considerable amount of water must have been present long ago to form the channels that have been observed on Mars. The question is, where did the water go? There is frozen water in the polar caps, but is there also water below the surface? The current atmospheric density is so low that liquid water cannot exist on the surface of Mars. Evidence from the Mars Pathfinder mission in 1997 showed that Ares Vallis was probably once a floodplain and that sand exists on Mars, suggesting free-flowing water at one time in the planet's history. Questions surrounding the history and evolution of water—and therefore perhaps life—on Mars have not yet been answered. However, the Mars Exploration Rovers and the European Space Agency Mars Express spacecraft in 2004 provided tantalizing evidence strongly suggesting that portions of the Martian surface had once been covered with water, and that layers of sedimentary rocks might well have resulted from that circumstance. Questions of life on Mars remained unanswered, but future probes were under design to investigate such questions, building upon the rich scientific harvest of the Viking spacecraft.

See also: Hubble Space Telescope: Science; Langley Research Center; Mariner 8 and 9; Mars Observer; Planetary Exploration; Search for Extraterrestrial Life; Vanguard Program; Viking 1 and 2.

Further Reading

Baker, Victor R. *The Channels of Mars.* Austin: University of Texas Press, 1982. Includes many illustrations obtained by the Viking orbiters and landers along with a few obtained by Mariners 4, 6, 7, and 9. This text includes a chapter about the existence of water, or water ice, on Mars.

Burrows, William E. *This New Ocean: The Story of the First Space Age.* New York: Random House, 1998. This is a comprehensive history of the human conquest of space, covering everything from the earliest attempts at spaceflight through the voyages near the end of the twentieth century. Burrows is an experienced journalist who has reported for *The New York Times, The Washington Post,* and *The Wall Street Journal.* There are many photographs and an extensive source list. Interviewees in the book include Isaac Asimov, Alexei Leonov, Sally K. Ride, and James A. Van Allen.

Corliss, William R. *The Viking Mission to Mars.* NASA SP-334. Washington, D.C.: Government Printing Office, 1975. Suitable for high school and college levels. Contains a description of the spacecraft and launch vehicle used, along with information on the scientific exploration accomplished during the mission.

Ezell, Edward Clinton, and Linda Neuman Ezell. *On Mars: Exploration of the Red Planet, 1958-1978*. NASA SP-4212. Washington, D.C.: Government Printing Office, 1984. The official history of the Viking missions. Written by historians, this work is suitable for high school and college levels. Contains substantial background information on earlier missions to Mars along with numerous illustrations, photographs of the Viking staff and spacecraft, and pictures returned from the mission.

Hamilton, John. *The Viking Missions to Mars*. Edina, Minn.: ABDO and Daughters Publishing, 1998. Although this is a juvenile book, it does give a good overview of the Viking missions to the Red Planet.

Hartmann, William K. *Moons and Planets*. 5th ed. Belmont, Calif.: Thomson Brooks/Cole Publishing, 2005. Provides detailed information about all objects in the solar system. Suitable on three separate levels: high school student, general reader, and college undergraduate studying planetary geology.

Heppenheimer, T. A. *Countdown: A History of Space Flight*. New York: John Wiley, 1997. A detailed historical narrative of the human conquest of space. Heppenheimer traces the development of piloted flight through the military rocketry programs of the era preceding World War II. Covers both the American and the Soviet attempts to place vehicles, spacecraft, and humans into the hostile environment of space. More than a dozen pages are devoted to bibliographic references.

Lee, Wayne. *To Rise from Earth: An Easy to Understand Guide to Spaceflight*. New York: Checkmark Books, 1996. This is a good introduction to the science of spaceflight. Although written by an engineer with the NASA Jet Propulsion Laboratory, it is presented in easy-to-understand language. In addition to the theory of spaceflight, it gives some of the history of the human endeavor to explore space.

Mari, Christopher, ed. *Space Exploration*. New York: H. W. Wilson, 1999. Twenty-five articles (reprinted from magazines), covering the state of the space program at the time of publication, are divided into five sections: John H. Glenn, Jr.'s return to space, the exploration of Mars, the International Space Station, recent mining efforts by commercial industries, and new types of space vehicles and propulsion systems.

Mishkin, Andrew. *Sojourner: An Insider's View of the Mars Pathfinder Mission*. New York: Berkley Books, 2003. A thorough exploration of the Mars Pathfinder mission that also provides a detailed historical perspective on the exploration of the surface of Mars (begun by Viking).

Pollack, James B. "Mars." *Scientific American* 233 (September, 1975): 16, 106-117. Contains an excellent overview of pre-Viking knowledge of the evolution of Mars and its atmospheric and surface characteristics.

Sagan, Carl. "The Solar System." *Scientific American* 233 (September, 1975): 16, 22-31. Suitable for both high school and college audiences. Describes each of the planets of the solar system and briefly discusses the then-current or planned missions for planetary exploration.

Henry W. Norris

Viking 1 and 2

Date: August 20, 1975, to November 13, 1982
Type of spacecraft: Planetary exploration

The Viking mission to Mars was the first long-duration, intensive exploration of the surface of another planet. Viking 1 and 2 each consisted of a soft lander, which examined the physical and chemical properties of the Martian surface and atmosphere, and an orbiter, which extensively surveyed Mars from orbit and served as an Earth-Mars transport vehicle and communications relay for its lander.

Key Figures

James S. Martin, Jr. (1920-2002), project manager
A. Thomas Young, mission director
Gerald A. Soffen (1926-2000), project scientist
B. Gentry Lee, director of Science Analysis and Mission Planning
Conway W. Snyder, orbiter scientist, Primary Mission, and project scientist, Extended Mission
Harold Masursky (1923-1990), team leader, Landing Site Certification
Thomas A. Mutch (1931-1980), team leader, Lander Imaging
Michael H. Carr, team leader, Orbiter Imaging
Robert B. Hargraves, Magnetic Properties Team
Seymour Hess, Meteorology Team leader

Summary of the Missions

Vikings 1 and 2 each consisted of a robotic soft lander and an orbiter that carried the lander to Mars and served as a communications relay station from Mars orbit. The Viking mission's primary objectives were to investigate the physical characteristics of Mars and to search for Martian life. To this end, the Viking spacecraft carried the experiments of thirteen teams of scientists who, throughout the mission, maintained remote control over their experiments from the Mission Control center at the Jet Propulsion Laboratory (JPL) in Pasadena, California.

Because Viking was to be the first attempt by the National Aeronautics and Space Administration (NASA) at soft-landing a craft on another planet, the spacecraft designs required tremendous technological advances over previous vehicles. In addition to the usual scientific tasks, the orbiters would carry the landers to Mars, hold them in orbit while certifying the preselected landing sites as safe, position and release the landers for descent to the surface, and serve as relay stations for communications between the landers and Earth. To meet these ends, the highly successful Mariner design—a flat, octagonal body with four protruding rectangular solar panels—was adopted, enlarged, and considerably modified for the task. Compared to their Mariner predecessors, the resulting Viking orbiters were larger (9.75 meters across the extended solar panels) and heavier (2,328 kilograms launch weight), and they carried vastly superior computer command systems (two redundant 4,096-word computers, either of which could operate the craft independently from programmed instructions or from commands sent from Earth).

The Viking Orbiter and Lander in outer space. (NASA)

Launch Pad Complex 41, the first scheduled for August 11, 1975, and the second ten days later. As the launch dates approached, a faulty thrust vector control valve (one of twenty-four valves that give fine guidance control over the thrust direction), followed by an accidental battery draining, caused the two craft to be switched. The first launch, designated Viking 1, occurred nine days late on August 20 at 2:22 P.M., Pacific daylight time, or 21:22 Coordinated Universal Time (UTC). The original craft was then repaired and prepared for launch as Viking 2. After the thrust vector control valve had been repaired, further repair work on its S-band (low-frequency) radio receiver was required. Viking 2 finally lifted off on September 9 at 11:39 A.M. Pacific daylight time (18:39 UTC), only three minutes before an approaching storm would have canceled the launch. To compensate for these launch delays, minor course corrections were made so that Viking 1 would reach Mars only one day behind its original schedule and Viking 2 would arrive on schedule.

On May 1, 1976, Viking 1's sensors first detected Mars—a calibration picture was taken from a distance of 11 million kilometers. Beginning six weeks later, a series of color photographs taken of Mars during approach showed that the dust storms that had hampered Mariner 9 were absent: The 4,500-kilometer-long Valles Marineris (Valley of the Mariners) and the 27-kilometer-high volcano Olympus Mons (Mount Olympus) were clearly visible, as were water-ice fogs and various surface and atmospheric brightenings.

Even as these images were being planned and taken, a leaking pressure regulator was found to be threatening to overpressurize the Viking Orbiter 1 rocket propellant tanks. The pressure was relieved by ordering two unscheduled engine burns (on June 10 and 15), which slowed Viking 1 by 4,000 ki-

The landers were essentially a new design. Hidden beneath a panoply of protruding scientific instruments, cameras, and antennae was a flat (0.457-meter-high), hexagonal body (1.494 meters across) with alternately long (1.1-meter) and short (0.56-meter) sides. A shock-absorbing landing leg extended from each of the three short sides, while inside the body were self-contained, miniaturized laboratories for analyzing the Martian soil and looking for signs of microbial life in it.

To avoid bringing to Mars any terrestrial organisms that might interfere with the search for life there (or, worse yet, to inadvertently contaminate Mars), each Viking lander was sterilized prior to launch and then sealed inside a contamination-proof bioshield spacecraft, where it would remain until safely outside Earth's atmosphere. Because the orbiters were not sterilized, the initial launch trajectories were set so the Vikings would miss Mars altogether (rather than crash-land there) if the craft proved uncontrollable after launch.

Both Vikings were to be launched on Titan III-E-Centaur launch vehicles from Cape Canaveral's

lometers per hour and delayed its arrival at Mars by about six hours. Finally, on June 19, a 38-minute engine burn—the longest burn to date in deep space—expelled 1,063 kilograms of propellant and placed Viking 1 into Mars orbit.

An orbit trim maneuver two days later settled Viking 1 into a highly elongated, synchronous orbit (an orbit in which the craft's orbital period is identical to the rotation period of the planet below—in this case 24 hours, 39 minutes, and 36 seconds). With each orbit, Viking 1 dived to a periapsis (closest approach to the planet) of only 1,514 kilometers—directly above its intended landing site—and then ascended to an apoapsis (farthest point from the planet) of 32,800 kilometers. The initial periapsis passage (on June 19) was designated "P0," the next "P1," and the others followed in this sequence; the historic first Mars landing was scheduled for P15, which (not coincidentally) would occur on July 4, 1976—the American bicentennial.

The primary landing target (designated "A1" for "mission A, site 1") was at 19.5° north latitude and longitude 34° west in Chryse, a now-dry delta region in the outflow pattern of an ancient flood channel. This site was selected because it was at a low elevation (where the higher atmospheric pressure made landing easier), close to the equator (as required by the lander's approach angle), and relatively level and devoid of high winds (which made landing safer), and because it showed evidence of past water (which enhanced the chances of finding life).

The first detailed photographs of A1, taken on P3 and P4, showed spectacular image quality but frightening details. What Mariner 9's cameras had picked up as a smooth plain was revealed to be a confusion of craters, depressions, knobs, and islands—apparently too hazardous a terrain in which to attempt a landing. This assessment was complicated by the fact that, despite their impressive quality, the Viking orbiter photographs could not reveal objects less than 100 meters in size; the greatest hazard to the Viking landers would be from objects 0.1 to 1 meter in size.

In the interest of safety, the planned landing was postponed, and the Viking flight team began feverishly looking for an alternative landing site. Orbiter 1 was directed to photograph areas immediately to the south and northwest of A1 as well as the Viking 2 primary target site B1 (44° north latitude, longitude 10° west) in Cydonia and the alternative site C1 (6.5° south latitude, longitude 42.75° west) in Capri. Viking geologists, assisted by a tireless team of undergraduate interns, worked around the clock assessing the potential landing hazards suggested by these photographs. Leonard Tyler, a member of the Radio Science Team, analyzed data from the Arecibo and Goldstone radio observatories on Earth which, through the pattern of radar signals reflected from Mars, could be used to estimate the average surface roughness resulting from centimeter-sized objects scattered over large areas

This first photo of the Martian surface was taken by Viking 1. (NASA CORE/Lorain County JVS)

of the Martian surface. All this information was used by the Landing Site Certification Team, led by astrogeologist Harold Masursky, to form geologic models from which estimates of the landing hazards presented by the unobservable meter-sized objects could be made.

In the end, the leader of the Magnetic Properties Team, Robert B. Hargraves, suggested looking farther "downstream"—that is, northwest—from A1, where the ancient floods might have deposited fine-grained sediments and left a relatively smooth surface. On July 8, Viking 1's orbit was altered to bring its periapsis point over a newly designated A1-NW site, 300 kilometers northwest of A1. After further evaluation of the area, a site 240 kilometers due west of A1-NW was selected as the final landing target. This target lay at 22.4° north latitude and longitude 47.5° west, just within the western edge of Chryse Planitia (the Plains of Gold). The landing was set for July 20, 1976; coincidentally, this would be the seventh anniversary of the first piloted lunar landing, a date now known as Space Exploration Day.

On the morning of July 20, Mars was 360 million kilometers from Earth. At that distance, one-way radio communication took nineteen minutes—too long for flight controllers on Earth to intervene in the landing. Viking would have to land on its own. At 1:51:15 A.M., Pacific daylight time (08:51:15 UTC), Viking Orbiter 1 released Viking Lander 1; a twenty-minute burn of the lander's own engines then nudged it out of orbit to begin a long, looping descent toward Mars. At 5:03 A.M. Pacific daylight time (12:03 UTC), Lander 1 entered the top of the Martian atmosphere at a shallow 16° angle; it would need to travel nearly horizontally for 1,000 kilometers in that thin atmosphere before its speed decreased sufficiently for landing. Radar monitored the descent while other instruments collected data on the composition and physical characteristics of the atmosphere and radioed them back to Earth. At 5:10 A.M. Pacific daylight time (12:10 UTC), at an altitude of 5,906 meters, Lander 1 opened its parachute. Forty-five seconds later, at an altitude of 1,462 meters, the parachute was jettisoned, and

three terminal descent engines, which exhausted sterile propellants in an outward fan-pattern to avoid contaminating or disturbing the landing site below, immediately burst into life. They slowed the lander to a scant 2 meters per second (5 miles per hour), at which speed it soft-landed at 22.46° north latitude, longitude 47.82° west—within 20 kilometers of the targeted site.

Immediately upon landing, the rate at which Viking Lander 1 was transmitting data back to Earth automatically increased from 4,000 bits per second to 16,000 bits per second. Nineteen minutes later, at 5:12:07 A.M. Pacific daylight time (12:12:07 UTC), mission controllers at JPL received that increased bit rate as the first indication of Viking's safe landing. Their immediate shout of "Touchdown. We have touchdown," echoed to the cheers of eight hundred Viking team members at JPL.

On Mars, it was late afternoon (4:13:12 P.M. local lander time upon landing), and Lander 1 was busy. Some 25 seconds after touchdown, it began taking the first photograph—an image of the ground around its right front footpad. During that four-minute exposure, the lander also erected and aimed a high-gain antenna at Earth and deployed a weather station. Within fifteen minutes, a second photograph was completed; both were transmitted to Earth, where they were processed immediately.

By 5:54 A.M. Pacific daylight time (12:54 UTC), excitement mounted at JPL as the first of these photographs—the first successful picture ever taken from the surface of another planet—appeared on the laboratory monitors. The image was incredibly sharp and clear. Small rocks, up to 10 centimeters in size, littered the area. On the right, the circular footpad was clearly visible; it had barely penetrated the ground, indicating that the surface was solid. On the left, dust still settling from the landing had left dark streaks on the photograph. Undeniably, Viking Lander 1 was on Mars.

The second photograph showed a 300° panorama of the Martian landscape. A series of ridges and depressions led to a horizon 3 or 4 kilometers in the distance. Boulders, perhaps meters in size, and smaller rocks were strewn about everywhere. In all, the scene was remarkably Earth-like—reminiscent of a desert in the American Southwest—except that nowhere was there any visible sign of vegetation or any other form of life.

The first color picture, taken on sol 1, mistakenly showed a delightfully blue Earth-like sky. (A sol is a Martian day as defined by one rotation of Mars—equal to 24 hours, 40 minutes of Earth time. The day of the landing was designated sol 0 and the next sol 1.) Subsequent corrections to the color balance revealed a salmon-pink Martian sky (made so by dust particles suspended in the atmosphere) overlooking a rusty, reddish-orange landscape. Meteorology Team leader Seymour Hess made history by issuing the first weather report from another planet: "Light winds from the east in the late afternoon, changing to light winds from the southwest after midnight. . . . Temperature range from −122° Fahrenheit (188 kelvins), just after dawn, to −22° Fahrenheit (243 kelvins). . . . Pressure is steady at 7.70 millibars."

On sol 3, the remaining 60° of landscape was scanned, and a field of large sand dunes with a huge boulder, affectionately dubbed "Big Joe," was found. Viking scientists were struck by the realization that if Lander 1 had come down on Big Joe, it would never have survived the landing. On sol 12, a picture of the left front footpad revealed yet another surprise: The footpad was buried several centimeters deep in soft sediment.

Also on sol 3, an attempt was made to extend a 3-meter-long surface sampler arm intended to dig up and collect soil samples; the arm stalled upon retraction. Viking engineers, using a duplicate lander on Earth, concluded that a locking pin used to stow the sampler arm during flight had failed to release. They then "repaired" it by sending a sequence of extension and rotation commands that caused the pin to drop free. On sol 8, the arm successfully dug trenches and delivered soil to biology, inorganic chemistry, and molecular analysis instruments inside the lander body. Early responses from the biology instruments were suggestive of life having been found. In a climate of utter hopefulness

coupled with scientific caution, it gradually became clear that the responses were most likely the result not of biota but of chemical reactions in the Martian soil. The next week and a half was spent attempting to make further repairs to the sampler arm, the molecular analyzer, and a radio transmitter, while obtaining more surface samples, photographing the surroundings in detail, and puzzling over results.

Viking 2 entered Mars orbit flawlessly on August 7, immediately adopting an orbit that would allow its periapsis point to "walk" around the planet, passing over the entire 40° to 50° north latitude band every eight days. With assistance from Orbiter 1, Orbiter 2 carefully surveyed Lander 2's prime landing site (B1) in Cydonia, its backup landing site (B2) near Alba Patera, and a hastily chosen additional backup site (B3) in Utopia Planitia. After much analysis, and with some misgivings, the B3 site at 47.9° north latitude and longitude 225.8° west was picked by a Viking crew too exhausted to consider further searching.

On September 3, 1976, Viking Lander 2 headed for a suspenseful landing. When it separated from its orbiter, flight controllers temporarily lost contact with the orbiter. Unable to receive progress reports relayed from the lander, they could only sit and wait as Lander 2 landed automatically and, miraculously, sent home the high-transmission-rate direct signal that indicated a safe landing. Touchdown was recorded at 3:37:50 P.M. Pacific daylight time (22:37:50 UTC).

The scene at Utopia was remarkably reminiscent of that at Chryse: A desolate, red, rock-and-boulder-strewn landscape was seen. There were, however, differences in detail: Utopia was generally flatter than Chryse, and it had perhaps twice as many rocks. Most of them were irregularly broken and laced with pits and holes. Apparently, the unphotographable back leg of Lander 2 had, in fact, come down on top of one of those rocks: The Lander was tilted about 8° from the upright position. Lander 2 set about a science analysis sequence similar to that followed by Lander 1. It also set another precedent in planetary exploration: It used its sampler arm to push aside a rock and sample the soil beneath.

Throughout all this, scientific activities with Viking 1 continued unabated. The harried Viking flight team had four spacecraft—two orbiters and two landers—with which to conduct scientific studies simultaneously. Each of the thirteen teams of scientists had cooperated closely in designing their instrumentation to fit and operate in the close confines of the Viking craft. Now they cooperated in sharing the available resources of the craft (for example, data storage and transmission and electrical power capabilities) and in making sure that each of the various investigations complemented the efforts of the others. These goals were all amply met as science investigators followed a hectic routine.

Each day at noon, the scientists met to share the findings of the previous twenty-four hours, discuss their significance, and plan future activities. Every two days, on average, a new set of instructions for future operations was laid out and "uplinked" (sent by radio telemetry) to the craft on Mars. Thus, while the Vikings always had in their onboard computers complete instructions for operating automatically in case the uplink capability was lost, the instructions were continually updated to meet the ongoing needs of the project. This updating capability contributed immeasurably to the mission's success as experimental procedures were altered in response to unexpected findings. Also—and perhaps more important—problems with the craft themselves were continually analyzed and in some cases repaired by remote control.

In an unprecedented acknowledgment of public interest in Viking, Viking scientists reported their findings and progress daily to the more than one hundred members of an international press corps resident at JPL. So close was the cooperation with the press that on one occasion the Viking biology team actually performed an experimental sequence suggested by reporter Jonathan Eberhart of *Science News*. From the seeming chaos of these day-to-day operations, a new picture of Mars gradually began to emerge.

The routine continued until November 15, 1976, when the orbit of Mars was about to take it behind the Sun. Radio communications with the Vikings would then be impossible, so the primary mission was declared ended. In mid-December, Mars reemerged from behind the Sun, and an extended mission began, mostly with replacement personnel.

On February 12, 1977, Orbiter 1's orbit was altered to bring it within 90 kilometers of the Martian moon Phobos for high-resolution photography. On March 11, 1977, its periapsis was lowered to 300 kilometers, allowing photographic identification of features on Mars as small as 20 meters across. On September 25, 1977, Orbiter 2's orbit was altered to bring it within 22 kilometers of the other moon, Deimos; on October 23, 1977, its periapsis was also lowered to 300 kilometers. Both craft continued observing and mapping Mars from orbit until an entire Martian year had elapsed. Then, on July 25, 1978, with nearly sixteen thousand photographs to its credit, Orbiter 2 was powered down when a series of leaks exhausted its steering gas. Orbiter 1 continued observing Mars until July 14, 1980, when it took the last of its more than thirty-four thousand photographs. Almost out of steering propellant, the orbiter was then used in a series of tests to determine exactly how close to empty the tanks could get before control of the craft was lost. This provided information crucial to the design of future space missions. After these tests, Orbiter 1 was finally deactivated on August 7, 1980.

On the surface, Lander 2 had observed ground frost in mid-August of 1977, late in the Martian winter. It continued monitoring its surroundings until April 12, 1980, when, apparently because of battery failure, it stopped transmitting information to Earth. Meanwhile, Lander 1 was programmed to take photographs automatically and to monitor the Martian weather. On January 7, 1981, it was designated by NASA as the Thomas A. Mutch Memorial Station, dedicated to the Viking Imaging Team leader after his untimely death in a tragic mountain-climbing accident. The Mutch Station continued returning photographs from Mars until November 13, 1982, when its radio transmitter unexpectedly fell silent. A heroic effort to revive the craft was mounted, but after five unsuccessful months the effort was abandoned and the Viking mission was at long last officially terminated. Lander 1's career, however, was not yet finished. On May 18, 1984, ownership of the Mutch Memorial Station was transferred to the National Air and Space Museum of the Smithsonian Institution to begin a new career as the most distant landed historical marker of human civilization.

Contributions

Each Viking orbiter carried two vidicon cameras (similar to television cameras), with a 475-millimeter telephoto lens and filters allowing color photography. In total, 51,539 pictures were returned, covering the entire planet at a resolution of 200 meters and much of it at resolutions as small as 8 meters. Pictures of the Martian moons, Phobos and Deimos, were also taken; those of Deimos were the highest-resolution pictures ever taken of a planetary body from a flyby or orbiting spacecraft.

Mars was revealed as a planet with a tremendous variety of land features. Of greatest interest were those that gave clear evidence of past water flow: Broad, dry flood channels, apparently formed by episodes of catastrophic flooding two or three billion years ago, and smaller, dry networks of runoff channels generally more than three and a half billion years old were commonly seen. Such features were surprising because the Martian atmosphere is too thin and the climate too cold for liquid water to exist. Indeed, no signs of liquid water, or any past or present ocean or lake basins, were seen. The water was apparently underground, frozen in a 1- to 3-kilometer-thick layer of permafrost. As evidence, some meteorite impact craters resembled giant mud splats, apparently created when heat from the impact had melted subsurface ice. Other areas showed large-scale polygonal features or fretted terrain (smooth, flat lowlands bounded by abrupt cliffs) perhaps caused by the activity of ground ice.

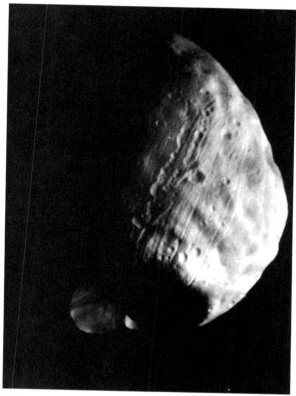

This photo of Phobos, Mars's nearest satellite, was taken during the Viking 1 mission. Phobos moves in an almost circular orbit 6,000 kilometers above the Martian surface. (NASA)

The polar regions consisted of alternating layers of water ice (as verified below) and entrapped dust; apparently, these were the regions where the underlying ice breached the surface.

Volcanism was also apparent. Large shield volcanoes (in which the lava flows out through cracks), including Olympus Mons and others in the Tharsis region, were seen, as were composite volcanoes (which sometimes erupt explosively). Unique to Mars were the low, broad volcanic vents known as pateras. Pedestal craters (craters sitting atop raised plateaus) were revealed by Viking to be similar to Icelandic table mountains (shield volcanoes erupting beneath thick ice sheets). Some volcanoes showed evidence of having been active within the last few hundred million years.

The Viking Lander captured this image, relayed by the Orbiter to Earth on May 18, 1979, of a thin layer of ice on the surface of Mars. (NASA)

Valles Marineris, a 4,500-kilometer-long system of steep-walled canyons up to 9 kilometers deep, was apparently formed by large-scale faults. Subsequent slumping (in-falling or landsliding) of the canyon walls was easily visible, as were horizontal sedimentary rock layers in the canyon walls.

Other features seen included numerous meteorite impact craters and basins, bright streaks indicative of wind-blown dust, dark streaks interpreted as erosion scars, atmospheric haze caused by airborne dust, and carbon-dioxide condensate clouds and fogs. Spectacular clouds routinely formed around Olympus Mons and other large volcanoes in the late morning. Two global dust storms and several dozen localized dust storms were observed during the extended mission.

High-resolution pictures of the Martian moon Phobos revealed sharp, fresh-looking impact craters and a peculiar system of parallel linear grooves, apparently formed during the impact that created Phobos's large crater Stickney. The other moon, Deimos, was seen to be saturated with impact craters and covered, apparently, with a layer of dust.

An infrared spectrometer on board each orbiter measured the concentration of water vapor suspended in the Martian atmosphere. The amounts found were highly variable, ranging from zero parts per million during the Martian winter to eighty-five parts per million near the poles during the Martian summer. Over the northern polar cap in mid-Martian summer, the atmosphere was saturated (held as much water vapor as possible without condensation occurring), strongly suggesting that the polar cap itself contained water ice. In total, the equivalent of 1.3 cubic kilometers of water ice was found in vapor form in the atmosphere.

A radiometer on each orbiter mapped the surface of Mars in infrared light (heat rays); from the amount of infrared radiation being given off, the temperature of the surface could be determined. The most significant finding was that the temperature of the permanent northern polar cap was 200 to 215 kelvins—too warm to be frozen carbon dioxide (dry ice); it therefore had to be made of water ice. A layer of carbon-dioxide ice apparently settled over the water ice in the winter but evaporated in the summer. The southern polar cap was much colder, suggesting that it was a mix of water and carbon-dioxide ice with a year-round cover of carbon-dioxide frost. The global temperature extremes ranged from 133 kelvins at the winter poles to 293 kelvins at noon on the equator.

The radio communications and radar instruments on board Viking were used for a number of experiments auxiliary to their primary purposes. When Mars went behind the Sun, the Sun's gravity was observed to slow the transit time of radio signals passing near it. This was in accord with predictions made by Albert Einstein's general theory of relativity, and it provided new confirmation of that theory. Also at that time, radio signals passing through the solar corona yielded information on the small-scale structure of the corona. Other radio data led to improved measurements of the planet's size, orientation, spin rate, orbital characteristics, and distance from Earth, finding the exact location of the landers on Mars (to 11-kilometer accuracy), and information on the electrical properties of the Martian surface.

During the descent to the surface, instruments on board the Viking landers examined the properties of the atmosphere. The upper atmosphere was found to be mostly carbon dioxide, with small amounts of nitrogen, argon, carbon monoxide, oxygen, and nitrogen monoxide. The concentration of nitrogen 14 (normal nitrogen) relative to that of nitrogen 15 (an isotope of nitrogen having one extra neutron) was found to be lower than it is expected to have been early in the planet's history. Because nitrogen 14 escapes from the Martian atmosphere more rapidly than does nitrogen 15, there must have been much more nitrogen 14 present in the past. This suggests that Mars had a much denser atmosphere sometime in the past—perhaps one dense enough to have allowed the flow of liquid water.

Each lander carried two facsimile cameras—cameras in which the field of view is divided into a grid of small squares, which are scanned one at a time to produce the whole picture. The horizon around Lander 1 showed meter-sized boulders and a crater 500 kilometers in diameter at a distance of 2.5 kilometers fronting a ridge 10 kilometers away. The near surface was duricrust (a coinage for the hard, cemented Martian soil) littered with small rocks. Throughout the field were numerous ventifacts (angular, multifaced blocks shaped by winds), a ubiquitous litter of blocks thought to be either debris from impact craters or deposition from past floods, sediment drifts (probably of silt and clay), and outcrops of bedrock. The drifts were stable; in fact, the only sign of movement observed was a small slumping of soil around Big Joe (significant, nevertheless, as the first observation of geologic change ever made on another planet). During the extended mission, darkening of the sky and softening of shadows were observed during two global dust storms.

The horizon around Lander 2 showed gently undulating plains with much less relief than at the Lander 1 site; the terrain was almost undoubtedly formed by ejecta (debris) from the impact crater Mie, 160 kilometers to the east. Like the terrain around Lander 1, there were a duricrust surface, ventifacts, and a litter of blocks and boulders. Most of the Lander 2 blocks, however, were vesicular (pitted and porous-looking), probably the result of gas bubbles trapped in cooling volcanic lava, but some were smoother, typical of finer-grained volcanic rock. Signs of wind erosion included the ventifacts, wind-scalloped blocks, and wind-sculpted pedestals under some boulders. A linear trough, 10 to 15 centimeters deep and more than 10 meters long, was seen as part of a polygonal network near the lander; it may have been formed by seasonal freezing and thawing of groundwater. The frost observed was probably water ice or clathrate (a mixture of carbon dioxide and water); the ambient temperatures were too high for the existence of frozen carbon dioxide alone.

Other activities of lander imaging included monitoring the sampler arm, the soil sample collection, and the magnetic properties experiment. Lander imagery also proved invaluable as an aid to diagnosing problems with other parts of the spacecraft. Some astronomy—photographs of Phobos, the Sun, and the shadow of Phobos during a solar eclipse—was also done. A few pictures, particularly those of the American flags on the landers and of the spectacular sunsets and sunrises, were taken primarily for their aesthetic value. In total, more than forty-five hundred lander photographs were taken.

Data compiled from photographs, the forces exerted by the ground on the landing legs and sampler arm, and other such clues were used to understand properties of the Martian surface material. The surface was generally firm (with the exception of the area under Lander 1's left footpad) and adhesive (meaning that it stuck to things, such as parts of the lander), but only weakly cohesive (meaning that it did not stick well to itself; instead, it crumbled like a clod of Earthly dirt).

An array of magnets attached to the collector head of each sampler arm picked up any magnetic materials in the soil. Photographs of the magnets then revealed the material and allowed determination of its abundance and magnetic properties. Roughly 10 percent of the soil was found to be magnetic, most likely maghemite (a compound of iron and oxygen similar to rust). If so, then the planet's surface is red because it is, quite literally, rusty.

Each lander carried a seismometer designed to detect "Marsquakes." During the first five months at Utopia Planitia, no major disturbances were detected, although data believed to indicate a small quake (2.8 on the Richter scale) were recorded on sol 80. Analysis of the underground reflections of that disturbance indicated that the crust in the region near Lander 2 was 15 kilometers thick. An analysis of the structure of the planet's deep interior, the primary goal of this investigation, was not possible because the seismometer on Lander 1 failed to unlock from its stowed position after landing. The Lander 2 seismometer routinely picked up spacecraft vibrations resulting from winds and the activities of other instruments on board the craft. It was unexpectedly pressed into use as a wind monitor to supplement the meteorology instruments.

A complete weather station, including air temperature, pressure, and wind condition sensors, occupied the end of a boom extending 1 meter from each lander. Subsequent reports collected during six years on Mars provided data important for understanding both global and local atmospheric circulation patterns. Lander 1 recorded temperatures ranging from 185 to 261 kelvins. Winds were mild—generally only a few meters per second—although they did exceed 50 meters per second during violent storms. The atmospheric pressure was observed to vary by some 30 percent as carbon dioxide from the atmosphere froze onto or sublimated from the polar caps with the seasons.

The chemical composition of Martian soil samples was analyzed by an x-ray fluorescence spectrometer, a device that identifies various chemicals through their absorption and reemission of x rays. The soil was basically an iron-rich clay. Its composition was 5 percent magnesium, 3 percent aluminum, 20.9 percent silicon, 3.1 percent sulfur, 0.7 percent chlorine, less than 0.25 percent potassium, 4 percent calcium, 0.51 percent titanium, and 12.7 percent iron, with traces of rubidium, strontium, ytterbium, and zirconium. The remainder consisted of elements that could not be identified by the spectrometer.

A gas chromatograph mass spectrometer on board each lander identified organic molecules by breaking them down into their constituent elements. Their primary finding was that the Martian soil contained no organic compounds in quantities that could be detected—making the probability of finding life there unlikely. This instrument was also used to analyze the chemical composition of the Martian atmosphere. It corroborated the atmospheric composition found by the entry science experiments (95 percent carbon dioxide, 2.7 percent nitrogen, 1.6 percent argon, 0.13 percent oxygen, and smaller amounts of other constituents). It also revealed an underabundance of argon 36 (argon with four fewer neutrons than normal) relative to argon 40 (normal argon) of 10 percent, compared with Earth's atmosphere. This observation provided vital clues to models suggesting that early in its history, Mars had a substantially warmer and denser atmosphere, which might have allowed liquid water to exist on its surface.

Three experiments designed to detect microorganisms in the Martian soil were on each lander.

Controversy over their results continued throughout the mission. In the gas exchange experiment, a soil sample was placed in an artificial atmosphere of helium, krypton, and carbon dioxide humidified with a liquid nutrient. The presumption was that if microorganisms were present, they would reveal themselves by metabolizing the nutrient and then emitting hydrogen, nitrogen, oxygen, methane, or carbon dioxide. Surprisingly, the first run on Lander 1 produced copious amounts of oxygen, suggesting the presence of life. Subsequent experiments—particularly the demonstration that heat sterilization did not prevent the release of oxygen—indicated that the reaction was not indicative of life but of a peculiar physical chemistry process in the Martian soil. Apparently, ultraviolet light from the Sun had bound oxygen to material in the soil, creating compounds known as peroxides (or perhaps variants known as superoxides or ozonides). Water vapor from the nutrient apparently released that bound oxygen, mimicking a biological response. Similar results were obtained by Lander 2.

In the labeled release experiment, a soil sample was moistened with a liquid nutrient seeded with radioactive carbon 14 (which served as an easily monitored "label" or tracer of the nutrient). The presumption was that if any microorganisms were present, they would take in the nutrient and expel carbon dioxide or other gases containing the carbon 14 labels. In fact, labeled carbon dioxide was released, again mimicking a biological reaction. When more nutrient was added to the sample, however, the carbon dioxide release did not increase (as it would if organisms were releasing it); instead, it decreased, consistent with the premise that the nutrient had been decomposing peroxides to create the carbon dioxide. Heat sterilization here prevented carbon dioxide production completely, a result consistent with either biological or physical chemistry explanations.

In the pyrolytic release experiment, a soil sample was exposed to normal Martian air that had been labeled with radioactive carbon monoxide and carbon dioxide. The sample was then incubated under simulated sunlight, with the presumption that if any plantlike microorganisms were present, they would take in the radioactive gases. After prolonged incubation, the sample was vaporized (by heating, hence the term "pyrolytic") and the vapors examined for radioactive carbon. In the first run on Lander 1, a substantial amount of radioactive carbon was found—again suggesting life. Subsequent runs, however, failed to confirm that result.

Experiment designer Norman H. Horowitz concluded that the chance of the results being caused by biological activity was negligible in that the first high carbon reading could not be reproduced, adding water vapor had no effect on the reactions, turning off the light had no substantial effect on the reactions, and laboratory simulations on Earth could reproduce the results through the interaction of the labeled gases with iron oxide compounds such as maghemite (which the magnetic properties experiment had suggested was prevalent on Mars).

In the final analysis, virtually all the Viking scientists agreed that although a biological basis for the results could not be completely ruled out, it was extremely unlikely. The lack of organic compounds in the soil argued strongly against a biological explanation, and scientists had shown that all the results could be explained by purely physical chemistry processes. Labeled release experiment designer Gilbert V. Levin continued to argue for the minute possibility that life might nevertheless have been found.

Context

Before the Space Age, Mars was a fantasy place. From Edgar Rice Burroughs's tales of John Carter on Mars to Percival Lowell's scientific speculations about Mars as an "abode of life," the Western world equated "Martians" with "extraterrestrial life." The Mariner missions replaced those fantasies with orbital photographs of Moon-like craters and Earth-like volcanoes, valleys, and flood channels; the surface landscape itself, however, remained a matter of speculation. When Viking 1 touched down on Mars, all that changed: The entire world saw the barren, red, rock-strewn desert landscape of Chryse Planitia. Instantly, Mars became real.

Project Scientist Gerald Soffen has often said that in Viking's first month on Mars, more was learned about Mars than had been previously learned in the entire history of humanity. The complete absence of detectable organic compounds in the soil and the failure to find confirmable signs of life there was disappointing, even as the copious evidence for a history of flowing water was tantalizing. What allowed that water to flow in the past? Was there once a warm, dense atmosphere on Mars? Where has that water gone? Theorists analyzing the Viking data now suspect that there indeed was once such an atmosphere, and that the water is now locked in a kilometer-thick layer of ice-laden permafrost, which undergirds the entire surface of the planet and emerges at the polar caps. This planetary picture—a once hospitable, now dry, cold, and lifeless world—contrasts with the sulfuric acid-laden hothouse of Venus and leaves scientists astounded at the temperate, water-rich world called Earth.

It is easy to argue that Viking was the pinnacle of robotic exploration of the solar system. Certainly, the task of simultaneously operating four craft in the vicinity of Mars while landing two of them successfully on the surface was an unprecedented challenge for mission controllers. At the time, the ability of the Vikings to operate either completely automatically or upon continually updated instructions from Earth made them the most versatile and sophisticated deep-space craft ever flown. The spectacular operational lifetime of Viking Lander 1 of six years, three months, and twenty-four days on Mars set a long-lasting record for continual scientific operations on the surface of another planet.

Subsequent exploration of Mars did not resume until July, 1988, when the Soviet Union launched Project Phobos, two craft intended to examine Phobos at close range and land probes upon the Martian moon for which they are named. The American Mars Observer mission of 1992-1993 failed, as did the Mars Climate Orbiter and Mars Polar Lander missions of 1998-1999, but the Mars Global Surveyor of 1996 and the Mars Pathfinder mission of 1997 both returned spectacular scientific results, including live images of the planet's surface. The disappointments in 1988 and 1999 were overshadowed by the tremendous success of the Mars Exploration Rovers in 2004. Launched in mid-2003, the Spirit and Opportunity Rovers, each the size of a golf cart, safely touched down about two weeks apart in two widely spaced areas of Mars in January, 2004.

The call for expanded human exploration of space beyond low-Earth orbit made by President George W. Bush in the aftermath of the February 1, 2003, *Columbia* accident foreshadowed the need for renewed robust robotic exploration of Mars in advance of the arrival of human beings on the Red Planet.

See also: Hubble Space Telescope: Science; Jet Propulsion Laboratory; Mariner 6 and 7; Mariner 8 and 9; Mars Global Surveyor; Mars Observer; Planetary Exploration; Search for Extraterrestrial Life; Vanguard Program; Viking Program.

Further Reading

Baker, Victor R. *The Channels of Mars.* Austin: University of Texas Press, 1982. A detailed analysis and summary of Martian geomorphology as revealed by Viking, focusing on water-cut channels and other evidence of water on Mars. Extensive references and numerous illustrations. Advanced college level.

Burgess, Eric. *To the Red Planet.* New York: Columbia University Press, 1978. A good chronological summary of the Viking missions, beginning with historical perceptions of Mars and continuing through the early Viking results. Some illustrations; written for general audiences.

Burrows, William E. *This New Ocean: The Story of the First Space Age.* New York: Random House, 1998. This is a comprehensive history of the human conquest of space, covering everything from the earliest attempts at spaceflight through the voyages near the end of the twentieth century. Burrows is an experienced journalist who has reported for *The New York Times*, *The Washington Post*, and *The Wall Street Journal.* There are many photographs and an extensive source list. Interviewees in the book include Isaac Asimov, Alexei Leonov, Sally K. Ride, and James A. Van Allen.

Carr, Michael H. *The Surface of Mars.* New Haven, Conn.: Yale University Press, 1981. A comprehensive survey by the leader of the Orbiter Imaging Team of physical processes affecting the surface of Mars. Also includes illustrations, a chapter on the search for life on Mars, and a chapter on the Martian moons. Advanced college level.

Cooper, Henry S. F., Jr. *The Search for Life on Mars: Evolution of an Idea.* New York: Holt, Rinehart and Winston, 1980. A masterful account by *The New Yorker*'s premier science writer. Covers the history and chronology of the Viking biology experiments in exquisite detail, with much attention to the human side of the experiments. Accurate and easily accessible for general audiences.

Eberhart, Jonathan. "Operation Red Planet." *Science News* 109 (June 5/12, 1976): 362. This article and its accompanying special reports began an extensive series of articles on Viking that continued in subsequent issues throughout the mission. Eberhart, the undisputed dean of space science reporters, is lucid, insightful, and accessible to all audiences.

Ezell, Edward Clinton, and Linda Neuman Ezell. *On Mars: Exploration of the Red Planet, 1958-1978.* NASA SP-4212. Washington, D.C.: Government Printing Office, 1984. NASA's official history of the Viking program. Includes extensive administrative, political, financial, and scientific background on the development of the U.S. Martian exploration program. Also contains an in-depth review of Viking landing-site selection and certification procedures and an overview of Viking science results. Detailed appendices and references; suitable for general audiences.

Hamilton, John. *The Viking Missions to Mars.* Edina, Minn.: ABDO and Daughters Publishing, 1998. Although this is a juvenile book, it does give a good overview of the Viking missions to the Red Planet.

Hartmann, William K. *Moons and Planets.* 5th ed. Belmont, Calif.: Thomson Brooks/Cole Publishing, 2005. Provides detailed information about all objects in the solar system. Suitable on three separate levels: high school student, general reader, and college undergraduate studying planetary geology.

Heppenheimer, T. A. *Countdown: A History of Space Flight.* New York: John Wiley, 1997. A detailed historical narrative of the human conquest of space. Heppenheimer traces the development of piloted flight through the military rocketry programs of the era preceding World War II. Covers both the American and the Soviet attempts to place vehicles, spacecraft, and humans into the hostile environment of space. More than a dozen pages are devoted to bibliographic references.

Horowitz, Norman H. *To Utopia and Back: The Search for Life in the Solar System.* New York: W. H. Freeman, 1986. An authoritative, firsthand analysis of the search for life in the solar system, written by one of the principal Viking biologists. Covers biological background, development of Viking biology experiments, and the Viking mission itself. Clear and accurate; college level.

Lee, Wayne. *To Rise from Earth: An Easy to Understand Guide to Spaceflight.* New York: Check-mark Books, 1996. This is a good introduction to the science of spaceflight. Although written by an engineer with the NASA Jet Propulsion Laboratory, it is presented in easy-to-understand language. In addition to the theory of spaceflight, it gives some of the history of the human endeavor to explore space.

Mari, Christopher, ed. *Space Exploration.* New York: H. W. Wilson, 1999. Twenty-five articles (reprinted from magazines), covering the state of the space program at the time of publication, are divided into five sections: John H. Glenn, Jr.'s return to space, the exploration of Mars, the International Space Station, recent mining efforts by commercial industries, and new types of space vehicles and propulsion systems.

Mishkin, Andrew. *Sojourner: An Insider's View of the Mars Pathfinder Mission.* New York: Berk-ley Books, 2003. A thorough exploration of the Mars Pathfinder mission that also provides a detailed historical perspective on the exploration of the surface of Mars (begun by Viking).

Moore, Patrick. *Guide to Mars.* New York: W. W. Norton, 1977. Great Britain's celebrated astronomer and writer presents a tourist's guide to Mars, beginning with basic introductions to the solar system and telescopes, continuing with a history of Mars exploration, and culminating with Viking and a look toward future exploration. Highly accessible and informative for all audiences.

National Aeronautics and Space Administration, Viking Lander Imaging Team. *The Martian Landscape.* NASA SP-425. Washington, D.C.: Government Printing Office, 1978. Viking Lander Imaging Team leader Thomas Mutch's anecdotal account of the conception, design, building, and operation on Mars of the lander cameras, followed by more than two hundred of the best and most representative photographs taken by the Viking landers—some in color, some in stereo, all with explanatory text. Stereo viewer included. For all audiences.

Science 193/194 (August 27, October 1, and December 17, 1976). These three issues are dedicated to the Viking missions; the overview was written by Gerald Soffen and Conway Snyder, two members of the various Viking teams (see entry below).

Soffen, Gerald A., and Conway W. Snyder. "The First Viking Mission to Mars." *Science* 193 (August 27, 1976): 759. This overview begins the first of three special issues of *Science* dedicated to the Viking missions (see entry above). Advanced college level.

Spitzer, Cary R., ed. *Viking Orbiter Views of Mars.* NASA SP-441. Washington, D.C.: Government Printing Office, 1980. The Viking Orbiter Imaging Team's public report on its findings. Contains brief introductions to Viking and Mars, followed by spectacular photographs—some in stereo (viewer provided)—illustrating all the major landforms and atmospheric phenomena observed by the Viking orbiters. Includes detailed photographs of the Viking landing sites. The extensive captions are college level; the photographs are enthralling to all audiences.

Philip J. Sakimoto

Voyager Program

Date: Beginning August 20, 1977
Type of program: Planetary exploration

The Voyager probes executed the first Grand Tour in planetary exploration by successively encountering Jupiter, Saturn, Uranus, and Neptune. Such a tour, using the "planetary-gravity-assist" technique to travel from planet to planet, is possible only once every 175 years.

Key Figures

Gary A. Flandro, discoverer of Grand Tour alignments of the outer planets

Charles E. Kohlhase, Principal Mission designer

Harris M. Schurmeier, Voyager project manager through development phase

John R. Casani (b. 1932), Voyager project manager from launch to Jupiter encounters

Raymond L. Heacock (b. 1928), Voyager project manager for Jupiter and Saturn encounters

Richard P. Laeser, Voyager project manager for Uranus encounter

Norman Ray Haynes (b. 1936), Voyager project manager for Neptune encounter

Edward C. Stone, Jr., Voyager project scientist

Ellis D. Miner, assistant project scientist

Andrei B. Sergeyevsky, principal trajectory designer for the Neptune encounter

Bradford A. Smith, principal investigator, imaging science experiment

G. Leonard Tyler, principal investigator, radio science experiment

Summary of the Program

Two Voyager spacecraft were launched from Earth in 1977; Voyager 1 encountered Jupiter in 1979 and Saturn in 1980, and Voyager 2 encountered Jupiter in 1979, Saturn in 1981, Uranus in 1986, and Neptune in 1989. It is this latter sequence of planetary encounters that is called the Grand Tour.

The Voyager spacecraft used the "planetary-gravity-assist" technique to move from one planet to the next. Concepts of using gravity to propel a spacecraft from one body to another have existed since the 1920's. The actual technique of executing a gravity assist was not well understood, however, until Michael Minovitch developed it in the early 1960's.

It is not possible to accomplish a Grand Tour unless the gravity-assist technique is used. The fuel re-

quirements of a nongravity-assist tour are vastly beyond the capability of present-day chemical propulsion systems. To illustrate, Voyager saved 1.5 million kilograms of fuel by using Jupiter's gravity to propel it toward Saturn. Similar amounts of fuel were saved at Saturn (using its gravity to assist the spacecraft to reach Uranus) and again at Uranus (using another gravity assist to proceed to Neptune).

The outer planets must be properly aligned or such a Grand Tour is not possible. In 1966, Gary Flandro discovered that this alignment occurs only every 175 years, that the last alignment had occurred in 1802, and that the next time the planets would be properly aligned for a Grand Tour launch would be in 1977. The Voyager mission was designed to take advantage of that opportunity.

The National Aeronautics and Space Administration (NASA) authorized the Jet Propulsion Laboratory (JPL) in 1972 to start the Voyager project. Initially, only a four-year mission to Jupiter and Saturn, with a launch date in 1977, was funded. This circumstance led to the original name of the project: Mariner Jupiter Saturn 77 (MJS77). In 1977, the name of the project was changed to Voyager. Although only the Jupiter and Saturn encounters were originally authorized, the spacecraft was built with capabilities to extend its mission and fulfill the Grand Tour if subsequent budget authorizations permitted.

Two spacecraft were to visit Jupiter, then Saturn. If the Voyager 1 encounter with Saturn was successful, then Voyager 2 would be permitted to go on to

Uranus and then Neptune. The Uranus option was authorized in 1981, the Neptune option in 1986.

The Voyager spacecraft, which were largely identical, used the Mariner spacecraft's decagonal shape. The ten bays housed electronics boxes. In the middle was a fuel tank for the propulsion system. Attached to one end was a large communication antenna for receiving spacecraft commands from Earth and for transmitting data back to it. Also attached were two deployable booms. At the end of one lay the three nuclear power plants that provided the electrical power to run the spacecraft. Along the sides and at the end of the other boom lay various scientific instruments.

The Voyager Grand Tour mission was one of scientific exploration. Each spacecraft carried the same eleven instruments; the complete package included sensors that point at an object (target body instruments) and ones that make in situ measurements (field, particle, and wave instruments). The target body sensors included wide-angle and telephoto cameras, an infrared telescope, an ultraviolet telescope, an instrument to measure certain characteristics of light, and a radio transmitter to send radio signals through planetary atmospheres and rings back to Earth. The in situ sensors included a magnetic field sensor, a radio receiver, a plasma wave sensor, and three particle detectors.

The two spacecraft were to be launched from Titan-Centaur launch vehicles, approximately three weeks apart. The first spacecraft would be on a slower path to Jupiter and Saturn; the second would be on a faster path to the planets and thus would arrive first. As the order in which the spacecraft would encounter each planet was far more important than the

The Voyager spacecraft during vibration testing. (NASA-JPL)

A model of the trajectory of the Voyager spacecraft from Earth through the solar system. (NASA CORE/Lorain County JVS)

Voyager 1 took exactly eighteen months to reach Jupiter, making its closest approach on March 5, 1979. Voyager 2 made its closest approach on July 9, 1979. Jupiter was known to have four huge moons (discovered by Galileo in 1610 and named "the Galilean satellites" in his honor) and nine smaller ones. The planet itself and the four large Galilean satellites were the main targets of interest; thus, the Voyager trajectories were designed to permit a close encounter with each of these five main bodies.

The planetary-gravity-assist technique was used by both Voyagers at Jupiter to propel the spacecraft on to Saturn. Before the Voyagers executed the gravity assist, they had been in an elliptical orbit about the Sun and Jupiter. After the maneuver, the spacecraft had enough energy to escape the gravitational pull of the Sun permanently. It was the gravity assist at Jupiter that started Voyager 2 on the Grand Tour.

The Voyagers' second planetary encounter was with Saturn. One spacecraft had preceded Voyager to Saturn: Pioneer 11, which had encountered Saturn on September 1, 1979. After a little more than twenty months of interplanetary cruising from Jupiter, Voyager 1 made its closest approach to Saturn on November 12, 1980. Voyager 2 followed nine months later, making its closest approach on August 25, 1981.

Saturn was known to have nine moons of varying sizes. Between the two spacecraft, close encounters with seven of the nine were made. One of the moons, Titan, which is much larger than the others, was known to have an atmosphere. Thus, there was the possibility of an environment capable of supporting life. For these reasons, Voyager 1 was targeted to pass within 4,000 kilometers of Titan's surface. Unfortunately, this close passage flung the spacecraft out of the plane of the ecliptic (the plane in which the planets orbit the Sun) at about a

order of launch, Voyager 2 was launched first, at 10:29:45 A.M., eastern daylight time or 14:29:45 Coordinated Universal Time (UTC), on August 20, 1977, and Voyager 1 was launched second, at 8:56:01 A.M. (12:56:01 UTC) on September 5, 1977.

Each spacecraft orbited Earth several times and then was injected onto a Jupiter-encounter path by the high-energy liquid-fueled Centaur upper stage. After burnout, all deployable booms were extended, and the spacecraft were thoroughly checked. Thirteen days after launch, Voyager 1 turned its telephoto camera toward Earth and shuttered the first image in history to contain both Earth and the Moon.

The Voyagers' first planetary encounter was with Jupiter. Two spacecraft had preceded the Voyagers there: Pioneers 10 and 11. Pioneer 10 had encountered the planet in 1973, and Pioneer 11 a year later. Neither spacecraft had an imaging camera, although each had an imaging photopolarimeter which could (and did) take rather crude photographs. It was left to Voyager to provide the first high-quality images of the solar system's largest planet.

35° angle, costing Voyager 1 the opportunity to encounter Uranus, Neptune, and Pluto.

Voyager 2's third planetary encounter was with Uranus, which had not yet been encountered by any spacecraft. After a four-year, four-month interplanetary cruise, Voyager 2 made its closest approach to Uranus on January 24, 1986. This planet is tilted on its side. As it orbits the Sun, first its northern pole, then its equator, then its southern pole, and then its equator points toward the Sun. When Voyager 2 encountered Uranus the planet's southern pole was pointed at the Sun.

At this time, five moons were known to orbit Uranus about its equator. As Voyager 2 approached the southern pole, the five moons appeared to orbit the planet in concentric circles, creating a bull's-eye effect. Because of this geometric pattern, Voyager 2 could pass close to only one moon. Fortunately, the gravity assist required at Uranus for the trip to Neptune allowed the spacecraft to pass close to Miranda, the moon closest to Uranus.

The final Voyager 2 planetary encounter was with Neptune, and the spacecraft approached the planet in 1989. Neptune is known to have one large moon, Triton, and one small one, Nereid. Of the two, Triton is known to have an atmosphere, and thus scientific interest in it is high. Voyager 2 came close to both bodies, receiving a final gravity assist at Neptune to bend the spacecraft's path by almost 45° so that it could pass within 40,000 kilometers of the surface of Triton. This dual encounter marked the end of the Grand Tour and began Voyager's interstellar mission.

Contributions

The Voyager mission provided the first high-quality visual study and the most comprehensive scientific investigation of the Jovian system until that time. The outer atmosphere of Jupiter is now known to be made up of about 89 percent hydrogen, about 11 percent helium, and trace amounts of many elements. Cloud-top lightning was observed at all latitudes on the dark side of the planet. At the place in the atmosphere where the pressure

is the same as that at Earth's surface, the temperature is about 165 kelvins. Voyager discovered a set of rings, no more than 30 kilometers thick, composed of fine particles.

For the first time, humanity saw what the surfaces of Jupiter's four major moons look like. On Io, nine volcanoes in the process of eruption were observed. Europa was observed to have the smoothest known surface in the solar system, with a difference in altitude between the highest peak and the lowest valley of 200 meters or less. Ganymede proved to be the largest known moon in the solar system. Three new Jovian moons were discovered.

Jupiter's magnetic field is the largest of all the planets in the solar system. (The magnetic field of the Sun is larger.) At times Jupiter's magnetic field stretches beyond the orbit of Saturn. An electric current of more than five million amperes flows between Io and Jupiter.

Voyager also provided the first high-quality visual images and a highly comprehensive scientific study of the Saturnian system. The outer atmosphere of Saturn is now known to be made up of about 94 percent hydrogen, about 6 percent helium, and trace amounts of many other elements. Saturn radiates about 80 percent more energy than it receives from the Sun. Lightning was observed at low latitudes on the dark side of the planet. At the place in the atmosphere where the pressure is the same as that at Earth's surface, the temperature is about 134 kelvins. Saturn's day is 10 hours, 39 minutes, and 15 seconds long.

Each of Saturn's three great rings (the A-, B-, and C-rings) actually contains many hundreds of thousands of ringlets. Pioneer 11's discovery of the thin F-ring was confirmed, and Voyager discovered two new rings: the diffuse D- and G-rings. Two small moons, one on either side of the thin F-ring, were discovered; the existence of these moons had been predicted before Voyager 1's encounter with Saturn.

An enormous crater was discovered on Saturn's innermost major moon, Mimas; the crater was one-third of that moon's diameter. An even larger

A record carried by both Voyager 1 and Voyager 2, "The Sounds of Earth," made of 12-inch discs of gold-plated copper with greetings in 60 languages, samples of music, and natural sounds of Earth. (NASA-JPL)

the two moons on either side of the F-ring and a moon just outside the A-ring.

Saturn's magnetic field is aligned with its north pole to within one degree. A torus of hydrogen and oxygen ions that orbits Saturn is probably provided by the moons Tethys and Dione. Titan provides its own orbiting torus of neutral hydrogen atoms.

In another first, Voyager provided the first encounter of any kind with the Uranian system. The quality of the visual imagery obtained and the completeness of its scientific investigation there were comparable to those attained at Jupiter and Saturn. For the first time, the appearance of the planet Uranus, its eleven rings, and its five major moons became known. The outer atmosphere of Uranus is now known to be made up of 85 percent hydrogen, about 15 percent helium, and trace amounts of many hydrocarbon compounds. Unlike Saturn, Uranus radiates only about one-third as much energy as it receives from the Sun. Lightning was also detected in Uranus's atmosphere. The Uranian day is 17 hours, 14 minutes, and 40 seconds long.

crater, 400 kilometers in diameter, was observed on Saturn's moon Tethys. The moon Enceladus has the second smoothest surface in the solar system. The moon Iapetus is half light and half dark, giving it the largest light-to-dark ratio of any body in the solar system. Voyager 1 established that Saturn's largest moon, Titan, is the second largest in the solar system. Titan's atmosphere has a near-surface pressure 1.5 times that of Earth, has a temperature of 94 kelvins, and is composed of 90 percent nitrogen, with methane, argon, and trace carbon and hydrocarbon compounds making up the remaining 10 percent. Three new moons were discovered:

Voyager 2 discovered two new rings, both of them very thin. All eleven of Uranus's rings are thin (no more than 100 kilometers wide). Two moons, one on either side of the outermost (and thickest) Uranian ring, the epsilon ring, were discovered. These moons help confine the epsilon ring particles. Ten new moons (including the two discussed above) were discovered. The surface of Uranus's innermost major moon, Miranda, revealed nearly every type of geological process observed anywhere else in the solar system.

Voyager 2 discovered that Uranus has a magnetic field. The field is tilted by 58.6° with respect to Uranus's north pole. In contrast to any other planet that has been explored, the magnetic field's center is offset from the center of Uranus by nearly one-third of the planet's radius.

Voyager 2 found Neptune to be a deeper blue color than Uranus, and to be even more energetic, with very high-speed winds and a large storm called the Great Dark Spot. Neptune's largest moon, Triton, appeared orange in color and seemed to have a tenuous atmosphere and an active ice geyser system.

Context

Voyagers 1 and 2 were the third and fourth spacecraft to encounter Jupiter, and the second and third to encounter Saturn. Voyager 2 was the first spacecraft to encounter Uranus and Neptune. It was also the first spacecraft to go on the Grand Tour of the outer solar system.

The solar system contains one star, nine planets, well over fifty moons, and thousands of asteroids and comets. Together, the two Voyager spacecraft used the planetary gravity assist a total of six times to encounter five of the nine planets and fifty-one moons. Voyager provided high-quality visual imagery of Jupiter and Saturn and the very first imagery of the moons of Jupiter, the rings and moons of Saturn, the planet Uranus and its rings and moons, and the planet Neptune and its rings and moons.

An extensive scientific investigation of the four gas giant planetary systems was conducted.

Voyager data contributed greatly to the understanding of the characteristics of the solar system's parts that is necessary before a full-fledged theory of the solar system can be developed. Such a theory would explain the creation and evolution of the entire system and each of its parts. The theory might then permit accurate predictions regarding the system's future evolution. The inhabitants of Earth have a vested interest in knowing what will happen to their planet. A well-supported theory of the solar system might also allow accurate predictions to be made of the density and characteristics of solar systems in the Milky Way galaxy and beyond.

As of 2005, the two Voyager spacecraft were continuing their scientific exploration of interstellar space. Voyager 1 was 13.5 billion kilometers from Earth, twenty-seven years after launch, and was traveling away from the solar system at 21 kilometers per second. Voyager 2 was 11 billion kilometers from Earth, traveling at 29 kilometers per second.

See also: Cassini: Saturn; Deep Space Network; Galileo: Jupiter; Huygens Lander; Magellan: Venus; Mariner 10; Nuclear Energy in Space; Pioneer 10; Pioneer 11; Pioneer Venus 2; Planetary Exploration; Ulysses: Solar Polar Mission; Viking Program; Voyager 1: Jupiter; Voyager 1: Saturn; Voyager 2: Jupiter; Voyager 2: Saturn; Voyager 2: Uranus; Voyager 2: Neptune.

Further Reading

Beatty, J. Kelly, Carolyn Collins Petersen, and Andrew Chaikin, eds. *The New Solar System*. 4th ed. New York: Sky Publishing, 1999. Gives a comprehensive description of the solar system, using the results of planetary exploration missions from all countries. Each of the twenty-one chapters is written by a pioneer in planetary exploration. Contains many illustrations and reproductions of images returned by planetary spacecraft. Suitable for general audiences.

Fimmel, Richard O., William Swindell, and Eric Burgess. *Pioneer Odyssey*. NASA SP-396. Washington, D.C.: Government Printing Office, 1977. Discusses the state of knowledge of the Jovian system before the Pioneer 10 and 11 encounters, the history of the Pioneer 10 and 11 missions, and the knowledge gained from the two spacecraft. Contains many illustrations, a list of project participants, and recommendations for further reading.

Fimmel, Richard O., James A. Van Allen, and Eric Burgess. *Pioneer: First to Jupiter, Saturn, and Beyond.* NASA SP-446. Washington, D.C.: Government Printing Office, 1980. This book is an update to *Pioneer Odyssey,* covering the Pioneer 11 encounter with Saturn and the mission after the planetary encounters. A more mature discussion of the knowledge gained from the Jupiter encounters is also provided. Contains many illustrations, images from both Jupiter and Saturn, a listing of project participants, and a bibliography.

Harland, David M. *Jupiter Odyssey: The Story of NASA's Galileo Mission.* London: Springer-Praxis, 2000. A detailed scientific and engineering history of the Galileo program, but also includes extensive discussion of Voyager program events and results.

_____. *Mission to Saturn: Cassini and the Huygens Probe.* London: Springer-Praxis, 2002. A detailed scientific and engineering history of the Cassini program, but also includes extensive discussion of Voyager program events and results.

Hartmann, William K. *Moons and Planets.* 5th ed. Belmont, Calif.: Thomson Brooks/Cole Publishing, 2005. Provides detailed information about all objects in the solar system. Suitable on three separate levels: high school student, general reader, and college undergraduate studying planetary geology.

Irwin, Patrick G. J. *Giant Planets of Our Solar System: Atmospheres, Composition, and Structure.* London: Springer-Praxis, 2003. Provides an in-depth comparison of Jupiter, Saturn, Uranus, and Neptune, incorporating data obtained from astronomical observations and planetary spacecraft encounters.

Lee, Wayne. *To Rise from Earth: An Easy to Understand Guide to Spaceflight.* New York: Checkmark Books, 1996. This is a good introduction to the science of spaceflight. Although written by an engineer with the NASA Jet Propulsion Laboratory, it is presented in easy-to-understand language. In addition to the theory of spaceflight, it gives some of the history of the human endeavor to explore space.

Mari, Christopher, ed. *Space Exploration.* New York: H. W. Wilson, 1999. Twenty-five articles (reprinted from magazines), covering the state of the space program at the time of publication, are divided into five sections: John H. Glenn, Jr.'s return to space, the exploration of Mars, the International Space Station, recent mining efforts by commercial industries, and new types of space vehicles and propulsion systems.

Miner, Ellis D., and Randii R. Wessen. *Neptune: The Planet, Rings, and Satellites.* London: Springer-Praxis, 2002. The assistant project scientist for the Voyager 2 Neptune encounter composed this thorough review of our contemporary understanding of the Neptune system.

Morrison, David. *Voyages to Saturn.* NASA SP-451. Washington, D.C.: National Aeronautics and Space Administration, 1982. Examines the state of knowledge of the Saturnian system before Pioneer 11 and gives an account of the Pioneer 11 encounter, the history of the Voyager mission, and the Voyager encounters. Contains many images returned by the Pioneer 11 and Voyager spacecraft, a list of Voyager project personnel, and suggestions for additional reading.

Morrison, David, and Tobias Owen. *The Planetary System.* 3d ed. San Francisco: Addison-Wesley, 2003. Organized by planetary object, this work provides contemporary data on all planetary bodies visited by spacecraft since the early days of the Space Age. Suitable for high school and college students and for the general reader.

Morrison, David, and Jane Samz. *Voyage to Jupiter.* NASA SP-439. Washington, D.C.: Government Printing Office, 1980. Discusses the state of knowledge of the Jovian system before any of the spacecraft encounters; gives an account of the Pioneer encounters, the history of the Voyager project, the Voyager encounters, and the prospects for a return to Jupiter. Reproduces many images returned by the Pioneer and Voyager spacecraft and includes a list of the Voyager project personnel and suggestions for additional reading.

William J. Kosmann

Voyager 1: Jupiter

Date: March 5 to April 15, 1979
Type of spacecraft: Planetary exploration

Voyager 1 collected detailed information on the planet Jupiter and its rings, satellites, and surrounding environment, including detailed photographs of the four Galilean satellites: Io, Europa, Ganymede, and Callisto. Voyager 1 demonstrated the viability of building a complex, semiautonomous spacecraft capable of lasting more than a decade.

Key Figures

Harris M. Schurmeier, first Voyager project manager
Raymond L. Heacock (b. 1928), Voyager project manager for the Jupiter encounter
Edward C. Stone, Jr., Voyager project scientist
Bradford A. Smith, principal investigator, imaging experiment
G. Leonard Tyler, principal investigator, radio science experiment
Norman F. Ness, principal investigator, magnetometer experiment
James W. Warwick, principal investigator, planetary radio astronomy experiment
Linda Kelly Morabito, discoverer of active volcanism on Io

Summary of the Mission

The story of Voyager 1 began in 1966 at the Jet Propulsion Laboratory (JPL) in Pasadena, California, where a team of scientists and engineers conceived the idea of a Grand Tour of the outer planets. A spacecraft can use the gravity of one planet to speed up and to deflect its trajectory toward another planet—a technique called gravity assist. In the late 1970's, Jupiter, Saturn, Uranus, and Neptune were all positioned in an arc on the same side of the Sun, making possible a gravity-assist trajectory from one planet to the next. This special alignment of planets occurs only once every 175 years.

To study the feasibility of a mission to the outer planets, the Thermoelectric Outer Planet Spacecraft (TOPS) group was formed. The TOPS group considered many problems posed in designing a spacecraft for the outer solar system. For example, the spacecraft would have to function at greater distances from the Sun and Earth than had any other spacecraft. Because a mission to the outer

planets would take about ten years to complete, all parts in the spacecraft had to be designed to last that number of years or have failproof backup systems. The spacecraft needed to be more automatic and more independent than any previous spacecraft. Because the outer planets are so distant from Earth, it would take hours for engineers to correct a spacecraft malfunction.

When some of the more significant questions were answered, the Space Science Board, a group of appointed scientists, carefully studied the recommendations of the TOPS group. In 1969, a series of five separate outer planet missions to visit Jupiter, Saturn, Uranus, Neptune, and Pluto were recommended. At the same time, many other missions were competing for funds within the National Aeronautics and Space Administration (NASA) in Washington, D.C. Because of budget constraints in 1972, Congress approved a revised plan to build only three spacecraft. Voyager 1 would fly by both

Jupiter and Saturn. If the Voyager 1 mission was a success, then Voyager 2 would be targeted to fly by not only Jupiter and Saturn but Uranus and Neptune as well. The third spacecraft would be built as a ground spare. To provide advanced knowledge about the environments around both Jupiter and Saturn, two additional, less complex spacecraft, Pioneer 10 and Pioneer 11, were separately funded and launched in 1972.

JPL was selected to implement the mission. The Mariner Jupiter Saturn 77 (MJS77) project, later renamed Voyager, began on July 1, 1972, under the management of Harris M. Schurmeier. When funds were authorized for the mission, NASA issued an "announcement of flight opportunity" to select the scientific instruments for the craft. Eventually, eleven instruments were built for Voyager 1. Edward C. Stone, Jr. was selected as project scientist

A Titan-Centaur launch vehicle boosts Voyager 1 into space and on its way to a meeting with Jupiter and Saturn. (NASA)

and charged with coordinating scientific activity. For Voyager 1's Jupiter flyby, Raymond L. Heacock was project manager.

The Voyager 1 spacecraft was modeled on the Mariner spacecraft series, which had flown earlier to Venus and Mars. The spacecraft was about the size of a compact car. It weighed 825 kilograms, including 117 kilograms of scientific instruments. Voyager 1 was not a spinning spacecraft; it was stabilized on all three axes, using one sensor locked on the Sun and a second sensor locked on a star. Voyager 1 was a ten-sided aluminum structure, containing its key electronic elements inside its inner walls. The center of the structure contained a spherical propellant tank filled with hydrazine fuel. The fuel was used for trajectory corrections and to control the orientation of the spacecraft so that the high-gain antenna, 3.66 meters in diameter, pointed toward Earth.

The spacecraft was powered by three nuclear power sources, radioisotope thermal generators (RTGs) that produced about 400 watts of electrical power. A digital tape recorder could store about 500 million bits of information—equivalent to about one hundred images. The spacecraft was controlled by six onboard computers (two of each kind): the attitude and articulation control subsystem, the flight data subsystem, and the computer command subsystem. The attitude control subsystem controlled the stability and orientation of the spacecraft and the scan platform. The flight data subsystem provided instrument control for the scientific instruments and digital tape recorder and formatted the scientific and engineering data before they were sent to the ground. The computer command subsystem provided primary control of the spacecraft. These Voyager 1 computers could accept precoded sets of instructions that could provide autonomous operation for days or even weeks. These systems also included detailed instructions to detect and correct problems without human intervention.

On Labor Day, September 5, 1977, Voyager 1 was launched from Cape Canaveral, Florida, at 8:56 A.M. eastern daylight time or 12:56 Coordinated Universal Time (UTC), five days after the launch window had opened. The launch vehicle was a Titan III-E-Centaur rocket. Unexpectedly, the Titan main engine shut down early during the launch, and the Centaur stage had to make up the difference during the trajectory-insertion burn. After completing the insertion burn, the Centaur stage shut down with less than five seconds' worth of fuel left in its tank. If the launch had proceeded five days before, as scheduled, the thrust from the remaining fuel would not have been enough to allow Voyager 1 to reach Jupiter. With a little bit of luck, Voyager 1 was on its way to Jupiter.

During the autumn of 1977, a series of small problems challenged the JPL engineers. Attitude control thrusters fired at the wrong times, and sometimes the computer control systems overrode the commands from the ground. The onboard computers had been programmed to be too sensitive to slight changes on the spacecraft, and some reprogramming of the computers was necessary.

On February 23, 1978, the scan platform malfunctioned and prematurely stopped. This platform contained important remote-sensing instrument, and full mobility of the platform during the planetary flybys was essential. At JPL, tests were run on an exact copy of the scan platform. Slowly and carefully, the spacecraft platform was commanded to move, and normal operation was resumed. Engineers suspected that some material caught in the platform gears had been moved out of the way or crushed.

On January 4, 1979, the science-intensive Jupiter encounter began. Voyager 1 was now transmitting information not obtainable from Earth. For the next three months, Voyager 1 carried out a scientific survey of Jupiter, its satellites, and its magnetosphere. More than thirty thousand images were transmitted to Earth during the encounter with Jupiter. Throughout the month of January, the Voyager 1 cameras sent back a series of images every two hours. The images were then turned into a color "motion picture" of Jupiter's weather patterns.

On February 28, Voyager 1 crossed Jupiter's bow shock—the boundary between the solar plasma that flows from the Sun (solar wind) and the planet's magnetosphere. Six hours later, the solar wind had pushed the magnetosphere back toward Jupiter, and Voyager 1 was once again in the solar wind. Over the next several days, variations in the solar wind pressure allowed Voyager 1 to cross the bow shock five times in all, as the Jovian magnetosphere repeatedly expanded and contracted.

A single eleven-minute imaging exposure of space, just above the equatorial cloud tops, was taken as the spacecraft passed through the plane of Jupiter's equator on March 4. Faint rings circling the planet were discovered. Close-range observations of the four Galilean satellites—Io, Europa, Ganymede, and Callisto—were made between March 4 and March 6. The closest flyby distances from each satellite were Io, 22,000 kilometers; Europa, 734,000 kilometers; Ganymede, 115,000 kilometers; and Callisto, 126,000 kilometers.

Voyager 1's closest approach to Jupiter took place on March 5, 1979, at 4:05 A.M. Pacific stan-

Jupiter, the largest planet in the solar system and one of four "Gas Giant" planets, is composed primarily of light-weight elements such as hydrogen and helium. (NASA CORE/Lorain County JVS)

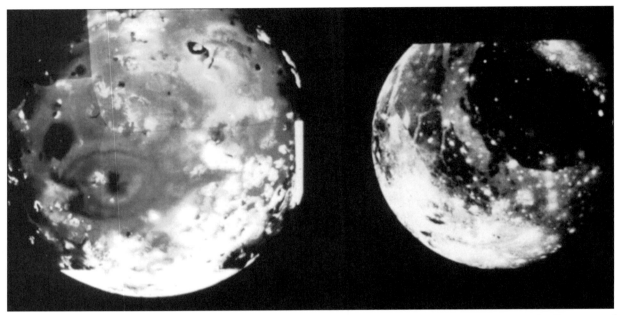

Images of two of the four large Galilean satellites of Jupiter taken by Voyager 1 (from left): Io, Ganymede. Europa and Callisto are not pictured. (NASA)

dard time (12:05 UTC). Thirty-seven minutes later, at 4:42 (12:42 UTC), the signals from the spacecraft reached Earth. At 8:14 (16:14 UTC), the spacecraft passed out of sight behind Jupiter, and the radio occultation of the Jovian atmosphere began. As the spacecraft flew behind Jupiter, the varying strength of the radio signal was used to probe the cloud structure in the atmosphere. Two hours later, Voyager 1 safely reappeared.

On March 5, during the period when it was closest to Jupiter, harsh radiation from the planet caused problems on the spacecraft. The main spacecraft clock slowed a total of eight seconds, and two computers were out of synchronization with the flight data subsystem computer. Some of the best images of Io and Ganymede were out of focus because the spacecraft started to move before the camera's shutter closed. Once the spacecraft moved farther away from Jupiter, this problem was corrected. On March 20, Voyager 1 crossed the Jovian bow shock, leaving Jupiter's magnetosphere and once again entering the solar wind. After a spectacular encounter with Jupiter, Voyager 1 was

on its way for an encounter with Saturn on November 12, 1980.

Contributions

Voyager 1's pictures of Jupiter provide details of a turbulent, colorful atmosphere unlike any seen before. Images of Jupiter's Great Red Spot, a feature whose diameter is three times that of Earth's and that is more than three hundred years old, reveal a huge hurricane-like storm towering above the surrounding clouds. White oval-shaped features about the size of Earth are other storms similar to the Great Red Spot.

Images of Jupiter reveal a stable zonal pattern of east-west winds. This planet-wide flow is more fundamental than the shifting cloud patterns within the east-west alternating belts (dark, deeper atmospheric regions) and zones (light, higher atmospheric regions). Within the belts and zones reside dark, brownish regions known as hot spots, holes in the uppermost cloud tops. These regions are warmer than the surrounding atmosphere, and both water vapor and germanium were discovered there. The minimum temperature on Jupiter was 110 kelvins at 0.1 bar (on Earth, the surface pressure is typically 1.0 bar).

Cloud-top lightning bolts, similar to superbolts on Earth, were photographed, and radio-frequency emissions associated with the lightning were observed by the instrument for the planetary radio astronomy experiment. Auroral emissions in the polar region were seen in both the ultraviolet and the visible spectra.

With the imaging system, eight active sulfur volcanoes were discovered on the surface of Io, with plumes extending 250 kilometers above the surface. Tidal heating as a result of interactions with the other satellites and with Jupiter's powerful gravity melts the interior of Io, producing spectacular volcanoes. The moon's surface is uncratered and young because the volcanoes bring about continual resurfacing. The infrared instrument discovered hot spots on the surface of the satellite, and infrared scientists independently concluded that volcanoes existed there. Infrared measurements identified sulfur dioxide gas over the volcano named Loki.

Imaging observations showed numerous intersecting, linear features on the surface of Europa. Two distinct types of terrain, craters and grooves, characterize the surface of Ganymede. Ganymede was the first body other than Earth to display evidence of tectonic activity. Callisto displays a heavily cratered surface.

Voyager discovered rings of material orbiting Jupiter, with an outer edge about 128,000 kilometers from the center of the planet. The rings consist of small, dusty particles. Two newly discovered satellites, Metis and Adrastea, are embedded in the rings and are probably the source of the tiny ring particles.

An electric current of more than one million amperes flows in a magnetic flux tube linking Jupiter and Io. An Io torus—an invisible "doughnut" containing ionized sulfur and oxygen circling Jupiter at Io's orbit—was discovered. Jupiter has unusual radio emissions at the kilometer wavelength, which may be generated by plasma interactions with the Io torus.

Pioneer had shown that the magnetic field of Jupiter was dipolar, with a tilt of 11°, and Voyager verified it. Both Pioneer and Voyager measurements showed that the magnetosphere of Jupiter is large.

Context

Voyager 1 was the third spacecraft to fly through the Jovian system and the first to take high-resolution pictures of all four Galilean moons. Voyager 2 followed four months later. The flybys five years earlier had both been U.S. missions as well, Pioneer 10 and Pioneer 11.

In the late 1960's, Earth observations had established that Jupiter had an internal heat source. Pioneer investigations had confirmed this finding: Jupiter is still cooling from its initial collapse and formation. Infrared instruments aboard Pioneer had provided the first measurement of the ratio of hydrogen to helium on Jupiter, and Voyager was able to refine this value further. This ratio, roughly ten hydrogen atoms for every helium atom, is comparable to the value for the Sun, supporting the idea that Jupiter and the Sun have similar compositions.

Observations of Jupiter are important for the understanding of the origin and evolution of the solar system. Jupiter has an extremely large mass. Had it been roughly one hundred times more massive it would have formed a star. With the same basic composition as the Sun, Jupiter constitutes a sample of the original material from which the solar system formed.

The four large Galilean satellites form a miniature planetary system. Ranging in size from just larger than the planet Mercury (Ganymede and Callisto) to just smaller than the Moon (Io and Europa), these satellites decrease in density with increasing distance from Jupiter, as do the planets in the solar system with increasing distance from the Sun. With the exception of Callisto's, the surfaces on the Galilean satellites are in general much younger than expected. Cratering records showed multiple periods of bombardment, interspersed with resurfacing.

Plans for a return to Jupiter focused on the Galileo spacecraft, a combined orbiter and probe. Galileo was launched in October of 1989; it reached Jupiter in July of 1995, when the probe and orbiter separated. The probe was sent into the atmosphere of Jupiter, and the orbiter circled the planet, recording data, thereafter encountering all the major moons, some several times. Far outliving the initially planned mission, Galileo's mission was extended beyond the year 2000. Galileo was purposely directed into Jupiter's crushing atmosphere on September 21, 2003, to prevent it from colliding with and contaminating any of the Galilean moons, especially Europa.

Another opportunity for a Jupiter flyby was to come with the Cassini mission to Saturn. The mission was launched on October 15, 1997, and reached Jupiter, flying by the planet on December 30, 2000. The gravity of Jupiter was used to speed up and deflect the Cassini spacecraft toward Saturn for an orbital insertion in early summer, 2004.

Voyager established the viability of building a spacecraft to last a decade or more. Scientists expected that this technology would play a major role in the design of spacecraft for extended missions to the outer planets, craft that would relay information to Earth for many years.

Both Voyager and Pioneer expanded the frontier of the outer solar system. With the pictures and information sent back by Voyager, the Galilean satellites were transformed from pinpoints of light into tiny worlds. Jupiter's turbulent atmosphere provides a model against which Earth's circulation patterns can be compared. Still, Voyager 1 only began to address fundamental questions about the formation and evolution of Jupiter and planets in the outer solar system.

As of 2005, the Voyager 1 spacecraft was continuing its scientific exploration of interstellar space. It had passed nearly 22 billion kilometers (13.5 billion miles), traveling away from the solar system at more than 21 kilometers per second.

See also: Cassini: Saturn; Deep Space Network; Galileo: Jupiter; Hubble Space Telescope; Huygens Lander; Jet Propulsion Laboratory; Planetary Exploration; Voyager Program; Voyager 1: Saturn; Voyager 2: Jupiter; Voyager 2: Saturn; Voyager 2: Uranus; Voyager 2: Neptune.

Further Reading

Beatty, J. Kelly, Carolyn Collins Petersen, and Andrew Chaikin, eds. *The New Solar System.* 4th ed. New York: Sky Publishing, 1999. Twenty chapters by distinguished researchers synthesize knowledge of the solar system. Findings on the Sun, the planets, the satellites, and the medium between are clearly discussed, with an emphasis on the discoveries of space probes such as Voyager.

Eberhart, Jonathan. "Jupiter and Family." *Science News* 115 (March 17, 1979): 164-165, 172-173. Highlights of Voyager 1's Jupiter encounter, particularly the discovery of volcanoes on Jupiter's moon Io, are described in this article, published less than two weeks after the closest approach to Jupiter. Color photographs of Jupiter and its moons are included.

Gore, Rick. "Voyager Views Jupiter's Dazzling Realm." *National Geographic* 157 (January, 1980): 2-29. This readable article describes the findings of the Voyager Jupiter flybys. Includes quotes from key individuals involved in the mission. Beautiful color images of Jupiter and the Galilean satellites are included.

Harland, David M. *Jupiter Odyssey: The Story of NASA's Galileo Mission.* London: Springer-Praxis, 2000. A detailed scientific and engineering history of the Galileo program, but also includes extensive discussion of Voyager program events and results.

_____. *Mission to Saturn: Cassini and the Huygens Probe*. London: Springer-Praxis, 2002. A detailed scientific and engineering history of the Cassini program, but also includes extensive discussion of Voyager program events and results.

Hartmann, William K. *Moons and Planets*. 5th ed. Belmont, Calif.: Thomson Brooks/Cole Publishing, 2005. Provides detailed information about all objects in the solar system. Suitable on three separate levels: high school student, general reader, and college undergraduate studying planetary geology.

Lee, Wayne. *To Rise from Earth: An Easy to Understand Guide to Spaceflight*. New York: Checkmark Books, 1996. This is a good introduction to the science of spaceflight. Although written by an engineer with the NASA Jet Propulsion Laboratory, it is presented in easy-to-understand language. In addition to the theory of spaceflight, it gives some of the history of the human endeavor to explore space.

Mari, Christopher, ed. *Space Exploration*. New York: H. W. Wilson, 1999. Twenty-five articles (reprinted from magazines), covering the state of the space program at the time of publication, are divided into five sections: John H. Glenn, Jr.'s return to space, the exploration of Mars, the International Space Station, recent mining efforts by commercial industries, and new types of space vehicles and propulsion systems.

Nicks, Oran W. *Far Travelers: The Exploring Machines*. NASA SP-480. Washington, D.C.: Scientific and Technical Information Branch, 1985. Discusses all major NASA planetary spacecraft during NASA's first quarter-century. Written by a senior NASA official involved with lunar and planetary programs during that era.

Simon, Seymour. *Jupiter*. New York: William Morrow, 1985. This well-written book summarizes knowledge of the planet Jupiter in a format suitable for elementary levels and older. Twenty color photographs taken during the Voyager encounters highlight this introduction to Jupiter.

Soderblom, Lawrence. "The Galilean Moons of Jupiter." *Scientific American* 242 (January, 1980): 88-100. The four Galilean satellites, Io, Europa, Ganymede, and Callisto, are discussed in detail in this journal article. Physical characteristics of the satellites and the parameters of the Voyager flybys of each satellite are detailed. Possible evolution scenarios and cratering rates are outlined. Suitable for high school and college students.

Linda J. Horn

Voyager 1: Saturn

Date: November 12 to December 15, 1980
Type of spacecraft: Planetary exploration

On its second planetary flyby, Voyager 1 encountered Saturn and sent back to Earth information on the planet's rings, satellites, and atmosphere. In the process, the probe helped demonstrate that a complex spacecraft could last more than a decade while operating in space semiautonomously.

Key Figures

Harris M. Schurmeier, first Voyager project manager
Raymond L. Heacock (b. 1928), Voyager project manager for the Saturn encounter
Edward C. Stone, Jr., Voyager project scientist
Bradford A. Smith, principal investigator, imaging experiment
Norman F. Ness, principal investigator, magnetometer experiment
G. Leonard Tyler, principal investigator, radio science experiment
James W. Warwick, principal investigator, planetary radio astronomy experiment

Summary of the Mission

The Grand Tour of the outer planets in the solar system was conceived by a team of engineers and scientists working at the Jet Propulsion Laboratory (JPL) in Pasadena, California. A plan to build three spacecraft for a mission to the outer planets was approved by Congress in 1972. One of those spacecraft was Voyager 1; its trajectory took it past both Jupiter and Saturn. Jupiter's gravity was used to increase the spacecraft's velocity and divert its trajectory toward Saturn, shortening the travel time to Saturn from 6 years to 3.3 years.

JPL was chosen to implement the mission. The Mariner Jupiter Saturn 77 (MJS77) project, later renamed Voyager, began on July 1, 1972, under the management of Harris M. Schurmeier. When funds were authorized for the mission, the National Aeronautics and Space Administration (NASA) began the process of selecting the scientific instruments that would be installed on the craft. Eventually, eleven instruments were built for Voyager 1. To coordinate the scientific activity, Edward C. Stone, Jr., was selected as project scientist,

and Raymond L. Heacock was project manager for the Saturn flyby.

There were two groups of Voyager instruments. The first group comprised the remote-sensing instruments. Most of these were mounted on a movable scan platform and obtained data on remote targets. The second group included in situ instruments, those that made direct measurements of the surrounding charged particles, magnetic field, and plasma waves.

The remote-sensing instruments were mounted on a movable scan platform at the end of a 2.3-meter science boom. The scan platform could move in two axes, scanning all the sky except for the region blocked by the spacecraft itself. The instruments mounted on the scan platform included the infrared interferometer spectrometer and radiometer, two imaging cameras, a photopolarimeter, and an ultraviolet spectrometer. Together, these instruments measured the properties of the objects they detected in wavelengths from the infrared through the visible to the ultraviolet.

Attached at the midpoint of the science boom were two in situ instruments, the cosmic-ray experiment and the low-energy charged particle experiment. The plasma experiment was an in situ instrument farther out on the science boom, near the scan platform. These instruments measured the distribution of energetic charged particles such as electrons, protons, and ions.

The planetary radio astronomy and plasma wave experiments shared an antenna. The antenna consisted of two thin metal rods, each 10 meters long, set at right angles to each other. These rods were extended from the spacecraft after launch. The planetary radio astronomy experiment was a remote-sensing instrument, and the plasma wave experiment took samples in situ. The in situ magnetic field experiment consisted of four three-axis magnetometers, two high-field sensors attached to the spacecraft, and two low-field sensors mounted at the end of a 13-meter-long boom. This boom was packed tightly in a canister during launch; once in space, the canister opened, and the boom extended automatically.

The final Voyager instrument, the radio science experiment, used the dish-shaped high-gain antenna. The high-gain antenna was the radio communications link between Earth and Voyager, relaying data from the outer science instruments as well as making remote-sensing measurements of various atmospheres as the spacecraft passed behind planetary bodies.

On Labor Day, September 5, 1977, Voyager 1 was launched from Cape Canaveral, Florida, at 8:56 A.M. eastern daylight time or 12:56 Coordinated Universal Time (UTC), five days after the launch window opened. The launch vehicle was a Titan III-E-Centaur rocket. The Titan main engine cut out early, forcing the Centaur stage to make up the difference during the trajectory insertion burn. After completing its burn, the Centaur stage shut down with very little fuel left in its tank. If the launch had not been delayed, the thrust from the remaining fuel would have been insufficient to allow Voyager 1 to reach first Jupiter and then Saturn.

During the one-and-one-half-year cruise period between the Jupiter and Saturn encounters, Voyager 1 continued to measure the solar wind. On January 1, ten months before the Voyager 1 encounter with Saturn, the planetary radio astronomy instrument discovered very long wavelength radio bursts from Saturn. These bursts were highly regular, allowing calculation of the rotation rate for Saturn's interior, which proved to be 10 hours and 39.4 minutes.

In the fall of 1980, the science-intensive Saturn encounter began. Voyager 1 performed a detailed survey of Saturn, its ring system, and its satellites—including a close flyby of the large moon Titan. More than eighteen thousand images were transmitted to Earth during the encounter with Saturn.

On October 6, an extensive set of images of Saturn's rings revealed unexpected and detailed structure in the rings. As a result of this observed structure, the spacecraft was reprogrammed, and on October 25 the spacecraft cameras pointed toward one ansae (end) of the rings and imaged the rings every five minutes for ten hours. These images were used to produce a "film" of ring activity that highlighted the "spokes," dark fingers of material extending radially outward over a portion of the rings. In images from this film, two tiny satellites, Prometheus and Pandora, were first discovered. They orbit on each side of the narrow F-ring.

On November 11, Voyager 1 crossed the Saturn bow shock, the boundary between the solar plasma, which flows from the Sun (solar wind), and Saturn's magnetosphere. Just inside the bow shock, a boundary called the magnetopause separates the turbulent area between the bow shock and the actual magnetosphere. The location of the magnetopause changes dynamically with variations in solar wind pressure. Five magnetopause crossings occurred during the period of one hour before Voyager 1 entered the magnetosphere for the final time.

The closest approach to Titan took place on November 11, within 4,000 kilometers. Infrared measurements of varying atmospheric levels were performed from the edge of Titan's north pole.

As Voyager 1 passed behind Titan, the varying strength of its radio signal probed Titan's atmosphere (a technique known as radio occultation), measuring its pressure and temperature as a function of distance above the surface. The ultraviolet instrument on board the spacecraft observed the sunset on Titan (solar occultation), and another measurement of Titan's atmosphere was made. About fifteen minutes later, the spacecraft safely reappeared from the shadow of Titan. As Voyager 1 was passing close to Titan, it simultaneously passed through the Saturn ring plane, which passes through Saturn's equator and through the known rings. No damaging particles hit the spacecraft during this passage.

November 12, 1980, was encounter day. Voyager 1's closest approach to Saturn took place at 6:45 P.M. eastern standard time (23:45 UTC) only 124,000 kilometers above the cloud tops. At 7:08 P.M. (00:08 UTC, the next day), the spacecraft passed behind Saturn, and the structure of the clouds in the planet's atmosphere was probed by measuring the varying strength of the radio signals sent to Earth. A solar occultation by the atmosphere was simultaneously observed. The spacecraft reappeared briefly as it reached the other side of Saturn, and then it passed behind Saturn's rings, where it measured the distribution of ring material in the main rings by using the variation in the strength of the radio signal. Almost forty-five minutes after the occultations ended, Voyager 1 crossed the ring plane again, at the orbit of the satellite Dione, called the Dione clear zone, where the risk of collision with a ring particle was less likely. Observing the sunlight shining through them, Voyager 1 spent twenty-three hours underneath the rings photographing views never before seen from Earth.

Close-up views of five of the major Saturnian satellites were obtained on November 12. The closest flyby distances to each satellite were Mimas, 88,000 kilometers; Enceladus, 202,000 kilometers; Tethys, 416,000 kilometers; Dione, 161,000 kilometers; and Rhea, 74,000 kilometers. The images with the highest resolution were of Rhea. To obtain sharp images, the entire spacecraft was turned to keep Rhea motionless in the cameras.

By the end of November, Voyager 1 had crossed the Saturn bow shock to leave Saturn's magnetosphere and enter the solar wind for the final time. No further planetary encounters were possible for Voyager 1. In order to bring it close to the satellite Titan, the spacecraft trajectory, using Saturn's gravity, was bent out of the plane containing the orbits of the major planets. Voyager 1 was therefore outward bound on a path leaving the solar system. Perhaps in several hundred thousand years Voyager 1 will fly close to another star.

Mounted on the spacecraft is a gold-plated recording of the sights and sounds of planet Earth. Should a distant civilization find Voyager 1 and de-

A montage of Voyager 1's images of Saturn and its moons. (NASA-JPL)

code the recording, greetings from Earth spoken in fifty-three languages and various sounds of the world will be heard. One hundred fifteen images will display the diversity of life and culture on Earth. With its departure from Saturn, Voyager 1 was beginning a journey of epic proportions.

Contributions

Images of Saturn reveal alternating east-west belts (darker bands) and zones (lighter bands) similar to those in the Jovian atmosphere, although on Saturn they are considerably more muted. Saturn's equatorial jet stream blows four times harder (about 0.5 kilometer per second) around Saturn's equator than Jupiter's winds blow around Jupiter. Measured temperatures in Saturn's atmosphere include a minimum temperature of 80 kelvins at 0.1 bar (on the surface of Earth, the average atmospheric pressure is 1.0 bar).

Voyager 1's data revealed that Titan is hidden by an atmosphere thicker than Earth's, and that it may possess hydrocarbon oceans. A thick, smoglike haze covers Titan and creates a greenhouse effect, warming the surface. Voyager 1 provided the first measurements of the near-surface atmospheric pressure (1.6 bars) and temperature (95 kelvins) on Titan. The main constituent of Titan's atmosphere was found to be nitrogen. Infrared measurements detected trace amounts of various hydrocarbons in the atmosphere. Nitriles—molecules composed of hydrogen, nitrogen, and carbon—were discovered also.

Images from the Voyager 1 cameras revealed three new, tiny satellites. Atlas orbits just outside the outer edge of the main ring system; Prometheus and Pandora fall on either side of the narrow F-ring, located 3,000 kilometers outside the main ring system. The F-ring appears braided and clumpy as a result of gravitational interactions with these satellites.

Using the imaging system, the sizes of the seventeen known satellites were determined for the first time, thus permitting a more accurate estimate of their densities. On Mimas, a huge crater, 130 kilometers in diameter and one-third the diameter

of Mimas itself, was discovered. Enceladus is the brightest satellite in the solar system; some regions of its bright surface are almost devoid of craters, indicating a young surface. An enormous canyon, covering nearly three-quarters of its circumference, engulfs Tethys. Both Dione and Rhea display a dark surface overlaid with bright, wispy terrain, possibly a product of internal processing. Iapetus has an unusual distribution of light and dark material on its surface: It is bright on one side and completely dark on the other. The light material is approximately ten times brighter than the dark material.

Voyager 1 revealed a ring system of structural complexity and variety. The rings are not bland sheets of material, as thought prior to the Voyager 1 flyby, but possess detailed structures on scales smaller than a kilometer. The rings are composed of particles in a wide range of sizes, from tiny pebbles to giant boulders, including a sprinkling of fine dust. Spokes and elliptical as well as discontinuous ringlets were discovered in the main ring system by Voyager 1. A new ring, the tenuous G-ring, was discovered between the orbits of Mimas and two co-orbiting satellites.

Voyager 1's instruments confirmed that, unlike Earth, Saturn has a magnetic dipole axis that is closely aligned with the planet's spin axis. The relative tilt between the magnetic dipole axis and the spin axis is only 0.7 degree for Saturn, compared to a tilt of 11.5° for Earth. The number of charged particles in Saturn's magnetosphere is much smaller than the number measured at Jupiter. The rings are effective particle absorbers in the inner magnetosphere.

Context

Voyager 1 was the second spacecraft to fly through the Saturn system and the first to take high-resolution images of Saturn, Titan, the icy satellites, and the rings. Voyager 2 followed, nine months later. One year earlier, Pioneer 11, another U.S. mission, had also conducted a flyby.

Saturn, the sixth planet from the Sun, is second in size only to Jupiter. Studies of the Saturn system have contributed to the understanding of the origin and evolution of the solar system. Earth-based observations established that Saturn has an internal heat source, and infrared measurements by both Pioneer 11 and Voyager 1 confirmed it. Primordial cooling from Saturn's initial formation and collapse should be complete; thus, an excess of heat was unexpected. Voyager 1 also measured the hydrogen-to-helium ratio of Saturn's upper atmosphere: roughly thirty hydrogen molecules for every helium molecule. Saturn has suffered a threefold depletion of helium relative to the solar abundance value. The mechanism responsible for this depletion in the upper atmosphere may also generate the excess heat.

With the exception of the Jovian satellite Ganymede, Titan is the largest satellite in the solar system, and it is the only one known to possess a substantial atmosphere. At a greater distance from the Sun than Ganymede, it is much colder and richer in ices. Some of the chemical reactions occurring in Titan's atmosphere provide possible analogues to some of the prebiotic chemistry that took place on primitive Earth to form the nucleic acids found in living organisms.

An interesting puzzle is the apparent youth of the ring system, only ten to one hundred million years, much shorter than the solar system age of 4.5 billion years. Understanding the evolution of planetary rings may also lead to a better understanding of planetary accretion from the disk of material originally surrounding the Sun.

Voyager 1 established the possibility of building a spacecraft to last for more than a decade. Sophisticated engineering for Voyager 1's Saturn flyby utilized complex maneuvers to track the limb of the planet during the radio occultation and to compensate for the rapid motion of Rhea during the close flyby. Both Voyager 1 and Pioneer 11 extended the frontier of the outer solar system. With images and information relayed to Earth, a wealth of new knowledge was provided about the Saturnian system.

As of 2005, the Voyager 1 spacecraft was continuing its scientific exploration of interstellar space. Having traversed over 13.5 billion kilometers from Earth, twenty-seven years after launch, Voyager 1 was heading out of the solar system at more than 21 kilometers per second.

See also: Cassini: Saturn; Deep Space Network; Galileo: Jupiter; Hubble Space Telescope; Huygens Lander; Jet Propulsion Laboratory; Planetary Exploration; Voyager Program; Voyager 1: Jupiter; Voyager 2: Jupiter; Voyager 2: Saturn; Voyager 2: Uranus; Voyager 2: Neptune.

Further Reading

Beatty, J. Kelly, Carolyn Collins Petersen, and Andrew Chaikin, eds. *The New Solar System.* 4th ed. New York: Sky Publishing, 1999. Well-known scientists have provided a useful overview of the solar system, including discussions of the Sun, the planets and their satellites, and the medium between the various bodies.

Cooper, H. *Imaging Saturn.* New York: Holt, Rinehart and Winston, 1981. The Voyager mission is chronicled in a day-by-day description of events that occurred during the Saturn encounters. Accounts of key individuals, in particular the Voyager Imaging Team members, are detailed. Suitable for general audiences.

Eberhart, Jonathan. "Secrets of Saturn: Anything but Elementary." *Science News* 120 (September 5, 1981): 148-158. Highlights of the Voyager Saturn encounters are described in this article, with a focus on Saturn's rings. Several photographs from the encounter are included. Suitable for general audiences.

Gore, Rick. "Voyager 1 at Saturn." *National Geographic* 160 (July, 1981): 2-31. This readable article describes the results of Voyager 1's Saturn flyby and includes statements from key individuals involved in the mission. Beautiful color images of Saturn, Titan, the icy satellites, and the rings are included.

Harland, David M. *Jupiter Odyssey: The Story of NASA's Galileo Mission.* London: Springer-Praxis, 2000. A detailed scientific and engineering history of the Galileo program, but also includes extensive discussion of Voyager program events and results.

_____. *Mission to Saturn: Cassini and the Huygens Probe.* London: Springer-Praxis, 2002. A detailed scientific and engineering history of the Cassini program, but also includes extensive discussion of Voyager program events and results.

Hartmann, William K. *Moons and Planets.* 5th ed. Belmont, Calif.: Thomson Brooks/Cole Publishing, 2005. Provides detailed information about all objects in the solar system. Suitable on three separate levels: high school student, general reader, and college undergraduate studying planetary geology.

Lauber, Patricia. *Journey to the Planets.* Rev. ed. New York: Crown, 1987. Lauber describes the history and nature of the planets, noting their resemblances and differences. Includes photographs and information from the Voyager missions.

Mari, Christopher, ed. *Space Exploration.* New York: H. W. Wilson, 1999. Twenty-five articles (reprinted from magazines), covering the state of the space program at the time of publication, are divided into five sections: John H. Glenn, Jr.'s return to space, the exploration of Mars, the International Space Station, recent mining efforts by commercial industries, and new types of space vehicles and propulsion systems.

Morrison, David, and Tobias Owen. *The Planetary System.* 3d ed. San Francisco: Addison-Wesley, 2003. Organized by planetary object, this work provides contemporary data on all planetary bodies visited by spacecraft since the early days of the Space Age. Suitable for high school and college students and for the general reader.

Morrison, David, and Jane Samz. *Voyages to Saturn.* NASA SP-451. Washington, D.C.: Government Printing Office, 1982. The official account of the Voyager encounters with Saturn. Well illustrated. Includes appendices of information about the personnel involved in the Voyager missions as well as a list of suggested reading material.

Poynter, Margaret, and Arthur L. Lane. *Voyager: The Story of a Space Mission.* New York: Macmillan, 1981. Provides a behind-the-scenes account of the missions during all phases, from planning to the planetary encounters.

Simon, Seymour. *Saturn.* New York: William Morrow, 1985. This easy-to-read book surveys knowledge of the planet Saturn. Includes twenty large color photographs taken during the Voyager encounters.

Linda J. Horn

Voyager 2: Jupiter

Date: August 20, 1977, to July 11, 1979
Type of spacecraft: Planetary exploration

The Voyager 2 flyby of Jupiter provided vital information about the Jovian system, in spite of a number of technical problems and equipment failures. Complementing the Voyager 1 flyby, this mission helped map Jupiter's moons, collected valuable data on the magnetic and radiation fields surrounding Jupiter, and monitored the planet's atmospheric phenomena.

Key Figures

Raymond L. Heacock (b. 1928), Voyager project manager

Edward C. Stone, Jr., Voyager project scientist

Bradford A. Smith, principal investigator, imaging experiment

Norman F. Ness, principal investigator, magnetometer experiment

Von R. Eshleman, radio science team leader

James W. Warwick, principal investigator, planetary radio astronomy experiment

Summary of the Mission

Voyager 2 was launched on August 20, 1977, sixteen days before the launch date of its twin, Voyager 1. The flight paths of the two spacecraft were such that between the orbits of Mars and Jupiter, in the asteroid belt, Voyager 1 overtook Voyager 2 and arrived at Jupiter four months and four days before Voyager 2. The differences in these two flight paths allowed the two spacecraft to complement each other, so that Voyager 2 obtained data on features of the Jovian, and later Saturnian, systems that Voyager 1 was unable to probe. Voyager 2's flight path was designed to carry it past Uranus and Neptune, while Voyager 1 would leave the plane of the solar system after its encounter with Saturn.

The data and pictures sent back to Earth by Voyager 2 are remarkable not only for their quality and uniqueness but also because of the number of problems that were overcome to obtain them. Even before launch, failures in two of the computer subsystems of the VGR77-2 spacecraft (which later became Voyager 1) delayed its launch and forced the substitution of the identical VGR77-3 spacecraft,

now known as Voyager 2. Also just before the August 20, launch, the low-energy charged particle instrument had to be replaced. During launch, Voyager 2 behaved as if it had been jolted or bumped, switching to backup systems and losing telemetry signals. Flight engineers later determined that the attitude and articulation control subsystem (AACS) had experienced some electronic gyrations. Just after launch, the lock on the scientific instrument boom failed to signal that the boom was extended, even though it was. During the fall of 1977, Voyager 2 continued its erratic behavior. Finally, it was determined that the spacecraft's systems had been programmed to be too sensitive to environmental changes; after reprogramming, the erratic behavior subsided.

A more serious problem soon developed. The Voyager spacecraft are equipped with two receivers, a primary and a secondary or backup, through which commands are received from Earth. In late November, 1977, Voyager 2's primary receiver began losing power; it failed in late March, 1978. The failure would have caused no major difficulty if the backup receiver had been working properly, but when the spacecraft's computer command subsystem (CCS) switched to the backup receiver seven days after it received its last communication from Earth, Voyager 2 still did not respond properly. Twelve hours later, on April 5, 1978, the CCS switched back to the primary receiver, which worked for half an hour before a power surge blew its fuses and permanently disabled it. During the next seven days, flight engineers devised a way of communicating through the faulty secondary receiver. The problem with the backup receiver was that it had lost its ability to compensate for slight frequency changes in signals from Earth, so that now only the most accurate signals could be recognized. Slight changes in frequency and receiver response result from Earth's rotation (Doppler shift), temperature fluctuations caused by electronic components switching on and off, and environmental factors, such as the magnetic and electric fields near planets and the associated radiation to which the spacecraft is subjected. Now all these fluctuations had to be accounted for when a signal was beamed from Earth, which made programming Voyager 2 from Earth much more difficult.

On June 23 and October 12, 1978, Voyager 2 was programmed for backup automatic missions at Saturn and Jupiter in the event that the faulty backup receiver should fail completely. In August, 1978, it was again reprogrammed to ensure better scientific results during its Jupiter encounter. In particular, Voyager 2 was instructed to compensate for motions caused by its tape recorder. These motions would cause time-exposed television images to blur and lose detail. Throughout the Voyager 2 mission, the faulty backup receiver was used repeatedly to reprogram the spacecraft. This reprogramming was crucial to the amount and quality of information sent back to Earth.

A serious fuel shortage also threatened the mission. Course corrections took 15 to 20 percent

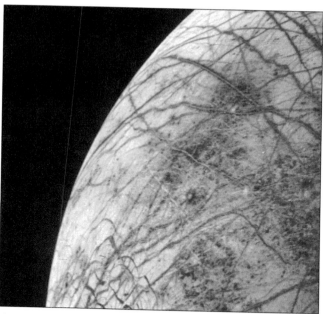

Another of Jupiter's moons, Europa, showing its fractured surface. (NASA CORE/Lorain County JVS)

more fuel than had been expected because the maneuvering jets were partially blocked by the struts that had connected the spacecraft to its last booster rocket. To reduce the number and duration of course corrections, Voyager 2 was tipped upside down. In this attitude the spacecraft was not as easily blown off course by the solar wind (charged particles streaming from the Sun), but its attitude control system had to be reprogrammed to steer by different guide stars. Another major fuel-saving maneuver was accomplished only two hours after Voyager 2's closest approach to Jupiter. Flight engineers determined that about 10 percent of the original fuel load could be saved by rescheduling a major course correction to that time. Executing this course correction so close to Jupiter made it much more difficult to monitor. In fact, just as the spacecraft began the 76-minute thruster firing that would send it on to Saturn, communication with Earth was lost. Afterward, once the interference from the Jovian magnetosphere (the region dominated by Jupiter's magnetic field instead of the Sun's) was penetrated, the flight controllers found that Voyager 2 had executed its new programming perfectly and was headed toward Saturn, with enough fuel to redirect it toward Uranus and beyond.

In spite of all these problems, Voyager 2 began sending back information about interplanetary space days after launch. During its cruise to Jupiter, the spacecraft's instruments were being calibrated and tested. Not only did this provide ground controllers and investigators with a better understanding of the instruments' behavior, but it also provided a chance to study infrared and ultraviolet radiation, magnetic fields, solar flares, and the solar wind far from Earth's influences. Unlike the Pioneer missions, Voyager 2 was not equipped to analyze the particulates of matter in the asteroid belt. Fortunately, the passage through this "shooting gallery" was uneventful, and by October, 1978, Voyager 2 had passed through the asteroid belt and was slightly more than halfway to Jupiter.

The July 9, 1979, encounter with Jupiter was planned so that Voyager 2 could photograph the unseen sides of Jupiter's largest moons. (Because of Jupiter's tremendous gravitational and tidal pull on its satellites, one side of these bodies always faces Jupiter and the other side is turned out toward space.) When Voyager 2 passed through the Jovian system, the moons were encountered first, so that their spaceward sides were facing the Sun and Voyager 2's cameras. The flight path followed by Voyager 2 brought it near enough to Europa to resolve features as small as 4 kilometers across. In contrast, the best Voyager 1 pictures of Europa revealed only features 33 kilometers across or larger. Even though Io was always more than 1 million kilometers away from Voyager 2, the volcanoes discovered with the help of Voyager 1 data had so fascinated project scientists that a ten-hour "Io volcano watch" was planned. Just after its closest approach to Jupiter, Voyager 2 turned its cameras and instruments toward Io and the glowing gases that surround its orbit about Jupiter.

Because Voyager 2 was scheduled to go to Uranus and Neptune, it could not come as close to Jupiter as Voyager 1 had. In spite of this, Voyager 2 data revealed a moon (later dubbed Adrastea), closer to Jupiter than Amalthea, and took spectacular pictures of Jupiter's rings. Voyager 2 data also showed that the magnetosphere and radiation belts surrounding Jupiter had enlarged and intensified since Voyager 1's flyby. Jupiter's radiation belts resemble Earth's Van Allen belts, except that they can be ten thousand times stronger. Their radiation is generated by energetic charged particles that are trapped by Jupiter's magnetic field. Voyager 2's closest distance from Jupiter was nearly twice Voyager 1's closest distance, yet Voyager 2 instruments detected radiation levels that were three times stronger than those detected by Voyager 1 (a circumstance that may explain the communication interruption at closest approach). Furthermore, Voyager 2 first encountered Jupiter's bow shock (where the solar wind collides with a planet's magnetosphere) at a greater distance than Voyager 1 had four months earlier. Jupiter's bow shock flutters in the solar wind, however, so that in three days Voyager 2 passed through the bow shock ten times.

Another important aspect of the Voyager 2 flyby of Jupiter was the information it returned on Jupiter's meteorology. Six weeks after the first Voyager flyby, Voyager 2 turned its cameras toward Jupiter and began to document the cloud movements of Jupiter. A series of pictures were taken and were combined into a motion picture of the dynamics of Jupiter's atmosphere. Throughout the encounter, the spacecraft took pictures of Jupiter itself; scientists hoped thereby to detect changes in cloud patterns, lightning flashes, or auroras.

Contributions

The information returned by Voyager 2 dealt with three basic aspects of the Jovian system: Jupiter's atmosphere, satellites and rings, and magnetosphere. This information permitted important comparisons with Voyager 1's data or provided detailed images and measurements of phenomena documented only briefly by Voyager 1. Not only did Voyager 2 confirm many of Voyager 1's sightings, but it also supplied significant new information.

The motions of Jupiter's upper atmosphere were revealed in a series of photographs taken by Voyager 2. High-altitude jet streams were shown to form bands at constant latitudes and alternate with Jupiter's belts (dark bands) and zones (light bands). The jet stream velocities varied slightly from those measured by Voyager 1. At the boundaries of Jupiter's belts and zones can be found turbulence and storms. Some of these storms are very persistent; the Great Red Spot, for example, has lasted more than four centuries, and three white ovals have been studied since 1939. Voyager 2 revealed that these ovals, like the Great Red Spot, are all anticyclonic (rotating counterclockwise in the southern hemisphere and clockwise in the north). This finding indicates that these ovals, and four other white spots identified by Voyager 2, are sites of upsurging material. In the four months between the Voyager flybys, a protrusion formed to the east of the Great Red Spot, blocking the circulation of small structures about it. During the Voyager 2 flyby, a white region covered a brown oval in the north, showing that the brown ovals are actually breaks in the higher-altitude clouds of white ammonia crystals. Ultraviolet studies of Jupiter by Voyager 2 revealed an absorbent layer of haze above the cloudtops, precipitation of charged particles from the magnetosphere, and auroras near the poles. The high-latitude auroras are induced by charged particles from Io and play a part in the atmospheric chemistry of Jupiter, as does lightning, which produced eight flashes detected by Voyager 2.

Jupiter's moons and main ring were also scrutinized by Voyager 2's cameras and instruments. The smooth surface of Europa was photographed with unprecedented resolution, so that the moon was shown to have uniform, bright terrain crisscrossed by dark lines and ridges and almost no features resembling craters. In contrast, Callisto's outward face, like its inward face (which is always turned toward Jupiter), is heavily cratered and very ancient. Ganymede, between Callisto and Europa, exhibits a variety of surface features such as the old and cratered surface of Regio Galileo, discovered by Voyager 2, and ancient parallel mountain ridges, nearly the size of Earth's Appalachian Mountains. Six of Io's volcanoes were still active at the time of the Io volcano watch. The Pele volcano, which had been the most violent, was now quiet, but its surrounding terrain had been visibly altered since the time of the Voyager 1 images; the plume of the volcano Loki was much larger than it had been. High-resolution images of Jupiter's ring were obtained, revealing a bright, narrow segment with a slightly brighter center surrounding a broader, dimmer disk—all surrounded by a halo of very fine particles. The outer edge of the ring is sharp, and two moons orbit just outside the edge. The inner disk of the ring probably extends all the way down to the cloud tops of Jupiter.

Jupiter has at least sixteen moons. Four are as large as or larger than Earth's moon. (NASA CORE/Lorain County JVS)

The outer edge of Jupiter's magnetosphere was observed to fluctuate considerably as Voyager 2 crossed it several times. In spite of the boundary's instability, it was clear that the magnetosphere had changed shape since the first Voyager probed it, protruding more toward the Sun and narrowing in its long tail. Inside the magnetosphere, a hot plasma (a gas so hot that its atoms cannot keep their electrons) of hydrogen, oxygen, and sulfur was slowly spiraling outward. The amount of oxygen and sulfur had decreased, though, since the Voyager 1 passage, suggesting that these heavier atoms were slowly settling back toward Jupiter and that Io was producing less of them. These heavier atoms of sulfur and oxygen are most likely injected into the magnetosphere by the volcanoes on Io and first collect in a plasma cloud that surrounds Io's orbit. During the Voyager 2 encounter, this plasma cloud was glowing twice as brightly with ultraviolet radiation as it had four months before, yet its temperature had decreased. It was also determined that the auroras observed on Jupiter were caused by Io's plasma cloud. Data from Voyager 2 also revealed that Ganymede swept up some of the charged particles from the magnetosphere's plasma, producing a plasma wake similar to the wake of a speedboat on a calm lake. A similar effect had been detected near Io by Voyager 1.

Context

Voyager 2 was the fourth spacecraft to encounter Jupiter. The first two, Pioneers 10 and 11, principally showed that the later Voyager missions were possible. Voyager 2's design and flight path were altered after these two Pioneer missions supplied measurements of Jupiter and its moons, permitted identification of certain problems—such as the intense radiation surrounding the planet—and suggested interesting phenomena for study. Some Voyager discoveries had been hinted at by Earth-based observations. Radio signals from Jupiter suggested that electrical current was flowing between Io and Jupiter, but the current proved much stronger than had been imagined.

Data from Voyager 1 provided the most guid-

ance for the second Voyager flyby of Jupiter. For example, theories had predicted that Jupiter could not have a ring, because gravity would pull the material into Jupiter's atmosphere; yet Voyager 1 found a ring. The ring could not be studied carefully at the time, because the discovery was unexpected. The presence of the ring could be explained by Io's volcanoes, which were subsequently detected by Voyager 1. Some of the material ejected by Io may find its way down toward Jupiter, where it could replace precipitating material. Another possible explanation is that the rings are renewed by collisions between high-velocity particles and Jupiter's nearest moons, Adrastea and Metis, which undoubtedly shape the ring even if they do not regenerate it. In any event, much of the information Voyager 2 collected about the rings, Io's volcanoes, and Adrastea would have been undiscovered if data from Voyager 1 had not pointed the way.

Another surprise of the Voyager mission to Jupiter was the smooth surface of Europa. The lack of large impact craters suggests a relatively new surface, and the smooth regions between intersecting lines resemble an aerial photograph of Arctic ice crossed by pressure ridges. If this interpretation is correct, Europa may be covered by a frozen ocean. How thick the ice is and whether liquid water lies underneath is still speculative; more information is needed. Without the Voyager 2 images, however, the idea of a recently frozen surface would seem very unlikely.

The information from Voyager 2's flyby of Jupiter has taken years to analyze, and some of the mysteries cannot be solved without another mission. The Galileo mission, launched October 18, 1989, reached Jupiter in 1995 and proceeded to send its probe to the surface. The orbiter continued its encounters with Jupiter and its major moons, proving so successful that its mission was extended into the 2000's.

As of 2005, the Voyager 2 spacecraft was continuing its scientific exploration of interstellar space. Twenty-seven years after its launch, Voyager 2 was more than 11 billion kilometers from Earth, the equivalent of seventy times the distance from Earth to the Sun, and was heading out of the solar system at approximately 29 kilometers per second.

See also: Cassini: Saturn; Deep Space Network; Galileo: Jupiter; Hubble Space Telescope; Hubble Space Telescope: Science; Jet Propulsion Laboratory; Planetary Exploration; Voyager Program; Voyager 1: Jupiter; Voyager 1: Saturn; Voyager 2: Saturn; Voyager 2: Uranus; Voyager 2: Neptune.

Further Reading

Beatty, J. Kelly, Carolyn Collins Petersen, and Andrew Chaikin, eds. *The New Solar System.* 4th ed. New York: Sky Publishing, 1999. A collection of articles by noted experts. Chapters 11, 12, 13, 14, and 19 are particularly relevant for readers who wish to learn more about Voyager 2's flyby of Jupiter.

Harland, David M. *Jupiter Odyssey: The Story of NASA's Galileo Mission.* London: Springer-Praxis, 2000. A detailed scientific and engineering history of the Galileo program, but also includes extensive discussion of Voyager program events and results.

_____. *Mission to Saturn: Cassini and the Huygens Probe.* London: Springer-Praxis, 2002. A detailed scientific and engineering history of the Cassini program, but also includes extensive discussion of Voyager program events and results.

Hartmann, William K. *Moons and Planets.* 5th ed. Belmont, Calif.: Thomson Brooks/Cole Publishing, 2005. Provides detailed information about all objects in the solar system. Suitable on three separate levels: high school student, general reader, and college undergraduate studying planetary geology.

Hunt, Garry E., and Patrick Moore. *Jupiter.* New York: Rand McNally, 1981. A succinct yet complete treatment of almost all aspects of Jupiter and the Voyager missions to Jupiter. Its maps of the Galilean satellites are notable.

Lee, Wayne. *To Rise from Earth: An Easy to Understand Guide to Spaceflight.* New York: Checkmark Books, 1996. This is a good introduction to the science of spaceflight. Although written by an engineer with the NASA Jet Propulsion Laboratory, it is presented in easy-to-understand language. In addition to the theory of spaceflight, it gives some of the history of the human endeavor to explore space.

Mari, Christopher, ed. *Space Exploration.* New York: H. W. Wilson, 1999. Twenty-five articles (reprinted from magazines), covering the state of the space program at the time of publication, are divided into five sections: John H. Glenn, Jr.'s return to space, the exploration of Mars, the International Space Station, recent mining efforts by commercial industries, and new types of space vehicles and propulsion systems.

Morrison, David, and Tobias Owen. *The Planetary System.* 3d ed. San Francisco: Addison-Wesley, 2003. Organized by planetary object, this work provides contemporary data on all planetary bodies visited by spacecraft since the early days of the Space Age. Suitable for high school and college students and for the general reader.

Morrison, David, and Jane Samz. *Voyage to Jupiter.* NASA SP-439. Washington, D.C.: Government Printing Office, 1980. The official summary of the Voyager missions to Jupiter. This volume is notable for its chronological approach to both Voyager flybys of Jupiter, its summaries, its photographs and tables, and its maps of Jupiter's moons.

Murray, Bruce C., ed. *The Planets.* New York: W. H. Freeman, 1983. This collection of reprints of *Scientific American* articles includes articles on Jupiter, its Galilean moons, and its planetary rings.

Snow, Theodore P. *Essentials of the Dynamic Universe: An Introduction.* 2d ed. St. Paul, Minn.: West Publishing, 1987. A well-written introduction to astronomy and astrophysics. Intended for nonscientists. The chapter on Jupiter contains a summary of the Voyager missions.

Time-Life Books. *The Far Planets.* Alexandria, Va.: Author, 1988. This volume is notable not only for its photographs and informative illustrations but also for its lively and complete account of the Voyager 2 mission. This is a volume in the Voyage through the Universe series.

Larry M. Browning

Voyager 2: Saturn

Date: June 5 to September 4, 1981
Type of spacecraft: Planetary exploration

The Voyager 2 flyby of Saturn produced high-resolution images of Saturn's ring system and of the satellites Iapetus, Hyperion, Enceladus, and Tethys. In addition, this second Voyager mission to Saturn collected valuable data on the magnetic and radiation fields surrounding that planet and observed atmospheric phenomena on Saturn.

Key Figures

Esker Davis, Voyager project manager
Edward C. Stone, Jr., Voyager project scientist
Bradford A. Smith, principal investigator, imaging experiment
Norman F. Ness, principal investigator, magnetometer experiment
Von R. Eshleman, radio science team leader
James W. Warwick, principal investigator, planetary radio astronomy experiment

Summary of the Mission

The Voyager 2 mission to Saturn was the last formal objective for the Voyager program, but the flight controllers knew that with careful planning and a little luck, Voyager 2 could continue to Uranus and Neptune on a Grand Tour of the outer solar system. Voyager 2's flight path took the probe only 32,000 kilometers from the edge of Saturn's F-ring, through the region where Pioneer 11 had a close encounter with the satellite now known as Janus. This trajectory was chosen so that Saturn's gravitational pull would send Voyager 2 toward a 1986 encounter with Uranus. The timing of Voyager 2's passage through the Saturnian system was also planned to optimize measurements of the moons Iapetus, Enceladus, and Tethys, as well as the ring system, which was better illuminated by the Sun than it had been during Voyager 1's flyby. Other satellites would also be photographed and compared with data returned by Voyager 1, which had achieved very close encounters with the moons Mimas, Rhea, and Titan.

During its approach to Saturn, Voyager 2 monitored the solar wind, the charged particles that stream from holes in the Sun's corona. This information warned Voyager 1 flight controllers and scientists of gusts or other changes in the solar wind that would affect the boundary of Saturn's magnetosphere (the region surrounding a planet that is dominated by that planet's magnetic field). The data were useful in interpreting information sent from Voyager 1 as it crossed Saturn's bow shock. (A planet's bow shock is the region where the solar wind first encounters the planet's magnetic field and loses most of its energy.)

In February of 1981, Voyager 2 passed through the tail of Jupiter's magnetosphere. It was expected that in August of 1981, during Voyager 2's Saturn encounter, Jupiter's magnetosphere would extend over Saturn, shading Saturn from the solar wind. That did not happen, however, and Saturn's magnetosphere was subjected to the solar wind's full fury, robbing Voyager 2 of the opportunity to study

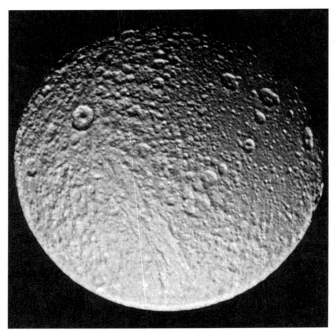

The Saturn moon Tethys has many craters and trenches. (NASA CORE/Lorain County JVS)

the interaction of Jupiter and Saturn's magnetospheres.

Ten weeks before periapsis (closest approach), Voyager 2 began taking a series of pictures that were later combined into films showing the motion of Saturn's atmosphere and rings. The atmospheric films showed banding and turbulence similar to those on Jupiter; wind speeds, however, were an amazing 400 to 500 meters per second, and cloud layers were deeper, giving Saturn a more uniform appearance than Jupiter. In the nine months since Voyager 1 had visited Saturn, atmospheric activity had increased so that more storms, spots, and waves could be seen in the cloudtops. The ring images showed the three major rings easily visible from Earth, which are named A, B, and C from the outermost inward. In the B-ring, radial "spokes" that rotate with the planet were clearly visible. The detection of these spokes by Voyager 1 had been a surprise, because such structures cannot orbit Saturn as a result of the gravitational forces that shape the concentric, nearly circular rings. The Voyager 2

pictures helped to establish that the spokes are very tiny dust or haze layers suspended over larger ring particles by Saturn's magnetic field and, consequently, rotate with the field as the planet rotates.

Between the orbits of the moons Iapetus and Titan, Voyager 2 encountered Saturn's bow shock. As on its flyby of Jupiter, Voyager 2 crossed the bow shock many times, because the solar wind would alternately gust, compressing the magnetosphere, and relax, allowing the magnetosphere to expand. Because the solar wind's strength had increased since Voyager 1's encounter, Voyager 2 finally passed into Saturn's magnetopause, or the magnetosphere's boundary, just inside Titan's orbit. The spacecraft then proceeded to pass through three distinct regions of Saturn's magnetosphere. The first is dominated by Titan, whose orbit is surrounded by a cloud of hydrogen gas emanating from the moon's atmosphere and extending toward Saturn up to Rhea's orbit. Between Rhea and Mimas are intense radiation belts composed of charged particles that rotate with Saturn's magnetic field. Closer to the planet, the rings almost completely neutralize the charged particles, making Saturn's rings one of the most radiation-free regions in the solar system.

The rings themselves were also very carefully studied by Voyager 2 scientists. Almost two and a half hours on the day of periapsis were devoted to a very careful photopolarimeter scan of the rings. This scan resolved objects as small as 100 meters across and was done by measuring the change in intensity of starlight as the ring passed between the probe and the star Delta Scorpii.

On August 22, 1981, Voyager 2 began its survey of Saturn's moons as it flew by Iapetus. The moons, except for Phoebe and possibly Hyperion, have synchronous orbits around Saturn, so that one side of the moon always faces Saturn. Put another way, one side of the moon always faces ahead, leading the moon in its orbit, and the other side always faces behind, trailing the moon as it circles. In the case of Iapetus, the leading edge is darker than the trailing edge; the moons Dione and Rhea, however, have brighter leading faces and darker trailing faces crossed by bright streaks. After passing Iapetus, Voyager 2 flew by Hyperion. The probe returned pictures that revealed this moon to be irregularly shaped, with its long axis pointing out toward space instead of toward Saturn, as had been expected. Just before the closest approach to Saturn, Voyager 2 turned its cameras toward the small satellites Telesto, Calypso, and Helene, and toward the larger moons Enceladus and Tethys, even though the closest approach to the larger moons would occur after periapsis.

The preliminary studies of Enceladus and Tethys proved fortunate, because fifty-five minutes after Voyager 2 crossed the plane of Saturn's rings, its camera platform's azimuth control became stuck. When the scan platform's back-and-forth motion stopped, Voyager 2 was behind Saturn and out of contact with Earth. The spacecraft was on automatic control and was recording its cameras' images for later relaying. Reviewing the images hours later, flight controllers and scientists watched as the cameras moved progressively off target so that first the high-resolution pictures were lost and then even the wide-angle pictures were blank.

As soon as contact was reestablished with Voyager 2 as it moved out of Saturn's shadow, flight controllers realized that the cameras and other sensitive instruments were pointed toward the Sun. If left in this position, the instruments would be destroyed, effectively blinding the spacecraft and making the rest of the mission practically useless. Quickly, commands were sent to rotate the entire spacecraft. Voyager 2 was so far from Earth, however, that an hour and a half would pass before these commands would be received. Adding to the danger was the possibility that Voyager 2 would not be able to decode the commands, as its primary receiver had failed soon after launch, and its backup receiver was faulty. If the flight engineers had not taken into account the fact that the spacecraft had cooled in the shadow of Saturn, and adjusted the commands accordingly, Voyager 2's last images would have recorded its cameras burning out in the Sun's glare.

Three days passed before the instrument platform was partially freed and a few final images of Saturn and the moon Phoebe were sent back to Earth. Unfortunately, the highest-resolution images of Enceladus and Tethys had been lost, along with three-dimensional pictures of the F-ring, a photopolarimeter scan of the F- and A-rings, and images of the night side of Saturn with backlit rings. It was also clear that for future flybys, the entire spacecraft would have to be slowly rotated to keep time-exposed pictures from blurring.

This photo of Saturn and its rings was taken on July 12, 1981, by Voyager 2. Saturn is the second largest planet in the solar system. Its rings vary in size, ranging from a few ten thousandths of an inch to more than 300 feet in diameter. (NASA CORE/ Lorain County JVS)

Contributions

During its flyby of Saturn, Voyager 2 conducted the first high-resolution reconnaissance of the moons Iapetus and Hyperion and of the two previously scanned moons, Tethys and Enceladus. The images of Iapetus revealed a dark leading edge surrounded by a concentric, dark circle. The trailing edge was much brighter, but the deepest craters showed dark bottoms. Voyager 2 flew so close to Iapetus that it was able to make the first direct measurement of this moon's mass as its gravity very slightly bent the spacecraft's trajectory.

The Hyperion pictures revealed an elongated moon, pockmarked by meteoric impacts, with its longest axis pointing away from Saturn. This dark, icy moon's orientation was seen to be so unusual that mission scientists suspected that Hyperion is not synchronously rotating about Saturn, as are all the other moons but Phoebe. Unfortunately, Voyager 2 could not observe Hyperion long enough to determine its exact orientation and rotation period.

Voyager 2's images of Tethys revealed a huge and ancient impact crater 400 kilometers wide—bigger than the moon Mimas. This crater, named Odysseus, is relatively shallow, because its floor re-

bounded after the initial meteoric impact; it now has nearly the same curvature as the rest of the moon. It was also discovered that the huge trench Ithaca Chasma, first photographed by Voyager 1, extends three-fourths of the way around Tethys.

The surface of Enceladus is incredibly bright, reflecting nearly all the light that reaches it. This extreme brightness suggested to the Voyager mission scientists that Enceladus's surface is very new and perhaps frequently restored by ice volcanoes. Voyager 2 detected no such ice volcanoes, but it did find a variety of terrain, which indicates recent geological activity. Such activity was unexpected in such a small moon and is unknown among Enceladus's neighbors.

Voyager 2's high-resolution images of Saturn's rings showed much more detail than had ever been seen before. When various sections of the rings were compared, it was realized that the rings are very dynamic; their fine structure is constantly shifting and changing. The Cassini Division and the Encke Division, which from Earth had appeared devoid of material, were shown to have small, dark ringlets. Scientists had expected to find small moons in the divisions sweeping away material, but none were detected. Voyager 2 also measured the rings' thickness and found it to be less than 300 meters; corrugations in the rings make them appear ten times thicker from Earth.

Weather patterns on Saturn proved to be remarkably stable. Storms identified by Voyager 2 as it approached Saturn were seen six weeks later at the same latitude and traveling with the same speed. Even more remarkable was that many storms appeared just where Voyager 1 data had predicted they would.

Context

The third probe to fly by Saturn, Voyager 2 provided much detailed information that Pioneer 11 and Voyager 1 could not. Pioneer 11 had made several important discoveries, including Saturn's F-ring, but its imaging systems and other instruments could not match the resolution capabilities of the Voyager spacecraft. Voyager 1 had observed

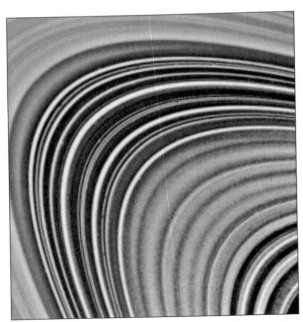

An image of Saturn's C ring returned by Voyager 2 in 1981. (NASA-JPL)

most of the previously unknown and in many cases unexpected aspects of the Saturnian system nine months before Voyager 2's passage, but a single flight through the Saturnian system could not capture every aspect for scrutiny.

Also, to have a close look at Titan, which was the only moon in the solar system known to have a substantial atmosphere, Voyager 1 mission scientists had to give up the chance to send the probe on to Uranus and Neptune. The second Voyager craft's trajectory was calculated to take it past those two planets. Voyager 2 had slightly better instruments, as well; its imaging system had about 50 percent more sensitivity and produced sharper pictures. Finally, Voyager 1's photopolarimeter was destroyed by the intense radiation near Jupiter, and important details about Saturn's rings and the polarization of light scattered from Titan were left for Voyager 2 to record.

Voyager 1 provided information that greatly influenced the planning of Voyager 2's Saturn flyby. For example, Voyager 1's data on the complexity of Saturn's rings made scientists and flight controllers realize how important a high-resolution scan of the rings would be. Nevertheless, not all the guidance for the Voyager 2 mission was provided by Voyager 1. During 1966 and 1980, Saturn's rings' edges were facing Earth. Without the brighter rings to obscure them, a number of new satellites and rings were observed in the Saturnian system by several teams of astronomers. Voyagers 1 and 2 confirmed their existence, and these objects are now known as the two co-orbital satellites Epimetheus and Janus; the satellite Helene, which shares an orbit with Dione; and the E-ring, near Enceladus. Another, much older, Earth observation was confirmed by Voyager 2 when it sent back images of Iapetus showing a dark leading surface and a bright trailing surface. In 1671, the astronomer Gian Domenico Cassini, who discovered Iapetus and the gap between the A- and B-rings that now bears his name, observed that Iapetus was easy to see when it was west of Saturn but could barely be seen when it was east of Saturn. These observations are consistent with a synchronously rotating satellite with a dark leading edge, which the Voyager 2 images revealed Iapetus to be.

Despite Voyager 2's successes, there was some sadness at the press conference held after the Saturn flyby. Everyone there knew that it would be the last such conference for quite some time. Even though Voyager 2 would eventually encounter Uranus, in 1986, that was five years in the future, and, more important, no new probes had been launched to investigate the outer planets. The Voyager 2 flyby of Saturn not only marked the end of the Voyager program's formal objectives but also marked the end of a period of intense planetary exploration.

As of 2005, the Voyager 2 spacecraft was continuing its scientific exploration of the far reaches of the solar system. Having traveled more than 11 billion kilometers from Earth (the equivalent of seventy times the distance from Earth to the Sun) in the twenty-seven years since its launch, Voyager 2 continued heading out of the solar system at approximately 29 kilometers per second.

See also: Cassini: Saturn; Deep Space Network; Galileo: Jupiter; Hubble Space Telescope; Hubble Space Telescope: Science; Jet Propulsion Laboratory; Mariner 10; Planetary Exploration; Voyager Program; Voyager 1: Jupiter; Voyager 1: Saturn; Voyager 2: Jupiter; Voyager 2: Uranus; Voyager 2: Neptune.

Further Reading

Beatty, J. Kelly, Carolyn Collins Petersen, and Andrew Chaikin, eds. *The New Solar System.* 4th ed. New York: Sky Publishing, 1999. A collection of articles by noted experts and authors. Several chapters are particularly relevant to Voyager 2's flyby of Saturn.
Cooper, Henry S. F., Jr. *Imaging Saturn: The Voyager Flights to Saturn.* New York: H. Holt, 1985. A very readable, chronological account of the Voyager missions to Saturn.

Couper, Heather, and Nigel Henbest. *New Worlds: In Search of the Planets*. Reading, Mass.: Addison-Wesley, 1986. A summary of planetary exploration, including the exploration of Saturn. Also contains information about how to find and observe Saturn. Accessible to all readers.

Frazier, Kendrick. *Solar Systems*. Alexandria, Va.: Time-Life Books, 1985. A volume in the series Planet Earth, this work is attractively illustrated and understandable to the layperson.

Harland, David M. *Jupiter Odyssey: The Story of NASA's Galileo Mission*. London: Springer-Praxis, 2000. A detailed scientific and engineering history of the Galileo program, but also includes extensive discussion of Voyager program events and results.

_____. *Mission to Saturn: Cassini and the Huygens Probe*. London: Springer-Praxis, 2002. A detailed scientific and engineering history of the Cassini program, but also includes extensive discussion of Voyager program events and results.

Hartmann, William K. *Moons and Planets*. 5th ed. Belmont, Calif.: Thomson Brooks/Cole Publishing, 2005. Provides detailed information about all objects in the solar system. Suitable on three separate levels: high school student, general reader, and college undergraduate studying planetary geology.

Hunt, Garry E., and Patrick Moore. *Saturn*. New York: Rand McNally, 1981. A succinct and complete treatment of almost all aspects of Saturn and the Voyager missions to that planet. Maps of some of the satellites and diagrams of the Voyager spacecraft and its instruments are included.

Lee, Wayne. *To Rise from Earth: An Easy to Understand Guide to Spaceflight*. New York: Checkmark Books, 1996. This is a good introduction to the science of spaceflight. Although written by an engineer with the NASA Jet Propulsion Laboratory, it is presented in easy-to-understand language. In addition to the theory of spaceflight, it gives some of the history of the human endeavor to explore space.

Mari, Christopher, ed. *Space Exploration*. New York: H. W. Wilson, 1999. Twenty-five articles (reprinted from magazines), covering the state of the space program at the time of publication, are divided into five sections: John H. Glenn, Jr.'s return to space, the exploration of Mars, the International Space Station, recent mining efforts by commercial industries, and new types of space vehicles and propulsion systems.

Morrison, David. *Voyages to Saturn*. NASA SP-451. Washington, D.C.: National Aeronautics and Space Administration, 1982. A lovely book with many color photographs from space probes, providing richly detailed background information. Morrison summarizes the history of Saturn observations and recounts the Pioneer and Voyager missions to Saturn. He explains theories about the planet's development and that of its rings and moon clearly for general readers.

Morrison, David, and Tobias Owen. *The Planetary System*. 3d ed. San Francisco: Addison-Wesley, 2003. Organized by planetary object, this work provides contemporary data on all planetary bodies visited by spacecraft since the early days of the Space Age. Suitable for high school and college students and for the general reader.

Murray, Bruce, ed. *The Planets*. New York: W. H. Freeman, 1983. A collection of reprints of *Scientific American* articles. Includes discussions of Saturn, Saturn's moons, and planetary rings.

Smoluchowski, Roman. *The Solar System*. New York: W. H. Freeman, 1983. A well-written and well-illustrated summary of humankind's understanding of the solar system.

Snow, Theodore P. *The Dynamic Universe: An Introduction to Astronomy.* 3d ed. St. Paul, Minn.: West Publishing, 1988. A well-written introduction to astronomy and astrophysics intended for nonscientists. The chapter on Jupiter contains a summary of the Voyager missions, and the chapter on Saturn discusses the use of Voyager 2's photopolarimeter to resolve Saturn's rings.

Time-Life Books. *The Far Planets.* Alexandria, Va.: Author, 1988. A volume in the Voyage through the Universe series, this source is notable not only for its pictures and informative illustrations but also for its lively and complete account of the Voyager 2 mission.

Larry M. Browning

Voyager 2: Uranus

Date: November 4, 1985, to February 25, 1986
Type of spacecraft: Planetary exploration

Voyager 2 was the first spacecraft to collect and return data from the planet Uranus. This encounter was the third of four potential encounters made possible by a planetary alignment that occurs only once every 175 years.

Key Figures

Richard P. Laeser, Voyager project manager

Edward C. Stone, Jr., Voyager project scientist

Ellis D. Miner, assistant project scientist

Bradford A. Smith, principal investigator, imaging experiment

Norman F. Ness, principal investigator, magnetometer experiment

Von R. Eshleman, radio science team leader

James W. Warwick, principal investigator, planetary radio astronomy experiment

Summary of the Mission

The Voyager 2 spacecraft first encountered the planet Uranus while the probe was on a trajectory that would ulimately take it out of the solar system. This flyby of Uranus was the third of four planetary encounters made possible by an alignment of the outer planets—Jupiter, Saturn, Uranus, and Neptune—that occurs only once every 175 years. This alignment allowed Voyager 2 to arrive at Uranus in nine years instead of sixteen by using the gravity of each planet to boost it on to the next.

In 1981, prior to Voyager 2's encounter with Saturn (but after Voyager 1's Saturn encounter), the National Aeronautics and Space Administration (NASA) approved the Voyager 2 Uranus mission. The gravity of Saturn would be used to direct the spacecraft's trajectory toward Uranus. The journey between these worlds would take more than another four years and cover a further distance of 1.5 billion kilometers.

The time between the Saturn and Uranus encounters was used by Voyager engineers to modify the craft's onboard computer programs. These modifications were needed to enable the probe to overcome the problems associated with visiting a planet that is twice as far from the Sun as is Saturn. One such problem was the decreasing light levels. At a distance of 3.2 billion kilometers from the Sun, the light at Uranus would be only one four-hundredth the level at Earth. This level of light made necessary longer photographic exposures, which would lead to blurred pictures if the spacecraft could not keep the target in the camera's field of view during the exposure. To address this need, a computer algorithm known as target motion compensation was written and sent to the spacecraft. This algorithm allowed Voyager to drift in such a way as to keep the target in the camera's sights. In essence, this routine allowed the spacecraft to "pan" the cameras, like a human photographer does to photograph a moving object.

The greater distance also forced ground engineers to reduce the rate at which the spacecraft transmitted data. This reduction compensated for the ever-decreasing signal strength as the spacecraft moved away from Earth. It also meant that as the spacecraft got farther away, more time would be required to send the same amount of information. At Jupiter, for example, one picture could be transmitted every 96 seconds; at Saturn, the rate had fallen to one picture every 3.2 minutes, and at Uranus, it would be one every 8.8 minutes. To reduce the impact of this problem during the cruise toward Uranus, project managers made a bold decision. Instead of using one flight data subsystem computer to format the data to be sent and the other as a backup, the two would be used together with no backup. The primary subsystem would still format the data; the secondary subsystem, however, would be used to combine the data more efficiently. This combination routine, known as image data compression, could send a complete photograph using less than 40 percent of the information bits normally required. Thus, even with the slower transmission rates from Uranus, one picture could be transmitted every 4.8 minutes.

Many more changes were made to Voyager 2 to compensate for the greater distance from the Sun and Earth and the lack of knowledge of the outer solar system. These changes involved not only the spacecraft program but also procedures used to operate the spacecraft. The spacecraft that finally arrived at Uranus was far superior to the one that had passed by Jupiter and Saturn.

On November 4, 1985, the first phase of the Uranus encounter began. Known as the observatory phase, it started when the quality of data from the spacecraft instruments surpassed the quality of data from ground-based instruments. During this phase, systematic searches were made of the Uranian system for new satellites and rings. In addition, atmospheric measurements began to be taken to gain information regarding atmospheric structure and composition and wind patterns.

Twenty-five days prior to the spacecraft's closest approach to Uranus, on December 31, 1985, Voyager 2 discovered its first Ura-

nian satellite, which was given the name Puck. Puck is 170 kilometers in diameter and is located between the outer Uranian ring, known as the epsilon ring, and the innermost satellite, known as Miranda. Very little information about Puck's surface was gained from the discovery pictures because of its small size and great distance from the spacecraft. Because of the importance of surface geology, however, the various Voyager teams worked quickly to modify the spacecraft's program to photograph Puck immediately before the busy near-encounter phase.

To modify the program, ground controllers decided to eliminate a planned observation of Miranda and replace it with an observation of Puck, then known as 1985U1. The spacecraft photographed Puck successfully, but the receiving ground antenna station started to drift, producing a poor alignment of the ground antenna with the spacecraft and resulting in the loss of the spacecraft signal that contained the Puck photograph. Fortunately, the photograph had been recorded on the

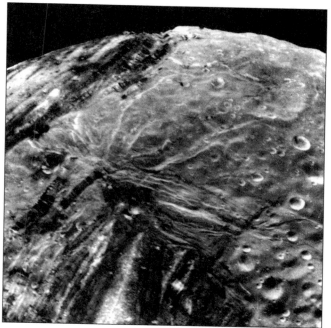

Uranus's moon Miranda captured as Voyager 2 flew by. (NASA CORE/Lorain County JVS)

spacecraft's tape recorder and could be played again. The commands to replay the data had to be sent quickly, however, before new information was recorded over the Puck data. The second playback was successful, revealing the surface of the moon discovered only days earlier.

The next phase of the encounter was known as the far encounter. This phase started on January 10, 1986, when Uranus and its rings were too large to fit comfortably into one narrow-angle photograph. To capture the entire planet and its rings required that the observation be designed as a mosaic of four pictures. Each picture in the mosaic would contain one quarter view of the planet and its rings.

During the far encounter, the spacecraft experienced its first hardware failure since its encounter with Saturn. On January 18, 1986, six days prior to Voyager 2's closest approach to Uranus, unexpected gaps started appearing in the photographs. Engineers quickly reviewed both the ground data system and the multimission image processing laboratory for hardware failures or software problems. Two days later, convinced that the problem was on the spacecraft and not on the ground, spacecraft controllers sent commands to the spacecraft instructing it to transmit the contents of its flight data subsystem memory. After reviewing the information, analysts found that one memory location had an incorrect value. The following day, on January 21, a command was sent to the spacecraft to use one of the few remaining spare memory locations instead of the one that appeared to have failed. After the craft received the command, the gaps in the spacecraft photographs disappeared, proving that a memory location had indeed failed.

On January 22, 1986, the near-encounter phase began. During this phase, which lasted only four days, the spacecraft would be closest to the planet and would be gathering most of the important scientific data of this part of its mission. Timing was critical during this phase. The spacecraft had to be in the correct place at the correct time if its commands were to execute properly. The Voyager project teams worked throughout the night of Jan-

A model of Voyager 2 flying by Uranus and its nine known rings. (NASA CORE/Lorain County JVS)

uary 23 and early into the next morning, adjusting the spacecraft commands. The final commands were transmitted to the spacecraft less than eight hours prior to their execution.

On January 24, at 17:59 Coordinated Universal Time (UTC), Voyager 2 passed 107,000 kilometers from the center of Uranus. The navigation team had done a superb job of directing the spacecraft to its destination. The spacecraft was off its schedule by only sixty-one seconds, and its placement was such that Voyager 2 would pass by Neptune several years later.

The last phase of the encounter, which began on January 26, 1986, was the postencounter. During this time, measurements resembled those that had been executed during the observation phase. Pictures showed Uranus as a crescent as Voyager 2 headed toward Neptune.

Contributions

The visible atmosphere of Uranus was found to be composed predominantly of hydrogen and helium, with concentrations very similar to those found in the Sun. Carbon, however, was found to be twenty times more abundant than in the Sun. This relatively large amount of carbon, which is combined with hydrogen to form methane, gives Uranus its blue-green color. Methane absorbs mainly in the red wavelengths, reflecting the blue light to the observer.

The infrared experiment found that almost the same amount of energy is emitted by the poles of Uranus as by the equator. This was surprising, for the poles are exposed to more sunlight during the eighty-four-year orbit of Uranus and thus should radiate more energy. In addition, the spin axis of Uranus was found to be tilted 98° relative to its orbital plane. Thus, even though Uranus spins on its axis once every 17.24 hours, the south pole faced the Sun during the entire Voyager 2 encounter. It takes forty-two years (one-half of a Uranian orbit) for the Sun to rise and set at the planet's poles.

Few discrete clouds were observed in the atmosphere of Uranus. The motion of those few clouds seen, however, indicated that winds rotated with the planet at a maximum speed of about 200 meters per second at a southern latitude of 60°. The winds slowed on either side of this region until they became almost nonexistent at the pole and at 20° south latitude. Farther north at the equator, however, the radio science experiment found the winds to flow in a direction opposite the rotation, with speeds up to 100 meters per second.

The upper atmosphere of Uranus was found to extend far above the planet. This part of the atmosphere has an extremely high temperature of 500° Celsius (773 kelvins) and may produce drag forces on the particles located in the rings. These forces may be responsible for removing dust-sized particles from the Uranian ring system. The extended atmosphere also interacts with sunlight, giving off emissions that were detected by the ultraviolet experiment.

The nine previously known rings were photographed by Voyager's cameras. They are very dark and contain few particles in the 1- to 10-centimeter range. The darkness may result from proton bombardment of the methane-water ice particles in the Uranian magnetosphere. The result of this bombardment is a carbon residue on the ice, which makes the ring particles as dark as coal. Two additional rings were found as Voyager 2 approached Uranus. As the probe left, one single long-exposure photograph indicated that micrometer-sized dust exists throughout the entire ring system.

Voyager 2 discovered ten new satellites in all, increasing the total number of known Uranian satellites from five to fifteen. The new satellites range from 40 to 170 kilometers in diameter and are located between the outer part of the ring system and the orbit of Miranda. Compared to Saturn's icy moons, the five previously known moons of Uranus were found to be relatively dark. Ariel, the second moon out from Uranus, has a fractured surface, which may indicate that ice once flowed across it. It also possesses one of the most geologically active surfaces in the Uranian system. Yet the most distinctive surface was found on Miranda. This satellite may have broken apart and re-formed during its early history.

This photo of Uranus's moon Ariel was taken by Voyager 2. (NASA CORE/Lorain County JVS)

The magnetic field experiment found that the axis of the Uranian magnetosphere was tilted at 59° to the planet's rotational axis. In addition, instead of being generated at the planet's center, as on Jupiter and Saturn, the magnetic field was found to be offset one-third of a Uranian radius.

Context

On March 13, 1781, Sir William Herschel discovered Uranus, the seventh planet in the solar system, from his home in Bath, England. At that time, Uranus was the first planet to be observed by modern astronomers that had not been known to the ancients; its discovery showed that the solar system's outer boundary was much farther than had been previously believed.

Since its discovery, many telescopes have been pointed at Uranus. Its extreme distance from the Sun, however, has made unraveling its secrets very difficult. If one counted all the photons of light collected from Uranus by all the telescopes in the years from its discovery up to the Voyager 2 encounter, the amount of light would equal that given off by a flashlight in one second. Thus, the Voyager encounter with Uranus greatly increased knowledge of the Uranian system.

As a result of the encounter, planetary scientists now have the ability to compare the system of Uranus with those of Jupiter and Saturn. This study, known as comparative planetology, will allow scientists to understand the physical characteristics of these worlds better. This knowledge can then be applied to Earth. For example, atmospheric studies of the outer planets can improve meteorologists' understanding of terrestrial atmospheric dynamics.

Study of the Uranian ring system showed that it had many similarities to the ring systems of both Jupiter and Saturn; nevertheless, models of ring dynamics cannot yet completely explain the thinness of the nine Uranian rings. In addition, there is much to learn from the interaction of the Uranian magnetosphere with the ring system. Better models need to be developed to improve the understanding of the Uranian magnetosphere as well as of the magnetic fields of Mercury, Earth, Mars, Jupiter, and Saturn.

The Voyager 2 mission, sometimes referred to as the Grand Tour (a consecutive set of flybys past Jupiter, Saturn, Uranus, and Neptune), is truly one of the most remarkable missions of all time. At its completion, it had observed all the gaseous giant outer planets and most of the major moons in the solar system. As of 2005, it was continuing its scientific exploration of interstellar space. Having traveled more than 11 billion kilometers (the equivalent of seventy times the distance between Earth and the Sun) from Earth in the twenty-seven years since its launch, Voyager 2 continued on a trajectory that would eventually take it beyond the distant edge of the solar system at approximately 29 kilometers per second.

See also: Cassini: Saturn; Deep Space Network; Galileo: Jupiter; Hubble Space Telescope; Hubble Space Telescope: Science; Jet Propulsion Laboratory; Mariner 10; Planetary Exploration; Voyager Program; Voyager 1: Jupiter; Voyager 1: Saturn; Voyager 2: Jupiter; Voyager 2: Saturn; Voyager 2: Neptune.

Further Reading

Ahrens, C. Donald. *Essentials of Meteorology: An Invitation to the Atmosphere.* 4th ed. Pacific Grove, Calif.: Thomson Brooks/Cole, 2005. This is a text suitable for an introductory course in meteorology. Comes complete with a CD-ROM to help explain concepts and demonstrate the atmosphere's dynamic nature.

Andrade, Alessandra A. L. *The Global Navigation Satellite System: Navigating into the New Millennium.* Montreal: Ashgate, 2001. Provides an international view of issues of availability, cooperation, and reliability of air navigation services. Attention is specifically paid to the American GPS (Global Positioning System) and Russian GLONASS systems, although the development of the Galileo civilian system in Europe is also presented.

Davis, Joel. *Flyby: The Interplanetary Odyssey of Voyager 2*. New York: Atheneum, 1987. This book, intended for general audiences, gives a behind-the-scenes look at the Voyager 2 encounter with Uranus. It describes the individuals responsible for making the encounter as successful as it was and the events in which they took part.

Gore, Rick. "Uranus: Voyager Visits a Dark Planet." *National Geographic* 170 (August, 1986): 178-195. This article, intended for general audiences, gives an overview of the Voyager 2 encounter of Uranus. It contains many photographs and drawings that are helpful in elucidating theories and physical characteristics of the Uranian system.

Hartmann, William K. *Moons and Planets*. 5th ed. Belmont, Calif.: Thomson Brooks/Cole Publishing, 2005. Provides detailed information about all objects in the solar system. Suitable on three separate levels: high school student, general reader, and college undergraduate studying planetary geology.

Heppenheimer, T. A. *Countdown: A History of Space Flight*. New York: John Wiley, 1997. A detailed historical narrative of the human conquest of space. Heppenheimer traces the development of piloted flight through the military rocketry programs of the era preceding World War II. Covers both the American and the Soviet attempts to place vehicles, spacecraft, and humans into the hostile environment of space. More than a dozen pages are devoted to bibliographic references.

Hunt, Garry E., ed. *Uranus and the Outer Planets*. New York: Cambridge University Press, 1982. This collection of papers, intended for college students, describes the history of the discovery of Uranus and the knowledge of the Uranian system prior to Voyager 2's swing-by.

Irwin, Patrick G. J. *Giant Planets of Our Solar System: Atmospheres, Composition, and Structure*. London: Springer-Praxis, 2003. Provides an in-depth comparison of Jupiter, Saturn, Uranus, and Neptune, incorporating data obtained from astronomical observations and planetary spacecraft encounters.

Laeser, Richard P., et al. "Engineering Voyager 2's Encounter with Uranus." *Scientific American* 225 (November, 1986): 36-45. This article, written by the Voyager project manager, the manager, and the deputy manager for the Flight Engineering Office, is intended for high school and college students. It explains how Voyager 2 had to be modified in preparation for the Uranus encounter. Contains photographs and illustrations.

Lambright, W. Henry, ed. *Space Policy in the Twenty-First Century*. Baltimore: Johns Hopkins University Press, 2003. This book addresses a number of important questions: What will replace the space shuttle? Can the International Space Station justify its cost? Will Earth be threatened by asteroid impact? When and how will humans explore Mars?

Lee, Wayne. *To Rise from Earth: An Easy to Understand Guide to Spaceflight*. New York: Checkmark Books, 1996. This is a good introduction to the science of spaceflight. Although written by an engineer with the NASA Jet Propulsion Laboratory, it is presented in easy-to-understand language. In addition to the theory of spaceflight, it gives some of the history of the human endeavor to explore space.

Leverington, David. *New Cosmic Horizons: Space Astronomy from the V2 to the Hubble Space Telescope*. New York: Cambridge University Press, 2001. This is a broad treatise exploring the development of space-based astronomical observations from the end of World War II to the Hubble Space Telescope and other major NASA space-based observatories.

Mari, Christopher, ed. *Space Exploration*. New York: H. W. Wilson, 1999. Twenty-five articles (reprinted from magazines), covering the state of the space program at the time of publication, are divided into five sections: John H. Glenn, Jr.'s return to space, the exploration of Mars, the International Space Station, recent mining efforts by commercial industries, and new types of space vehicles and propulsion systems.

Morrison, David, and Tobias Owen. *The Planetary System*. 3d ed. San Francisco: Addison-Wesley, 2003. Organized by planetary object, this work provides contemporary data on all planetary bodies visited by spacecraft since the early days of the Space Age. Suitable for high school and college students and for the general reader.

National Aeronautics and Space Administration. *Space Network Users' Guide (SNUG)*. Washington, D.C.: Government Printing Office, 2002. This users' guide emphasizes the interface between the user ground facilities and the Space Network, providing the radio frequency interface between user spacecraft and NASA's Tracking and Data-Relay Satellite System, and the procedures for working with Goddard Space Flight Center's Space Communication program.

Ordway, Frederick I., III and Mitchell Sharpe. *The Rocket Team*. Burlington, Ont.: Apogee Books, 2003. A revised edition of the acclaimed thorough history of rocketry from early amateurs to present day rocket technology. Includes a disc containing videos and images of rocket programs.

Radlauer, Ruth, and Carolyn Young. *Voyager 1 and 2: Robots in Space*. Chicago: Children's Press, 1987. This book, intended for elementary and junior high school students, describes the entire Voyager 1 and 2 program. It contains many photographs and includes Voyager 2 data from Uranus.

Siddiqi, Asif A. *Sputnik and the Soviet Space Challenge*. Gainesville: University Press of Florida, 2003. This two-volume set provides a comprehensive history of Soviet space efforts at the dawn of the Space Age.

Tassoul, Jean-Louis, and Monique Tassoul. *A Concise History of Solar and Stellar Physics*. Princeton, N.J.: Princeton University Press, 2004. A comprehensive study of the historical development of humanity's understanding of the Sun and the cosmos, written in easy-to-understand language by a pair of theoretical astrophysicists. The perspective of the astronomer and physicist are presented.

Zimmerman, Robert. *The Chronological Encyclopedia of Discoveries in Space*. Westport, Conn.: Oryx Press, 2000. Provides a complete chronological history of all crewed and robotic spacecraft and explains flight events and scientific results. Suitable for all levels of research.

Randii R. Wessen

Voyager 2: Neptune

Date: Beginning August 20, 1977
Type of spacecraft: Planetary exploration

Voyager 2 essentially completed the mission of the originally proposed Grand Tour of the outer solar system, flying by Jupiter, Saturn, Uranus, and Neptune, on a trajectory that will ultimately take it beyond the solar system.

Key Figures

Norman Ray Haynes (b. 1936), Voyager project manager
Edward C. Stone, Jr., Voyager project scientist
Ellis D. Miner, assistant project scientist
Bradford A. Smith, principal investigator, imaging experiment
Norman F. Ness, principal investigator, magnetometer experiment
Von R. Eshleman, radio science team leader
James W. Warwick, principal investigator, planetary radio astronomy experiment

Summary of the Mission

Voyager 2 departed from Cape Canaveral's Pad 41 on August 20, 1977, atop a Titan III-E-Centaur D1 booster. The spacecraft was inserted into a coasting Earth orbit. At the proper point in the orbit, the Centaur upper stage ignited to boost Voyager 2's speed by 12 kilometers per second above orbital speed. Its fuel exhausted, the Centaur was separated from Voyager 2. A solid rocket motor attached to Voyager 2 then fired to provide an additional 2 kilometers per second to send the spacecraft out of Earth's gravitational influence on the proper heading for it to pass through the asteroid belt and fly past Jupiter.

Because of the trajectories chosen for the two spacecraft, Voyager 2 was launched sixteen days earlier than its sister Voyager 1. The latter overtook Voyager 2 on a faster trajectory and encountered Jupiter and then Saturn before Voyager 2's flybys of those same two planets.

With launch and Earth escape behind them, Voyager spacecraft controllers began configuring the spacecraft for the long cruise ahead. Several booms and the science scan platform had to be properly deployed. A few hours into the mission, hope for a successful Grand Tour of the outer solar system began to look bleak. Voyager 2's science scan platform apparently failed to lock into the proper position. Without that lock, it appeared that there would be no way to steer television cameras and other scientific equipment to desired targets during closest approach to the outer planets. After detailed analysis, it was determined that the platform had indeed extended to within less than 1° of the proper position and would support movements required for scientific observations.

With the scan platform problem overcome, technicians began determining the health of Voyager 2's science instruments. All were turned on and found to be in working order by September 2.

A much more serious problem developed in April, 1978. Sidetracked by problems with Voyager 1, controllers forgot to check in with the spacecraft. As part of its computer code, Voyager 2, believing itself to be malfunctioning because it had not heard from Earth within a predetermined period, switched from its primary radio receiver to the backup receiver. Unfortunately, the backup receiver had suffered a malfunction in its tracking-loop capacitor while not in use. This left Voyager 2 unable to communicate with Earth. Also, as part of its computer code, the spacecraft switched back to the primary receiver after a predetermined period, but this time the primary receiver suffered a power-supply short circuit and was permanently lost. One week later, the spacecraft, as required by computer programming, switched again to the backup receiver. As a result of the tracking-loop capacitor problem, the frequency at which the backup receiver could acquire commands drifted. Controllers on the ground sent commands to Voyager 2 at a variety of frequencies, one of which reestablished contact with the spacecraft. For the remainder of the mission, controllers had to monitor the frequency drift, eventually developing a technique that allowed them to properly determine the drift behavior and maintain command of Voyager 2. The spacecraft passed through the asteroid belt without any significant damage.

Voyager 2 flew by Jupiter on July 9, 1979, coming within 645,000 kilometers of the giant planet. It provided enough high-resolution images of Jupiter to produce time-lapse motion pictures of Jupiter's atmospheric circulation patterns, including the interaction of high-speed winds with the planet's Great Red Spot, a feature known through telescopic observations from the time of Galileo. In traversing the Jovian system, Voyager 2 made a close pass by Callisto, the outermost Galilean moon; Ganymede (at only 62,000 kilometers); and Europa. Voyager 2 did not pass as close to Io, the innermost Galilean moon, as had Voyager 1, but the spacecraft monitored its volcanic activity. Only two hours after closest approach to Jupiter, Voyager 2 fired its propulsion system for 76 minutes, putting the spacecraft on course for its next target—the ringed planet Saturn.

Voyager 2 flew by Saturn on August 26, 1981, coming within 101,000 kilometers of the mysterious ringed planet. Although Saturn's numerous and diverse moons were objects of study on the inward and outbound portions of the closest approach, the complex ring system was scrutinized with tremendous intensity. Unlike Voyager 1, Voyager 2 did not make a close pass by Saturn's largest moon, Titan, one of the few moons in the solar system to hold its own atmosphere. A Titan encounter was sacrificed on Voyager 2 in order to target the spacecraft for a gravitational assist that would send it on its way to Uranus.

It took Voyager 2 four and one-half years to blaze the previously untraveled trail to Uranus, a ringed gas giant that rotates on its side. Whereas Voyager 2's Jupiter and Saturn encounters provided sensational photographs that received significant attention in the press, the spacecraft's Uranian encounter preceded the *Challenger* accident (January 28, 1986). The deaths of seven astronauts overshadowed the results being sent by Voyager 2 from Uranus.

The spacecraft passed within 71,000 kilometers of Uranus on January 24, 1986. Although Uranus was a rather bland looking world, high-resolution photographs of its five larger moons provided striking evidence of unusual surface features, particularly on Miranda. Because of the spacecraft's tremendous speed, special motion compensation techniques were developed to prevent images from blurring. Because the scan platform could not be skewed enough to accomplish that, the entire spacecraft was carefully moved using short, precise thruster firings. Ten new moons and several more rings were discovered during the Uranian encounter.

On February 14, 1986, Voyager 2 burned 12 kilograms of propellant during a 2.5-hour-long course adjustment that provided 21.1 meters per second more to the spacecraft's speed, placing Voyager 2 on a trajectory aimed to take it to within only 1,300 kilometers of Neptune's outer atmosphere and allow it to pass by Neptune's largest moon, Triton, at a distance of 6,000 kilometers. Although the

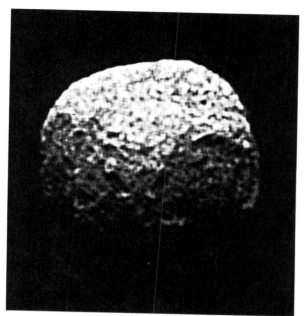

Voyager 2 discovered Neptune's Satellite 1989N1. (NASA CORE/Lorain County JVS)

spacecraft was in reasonably good health and flight controllers had an excellent record of maintaining contact with Voyager 2's pesky radio receiver, in September, 1986, a backup mission load was transmitted to the spacecraft's computer, programming it to carry out the entire Neptune encounter autonomously and send its data back to Earth even if contact was lost during the 3.5-year cruise to Neptune.

Voyager 2 provided the first close-up pictures of the blue planet Neptune twelve years after launch and at a distance of 4.4 billion kilometers from Earth, generating a wealth of information about Neptune's atmospheric circulation patterns; its frigid, light-pink and white moon Triton; and its asymmetrical ring systems.

A deep blue storm system, almost the size of Earth, referred to as the Great Dark Spot, located 22° south of the planet's equator, completed a rotation in 18 hours, 18 minutes, while spinning in a counterclockwise direction. The Great Dark Spot changed shape as it rotated, interacting with white cirrus clouds passing around its boundaries.

Another, smaller dark spot was found at 54° south latitude. This one contained bright white clouds near its center, indicative of atmospheric upwelling. Between this and the Great Dark Spot was a feature referred to as the Scooter, a rapidly moving bright cirrus cloud, possibly composed of frozen methane.

Prior to Voyager 2, segments referred to as ring arcs were known to orbit Neptune, which was verified by the first ring photographs imaged prior to close encounter. Long exposures, some up to ten minutes in duration and backlit by sunlight, clearly revealed a total of four complete diffuse rings of nonuniform density and a broad sheet of dust. Brighter clumps were the sources of ring arcs detected from Earth. To determine particle density and size, Voyager 2's photopolarimeter recorded the brightness of the star Sigma Sagitarii as it was occulted by the rings and by Neptune itself.

Voyager 2 recorded fewer charged particles (hydrogen, helium, and nitrogen) in the magnetosphere of Neptune than in those of the other gas giants. Not only was Neptune's magnetic field tilted from the rotation axis, but its structure also was quite complex, with a repetition rate of 16 hours and 3 minutes. Triton moves through the magnetic field equator, contorting the field's shape. (Triton was the source of the nitrogen ions observed in Neptune's magnetosphere.) The field was displaced from Neptune's center, suggesting an electric dynamo effect occurring in conductive layers surrounding Neptune's core rather than in the core itself (as on Earth).

Voyager 2 approached Triton from the south, providing high-resolution views of the polar cap, a white cover of frozen nitrogen, and much of the southern hemisphere then illuminated by solar conditions experienced during the moon's summer. It appeared that the polar cap was in the process of sublimating as the height of season approached. Spacecraft images revealed several distinct surface features, of various ages, but no heavily cratered areas indicative of an age greater than three billion years. Near the equator was a relatively young region marked with linear regions reminiscent of a cantaloupe's skin. In the northern hemisphere were calderas and frozen lakes suggesting a volcanic flow of water, ammonia, methane, and nitrogen mixtures. Voyager 2 discovered evidence of active volcanism—nearly a dozen dark plumes (blowing northeasterly) depositing radiation-darkened methane on the surface. The spacecraft probed Triton's tenuous atmosphere, using an occultation of the star Beta Majoris.

After the spacecraft headed out of the Neptunian system, it looked back and transmitted images of both a crescent Neptune and Triton. Voyager 2's trajectory was bent by Neptune's gravity in such a way that it headed 48° below the plane of the solar system. This portion of the spacecraft's mission was referred to as the Voyager Interstellar Mission (VIM).

Contributions

Voyager 2 refined measures of Neptune's mass (17.135 times that of Earth), diameter (24,764 kilometers), density (1.64 grams per cubic centime-

ter), rotation rate (16.05 hours), atmospheric wind speed (up to 1,000 kilometers per hour from the west), and magnetic field (inclined 50° to the planet's rotation axis).

Prior to Voyager 2's encounter, Neptune was known to have only two moons: Triton and Nereid. Voyager 2 discovered six additional relatively small moons, initially given designations 1989N1 through 1989N6. Although Voyager 2 came no closer than 4.7 million kilometers to Nereid, it was determined that Nereid, in actuality, was smaller than 1989N1 (400 kilometers in diameter). All the other new moons were less than 210 kilometers in diameter and quite irregular in shape. With the exception of 1989N6, which was inclined 4.5°, the new moons orbited in the plane of Neptune's rings. Two of the small moons exercised shepherding roles close to rings.

Triton was a curious world, almost certainly formed independently from Neptune and later captured by the gas giant when it strayed too close. This would account for Triton's unusual orbital inclination: 20° relative to the ring plane and the plane of most of the other small Neptunian moons. Voyager 2 refined measures of Triton's mass (2.13×10^{22} kilograms), diameter (2,720 kilometers), and density (2.03 grams per cubic centimeter). Triton's atmosphere consisted of nitrogen, with traces of methane. Surface pressure was a mere 10 microbars. Nitrogen extended as far as 800 kilometers from the surface, with a haze layer existing in the closest 25 kilometers. Methane, a heavier molecule, remained much closer to the surface.

Although Voyager 2 could not be diverted to intercept Pluto, the spacecraft may have provided clues to the nature of the Pluto-Charon system through its examination of Triton, revealing the presence of a combination of rock and frozen nitrogen, methane, and water. It is expected that Triton undergoes seasonal changes, and its atmosphere of nitrogen and methane may well freeze out during winter. Triton's aver-

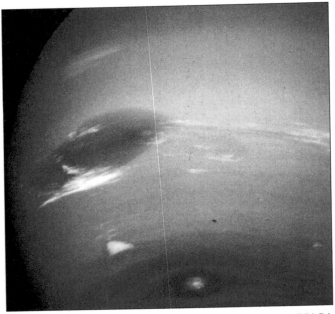

Neptune's Great Dark Spot and high-altitude clouds. (NASA CORE/Lorain County JVS)

age temperature, 37 kelvins, is typical of Pluto as well, so Pluto may also have a similar behavior and composition.

Context

In 1966, Gary Flandro, then an aeronautics graduate student at JPL, published a paper describing a unique alignment of the outer planets that would permit spacecraft to use energy provided by the planets themselves to alter the spacecraft's trajectory and allow a Grand Tour of the outer solar system in an amount of time far less than would be required by least-energy-transfer orbits. In 1969, NASA began planning for an ambitious program that it entitled the Grand Tour, in which four spacecraft would be dispatched from Earth to encounter Jupiter, Saturn, Uranus, Neptune, and Pluto. Two would be sent to Jupiter, Saturn, and Pluto, and the other two would fly by Jupiter, Uranus, and Neptune. Budget cutbacks forced the cancellation of this ambitious program, but NASA was able to obtain funding for what it initially called the Mariner-Jupiter-Saturn (MJS) project. The name was later changed to Voyager, and although the scope of the scientific investigation was scaled back, the opportunity still remained for at least one of the Voyager spacecraft to use a gravitational assist at Saturn to then send it on to Uranus and possibly also Neptune.

Without gravitational assists, the Voyager 2 mission to visit four planets would have been impossible. That navigational technique was first demonstrated on the Mariner 10 mission (1973-1975) to Venus and Mercury. Part of Voyager 2's journey was flown in the wake of previous explorers. The pathway to Jupiter and Saturn had been blazed by the successful Pioneers 10 (1973 Jupiter flyby) and 11 (1974 Jupiter flyby and 1979 Saturn flyby), and the Voyager 1 spacecraft. Voyager 2, however, entered virgin territory when it investigated the Uranian and Neptunian systems in 1986 and 1989, respectively. It then joined the Pioneers 10 and 11, and its fellow Voyager 1 spacecraft, to form a quartet of interplanetary travelers leaving our solar system and heading away in different directions, each carrying a message from the people of Earth and investigating the outer solar system boundaries before eventually running out of energy.

As of 2005, the Voyager 2 spacecraft was continuing its scientific exploration of interstellar space. Having traveled more than 11 billion kilometers from Earth during the twenty-seven years since its launch, Voyager 2 was heading out of the solar system at approximately 29 kilometers per second. After about eighty-five hundred years in flight, Voyager 2 should pass within four light-years of Barnard's Star. Its closest approach to a stellar system will occur in 40,457 years, when it should breeze within 1.5 light-years of Ross 248. Those last two encounters presuppose that, in the intervening time, humanity will fail to develop interstellar travel and seek out the intrepid Voyager 2, returning it to Earth for an honored resting place in an appropriate museum.

In 2004, NASA sent forth requests for proposals from industry for a Neptune Orbiter Probe that would extend the knowledge of the Neptunian system provided by Voyager 2's flyby. However, a year later, the Neptune Orbiter Probe, as well as other new projects to study the outer solar system (with the exception of the Jupiter Icy Moons Orbiter), were cut from the NASA budget.

See also: Cassini: Saturn; Deep Space Network; Galileo: Jupiter; Hubble Space Telescope; Hubble Space Telescope: Science; Jet Propulsion Laboratory; Planetary Exploration; Voyager Program; Voyager 1: Jupiter; Voyager 1: Saturn; Voyager 2: Jupiter; Voyager 2: Saturn; Voyager 2: Uranus.

Further Reading

Beatty, J. Kelly, Carolyn Collins Petersen, and Andrew Chaikin, eds. *The New Solar System.* 4th ed. New York: Sky Publishing, 1999. Contains detailed information about all major solar system objects, including photographs and data returned by robotic spacecraft from Mariner 2 through Voyager 2. Many chapters were written by principal investigators from the Voyager program.

Berry, Richard. "Neptune Revealed." *Astronomy* 17 (December, 1989): 22-34. Continues the presentation begun in the November, 1989, issue. Results are presented after further reflection. Makes comparisons of Neptune data with that from Voyager 2's Uranus encounter.

_____. "Triumph at Neptune." *Astronomy* 17 (November, 1989): 20-28. Contains excellent color images returned by Voyager 2 from Neptune, its rings, and its moons. Thorough presentation of preliminary results, primarily on Neptune and Triton.

Burgess, Eric. *Far Encounter: The Neptune System.* New York: Columbia University Press, 1991. Provides a full account of the Voyager 2 journey from launch through the Neptune encounter.

Davis, Joel. *Flyby: The Interplanetary Odyssey of Voyager 2.* New York: Atheneum, 1987. Provides a readable account of the journey of Voyager 2 from launch through the Uranus encounter. Provides numerous discussions with scientists involved with the program, and details important discoveries made at Jupiter, Saturn, and Uranus. Profiles the expected Neptune encounter.

Hartmann, William K. *Moons and Planets.* 5th ed. Belmont, Calif.: Thomson Brooks/Cole Publishing, 2005. Provides detailed information about all objects in the solar system. Suitable on three separate levels: high school student, general reader, and college undergraduate studying planetary geology.

Irwin, Patrick G. J. *Giant Planets of Our Solar System: Atmospheres, Composition, and Structure.* London: Springer-Praxis, 2003. Provides an in-depth comparison of Jupiter, Saturn, Uranus, and Neptune, incorporating data obtained from astronomical observations and planetary spacecraft encounters.

Kohlhase, Charles, ed. *The Voyager Neptune Travel Guide.* JPL Publication 89-24. Washington, D.C.: Superintendent of Documents, 1989. An excellent pre-encounter publication describing the objectives and maneuvers of Voyager 2 at its Neptune encounter. Provides numerous detailed charts and diagrams. Some parts are rather technical, but most are accessible to the general reader.

Lee, Wayne. *To Rise from Earth: An Easy to Understand Guide to Spaceflight.* New York: Checkmark Books, 1996. This is a good introduction to the science of spaceflight. Although written by an engineer with the NASA Jet Propulsion Laboratory, it is presented in easy-to-understand language. In addition to the theory of spaceflight, it gives some of the history of the human endeavor to explore space.

Miner, Ellis D., and Randii R Wessen. *Neptune: The Planet, Rings, and Satellites.* London: Springer-Praxis, 2002. The assistant project scientist for the Voyager 2 Neptune encounter composed this thorough review of our contemporary understanding of the Neptune system.

Moore, Patrick. *The Planet Neptune.* Chichester, West Sussex, England: Ellis Horwood, 1988. Provides a historical context of the investigation of the planet Neptune, from early searches for its existence up to the advent of spacecraft. Does not include data from Voyager 2.

United States Congress. House Committee on Science, Space, and Technology. *Voyager 2 Flyby of Neptune.* Washington, D.C.: Government Printing Office, 1990. A report to the

Congressional Subcommittee on Space Sciences and Applications by Drs. Lennard A. Fisk, Lew Allen, and Edward Stone (October 4, 1989). Includes black-and-white Voyager 2 Neptune images as well as descriptions of the scientific results of the mission, its importance, and implications for future planetary science programs.

"Voyager's Last Picture Show." *Sky and Telescope* 78 (November, 1989): 463-470. A pictorial review of the images returned by Voyager 2 at Neptune. Includes close-ups of atmospheric features such as the Great Dark Spot and Scooter, and surface features on Triton.

David G. Fisher

Wilkinson Microwave Anisotropy Probe

Date: Beginning June 30, 2001
Type of program: Scientific platform

The Wilkinson Microwave Anisotropy Probe (WMAP), a space-based astronomical observatory designed to measure the "echo" or "afterglow" of the Big Bang known as cosmic microwave background (CMB), provided the first accurate measure of the age of the universe, changed the way astronomers think about the earliest star formation, and supported some of the leading cosmological theories.

Key Figures

Charles L. Bennett, principal investigator at Goddard Space Flight Center
David T. Wilkinson (1935-2002), project scientist, Princeton University

Summary of the Mission

The Wilkinson Microwave Anisotropy Probe (WMAP) is a space-based observatory and the most successful observational cosmology project to date. Launched on June 30, 2001, atop a Delta II rocket, this unique orbiting radio telescope began a journey that would answer many fundamental questions about the evolution of the universe.

The concept for WMAP, originally referred to simply as MAP, came from David Wilkinson of Princeton University, one of the first astronomers to study the cosmic microwave background (CMB). The CMB is the nearly uniform electromagnetic radiation from an epoch when the universe was only 380,000 years old, according to WMAP's findings. This "baby picture" can reveal much about the structure and evolution of the universe. Wilkinson spearheaded another orbiting observatory, the Cosmic Background Explorer (COBE), which in 1992 discovered the anisotropies, or irregularities, in the otherwise smooth CMB. He died before any MAP results were made public, and the observatory was promptly renamed in his honor. The current principal investigator is Charles L. Bennett of the National Aeronautics and Space Administration's (NASA's) Goddard Space Flight Center (GSFC), and collaborators come from GSFC, Princeton, the University of Chicago, the University of California at Los Angeles, the University of British Columbia, and Brown University.

Astronomer George Gamow speculated in 1948 on the existence of an "echo" of the early events in the history of the universe. After the Big Bang, or moment of creation, light and matter were bound together in such a way that the universe was opaque to light. When the temperatures from the hot Big Bang cooled to approximately 300 kelvins (K), matter and light separated in what is called "last scattering." The light from that moment is still reaching Earth today but has been redshifted to 2.73 K given the large distance. (The Kelvin scale measures degrees above absolute zero temperature.) This radiation was accidentally discovered in 1965 by astronomers and Bell Telephone Laboratory scientists Arno A. Penzias and Robert W. Wilson. They discovered this seemingly isotropic radiation while studying emissions from the Milky Way. This discovery earned them the Nobel Prize in Physics in 1978. Only a few miles away, at Princeton University, David Wilkinson had been working with Robert Dicke on a receiver to detect the CMB.

Astronomers continued to study the CMB for decades. However, it puzzled them in that the measured temperature was perfectly smooth in all directions. After all, if the universe was perfectly smooth, then clumps of matter like galaxies, stars, and planets could not have formed. Hence, although the existence of the CMB supported the Big Bang theory, its uniformity did not fit in with observations of the modern universe. Finally, in 1992, the COBE satellite measured irregularities in the CMB temperature on small scales. Although this fuzzy picture did not provide much scientific detail, the seeds of the early galaxies had finally been detected. WMAP was proposed to NASA in 1995 to follow up on this remarkable discovery and was authorized in 1997.

WMAP features two radio telescopes 140° apart in order to map the temperature of the sky in all directions. This design allows for differential mapping—or subtracting the temperature in one area of the sky from the temperature at another point—which allows for subtraction of false signals. In other words, the relative temperatures of different regions are measured. In this way, WMAP can achieve a sensitivity of 0.000020 K. This is necessary in order to determine the tiny density variations in a radiation field that is only 2.73 K. The temperature of an object or energy field is related to the peak wavelength of the emission. Since the CMB temperature is so low, it has a low energy and therefore peaks at long wavelengths, specifically in the microwave region of the electromagnetic spectrum.

The cosmic background radiation is easily washed out by "foreground" objects, mainly galaxies, gas clouds, or human-made signals. So, WMAP operates at five frequencies: 22, 30, 40, 60, and 90 gigahertz (1 gigahertz equals one billion cycles per second). Because many ground-based telescopes operate at these frequencies, astronomers know much about objects in the radio sky and can subtract that radiation from the CMB radiation in the WMAP data with great precision. The result is a spectacular map of the early universe and the very first anisotropies that grew into the stars and galaxies that exist today. Angular resolutions of the maps range from 0.93° at 22 gigahertz to 0.23° at 90 giga-

hertz. (For comparison, the angular size of the full Moon is 0.5°.)

The observatory does not orbit Earth but a point in space known as L2, or the second Lagrange point. This is a gravitationally stable point between the Earth and the Sun where the Earth and Moon do not obstruct the telescope's view or provide background noise. Here, the telescope is 1.5 million kilometers away from Earth. A large sail acts as a shield against the Sun, and radiators act to keep the telescope cool. Radio astronomy receivers must be kept at very low temperatures in order to detect such faint radiation.

Having used the Moon for a small gravity assist, WMAP took three months to reach the L2 point. Initial results were announced on February 11, 2003. The map of the universe that was produced from the first-year results has become a mainstay of science media. The map is color-coded to show tiny temperature variations. The warmer temperature spots indicate density clumps, and the cooler spots indicate empty space. These clumps vary in temperature by 0.0002 K. From these data, many conclusions could be made about the nature, history, and future of the universe.

Contributions

Detecting tiny anisotropies was not WMAP's greatest achievement. In fact, the data gleaned from the mission helped astronomers pin down the most important cosmological parameters. Using sophisticated modeling techniques, astronomers are now able to start with a clear early picture of the universe and test different evolutionary models until the result resembles the current universe.

For the first time, the age of the universe has been pinpointed with incredible accuracy at 13.7 billion years, because the Hubble constant, or a measure of the expansion rate of the universe, has been determined to be 71 (km/sec)/Mpc. (A megaparsec, or Mpc, is approximately 3.26 million light-years; kilometers per second is denoted as km/sec.) Also, the first stars seem to have turned on, or have begun nuclear fusion, 200 million years

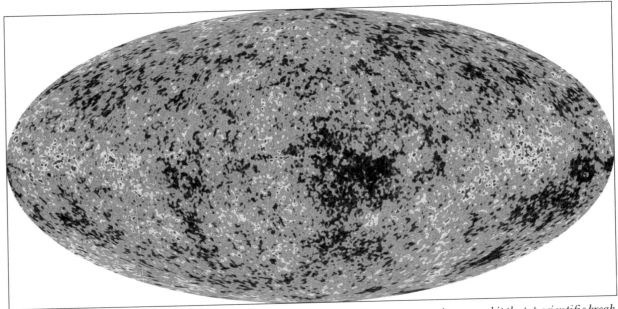

When WMAP provided data for this image of the microwave universe, Science *magazine named it the top scientific breakthrough of 2003. The mission made it possible to peg the age of the universe at 13.7 billion years and revealed that only 4 percent of the universe is normal matter; up to 23 percent is "dark" matter, and 73 percent is dark energy.* (NASA/ WMAP Science Team)

after the Big Bang—long before anyone had originally thought. WMAP also confirms that the geometry of the universe is flat. That is, the Euclidean geometry that is learned in high school applies over large scales. Other theories had surmised that the universe could be curved. In these strange geometries, parallel lines could eventually intersect or diverge over long distances.

Our universe has its own strange qualities, nevertheless. This flat geometry, along with other cosmological data to date, suggest that only 4 percent of the matter in the universe is the matter that makes up stars, planets, and humans, or baryonic matter, while 23 percent of the matter in the universe is known as dark matter—that which has not yet been detected but exerts gravitational forces on nearby objects. An even stranger thing, known only as dark energy, forms 73 percent of the universe. Theorists do not yet know of what this "energy" may consist. However, they propose that it exerts a long-distance repulsive force. This dark energy is also known as Albert Einstein's "cosmological con-

stant," or a constant used by Einstein to complete his theory of gravitation. Einstein had originally called it his "greatest blunder," but this constant was resurrected in 1998 when astronomers discovered the recent acceleration of the expansion of the universe. If this is the case, the universe will continue to expand forever, long after all of the stars have grown dark and cold.

WMAP data also conform with the inflationary model for the universe. In this model, the first few seconds after the Big Bang involved a very rapid, energetic expansion of space. In order to confirm this model further, even more sensitive maps of the CMB are needed. Inflation should have created gravitational waves that would be imprinted on the CMB. Until these can be detected, other theories of early universe formation cannot be entirely ruled out.

Context

Astronomers truly began to unravel the mysteries of the universe during the twentieth century. In

1927, Georges Lemaître proposed that the universe began as a tiny atom that exploded and expanded into the modern universe. Edwin Hubble gave evidence to this in 1936 when he discovered that there were many galaxies like our own and that they were rapidly moving away from one another. The term "Big Bang" was later coined by Fred Hoyle in 1950 as a derogatory term for the theory. Penzias and Wilson's discovery of the CMB, however, put the final nail in the coffin of the rival theory of which Hoyle was a proponent. By the time of WMAP's findings, cosmology had grown from speculation and theory to an observational science. Hubble's constant has finally been pinned down with great accuracy after being estimated at ranging values for decades. As a result, we now know the age of the universe, 13.7 billion years.

WMAP data are now available to all researchers in the Legacy Archive for Microwave Background Data Analysis (LAMBDA)—appropriately named after Einstein's cosmological constant, which is symbolized by the capital Greek letter lambda. The continued study of the CMB from orbiting and ground-based observatories will certainly surprise and inform astrophysicists and cosmologists for decades to come.

See also: Ulysses: Solar Polar Mission.

Further Reading

National Aeronautics and Space Administration. "Wilkinson Microwave Anisotropy Probe." http://map.gsfc.nasa.gov. The official Web site of the WMAP mission, including technical information, published papers, an introduction to cosmology, summary of findings, and images. Accessed February 21, 2005.

Plionis, Manolis, ed. *Multiwavelength Cosmology*. New York: Springer, 2004. Recent achievements in observational cosmology were presented and discussed at the "Multiwavelength Cosmology" conference that took place from June 17-20, 2003. Papers announce discoveries that range over the entire electromagnetic spectrum, including the newest WMAP results. Recommended for those with a scientific background.

Rowan-Robinson, Michael. *Cosmology*. London: Oxford University Press, 2003. This introductory text is well suited to undergraduates or the general public. It explains the basic concepts of Big Bang theory and observational evidence in support of it. It also delves into other cosmological theories and includes the first release of WMAP data.

Nicole E. Gugliucci

X-15

Date: June 8, 1959, to December 31, 1968
Type of mission: Piloted spaceflight

The first winged reusable spacecraft, the X-15 was designed to provide data on aerodynamic structures, flight controls in reduced atmosphere, and the physiological effects of high-altitude, hypersonic piloted flight. Many of the lessons learned from these experiments were incorporated into the piloted space program, with specific applications to the Apollo and space shuttle efforts.

Key Figures

A. Scott Crossfield (b. 1921), pilot
Joseph A. Walker (1921-1966), pilot
Robert M. White (b. 1924), pilot
Forrest Petersen (1922-1990), pilot
John B. McKay (1922-1975), pilot
Robert A. Rushworth (1924-1993), pilot
Neil A. Armstrong (b. 1930), pilot
Joseph H. Engle (b. 1932), pilot
Milton O. Thompson, pilot
William J. "Pete" Knight (1929-2004), pilot
William H. Dana (b. 1930), pilot
Michael J. Adams (1930-1967), pilot

Summary of the Program

Initiated a mere six years after Chuck Yeager broke the sound barrier in the Bell X-1, the X-15 program was originally intended to answer some of the questions that arose with the X-1 and X-2 flights, specifically the problems of aerodynamic heating and stability at high speeds. The X-15 project was a joint effort by the National Advisory Council for Aeronautics (NACA, later renamed the National Aeronautics and Space Administration, or NASA), the U.S. Air Force, and the U.S. Navy. The project was initiated in response to a 1952 NACA recommendation concerning the analysis of potential problems involved in spaceflight. This interest evolved into a recommendation for a high-altitude, hypersonic research aircraft that could be used to study high-temperature structures, hypersonic stability and control, and the physiology of flight in a reduced-gravity environment.

On June 11, 1956, North American Aviation was awarded a contract to construct three research aircraft to be designated the X-15. Much of the original design work for the vehicles was conducted at the Langley Research Center in Virginia. The final design resulted in a fixed-wing, rocket-powered, single-pilot aircraft of substantially conventional design. The structure was to be a semi-monocoque design constructed of titanium with a skin of Inconel-X nickel chromium alloy. The need for the exotic materials grew out of the requirement to control aerodynamic heating at hypersonic speeds. It was estimated that the aircraft would have to withstand localized temperatures of up to 1093.3° Celsius upon atmospheric reentry. The X-15 was to be capable of speeds in excess of Mach 6 at altitudes exceeding 76,200 meters. Eventually the aircraft flew at a speed of Mach 6.7 at an altitude of 107,960 meters.

Operations at such extreme altitudes required a control system that would be effective in the reduced atmospheric pressure of space as well as a conventional control system for lower-speed operation within the atmosphere. Twelve hydrogen peroxide rockets, installed in the nose and the wings, were used for a Reaction Control System (RCS) for out-of-atmosphere operations. The aircraft actually had three control sticks in the cockpit. With a mid-wing design with a length of 15.24 meters and

a wingspan of 6.706 meters, the aircraft weighed 6350.3 kilograms empty and 15,422.1 kilograms at maximum weight. A single Reaction Motors YLR-99 rocket powered it. This was the first rocket engine with controllable thrust and the capability of being shut down and restarted in flight by the pilot.

Because of the extremely high fuel consumption of the 260-kilonewton-thrust engine, the X-15 was designed to be launched from a B-52 mother ship. Launch was to occur at approximately 11,582.4 meters at Mach 6. Upon launch, the rocket engine would operate for approximately 90 seconds with the remainder of the flight being powerless. The descent, beginning at speeds in excess of 6437.4 kilometers per hour, was followed by a 321.9-

The X-15 in 1961 on the dry lake bed near Dryden Flight Research Center in Southern California. A B-52 flies overhead. (NASA)

kilometer-per-hour, dead-stick landing on one of the dry lakebeds in or around Edwards Air Force Base, California. The aircraft was designed to utilize the friction of the atmosphere to slow it down, a procedure that continues in shuttle flights.

The first flight of the X-15 took place on June 8, 1959. This was a powerless flight piloted by North American Aviation test pilot A. Scott Crossfield. The first powered flight occurred on September 17, 1959. Over the next nine years, 199 flights were completed, with the average flight lasting approximately 10 minutes, for a total flight time of 30 hours, 13 minutes, and 49.4 seconds.

Initially the missions examined the primary design parameters of hypersonic heating and stability and control, as well as confirming the research findings from tests conducted in cold-flow hypersonic wind tunnels. These goals were soon reached, and the project focus shifted to operations at the edge of space. Space operations had not received full support during the initial design of the project. Many believed that spaceflight was still far in the future. The success of the X-15 quickly convinced skeptics of the value of flights outside the Earth's atmosphere. By the end of 1962, the X-15 was routinely operating at Mach 6.4 at altitudes above 91,440 meters. Each flight was preceded by fifteen to twenty hours in an aircraft-specific simulator where flight operations and experiments were practiced.

The X-15 research was so successful in examining the two principal design questions that by 1961, the X-15 became a test bed for additional experiments. By 1965, more than 65 percent of the data being retrieved concerned follow-on experiments not involving the original design questions. By the end of the program, twenty-eight additional experiments had been conducted by the X-15 pilots. These included experiments in astronomy, micrometeorite collection, testing of various insulation materials, and photography. Tragically, it was a flight conducting an experiment in horizon identification that resulted in the only fatality of the program. On November 15, 1967, aircraft X-15-3, piloted by Major Michael J. Adams, entered a Mach 5 spin upon reentry into the atmosphere. It broke apart, killing the pilot and destroying the aircraft.

The research program continued through 1968. The X-15 exceeded all of its performance goals and is considered the most successful research aircraft ever designed. In addition to providing technical information about aerodynamic heating and stability at hypersonic speeds, the program provided valuable information about operations in the space environment, pilot physiology when subject to gravitational loads in excess of 5g, pilot-guided reentry to the atmosphere, and the feasibility of reusable space vehicles. The X-15 was the last high-speed research aircraft to fly as part of the research airplane program, and it held the winged vehicle speed record of Mach 6.7 and the altitude record of 107,960 meters until the orbital flights of the space shuttle *Columbia* in 1981.

The last flight of the X-15 took place on October 24, 1968. Repeated attempts to conduct flight number 200 were unsuccessful, and the program was terminated on December 31, 1968. The total cost of the X-15 project was $300 million. While the cost was more than thirty times the original estimate, the X-15 is still considered the most successful research airplane in history. The two surviving X-15 aircraft are on display at the National Museum of the United States Air Force in Dayton, Ohio, and at the Air and Space Museum of the Smithsonian Institution in Washington, D.C.

Contributions

Although the X-15 project was originally designed to examine specific questions regarding aerodynamic heating and stability, as well as control at hypersonic speed, the project contributed much more. The project added to the knowledge gained from the X-1 and X-2 flights and dispelled the theory of the "stability barrier" that had been experienced by the earlier aircraft operating above Mach 1. The project also demonstrated the feasibility of a piloted reentry into the atmosphere, with the pilot controlling the vehicle as opposed to the technique used in rocket-launched piloted space capsules.

Engine failure forced pilot John "Jack" McKay to make an emergency landing near Edwards Air Force Base in November, 1962. He survived, with injuries. Many pilots have risked their lives to advance supersonic and space flight. (NASA)

The techniques and procedures developed, as well as the knowledge gained about the stresses and temperatures involved in reentry into the atmosphere, have been directly transferred to the space shuttle program. The X-15 was the first vehicle to demonstrate that the concept of a reusable space vehicle was practical and cost-effective. This feature permitted a full recovery of test equipment and proved to be the least costly method available for achieving the desired results. It also provided the capability of repeating experiments to confirm results.

The X-15 program also examined the physiology of high-speed, high-altitude flight. It answered many questions concerning the ability of pilots to function in a zero gravity and conditions of high gravitational load. Data on the heart rates of X-15 pilots involved in missions were later used as a baseline for astronauts prior to launch. The ability to function and conduct scientific experiments outside Earth's atmosphere was demonstrated repeatedly during X-15 missions.

One of the primary accomplishments of the program was the confirmation of data gathered during cold-flow hypersonic wind-tunnel testing. Prior to the X-15, there was no way to confirm these data. The X-15 proved the accuracy of these types of data generation, which, with the exception of drag information, were extremely accurate. The drag error was found to be caused by the mounting technique, and once that was corrected, the error was eliminated. Wind-tunnel testing, combined with the development of a complex flight simulator, allowed scientists as well as pilots to prepare for every aspect of a mission prior to leaving the ground. This practice continues today in the space shuttle program.

Context

The X-15 project was a continuation of the X-plane program, which began with the X-1. It was originally intended to answer questions that arose from X-1 and X-2 flights. The program was so successful that it expanded and eventually included many additional experiments. Although the X-15 was the last of the high-speed experimental research aircraft to fly, many of the lessons learned contributed to modern space operations. The value of the program does not lie in the hardware utilized on the X-15, as most of it is obsolescent today, but rather in the concepts and techniques developed during the program. Piloted reentry procedures and demonstration of temperatures and stresses experienced during reentry have proven invaluable to the space shuttle program, as did the evaluation of heat-dissipation methods and materials.

The development of a successful pressure suit, which not only protected the pilot but also allowed him to operate the aircraft and conduct complex scientific experiments, was a major contribution to space operations. The suit also monitored and transmitted physiological data to the ground and evolved into the pressure suits worn by astronauts today.

While eight X-15 pilots completed a total of thirteen astronaut-qualification flights, neither the X-15 program nor its pilots received the recognition awarded the original Mercury astronauts until nearly forty years later. On August 23, 2005, NASA awarded astronaut wings to three of the X-15 test pilots. Retired NASA pilot Bill Dana and family members representing deceased pilots John McKay and Joseph Walker received the civilian astronaut wings. The wings acknowledged the fact that the pilots flew the X-15 at altitudes of 50 miles or higher. The men were honored in a ceremony at NASA's Dryden Flight Research Center, Edwards, California, the site of their achievements. Dana was philosophical about it: "NASA pilots didn't wear wings anyway, and the concept of winning special wings was probably more crucial to a military pilot's career ladder," he explained.

Largely in response to the perceived Soviet threat following the launch of the first artificial satellite, Sputnik, in 1957 and successful manned orbital spaceflights beginning with that of Yuri A. Gagarin on April 12, 1961, the United States' space program took the direction of the space capsule rather than the controllable, reusable space plane—at least until the advent of the space shuttle. The X-15 carried on until it was quietly retired on the last day of December, 1968.

See also: National AeroSpace Plane; Pegasus Launch Vehicles; Space Shuttle; Space Shuttle: Ancestors; Space Shuttle: Approach and Landing Test Flights; Space Task Group.

Further Reading

Crossfield, A. Scott, with Clay Blair, Jr. *Always Another Dawn: The Story of a Rocket Test Pilot.* New York: Arno Press, 1972. While technically not a book about the X-15, this work provides insight into the people involved and their dedication to the program. It is a good introduction suitable for secondary as well as college students.

Goodwin, Robert, ed. *X-15: The NASA Mission Reports.* Burlington, Ont.: Apogee Books, 2001. Developed from the archives of the National Aeronautics and Space Administration and the U.S. Air Force, this volume offers the most complete and detailed coverage of the X-15 and its accomplishments. It includes a complete review of the program, a biography of each pilot, flight reports for every flight, and a CD-ROM with a number of exciting video clips. Essential to anyone interested in the X-15 program.

Jenkins, Dennis R. *Hypersonic: The Story of the North American X-15.* North Branch, Minn.: Specialty Press, 2003. A detailed history of the X-15 project from its genesis in 1952 to the retirement of the aircraft in 1968. Included are a number of interesting photographs and illustrations. The book tends toward technical details and may not be suitable for some high school students.

Thompson, Milton O. *At the Edge of Space: The X-15 Flight Program.* Washington, D.C.: Smithsonian Institution Press, 2003. Covers the personal insights and experiences of an X-15 pilot. Thompson completed fourteen flights in the X-15. In this volume he recalls both the excitement of piloting the space plane and the tragedy of losing a colleague to the program's only fatal accident. A well-written book from the perspective of an insider.

Ronald J. Ferrara

X-20 Dyna-Soar

Date: 1957 to 1963
Type of mission: Piloted spaceflight

Although canceled before a prototype was built, the Dyna-Soar project became a model for other efforts to build a space plane, including the space shuttle, by pioneering technologies and fabrication methods.

Key Figures

Eugen Sänger (1905-1964), Austrian engineer who proposed an antipodal bomber
James Wood (1924-1990), major, USAF, astronaut-engineer
Henry C. Gordon (1925-1996), major, USAF, astronaut-engineer
Russell L. Rogers (1928-1967), major, USAF, astronaut-engineer
Milton O. Thompson, NASA test pilot
William J. "Pete" Knight (1929-2004), captain, USAF, astronaut-engineer
Albert Crews, Jr. (b. 1929), captain, USAF, astronaut-engineer

Summary of the Program

The convoluted saga of the X-20 Dyna-Soar project has caused passionate controversy among historians, aerospace scientists, and science journalists ever since its cancellation in 1963. Milton O. Thompson, a Dyna-Soar test pilot, called the cancellation a crime. Historian Clarence Geiger complained that no research project was ever so callously treated by nonscience bureaucrats. Robert Godwin (editor of *Dyna-Soar: Hypersonic Strategic Weapons System*) lamented that no other project was ever subjected to so many changes in direction. Martin Caidin, an eminent science writer, charged that the program's demise left the United States without a vehicle it still badly needed, a lightweight, one-person rocket-glider capable of orbiting the Earth for a variety of missions and returning to landing fields for rapid reuse.

The space shuttle is the heir to Dyna-Soar's technology, but the connection between the two crewed space plane programs is complex, because the demise of Dyna-Soar itself was complex. It failed to produce a prototype vehicle because of Cold War politics, competition among space-

flight programs, U.S. Air Force ambitions for space, and budgetary restraints, but the overriding problem was that the program never fixed on a single, clear purpose for the vehicle that it was to develop.

The origin of Dyna-Soar, in concept, dates to 1934, when Austrian engineer Eugen Sänger proposed his "antipodal bomber." His idea was to build a rocket-powered plane that could boost into suborbital space and then skip off the atmosphere like a stone skipping over water until it reached its objective, where it would descend by gliding and drop its bomb payload. Sänger's concept intrigued Germany's high command during World War II, including the commander of rocket research, General Walter Dornberger. The Germans never progressed past the design stage, but Dornberger brought up the idea of a military space plane again while he was working as a consultant and lobbyist for a U.S. aerospace company. Beginning in 1952, together with his wartime colleague Wernher von Braun, he urged the Air Force to conduct feasibility studies.

Faith in technological progress during the 1950's and a burgeoning enthusiasm for space-flight meant that ideas such as the space plane of von Braun and Dornberger found a wide, receptive audience, both civilian and military. The Air Force was particularly interested because it wanted to exploit the tactical possibilities of outer space. Specifically, air-war theorists saw a space plane as an unstoppable method of taking nuclear weapons into the Soviet Union. This interest suddenly gelled into action after the Soviets launched the world's first satellite, Sputnik 1, on October 4, 1957, and the Space Race intensified. Consolidating three preliminary research programs, the Air Force opened bidding on October 10 for primary contractors in its three-phase Dyna-Soar (a contraction of "dynamic soaring") program. For some aspects of design and technological development, the Air Force enlisted the help of the National Advisory Committee for Aeronautics (NACA, later the National Aeronautics and Space Administration, NASA). From the nine contractor teams bidding, the Air Force chose Boeing to build the spacecraft and Martin to develop a launch vehicle from a modified Titan intercontinental ballistic missile (ICBM).

From the first, there was disagreement over Dyna-Soar's design. Some planners favored a purely nonballistic space capsule—the seed idea for what later became the Mercury spacecraft. Others wanted a lifting body. It was the winged design in the spirit of Sänger that won.

Step I, as envisioned in the Air Force's first system development directive, called for a suborbital test vehicle designed for a single pilot. With a length of 10.8 meters, it would have delta-shaped wings of 6.4-meter wingspan on the bottom of the fuselage and would weigh about 5,170 kilograms at launch. To withstand an expected maximum heating of 2,000° to 2,400° Celsius during reentry, it would have a rounded, tilted nose of graphite, reinforced with zirconia rods, a frame of René 41 nickel super-alloy, an upper surface composed of René 41 panels, and a heatshield and leading edges made of zirconia and molybdenum alloys. The vehicle was to have four objectives: It would test the heating to its skin during flight, maneuverability while reentering the atmosphere from space, conventional glider landing, and human tolerance for hypersonic flight (more than five times the speed of sound). The first crewed demonstration flight of the vehicle was scheduled for 1964.

Step IIA would conduct a variety of tests during orbital flights, beginning in 1966, including orbital velocities, military radar and sensors, bombing and navigation systems, air-to-surface missiles, docking equipment and techniques, and guidance and control systems. Step IIB was to supply the Air Force in 1967 with a vehicle able to perform reconnaissance from space and to inspect satellites. Step III was to be a fully operational bomber-reconnaissance spacecraft. The Air Force asked for nearly $500 million to finance the first step.

While the Air Force general staff was enthusiastic about the Dyna-Soar, the civilian leaders at the Department of Defense were not. The Air Force's funding requests were routinely scaled back or canceled, only to be partially restored, and doubts arose about the purpose of the project, especially its military goals. Secretary of Defense Robert S. McNamara questioned whether it was a wise use of national resources at all and thought that tests of military equipment might be better conducted aboard the Gemini spacecraft then under development.

In any case, the Dyna-Soar program was abbreviated. In a December, 1961, "redirection," Steps I and IIA were merged and the suborbital flight dropped. The new plan was to launch a vehicle into orbit in 1965 to test the technology for crewed, maneuverable orbital systems and demonstrate its military potentiality. To emphasize its experimental rather than operational purpose, Secretary McNamara directed that the program's name be changed to X-20 in June, 1962, even though the Air Force preferred its original name. Thereafter, it was referred to as the X-20 Dyna-Soar.

Meanwhile, an important problem remained unsettled: how to boost the craft into space. Redirection from a suborbital to orbital mission compli-

Test pilot Neil Armstrong, who would later be the first human being to step on the Moon, prepares to fly an FSD-1 Skylancer for a Dyna-Soar abort simulation. The FSD-1 had a wing planform similar to that of the Dyna-Soar. (NASA)

cated the problem. The original design called for a modified Titan I ICBM, but that was determined to be underpowered. Next the Titan II, then undergoing testing, was proposed; it was 3 meters longer and had a wider upper stage that would accommodate the Dyna-Soar more readily. It was eventually dropped as too limited, and for a while the massive Saturn C-1 was considered until it was eliminated because of its high cost and complexity. Finally, in mid-1962 planners chose a Titan II with two solid boosters attached for extra lift; this configuration was designated Titan III-C. However, delays in its development threatened to set back the whole X-20 Dyna-Soar program as well.

On September 20, 1962, at the Las Vegas Convention Center, a mock-up of the Dyna-Soar was unveiled to the public. General Bernard Schriever, commander of the Air Force Systems Command, introduced the program's six astronaut-engineers, of whom five were Air Force officers: Major James

Wood, Major Henry C. Gordon, Major Russell Rogers, NASA test pilot Milton O. Thompson, Captain William J. Knight, and Captain Albert Crews, Jr. For a while X-15 test pilot Neil A. Armstrong, who later became the first person to step onto the Moon, served as a consultant. These men tested the new technology for Dyna-Soar, including space suits, piloting controls, and cockpit arrangements, while hundreds of engineers and fabricators at Boeing, Martin, and their subcontractors, sometimes exploiting the technology of programs such as the X-15 and at other times inventing materials and systems, labored to put together a vehicle even as its goal changed from a suborbital to an orbital military glider and finally to an orbital test platform.

These changes, coupled with competition for funds with other NASA and Air Force projects and lukewarm support from the Department of Defense, eventually doomed the X-20 Dyna-Soar. In a December 10, 1963, press release, Secretary McNamara announced its cancellation in order to divert its funding to a new Air Force project, a near-Earth Manned Orbiting Laboratory, which in turn was canceled in 1969.

Contributions

Even though the X-20 Dyna-Soar program ended before it could produce a prototype spacecraft, it advanced aerospace technology considerably and provided crucial data incorporated by space shuttle designers. In fact, testing of some components created for the Dyna-Soar continued after its cancellation.

More than fourteen thousand hours of tests in wind tunnels were conducted before engineers settled on the final design for the slender glider, the first such spacecraft to receive such scrutiny. During that time, the configuration of its wings and tail, its handling and maneuverability at supersonic and hypersonic velocities, and the heating rates of its surfaces were assessed. These tests provided data of use to later experimental space plane designs. Among the special materials tested were super alloys (such as René 41), ceramics, and high-temperature bearings and insulation.

Several devices to aid and protect pilots were developed for the program. The energy management display indicator was a cathode-ray display helping the pilot keep within the craft's structural thermal limits and reach the landing site by indicating its progress in flight as a function of velocity. Computer-assisted adaptive controls were designed to control the craft within and beyond the atmosphere, for normal gliding, and for an aborted takeoff. A landing gear and skids that would absorb the strain of landing after being heated hundreds of degrees was also developed, as well as various types of heatshields. Specifically, the proper design of the nose cap, as the point of greatest heating upon reentry, was a crucial problem, and two designs received independent testing, both of which studied the heat-shedding properties of zirconia. A significant achievement of the project was the water-wall designed to protect the pilot, equipment bay, and secondary power bay from heat during reentry. Other advances were made in hydraulic control systems, control surfaces, and instrumentation.

Context

The Air Force wanted an unstoppable bomber capable of attacking the Soviet Union with nuclear weapons, and Dyna-Soar, which would strike from orbit, was a prime candidate. That placed the program squarely in the context of Cold War politics. Both U.S. presidents ultimately responsible for the X-20 Dyna-Soar, Dwight D. Eisenhower and later John F. Kennedy, had serious reservations about it.

First was their concern about adding another dimension to the arms race. If the United States built an orbital bomber, the Soviets would either build one of their own or, worse, invent a means to shoot it down, which would also endanger American satellites. Those satellites, in fact, posed a second, greater concern. Presidents Eisenhower and Kennedy favored an international open-space policy that would forbid any nation claiming the extra-atmospheric area above it as its territory and controlling access. That way, satellites could pass over any portion of the Earth. If Russia were entitled to restrict outer space, America's highly successful intelligence-gathering satellites would be illegal, liable to defensive measures, and possibly a cause for war. The Dyna-Soar threatened to disrupt the delicate American-Soviet negotiations concerning outer space, and neither president wanted that.

Moreover, international tensions were particularly high following the Cuban Missile Crisis of 1962 and the Berlin crisis of 1963. Given Dyna-Soar's potential to heighten military rivalry, presidential preference for satellites, the capability of the Gemini Program to carry out military experiments, and Secretary McNamara's disinclination to let the Air Force run a space program independent of NASA, the X-20 Dyna-Soar was terminated.

See also: Launch Vehicles: Reusable; Manned Orbiting Laboratory; Mercury-Atlas 7; Space Shuttle; Space Shuttle: Ancestors; Space Shuttle: Approach and Landing Test Flights.

Further Reading

Geiger, Clarence J. "Strangled Infant: The Boeing X20A Dyna-Soar." In *The Hypersonic Revolution: Case Studies in the History of Hypersonic Technology*, edited by Richard P. Hallion. Vol. 1, *From Max Valier to Project PRIME*. Bolling Air Force Base, D.C.: Air Force History and Museum Program, 1998. A series of articles that explains (at a moderately technical level) the history, systems development, phases, cancellation, and technical legacy of the Dyna-Soar project, with diagrams.

Godwin, Robert. *Dyna-Soar: Hypersonic Strategic Weapons System*. Burlington, Ont.: Apogee Books, 2003. Following a thirty-eight-page introduction about the Dyna-Soar program, Godwin offers an omnium-gatherum of technical and administrative documents, articles, graphics, photographs, letters, and transcribed public testimony about it. An accompanying DVD contains vintage films about the vehicle, the project's astronaut corps, and construction methods.

Houchin, Roy F., II. "Hypersonic Technology and Aerospace Doctrine." *Air Power History* 46 (Fall, 1999): 4-18. Houchin reviews the Cold War political tensions that prompted the Defense Department to back away from a space bomber.

Roger Smith

Appendices

Chronology of the U.S. Space Program

Below, major developments in rocketry and space exploration are preceded by the titles of essays in this publication where more information can be found.

c. 360 B.C.E.	**Rocketry: Early History:** Archytas of Tarentum, in southern Italy, uses steam to provide thrust to propel a wooden pigeon traveling on wires around in a circle.
c. 206 B.C.E.-220 C.E.	**Rocketry: Early History:** Chinese use bamboo tubes filled with black powder ("fire arrows") during the Han Dynasty.
c. 100 B.C E.	**Rocketry: Early History:** Hero of Alexandria, Greece, creates an engine that uses steam to provide thrust that rotates a centrally mounted sphere. Hero seems to have used his engine for amusement rather than for practical applications.
c. 618-907 C.E.	**Rocketry: Early History:** Fireworks are popular in China during the Tang Dynasty.
1180	**Rocketry: Early History:** Chinese used bamboo tubes filled with black powder with a spearhead attached as a weapon of war.
1241	**Rocketry: Early History:** Chinese introduce rockets to Europe in the Battle of Sejo, against the Magyars, near what is now Budapest.
1268	**Rocketry: Early History:** Arabs are on the receiving end of Mongol rocket barrages during the capture of Baghdad (1258) and soon employ them against French Crusaders.
1429	**Rocketry: Early History:** The French army uses rockets against the English in the Siege of Orléans.
1730's	**Rocketry: Early History:** German artillery officer Colonel Friedrich von Geissler builds military rockets weighing between 25 and 54 kilograms, although these fail to make any impact on warfare of the day.
1800's	**Rocketry: Early History:** British Colonel Sir William Congreve designs the rocket system that would bear his name.
1806	**Rocketry: Early History:** Congreve rockets first see action in an attack on the French city of Boulogne. The two thousand rockets so terrify the city's defenders that they surrender without returning a shot.
1814	**Rocketry: Early History (September 13):** Congreve rockets are memorialized when British ships fire them at Fort McHenry (Baltimore, Maryland), inspiring Francis Scott Key to include "the rockets' red glare" in his poem "The Star-Spangled Banner."
1847	**Rocketry: Early History (March 24):** The first American rocketeer company goes on campaign during the Mexican War, accompanying Major General Winfield Scott's expedition against Mexico City. The company first sees action during the Siege of Vera Cruz, Mexico.
1855	**Rocketry: Early History:** Lieutenant Colonel Charles Boxer of the Royal Laboratory in Britain develops a two-stage rocket, greatly improving the distance over which life lines can be sent to rescue people from sinking ships.

1903   **Rocketry: Modern History:** Konstantin Tsiolkovsky details how rockets capable of traveling to outer space can be built. In his *Exploration of Cosmic Space by Rocket Devices*, he explains how liquid oxygen, when combined with either kerosene or liquid hydrogen in a combustion chamber, can provide the propulsive force needed, especially when "rocket trains" (multi-stage rockets) are used, to attain velocities necessary to escape from the Earth's gravity.

1906-1908   **Rocketry: Modern History:** Robert Goddard begins experiments with powder rockets.

1915   **National Aeronautics and Space Administration (March 3):** A rider to the Naval Appropriations Act establishes the National Advisory Committee for Aeronautics (NACA), the predecessor to NASA.

            Rocketry: Modern History: Robert Goddard proves experimentally that a rocket will provide thrust in a vacuum.

1917   **Langley Research Center (July 17):** Langley Memorial Aeronautical Laboratory, the site for the NACA's first experimental aeronautical facility, is founded. These facilities and their successors contribute to the advancement of aeronautics and spaceflight for the indefinite future.

1920-1925   **Rocketry: Modern History:** Robert Goddard develops the first rocket motor using liquid propellants (liquid oxygen and gasoline).

1926   **Rocketry: Modern History (March 16):** Robert Goddard launches the first liquid-fueled rocket at Auburn, Massachusetts. It is powered by the combustion of liquid oxygen and kerosene.

1929   **Rocketry: Modern History:** Hermann Oberth launches his first liquid-fueled rocket, named Kegeldüse (cone nozzle). His students at the Technical University of Berlin, including Wernher von Braun, help him in this experiment.

1935   **Rocketry: Modern History (March 8):** Robert Goddard becomes the first to launch a liquid-propellant rocket that attains a speed greater than that of sound (1,100 kilometers per hour).

1940   **Ames Research Center (April 18):** Ames Aeronautical Laboratory in Moffett Field, California (later renamed NASA Ames Research Center), is formed.

1941   **Lewis Field and Glenn Research Center (January 23):** NACA Aircraft Engine Laboratory in Cleveland, Ohio, is founded. In 1949, the center is renamed the Lewis Research Center in honor of George W. Lewis, the NACA's director of aeronautical research. On March 1, 1999, the center is officially renamed the NASA John H. Glenn Research Center at Lewis Field.

1942   **Rocketry: Modern History:** Adolf Hitler approves the production of the A-4 as a "vengeance weapon" to rain explosives on London. The first A-4 flies in March, flying about 1.6 kilometers and crashing into the water. The second launch reaches an altitude of 11 kilometers before exploding. The third rocket is launched on October 3 and follows its trajectory perfectly to land 193 kilometers away and become the first manufactured object to enter space.

1944   **Rocketry: Modern History (September 7):** The first combat A-4, now called the V-2 for Vergeltungswaffe 2 (retaliation weapon 2, a name invented by Josef Goebbels), is launched toward England.

1945	**Rocketry: Modern History (June 20):** U.S. Secretary of State Cordell Hull approves the transfer of Wernher von Braun's German rocket specialists. This transfer was known as Operation Paperclip because of the large number of Germans stationed at Army Ordnance; the paperwork of those selected to come to the United States was kept together with paperclips. Under British supervision, there are three V-2 rocket launches to demonstrate the launch capability in October. Soviet rocket pioneer Sergei Korolev is awarded the Badge of Honor, his first decoration, for his work on the development of rocket motors for military aircraft.

Space Centers and Launch Sites: The first testing centers away from populated areas become a necessity with the advent of the German V-2 rocket in World War II.

1946   **Space Shuttle: Ancestors (January 19):** The first glide flight of the NACA X-1 rocket research airplane occurs in California.

1947   **Rocketry: Modern History:** Soviet engineers, thanks in part to disassembled captured V-2 rockets, begin producing a working replica of the rocket. Designated the R-1, it is first tested in October. A total of eleven are launched, with five landing on target. This was comparable to the German success rate and demonstrated the unreliability of the rocket. The Russians continue to utilize the expertise of the Germans on their rocket designs until about 1952, when the first groups begin to return home.

Space Shuttle: Ancestors (October 14): Captain Charles E. Yeager (USAF) makes the world's first supersonic flight in level or climbing flight at Muroc, California, in a rocket-powered NACA-USAF research plane. During its fiftieth flight, the Bell X-1 No. 1 attains Mach 1.06 at 13,115 meters, approximately 1,126 kilometers per hour.

1948   **Space Shuttle: Ancestors:** The first flight of the Douglas D-558-II (No. 1) supersonic research aircraft occurs with John Martin of Douglas as pilot. D-558-II had both jet and rocket engines and was flown from ground takeoff.

1949   **Rocketry: Modern History:** The U.S. rocketry program makes use of the supply of unused V-2 rockets left from the war. Some of these are equipped with a WAC-rocket as a second stage and called Bumper. One such rocket reaches a then-record altitude of 400 kilometers at its launch from White Sands. The Bumper is the first rocket launched from Cape Canaveral, Florida.

1950   **Cape Canaveral and the Kennedy Space Center:** Cape Canaveral Air Force Station becomes the launch site for early missile tests and eventually becomes one of three principal facilities for launching satellites and piloted spacecraft.

1952   **Rocketry: Modern History:** Based on the V-2 and other rockets, the Redstone intermediate range ballistic missile (IRBM) is tested at Cape Canaveral, establishing its range of 320 kilometers and its usefulness as a support for army ground forces.

1953   **Space Shuttle: Ancestors:** D-558-2, flown by Scott Crossfield, is the first aircraft to break Mach 2, or twice the speed of sound. The achievement culminates a joint Navy/NACA high-speed flight research program.

1955   **Vanguard Program (1955-1959):** Remembered more for its failed attempts than its three successful launches of U.S. satellites, the Vanguard program generates important developments in rocket propulsion, satellite design, and satellite telemetry and tracking.

1956 **Soyuz Launch Vehicle:** The Soviet R-5, the world's first IRBM, completes a successful first flight.

1957 **Funding Procedures of Space Programs:** With the onset of the space race, the United States begins serious funding of the space program, peaking in the 1960's and declining thereafter.

Lifting Bodies (1957-1975): The research program on lifting bodies creates a fleet of wingless vehicles designed specifically to be able to bring human explorers back to Earth from space and to land on a specified runway, just like an airplane. It demonstrates that pilots can safely maneuver and land lifting bodies.

Soyuz Launch Vehicle: In connection with the International Geophysical Year (July, 1957, to December, 1958), both the United States and the Soviet Union plan to launch satellites to study the Earth's outer-space environment. The first true intercontinental ballistic missile (ICBM) is the Soviet R-7, developed by Sergei Korolev. It is a two-stage rocket with a maximum payload of 5.4 tons and a range of 7,000 kilometers. After several test failures, the R-7 successfully sends a dummy payload to Kamchatka in August. The R-7 places in orbit Sputnik 1, the world's first artificial satellite, on October 4. Sputnik 2, weighing six times the mass of Sputnik 1, is launched on November 3. It carries the dog Laika (Russian for "Barker"), the world's first live orbiting payload.

X-20 Dyna Soar (1957-1963): Although canceled before a prototype was built, the Dyna-Soar project becomes a model for other efforts to build a space plane, including the space shuttle, by pioneering technologies and fabrication methods.

1958 **Ames Research Center:** The Ames Research Center becomes part of NASA.

Deep Space Network (January): The forerunner of the Deep Space Network (DSN) is established when the Jet Propulsion Laboratory (JPL), then under contract to the U.S. Army, deploys portable radio tracking stations in Nigeria, Singapore, and California to receive telemetry and plot the orbit of Explorer 1.

Explorers 1-7 (1958-1961): The first successful U.S. program to launch a human-made satellite into orbit, with five of the first seven Explorers achieving Earth orbit.

Jet Propulsion Laboratory (December 3): JPL comes under NASA jurisdiction, and its emphasis is changed to lunar and planetary exploration.

Langley Research Center: The oldest aeronautics research and testing center, the Langley Memorial Aeronautical Laboratory, becomes part of NASA and now conducts programs on advanced aerodynamics and the future of piloted and robotic space travel.

Lewis Field and Glenn Research Center: Lewis Research Center joins NASA and performs basic and applied research to develop technology in aircraft propulsion, space propulsion, space power, microgravity science, and satellite communications.

Mercury Project (1958-1963): The first U.S. piloted orbital space program yields important data on human adaptability to space travel.

Mercury Spacecraft (1958-1963): The United States' first spacecraft, the Mercury capsule, tests technical and conceptual approaches, gives astronauts practical experience in outer space, and boosts the nation's sagging prestige early in the Space Race, all of which deeply influences subsequent piloted space programs.

National Aeronautics and Space Administration (October 1): The National Aeronautics and Space Administration is formally established. Thomas Keith Glennan is the first administrator, serving until January 20, 1961.

Pioneer Missions 1-5 (1958-1960): All five Pioneers make important discoveries about the radiation belts around Earth even though only five are unqualified successes, achieving heliocentric orbit between the paths of Earth and Venus.

Space Task Group (1958-1961): The first U.S. civilian agency for piloted spaceflight is responsible for the Mercury, Gemini, and Apollo programs, as well as the ancestor of the Manned Spacecraft Center, now the Johnson Space Center.

Telecommunications Satellites: Military: Communications satellites provide reliable, worldwide communications between troops and decision makers.

Vandenberg Air Force Base: Vandenberg becomes the first operational ICBM facility in the United States. Vandenberg will be the launch site for more than five hundred orbital and one thousand nonorbital launches of rockets and ballistic missiles, the launch complex for the Manned Orbiting Laboratory, a West Coast launch center for the space shuttle, and the Western Commercial Space Center.

1959

Apollo Program: Developmental Flights (1959-1969): During the initial phases of the Apollo Program, space scientists develop increasingly powerful launch vehicles, including Little Joe II, Saturn I, Saturn IB, and Saturn V. They also test escape mechanisms, should the boosters fail in flight, and the susceptibility of spacecraft to meteoroids. All these developments ultimately make possible several successful piloted missions to the Moon.

Astronauts and the U.S. Astronaut Program: NASA recruits the first seven astronauts and launches the astronaut-training program.

Atlas Launch Vehicles: Originally developed to launch and carry thermonuclear warheads a distance of 8,200 kilometers, the Atlas is first tested. The United States' first successful ICBM, it was a 1½-stage, liquid-fueled (LOX and RP-1) rocket, with three engines producing 1,590 kilonewtons of thrust.

Explorers: Astronomy: The Astronomy Explorers collect data on solar radiation, gamma rays, meteoroids, x rays, and radio waves.

Gemini Program (1959-1969): The Gemini spacecraft is developed to bridge the gap between the one-person Mercury vehicle and the much more ambitious three-person Apollo spacecraft.

Goddard Space Flight Center: Founded in 1959, Goddard Space Flight Center is called the intellectual brain trust of NASA.

Mercury Project Development (September 9, 1958): "Big Joe" (Atlas 10-D) launches a boilerplate Mercury capsule from Cape Canaveral. The objective of Big Joe is to test the Mercury spacecraft heatshield, which survives reentry and is in remarkably good condition when retrieved from the Atlantic.

Spy Satellites: Spy satellites are operational by 1959 to provide the United States with sophisticated information in a more effective and less dangerous way than other reconnaissance methods.

Titan Launch Vehicles (February 6): The first flight of the Titan I occurs. It is the United States' first true multistage ICBM and the first in a series of Titan rockets.

X-15 (1959-1969): Three X-15 aircraft make 199 flights. On August 22, 1963, flight reaches an all-time altitude record of 108 kilometers. On the flight of October 3, 1967, a record speed of Mach 6.7 is reached. This program produces invaluable data on aerodynamic heating, high-temperature materials, reaction controls, and space suits.

1960

Apollo Program: Developmental Flights: As part of the Mercury Program and using a "Little Joe" rocket as a booster, tests begin on abort maneuvers and escape technologies. A Little Joe booster from Wallops Island, Virginia, launches a monkey in a spacecraft topped by an escape tower. The test is a success and the monkey is returned safely to Earth.

Apollo Program: Lunar Lander Training Vehicles (1960-1971): When NASA undertakes the challenge set forth by President Kennedy to put an American on the Moon before 1970, the need for an inflight lunar landing trainer becomes apparent. The Lunar Landing Research Vehicle (LLRV) and the production-version Lunar Landing Test Vehicle (LLTV) fill that need. Built of tubular steel like a giant four-legged bedstead, the LLRV (nicknamed the "flying bedstead") has a jet engine with 19,000 newtons of thrust to simulate a lunar-landing profile.

Early-Warning Satellites: Products of the Cold War, these satellites are designed to detect firings of intercontinental ballistic missiles and transmit warnings to Earth.

Electronic Intelligence Satellites: ELINT satellites receive various electronic signals, providing a major portion of the intelligence on which the United States and Soviet Union rely.

Explorers: Ionosphere: These Explorer satellites conduct the first comprehensive direct measurements of Earth's ionosphere.

Global Positioning System: This network of satellites is designed for military application; it will evolve into a navigation system that allows users to determine their positions anywhere in the world.

Marshall Space Flight Center: Ground is broken for the new NASA facility, built around the core of German scientists and engineers under the direction of Wernher von Braun. President Eisenhower directs the transfer of the Army Ballistic Missile Agency's Development Operations Division, headed by von Braun, to NASA on January 14.

Mercury Project Development (January 21): Little Joe 1B launches from Wallops Island, Virginia, with Rhesus monkey "Miss Sam" on board. This is a test of launching and abort systems of the Mercury spacecraft.

Meteorological Satellites: Since 1960, meteorological satellites have helped predict the weather and have revolutionized meteorological science.

Meteorological Satellites: Military: Since 1960, military meteorological satellites have provided cloud-cover photographs and other weather data from orbit to help schedule reconnaissance satellite launches and military operations.

Navigation Satellites: These military satellites, made accessible to civilians, allow ships, aircraft, and land vehicles to pinpoint their positions on Earth.

Telecommunications Satellites: Passive Relay (1960-1969): The U.S. Echo satellite project includes the first passive relay communications satellite launched into space and the first cooperative space venture between the United States and the Soviet Union.

Telecommunications Satellites: Private and Commercial: Since 1960, communications satellites have been designed and built by private corporations to serve the needs of their customers.

TIROS Meteorological Satellites (1960-1986): These satellites have provided high-altitude views that have increased meteorologists' capability to forecast weather.

1961   **Amateur Radio Satellites (December 12):** The first amateur radio satellite, OSCAR 1, is launched. It is the first privately owned, nongovernmental satellite launched. Today, amateur radio satellites allow residents of different countries to communicate with one another, bring space exploration into the classroom, and assist in emergency relief projects.

Explorers: Air Density (1961-1970): Data from the four Explorers determine the effect on Earth's atmosphere of solar heating over a sunspot cycle, provide the first evidence of the winter helium and summer atomic oxygen bulges, and show that the atmospheric constituents above 300 kilometers exist in layers and maximize at points of maximum temperature.

Explorers: Micrometeoroid (1961-1972): These Explorers perform direct measurements of the micrometeoroid environment in near-Earth space.

Gemini Program (December 7): Project Gemini begins. During a five-year period, it places humans into Earth orbit and teaches them how to track, maneuver, and control orbiting spacecraft; dock with other orbiting vehicles; and reenter Earth's atmosphere and land at specified locations. Later missions provide astronauts with experience in long-duration flights and extravehicular activity (EVA).

Johnson Space Center: The Manned Spacecraft Center (later the Johnson Space Center) becomes an independent entity, and a suburb of Houston is selected as the site for this piloted spacecraft headquarters.

Mercury Project (January 31): Ham the chimpanzee flies a suborbital lob on MR-2, experiencing six minutes of weightlessness. Despite problems with electronics aboard the Mercury spacecraft and much joking from their test-pilot friends about astronauts being no better than apes, the Mercury Seven astronauts are heartened by Ham's success. Mercury-Atlas tests 2, 3, and 4, all robotic, are completed by September, 1961.

Mercury-Atlas 6 (November 29): Enos the chimpanzee flies a two-orbit Mercury mission. This event prompts one cartoonist to depict Enos walking away from the spacecraft, helmet in hand, saying, "We're a little behind the Russians and a little ahead of the Americans."

Mercury-Redstone 3 (May 5): Alan Shepard becomes the first American to reach space on Mercury-Redstone 3, a fifteen-minute suborbital mission.

Mercury-Redstone 4 (July 21): Gus Grissom duplicates the successful suborbital flight of Alan Shepard.

Mercury-Redstone 4 (August 6-7): Vostok 2 takes Gherman Titov for the first daylong orbital spaceflight.

National Aeronautics and Space Administration (February 14, 1961): James Webb becomes the second NASA administrator, serving until October 7, 1968.

Ranger Program (1961-1965): The Ranger Program is the first part of NASA's three-stage plan leading to the piloted exploration of the Moon. It scores an impact on the Moon with Ranger 4, but the mission is a failure as its sequencer fails and the spacecraft returns no data. Ranger 5 launches on October 18, 1962, loses contact with Earth, and flies past the

Moon at a distance of 725 kilometers. Ranger 6 follows the proper trajectory to its selected target, and its instruments respond to command. However, when the spacecraft is ready to begin sending images during final approach to impact on February 2, 1964, its television cameras unexpectedly fail. All program goals are accomplished with highly successful Ranger 7, 8, and 9 missions.

Saturn Launch Vehicles (October 27): The first flight test of the Saturn I (Block I) launch vehicle uses a Saturn I first stage to carry water-filled dummy upper stages to an altitude of 136.5 kilometers and a downrange distance of 345.7 kilometers. The flight accomplishes its objective of verifying the aerodynamic and structural design of the Saturn I booster.

Soyuz Launch Vehicle (April 12): Vostok 1, carrying Cosmonaut Yuri A. Gagarin, is launched. On his one-orbit mission, Gagarin becomes the first human to travel into space.

Space Suit Development: Russian cosmonauts and American astronauts first wear full-pressure space suits in space.

1962

Apollo Program: Orbital and Lunar Surface Experiments: NASA begins soliciting scientific experiments for the Apollo flights. Chemists, physicists, geologists, geophysicists, and astrobiologists are brought together for conferences during the summers of 1962 and 1965 to identify three general types of scientific work: astronaut observations, sample collection, and instrument placement to return data to Earth. At a conference during the summer of 1967, NASA establishes an advisory body of scientists who will assist the Manned Spacecraft Center (Houston, Texas) in planning the scientific work for each mission.

Mariner 1 and 2 (1962-1963): The first interplanetary spacecraft directed to study Venus. Mariner 1 suffers a launch failure, and Mariner 2's flyby revolutionizes knowledge of the conditions on Venus.

Mercury-Atlas 6 (February 20): John Glenn proves that humans can work in a microgravity environment and that the Mercury spacecraft is sound during the first U.S. piloted Earth-orbiting flight.

Mercury-Atlas 7 (May 24): This is the first U.S. space mission devoted to piloted scientific research in space.

Mercury-Atlas 8 (October 3): Wally Schirra demonstrates that humans can work in a microgravity environment for an extended period of time.

Orbiting Solar Observatories (1962-1978): These observatories are designed to study the structure of the Sun and its outward flow of high-energy particles.

X-20 Dyna Soar (September 20): A mock-up of the Dyna-Soar is unveiled to the public at the Las Vegas Convention Center. General Bernard Schreiver, commander of the Air Force Systems Command, introduces the program's six astronaut-engineers, of whom five are Air Force officers. For a while X-15 test pilot Neil Armstrong serves as a consultant.

1963

Apollo Program: Developmental Flights (August 28): The first Little Joe II "qualification test vehicle" is launched from White Sands Missile Range in New Mexico and successfully accomplishes its objective of gathering data on temperatures within and outside the spacecraft and pressures on it.

Ethnic and Gender Diversity in the Space Program (June 16): Soviet cosmonaut Valentina Tereshkova pilots the Vostok 6 spacecraft and becomes the first woman and first civilian to fly into space.

Explorers: Atmosphere (1963-1981): These satellites gather invaluable data on Earth's atmosphere and ionosphere.

Interplanetary Monitoring Platform Satellites (1963-1973): The ten spacecraft in this part of the Explorer program measure cosmic radiation levels, magnetic field intensities, and solar wind properties in the near-Earth and interplanetary environment.

Lifting Bodies: The M2-F1, known affectionately as the "flying bathtub," is completed in 1963. It has elevons, attached to each of the two rudders, instead of ailerons (which control roll). A large flap on the trailing edge of the body acts as an elevator. The M2-F1 is tested by towing it behind a Pontiac convertible at speeds of 195 kilometers per hour. Hundreds of such tests are conducted. Following the successful tests, the craft is towed behind a NASA R4D tow plane at higher altitudes (3,600 meters). Pilot Milt Thompson glides the M2-F1 back to the base.

Manned Orbiting Laboratory (1963-1969): The first U.S. space station project evolves beyond the study stage.

Mercury Spacecraft: Following one failed launch attempt, there are five Mercury-Redstone missions (two robotic, one with a chimpanzee, and two piloted, all suborbital) and nine Mercury-Atlas missions (four robotic, one with a chimpanzee, and four piloted, the last six orbital). The final splashdown occurs on May 16, 1963.

Mercury-Atlas 9 (May 15-16): Gordon Cooper demonstrates that humans can work in a microgravity environment for more than one day.

Nuclear Detection Satellites: The Nuclear Test-Ban Treaty is signed. Satellites are used to detect secret nuclear explosions in space and within Earth's atmosphere.

X-20 Dyna Soar (December 10): In a press release, Secretary of Defense Robert McNamara announces cancellation of the X-20 Dyna Soar in order to divert its funding to a new Air Force project, a near-Earth Manned Orbiting Laboratory, which in turn is canceled in 1969.

1964

Apollo Program (1964-1972): The Apollo Program is the United States' bid for international leadership in space exploration; this leadership is demonstrated by landing humans on the Moon and returning them safely to Earth.

Apollo Program: Developmental Flights (May 13): The first Little Joe II to fly with an Apollo CM is launched. On December 8, a Little Joe II launch vehicle successfully demonstrates Apollo's escape system.

Apollo Program: Lunar Lander Training Vehicles (October 30): On the day of the first flight of the Lunar Landing Research Vehicle (LLRV), research pilot Joe Walker flies it three times for a total of just under 60 seconds to a peak altitude of 3 meters.

Lifting Bodies: The M2-F2 is designed to remedy control problems that are identified during the testing of the M2-F1. A major design problem results in airflow separation, which causes aerodynamic instability. The opening footage of the 1970's television program *The Six Million Dollar Man* shows the M2-F2, piloted by Bruce Peterson, crashing and tumbling violently along the runway. Peterson survived to fly again and the craft was rebuilt as the Northrop M2-F3 (1967) with a central rudder to correct the problem.

Mariner 3 and 4 (1964-1965): Mariner 4 is the first U.S. robotic probe to Mars, providing data that greatly increases the understanding of Earth's neighboring planet.

Nimbus Meteorological Satellites (1964-1986): By remote sensing from orbit, the Nimbus satellites are used to develop new techniques for observing Earth, especially its atmosphere and oceans.

Orbiting Geophysical Observatories (1964-1972): These six spacecraft return significant data on various geophysical phenomena.

Soyuz Launch Vehicle (October 12-14): Voskhod 1 is the first spaceflight to carry more than one person into space and the first flight without space suits.

1965

Apollo Program: Developmental Flights: The final three Saturn I Block II vehicles—SA-9, SA-8, and SA-10—launched Pegasus satellites in February, May, and July of 1965, providing the most information ever gathered on micrometeoroids. However, because the Saturn I is unable to launch the Apollo Command and Service Module (CSM) and the Lunar Module (LM) together, NASA officials cancel the Saturn I program after ten successful missions, and in its place, the Saturn IB program accelerates.

Explorers: Solar: The information obtained from the Solar Explorers increases understanding of the effects of solar activity on piloted space activity and radio communications.

Gemini 3 (March 23): The first U.S. two-person orbital spaceflight completes three Earth orbits with a total flight time of 4 hours and 53 minutes.

Gemini IV (June 3-7): The second piloted Gemini mission is flown. It is the first U.S. mission to include an extravehicular activity.

Gemini V (August 21-29): This flight demonstrates the human ability to survive eight days in space, rudimentary rendezvous techniques, and fuel-cell electrical power generation systems.

Gemini VII and VI-A (December 4-18): Gemini VI-A tests the ability to rendezvous in space and Gemini VII's flight of fourteen days demonstrates that humans can endure long-duration flights.

Intelsat Communications Satellites: Intelsat is created to develop and operate a global satellite communications system that will guarantee access to international satellite communications for all member nations.

Pioneer Missions 6-E: Pioneer 6 through Pioneer E are the first spacecraft specifically prepared to obtain synoptic information on the effects in planetary space of solar activity.

Skylab Program (1965-1974): Three crews spend a total of 171 days in space on the first American space station.

Skylab Space Station (1965-1979): Skylab, the United States' first space station, is a "monolithic" station, intended to be constructed and launched in one piece and then occupied by a crew later. The 91-metric-ton Skylab is in Earth orbit from 1973 to 1979 and occupied by crews three times in 1973 and 1974.

Soyuz Launch Vehicle (March 18-19): Aleksei Leonov becomes the first person to perform a spacewalk when he ventures outside the Voskhod 2 spacecraft for twelve minutes at the end of a tether.

1966

Apollo Program: Developmental Flights (February 26): The Saturn IB, an improved Saturn I with an upgraded first stage, is first launched from Cape Kennedy, primarily to test the booster and the attached Apollo spacecraft. The flight is designated Apollo-Saturn 201 (AS-201).

Apollo Program: Lunar Lander Training Vehicles (April): The Lunar Landing Research Vehicle (LLRV) has performed more than 100 successful flights and NASA has accumulated enough data from the LLRV flight program to have three LLTVs built at a cost of $2.5 million each. The LLTV is similar to the LLRV, but the pilot has a three-axis side control stick and a more restrictive cockpit view, both features of the real Lunar Module.

Applications Technology Satellites (1966-1979): This satellite program is designed to demonstrate the promise of artificial satellites for direct broadcast communications and meteorological monitoring of Earth's surface.

Biosatellites (1966-1970): A series of spacecraft are launched to investigate the effects of spaceflight on basic life processes.

Environmental Science Services Administration Satellites (1966-1976): The first true weather satellite system creates a base for meteorological satellites used by scientists on Earth.

Gemini VIII (March 16-17): Gemini VIII becomes the first piloted spacecraft to rendezvous and dock with the Agena target vehicle. The Agena has an engine that can be restarted either from the Gemini spacecraft or by ground control.

Gemini IX-A and X (June 3-6, July 18-21): Two Gemini missions conclude three-day flights designed to demonstrate rendezvous and docking techniques and to evaluate extravehicular activity.

Gemini XI and XII (September 12-15, November 11-15): The final two missions of the Gemini Program share the primary objectives of performing rendezvous and docking with the Agena target vehicles and conducting extravehicular activities.

Lifting Bodies (July, 1966-November, 1975): The Northrop HL-10 is one of five heavyweight lifting body designs flown at NASA's Flight Research Center (later Dryden Flight Research Center), Edwards, California, to study and validate the concept of safely maneuvering and landing a low lift-over-drag vehicle designed for reentry from space. It is a NASA design and built to evaluate "inverted airfoil" lifting body and delta planform.

Lunar Orbiters (1966-1968): The Lunar Orbiter becomes one of three robotic spacecraft programs designed to help scientists select safe landing sites on the Moon for the Apollo Program.

Orbiting Astronomical Observatories (1966-1972): These observatories provide astronomers with an opportunity to conduct observations at specific wavelengths above Earth's atmosphere.

Surveyor Program (1966-1968): The Surveyor Program develops the technology for soft-landing on the Moon.

Tethered Satellite System: Since Gemini XI and XII, scientists have experimented with tethered satellite systems with the potential to provide electrical power to the space shuttle or to recharge failing satellite batteries.

1967   **Apollo 1 (January 27):** During a ground test of the Apollo spacecraft, the prime crew of Apollo 1 is killed when an oxygen-fed fire sweeps through their sealed spacecraft.

Apollo Program: Developmental Flights (November 9): The Saturn V, biggest of the Saturn family of boosters, is launched on its first robotic flight, known as Apollo 4 (Apollo-Saturn 501). The rocket, which generates 33 meganewtons of thrust in the first stage alone, reaches an altitude of 16,100 kilometers. The spacecraft's engines propel Apollo at an angle into the atmosphere at such high speeds that the heatshield reaches a temperature of about 3,000 kelvins.

Apollo Program: Developmental Flights (1967-1968): The early Apollo missions are robotic test flights that further spaceflight technology for the eventual safe transport of astronauts to and from the Moon.

Mariner 5 (1967-1968): The primary scientific mission of Mariner 5 is to gather data on the atmosphere and the ionosphere of Venus and to study the interaction of the solar plasma with Venus's environment.

Ocean Surveillance Satellites: Since 1967, these satellites have been used to locate ships, identify them, and determine their speed and course.

Soyuz and Progress Spacecraft (April 24-24): First piloted launch (Soyuz 1) of the Soyuz 11A511 launch vehicle and Soyuz 7K-OK spacecraft in a planned rendezvous mission. The spacecraft fails to deploy one of its solar panels, and the decision is made to bring Cosmonaut Vladimir Komarov back early. Reentry is successful, but a failed pressure sensor prevents the main parachute from deploying. Komarov releases the reserve chute, but it tangles with the drag chute. The descent module crashes into a field near Orenburg, Russia, killing Komarov instantly.

1968

Apollo Program: Developmental Flights (January 22): Apollo 5 is the first test flight of the Lunar Module (LM-1). Its mission objectives include verification of the ascent and descent stages, the propulsion systems, and the restart operations. It also evaluates the spacecraft structure, LM staging, second stage (S-IVB), and instrument unit orbital performance.

Apollo Program: Lunar Lander Training Vehicles: Neil Armstrong is flying LLRV-1 on May 6, 1968, when the helium pressurization system for the steering jets on the LLRV fail, leaving Armstrong no way to control the vehicle. He safely ejects before the LLRV becomes a pile of crumpled metal.

Apollo 7 (October 11-22): The first piloted flight of the Apollo spacecraft, the first piloted launch of the Saturn IB launch vehicle, and the first three-person American space mission is successfully conducted.

Apollo 8 (December 21-28): The first piloted mission to the Moon is also the first piloted launch of the Saturn V rocket.

Explorers: Radio Astronomy (1968-1975): RAE-A and RAE-B are orbited to detect and measure extraterrestrial radio noise.

1969

Apollo 9 (March 3-13): The first piloted mission of the Lunar Module, an Earth-orbital flight to test the Command and Service Module (CSM) and the Lunar Module (LM) for both docking and rendezvous.

Apollo 10 (May 18-26): Apollo 10 successfully orbits the Moon and provides the final testing of all systems needed for an actual landing.

Apollo 11 (July 16-24): Humans make their first landing on the Moon, exiting the Lunar Module to walk on the surface and successfully returning to Earth.

Apollo 12 (November 14-24): The second successful lunar landing, Apollo 12 makes a pinpoint landing in the Ocean of Storms, within walking distance of the Surveyor 3 probe, which had landed there in April, 1967.

Apollo Program: Orbital and Lunar Surface Experiments (July): The first lunar scientific package, the Early Apollo Scientific Experiment Package (EASEP), is deployed by the Apollo 11 crew. EASEP consists of a passive seismograph experiment (PSEP) and a Laser Ranging Retroreflector (LRR).

Lifting Bodies (April 17): The X-24A—a fat, short, teardrop-shaped aircraft with vertical fins for control—makes its first, unpowered, glide flight, It will be flown twenty-eight times at speeds up to 1,667 kilometers per hour and altitudes up to 21.8 kilometers.

Mariner 6 and 7 (1969-1970): The second and third robotic missions to Mars provide valuable photographic information on the surface characteristics of the planet.

National Aeronautics and Space Administration (March 21): Dr. Thomas O. Paine becomes NASA administrator; he will serve until September 15, 1970.

Soyuz and Progress Spacecraft (January 14-18): Soyuz 4 and 5 perform the first docking between two piloted spacecraft. During the second-ever Soviet spacewalk and the first two-person spacewalk, Cosmonauts Aleksei Yeliseyev and Yevgeni Khrunov leave Soyuz 5 and float over to Soyuz 4, completing the first inflight crew exchange.

Soyuz and Progress Spacecraft (October 11-18): Soyuz 6, 7, and 8 complete the first missions that involve three spacecraft in orbit together at the same time. They set a record by carrying seven cosmonauts.

1970

Apollo 13 (April 11-17): During the third scheduled lunar landing mission, an explosion on board causes the landing to be aborted and requires the crew to use the Lunar Module as a lifeboat.

ITOS and NOAA Meteorological Satellites (1970-1979): A series of meteorological satellites is developed by the United States to maintain constant surveillance of weather conditions around the world. NOAA is the federal agency that operates the ITOS satellites during the 1970's.

1971

Apollo 14 (January 31-February 9): The third successful landing of an American scientific team on the lunar surface is the first to use an equipment cart on the Moon. Veteran astronaut Alan Shepard becomes the first lunar "golfer" when he uses a makeshift six iron to whack two golf balls for "miles and miles."

Apollo 15 (July 26-August 7): The fourth landing of an American scientific team on the lunar surface. Astronauts explore and return samples from the Hadley-Apennine region.

Apollo 15's Lunar Rover (July 31-August 2): The battery-powered Lunar Rover greatly expanded the area the astronauts could explore.

Apollo Program: Orbital and Lunar Surface Experiments (1971-1972): Sector 1 of the Apollo Service Module (SM), usually filled with ballast to maintain the SM's center of gravity, houses a Scientific Instrument Module (SIM) for lunar study on Apollo missons 15-17. The equipment includes camera systems, a lunar subsatellite (Apollo 15 and 16), a laser altimeter, an S-band transponder, a lunar sounder (Apollo 17), an ultraviolet spectrometer, and an infrared radiometer. Prior to their return to Earth, the Apollo 15 and 16 crews deploy small, 38-kilogram subsatellites into lunar orbit. These subsatellites perform three experiments that continue the study of the lunar environment. Accurate tracking of the spacecraft from Earth using the S-band transponder allows details of the Moon's gravity field to be mapped, which provides information about the distribution of mass in the Moon's interior.

Mariner 8 and 9 (1971-1972): As the first artificial satellite of another planet, Mariner 9 does the work planned for both Mariners.

Materials Processing in Space: The lack of gravitational effects in space promises great technological advances for processing materials.

National Aeronautics and Space Administration (April 27): Dr. James C. Fletcher becomes NASA administrator and will serve until May 1, 1977.

Space Stations: Origins and Development (April 19-October 10): Soviet Salyut 1 becomes the first space station placed into orbit. Soyuz 10 fails to make first docking to the station.

1972   **Apollo 16 (April 16-27):** The second-to-last of the Apollo lunar flights, Apollo 16 allowed the astronauts to explore more difficult terrain than previous crews. This is the only landing in the lunar highlands during the Apollo Program; the landing site is in the Descartes region.

Apollo 17 (December 7-19): The last and perhaps most ambitious of the Apollo flights confirms findings of earlier flights and returns the largest number of lunar rocks for study.

Landsat 1, 2, and 3 (1972-1982): These satellites collect data about Earth's agriculture, forests, flatlands, minerals, waters, and environment.

Pioneer 10: Pioneer 10 becomes the first spacecraft to provide close-up reconnaissance of Jupiter and to sample directly its magnetic and particle environment.

Space Shuttle: The Space Transportation System is established to develop an economic and reusable system that can transport humans, satellites, and equipment to and from Earth orbit on a regular basis.

Space Stations: Origins and Development (April 3-May 28): The Soviet Union's Almaz space station, the world's first military station, is orbited under the name Salyut 2 to hide its true purpose. It depressurizes before a crew can arrive and is never occupied.

Spaceflight Tracking and Data Network: This network of ground communications and tracking stations begins to provide data relay, data processing, communications, and command support to the U.S. space shuttle program and other orbital and suborbital spaceflights.

1973   **Lifting Bodies:** The X-24B's design evolves from a family of potential reentry shapes, each with higher lift-to-drag ratios, proposed by the Air Force Flight Dynamics Laboratory. To reduce the costs of constructing a research vehicle, the Air Force returns the X-24A to the Martin Marietta Corporation for modifications that convert its bulbous shape into one resembling a "flying flatiron." John Manke makes the first glide flight on August 1, 1973, and pilots the first powered mission on November 15, 1973.

Mariner 10 (1973-1975): Mariner 10 collects vital data on the inner solar system, including detailed photographs of Venus and Mercury.

Pioneer 11: Pioneer 11 collects critical data on the outer solar system.

Skylab 2 (May 25-June 26): The crew establishes a record for the longest Earth-orbiting flight with their twenty-eight-day mission in space.

Skylab 3 (July 28-September 25): Skylab 3's crew exceeds the objectives of their mission in fifty-nine days in the orbital workshop.

Skylab 4 (November 16, 1973-February 8, 1974): In the last flight to the first U.S. space station, the three astronauts spend eighty-four days in Earth orbit, making it the longest piloted U.S. spaceflight up to that time.

Skylab Space Station (1973-1979): The United States' first space station, the 75-metric-ton Skylab, is launched on May 14 and visited by crews three times in 1973 and 1974.

1974

SMS and GOES Meteorological Satellites: Beginning in 1974, these satellites begin to provide continuous coverage of weather conditions on Earth.

Space Stations: Origins and Development (June, 1974-January, 1975): First successful Almaz military manned space station flight, launched as Salyut 3. Following the successful Soyuz 14 and unsuccessful Soyuz 15 missions, on September 23, the station ejects a film return capsule, which is successfully recovered.

Space Stations: Origins and Development (December, 1974-February, 1977): Two Soyuz crews visit the first Soviet non-military space station, Salyut 4.

1975

Apollo-Soyuz Test Project (July 15-24): The first U.S.-Soviet cooperative piloted spaceflight.

Viking Program (1975-1982): The Viking missions to Mars are the first long-duration intensive explorations of the surface of another planet.

Viking 1 and 2 (1975-1982): The Viking mission to Mars is the first long-duration intensive exploration of the surface of another planet.

1976

Get-Away Special Experiments: A program to allow a wide variety of users to have their space experiment packages launched inside small canisters aboard space shuttle missions at relatively low cost.

Space Stations: Origins and Development (June, 1976-August, 1977): The third and most successful Almaz military space station, called Salyut 5, is occupied by two Soyuz crews.

Telecommunications Satellites: Maritime: A network of satellites designed to upgrade the communications capabilities of commercial and military maritime vessels was first proposed in 1972; satellites are launched between 1976 and 1979, when the International Maritime Satellite Organization was formed.

1977

High-Energy Astronomical Observatories: These observatories provide a detailed survey of the celestial sphere, studying x-ray sources and detecting gamma-ray and cosmic radiation.

International Sun-Earth Explorers (1977-1985): Three ISEEs perform observations and experiments in deep space and Earth's magnetosphere. ISEE 3 is sent around the Moon to make the first spacecraft contact with a comet.

National Aeronautics and Space Administration: Dr. Robert A. Frosch is NASA administrator from June 21, 1977, to January 20, 1981.

Space Shuttle: Approach and Landing Test Flights: *Enterprise*, the first space shuttle orbiter, tests the approach and landing techniques of the Space Transportation System.

Space Stations: Origins and Development (September, 1977-July, 1982): Salyut 6, the Soviet Union's most successful space station to date, features several revolutionary advances including a second docking port where the Progress cargo spacecraft could dock and refuel the station. With Salyut 6, the Soviet space station program evolves from short-duration to long-duration stays. Five long-duration and eleven short-duration Soyuz crews occupy Salyut 6 for a total of 683 days.

Voyager Program: The Voyager probes execute the first Grand Tour in planetary exploration by encountering Jupiter, Saturn, Uranus, and Neptune.

Voyager 2: Jupiter (1977-1979): The Voyager 2 flyby of Jupiter provides vital information about the Jovian system, in spite of a number of technical problems and equipment failures.

1978

Air Traffic Control Satellites: Beginning with the earliest satellite program, the United States has conducted experiments using artificial satellites to aid in air traffic control operations. These efforts led to Navstar, a network of navigational satellites, and to Nusat, the first satellite designed exclusively for air traffic control.

Dynamics Explorers: The first satellites placed in Earth orbit to provide data on the energy and momentum of charged particles in Earth's upper atmosphere.

Heat Capacity Mapping Mission: HCMM is the first satellite designed to measure thermal inertia.

International Ultraviolet Explorer: Conceived as an orbiting observatory designed to observe the ultraviolet portion of the spectrum, IUE proves one of the most productive and oldest functioning satellites in the history of the space program.

Pioneer Venus 1: The first Venus mission to map the planet's surface and investigate its atmosphere and ionosphere.

Pioneer Venus 2: Called the multiprobe, this cluster of five spacecraft is designed to penetrate Venus's cloud cover and gather information about its atmosphere at separate locations.

Seasat: This satellite performs a variety of experiments, including photographic and remote-sensing procedures, above the world's oceans.

1979

Stratospheric Aerosol and Gas Experiment: The SAGE instrument was designed to measure the concentration of some of the constituents of Earth's atmosphere.

Voyager 1: Jupiter: Voyager 1 collected detailed information on the planet Jupiter, its rings, satellites, and surrounding environment, including detailed photographs of the four Galilean satellites.

1980

Solar Maximum Mission (1980-1989): Solar Max is a mission designed to study solar phenomena, particularly solar flares, during the 1980 peak of the eleven-year sunspot cycle. In 1984, the Space Shuttle *Challenger* intercepts it and maneuvers the satellite into the shuttle's payload bay for maintenance and repairs.

Soyuz and Progress Spacecraft (January 20): Progress 1, the first autonomous, robotic supply vessel, delivers fuel, consumable materials, and equipment to the Salyut 6 station. It is the first of many resupply missions to Salyut 6, Salyut 7, the Mir Complex, and the International Space Station.

Soyuz and Progress Spacecraft (June 5-9): The Soyuz T-2 flight brings the sixth visiting crew to Salyut 6 and introduces the second-generation Soyuz spacecraft. The Soyuz-T (Transport) features larger solar panels for longer independent flights to the Salyut space stations and carries a crew of two. It would act as a crew ferry for the next six years.

Voyager 1: Saturn: On its second planetary flyby, Voyager 1 encounters Saturn and sends back to Earth information on the planet's rings, satellites, and atmosphere.

1981

National Aeronautics and Space Administration: James M. Beggs is NASA administrator from July 10, 1981, to December 4, 1985.

Space Shuttle: Living Accommodations (April 12ff.): Every effort has been made to provide safe and comfortable living conditions in a microgravity environment for space shuttle astronauts since STS-1 in 1981.

Space Shuttle Mission STS-1 (April 12-14): The objectives of the first launch of the space shuttle are to demonstrate a safe launch to orbit, test basic systems in space, achieve reentry, and land safely.

Space Shuttle Mission STS-2 (November 12-14): The second flight of the shuttle *Columbia* carries the first scientific experiments and tests the remote manipulator system for the first time.

Voyager 2: Saturn: The Voyager 2 flyby of Saturn produces high-resolution images of Saturn's ring system and of the satellites Iapetus, Hyperion, Enceladus, and Tethys.

1982

High-Energy Astronomical Observatories: These observatories provide a detailed survey of the celestial sphere, studying x-ray sources and detecting gamma and cosmic radiation.

Landsat 4 and 5: These satellites collect data about the Earth's agriculture, forests, flatlands, minerals, waters, and environment.

Space Shuttle Flights, 1982: Two additional test flights (STS-3, in March, and STS-4, in June) determine the flightworthiness of the space shuttle and lead to a flight carrying the first commercial payload.

Space Shuttle Flights, 1982 (November 11-16): STS-5, the first so-called operational flight of the Space Transportation System, deploys the first American commercial satellites, ANIK C-3 for TELESAT Canada and SBS-C for Satellite Business Systems.

Space Stations: Origins and Development (1982-1991): The last Salyut space station, Salyut 7, is aloft for four years and two months, during which time it is visited by ten crews constituting six main expeditions and four secondary flights (including French and Indian cosmonauts).

1983

Infrared Astronomical Satellite: From its vantage point in Earth orbit, this satellite enables scientists to study in detail, and without atmospheric interference, the heat emitted by astronomical sources.

Shuttle Amateur Radio Experiment: Since 1983, amateur radio operators have communicated with space shuttle astronauts orbiting the Earth.

Space Shuttle Flights, 1983: This year of firsts sees the introduction of the second orbiter (*Challenger*), the first American woman astronaut (Sally Ride), the first African American astronaut (Guy Bluford), the first non-American astronaut on an American spacecraft (ESA astronaut Ulf Merbold of Germany), and the first four-, five-, and six-person crews.

Space Shuttle Mission STS-6 (April 4-9): The first voyage of *Challenger* deploys the first Tracking and Data Relay Satellite and permits two astronauts to work in space thanks to the extravehicular mobility unit.

SPACEHAB: This commercial mini-laboratory fits into the cargo bay of the space shuttle and since 1983 has been leased to private corporate interests.

Strategic Defense Initiative: This project, which proposed using space- and Earth-based technology to counter nuclear missile attacks, is essentially canceled following the collapse of the Soviet Union, the election of Bill Clinton to the presidency, and the common-sense fact that it does not work.

Tracking and Data-Relay Communications Satellites: Both the United States and the Soviet Union begin implementation of separate communications satellite systems, designed to improve tracking and data relay capabilities from low-Earth orbiting spacecraft.

1984   **Extravehicular Activity:** In February, 1984, Bruce McCandless becomes the first free-flying astronaut who, by using the Manned Maneuvering Unit (MMU), is able to propel himself in space during STS 41-B.

International Space Station: Design and Uses: This planned Earth-orbiting facility is designed to house experimental payloads, distribute resource utilities, and support permanent human habitation for conducting research in a microgravity environment.

International Space Station: Development: Completion of an international space station is planned for the early 2000's.

International Space Station: Living and Working Accommodations: Plans call for a crew of six persons to live and work both in and out of a habitation module on the International Space Station.

International Space Station: Modules and Nodes: The International Space Station will include the pressurized Habitation Module; three laboratory modules, one built by the United States, one by Europe, and one by Japan; and propulsion modules from Russia.

International Space Station: U.S. Contributions: The International Space Station will be the largest international scientific program in history, drawing on the resources and experience of thirteen nations, led by the United States.

International Sun-Earth Explorers: First spacecraft to perform active experiments on the Sun's effects on Earth's magnetosphere and radiation belts.

National Commission on Space (1984-1986): The group that prepares a report on the long-term space goals, including a "Declaration for Space," is established.

Space Shuttle Flights, 1984: In 1984, the space shuttle program achieves a number of firsts, overcame nagging problems, and displays the ability of humans to play an important role in spaceflight with the capture and return of two failed satellites. The third orbiter, *Discovery*, joins the fleet during STS 41-D, the first flight to include two women.

Space Shuttle Mission STS 41-B (February 3-11): The tenth mission of the space shuttle program is the first to return to Earth from where it was launched, at Kennedy Space Center, and demonstrates the ability of astronauts to maneuver untethered in space.

Space Shuttle Mission STS 41-C (April 6-13): STS 41-C deploys into low-Earth orbit the Long-Duration Exposure Facility to study materials degradation and sees the crew repair the Solar Maximum Mission satellite.

Space Shuttle Mission STS 51-A (November 8-16): During STS 51-A, the crew captures and returns to Earth two satellites that otherwise would be left in a useless orbit.

1985

National Aeronautics and Space Administration: Dr. William R. Graham is acting NASA administrator from December 4, 1985, to May 11, 1986.

Space Shuttle Flights, January-June, 1985: STS 51-C is the first American mission dedicated to a classified Department of Defense payload; STS 51-D carries the first nonpilot, nonscientist astronaut (U.S. Senator Jake Garn, R-Utah). STS 51-B carries the Spacelab 3 science payload into orbit. STS 51-G releases four satellites, retrieves one, and completes important technological and scientific investigations.

Space Shuttle Flights, July-December, 1985: During space shuttle missions STS 51-F, 51-I, 51-J, 61-A, and 61-B many experiments are conducted, satellites are launched, inflight maintenance is completed on a malfunctioning satellite, and astronauts practice assembling large structures in space in preparation for the construction of the International Space Station.

Space Shuttle Mission STS 51-I (August 27-September 3): The twentieth flight in the space shuttle program deploys three communications satellites and crew members capture and repair a malfunctioning satellite.

United States Space Command: Beginning in 1985, the U.S. Space Command begins to provide the needed operational focus across all armed services, consolidating control of space assets and activities in support of nonspace missions.

Voyager 2: Uranus (1985-1986): Voyager 2 is the first spacecraft to collect and return data from the planet Uranus.

1986

National Aeronautics and Space Administration: Dr. James C. Fletcher is NASA administrator from May 12, 1986, to April 8, 1989.

National AeroSpace Plane: The National AeroSpace Plane is a plan to design and build a single-stage-to-orbit vehicle that will land and take off on conventional runways. Terminated in 1994, the NASP program never builds any test devices to demonstrate the capabilities of the various technologies. Large funding requests delay obtaining the required budget from Congress.

Space Shuttle Activity, 1986-1988: Following nine shuttle flights in 1985, fifteen launches were scheduled for 1986 as the United States moves toward its aim of eventually having twenty-four launches annually. When *Challenger* is destroyed during launch early in 1986, killing all on board, the entire space program is reevaluated, causing the cancellation of launches planned for 1986, 1987, and early 1988.

Space Shuttle Mission STS 61-C (January 12-18): Astronauts aboard *Columbia* launch a communications satellite, test a new payload carrier system, and photograph Halley's comet.

Space Shuttle Mission STS 51-L (January 28): STS 51-L experiences a structural breakup caused by a faulty joint seal in its right-hand solid rocket booster 73 seconds after launch, killing its crew of seven and completely destroying the shuttle *Challenger* and its satellite cargo. After analyzing the cause of the accident, the Rogers Commission makes a number of recommendations to NASA. These recommendations fall into several categories, including the design of the SRBs and shuttle management. The goal of the recommendations is to improve the reliability of the entire shuttle. The shuttle program flies 87 successful missions between 1988 and 2003 before experiencing another such catastrophic accident.

Space Stations: Origins and Development (February 20, 1986-March 23, 2001): The Mir Complex was a highly successful Soviet (and later Russian) space station. It was humanity's first permanently inhabited long-term research station in space. Through a number of collaborations, it was made internationally accessible to cosmonauts and astronauts of many different countries. Mir is constructed by connecting several modules, each placed into orbit separately. The Mir Core Module (launched in 1986) provided living quarters and station control. Kvant I (1987) and Kvant II (1989) contained scientific instruments and the crew's shower. Kristall (1990) extended Mir's scientific capabilities. Spektr (1995) served as the living and working space for American astronauts. Priroda (1996) conducted earth remote sensing. The Docking Module (1996) provided a safe and stable port for the space shuttle. It commonly housed three crew members, but it sometimes supported as many as six, for up to a month. Except for two short periods, Mir was continuously occupied until August, 1999. More than 100 cosmonauts, astronauts, and international crew members visited and lived on board the space station. These included visitors from Afghanistan, Austria, Bulgaria, Canada, France, Germany, Great Britain, Japan, and Syria. More than 75 spacewalks were performed, totaling more than 350 hours. The longest-duration spaceflight records are held by Russian Mir residents, including Yuri Romanenko (1987, 326 days); Vladimir Titov and Musa Manarov (1995, 366 days); and Valery Polyakov (1995, 438 days). Polyakov holds the overall endurance record of 679 days in two flights. Russian Elena Kondakova (1995, 169 days) and the American Shannon Lucid (1996, 188 days) set women's spaceflight endurance records.

1987

Lifting Bodies: After more or less being forgotten, the lifting body reappears as the HL-20, also known as ACRV (Assured Crew Return Vehicle), CERV (Crew Emergency Return Vehicle), and PLS (Personnel Launch System). NASA Langley Research Center designs the piloted space plane as a backup to the space shuttle (in case it is abandoned or grounded) and as a CERV from the *Freedom* space station. It is designed for two flight crewpersons and eight passengers to make a piloted landing at an airfield on landing gear. Although it is studied by contractors and a full-size mock-up is built, the design is not selected for further development. The Russian Soyuz spacecraft is designated as the International Space Station CERV.

Mobile Satellite System: In 1987, Motorola engineers Ray Leopold, Ken Peterson, and Barry Bertiger propose the Iridium system, consisting of a constellation of low-orbiting satellites interfacing with existing terrestrial telephone systems around the world through a number of ground station gateways.

Soyuz and Progress Spacecraft (February-July): Soyuz TM-2 travels to the Mir Space Complex, carrying its second permanent expedition crew. It introduces the Soyuz TM (Transport Modification) spacecraft, which features multiple improvements in the design, including the introduction of a new weight-saving computerized flight-control system and improved emergency escape system. These upgrades allow for three crew members wearing pressure suits.

Space Shuttle Mission STS-26: Testing of the new solid rocket booster joints begins as NASA begins the recovery from the *Challenger* accident.

Space Stations: Origins and Development (March 31): The Kvant-1 (meaning "quantum") Astrophysics Module is launched and docks to the Mir Complex. It will be used to measure electromagnetic spectra and x-ray emissions. Two permanent expedition crews, two visiting crews, and seven Progress resupply missions fly to Mir during the year.

1988

Space Shuttle Activity, 1986-1988: The *Challenger* accident puts future launches on hold while a presidential commission headed by William Rogers investigated both the accident and the administration of NASA. No shuttles are launched between January 29, 1986 and September 29, 1988, when *Discovery* (STS-26) is sent aloft to put a communications satellite into geosynchronous orbit. This launch is followed on December 2 by the launching of *Atlantis* (STS-27), which carries a Department of Defense payload into orbit.

Space Shuttle Mission STS-26: The first shuttle flight after the *Challenger* accident, STS-26 launches a vital communications satellite into geosynchronous orbit and returns the U.S. piloted space program to active flight status.

Space Stations: Origins and Development: One permanent expedition crew, three visiting crews, and six Progress resupply missions fly to Mir during the year.

1989

Galileo: Jupiter: Galileo has provided more extensive sampling of Jupiter's outer atmosphere, and far closer encounters of its principal satellites, than previous missions.

Magellan: Venus (1989-1994): Magellan mapped 99 percent of Venus's surface using powerful radar imaging instruments.

National Aeronautics and Space Administration: Richard H. Truly is NASA administrator from May 14, 1989, to March 31, 1992.

Nuclear Energy in Space (October 18): The Galileo probe, carrying 49 pounds of Pu-238, becomes the first radioisotopic thermoelectric generator (RTG)-powered mission to be launched by a space shuttle, *Atlantis*.

Space Shuttle Flights, 1989: Space shuttle crews orbit the Earth for more than four weeks during 1989 as NASA successfully launches and lands five space shuttle missions.

Space Stations: Origins and Development (November 26): The Kvant-2 ("quantum") Module is launched and docks to the Mir Complex. Designed to host life science, materials science, and Earth observation experiments, it also carries a substantial amount of equipment to improve living conditions and operations of the Mir complex, including the Elektron electrolysis system to provide oxygen from recycled water, a water supply system, two water regeneration systems, sanitation facilities, a shower, and an airlock compartment. One permanent expedition crew and four Progress resupply missions fly to Mir during the year.

Voyager 2: Neptune: In 1989, Voyager 2 completes the mission of the originally proposed Grand Tour of the outer solar system, having flown by Jupiter, Saturn, Uranus, and Neptune, on a trajectory that would ultimately take it beyond the solar system.

1990

Hubble Space Telescope: The largest optical astronomical observatory to orbit the Earth is deployed.

Mission to Planet Earth: This long-term NASA program studies how the global environment is changing.

Mobile Satellite System: In 1990, the Iridium system is announced at simultaneous press conferences in Beijing, London, Melbourne, and New York City. At these conferences, Iridium unveils its revolutionary concept for global personal communications: to link existing terrestrial telephone networks, using the Iridium satellite constellation as a base.

Nuclear Energy in Space (October): The Ulysses mission is launched. For this mission, a single radioisotopic thermoelectric generator (RTG) provides all the power for instruments and other equipment, and will do so for several years, as the Sun and interstellar space are explored.

Pegasus Launch Vehicles: Pegasus is a satellite launch vehicle, developed by the Orbital Sciences Corporation, which uses a novel air-launching system. Pegasus is dropped from an aircraft, flying at about 12 kilometers, and serves as a low-cost, recoverable first stage, reducing the cost of satellite launching compared to that of conventional, ground-launched rockets.

Space Shuttle Flights, 1990: NASA flies six space shuttle flights in 1990, highlighted by STS-31, the flight deploying the Hubble Space Telescope in orbit.

Space Stations: Origins and Development (May 31): The Kristall ("crystal") Module is launched to the Mir Complex to develop biological and materials production technologies in the space environment. One component of the module is a radial docking port. Three permanent expedition crews and four Progress resupply missions fly to Mir during the year.

Ulysses: Solar-Polar Mission: This joint mission between NASA and the European Space Agency is the first spacecraft to explore interplanetary space out of the plane of the ecliptic.

1991

Asteroid and Comet Exploration (October 29): The Galileo spacecraft flies within 1,600 kilometers of Gaspra at a relative speed of 8 kilometers per hour. While crossing the asteroid belt on its way to enter orbit around Jupiter, Galileo became the very first robotic probe to encounter an asteroid at relatively close range and return images of the pockmarked surface of the body. This asteroid was very irregular in shape (20 by 12 by 11 kilometers), heavily cratered, and covered with a thin layer of dust and rubble.

Compton Gamma Ray Observatory: The most massive robotic civilian spacecraft the United States has ever built provides dramatic new insights into some of the highest-energy phenomena in the universe.

Dynamics Explorers: NASA officially retires the Dynamics Explorer-1 (DE-1) satellite, which acquired the first global images of the aurora, on February 28, after nine years of collecting scientific data. Designed to operate for three years, DE-1 performed for nearly a decade in space.

Space Shuttle Flights, 1991: Three missions are flown during which satellites are deployed; astronauts conduct experiments to determine the physiological effect of spaceflight and demonstrate techniques for constructing the space station.

Space Shuttle: Life Science Laboratories: STS-40 is launched on June 5. This is the fifth dedicated Spacelab mission, carrying Spacelab Life Sciences-1 (SLS-1), and the first Spacelab mission dedicated solely to life sciences. The mission features the most detailed and interrelated physiological measurements in space since the 1973-1974 Skylab missions. The subjects are humans, thirty rodents, and thousands of tiny jellyfish. The primary SLS-1 experiments study six body systems. Of eighteen investigations, ten experiments involve humans, seven involve rodents, and one uses jellyfish.

Space Stations: Origins and Development: Two permanent expedition crews and five Progress resupply missions fly to Mir during the year.

1992

Atmospheric Laboratory for Applications and Science (1992-1994): A package of instruments to record data relating to the upper atmosphere, especially ozone depletion, and irradiance from the Sun and outer space is deployed from the space shuttle.

Space Shuttle Flights, 1992: The eight missions of 1992 see the longest shuttle mission and the most spacewalks to date; *Endeavour* replaces *Challenger.*

Extreme Ultraviolet Explorer (June 7): The Extreme Ultraviolet Explorer (EUVE) unveils the universe in the last remaining portion of the electromagnetic spectrum, the shortest ultraviolet wavelengths. Since its 1992 launch, EUVE has studied a wide range of astronomical objects at these wavelengths. This work is impossible to conduct from Earth's surface, and prior to the mission, many astronomers thought that extreme ultraviolet astronomy was not possible at all.

Mars Observer (1992-1993): Mars Observer was to have photographed the Martian surface and to have carried out scientific experiments on the Martian atmosphere, but it was unsuccessful.

National Aeronautics and Space Administration: Daniel S. Goldin is NASA administrator from April 1, 1992, to November 17, 2001.

Pioneer Venus 1 (October 8): Pioneer Venus 1 ceases operation when radio contact is lost. During its fourteen-year life, it obtained important information on Venus's topography and atmosphere.

Small Explorer Program: The first satellite in the SMEX program is launched on July 3, only three years after the program was announced. The Solar Anomalous and Magnetospheric Particle Explorer (SAMPEX) is placed into orbit by a Scout rocket fired from Vandenberg Air Force Base in California. The SAMPEX spacecraft weighs 157 kilograms and carries 40 kilograms of instruments that consume an average of 22 watts of electric power. The SAMPEX satellite carries four instruments that monitor the charged particles emitted by the Sun, in order to investigate the isotopic composition of these particles. SAMPEX uses a novel, momentum reaction system to keep its instruments pointed at the Sun.

Space Shuttle Mission STS-49: The newest space shuttle orbiter, *Endeavour,* sees its maiden voyage on May 7.

Space Shuttle: Life Science Laboratories: STS-42 is launched on January 22. The primary payload is the International Microgravity Laboratory-1 (IML-1). The IML-1 is making its first flight and used the pressurized Spacelab module. The international crew is divided into two teams for around-the-clock research on the human nervous system's adaptation to low gravity and the effects of microgravity on other lifeforms, such as shrimp eggs, lentil seedlings, fruit fly eggs, and bacteria.

Space Shuttle: Microgravity Laboratories and Payloads: STS-42 is launched on January 22. The primary payload of STS-42 is the International Microgravity Laboratory-1 (IML-1), making its first flight and using the pressurized Spacelab module. Materials processing experiments are also conducted. These include crystal growth from a variety of substances such as enzymes, mercury iodide, and a virus.

Space Shuttle: Microgravity Laboratories and Payloads: STS-50 begins its flight on June 25. The primary payload is the United States Microgravity Laboratory-1 (USML-1) making its first flight. The experiments conducted include the Crystal Growth Furnace, the Drop Physics Module, Surface-Tension-Driven Convection Experiments, Zeolite Crystal Growth, Protein Crystal Growth, and the Space Acceleration Measurement System (SAMS). The SAMS is designed to provide information about the acceleration environment in which the other USML-1 experiments are being conducted.

Space Shuttle: Microgravity Laboratories and Payloads: STS-52 is launched on October 22. One of the two primary objectives of STS-52 is the operation of the United States Microgravity Payload-1 (USMP-1). USMP-1 includes three experiments mounted on two connected structures that are mounted in the cargo bay. USMP-1 experiments include the Lambda Point Experiment and Material for the Study of Interesting Solidification Phenomena on Earth and in Orbit (MEPHISTO). MEPHISTO is a cooperative program between NASA and the French space agency, CNES, which also built the payload.

Space Stations: Origins and Development: Two permanent expedition crews and five Progress resupply missions fly to Mir during the year.

1993   **Asteroid and Comet Exploration (August 28):** The Galileo spacecraft encounters a large asteroid named Ida (longest dimension, 55 kilometers), which shows an older surface than the previously encountered asteroid Gaspra and which sports a 1.5-kilometer-diameter small moon, Dactyl. Galileo detects magnetic fields originating from both these asteroids.

Space Shuttle Flights, 1993: The seven space shuttle missions of 1993 include science missions, a mission to retrieve a satellite left in space a year earlier, missions to deploy new satellites, and the mission to service and repair the Hubble Space Telescope.

Space Shuttle: Life Science Laboratories: STS-58 goes into orbit on October 18. This is the second dedicated Spacelab Life Sciences mission (SLS-2). Fourteen experiments are conducted in four areas: regulatory physiology, cardiovascular/cardiopulmonary, musculoskeletal, and neuroscience. Eight experiments focus on the crew. Six other experiments focus on forty-eight rodents. The crew collected more than 650 different samples from themselves and rodents, thus increasing the statistical base for life sciences research.

Space Shuttle Mission STS-61: This important mission repairs the Hubble Space Telescope, demonstrating the value of piloted spaceflight.

Space Stations: Origins and Development: Two permanent expedition crews and five Progress resupply missions fly to Mir during the year.

Strategic Defense Initiative: The Strategic Defense Initiative comes to an inauspicious conclusion with the inauguration of Bill Clinton as president of the United States. The cancellation is made possible by the collapse of the Soviet Union, by a variety of strategic arms limitations treaties, and by the common-sense fact that it did not work.

1994   **Clementine Mission to the Moon:** A new generation of spacecraft, designed to employ new technologies within a low-cost budget, accomplishes the first complete mapping of the Moon.

Manned Maneuvering Unit: The Simplified Aid For Extravehicular Activity Rescue (SAFER) is first tested on STS-64 in September, ten years after the last Manned Maneuvering Unit mission. Astronauts Mark Lee and Carl Meade perform an engineering evaluation, an EVA self-rescue demonstration, and an overall flight quality evaluation, which included a demonstration of precision flying by tracking the Remote Manipulator System arm.

National AeroSpace Plane: The National AeroSpace Plane program fails to build any test devices to demonstrate the capabilities of the various technologies. Large funding requests delay obtaining the required budget from Congress. The NASP study is terminated when it is concluded that the high-temperature materials and air-breathing propulsion technology required for such prolonged high speeds within Earth's atmosphere would take many more years to mature than had originally been estimated.

Space Shuttle Flights, 1994: Seven space shuttle missions, all dedicated to science, are launched.

Space Shuttle: Microgravity Laboratories and Payloads: STS-62, the first Extended Duration Orbiter, is launched on March 4. Two of the USMP-2 experiments focus on directional solidification, a well-known industrial process for making semiconductors and metals. The goal of the Advanced Automated Directional Solidification Furnace was to exploit the microgravity environment of space to gain understanding of the effects of gravitational forces on the material properties of semiconductors. The other directional solidification experiment, Material for the Study of Interesting Solidification Phenomena on Earth and in Orbit (MEPHISTO), flew on the first USMP as well.

Space Shuttle: Radar Imaging Laboratories: The Space Radar Imaging Laboratory-1 (SRL-1) is launched aboard the space shuttle *Endeavour* (STS-59) on April 9. The SRL-1 comprises two elements: a suite of radar instruments called Spaceborne Imaging Radar-C/X-Band Synthetic Aperture Radar (SIR-C/X-SAR) and the Measurement of Air Pollution from Satellite (MAPS) instrument.

Space Shuttle: Radar Imaging Laboratories: The Space Radar Imaging Laboratory-2 (SRL-2) is launched on the space shuttle *Endeavour* (STS-68) on September 30. Flying SRL-2 during different seasons allowed comparison of changes between first and second flights.

Space Stations: Origins and Development: Three permanent expedition crews and five Progress resupply missions fly to Mir during the year.

1995

Explorers: Astronomy: The Rossi X-Ray Timing Explorer (RXTE) is launched on December 30. It is designed to look at cosmic x-ray sources at short timescales over a broad energy range. Astronomers can learn about very fast phenomena such as the flickering of matter as it falls into a black hole or the rotation of disks of matter around a neutron star. They can also look at the longer-term variability of sources, such as the precession of orbiting neutron stars or accretion disks. RXTE was designed for a required lifetime of two years, with a goal of five years.

Hubble Space Telescope: Science: The Hubble Space Telescope finds evidence of a primordial reservoir of 200 million comets at the edge of the solar system, proving the existence of the Kuiper Belt, which was proposed in the early 1960's. Here, astronomers see forty large-sized proto-comets that were not previously visible. This region will help scientists understand the formation of the solar system.

International Space Station: Crew Return Vehicles: A lifeboat in space, the X-38 crew return vehicle affords astronauts aboard the International Space Station an escape route in case an accident renders the station uninhabitable and can serve as an ambulance to carry injured astronauts to an Earthside hospital. On April 29, 2002, NASA announces the cancellation of the X-38 program due to budget pressures associated with the International Space Station. The X-38 was two years short of completing its flight test phase.

Lifting Bodies: When doubts about the availability of the Soyuz spacecraft developed in 1995, NASA proceeds with the design of the X-38. A NASA Johnson concept, it is a smaller version of the X-24 lifting body with a parafoil to assist in landing. The one key difference is that the X-38 does not land conventionally on wheels on a runway. Instead, the X-38 deploys a rectangular parachute as it nears the ground and floats to a landing. This method is chosen because NASA learned during its earlier lifting body research that lifting bodies have unsafe landing characteristics.

Pioneer 11: The last communication from Pioneer 11 is received in November, shortly before the Earth's motion carries it out of view of the spacecraft antenna. The spacecraft is headed toward the constellation of Aquila (The Eagle), northwest of the constellation of Sagittarius. Pioneer 11 may pass near one of the stars in the constellation in about 4 million years.

Solar and Heliospheric Observatory: Armed with twelve instruments to study the entire Sun in detail, the Solar and Heliospheric Observatory (SOHO) travels to an orbit 1.5 million kilometers inside Earth's orbit. This stable vantage point allows SOHO to study the Sun continuously, without the regular interruptions plaguing Earth-orbiting satellites as they pass through Earth's shadow.

Space Shuttle Flights, 1995: Of the year's seven missions, three involve a rendezvous with the Russian space station Mir.

Space Shuttle Mission STS-63: STS-63 begins a phase of cooperation between NASA and the Russian Space Agency, in which techniques for joint operations between the space shuttle and Mir demonstrated the feasibility of an International Space Station.

Space Shuttle-Mir: Joint Missions (1995-1998): Between March, 1995, and May, 1998, the Russian space station Mir hosts a series of NASA astronauts as crew members. The Phase I program operates under a complicated logistical scheme. In the history of human spaceflight, no previous program has required so many transport vehicles, so much interdependent operation between organizations, and so much good timing. Shuttle-Mir experience gave participants an opportunity to gear up for the formidable cooperative effort the International Space Station requires. Norman Thagard becomes the first American to ride into space aboard the Russian Soyuz spacecraft. He and fellow Soyuz TM-21 crewmates Vladimir Dezhurov and Gennady Strekalov become the first persons to be launched in one type of spacecraft and return in another when they ride the space shuttle back to Earth. Shannon Lucid sets a record for women and for Americans of either gender for time spent in space on a single mission: eighty-eight days.

Space Shuttle Mission STS-71/Mir Primary Expedition 18: STS-71 conducts the first U.S. docking with Mir, bringing supplies and a new crew from Earth, returning materials and the old crew, and gaining experience in joint operations in space.

Space Shuttle: Microgravity Laboratories and Payloads: STS-73 is launched on October 20. It is the second flight of the USML and builds on the foundation of its predecessor, which flew on *Columbia* during mission STS-50 in 1992. Research of the USML-2 is concentrated within the same overall areas of USML-1, with many experiments flying for the second time.

Space Stations: Origins and Development (June 1): The Spektr ("spectrum") Module, the fifth Mir component to attain orbit, was primarily designed for Earth observation (specifically, natural resources and atmospheric studies). The module also carried US/Russian equipment for material science, biotechnology, life sciences, and space technology studies. A small airlock and manipulator arm were available to attach small external experiments and deploy small satellites. The module's total pressurized volume was 62 cubic meters. Two permanent expedition crews, two space shuttle visiting crews, and five Progress resupply missions fly to Mir during the year.

1996

Asteroid and Comet Exploration: The Near-Earth Asteroid Rendezvous (NEAR) mission is launched on February 17. NEAR's mission is to help answer many fundamental questions about the nature and origin of asteroids and comets.

Hubble Space Telescope: Science: HST's Deep Field Imaging Telescope takes pictures of galaxies that had never before been seen. Some galaxies "seen" may be between ten and fourteen billion years old, around the commonly accepted time that the universe began. Astronomers were shocked that galaxies were formed so early in the history of the universe.

International Ultraviolet Explorer: The International Ultraviolet Explorer is shut down on September 30. It produces more published scientific papers than any other satellite. It provides information about physical conditions in the central regions of distant galaxies that may contain black holes. It also provides scientists with more knowledge of the physical conditions in very hot stars, the effect of solar winds on the atmospheres of the planets in our solar system, and the loss of mass from stars when stellar winds and flares occur.

Manned Maneuvering Unit: While docked to the Mir Space Station in March, astronauts Linda Godwin and Michael Clifford attach four experiments, known collectively as the Mir Environmental Effects Payload (MEEP), on the outside of the Mir Docking Module. As a precaution, each wore a Simplified Aid for Extravehicular Activity Rescue (SAFER) pack.

Space Shuttle Flights, 1996: The space shuttle flies seven times during the year, continuing its mission to the Russian space station Mir. Astronaut Shannon Lucid sets a record for women living in space with her 181 days on Mir. Other flights are devoted to microgravity studies, life sciences, a reflight of the tethered satellite, and ultraviolet spectrographic observations.

Space Shuttle: Microgravity Laboratories and Payloads: STS-75 sets off on February 22. The USMP-3, flying on the shuttle for the third time, includes U.S. and international experiments, all of which have flown at least once before. The experiments are the Advanced Automated Directional Solidification Furnace, Critical Fluid Light Scattering Experiment, Isothermal Dendritic Growth Experiment, and Material for the Study of Interesting Solidification Phenomena on Earth and in Orbit (MEPHISTO).

Space Shuttle: Life Science Laboratories: STS-78 lifts off on June 20. Five space agencies, NASA, the European Space Agency, the French Space Agency, the Canadian Space Agency, and the Italian Space Agency, and research scientists from ten countries worked together on the primary payload of STS-78, the Life and Microgravity Spacelab (LMS).

Space Stations: Origins and Development (April 23): Priroda ("nature") Module, designed to study the atmosphere and oceans, with an emphasis on pollution and other environmental impact of human activities on them, is launched to the Mir Complex. Two permanent expedition crews, two space shuttle visiting crews, and three Progress resupply missions fly to Mir during the year.

Tethered Satellite System: NASA again flies the Tethered Satellite System on STS-75. The mission is launched on February 22, and lasts approximately 16 days. The 20.6-kilometer tether is deployed from *Columbia*'s cargo bay on the third day of the flight. Approximately five hours into the deployment at a length of 19.7 kilometers, the tether unexplainably snaps and separates rapidly from the shuttle. A rendezvous and recovery operation is planned, but insufficient quantities of propellant onboard *Columbia* prevent it. The satellite reenters the atmosphere on March 19, over the Atlantic Ocean near Africa. There are no plans to attempt another satellite flight in the near future.

United Space Alliance: In 1996, the National Aeronautics and Space Administration (NASA) cedes the responsibility for launching space shuttles and training astronauts to the United Space Alliance (USA) under a six-year, seven-billion-dollar contract. This results in cutting costs and freeing NASA to pursue more vigorously research and development relating to its shuttle program in preparation for constructing the International Space Station.

1996-1997   **Mars Pathfinder:** Mars Pathfinder is a spectacularly successful mission to Mars, which sends back tens of thousands of close-up pictures of the Martian surface. It consists of a lander (Pathfinder) and a rover (Sojourner), which analyze the surface rocks and the soil, and provide extensive data on the thin Martian atmosphere and weather patterns.

1997   **Asteroid and Comet Exploration:** On June 27, 1997, Near Earth Asteroid Rendezvous (NEAR) makes a 1,200-kilometer flyby of asteroid 243 Mathilde, the closest ever of an asteroid, and radios back more than five hundred pictures. On December 23, 1998, NEAR makes a similar flyby of asteroid 433 Eros.

Cassini: Saturn (1997-2008): The Cassini mission begins on October 15, during a one-month launch window with the most favorable arrangement of planets for the trip to Saturn. After a twenty-minute burn, the Titan IV-Centaur rocket inserts Cassini-Huygens into a 445-kilometer-high parking orbit. After separation from the Centaur upper stage, the spacecraft establishes communications with the Deep Space Network, receives thorough systems checks, and fires its main engines to leave orbit on a course for Saturn.

Hubble Telescope Servicing Missions: STS-82, representing the first HST-dedicated servicing mission, is launched. Scientific equipment that had gathered highly valuable data since installation either preflight or on STS-61 is replaced with new detectors. One particularly is meant to expand HST's investigations into the infrared region of the electromagnetic spectrum. Some of HST's systems are upgraded, and some temporary patchwork repairs are made to the telescope's thermal blankets.

Huygens Lander: The combined Cassini-Huygens spacecraft is launched from Earth on October 15, 1997. Titan, the largest moon in the Saturn system and the second largest moon in the solar system, is shrouded from telescopic observation in visible wavelengths. The only ways to determine surface information are an imaging radar or a probe that enters the atmosphere and proceeds to the surface. The former is the job of the Cassini orbiter, the latter is the job of the Huygens probe that Cassini transported to the Saturn system.

Mars Global Surveyor: The Mars Global Surveyor (MGS) was designed to spend more than a full Martian year (687 Earth days) mapping the surface of Mars with unprecedented accuracy, making high-resolution images, observing short-term and seasonal changes on the surface and in the atmosphere, and measuring gravitational and magnetic properties of Mars.

Mobile Satellite System: Iridium successfully places forty-seven satellites into orbit. Its board members receive the first Iridium pager message delivered by orbiting satellites. Iridium selected AlliedSignal as its aeronautical strategic partner to develop global wireless telecommunications for aircraft passengers and crew.

Nuclear Energy in Space: The Cassini mission to Saturn, carrying three radioisotopic thermoelectric generators (RTGs) with a total of 72 pounds of Pu-238, is launched.

Space Shuttle Flights, 1997: In 1997, NASA conducts eight piloted Earth-orbiting spaceflights. Each mission obtains important information relating to possible advancements in the quality of human life on Earth, including the efficacy of modern medicine and technology and the development of an international space station.

Space Shuttle: Microgravity Laboratories and Payloads: STS-83 is launched on April 4. The first flight of the Microgravity Science Laboratory-1 (MSL-1) is cut short due to concerns about one of three fuel cells, marking only the third time in shuttle program history that a mission ends early. The crew is able to conduct some experiments in the MSL-1 Spacelab Module despite the early return. Work is performed in the German electromagnetic levitation furnace facility on an experiment that studied the amount of undercooling that can be achieved before solidification occurs. Another experiment conducted is the Liquid-Phase Sintering II experiment in the large isothermal furnace. This investigation uses heat and pressure to test theories about how a liquefied substance bonds with the solid particles of a mixture without reaching the melting point at which a new alloy combination is created.

Space Shuttle: Microgravity Laboratories and Payloads: STS-94 is launched on July 1 and marks the first reflight of the same vehicle, crew, and payloads. The repeat mission is due to the shortened STS-83 mission in April, 1997. The crew maintains twenty-four-hour, two-shift operations.

Space Shuttle: Microgravity Laboratories and Payloads: The fourth flight of the USMP focuses on materials science, combustion science, and fundamental physics and is launched on November 19. With Material for the Study of Interesting Solidification Phenomena on Earth and in Orbit (MEPHISTO), researchers are able to separate for the first time two separate processes of solidification. The Particle Engulfment and Pushing by a Solid/Liquid Interface (PEP) experiment, conducted with the glovebox facility, examines the solidification of liquid metal alloys. For the first time, researchers observe large clusters of particles being pushed, forcing them to reassess theories of how alloys solidify.

Space Stations: Origins and Development (February 23 and June 25): The Mir Complex is evacuated twice during the year, because of an onboard fire and a collision between a Progress resupply vessel and the Spektr Module. Two permanent expedition crews, three space shuttle visiting crews, and four Progress resupply missions fly to Mir during the year.

Spacelab Program: The Spacelab Program ends when NASA redirects its efforts toward the International Space Station. Between 1983 and 1997, twenty-four Spacelab missions were flown successfully.

1998

Astronauts and the U.S. Astronaut Program: Barbara Morgan, Backup Teacher-in-Space to Christa McAuliffe, joins twenty-four men and women in the astronaut class of 1998. This year's class consists of eight pilot and seventeen mission specialist candidates. Of the twenty-five class members, twenty-one are male and four are female. In January, Morgan is named the first Educator Mission Specialist.

International Space Station: Living and Working Accommodations: The 19,323-kilogram pressurized module Zarya ("sunrise" in Russian) is launched atop a Russian Proton 8K82K vehicle on November 20, 1998. The Zarya Control Module, known in Russia as the Functional Cargo Block (in Russian, *Funktsionalya-gruzovod blokor*, or FGB) was funded by NASA and built by Khrunichev in Moscow under subcontract from Boeing for NASA. In early December, 1998, the STS-88 (ISS-2A) mission sees space shuttle *Endeavour* attach the Unity Module to Zarya, initiating the first ISS assembly sequence. Unity was launched passive with two Pressurized Mating Adapters (PMAs) attached and one stowage rack installed inside.

International Space Station: 1998: Despite financial problems, in 1998 the Russian Space Agency launches the Zarya Control Module, the first component of the International Space Station, and soon afterward NASA orbits the Unity connecting module. Space shuttle *Endeavour* crew members join the two components during the STS-88 mission.

Manned Maneuvering Unit: Astronaut Scott Parazynski and Russian cosmonaut Vladimir Titov test the first flight production model of the Simplified Aid for Extravehicular Activity Rescue (SAFER) on STS-86 during the September mission. During the third STS-88 spacewalk to assemble the International Space Station in December, astronaut Jerry Ross achieves only 50 percent of the evaluation objectives for SAFER. Still, the tests are considered successful, and SAFER will be worn by astronauts during the station's construction.

Mobile Satellite System: Iridium completes the constellation of sixty-six low-Earth orbit (LEO) satellites with 100 percent launch success, and following extensive testing, the Iridium system enters commercial service on November 1.

Space Shuttle Flights, 1998: Space shuttle flights in 1998 consist of missions STS-89, STS-90, STS-91, STS-95, and STS-88. Missions to Mir, which include studies relating to solar winds, and experiments on the effects of microgravity environments on biological systems, continue. Astronaut John Glenn returns to space. The first Space Station Assembly Flight is flown.

Space Shuttle: Life Science Laboratories: STS-90 (Neurolab) is launched on April 17. The Neurolab's twenty-six experiments target one of the most complex and least understood parts of the human body: the nervous system. Primary goals are to conduct basic research in space neurosciences and expand our understanding of how the nervous system develops and functions in space. The test subjects are crew members, rats, mice, crickets, snails, and two kinds of fish. The work is a cooperative effort of NASA, several domestic partners, and the space agencies of Canada, France, and Germany as well as the European Space Agency and the National Space Development Agency of Japan. Most experiments are conducted in the pressurized Spacelab long module located in *Columbia*'s payload bay, and many are the same as those carried out during the flights of STS-58 and STS-78.

Space Shuttle Mission STS-95: John H. Glenn, Jr., the first American to orbit the Earth, left NASA in the mid-1960's to enter the worlds of business and politics, eventually becoming a U.S. senator from his native Ohio. Yet he always desired to return to orbit. At age seventy-seven, Glenn fulfilled that dream on shuttle mission STS-95.

Space Stations: Origins and Development: Two permanent expedition crews, two space shuttle visiting crews, and three Progress resupply missions fly to Mir during the year.

1998-1999   **Lunar Prospector:** While collecting data from orbit around the Moon, the Lunar Prospector's instruments detect high levels of hydrogen. The hydrogen seems to be located in dark craters caused by the impact of comets in years past. NASA scientists, along with engineers and astronomers from the University of Texas, decide to perform a controlled crash with the Prospector to investigate whether the detected hydrogen exists in the form of ice or in some other form.

Mars Climate Orbiter and Mars Polar Lander: As part of the Mars Surveyor Program, NASA's long-term Mars exploration program, the Mars Climate Orbiter and the Mars Polar Lander are sent to arrive at Mars in late 1999. In order to provide evidence for or against the existence of life on Mars they were to study Mars's climate and history as well as the nature and extent of its resources. Unfortunately, both missions fail during their encounters with the Red Planet.

Space Centers, Spaceports, and Launch Sites (January, 1998): The Alaska Aerospace Development Corporation begins building a commercial spaceport at Narrow Cape on Kodiak Island, about 400 kilometers south of Anchorage and 40 kilometers southwest of the City of Kodiak. The location, combined with low-cost operations, was promoted for launching telecommunications, remote sensing, and space science payloads of up to 3,500 kg into low-Earth polar orbits. Through an agreement with the State Division of Land, AADC was granted a thirty-year lease of 1,200 hectares at Narrow Cape with an option for a second thirty-year term. On November 5, 1998, the USAF conducts the launch of the AIT (atmospheric interceptor technology) suborbital rocket, marking the first launch from the Kodiak Launch Complex (KLC).

1999   **Asteroid and Comet Exploration:** The Stardust mission is launched in February 7, 1999, to encounter the comet Wild 2 in January, 2004, at a distance of about 400 million miles from Earth.

Chandra X-Ray Observatory and Space Shuttle Mission STS-93: The Chandra X-Ray Observatory, the third space-based telescope in NASA's Great Observatories Program, is deployed from the space shuttle *Columbia* on July 23, during the STS-93 mission. This flight is the first shuttle mission commanded by a woman, Eileen Collins.

Earth Observing System Satellites: The flagship of the EOS series, Terra (originally called EOS AM-1), is launched from Vandenberg Air Force Base in California on December 18, 1999. It follows a Sun-synchronous, polar orbit (that is, one that passes near the Earth's poles and carries the satellite over any given point on the planet's surface at the same local time on each pass) approximately 705 kilometers above the Earth's surface.

Earth Observing System Satellites: The ACRIMSAT spacecraft is launched December 20, 1999, and carries the Active Cavity Radiometer Irradiance Monitor III (ACRIM III) instrument. ACRIM III provided precise measurements of the total amount of the Sun's energy that falls on Earth's land surface, oceans, and atmosphere.

Far Ultraviolet Spectroscopic Explorer: The Far Ultraviolet Spectroscopic Explorer (FUSE) is launched on June 24, 1999, atop a Boeing Delta II rocket from Cape Canaveral, Florida, to an orbit about 768 kilometers above Earth's surface. FUSE is an orbital telescope designed to study wavelengths of ultraviolet light that are inaccessible from either the ground or the Hubble Space Telescope. FUSE is approximately ten-thousand times more sensitive than the only previous satellite to study these wavelengths. It investigates questions related to our origins: What were the conditions just after the Big Bang, what are the properties of interstellar clouds that will eventually form solar systems, and how do galaxies evolve?

Hubble Telescope Servicing Missions STS-103: The Third HST Servicing Mission is launched. It had been placed in the shuttle schedule as an opportunity to change out scientific equipment and make some repairs, but its urgency increased when HST suffered unexpected failures in its gyroscope pointing system.

International Space Station: 1999: Delays in assembling and launching components disrupt the International Space Station's construction schedule and heighten the controversy surrounding it. Nevertheless, STS-96 crew members of the space shuttle *Discovery* add new equipment, repair malfunctions, and supply the station; ground controllers resolve a series of minor threats to the station.

International Space Station: Living and Working Accommodations: Two shuttle missions—STS-96 (ISS-2A.1) in June, 1999, and STS-101 (ISS-2A.2a) in May, 2000—supply the Zarya and Unity modules with tools and cranes, and deliver provisions in preparation for the arrival of the Zvezda Service Module and the station's first permanent crew. Zvezda (Russian for "star"), built and financed by Russia, docks with the ISS on July 26 and becomes the third major component of the station.

Landsat 7: The Landsat 7 mission is a continuation of the Landsat Program, which since 1972 has provided digital images of the Earth's land surface and coastal areas to a diverse group of users. Landsat 7 is the first of these platforms to feature the Enhanced Thematic Mapper Plus (ETM+) instrument.

New Millennium Program: On July 28, 1999, Deep Space 1 flies to within 26 kilometers of asteroid 9969 Braille. Its infrared sensor indicates that Braille is similar to Vesta, one of the largest members of the asteroid belt between Mars and Jupiter. The Braille flyby uses a new space autopilot system called AutoNav, and its successful piloting of DS1 further validates the technologies on board the spacecraft.

New Millennium Program: The two miniature probes, carrying ten experimental technologies each, fail to respond to communication efforts by NASA engineers. The probes, which piggybacked aboard the Mars Polar Lander, slam into Martian soil December 3, 1999.

Space Shuttle Flights, 1999: The year 1999 witnesses many advances in space exploration along with several setbacks, the most notable being a decrease in the number of missions undertaken. Advances include learning about working rapidly in a weightless environment, successfully launching the Chandra X-Ray Observatory, and deploying a base for the Russian construction crane on the International Space Station. During this time NASA's focus shifts from saving money to making safety its highest priority.

Space Shuttle Mission STS-93: STS-93 achieves two landmark distinctions: (1) the first flight of a female space shuttle commander and (2) orbital deployment of the Advanced X-Ray Astrophysics Facility (AXAF), the third space-based telescope in the National Aeronautics and Space Administration's (NASA's) Great Observatory Program.

Space Stations: Origins and Development: One permanent expedition crew and two Progress resupply missions fly to Mir during the year, as activities switch to the International Space Station.

Stardust Project (1999-2006): Stardust, an interplanetary spacecraft, is launched on February 7, 1999. Its purpose is to investigate the makeup of the comet Wild-2 and its coma. Stardust is the first United States space mission designed from the beginning to study a comet. It is only the second sample return mission since the Apollo missions to the Moon. Stardust will return samples of Comet Wild-2 and samples of an interplanetary dust stream, making these the first samples returned from beyond the region of the inner solar system near Earth as well as possibly the first samples originating from beyond the solar system.

2000

Asteroid and Comet Exploration: On February 14, 2000, NEAR successfully rendezvoused with 433 Eros and begins a year-long orbit to map the asteroid's surface and determine the asteroid's mass, structure, geology, chemical composition, rotational characteristics, gravity, and magnetic field. In 2000, the NEAR project acquires a more personal surname: Shoemaker. Gene Shoemaker was a pioneer in the science of astrogeology. He had trained astronauts who landed on the Moon in lunar geology and fieldwork to maximize the scientific returns from the Apollo lunar landings. Later he and his wife Carolyn championed the search for near-Earth objects in order to alert the world to hazards from potential impacts by these bodies. In 1997, he died in a road accident while on a research trip in Australia. The new name for the project, NEAR Shoemaker, is in his honor.

Compton Gamma Ray Observatory: The Compton Gamma Ray Observatory reenters on June 4. It breaks up, as planned, and the pieces splash into the Pacific Ocean. In the first deliberate and controlled crash of a satellite, NASA engineers direct the Compton through a series of rocket firings that drop it from a high orbit and send it plunging to Earth.

Galileo: Jupiter: Galileo continues to gather data from the Jovian system nearly five years after its arrival. A large maneuver in March, 1996, raised the inner end of Galileo's orbit away from Jupiter's hazardous radiation belts. The first encounters of the Jovian moons Ganymede and Io began June 27, 1996, and a second Ganymede encounter on September 6, 1996. The first encounter with Callisto occurred on November 2, 1996, and with Europa on

December 19, 1996. Each encounter involved a one-week, high-rate observation of Jupiter and at least one satellite. Each flyby brought Galileo to within a few hundred kilometers of the satellites and gave it a gravity assist into the next orbit. In January, 1997, Galileo and Jupiter entered another superior conjunction, after which the orbiter continued its close flybys for another year. On February 20, 1997, the Galileo orbiter encountered Europa for a second time. It encountered Ganymede on April 5, 1997, at a distance of only 3,095 kilometers, nineteen times closer than Voyager 2 and, again, on May 7, 1997. This time it got within 1,600 kilometers of the satellite—thirty-seven times closer than Voyager 2. On June 25, 1997, the probe glided to within 415 kilometers of Callisto. The encounter was repeated on September 17, 1997. Between November 6, 1997, and February 1, 1999, Galileo played tag with Europa nine times, swooping down on its icy surface. In May 5, 1999, Galileo began another four-visit tour of Callisto, ending on September 16, 1999. In late 1999 and early 2000, near the end of its two-year mission extension known as the Galileo Europa Mission, the Galileo spacecraft dipped closer to Jupiter than it had been since it first went into orbit around the giant planet in 1995. These maneuvers allowed Galileo to make three flybys of the volcanically active moon Io and also made possible new high-quality images of Thebe, Amalthea, and Metis, which lie very close to Jupiter, inside the orbit of Io. Volcanic calderas, lava flows, and cliffs could be seen in a false-color image of a region near the south pole of Jupiter's volcanic moon Io. Combining a black and white image taken by the Galileo spacecraft on February 22, 2000, with lower-resolution color images taken by Galileo on July 3, 1999, JPL scientists created the image. Included in the image are three small volcanic calderas about 10-20 kilometers in diameter.

Hubble Space Telescope: Science: In April, 2000, HST's Wide Field/Planetary Camera II witnesses an unusual planetary nebula, NGC, 5,400 light-years from Earth. This nebula, which glows like a big eye, was ejected several thousand years ago from a hot star in the constellation Aquila.

International Space Station: 2000: The first crew for the International Space Station (ISS) takes up residence in November, a milestone for the massive program.

International Space Station: Living and Working Accommodations: STS-106 (ISS-2A.2b) and the crew of *Atlantis* visit the station in September, 2000, to deliver supplies and outfit Zvezda in preparation for the first permanent crew (Expedition 1). Prior to the Expedition One crew's arrival, STS-92 (ISS-3A) delivers the Integrated Truss Structure (ITS) Z1, Pressurized Mating Adapter 3, and four Control Moment Gyros (CMGs) in October. Soyuz TM-31 (ISS-2R) launches from the Baikonur Cosmodrome on October 31, 2000. When it docks with the ISS on November 2, the first human presence on the space station is established. The Soyuz spacecraft, which remains docked to the station after the crew returns to Earth, provides assured crew return capability without the space shuttle present.

International Space Station: Modules and Nodes: Years behind schedule, the Zvezda ("Star") Service Module of the International Space Station, built and financed by Russia, finally reaches orbit. The Z1 truss (containing four large gyroscopic devices) is mounted on the upper (zenith) port of Unity Module. The Z1 truss and the Pressured Mating Adapter (PMA) are delivered on the space shuttle *Discovery* (flight ISS-3A) and mounted on Unity. The P6 module (a set of large solar arrays) is delivered to the ISS aboard *Endeavour* (flight ISS-4A) and mounted on the Z1 truss.

New Millennium Program: The Earth Observing 1 (EO-1) satellite of the New Millennium Program is launched on a Delta II booster from Vandenberg Air Force Base on November 21, 2000, and following deployment from the second stage, enters a 705-kilometer circular, Sun-synchronous orbit with a 98.7° inclination with respect to the equator. This orbit matches within a minute the orbit of Landsat 7 and comparisons of identical images can be made during ground-based analysis.

Pioneer 10: As of March 2—the twenty-eighth anniversary of its launch—Pioneer 10 has traveled more than 11 billion kilometers (75 astronomical units), about twice the distance of Pluto from the Sun. It is still transmitting data about the far reaches of space beyond the solar system, despite its dwindling nuclear power supply. It will soon reach the extent of the heliosphere, an elliptical field of solar wind that radiates from the Sun.

Search and Rescue Satellites: There are seven COSPAS/SARSAT satellites in low-altitude Earth orbit and three satellites in geostationary orbit, eighteen ground segment providers, and eight user states. It is estimated that more than 800,000 distress beacons will be deployed. In 1998 alone, more than thirteen hundred persons had been rescued in 385 events with the aid of COSPAS/SARSAT.

SMS and GOES Meteorological Satellites: GOES-11 is launched on May 3, 2000, and renamed GOES-L when it achieves operational status. NASA turns it over to NOAA as an orbital backup to either GOES-10 (GOES-K, or GOES-West) or GOES-12 (GOES-East).

Space Shuttle Flights, 2000: Five space shuttle flights are made in the year 2000. These flights include only one research mission to study the Earth through radar imaging. The other four missions are construction missions to the International Space Station (ISS).

Space Shuttle: Radar Imaging Laboratories: The Shuttle Radar Topography Mission (SRTM) flies on the space shuttle *Endeavour* in February. The mission is a partnership between NASA and the National Imagery and Mapping Agency. In addition, the German and Italian space agencies contribute an experimental high-resolution imaging radar system. To acquire topographic (or elevation) data, the SRTM instrument is configured with two receiving antennae separated by a distance.

Space Stations: Origins and Development: One permanent expedition crew and three Progress resupply missions fly to Mir during the year.

Spaceflight Tracking and Data Network: TDRS-8 lifts off on June 30, 2000, from Cape Canaveral Air Force Station, Florida, aboard an Atlas IIA rocket. It is the first of three new satellites featuring improved multiple-access and S-band single-access performance, along with a new high-frequency Ka-band service.

Ulysses: Solar-Polar Mission: Ulysses continues to observe the Sun's polar regions and the surrounding heliosphere. Its instruments are functioning, despite the harsh environment the spacecraft had encountered during its ten-year journey. Although the spacecraft is expected to be capable of sending data through the end of 2007, funding for the project was slated to end in December, 2001.

Upper Atmosphere Research Satellite: In February, five years past the expected shutdown date for receiving transmissions from the UARS, six of the ten instruments aboard the satellite are still operational, sending data back to Earth. The observation time of several of the experiments is reduced to conserve battery power, but they continue to transmit scientifically useful data.

Voyager Program: The two Voyager spacecraft continue their scientific exploration of interstellar space. Voyager 1 is 11.5 billion kilometers from Earth, twenty-three years after launch, and is traveling away from the solar system at 21.7 kilometers per second. Voyager 2 is 9.0 billion kilometers from Earth, traveling at 29.3 kilometers per second.

2001

Asteroid and Comet Exploration: On February 12, 2001, the NEAR Shoemaker spacecraft is gently maneuvered to softly impact on the surface of the asteroid Eros. Images are returned through a successful touchdown in an area bordering a saddle-shaped depression named Himeros. Consideration is given of attempting to relaunch NEAR in order to look at the impact site, but the spacecraft does not respond and remains the first asteroid lander.

Earth Observing System Satellites: Jason-1, a joint U.S.-France oceanography mission, launches from the Vandenberg Air Force Base on December 7, 2001. It carries the Poseidon-2 and Jason Microwave Radiometer (JMR) instruments to study global ocean circulation. The Stratospheric Aerosol and Gas Experiment III (SAGE III), which monitors trace gases and other atmospheric parameters, is aboard Russia's METEOR 3M-1 spacecraft when it launches from the Tyuratam Cosmodrome on December 10, 2001.

Extreme Ultraviolet Explorer: In the summer of 2000, NASA decides that EUVE mission operations should cease within a few months. EUVE science operations end on January 26, 2001, and there follows several days of end-of-life mission engineering tests of the never-used backup high voltage supplies and checking of the remaining battery capacity. EUVE is stabilized pointing away from the Sun and sent into safe hold on January 31, 2001. The transmitters are finally commanded off on February 2, 2001.

International Space Station: 2001: During 2001, the International Space Station is a busy construction site and science platform, but the program nevertheless faced increasing criticism over its budgeting and scientific justification, and a decline in confidence in the National Aeronautics and Space Administration forces it to reduce its long-term goals for the station.

International Space Station: Living and Working Accommodations: STS-98 (ISS-5A) delivers and mates the U.S. Destiny Laboratory to the forward port of the Unity Module in February, 2001. Astronauts work inside the pressurized facility to conduct research in numerous scientific fields. Launched on STS-100 (ISS-6A) in April, the Canadarm2 is a bigger, improved, and computer upgraded version of the space shuttle's original robotic arm. Canadarm2 is 17.6 meters long when fully extended and has seven motorized joints. In July, the STS-104 (ISS-7A) crew delivers the Quest Airlock and installs it on the station's Unity Node. Expedition Two Flight Engineer Susan Helms uses the Canadarm2 to lift Quest from the orbiter's payload bay. In September, Progress M-SO1 (ISS-4) delivers the Pirs (Pier in Russian) Docking Compartment and docks it to the nadir port on Zvezda.

Mars Odyssey: Mars Odyssey is launched April 7, 2001, on a Delta II rocket from Cape Canaveral, Florida, and reaches Mars on October 24, 2001. It is part of a series of missions over several years to study the Red Planet. Following a series of Mars mission failures, Mars Odyssey was successful and a much-needed public relations boost for NASA. The Mars Odyssey mission discovered evidence suggesting that Mars may have had water in the distant past.

National Aeronautics and Space Administration: Dr. Daniel R. Mulville is Acting NASA administrator from November 19, 2001, to December 21, 2001. Sean O'Keefe is administrator from December 21, 2001, to February 11, 2005.

Shuttle Amateur Radio Experiment: With the advent of the International Space Station, SAREX has matured to become the Amateur Radio on the International Space Station (ARISS), sponsored by the American Radio Relay League (ARRL), the Radio Amateur Satellite Corporation (AMSAT), and NASA. The space shuttle *Atlantis* carries the initial ARISS equipment on mission STS-106 in September, 2000. Two new call signs are issued for U.S. Amateur Radio operations, NN1SS and NA1SS, to the International Space Station Amateur Radio Club on October 11, 2000. A technical team, called ISS Ham, has been officially established to serve as the interface to support hardware development, crew training, and on-orbit operations.

SMS and GOES Meteorological Satellites: The GOES-12 satellite is launched on July 23, 2001, positioned as the eastern satellite, and renamed GOES-M (or GOES-East).

Space Shuttle Flights, 2001: Six space shuttle flights are made in the year 2001, the most flights in one year since 1997. All six missions are to the International Space Station (ISS). This is the first year since the beginning of the space shuttle program that the National Aeronautics and Space Administration (NASA) has no shuttle missions planned primarily as science missions.

Space Stations: Origins and Development (March 23): After fourteen years of service to space research and Earth observation, the Mir Complex is deorbited to burn up in the atmosphere over the South Pacific.

Stratospheric Aerosol and Gas Experiment: SAGE III is launched on December 10, 2001, from the Tyuratam Cosmodrome on a Zenit 2 rocket. The instrument is aboard the Russian Meteor-3M spacecraft and represents a joint NASA, Ukrainian, and Russian project.

Wilkinson Microwave Anisotropy Probe: The Wilkinson Microwave Anisotropy Probe (WMAP) is launched by a Delta II rocket on June 30, 2001, from Cape Canaveral Air Force Station, Florida. A space-based astronomical observatory, it is designed to measure the "echo" or "afterglow" of the Big Bang, known as cosmic microwave background (CMB). WMAP provides the first accurate measure of the age of the universe, changing the way astronomers think about the earliest star formation, and supports some of the leading cosmological theories.

2002

Asteroid and Comet Exploration: The Comet Nucleus Tour (CONTOUR) is a NASA Discovery mission that launched on July 3, 2002. Its primary objectives were close flybys (within approximately 100 kilometers) of the comets Encke, Schwassmann-Wachmann-3, and d'Arrest to gather data concerning the makeup of comet nuclei. Unfortunately, the CONTOUR mission was lost on August 15, 2002, when it failed to contact Earth after attempting a firing of its main rocket motor.

Earth Observing System Satellites: The Aqua satellite (formerly EOS PM-1) launches from Vandenberg Air Force Base on May 24, 2002. It is designed to gather data concerning clouds, precipitation, the atmosphere's moisture content and temperature, terrestrial snow, sea ice, and sea-surface temperature.

Extreme Ultraviolet Explorer: EUVE destructively reenters the Earth's atmosphere on January 31, 2002.

International Space Station: 2002: The International Space Station continues to grow in size and sophistication during a year that also is crowded with scientific research, but budgeting problems and scaled-back plans for its future create conflict among the sixteen nations involved in the program.

International Space Station: Living and Working Accommodations: The ISS Truss forms the backbone of the Space Station, with mountings for unpressurized logistics carriers, radiators, solar arrays, and other equipment. The S0 (Segment Zero) Truss, the center segment of 11 integrated trusses, is attached to the top of the Destiny Laboratory on April 11, 2002, by the STS-110 (ISS-8A) crew. Through November, when the STS-113 crew delivers the P1 (P-One) Truss and sixth Expedition crew, the station has twelve major components in place. These include Zarya, Unity, Zvezda, Z1 Truss, P6 Integrated Truss, Destiny, Canadarm2, the Joint Airlock, Pirs, the S0 Truss, the S1 Truss, and the P1 Truss.

Lifting Bodies: On April 29, 2002, NASA announces the cancellation of the X-38 program due to budget pressures associated with the International Space Station. The X-38 is two years short of completing its flight test phase.

Small Explorer Program: The High Energy Solar Spectroscopic Imager (HESSI) is launched by a Pegasus XL rocket on February 5, 2002, off the coast of Florida from the Cape Canaveral area. Designed to observe the Sun, HESSI studies how particles are accelerated and how energy is released in solar flares. After being commissioned, it is renamed RHESSI, after Reuven Ramaty, a pioneer NASA scientist who worked in the areas of solar physics and gamma-ray astronomy until his death in 2001.

Soyuz and Progress Spacecraft (October, 2002-May, 2003): A fourth-generation Soyuz spacecraft, delivers the fourth visiting crew to the International Space Station on the Soyuz TMA-1 mission. Soyuz TMA (Transport Modification Anthropometric) contains redesigned seats and suspension to accommodate American astronauts, who on average are taller than cosmonauts, and a new set of computer displays.

Space Shuttle Flights, 2002: Five space shuttle flights are made in the year 2002. For the second year in a row, no shuttle flight has scientific research as a primary mission. The first mission is the last Hubble Space Telescope (HST) servicing mission, and the other missions are support missions for the International Space Station.

Spaceflight Tracking and Data Network: TDRS-9 is placed into orbit on March 8, 2002, by Atlas IIA AC-143. The third and final Advanced Tracking and Data Relay Satellite, TDRS-10, separates from the Centaur upper stage 30 minutes after its launch on December 4, 2002. This completes the $800 million, three-satellite system.

Stardust Project: Stardust flies within 3,300 kilometers of the asteroid 5535 Annefrank on November 2, 2002, and takes several photographs on its way to a rendezvous with comet Wild 2 in January, 2004.

2003

Earth Observing System Satellites: The Ice, Cloud, and Land Elevation Satellite (ICESat) mission launches on January 12, 2003, from Vandenberg Air Force Base. The ICESat platform follows a near-polar orbit at an altitude of 600 kilometers, carrying a single scientific instrument: the Geoscience Laser Altimeter System (GLAS). GLAS is designed to measure the elevation of the Earth's ice sheets, clouds, and land. The Solar Radiation and Climate Experiment (SORCE) launches from the Cape Canaveral Air Force Station on January 25, 2003. It includes the Solar/Stellar Irradiance Comparison Experiment (SOLSTICE) and Total Irradiance Monitor (TIM) to measure the solar ultraviolet, far ultraviolet, and total irradiance.

Ethnic and Gender Diversity in the Space Program: The first Chinese national and the first person to fly in a spacecraft launched by a country other than the United States or Russia is Liwei Yang, who rockets into space on October 15, 2003. China's Chang Zheng 2F booster places his spacecraft, Shenzhou 5, into a 200 by 343 kilometers orbit. He spends nearly a day in orbit, completing 14 revolutions of Earth.

Galaxy Evolution Explorer: The Galaxy Evolution Explorer (GALEX) is designed to observe ultraviolet emissions and produce a comprehensive survey of galaxies. GALEX data could prove crucial to understanding galaxy formation and evolution.

Galileo: Jupiter: To preclude the radioactive material in Galileo's power generation system from contaminating Europa or any of the other moons, the fourteen-year odyssey of the spacecraft concludes on September 21, 2003, with a controlled plunge into the outer atmosphere of Jupiter while the spacecraft is on its thirty-fifth orbit. The spacecraft passes into Jupiter's shadow and the Deep Space Network receives its final signal from Galileo. It hits the outer atmosphere just south of the gas giant's equator at a speed of 48.3 kilometers per second. Due to the time delay in receipt of radio signals, this message arrives 46 minutes after Galileo is crushed, vaporized, and dispersed into Jupiter's dense atmosphere.

International Space Station: 2003: Loss of the space shuttle *Columbia* in February causes the National Aeronautics and Space Administration to suspend construction of the International Space Station and reduce the permanent crew from three to two, yet ISS caretaker crews are still able to conduct some scientific research.

International Space Station: Crew Return Vehicles: The reliable Russian Soyuz spacecraft steps in to service the International Space Station in the wake of the *Columbia* accident and subsequent shuttle-fleet grounding. Originally designated as a backup to the CRV, the Soyuz TMA continues to serve in the role of transport and return vehicle.

Landsat 7: NASA handled Landsat 7's day-to-day operations until early into the twenty-first century, when the US Geological Survey (USGS) assumed management responsibilities. By 2003, the three ground stations managed by USGS had collected more than three-hundred thousand images for archiving.

Launch Vehicles: The final Titan II booster launched from Vandenberg in October, 2003, delivering a military weather satellite to orbit.

Pioneer 10: Pioneer 10's final transmitted signal is received through the Deep Space Network on January 22, 2003. At this point, signals from the spacecraft, 7.6 billion miles distant from Earth, take eleven hours and twenty minutes to arrive. It continued to send data concerning the nature of the far reaches of the outer solar system until its radioisotope nuclear power supply dwindled. The spacecraft will continue to head out of the solar system as a silent ambassador of humanity.

Space Shuttle Flights, 2003-2004: The only space shuttle flight in 2003 ends in tragedy when the shuttle *Columbia* disintegrates during reentry on February 1. All shuttle flights are postponed until the *Columbia* Accident Investigation Board can determine the cause of the accident and NASA fixes the design flaws leading to the accident. Space shuttle activity during 2003-2004 consists primarily of working to insure that the *Columbia* accident never reoccurs.

Space Shuttle Mission STS-107: The orbiter *Columbia* breaks apart during atmospheric reentry over Texas, killing all crewmembers. STS-107 is the second major loss to the space shuttle fleet and the third crew loss-of-life accident for the National Aeronautics and Space Administration (NASA).

Spitzer Space Telescope: On August 25, 2003, the Spitzer Space Telescope, originally designated the Space Infrared Facility (SIRTF), is launched. It completes the National Aeronautics and Space Administration's (NASA's) Great Observatory Program. Designed to detect celestial infrared emissions, it provides a different window on the universe for collecting data complementary to observations made by the other space-based Great Observatories.

Tethered Satellite System: NASA announces plans to study Momentum-eXchange/Electrodynamic Reboost (MXER) tether technology. If implemented MXER would station kilometers of cartwheeling cable in orbit around the Earth. Then, rotating like a giant sling, the cable would swoop down and pick up spacecraft in low orbits, and then hurl them to higher orbits or even lob them onward to other planets. The hope is to harness momentum while dramatically lowering the cost of launching space missions. The study is managed in the Office of Space Sciences at NASA Headquarters.

2004

Ansari X Prize: The Ansari X Prize competition is an attempt to duplicate the challenge that the Orteig Prize, which was eventually won by Charles A. Lindbergh in 1927, made to the aviation industry. This time the X Prize challenged the private sector to take up where various national space programs failed to go in developing less expensive reuseable spacecraft. If this challenge could be met, it would literally open up space to the private sector as well as the tourist trade. The first flight to qualify to win the Ansari X Prize takes place on September 29, 2004, as SpaceShipOne, with Mike Melvill at the controls, soars into the blue California sky over the Mojave Desert, and reaches an altitude of over 100 kilometers. Five days later on October 4, SpaceShipOne again heads toward space with pilot Brian Binnie at the controls. After an 80-second rocket burn SpaceShipOne continues to climb until Earth's gravity overpowers the craft's upward momentum. Upon landing, the Ansari X Prize is claimed.

Cassini: Saturn: On December 25, 2004, the Huygens probe separates from the Cassini orbiter. The probe reaches Saturn's moon Titan on January 14, 2005, where it makes an atmospheric descent to the surface and relays scientific information. It is the first spacecraft to orbit Saturn and just the fourth spacecraft to visit Saturn.

Delta Launch Vehicles: The initial Delta IV Heavy booster is launched on December 22, 2004.

Earth Observing System Satellites: Aura (formerly EOS Chem) launches on July 15, 2004, from the Vandenberg Air Force Base aboard a Delta booster. It is the third in the series of

EOS missions and follows a polar, Sun-synchronous orbit approximately 705 kilometers above the Earth's surface. The primary objective of the Aura mission is to study the chemistry and dynamics of the Earth's atmosphere, particularly in the upper troposphere and lower stratosphere (5 to 20 kilometers above Earth).

Far Ultraviolet Spectroscopic Explorer: In June, 2004, FUSE celebrates its fifth anniversary. During the first five years of operation, FUSE collected nearly fifty million seconds of scientific data on 2,200 different astronomical objects. These data have led to a few hundred significant scientific papers. As of late 2004 FUSE was still going strong, with no plans to end the mission. However, on December 27, 2004, FUSE suffers a major malfunction that threatens to halt science operations. One of the two remaining reaction wheels, the one for the roll axis, stalls. FUSE remained in a safe mode for nearly three months, but on March 29, 2005, the control team reopens the telescope's door and the following day achieves first light for the third time. A spectrum of a previously examined object, the central star in planetary nebula IC2448, is taken to compare present scientific capability with previous observations.

Funding Procedures of Space Programs: In 2004, political bravado and budgetary reality clash when the White House declares a goal of sending humans back to the Moon and on to Mars. Facing the worse federal deficit since the Hoover administration, funding for such projects would require NASA to cancel piloted and robotic missions already in the pipeline. In response to the announcement, NASA eliminates all space shuttle flights not directly supporting the International Space Station. This includes scheduled servicing missions to the Hubble Space Telescope. Without these essential repairs and upgrades, Hubble will wither and die a fiery death.

Galaxy Evolution Explorer: Late in December, 2004, GALEX produces surprising images of massive newborn galaxies relatively close to the neighborhood of the Milky Way. Models had predicted that formation of large galaxies had drastically diminished since the early era of galaxy formation, and that presently only small ones were forming. This finding by GALEX suggests that galaxy formation remained strong in various portions of the expanding universe.

Get-Away Special Experiments: During the first twenty years of flight aboard the space shuttle, 167 individual GAS missions are flown. Due to the International Space Station assembly sequence, no GAS opportunities will be available for some time. While no shuttle flights are currently available, GAS is considered by NASA to be an acceptable substitute for manifested station components.

Gravity Probe B: Gravity Probe B (GP-B) is launched aboard a Boeing Delta II rocket into polar orbit on April 20, 2004, from Vandenberg Air Force Base, California. It orbits 640 kilometers above the Earth. On August 27 the spacecraft began science data collection, and apart from some minor glitches, GP-B continues to work well. The spacecraft experienced some anomalous behavior when passing over the South Atlantic Anomaly, but spacecraft controllers were able to develop means of compensating for this. The mission of GP-B is to measure accurately two subtle effects predicted by Albert Einstein's general theory of relativity. If the results support it, GP-B will increase confidence in general relativity; otherwise, the results might point the way to a more comprehensive theory.

Hubble Space Telescope: Servicing Missions: In response to budgetary cuts in the wake of the STS-107 accident, NASA eliminates all space shuttle flights not directly supporting the International Space Station, including scheduled servicing missions to the Hubble Space Telescope. NASA had planned to visit Hubble one last time in 2006 to change out instruments and replace its gyroscopes with the intent of keeping the telescope in service until at least 2011, when its heir apparent, the James Webb Space Telescope, is expected to launch. Scrapping the final servicing mission raises the likelihood that Hubble will fail before Webb is in orbit.

International Space Station: 2004: Because the space shuttle fleet remains grounded throughout the year, the International Space Station sees no significant construction and the two-person crews spend most of their time in maintenance, although they do conduct scientific research.

Mars Exploration Rovers: The Mars Exploration Rovers (MER), Spirit (MER-A) and Opportunity (MER-B), are launched about a month apart in the summer of 2003 and reach Mars in early 2004. Both of these MER missions were timed to take advantage of the closest Mars approach to Earth in approximately 60,000 years. The twin robotic expeditions exploring the surface of Mars land on different locations on opposite sides of Mars to maximize the scientific return. With the ability to trek the length of a football field each day, they can explore a larger portion of the Martian surface than previous missions could. Each rover contains a package of scientific instruments designed to analyze Mars's surface geology to help trace the history of liquid water on Mars.

Mars Reconnaissance Orbiter: With its sophisticated scientific instruments, the Mars Reconnaissance Orbiter (MRO) is designed to return much more data on the atmosphere, surface, and subsurface of Mars than any previous mission. These data will provide significant support for future landers, rovers, and sample-return ventures.

MESSENGER: On August 3, the MErcury Surface, Space ENvironment, GEochemistry, and Ranging (MESSENGER) spacecraft is launched aboard a Delta 2 booster. For the spacecraft this begins a journey to Mercury that will cover 7.8 billion kilometers. Mercury, the innermost planet in the solar system, is both Earth-like and Moon-like; Earth-like in that it is a terrestrial planet with a magnetic field, and Moon-like in that it is heavily cratered. Only the Mariner 10 spacecraft in 1974 and 1975 prior to the MESSENGER program imaged this planet's surface from close proximity, and that was during three brief flyby encounters only.

Mission to Planet Earth: In June, 2004, NASA begins the transformation of its Earth and space science programs by combining them into an integrated Science Mission Directorate. One specific goal of the Vision for Space Exploration is the scientific investigation of the Earth, Moon, Mars, and beyond with emphasis on understanding the history of the solar system, searching for evidence of habitats for life on Mars, and preparing for future human exploration. Another goal is the search for Earth-like planets and habitable environments around other stars. A third goal is to explore the solar system for scientific purposes and to support human exploration in order to establish a sustained "presence" throughout the solar system.

Space Centers, Spaceports, and Launch Sites: The Mojave Spaceport, also known as the Mojave Airport and Civilian Flight Test Center, is the first facility licensed in the United

States for horizontal launches of reusable spacecraft. Certified as a spaceport by the Federal Aviation Administration on June 17, 2004, it is also the first inland spaceport in the United States. The Spaceport has been a test site for several teams in the Ansari X Prize. Most notably, it is the test launch site for SpaceShipOne, developed by Scaled Composites, which conducted the first privately funded human spaceflight on June 21, 2004. Other groups based at the Mojave Spaceport include XCOR Aerospace, Orbital Sciences Corporation, and Interorbital Systems. In December, 2004, the U.S. Congress passed the Commercial Space Launch Amendments Act of 2004. That legislation gave the Federal Aviation Administration (FAA), among other duties, the go-ahead to start shaping rules on medical requirements for a spaceship passenger—termed a "space flight participant"—an individual (who is not crew) carried within a launch vehicle or reentry vehicle. Along with Scaled Composites group and Virgin Galactic, XCOR, Rocketplane, SpaceX, Armadillo Aerospace, and others are working to complete reusable launch vehicles in the hopes of carrying passengers in the near future.

SpaceShipOne: On June 21 of 2004, Mike Melvill flies SpaceShipOne on the first successful entry into space by a private company solely financed by private investment. The spacecraft is significant because of the unique feathering system designed to allow a low speed and controlled re-entry into the atmosphere. The total cost of the program is significantly lower than the billions of dollars spent on government-sponsored space programs.

Stardust Project: Stardust flies by comet Wild 2 on January 2, 2004. During the flyby, it collects dust samples from the comet's coma and takes detailed pictures of its icy nucleus. Stardust is now on its way back to Earth and will arrive at Earth in 2006, to deliver the sample material in a capsule.

Swift Gamma-Ray Burst Mission: Swift, the first telescope of its kind designed to detect and study gamma-ray bursts in real time, is launched November 20, 2004, on top of a Delta 2 rocket. Gamma-ray bursts are high-energy explosions that occur daily throughout the sky. Swift will detect them and automatically move to observe them, giving astronomers their first chance to view a gamma-ray burst in the critical first few minutes. This will allow some theories on the origin and aftermath of gamma-ray bursts to be tested. It will also allow classifications of gamma-ray bursts and studies on how they affect their environments. The gamma-ray bursts will also give more information as to the history of the early universe.

Ulysses: Solar Polar Mission: Between November, 2003, and April, 2004, the Ulysses spacecraft was again in the vicinity of Jupiter and able to make observations of the Jovian environment. Solar observations continued, and in February, 2004, ESA's Science Programme Committee approved an extension of the Ulysses mission through March, 2008.

2005

X-15 (August 23): NASA awards astronaut wings to three X-15 test pilots in a ceremony at NASA's Dryden Flight Research Center, Edwards, California. Retired NASA pilot Bill Dana and family members representing agency deceased pilots John McKay and Joseph Walker receive civilian astronaut wings. The wings acknowledge the fact the pilots flew the X-15 at altitudes of 50 miles or higher.

Apollo Program: Orbital and Lunar Surface Experiments: The Laser Ranging Retrore-flectors from Apollo 11, 14, and 15 are still being used to increase the accuracy of the Earth-

Moon measurements to near millimeter accuracy. More than thirty years later, they are the only Apollo science experiments still functioning.

Asteroid and Comet Exploration: On April 3, 2003, the Rossi X-Ray Timing Explorer (RXTE) suffers a corruption of its telemetry interface, but, after rebooting, normal operation is restored. Like so many intrepid robotic craft, RXTE continues well beyond its design lifetime. As of 2005, RXTE-based research continues to routinely provide new and exciting published results concerning such exotic astrophysical objects as magnetars (a rare class of neutron stars with extremely intense magnetic fields), accretion disks illuminated by superbursts, spinning black holes, and the limitation of rotation rates on pulsars.

Atlas Launch Vehicles: Five decades after program inception the Atlas family of boosters continues to reliably place payloads in low-Earth orbit, geosynchronous positions, and on escape trajectories. The final Atlas II boosters were launched in 2004. The Atlas III family builds on the design of Atlas II with the use of a new single-stage Atlas main engine, the Russian RD-180. The Atlas IIIA uses a twin RD-180 configuration and has a single-engine Centaur atop it. The changes to Centaur for Atlas IIIB are a stretched tank and the addition of a second engine. The first Atlas III launched from Cape Canaveral's Pad 36 on May 24, 2000. This design was meant only as a transitional form from the Atlas II version to the more powerful Atlas V. The final Atlas III booster launched from Cape Canaveral on February 3, 2005. The Atlas V family is capable of lifting payloads up to 8,200 kilograms to GTO and over 5,940 kilograms directly to geosynchronous orbit (Atlas V-Heavy).

Chandra X-Ray Observatory: The data gathered by Chandra have greatly advanced the field of x-ray astronomy. The first light image, of supernova remnant Cassiopeia A, gives astronomers their first glimpse of the compact object at the center of the remnant, probably a neutron star or black hole. Chandra finds much cooler gas than expected spiraling into the center of the Andromeda Galaxy. Chandra shows for the first time the shadow of a small galaxy as it is being cannibalized by a larger one, in an image of Perseus A. A new type of black hole is discovered in galaxy M82, mid-mass objects purported to be the missing link between stellar-sized black holes and supermassive black holes. TWA 5B, a brown dwarf, is seen orbiting a binary system of Sun-like stars. The x-ray shadow of Titan is seen when it transits the Crab Nebula.

Global Positioning System: Air Force projections had assumed it would be necessary to launch two replacement GPS satellites in 1995, but in the year 2005, most of the original Navstar II satellites are still operational. Navstar IIR comprises twenty replacement satellites that incorporate autonomous navigation based on crosslink ranging.

Hubble Space Telescope: Servicing Missions: Flight hardware preparations for a fourth servicing mission resume under orders from new NASA Administrator Michael Griffin even though no decision was made to adopt a shuttle flight to accomplish that servicing.

Huygens Lander: Huygens separates from the Cassini orbiter on December 25, 2004, and lands on Titan on January 14, 2005, near the Xanadu region. It lands on land and continues to send data for about 90 minutes after reaching the surface.

International Space Station: Living and Working Accommodations: The International Space Station has been permanently occupied since November, 2000. Until the STS-107 accident grounded the shuttle fleet in February, 2003, three crew members spent six-month tours on the station. During the hiatus, the crew was reduced to two for safety considerations. After the successful STS-114 resupply mission, NASA announces that when STS-121 is launched in 2006 it will carry a third crew member. Once the station is completed, an international crew of up to seven will live and work in space between three and six months. Crew return vehicles will always be attached to the station to ensure the safe return of all crew members in the event of an emergency.

International Space Station: Modules and Nodes: Construction of the station is to resume in 2006 after the STS-114 and STS-121 resupply missions.

Launch Vehicles: The Titan IV is scheduled to be retired in 2006. The final launch (B-30) from Cape Canaveral occurs on April 29, 2005, and the final launch from Vandenberg AFB is set for 2005 or 2006.

Mars Exploration Rovers: On April 6, 2005, with both rovers suffering from only minor problems, and each covering greater and greater distances during translations across the Martian surface, NASA announces that the MER program is being extended for the third time. This latest extension will run for the coming eighteen months.

National Aeronautics and Space Administration: Frederick D. Gregory is acting NASA administrator from February 11, 2005, to April 14, 2005. Dr. Michael Griffin begins his duties as the eleventh NASA administrator on April 14.

Navigation Satellites: Russian Global Navigation Satellite System (GLONASS) satellites are in orbit and work in coordination with the American Global Positioning System. However, only eleven satellites are functional and an operational system requires eighteen. Russia has signed an agreement with India to expand on the diminished GLONASS system.

Pegasus Launch Vehicles: The Pegasus booster has been launched thirty-eight times with only six failures. Notable spacecraft placed into orbit are the Far Ultraviolet Spectroscopic Explorer, Galaxy Explorer, Reuven Ramaty High Energy Solar Spectroscopic Imager, High Energy Transient Explorer, Transition Region and Coronal Explorer, and the X-43A scramjet demonstration vehicles.

Search and Rescue Satellites: As of 2005, there are seven COSPAS/SARSAT satellites in low-altitude Earth orbit and three satellites in geostationary orbit, eighteen ground segment providers, and eight user states. One SARSAT satellite suffered malfunctions resulting in intermittent loss of service that could affect an entire or just partial satellite pass. More than 800,000 distress beacons will be deployed. By 2003, more than seventeen thousand people had been rescued in more than 4,800 responding incidents. NASA and others continue to seek new methods to cut costs and time in search and rescue operations while saving even more lives.

SMS and GOES Meteorological Satellites: By January, 2005, SOHO had discovered its nine-hundredth comet. One of these comets, Kudo-Fujikawa, was shown live on the Internet as it passed through SOHO's field of view. Another comet, named NEAT, was observed by SOHO as it was hit by a severe solar storm in 2003. That was the first recording of such an event.

Space Shuttle Mission STS-114: Originally scheduled to fly in March, 2003, on *Atlantis*, the flight is delayed by the February, 2003, *Columbia* reentry accident. The mission is launched on July 26, 2005. NASA's return to piloted spaceflight is successful, despite some minor damage to *Discovery*'s underside and the loss of insulating foam on the external tank. The Raffaello Multi-Purpose Logistics Module carries supplies and equipment for the International Space Station. The mission makes planned repairs to the Space Station, tests new safety features, and demonstrates inflight repair of the orbiter's reentry tiles.

SPACEHAB: Although supplanted by the larger Multi-Purpose Logistics Modules, SPACEHAB continues its contributions to the ISS by providing flexible-configuration pressurized and open cargo carriers. With the cancellation of all shuttle flights not related to the ISS, no plans have been made to replace the Research Double Module lost on STS-107. STS-114, the space shuttle's Return-to-Flight Mission, includes the Raffaello Multi-Purpose Logistics Module in its payload bay. The Italian-built module carries supplies and cargo to the International Space Station. *Discovery* also carries SPACEHAB's External Stowage Platform 2 (ESP2), a modified version of the company's Integrated Cargo Carrier (ICC) system. ESP2 carries replacement parts, known as Orbital Replacement Units (ORU), to the orbiting ISS. This resupply platform is attached to the Space Station's airlock and marks SPACEHAB's first permanent hardware residence on the ISS. Plans call for SPACEHAB to have payloads on at least three more shuttle flights to the ISS: STS-121, STS-116, and STS-118. STS-121, a logistics and supply mission, is scheduled to carry the ICC. STS-116 will be a crew rotation and logistics mission utilizing the SPACEHAB Single Module. STS-118 will feature the SPACEHAB Single Module and ICC as the shuttle resupplies the ISS.

Spitzer Space Telescope: The infrared eyes of NASA's Spitzer Space Telescope spot an enormous light echo etched in the sky by a fitful dead star. The surprising finding indicates Cassiopeia A, the remnant of a star that died in a supernova explosion 325 years ago, is not resting peacefully. Instead, this dead star likely shot out at least one burst of energy as recently as 50 years ago.

Swift Gamma-Ray Burst Mission: As of April, 2005, Swift has detected about two dozen gamma-ray bursts and x-ray afterglows for about fifteen of them, and optical afterglows for a few. Optical afterglows are important for providing distance estimates.

Tracking and Data-Relay Communications Satellites: There are ten TDRS satellites in the constellation. Due to increasing orbit inclination, TDRS-1 was the first satellite able to see both Poles. In cooperation with the National Science Foundation (NSF), an uplink/downlink station for TDRS-1 was installed in January, 1998, at the exact South Pole. TDRS-1 is located at 48.8° west longitude (W). TDRS-3, deployed during STS-26 in September, 1988, is in orbit at 84.8° east longitude (E), TDRS-4 (STS-29, March, 1989) is at 41.5°W, and TDRS-5 (STS-43, August, 1991) is at 171.4°W. TDRS-6 (STS-54, January, 1993) is located at 46.8°W and TDRS-7 (STS-70, July, 1995) is at 150.8°W. TDRS-8 and TDRS-10 are the primary satellites, with TDRS-9 serving as a spare. TDRS-8 is at 174.4°W, TDRS-9 is at 64.3°W, and TDRS-10 is at 42.2°W. TDRS 8, 9, and 19 were launched by Atlas 2A launch vehicles.

Voyager 1: Jupiter, Saturn: The Voyager 1 spacecraft continues its scientific exploration of interstellar space. It has passed 13.5 billion miles, the equivalent of ninety times the distance between the Earth and the Sun, traveling away from the solar system at well over twenty-one kilometers per second.

Voyager 2: Jupiter, Saturn, Uranus, and Neptune: The Voyager 2 spacecraft is continuing its scientific exploration of interstellar space. More than eleven billion kilometers from Earth, the equivalent of seventy times the distance from Earth to the Sun, twenty-seven years after launch, Voyager 2 is heading out of the solar system at approximately twenty-nine kilometers per second.

Wilkinson Microwave Anisotropy Probe: According to current models of the universe, WMAP data shows the universe is 13.7 billion years old (to within about 1 percent error). The universe is apparently composed of 4 percent ordinary matter, 23 percent of an unknown type of dark matter, and 73 percent of a mysterious postulated dark energy. This is a confirmation of the so-called concordance Lambda-CDM model. The cosmological scenarios of cosmic inflation agree with the observations, though there is an unexplained anomaly on large angular scales. The Hubble constant is 71 ± 4 kilometers per second per megaparsec (Mpc). Current theories applied to the WMAP data indicate that the universe will expand forever.

2006 and After
Dawn Mission (2006-2015): The Dawn Mission represents the first time that a spacecraft will orbit two separate planetary bodies as part of the same mission. In terms of its mission goals and technological innovations, the Dawn spacecraft will not only examine the two largest asteroids, but it will also represent the first fully scientific space mission to use ion propulsion to power the spacecraft throughout its long nine year journey.

Deep Impact: Deep Impact is the first space mission designed to study directly the interior of a primitive celestial body. This will be accomplished by slamming a heavy impactor into Comet Tempel-1, creating a fresh crater. Never before has any space mission tried to make an impact crater of this size in any object.

Gamma-Ray Large Area Space Telescope: The Gamma-ray Large Area Space Telescope (GLAST) will study the highest energy gamma rays from celestial objects. It will be about thirty to fifty times more sensitive, detect higher energies, see a larger area of the sky, and provide more accurate positional information than previous missions investigating gamma emissions could provide. By virtue of its sensitivity to higher energies, GLAST will provide new information about the most energetic violent events and objects in the universe, including gamma-ray bursts, the most active galactic nuclei, pulsars, black holes, and the Big Bang itself.

Hubble Space Telescope: Science: The James Webb Space Telescope (JWST) will be the successor to Hubble. It is a large, infrared-optimized space telescope scheduled for launch in August 2011. JWST is designed to study the earliest galaxies and some of the first stars formed after the Big Bang. These early objects have a high red shift from our vantage-point, meaning that the best observations for these objects are available in the infrared. JWST's instruments will be designed to work primarily in the infrared range of the electromagnetic spectrum, with some capability in the visible range.

International Space Station: Development: Russia will modify the backup Functional Cargo Block (FGB-2) into a Multipurpose Laboratory Module (MLM). It will replace the cancelled Universal Docking Module (UDM). The MLM should be ready for launch in November, 2006. Node 2 is the Space Station's "utility hub," containing eight racks that provide air, electrical power, water, and other systems essential to support life on the spacecraft. Node 2 is the second of three connectors between the ISS modules and is set for launch in

January, 2007, aboard STS-120. Columbus is a science laboratory contracted by ESA. The laboratory is a cylindrical module very similar in shape to the Multi-Purpose Logistics Module. Once launched on STS-123 in April, 2007, it will be attached at Node 2's starboard side, with the cylinder pointing outwards. The Special Purpose Dexterous Manipulator, or Canada Hand, is a smaller two-armed robot capable of handling the delicate assembly tasks currently handled by astronauts during spacewalks. It is scheduled to be attached to the ISS during STS-125 in November, 2007. The Japanese Experiment Module (JEM) named Kibo (Japanese for hope) consists of four components: the Pressurized Module (PM), the Exposed Facility (EF), the Experiment Logistics Module (ELM), and the Remote Manipulator System (JEMRMS). The Experiment Logistics Module will be launched in October, 2008, aboard STS-129. The Pressurized Module will arrive in January, 2009, aboard STS-130. The Science Power Platform (SPP) is a Russian element of the ISS that will be brought up by the space shuttle in July, 2009, on STS-132. The Centrifuge Accommodations Module (CAM) will provide controlled gravity for experiments and the capability to expose a variety of biological specimens to artificial gravity levels between 0.01g and 2g. The CAM will be attached to Node 2 of the ISS in October, 2009, during the STS-133 mission. The Cupola is a U.S. element of the ISS that will provide direct viewing for robotic operations and space shuttle payload bay viewing, as well as a spectacular observation point of Earth, for astronauts. It is scheduled for launch in April 2010.

New Horizons Pluto-Kuiper Belt Mission (2006-2026): The New Horizons spacecraft will be the first human-made probe to explore Pluto, its moon Charon, and bodies in the Kuiper Belt. Insights will be gained about the origin and evolution of the outer solar system, including the geology and geochemistry of Pluto, Charon, and icy bodies in the Kuiper Belt. Study of Kuiper Belt objects should provide increased understanding of how stars, planets, and solar systems form.

Nuclear Energy in Space: Development of advanced nuclear propulsion has been proposed under the name of Project Prometheus. The flagship mission of the Prometheus project involved the Jupiter Icy Moons Observer (JIMO), a vehicle that would steer through the Jovian system with a minimum of chemical propulsion.

Small Explorer Program: The Aeronomy of Ice in the Mesosphere (AIM) mission is slated for a launch in September, 2006. This mission will investigate Earth's highest-level clouds as part of a study of climate change. On January 26, 2005, NASA announced several new Small Explorer missions had been selected from over two dozen proposals. The Interstellar Boundary Explorer (IBEX) is intended to study galactic cosmic rays and indirectly detect the edge of the solar system. IBEX launch is tentatively scheduled for 2008. The Nuclear Spectroscopic Telescope Array (NuSTAR) is intended to become the first x-ray telescope specifically designed to detect black holes. NuSTAR is funded as a feasibility study; a decision will be made in early 2006 whether or not to develop NuSTAR and proceed to flight development.

Space Centers, Spaceports, and Launch Sites: The Oklahoma Space Industry Development Authority (OSIDA) was founded in 1999, with the signing of the Space Industry Tax Incentive Act by the Oklahoma state legislature. That legislation also directed that they be given municipal authority to develop a 168-square-mile "Spaceport Territory." The old Clinton-Sherman Air Force Base at Burns Flat is the site for the Oklahoma Spaceport, due to begin flight operations in 2006 or 2007. New Mexico is establishing the Southwest Regional Spaceport at Upham, an undeveloped location in Southern New Mexico approximately 72 kilometers north of Las Cruces and 48 kilometers east of Truth or Consequences. The proposed site is approximately 70 square kilometers of open, generally level, rangeland with an average elevation of 1,500 meters. The complete lack of conflicting operations, facilities, and environmental constraints provides a unique opportunity to design and develop a purpose-built launch complex that meets the needs of spaceport customers. The spaceport will include a launch complex; a 400-meter runway and aviation complex; a payload assembly complex; a support facilities complex; a system development complex; and site infrastructure.

Russell R. Tobias and David G. Fisher

Key Figures in the History of Space Exploration

Abbey, George W. S. (b. 1932): Deputy Associate Administrator for Space Flight, NASA Headquarters (1988-1991). Responsible for the development and operation of the Space Transportation System. Director, Johnson Space Center (JSC), Houston, Texas (1996-2001). Planned, organized, and directed all activities required to accomplish the missions assigned to JSC.

Abbott, Ira H. (b. 1906): A prominent aeronautical engineer in the early years of the American space program, Abbott joined the Langley Memorial Aeronautical Laboratory, Hampton, Virginia, in 1929. Assistant Chief of Research, Langley (1945-1948). National Advisory Committee for Aeronautics (NACA) Assistant Director of Aerodynamics Research (1948-1959). NASA Director of Advanced Research Programs (1959-1961). Director of Advanced Research and Technology (1961-1962). Supervised the X-15, the supersonic transport, the nuclear rocket, and the advanced reentry programs.

Abrahamson, James A., Jr. (b. 1933): Lieutenant General, United States Air Force (Retired). NASA Associate Administrator, Space Transportation System (1980-1984). Director, Strategic Defense Initiative Organization (1984-1989).

Adams, Michael J. (1930-1967): Major, United States Air Force. Chosen for the Manned Orbiting Laboratory (MOL) program and transferred to the X-15 program. He made seven flights in the X-15. On November 15, 1967, after attaining an altitude of 81 kilometers, the X-15 crashed, killing him. He was posthumously awarded Air Force astronaut wings, because his final flight exceeded 80 kilometers in altitude. Adams was the only pilot lost in the 199-flight X-15 program.

A'Hearn, Michael F.: Professor, University of Maryland (1982-present). He developed a variety of observational techniques to study the structure and composition of comets. Asteroid 3192 was named for A'Hearn for his contributions to the field of cometary science. As the Principal Investigator for the Deep Impact mission, Professor A'Hearn would be responsible for the mission's overall success in meeting its science objectives.

Aitken-Cade, Phillip: Colonel, United States Air Force. Deputy Program Director, X-30 National AeroSpace Plane (NASP), 1993-1994.

Aldrin, Edwin E. "Buzz," Jr. (b. 1930): Colonel, United States Air Force (Retired). NASA Astronaut (1963-1971), Gemini XII Pilot, Apollo 11 Lunar Module Pilot. Second human to walk on the Moon. Gemini IX Backup Pilot. Apollo 8 Backup Command Module Pilot. His books include *Return to Earth* (1974), an account of his Moon trip and his views on the United States' future in space, *Men from Earth* (1989), and a science-fiction novel, *Encounter with Tiber* (1996). A small crater on the Moon near the Apollo 11 landing site is named in his honor. He also has a star on the Hollywood Walk of Fame on Hollywood and Vine in Los Angeles.

Alexandrov, Alexander P. (b. 1943): Soviet/Russian Cosmonaut (1978-1993). Flight Engineer on Soyuz T-9 and Soyuz TM-3. Altogether, he spent 309 days, 18 hours, and 3 minutes in space. Chief of Russian NPO Energia Cosmonaut group (1993-1996). Chief Flight Test Director of S. P. Korolev Rocket and Space Corporation Energia, also known as RKK Energia (1996-present). Cochair for the Crew Training and Exchange Working Group, Shuttle-Mir Phase 1 Program (1995-1998).

Allen, H. Julian (1910-1977): Director, NASA Ames Research Center, Moffett Field, California (1965-1969). Senior Aeronautical Engineer at Ames and Chief of the High-Speed Research Division (1945-1959). Ames Assistant Director for Astronautics (1959-1965). In 1952, the National Advisory Committee for Aeronautics (NACA) was already thinking about aircraft that went very high and had to reenter the Earth's atmosphere at a high rate of speed, producing a great deal of heat. That year, Allen conceived the "blunt nose principle," which suggested that a blunt shape would absorb only a very small fraction of the heat generated by the reentry of a body into the Earth's atmosphere. The principle was later significant to the development of the intercontinental ballistic missile nose cone and NASA Mercury capsule. The H. Julian Allen Award is presented annually by NASA Ames Research Center for outstanding research.

Allen, Joseph P. (b. 1937): NASA Astronaut (1967-1985). Mission Specialist on STS-5 and STS 51-A. Flew the first space salvage attempt in history (STS 51-A), using the Manned Maneuvering Unit to retrieve the Palapa B2 and Westar 6 communications satellites and return them to Earth. NASA Assistant Administrator for Legislative Affairs in Washington, D.C. (1975-1978). Chief executive officer, Space Industries, Houston, Texas (1989-2004).

Allen, Lew, Jr. (b. 1925): General, United States Air Force (Retired). Vice President, California Institute of Technology and Director, Jet Propulsion Laboratory (JPL), Pasadena, California (1982-1990). Chair of the Charles Stark Draper Laboratory, Inc. Tenth Chief of Staff of the U.S. Air Force and a member of the Joint Chiefs of Staff, Washington, D.C. Allen also served as Director of the National Security Agency (1973-1977). Member, National Academy of Engineering. The Air Force remembers Allen with the General Lew Allen, Jr., Trophy, which is presented to a base-level officer and noncommissioned officer in recognition of outstanding performance in aircraft sortie-generation.

Ames, Joseph S. (1864-1943): President of Johns Hopkins University (1929-1935). One of the first members (1915) and chairperson of the National Advisory Committee for Aeronautics (NACA), 1927-1939. When Ames retired as NACA Chairperson, President Franklin Roosevelt cited him for his "inspiring leadership in the development of new research facilities and in the orderly prosecution of comprehensive research programs." NACA renamed the Moffett Field Laboratory the Ames Aeronautical Laboratory (now the Ames Research Center) in 1944.

Anders, William A. (b. 1933): Major General, United States Air Force (Retired). NASA Astronaut (1964-1969). Apollo 8 Lunar Module Pilot. Flew the first spacecraft to leave Earth's orbit, leave the influence of Earth's gravity, and orbit the Moon (Apollo 8). Vice President and General Manager of General Electric's Nuclear Products Division, San Jose, California (1977-1980). General Manager of the General Electric Aircraft Equipment Division, Utica, New York (1980-1984). Executive Vice President for Aerospace at Textron (1984-1986) and Senior Executive Vice President for Operations (1986-1990). In 1990, Anders became Vice Chairperson of General Dynamics. On January 1, 1991, he became its Chairperson and CEO. He retired in 1993 but remained Chairperson until May, 1994.

Anderson, Michael P. (1959-2003): NASA Astronaut (1994-2003). STS-89 Mission Specialist, STS-107 Payload Commander. He was killed when the space shuttle *Columbia* (STS-107) disintegrated during orbital reentry into the Earth's atmosphere.

Armstrong, Neil A. (b. 1930): Civilian Test Pilot for NASA at Edwards Air Force Base in California, flying the X-15 (1955-1962). He made a total of seven flights in the rocket plane, reaching an altitude of 63 kilometers in the X-15-3 and a Mach number of 5.74 (6,420 kilometers per hour) in the X-15-1. Pilot involved in the canceled U.S. Air Force Dyna-Soar orbital glider program

(1960-1962). NASA Astronaut (1962-1971), Apollo 8 Backup Mission Commander, Gemini XI Backup Command Pilot, Gemini V Backup Command Pilot, Gemini VIII Command Pilot, Gemini XIII Command Pilot, and Apollo 11 Commander. First human to walk on the Moon. There is a small crater on the Moon near the Apollo 11 landing site that is named in his honor. Armstrong joined the faculty of the University of Cincinnati in 1971 and remained there as a professor of aerospace engineering until 1979. Chairperson, Presidential Advisory Committee for the Peace Corps (1971-1973). Member of the National Commission on Space (1985-1986). Vice Chairperson of the Presidential Commission on the Space Shuttle *Challenger* Accident (1986).

Arvidson, Raymond E. (b. 1948): Chair of the Earth and Planetary Sciences Department and Director of the Earth and Planetary Remote Sensing Laboratory at Washington University, St. Louis, Missouri. He was the Team Leader for the Viking Lander Imaging System for the extended missions and a Member of the Venus Orbiter Magellan Science Team. In 2005, he was the Director, Geosciences Node of NASA's Planetary Data System (PDS), an Interdisciplinary Scientist on Mars Global Surveyor and Mars Odyssey Missions, the Deputy Principal Investigator for the Mars Surveyor Rover Mission, and a Coinvestigator for the hyperspectral imaging systems on the European Mars Express and the NASA Mars Reconnaissance orbiters.

Asrar, Ghassem R. (b. 1951): NASA Science Deputy Associate Administrator (as of 2004), NASA Earth Science Enterprise (ESE) Associate Administrator (1987-2004). Since Asrar became the Associate Administrator, ESE has successfully launched the first EOS satellites and developed a comprehensive data and information system for managing the wealth of information resulting from these missions.

Atkins, Kenneth L.: Stardust Project Manager, Jet Propulsion Laboratory (JPL), Pasadena, California (1995-1999).

Ausman, Neal E., Jr.: Galileo Mission Director, Jet Propulsion Laboratory (JPL), Pasadena, California.

Austin, Gene (b. 1928): X-33 Program Manager, Marshall Space Flight Center (MSFC), Huntsville, Alabama (1996-2001).

Babakin, Georgi Nikolaevich (1914-1971): From 1965 until his death in 1971, Babakin led the development of planetary spacecraft at the Lavochkin design bureau.

Bailey, James T.: Search and Rescue Satellite-Aided Tracking (SARSAT) Program Manager.

Barmin, Vladimir Pavlovich (1909-1993): Pioneer of the Soviet rocket program who led the development of launch infrastructure for Russian rocketry.

Baron, Bob: X-38 Project Manager at Dryden Flight Research Center (DFRC), Edwards, California (1995-2002).

Bassett, Charles A., II (1931-1966): Captain, United States Air Force. NASA Astronaut, (1931-1966), original Gemini IX Pilot. Died February 28, 1966, in St. Louis, Missouri, in the crash of his T-38 jet.

Bayer, David A.: Founder (in 1993) and former Chairperson of the Board of Leo One Worldwide, Inc., the company that builds, launches, and operates a forty-eight-satellite Little LEO (low-Earth-orbit) satellite system to provide wireless messaging services throughout the world. Bayer contributed to the FCC rule-making process for cellular licensing and participated in the coalition to persuade the FCC to devise a duopoly structure for licensing.

Bean, Alan L. (b. 1932): Captain, United States Navy (Retired). NASA Astronaut (1963-1981), Apollo 12 Lunar Module Pilot, Skylab 3 Commander. Fourth human to walk on the Moon (Apollo 12). Bean resigned from NASA in June, 1981, to devote his time to painting. Many of his paintings reside on the walls of space enthusiasts. He said his decision was based on the fact that, in his eighteen years as an astronaut, he was fortunate enough to visit worlds and see sights no artist's eye, past or present, has ever viewed firsthand and he hopes to express these experiences through the medium of art. Visit the Alan Bean Art Gallery at http://www.alanbeangallery .com.

Beerer, Joseph G. (b. 1943): Flight Operations Manager for the Mars Global Surveyor, Jet Propulsion Laboratory (JPL), Pasadena, California. Beerer joined JPL in 1969 as an engineer in the Mission Design Section and has worked on a variety of spacecraft missions, including the 1971 Mariner mission to Mars, the 1973 Mariner mission to Venus and Mercury, and the 1977 Voyager mission to the outer planets of the solar system.

Beggs, James M. (b. 1926): NASA Administrator (1981-1985). He was responsible for the early testing of the space shuttle, as well as the Voyager planetary probes. Since leaving NASA, Beggs has worked as a consultant from his offices in Bethesda, Maryland. Chairperson Emeritus, SPACEHAB, Inc.

Belew, Leland F.: Skylab Program Manager, Marshall Space Flight Center (MSFC), Huntsville, Alabama. Author of many definitive books on the Skylab program.

Bellman, Donald: Lunar Landing Research Vehicle Project Manager, Flight Research Center, Edwards, California, later the Dryden Flight Research Center (DFRC).

Berry, Charles A. (b. 1923): Chief, Center Medical Operations Office, Manned Spacecraft Center (MSC), later the Johnson Space Center (JSC), in Houston, Texas, during the preshuttle era. Gemini Program physician. Berry served as Director of Medical Research and Operations for MSC (JSC) and Director of Life Sciences at NASA Headquarters in Washington, D.C., until 1974. He is the recipient of more than forty national and international awards and was nominated for the Nobel Prize in Physiology or Medicine in 1979 and 1980.

Berry, Robert: Mars Odyssey Program Manager, Lockheed Martin Space Systems, Astronautics Operations.

Bikle, Paul (1916-1991): Director of NASA's Flight Research Center (FRC), Edwards, California, later the Dryden Flight Research Center (DFRC), 1959-1971. He approved a program to build and test a prototype based on the wingless concept in 1962.

Binder, Alan (b. 1940): Lunar Research Institute Founder, Director, and President of the Board. Principal Investigator on the 1976 Viking Mars Lander Missions. Principal Investigator for Lunar Prospector.

Binnie, Brian (b. 1953): SpaceShipOne Flight Crew Member and Astronaut. On December 17, 2003, the one hundredth anniversary of the Wright brothers' first powered flight, Binnie piloted the first powered test flight of SpaceShipOne (flight 11P), which reached a top speed of Mach 1.2 and a height of 20.7 kilometers. On October 4, 2004, he piloted SpaceShipOne's second Ansari X Prize flight (flight 17P), winning the X Prize and becoming the 434th person to go into space. Reaching a height of about 112 kilometers, Binnie become only the second person to earn his astronaut wings on a non-government spacecraft.

Black, Kent M. (b. 1939): Chief Executive Officer, United Space Alliance (USA), Houston, Texas. USA, jointly owned by the Boeing Company and Lockheed Martin Corporation, manages and conducts space-operations work involving the operation and maintenance of multipurpose space systems, including systems associated with NASA's human spaceflight program, space shuttle applications beyond those of NASA, and other reusable launch and orbital systems beyond the space shuttle and space station. As the prime contractor for NASA's space shuttle program, USA is responsible for the day-to-day operation and management of the U.S. space shuttle fleet.

Blagonravov, Anatoli A. (1895-1975): Soviet representative to the United Nations' Committee on the Peaceful Uses of Outer Space (COPUOS) in the early 1960's. He was also senior negotiator with NASA's Hugh L. Dryden for cooperative space projects at the height of the Cold War in the early 1960's. He worked in developing infantry and artillery weapons in World War II and on rockets.

Bluford, Guion S., Jr. (b. 1942): Colonel, United States Air Force (Retired). NASA Astronaut (1978-1993), STS-8 Mission Specialist, STS 61-A Mission Specialist, STS-39 Mission Specialist, STS-53 Mission Specialist. First African American to fly in space. Bluford left NASA in July, 1993, to take the post of Vice President/General Manager, Engineering Services Division of NYMA, Inc., a Greenbelt, Maryland, engineering and computer software company.

Boggess, Albert, Jr.: Principal Investigator, Orbiting Astronomical Observatory B. International Ultraviolet Explorer Project Scientist. Associate Director of Science for the Space Telescope Sciences Directorate (as of 1983).

Bogomolov, Alexei F. (b. 1913): Chief Designer, OKB MEI (Moscow Energy Institute), developer of flight control systems for Russian rocketry.

Borman, Frank (b. 1928): Colonel, United States Air Force (Retired). NASA Astronaut (1962-1970), Gemini VII Command Pilot, Apollo 8 Commander. Commanded the first orbital rendezvous of two American spacecraft (Gemini VII and Gemini VI-A). Commanded the first spacecraft to leave Earth orbit, leave the influence of Earth's gravity, and orbit the Moon (Apollo 8). He retired from NASA and the Air Force in 1970, becoming special adviser to Eastern Airlines. He rose in the ranks of Eastern, becoming CEO in December, 1975. The airline business underwent many changes in the late 1970's, and Eastern did well under Borman, reporting the four most profitable years in company history during his tenure. Borman retired from Eastern in 1986.

Bossart, Karel J. (1904-1975): Engineer with the Convair Corporation and Director of Project MX-774. In the 1950's, he was largely responsible for the design of the Atlas ICBM booster with a very thin, internally pressurized fuselage instead of massive struts and a thick metal skin.

Braun, Wernher von (1912-1977): German American rocket engineer and space-exploration visionary, responsible for the development of the Redstone, Jupiter, and Saturn family of launch vehicles. Technical Director, Army Ballistic Missile Agency (ABMA) and Director of the Marshall Space Flight Center (MSFC), Huntsville, Alabama (1960-1970).

Bressette, Walter E.: Designer and developer of the Air Density Explorer inflatable sphere, inflation system, and packaging techniques.

Bridges, Roy D., Jr. (b. 1943): Major General, United States Air Force (Retired). NASA Astronaut. STS 51-F Pilot. Director, Kennedy Space Center (KSC), Florida (1997-2003). Director, Langley Research Center (LaRC), Hampton, Virginia (as of 2002).

Briegleb, William "Gus" (1912-2002): Sailplane builder from Mirage Dry Lake, California, who hand-built the M2-F1 plywood prototype in 1962.

Brown, David M. (1956-2003): NASA Astronaut (1996-2003). STS-107 Mission Specialist. He was killed on his first spaceflight when the space shuttle *Columbia* (STS-107) disintegrated during orbital reentry into the Earth's atmosphere.

Carpenter, M. Scott (b. 1925): Commander, United States Navy (Retired). NASA Astronaut (1959-1967), Mercury-Atlas 7 Pilot. The second American in orbit. Fuel consumption was a problem during the flight and a mistimed reentry burn resulted in the craft overshooting the planned landing point by 400 kilometers. He was never chosen to fly in space again, and was given an extended leave of absence to work in the Navy's Man-in-the-Sea Project as an Aquanaut in the SEALAB II program off the coast of La Jolla, California. In the summer of 1965, during the forty-five-day experiment, Carpenter spent thirty days living and working on the ocean floor. He was team leader for two of the three ten-man teams of Navy and civilian divers who conducted deep-sea diving activities in a seafloor habitat at a depth of 62.5 meters. He returned to duties with NASA as Executive Assistant to the Director of the Manned Spaceflight Center and took an active role in the design of the Apollo Lunar Landing Module and in underwater extravehicular activity (EVA) crew training. In 1967, he returned to the Navy's Deep Submergence Systems Project (DSSP) as Director of Aquanaut Operations during the SEALAB III experiment. (The DSSP office was responsible for directing the Navy's Saturation Diving Program, which included development of deep-ocean search, rescue, salvage, ocean engineering, and Man-in-the-Sea capabilities.)

Casani, John R. (b. 1932): Manager, Mariner Spacecraft Systems and Voyager Project Manager from launch to Jupiter encounters, Jet Propulsion Laboratory (JPL), Pasadena, California.

Cassini, Gian Domenico (1625-1712): Professor of astronomy at the University of Bologna. He was the first to observe Saturn's four moons. In 1675, he discovered the gap in the ring system of Saturn now known as the Cassini Division.

Cernan, Eugene A. (b. 1934): Captain, United States Navy (Retired). NASA Astronaut (1963-1976), Gemini IX-A Pilot, Apollo 10 Lunar Module Pilot, Apollo 17 Commander. Second American to walk in space (Gemini IX-A). Piloted the first Lunar Module flight in lunar orbit (Apollo 10). Commanded the last piloted lunar landing (Apollo 17). Twelfth and last human to walk on the Moon (Apollo 17). Gemini XII Backup Pilot, Gemini IX Pilot, Apollo 7 Backup Pilot.

Chaffee, Roger B. (1935-1967): Lieutenant Commander, United States Navy. NASA Astronaut (1963-1967), Apollo 1 Senior Pilot. Chaffee died along with fellow Astronauts Virgil I. "Gus" Grissom and Edward H. White II in the Apollo 1 fire at Cape Kennedy, Florida. Chaffee Crater is located in the southern hemisphere on the far side of the Moon. It lies within the huge Apollo walled plain and is one of several craters in that formation named for astronauts and people associated with the Apollo Program. NASA has memorialized the Apollo 1 crew by dedicating the hills surrounding the Mars Exploration Rover (MER) Spirit's landing site in Gusev Crater to the late astronauts.

Chandrasekhar, Subrahmanyan (1910-1995): Nobel laureate known to the world as Chandra (which means "Moon" or "luminous" in Sanskrit). Widely regarded as one of the foremost astrophysicists of the twentieth century. He was the first person to theorize that a collapsing massive star would become an object so dense that not even light could escape it (black hole). The Chandra X-Ray Observatory is named for him.

Chawla, Kalpana (1961-2003): NASA Astronaut (1995-2003). STS-87 Mission Specialist, STS-107 Mission Specialist. She died aboard STS-107 when the space shuttle *Columbia* disintegrated during reentry into the Earth's atmosphere.

Chelomei, Vladimir Nikolaevich (1914-1984): Soviet mechanics scientist and rocket engineer. Chelomei designed, built, and tested the first Soviet cruise missile in 1944. Under the leadership of Chelomei, the USSR Special Design Bureau (OKB-51) designed the first pilotless aircraft. In 1955, Chelomei was appointed the Chief Constructor of the OKB-52, where he continued to work on cruise missiles. In 1959, Chelomei was appointed Chief Constructor of Aviation Equipment and OKB-52, and, in addition to designing ICBMs, he started work on spacecraft. In 1961, OKB-52 designed the UR-500, a rocket that could launch a small two-person craft on a lunar flyby mission. Before the first launch (on March 10, 1967), UR-500 was renamed Proton. Proton was used to launch Soviet satellites and the Salyut and Mir space stations. In the 1970's, Chelomei's OKB worked on the Almaz orbital complex, which became the basis for Salyut 2, Salyut 3, and Salyut 5, which were also designed by Chelomei and his OKB.

Cheng, Andrew Francis (b. 1951): Project Scientist for the NEAR (Near Earth Asteroid Rendezvous) Shoemaker Mission at the Johns Hopkins University. Co-investigator, New Horizons Project, Johns Hopkins University Applied Physics Laboratory.

Chertok, Boris E. (b. 1912): Pioneer of the Soviet rocket development program. He worked as deputy to Sergei Korolev at OKB-1, where he was responsible for flight control systems. OKB-1 was the classified Soviet name of Korolev's design bureau, which later became the Energia Rocket and Space Corporation (RSC Energia).

Christy, James W. (b. 1938): Astronomer at the U.S. Naval Observatory. On June 22, 1978, he discovered Pluto's moon and named it Charon (after the Greek mythological ferryman who carried souls across the river Styx to Pluto's underworld). Christy named it Charon, but pronounced it differently. The "ch" at the beginning of the moon's name is soft so it sounds like "Sharon," after the astronomer's wife Charlene, nicknamed Char, which both have soft "ch" sounds. The mythological figure's name is pronounced with a hard "ch" sound like the modern letter "k," like "ch" in Christy's name. The name "Charon" was officially accepted by the International Astronomical Union in 1985.

Citron, Robert (b. 1932): Kistler Aerospace Cofounder and President (1993-1996). From 1956-1976, Citron managed a series of space research projects under contracts to NASA for the Smithsonian Institution Astrophysical Observatory. Citron founded several successful companies in the fields of global communications, scientific field research, publishing, and commercial space development. Founded SPACEHAB, Inc. (1982), the only entrepreneurial company to develop a successful commercial-crewed spaceflight module.

Clark, Laurel B. (1961-2003): NASA Astronaut (1996-2003). STS-107 Mission Specialist. She died aboard STS-107 when the space shuttle *Columbia* disintegrated during reentry into the Earth's atmosphere.

Clarke, Sir Arthur C. (b. 1917): Science-fiction author. His most important contribution may be the concept that geostationary satellites are ideal telecommunications relays. He proposed this concept in a paper titled "Extra-Terrestrial Relays: Can Rocket Stations Give Worldwide Radio Coverage?" published in *Wireless World* in October, 1945. The geostationary orbit is now sometimes known as the Clarke orbit in his honor. However, it is not clear that his article was actually the inspiration for modern telecommunications satellites. He also wrote several novels, the most well known of which is *2001: A Space Odyssey,* based on a screenplay of the same name that he prepared for film director Stanley Kubrick.

Cobb, Geraldyn "Jerrie" M. (b. 1931): Aviator and the first female astronaut candidate in 1960. Set world records in aviation for speed, distance, and absolute altitude. Member of the Mercury 13 group and participated in the first female astronaut program in 1959. Though in the top 2 percent of her Mercury Program graduating class, she could not be an astronaut because NASA had determined that space travel was not safe for women. At the time, she had already flown sixty-four types of aircraft, including a jet fighter. She had set records for speed, distance, and absolute altitude.

Cochran, Jacqueline "Jackie" (b. Bessie Lee Pittman; 1906-1980): Aviation pioneer. In 1939, she set a new altitude and international speed record and received the Clifford Burke Harmon Trophy (given to the outstanding woman flier in the world) five times. Encouraged by her pilot friend Chuck Yeager, on May 18, 1953, at Rogers Dry Lake, California, she flew a Canadian F-86 Sabre jet at an average speed of 1,050 kilometers per hour, becoming the first woman to break the sound barrier. She was also the first woman to land and take off from an aircraft carrier.

Cohen, Aaron: NASA Acting Deputy Administrator (1992). Director, Johnson Space Center (JSC), Houston, Texas (1986-1993).

Collins, Eileen M. (b. 1956): Lieutenant Colonel, United States Air Force. NASA Astronaut (as of 1990), STS-63 Pilot, STS-84 Pilot, STS-93 Commander, STS-114 Commander. The first woman to pilot a space shuttle mission; the first woman to pilot the shuttle twice; the first woman to command a space shuttle mission.

Collins, Michael (b. 1930): Major General, United States Air Force (Retired). Experimental Test Pilot at the Air Force Flight Test Center, Edwards Air Force Base, California (1959-1963). NASA Astronaut (1963-1970), Gemini X Pilot, Apollo 11 Command Module Pilot. He wrote a book, *Carrying the Fire*, about his experiences as an astronaut. There is a small crater on the Moon near the Apollo 11 landing site that is named in his honor. He has a star on the Hollywood Walk of Fame, and Asteroid 6471 is named in his honor.

Compton, Arthur Holly (1892-1962): Chair of the physics department at Washington University in St. Louis. He and C. T. R. Wilson of England won the Nobel Prize in Physics for their discovery and explanation of the wavelength changes in diffused x rays when they collide with electrons in 1927. The Compton Gamma Ray Observatory is named after him.

Congreve, Sir William (1772-1828): British artillery officer and inventor best known for his work on black powder rockets used for bombardment of enemy fortifications. He based his rocketry on the pioneering work of Indian prince Hyder Ali, who had successfully used them against the British in 1792 and 1799 at Seringapatam. Congreve's rockets were used in the Napoleonic Wars and in the War of 1812.

Conrad, Charles "Pete," Jr. (1930-1999): Captain, United States Navy (Retired). NASA Astronaut (1962-1973), Gemini V Pilot, Gemini XI Command Pilot, Apollo 12 Commander, Skylab 2 Commander, Gemini VIII Backup Command Pilot, Backup Apollo 9 Mission Commander. Third human to walk on the Moon (Apollo 12). Commander of the first American space station mission (Skylab 2).

Cook, Richard: Project Manager, Mars Polar Lander and Deep Space 2, Mars Pathfinder Mission Manager, Jet Propulsion Laboratory (JPL), Pasadena, California (1997-2004). Project Manager, Mars Science Laboratory (as of 2004).

Cooper, L. Gordon, Jr. (1927-2004): Colonel, United States Air Force (Retired). NASA Astro-naut (1959-1970), Mercury-Atlas 9 Pilot, Gemini V Command Pilot. Flew the first American daylong orbital mission. Cooper served as a test pilot in the U.S. Air Force before being selected as a Mercury astronaut in April, 1959. Cooper is the last American astronaut to orbit the Earth alone for an entire mission. During his flight, he became the first astronaut to sleep in space. He was the first member of the active-duty military to address joint sessions of Congress twice. Vice President for Research and Development of EPCOT for Walter E. Disney Enterprises, Inc., the research and development subsidiary of Walt Disney Productions (1973-1975).

Coughlin, Thomas B.: Programs Manager for Johns Hopkins University Applied Physics Laboratory (APL) Space Department. NEAR (Near Earth Asteroid Rendezvous) Shoemaker Mission Project Manager at APL. New Horizons Project Manager.

Crews, Albert, Jr. (b. 1929): Captain, United States Air Force. Pilot Astronaut in the X-20 Dyna-Soar program (1960-1963). He retired as an astronaut when the Dyna-Soar program was canceled on December 10, 1963.

Crippen, Robert L., Jr. (b. 1937): Captain, United States Navy (Retired). NASA Astronaut (1969-1995), Pilot STS-1, Commander STS-7, Commander STS 41-C, Commander STS 41-G. Deputy Director, Shuttle Operations, Kennedy Space Center (KSC), Florida (1986-1989). Responsible for final shuttle preparation, mission execution, and return of the orbiter to Kennedy after its landing at Edwards Air Force Base, California. Director, space shuttle program, NASA, Washington, D.C. (1990-1992). Director, KSC (1992-1995). Piloted the first spaceflight of the space shuttle (STS-1).

Crisp, Joy: Mars Exploration Rovers Project Scientist, Jet Propulsion Laboratory (JPL), Pasadena, California (as of 2000).

Crossfield, A. Scott (b. 1921): Aeronautical Research Pilot with the National Advisory Committee for Aeronautics (NACA), 1950-1955, flying the X-1 and D-558-II rocket planes and other experimental jets. Chief Engineering Test Pilot for North American Aviation, Inc. (1955-1961), he flew nearly all of the experimental aircraft tested at Edwards, including the X-1, XF-92, X-4, X-5, D-558-I, and the Douglas D-558-II Skyrocket. The first man to fly at twice the speed of sound (Mach 2), in the D-558-II in 1953, Crossfield reached Mach 2.11 in the first powered flight of the X-15 in 1959.

Cunningham, Glenn E. (b. 1943): Project Manager, Mars Observer, Jet Propulsion Laboratory (JPL), Pasadena, California. Cunningham joined JPL in 1966 as an engineer in the Spacecraft System Design and Integration Section and has worked on a variety of spacecraft missions, including the 1969 Mariner spacecraft to Mars and the 1977 Voyager spacecraft to the outer planets of the solar system. He has also led studies for robotic precursor missions in support of eventual human exploration of Mars.

Dallas, Sam: Manager, Mars Observer Mission, Jet Propulsion Laboratory (JPL), Pasadena, California.

Dana, William H. (b. 1930): Chief Engineer, Dryden Flight Research Center (DFRC), Edwards, California (1993-1998). Chief Pilot at Dryden (1986-1993). Pilot Astronaut in the X-20 Dyna-Soar program (1960-1963). He was a Project Pilot on the hypersonic X-15 research aircraft and flew the rocket-powered vehicle sixteen times, reaching a top speed of 6,272 kilometers per hour and a peak altitude of 94 kilometers. He was the pilot on the final (199th) flight of the ten-year program. In the late 1960's and in the 1970's, Dana was a Project Pilot on the manned lifting body program and flew several versions of the wingless vehicles: M2-F1, HL-10, M2-F3, and X-24B flights.

DeBra, Daniel B.: Edward C. Wells Professor Emeritus in the Department of Aeronautics and Astronautics and the Department of Mechanical Engineering at Stanford University. Gravity Probe B Coprincipal Investigator.

Debus, Kurt H. (1908-1983): Director, Kennedy Space Center (KSC), Florida (1962-1974). He supervised the development and construction of rocket launch facilities at Cape Canaveral for the Redstone, Jupiter, Jupiter C, Juno, and Pershing military configurations beginning in 1952 and continuing through 1960. Beginning in 1961, he directed the design, development, and construction of NASA's Apollo/Saturn facilities on Cape Canaveral and the adjacent Kennedy Space Center.

DeFrance, Smith J. (1896-1985): Military aviator and aeronautical engineer. He worked in the flight research section at Langley Aeronautical Laboratory, Hampton, Virginia, and designed its 30-by-60-foot (9.14-by-18.29 meter) wind tunnel, which at the time was the largest ever built (1929-1931). Director, Ames Aeronautical Laboratory, Moffett Field, California (1940-1965). The center built nineteen major wind tunnels and conducted extensive flight research, including the blunt-body research necessary for safely returning spacecraft from orbit through the Earth's atmosphere.

Diamandis, Peter H.: Cofounder, Chairperson, and CEO of Zero Gravity Corporation, a privately held space entertainment and tourism company, and the first and only provider of weightless flight for the general public approved by the Federal Aviation Administration. Chairperson of the Ansari X Prize Foundation.

Diaz, Alphonso: NASA Associate Administrator of Science (as of 2004). Director, Goddard Space Flight Center (GSFC), Greenbelt, Maryland (1998-2004).

Dittemore, Ronald D. (b. 1952): Space shuttle Program Manager (1999-2003), responsible for the overall management, integration, and operations of the space shuttle program.

Donlan, Charles J.: Associate Director of the Space Task Group (1958-1961), which helped establish what would become the Johnson Space Center (JSC), Houston, Texas. Associate Director, Langley Research Center (LaRC), Hampton, Virginia (1961-1967), Deputy Director at Langley (1967-1968). Deputy Associate Administrator for Manned Space Flight (1968-1970). Acting Director of the space shuttle program (1970-1973). Consultant for the Institute for Defense Analyses, which studied military uses for the shuttle (1976-1988).

Dordain, Jean-Jacques (b. 1946): European Space Agency Director General (as of 2003).

Douglas, D. W.: Mariner-Venus (later Magellan) Missions Operation System Manager at the Jet Propulsion Laboratory (JPL), Pasadena, California.

Drake, Hubert: Engineer, NASA Flight Research Center (FRC), Edwards, California, later the Dryden Flight Research Center (DFRC), credited with the idea of a Lunar Landing Research Vehicle.

Draper, Charles S. (1901-1987): Physics professor at the Massachusetts Institute of Technology (MIT). Founded the Instrumentation Laboratory at MIT. Its first major achievement was the Mark 14 gyroscopic gun sight for Navy antiaircraft guns. Draper and the laboratory applied gyroscopic principles to the development of inertial guidance systems for airplanes, missiles, submarines, ships, satellites, and space vehicles, notably those used in the Apollo Moon landings.

Dryden, Hugh L. (1898-1965): Chairperson, National Advisory Committee for Aeronautics (NACA), 1947-1958. Deputy Administrator, NASA (1958-1965). He helped shape policy that led to development of the high-speed research program and its record-setting X-15 rocket aircraft. He established vertical and short takeoff and landing aircraft programs and sought solutions to the problem of atmospheric reentry for piloted spacecraft and ballistic missiles. The Dryden Flight Research Center (DFRC), Edwards, California, is named for him.

Duke, Charles M., Jr. (b. 1935): Brigadier General, United States Air Force (Retired). NASA Astronaut (1966-1975), Apollo 16 Lunar Module Pilot. Tenth human to walk on the Moon (Apollo 16). Duke served as CapCom for Apollo 11. As a result, his distinctive southern drawl became familiar to viewers around the world. He was

Backup Lunar Module Pilot on Apollo 13, but, shortly before the mission, he caught German measles from a friend's child. He had exposed it to the prime crew, which resulted in the replacement of Thomas K. "Ken" Mattingly by John L. "Jack" Swigert, as Mattingly had no natural immunity to the disease.

Duxbury, Thomas: Stardust Project Manager, Jet Propulsion Laboratory (JPL), Pasadena, California (as of 2000). Duxbury has actively participated in a wide variety of space missions, including the Science Imaging Teams of Mariners 6, 7, 9, and 10; Viking Orbiters 1 and 2; Viking Lander 1; Pioneers 10 and 11; Voyagers 1 and 2; the Soviet PHOBOS Mission (PHOBOS 88); Mars Observer (MO); Mars Global Surveyor (MGS); the DoD/NASA Clementine Mission; and the Russian Mars 96 Mission (Mars 96).

Earls, Julian M.: Director, Glenn Research Center (GRC), Cleveland, Ohio, as of 2003. Since beginning his career with NASA in 1965, Earls has written twenty-eight publications for technical and educational journals. He wrote the first health-physics guides used at NASA. He has been a Distinguished Honors Visiting Professor at numerous universities throughout the nation. On two occasions, he has been awarded NASA medals for exceptional achievement and outstanding leadership.

Eggers, Alfred J., Jr. (b. 1922): Assistant Director for Research and Development Analysis and Planning at what later became NASA's Ames Research Center, Moffett Field, California. Eggers conceived the original idea of lifting bodies in 1957.

Emme, Eugene M. (1919-1985): NASA Chief Historian (1959-1979). Prior to joining NASA, he was a historian with the Air University of the U.S. Air Force.

Engle, Joseph H. (b. 1932): Brigadier General, United States Air Force (Retired). NASA Astronaut (1966-1986). Test pilot in the X-15 research program at Edwards Air Force Base, California (1963-1966). Three of his sixteen flights in the X-15 exceeded an altitude of 80 kilometers (the altitude that qualifies a pilot for astronaut rating). Prior to that time, he was a test pilot in the Fighter Test Group at Edwards. Commander of STS-2 and STS 51-I. Deputy Associate Administrator for Manned Space Flight at NASA Headquarters (1982). Member of the Presidential Commission on the Space Shuttle *Challenger* Accident (1986).

Erickson, Jim: Mars Exploration Rovers Project Manager at the Jet Propulsion Laboratory (JPL), Pasadena, California (as of 2004).

Erickson, Kerry: Galaxy Evolution Explorer (GALEX) Project Manager at the Jet Propulsion Laboratory (JPL), Pasadena, California (as of 2004).

Estess, Roy S.: Acting Director, Johnson Space Center (JSC), Houston, Texas (2001-2002). Director, Stennis Space Center (SSC), Hancock County, Mississippi (1989-2001).

Everitt, C. W. Francis (b. 1924): Experimental physicist, Stanford University. Gravity Probe B Principal Investigator (as of 1981).

Faget, Maxime A. (1921-2004): Research Scientist, Langley Research Center (LaRC), Hampton, Virginia (1946-1958). Chief, Flight Systems Division, Space Task Group (1958-1961). Director of Engineering and Development, Manned Spacecraft Center, later the Johnson Space Center (JSC) in Houston, Texas (1961-1981). He was the designer of the Mercury spacecraft and a contributor to the Gemini and Apollo spacecraft and the space shuttle. After his retirement, Faget was among the founders of Space Industries, Inc., established in 1982. One of the company's projects was the Wake Shield Facility, a device to create a near-perfect vacuum in space.

Fanson, James: Galaxy Evolution Explorer (GALEX) Project Manager at the Jet Propulsion Laboratory (JPL), Pasadena, California (2003-2004).

Figueroa, Orlando (b. 1955): Director of the Mars Exploration Office at NASA Headquarters (as of 2001). Director of NASA's Office of Solar System Exploration (as of 2003). Mars Odyssey Program Director.

Fimmel, Richard O.: Pioneer Venus and Pioneer Jupiter Program Manager, Ames Research Center, Moffett Field, California.

Fisher, Anna L. (b. 1949): NASA Astronaut (1978-present). One of the six women named as NASA's first female astronaut candidates in January, 1978. Mission Specialist, STS-51-A. She participated in the first space salvage mission in history (STS 51-A) with the return to Earth of the Palapa B2 and Westar 6 satellites. She was a Deputy for the Mission Development Branch of the Astronaut Office (1986-1987) and joined the Astronaut Selection Board for the 1987 class. Fisher took an extended leave of absence to raise her family (1989-1996) but returned to the Operations Planning Branch of the Astronaut Office for the International Space Station (ISS) from 1996 to 1997. She was Branch Chief for the ISS Operations Planning Branch (1997-1998) and Deputy for Operations/Training, Space Station Branch (1998-1999). Beginning in 2000, she has been Astronaut Office representative on numerous Space Station Program Boards and Multilateral Boards.

Fisher, William F. (b. 1946): NASA Astronaut (1978-1991). STS 51-I Mission Specialist. Performed a successful on-orbit rendezvous with the ailing 7,000-kilogram Syncom IV-3 satellite, which was repaired in space and returned to orbit (STS 51-I).

Fisk, Lennard A.: Chair, Space Studies Board, National Research Council (2000-2005). NASA Associate Administrator, Office of Space Science and Applications (1987-1993).

Fletcher, James C. (1919-1991): NASA Administrator from April 27, 1971, to May 1, 1977, and from May 12, 1986, to April 8, 1989. Responsible for starting the Space Transportation System and the Viking-Mars Landing program. During his second term, he helped NASA recover from the *Challenger* accident.

Friedman, Peter: Galaxy Evolution Explorer (GALEX) Project Scientist at the Jet Propulsion

Laboratory (JPL), Pasadena, California (as of 1996).

Frosch, Robert A. (b. 1928): NASA Administrator (1977-1981). He was responsible for overseeing the continuation of the space shuttle development, as well as the early orbiter approach and landing test program.

Gagarin, Yuri A. (1934-1968): Soviet Cosmonaut who became the first human to travel into space on April 12, 1961. On March 27, 1968, he and his instructor were killed in a crash of a MiG-15 on a routine training flight near Moscow. It is uncertain what caused the crash, but a 1986 inquest suggests that the turbulence from a Su-11 using afterburners may have caused the craft to go out of control. Gagarin Crater is located in the southern hemisphere on the far side of the Moon.

Galileo Galilei (1564-1642): Italian astronomer. He made a series of telescopes whose optical performance was much better than any other of that period. In 1613 he discovered that, when seen in the telescope, the planet Venus showed phases like those of the Moon and therefore must orbit the Sun, not the Earth. He discovered the four small bodies orbiting Jupiter. With an eye on getting a job in Florence, he promptly named these "the Medicean stars," after the powerful Medici family. Later, they were confirmed to be the four large moons of Jupiter and were renamed the Galilean moons in his honor.

Gallagher, David: Space Infrared Telescope Facility (SIRTF) Project Manager, Jet Propulsion Laboratory (JPL), Pasadena, California (2001-2004). SIRTF was renamed the Spitzer Space Telescope.

Gardner, Trevor (1915-1963): Special Assistant to the Secretary of the Air Force (1953-1956). Presidential Space Task Force Commissioner (1960-1961). He was instrumental in securing approval for the creation of the intercontinental ballistic missile (ICBM) system.

Gavit, Sarah: Associate Manager, Interstellar and Solar Sail Technology Program (1999-2002).

Project Manager, Deep Space 2, Jet Propulsion Laboratory (JPL), Pasadena, California.

Geldzahler, Barry: Program Executive for Space Operations, NASA Deep Space Network. Stardust Program Executive.

Giberson, Walker E. "Gene": Surveyor Program Manager, Jet Propulsion Laboratory (JPL), Pasadena, California (1960-1965). Mariner Project Manager, JPL.

Gidzenko, Yuri Pavlovich (b. 1962): Russian Air Force Lieutenant Colonel, Soviet/Russian Cosmonaut (as of 1984). Mir Expedition Twenty Primary Crew Commander, Soyuz Commander and International Space Station (ISS) Expedition One Crew Member, Soyuz TM-34 (Soyuz Taxi 3) ISS Visiting Crew 2 Commander.

Gilruth, Robert R. (1913-2000): Engineer at the National Advisory Committee for Aeronautics (NACA). Worked at the Langley Aeronautical Laboratory, Hampton, Virginia (1937-1946). First Director of the Manned Spacecraft Center (MSC), later the Johnson Space Center (JSC), in Houston, Texas (1961-1972). Project Mercury Director (1959-1961).

Givens, Edward G., Jr. (1930-1967): NASA Astronaut (1966-1967). Assistant to the Commandant at the Air Force Experimental Flight Test Pilot School (1961-1962). He was responsible for developing the Astronaut Maneuvering Unit. He was killed in an automobile accident while driving his Volkswagen home from a meeting of the Society for Experimental Test Pilots.

Glahn, Carl W.: Mariner-Mars Program Manager at NASA Headquarters.

Glenn, John H., Jr. (b. 1921): U.S. Senator (D-Ohio), 1974-1999. Colonel, United States Marine Corps (Retired). NASA Astronaut (1959-1964). Mercury-Atlas 6 Pilot, STS-95 Payload Specialist. First American to fly in Earth orbit. Glenn is the oldest person to fly in space. The John H. Glenn Research Center (GRC), Cleveland, Ohio (formerly the Lewis Research Center), is named after him.

Glennan, T. Keith (1905-1995): First NASA Administrator (1958-1961). He incorporated into

NASA several organizations involved in space exploration projects from other federal agencies to ensure that a viable space exploration program could be reasonably conducted over the long term. He brought part of the Naval Research Laboratory into NASA and created the Goddard Space Flight Center (GSFC), in Greenbelt, Maryland. He also incorporated several disparate satellite programs, two lunar probes, and the important research effort to develop a million-pound-thrust, single-chamber rocket engine from the Air Force and the Department of Defense's Advanced Research Projects Agency. In December, 1958, Glennan also acquired control of the Jet Propulsion Laboratory, a contractor facility operated by the California Institute of Technology in Pasadena, California. In 1960, Glennan obtained the transfer to NASA of the Army Ballistic Missile Agency (ABMA), located at Huntsville, Alabama, and renamed it the Marshall Space Flight Center (MSFC).

Glushko, Valentin Petrovitch (1908-1989): Soviet rocket design pioneer. From 1946 until 1974, he was head of OKB-456, an independent design bureau that was a prominent developer of rocket engines within the Soviet Union. Among his designs was the powerful RD-170 liquid-propellant engine. In 1974, Glushko was selected to lead NPO Energia, a major Russian developer of piloted spacecraft. He had long criticized the N1-13 moon rocket, and one of his first steps was to request the cancellation of that program. He was an advocate of a new line of powerful launchers, which he wanted to use for the establishment of a Russian lunar base. However, the Apollo Program had come to an end, and the government wanted to build a craft to compete with the space shuttle. The Moon's Glushko Crater is named after him.

Goddard, Robert H. (1882-1945): Rocket scientist known as the "father of modern rocketry." He developed the first practical liquid-propellant rocket. Though his work in the field was revolutionary, he was often ridiculed for his theories, which were ahead of their time. He received little recognition during his own lifetime. By 1914, he was designing rocket motors with financial assistance from the Smithsonian Institution, and, by 1919, he was writing about the possibilities of Moon flight. Goddard launched the first liquid-fueled rocket on March 16, 1926, at Auburn, Massachusetts. The Goddard Space Flight Center (GSFC), in Greenbelt, Maryland, was named in his honor in 1959.

Goett, Harry J. (1910-2000): Director, Goddard Space Flight Center (GSFC), Greenbelt, Maryland (1959-1965). After holding a number of engineering posts with private firms, he became a project engineer at Langley Aeronautical Laboratory, in Hampton, Virginia, in 1936. He later moved to Ames Aeronautical Laboratory, at Moffett Field, California, where he was Chief of the Full-Scale and Flight Research Division (1948-1959). Special assistant to NASA Administrator James E. Webb (1965).

Goldin, Daniel S. (b. 1940): NASA Administrator (1992-2001). During his tenure, the Agency's civil service workforce was reduced by about a third, while the headquarters' civil service and contractor workforce was reduced by more than half. These reductions were accomplished without resorting to forced layoffs. At the same time, NASA's productivity gains climbed 40 percent. Goldin also cut the time required to develop Earth- and space-science spacecraft by 40 percent and reduced the cost by two-thirds, simultaneously increasing the average number of missions launched per year by a factor of four. During this time, space shuttle costs were reduced by about a third, while all safety indicators and mission capabilities achieved significant improvements. To expand opportunities for public and educational participation in the adventure of space exploration and research, Goldin directed NASA's program managers to incorporate Internet access into mission-outreach plans.

Gordon, Henry C. (1925-1996): Colonel, United States Air Force (Retired). Pilot Astronaut in the

X-20 Dyna-Soar program (1960-1963). Gordon was selected as an astronaut in the X-20 Dyna-Soar program in April, 1960, and began training at the Air Force Flight Test Center at Edwards Air Force Base in California. He retired as an astronaut when the Dyna-Soar program was canceled on December 10, 1963.

Gordon, Richard F., Jr. (b. 1929): Captain, United States Navy (Retired). NASA Astronaut (1963-1972). Apollo 12 Command Module Pilot, Gemini XI Pilot. Test Pilot and winner of the Bendix Trophy Race from Los Angeles to New York in May, 1961, he established a new speed record of 1,400 kilometers per hour and a transcontinental speed record of 2 hours and 47 minutes. He has logged more that 4,500 hours flying time, including 3,500 hours in jet aircraft.

Graf, James E.: Project Manager, Mars Reconnaissance Orbiter (MRO), Jet Propulsion Laboratory (JPL), Pasadena, California.

Graham, William R., Jr.(b. 1937): NASA Deputy Administrator (1985-1986). NASA Acting Administrator (1985-1986). Graham left NASA in October, 1986, to become Director of the White House Office of Science and Technology Policy. Science Advisor to the President (1986-1989).

Grammier, Rick: Deep Impact Project Manager at the Jet Propulsion Laboratory (JPL), Pasadena, California.

Gregory, Frederick D. (b. 1941): Colonel, United States Air Force (Retired). NASA Astronaut (as of 1978). NASA Acting Administrator (February, 2005-April, 2005). NASA Deputy Administrator (as of 2002). Pilot on STS 51-B. Commander of STS-33 and STS-44. NASA Associate Administrator for Space Flight (2001-2002). NASA Associate Administrator, Office of Safety and Mission Assurance (1992-2001).

Griffin, Michael (b. 1949): NASA Administrator (as of 2004). Griffin was President and Chief Operating Officer of In-Q-Tel, Inc., before joining Johns Hopkins University in April, 2004. He also served in several positions within Orbital Sciences Corporation of Dulles, Virginia, including Chief Executive Officer of Magellan Systems, Inc. Earlier in his career, Griffin served as Chief Engineer and Associate Administrator for Exploration at NASA headquarters and also worked at NASA's Jet Propulsion Laboratory. He also served as Deputy for Technology at the Strategic Defense Initiative Organization.

Griffith, Douglas G.: Magellan Project Manager, Jet Propulsion Laboratory (JPL), Pasadena, California.

Grissom, Virgil I. "Gus" (1926-1967): Lieutenant Colonel, United States Air Force. NASA Astronaut (1959-1967), Mercury-Redstone 4 Pilot, Gemini 3 Commander. The second American in space. Commander of first Gemini mission and Apollo 1. Grissom was killed along with fellow astronauts Edward H. White II and Roger B. Chaffee in the Apollo 1 fire at Cape Kennedy, Florida. He is buried at Arlington National Cemetery. Grissom Crater lies on the far side of the Moon. It is located just to the south of the huge Apollo impact basin and to the northeast of Cori Crater. NASA has memorialized the Apollo 1 crew by dedicating the hills surrounding the Mars Exploration Rover (MER) Spirit's landing site in Gusev Crater to the late astronauts.

Hagen, John P. (1908-1990): Director, Vanguard Program (1955-1960). NASA Assistant Director of Spaceflight Development (1958-1960).

Haglund, Howard H.: Surveyor Program Manager, Jet Propulsion Laboratory (1966-1968).

Haise, Fred W., Jr. (b. 1933): NASA Astronaut (1966-1979). Apollo 13 Lunar Module Pilot. Aeronautical Research Pilot, Lewis Research Center (now the Glenn Research Center), Cleveland, Ohio (1959-1963). Research Pilot at Dryden Flight Research Center (DFRC), Edwards, California (1963-1966). He flew five flights as the Commander of the space shuttle *Enterprise* in 1977 for the approach and landing test program at Edwards Air Force Base.

Hall, Charles F. (b. 1920): Pioneer Project Manager (1969-1980).

Halley, Sir Edmond (1656-1742): Astronomer. He proposed using transits of Mercury and Venus to determine the distance to the Sun and therefore the scale of the solar system using Kepler's third law. Beginning around 1695, Halley made a careful study of the orbits of comets. Using his theory of cometary orbits, he calculated that the comet of 1682 (now called Halley's Comet) was periodic and was the same object as the comets of 1531 and 1607.

Haynes, Norman Ray (b. 1936): Voyager Program Manager for the Neptune encounter, Jet Propulsion Laboratory (JPL), Pasadena, California. Director of Mars Exploration Directorate, JPL (1996-2001).

Heacock, Raymond L. (b. 1928): Voyager Program Manager for the Jupiter and Saturn encounters, Jet Propulsion Laboratory (JPL), Pasadena, California.

Holloway, Tommy W. (b. 1940): Space shuttle Program Manager (1995-1999). International Space Station Program Manager (1999-2002).

Holmes, D. Brainerd (b. 1921): NASA Associate Administrator for Manned Space Flight (1961-1963).

Howell, Jefferson D., Jr. (b. 1940): Lieutenant General, United States Marine Corps (Retired). Director, Johnson Space Center (JSC), Houston, Texas (as of March, 2002). Howell guided JSC through the difficult days after the February 1, 2003, loss of *Columbia* on STS-107, contributing to the investigation of the accident and beginning the process of return to shuttle flight.

Hubble, Edwin P. (1889-1953): Astronomer. He proved that galaxies are indeed "island universes" and outlined a classification system for galaxies that is still in use. His greatest discovery was the linear relationship between a galaxy's distance and the speed with which it is moving. The ratio of the two is known as the Hubble constant. The Hubble Space Telescope is named for him.

Husband, Rick D. (1957-2003): NASA Astronaut (1994-2003). STS-96 Pilot, STS-107 Commander. He was killed when the space shuttle *Columbia* (STS-107) disintegrated during orbital reentry into the Earth's atmosphere.

Huygens, Christiaan (1629-1695): Astronomer. Around 1654, he devised a new and better way of grinding and polishing lenses. Using one of his own lenses, he detected the first moon of Saturn in 1655.

Irwin, James B. (1930-1991): Colonel, United States Air Force (Retired). NASA Astronaut (1966-1972). Apollo 15 Lunar Module Pilot and eighth human to walk on the Moon (Apollo 15). He was the first to drive the Lunar Roving Vehicle on the Moon (Apollo 15). He left NASA and retired from the Air Force in 1972 and founded High Flight, a Christian ministry. Beginning in 1973, Irwin led several expeditions to Mount Ararat, Turkey, in search of the remains of Noah's Ark.

Isayev, Alexei (1908-1971): Soviet rocket designer. Isayev worked with Soviet space program Chief Designer Sergei Korolev on all early Soviet spacecraft and interplanetary probes as head of KB Isayev's design bureau. Developed numerous liquid-fuel rocket engines and gas turbines. Created first non-modular liquid-fuel rocket engine.

Jacobson, Richard K. "Jake" (d. 2001): President and Chief Operating Officer, SPACEHAB, Inc. (1987-1991).

Jansky, Karl G. (1905-1950): Researcher for Bell Laboratories in New Jersey. While studying the static that often disrupted radio communications, he discovered interstellar radio waves, thus creating the field of radio astronomy. He was the first to detect radio waves in the Milky Way in 1932. In 1973, the International Astronomical Union named the "jansky" in his honor. The jansky is the flux density of a radio wave in that it is power per unit area per interval of frequency; 1 jansky (Jy) = 10^{-26} watt/square meter/hertz (W/m^{2}Hz).

Jarvis, Gregory B. (1944-1986): Communications Subsystem Engineer, Hughes Aircraft Company (1973-1986). STS 51-L Payload Specialist. He died when the space shuttle *Challenger* was destroyed during launch on mission STS 51-L. Only one year before his death, Jarvis gave an inspirational commencement address at the 1985 graduation ceremony at the State University of New York at Buffalo. After his death, students nailed a sign reading "Jarvis Hall" to the Engineering East building. The name was made official in a 1987 dedication ceremony.

Johnson, Lyndon B. (1908-1973): President of the United States (1963-1969). The major presidential supporter of NASA funding. In 1973, congressional legislation renamed the Manned Spacecraft Center in Houston, Texas, the Lyndon Baines Johnson Space Center.

Kármán, Theodore von (1881-1963): Engineer and physicist who was primarily active in the fields of aeronautics during the 1940's and 1950's. He is responsible for many key advances in aerodynamics, notably in supersonic- and hypersonic-airflow characterization. In 1944, he helped found the Jet Propulsion Laboratory (JPL) in Pasadena, California, and in 1946 became the first Chairperson of the Scientific Advisory Group. The group studied aeronautical technologies for the Army Air Force. He also helped found AGARD, the NATO aerodynamics research oversight group (1951), the International Council of the Aeronautical Sciences (1956), the International Academy of Astronautics (1960), and the Von Kármán Institute in Brussels.

Kearney, Michael: President and Chief Executive Officer, SPACEHAB, Inc. (as of 2003).

Keathley, William C.: Project Manager for the Skylab Apollo Telescope Mount experiments. Chief, Skylab Optical Telescope Assembly Project. In 1977, he was named manager of the Space Telescope project (later named the Hubble Space Telescope).

Keldysh, Msitslav Vsevolodovich (1911-1978): Physicist and mathematician who became the chief theoretician of Soviet cosmonautics in the 1960's. He had previously served many years in a variety of positions at the Central Institute of Aerohydrodynamics at Moscow University and

the Steklov Mathematical Institute. Vice President of the Soviet Academy of Sciences (1960-1961), President of the Soviet Academy of Sciences (1961-1975).

Kennedy, James W.: Director, Kennedy Space Center (KSC), Florida (as of 2003). Prior to this appointment, he served as KSC's Deputy Director and earlier as the Deputy Director, Marshall Space Flight Center (MSFC), Huntsville, Alabama.

Kennedy, John F. (1917-1963): President of the United States, 1961-1963. Merritt Island, adjacent to Cape Canaveral, Florida, was renamed the John F. Kennedy Space Center (KSC), in November, 1963, shortly after he was assassinated. Cape Canaveral was also renamed Cape Kennedy, but this change was unpopular with the local people and the name reverted in 1973.

Kibalchich, Nikolai Ivanovitch (1854-1881): Russian revolutionary activist who is considered to be an author of one of the earliest proposals for a rocket-powered piloted flying apparatus. Kibalchich was executed for his role in the conspiracy to assassinate Czar Alexander II.

Kizim, Leonid (b. 1941): Russian Cosmonaut (1965-1987). He flew as Commander on Soyuz T-3, Soyuz T-10, and Soyuz T-15, the last mission to the Salyut 7 space station (Expedition Five), and the first to the Mir Complex (Expedition One).

Kleinknecht, Kenneth S. (b. 1919): Skylab Project Manager, Manned Spacecraft Center (MSC), later the Johnson Space Center (JSC), in Houston, Texas (1970-1974). Gemini Program Deputy Manager. Mercury Project Manager. He started his career in 1942 at the Lewis Research Center (now the Glenn Research Center). In 1951, Kleinknecht transferred to the Flight Research Center (FRC), Edwards, California, later the Dryden Flight Research Center (DFRC). After NASA was formed, he transferred to the MSC in 1959. Kleinknecht also served as the Advanced Projects Management Officer on the X-15 project and as the Technical Assistant to the Director of the MSC.

Knight, William J. "Pete" (1929-2004): Captain, United States Air Force. California State Senator (1996-2004). California State Assemblyman (1992-1996). Pilot Astronaut in the X-20 Dyna-Soar program (1960-1963). In 1960, he was one of six test pilots selected to fly the X-20 Dyna-Soar, which was slated to become the first winged orbital space vehicle capable of lifting reentries and conventional landings. After the X-20 program was canceled in 1963, he completed the astronaut-training curriculum at Edwards Air Force Base and was selected to fly the X-15. On October 3, 1967, he set a world aircraft speed record by piloting the X-15A-2 to 7,275 kilometers per hour (Mach 6.7), a record that still stands. During sixteen flights in the aircraft, Knight also became one of only five pilots to earn their astronaut's wings by flying an airplane in space, reaching an altitude of 85.5 kilometers.

Kohrs, Richard: Director, Space Station Freedom Program (1989-1993). Deputy Manager, National Space Transportation System (NSTS), 1985-1987. Deputy Director, NSTS (1987-1989).

Kondratyuk, Yuri Vasilievich (1897-1941): Visionary writer and pioneer of space exploration and interplanetary travel.

Koptev, Yuri Nikolaevich (b. 1940): Veteran of the Soviet space industry. Director General of the Russian Aviation and Space Agency, Rosaviakosmos (1992-2004).

Korolev, Sergei Pavlovich (1907-1966): Korolev is widely regarded as a founder of the Soviet space program. Involved in pre-World War II studies of rocketry in the Soviet Union, Korolev, like many of his colleagues, went through Joseph Stalin's prisons and later participated in the search for rocket technology in occupied Germany. His incredible energy, intelligence, belief in the prospects of rocket technology, managerial abilities, and almost mythical skills in decision making made him the head of the first Soviet rocket development center, known today as RKK Energia. He deserves credit for turning rocket weapons into instruments of space exploration and making the Soviet Union the world's first spacefaring nation.

Kraft, Christopher C., Jr. (b. 1924): Chief of Flight Operations, Space Task Group (1961-1972). Flight Director for all of the Project Mercury and Gemini Program flights. He directed the design of Mission Control at the Manned Spacecraft Center (MSC), later the Johnson Space Center (JSC), in Houston, Texas. Deputy Director, MSC (1970-1972). Director, MSC/JSC (1972-1982). Kraft personally chose and trained an entire generation of NASA Flight Directors, including Eugene F. Kranz.

Kranz, Eugene F. (b. 1933): Assistant Flight Director for Project Mercury. Flight Director for all Gemini Program missions. Flight Director for the Apollo Program. Led the "Tiger Team" for the successful return of the Apollo 13 crew. Flight Director and Flight Operations Director for the Skylab Program. Director, Mission Operations, Johnson Space Center (JSC), Houston, Texas (1983-1994).

Krikalev, Sergei Konstantinovich (b. 1958): Soviet/Russian Cosmonaut (as of 1985). International Space Station (ISS) Expedition Eleven Commander. ISS Expedition One Flight Engineer. Payload Specialist, STS-88 (the first International Space Station assembly mission). Payload Specialist, STS-60 (first joint U.S./Russian space shuttle mission). Soyuz TM-12 Flight Engineer. Soyuz TM-7 Flight Engineer. He has been dubbed by many as "the last Soviet Citizen": He spent 151 days aboard the Mir Space Station (Soyuz TM-12) while the Soviet Union collapsed during 1991 and 1992. In 1981, he joined NPO Energia, the Russian industrial organization responsible for piloted spaceflight activities. He tested spaceflight equipment, developed space operations methods, and participated in ground control operations. When the Salyut 7 space station failed in 1985, he worked on the rescue mission team, developing procedures for docking with the uncontrolled station and repairing the station's onboard system.

Laeser, Richard P.: Voyager Project Manager (1981-1986), Jet Propulsion Laboratory (JPL), Pasadena, California.

Landano, Matthew: Director of Office of Safety and Mission Success, Jet Propulsion Laboratory (JPL), Pasadena, California (as of 2002). Mars Odyssey Project Manager.

Langley, Samuel P. (1834-1906): American astronomer, physicist, and aeronautics pioneer who contributed to the knowledge of solar phenomena as related to meteorology and built the first heavier-than-air flying machine to achieve sustained flight. He was the third Secretary of the Smithsonian Institution and the founder of the Smithsonian Astrophysical Observatory. A number of things related to aviation have been named in his honor: the Langley medal (highest award of the Smithsonian Institution); NASA Langley X-43A Hypersonic scramjet research vehicle; NASA Langley Research Center (LaRC), Hampton, Virginia (formerly the Langley Memorial Aeronautical Laboratory); Langley Field (later Langley Air Force Base), Virginia; Langley unit of solar radiation; the Navy's first aircraft carrier, USS *Langley* (CV-1); and the aircraft carrier USS *Langley* (CVL-27).

La Piana, Lia: Spitzer Space Telescope Program Manager, NASA Headquarters.

Lebreton, Jean-Pierre: Huygens Mission Manager. Project scientist for the European Space Agency (ESA).

Lee, B. Gentry: Chief Engineer, Planetary Projects Directorate, Jet Propulsion Laboratory (JPL), Pasadena, California. JPL Director of Science Analysis and Mission Planning.

Lee, Chester M. "Chet" (1920-2000): Captain, United States Navy (Retired). NASA Director of Shuttle Operations. President and Chief Operating Officer, SPACHAB (1996-1998). Program Director of the Apollo-Soyuz Test Project (1973-1975). Assistant Apollo Mission Director (1966-1973). He was present in the blockhouse in 1967 when the three Apollo 1 astronauts died. Mission Director for Apollo 12 through Apollo 17. NASA Chief Mission Planning in Mission Operations, Office of Manned Space Flight (1965-1966).

Lehman, David: Project Manager, Deep Space 1, Jet Propulsion Laboratory (JPL), Pasadena, California (1995-2000).

Leibee, Jack: Mission Manager, Gamma-ray Large Area Space Telescope (GLAST), Goddard Space Flight Center (GSFC), Greenbelt, Maryland.

Leonov, Alexei (b. 1934): Soviet/Russian Cosmonaut (1960-1991). Pilot, Voskhod 2. Commander, Soyuz 19, Apollo-Soyuz Test Project. Commander of the Cosmonaut Team ("Chief Cosmonaut") and Deputy Director of the Yuri A. Gagarin Cosmonaut Training Center, where he oversaw crew training from 1976 to 1982. Leonov was one of the twenty air force pilots selected to be in the first cosmonaut group in 1960. His spacewalk (the first ever) was originally to have taken place on the Vostok 11 mission, but that mission was canceled and the historic moment happened on the Voskhod 2 flight instead.

Levy, David H. (b. 1948): Astronomer. Levy is one of the most successful comet discoverers in history. He has discovered more than twenty comets, many of them using his own backyard telescopes. With Eugene and Carolyn Shoemaker at the Palomar Observatory in California, he discovered Shoemaker-Levy 9, the comet that collided with Jupiter in 1994. That episode produced the most spectacular explosions ever witnessed in the solar system.

Lewis, George W. (1882-1948): Director of Aeronautical Research, National Advisory Committee for Aeronautics (NACA) until he retired in 1947. He taught at Swarthmore College from 1910 to 1970 and was appointed the first Executive Officer of the NACA in 1919. The Glenn Research Center at Lewis Field (GRC), Cleveland, Ohio (formerly the Lewis Research Center), is partly named for him.

Li, Fuk: Program Manager, New Millennium Program, Jet Propulsion Laboratory (JPL), Pasadena, California.

Lichtenberg, Byron K. (b. 1948): Astronaut. First space shuttle Payload Specialist. He flew on Spacelab 1 (STS-9) and STS-45. Founding Member, Association of Space Explorers. Cofounder and President, Payload Systems, Inc. Trustee and Cofounder, Ansari X Prize. President of Zero Gravity Corporation, a commercial private company providing parabolic flights to the entertainment, tourism, and scientific communities.

Lichti, Giselher: Coprincipal Investigator, Gamma-ray Large Area Space Telescope (GLAST) Burst Monitor team, Max-Planck-Institut für Informatik, Saarbrücken, Germany.

Lippershey, Hans (1570-1619): Dutch lens maker credited with creating and disseminating designs for the first practical telescope.

Lovelace, W. Randolph, II (1907-1965): Founding Director (1947-1965), Lovelace Foundation (founded by his uncle, William Lovelace, in 1922 and now called the Lovelace Respiratory Research Institute), best remembered today for his work with NASA to develop the extensive tests needed to measure the fitness of the first candidates for space travel. In 1959, the Lovelace Medical Foundation performed clinical examinations for the Project Mercury astronauts.

Lovell, James A., Jr. (b. 1928): Captain, United States Navy (Retired). NASA Astronaut (1962-1973), Gemini VII Pilot, Apollo 8 Command Module Pilot, Apollo 13 Commander. Flew the first orbital rendezvous of two American spacecraft (Gemini VII and Gemini VI-A). Flew the first spacecraft to leave Earth orbit, leave the influence of Earth's gravity, and orbit the Moon (Apollo 8). He retired from the Navy and the space program in 1973 and went to work at the Bay-Houston Towing Company in Houston, Texas, becoming CEO in 1975. He became President of Fisk Telephone Systems in 1977, and later worked for Centel, retiring as an Executive Vice President on January 1, 1991. In 1999, Lovell and his family opened Lovell's of Lake Forest, a fine-dining restaurant in Lake Forest, Illinois. The restaurant displays many artifacts

from Lovell's time with NASA, as well as from the filming of *Apollo 13*.

Low, George M. (1926-1984): NASA Director of Manned Spaceflight programs (1958-1964). Deputy Director of the Manned Spacecraft Center (MSC), later the Johnson Space Center (JSC), in Houston, Texas (1964-1967). Manager of the Apollo Spacecraft Project Office (1967-1969). NASA Deputy Administrator (1969-1976).

Lozino-Lozinskiy, Gleb E. (1909-2001): Key figure in the development of Buran, the Soviet equivalent of the space shuttle orbiter. In 1941, Lozino-Lozinskiy joined a design bureau led by Artem Mikoyan, the developer of the Soviet fighter jets, known around the world as MiGs. In 1976, Lozino-Lozinskiy was put in charge of NPO Molniya, a newly created design center on the outskirts of Moscow. The organization had the unprecedented task of developing a reusable orbiter with capabilities similar to or exceeding those of the U.S. space shuttle.

Lucid, Shannon W. (b. 1943): NASA Astronaut (as of 1979). One of six women named as NASA's first female astronaut candidates in January, 1978. STS 51-G Mission Specialist, STS-34 Mission Specialist, STS-43 Mission Specialist, STS-58 Mission Specialist, STS-76 Mission Specialist, STS-79 Mission Specialist, and Mir Permanent Expedition Crew 21 and 22 Astronaut-Investigator. She spent 188 days (March 22 to September 26, 1996) aboard the Mir Space Station and holds the record for the most flight hours in orbit by a woman.

McAuliffe, Sharon Christa Corrigan (1948-1986): Teacher and Astronaut. STS 51-L Payload Specialist, NASA Teacher in Space. She died when the space shuttle *Challenger* was destroyed during launch on mission STS 51-L. Asteroid 3352 McAuliffe is named in her memory, as is the Christa McAuliffe Planetarium in Concord, New Hampshire.

McCandless, Bruce M., II (b. 1937): Captain, United States Navy (Retired). NASA Astronaut (1966-1990), STS 41-B Mission Specialist, STS-31

Mission Specialist. Performed the first untethered spacewalk and first use of the Manned Maneuvering Unit (on STS 41-B).

McCool, William C. "Willie" (1961-2003): NASA Astronaut (1996-2003). STS-107 Pilot. He was killed when the space shuttle *Columbia* (STS-107) disintegrated during orbital reentry into the Earth's atmosphere.

McKay, John B. (1922-1975): Astronaut and X-15 Pilot. One of the first pilots assigned to the X-15 flight research program at NASA's Flight Research Center (FRC), Edwards, California, later the Dryden Flight Research Center (DFRC). As a civilian research pilot and aeronautical engineer, he made thirty flights in X-15s between 1960 and 1966. His peak altitude was 90 kilometers, a height that qualified him as an astronaut, and his highest speed was 6,217 kilometers per hour (Mach 5.64). McKay was with the National Advisory Committee for Aeronautics (NACA) and NASA from February 8, 1951, until October 5, 1971, and specialized in high-speed flight research programs. In addition to the X-15, he flew the D-558-1, D-558-2, X-lB, and the X-lE.

McKee, Daniel: Director of Development for the second-generation astronaut jet pack, called the Astronaut Maneuvering Unit (AMU), which was the first backpack-mounted maneuvering unit.

MacMillan, Logan T.: McDonnell Corporation Mercury Project Manager.

McNair, Ronald E. (1950-1986): NASA Astronaut (1978-1986). STS 41-B Mission Specialist, STS 51-L Mission Specialist. Second African American astronaut in space. He died when the space shuttle *Challenger* was destroyed during launch on mission STS 51-L. During the 1970's, Nichelle Nichols (of *Star Trek* fame) was employed by NASA to recruit minority candidates for the space program. McNair became one of these candidates and was selected for the astronaut program in 1978. McNair was a saxophonist; before the STS 51-L mission, he worked with composer Jean Michel Jarre on a piece of music called *Rendezvous VI*. McNair planned to record his saxophone solo on board *Challenger*, mak-

ing it the first piece of music played in space. After the accident, the piece was renamed *Ron's Piece*. After his death, members of Congress provided funding for the Ronald E. McNair Post-Baccalaureate Achievement Program to encourage college students who, like McNair, are low-income, first-generation college students of color to pursue graduate studies. McNair High School in Atlanta, Georgia, was named for him.

McNeill, Kevin: Lockheed Martin's Program Manager for the Mars Reconnaissance Orbiter (MRO).

Malina, Frank J. (1912-1981): Cofounder, Jet Propulsion Laboratory (JPL), Pasadena, California. Director, JPL (1944-1946).

Martin, Christopher: Galaxy Evolution Explorer (GALEX) Principal Investigator, California Institute of Technology, Pasadena, California.

Maryniak, Gregg E.: Executive Director, Ansari X Prize. In 1974, he founded Oak Research Enterprises, which designed, built, and installed electronic control systems for quadriplegics. In 1985, he developed the idea for the Lunar Prospector space probe mission.

Masursky, Harold (1923-1990): Geologist and astronomer. Beginning in 1962, he participated in the Mariner, Apollo, Viking, Pioneer, and Voyager programs. Masursky was responsible for surveying lunar and planetary surfaces and choosing landing sites for NASA. He was a member of the working groups that monitored and guided the Moon landing in 1969 and analyzed its data. He also led the team that monitored observations of Mars made by Mariner 9 (1971) and selected landing sites on Mars for the Viking probes. He was a member of the Venus Orbiter Imaging Radar Science Working Group (1978-1990). A crater on Mars and the asteroid 2685 Masursky were named in his honor. The Masursky Award and the Masursky Lecture are also named after him.

Mathews, Charles W. (1921-2002): Associate Administrator, Office of Applications (1971-1976). Deputy Associate Administrator, OMSF (1968-1971). Director, Apollo Applications Program,

Office of Manned Space Flight (OMSF), 1966-1968. Manager, Gemini Program Office (1963-1966). Member of NASA's Space Task Group, Langley Field, Virginia (1958-1962). Member of the National Advisory Committee for Aeronautics (1943-1958).

Matranga, Gene: Senior Engineer on the Lunar Landing Research Vehicle project at the Flight Research Center (FRC), Edwards, California, later the Dryden Flight Research Center (DFRC).

Mattingly, Thomas K. "Ken," II (b. 1936): Rear Admiral, United States Navy (Retired). NASA Astronaut (1966-1985). STS 51-C and STS-4 Commander, Apollo 16 Command Module Pilot. His first flight assignment was Command Module Pilot on the Apollo 13 mission, but he was removed from the mission three days before the scheduled launch because he had been exposed to German measles. He helped the crew conserve as much power as possible for reentry.

Medaris, John B. (1902-1990): Director of Army Ballistic Missile Development. Commander, ABMA Ordinance Corps. Worked with Wernher von Braun to launch Explorer I.

Meegan, Charles A.: Principal Investigator, Gamma-ray Large Area Space Telescope (GLAST) Burst Monitor team, Marshall Space Flight Center (MSFC), Huntsville, Alabama.

Melvill, Mike (b. 1941): SpaceShipOne Flight Crew and Astronaut. Test pilot for SpaceShipOne, the experimental space plane developed by Scaled Composites. Melvill piloted SpaceShipOne on its first flight past the edge of space (flight 15P) on June 21, 2004, and in doing so became the first commercial astronaut and the 433rd person to go into space. In a ceremony two hours after landing, Melvill was awarded his astronaut wings; he was the first person to earn them for a space-plane flight since the X-15 flights of the 1960's. He was also the pilot on SpaceShipOne's flight 16P, the first competitive flight in the Ansari X Prize competition.

Michelson, Peter F.: Principal Investigator, Gamma-ray Large Area Space Telescope (GLAST) Large Area Telescope, Stanford University.

Mishin, Vasili (1917-2001): Pioneer of Russian rocketry and a key leader of the Soviet Union's ill-fated effort to beat the United States to the Moon. Chief Designer, Korolev OKB-1 design bureau (1966-1974). Chief Designer of the Progress spacecraft.

Mitchell, Edgar D. (b. 1930): Captain, United States Navy (Retired). NASA Astronaut (1966-1972), Apollo 14 Lunar Module Pilot. Sixth human to walk on the Moon (Apollo 14). After retiring from the Navy in 1972, Mitchell founded the Institute of Noetic Sciences to sponsor research into the nature of consciousness as it relates to cosmology and causality. In 1984, he cofounded the Association of Space Explorers, an international organization of those who have experienced space travel.

Mitchell, Robert: Cassini Program Manager, Jet Propulsion Laboratory (JPL), Pasadena, California.

Mitchell, Willis B., Jr. (b. 1920): Manager of the Office of Vehicles and Missions, Gemini Program Office, Manned Spacecraft Center (MSC), later the Johnson Space Center (JSC), in Houston, Texas.

Moore, Jesse W.: Director, Johnson Space Center (JSC), Houston, Texas (January 23, 1986-October 2, 1986). Associate Administrator for Space Flight, NASA (1984-1985). The *Challenger* accident occurred just five days after Moore was named as JSC Director.

Moos, H. Warren: Principal Investigator, Far Ultraviolet Spectroscopic Explorer (FUSE), Johns Hopkins University, Baltimore, Maryland.

Morabito, Linda Kelly: Voyager Navigation Engineer at the Jet Propulsion Laboratory (JPL), Pasadena, California. She discovered active volcanism on the Jovian moon Io in 1979.

Morgan, Barbara R. (b. 1951): NASA Astronaut (as of 1998). First NASA Educator Astronaut. Backup candidate for NASA's Teacher in Space program (1985). Teacher in Space designee (1986). Morgan is assigned to the crew of STS-118, an assembly mission to the International Space Station.

Mueller, George E. (1918-2001): NASA Associate Administrator for Manned Space Flight (1963-1969). He was responsible for overseeing the completion of Project Apollo and beginning the development of the space shuttle. He moved to the General Dynamics Corporation, as Senior Vice President, in 1969 and remained until 1971. He then became president of the Systems Development Corporation (1971-1980) and its Chairperson and CEO (1981-1983).

Murray, Bruce C. (b. 1932): Cofounder and Chairperson of the Board of the Planetary Society (as of 1979). Director, Jet Propulsion Laboratory (JPL), Pasadena, California (1976-1982).

Myers, Dale DeHaven (b. 1922): NASA Deputy Administrator (1986-1989). Associate Administrator for Manned Space Flight (1970-1974). Vice President and Program Manager, space shuttle program, Rockwell International (1969-1970). He was also Vice President and Program Manager for the Apollo Command and Service Module Program at North American-Rockwell (1964-1969).

Naderi, Firouz (b. 1946): Associate Director for Programs, Project Formulation, and Strategy, Jet Propulsion Laboratory (JPL), Pasadena, California (as of 2005). Mars Exploration Program Manager at JPL (2000-2005).

Nelson, Clarens William "Bill" (b. 1950): U.S. Senator (D-Florida), as of 2000. Congressman (D-Florida), 1972-1978. STS 61-C Payload Specialist. Second congressional member in space (STS 61-C). First lawyer in space.

Nelson, George D. "Pinky" (b. 1950): NASA Astronaut (1978-1989), STS 41-C Mission Specialist, STS 61-C Mission Specialist, STS-26 Mission Specialist. Participated in on-orbit repair of Solar Maximum satellite using the Manned Maneuvering Unit during mission STS 41-C.

Newell, Homer E. (1915-1983): NASA Assistant Director of Space Flight Programs (1963-1973). Theoretical physicist and mathematician at the Naval Research Laboratory (1944-1958). While at the Naval Research Laboratory, he was Science Program Coordinator for Project Vanguard and was Acting Superintendent of the Atmosphere and Astrophysics Division. In 1958, he transferred to NASA to assume responsibility for planning and development of the new agency's space science program. Deputy Director of Space Flight Programs (1958-1961). Director of the Office of Space Sciences (1961-1963).

Newton, Sir Isaac (1643-1727): English physicist, mathematician, astronomer, philosopher, and alchemist. He was the first to promulgate a set of natural laws that could govern both terrestrial motion and celestial motion. He is associated with the scientific revolution and the advancement of heliocentrism. Newton is also credited with providing mathematical substantiation for Kepler's laws of planetary motion. He expanded these laws by arguing that orbits (such as those of comets) could also be hyperbolic and parabolic as well as elliptic. He also argued that light was composed of particles, and was the first to realize that the spectrum of colors observed when white light passed through a prism is inherent in the white light (and not added by the prism as Roger Bacon had claimed in the 13th century). Newton also developed a law of cooling, describing the rate of cooling of objects when exposed to air, and proposed the binomial theorem as well as the principles of conservation of momentum and angular momentum. He studied the speed of sound in air, proposed a theory on the origin of stars, and improved understanding of telescope optics.

Oberth, Hermann J. (1894-1989): One of the three recognized fathers of spaceflight. A Transylvanian by birth but of German ancestry, he was educated at the Universities of Klausenburg, Munich, Göttingen, and Heidelberg. His doctoral dissertation was rejected because it did not fit into any established scientific discipline, but he published it privately as *Die Rakete zu den Planetenräumen* (the rocket into interplanetary space) in 1923. This volume and its expanded version,

Ways to Spaceflight (1929), set forth the basic principles of spaceflight and directly inspired many subsequent spaceflight pioneers, including Wernher von Braun.

O'Connor, Bryan D. (b. 1946): Colonel, United States Marine Corps (Retired). NASA Astronaut (1980-1996). STS 61-B Pilot, STS-40 Commander. Assistant to the Shuttle Program Manager (1986-1988). Director, space shuttle program (1994-1996). Acting Space Station Program Director (1993-1994), responsible for transforming Space Station Freedom into the International Space Station.

O'Neil, William J.: Galileo Project Manager, Jet Propulsion Laboratory (JPL), Pasadena, California.

O'Neill, Gerard K. (1927-1992): Founder and President of the Space Studies Institute at Princeton University, Princeton, New Jersey (1977-1992). His research included the invention of particle storage rings and work on mass drivers for space propulsion, research, and design concepts for space stations, space colonization, solar power satellites, and lunar and asteroid mining. He authored the book *The High Frontier: Human Colonies in Space*, which inspired a generation of space-exploration advocates.

Onizuka, Ellison S. (1946-1986): Lieutenant Colonel, United States Air Force. NASA Astronaut (1978-1986). STS 51-C Mission Specialist, STS 51-L Mission Specialist. He died when the space shuttle *Challenger* was destroyed during launch on mission STS 51-L. Onizuka Air Force Station, originally known as the Air Force Satellite Test Center, in Sunnyvale, California, was renamed for Onizuka in 1986. A shuttlecraft from the *Star Trek: The Next Generation* television series was named *Onizuka* in his honor.

Opel, Fritz von (1889-1971): German industrialist and early rocketeer. He is remembered mostly for his spectacular demonstrations of rocket propulsion that earned him the nickname "Rocket Fritz." On March 15, 1928, von Opel tested his first rocket-powered car, the RAK.1, and achieved a top speed of 75 kilometers per hour, proving the validity of the concept. Less than two months later, he reached a speed of 230 kilometers per hour in the RAK.2, driven by twenty-four solid rockets.

O'Sullivan, William J., Jr.: An engineer at the National Advisory Committee for Aeronautics's (NACA) Langley Research Center (LaRC) in Hampton, Virginia. He proposed the use of a large balloon satellite to measure atmospheric drag as part of the International Geophysical Year (IGY). This idea inspired the development of the Echo satellite and the Air Density Explorer.

Paine, Thomas O. (1921-1992): American scientist. He was the third Administrator of NASA (1969-1970), and guided NASA through the early Apollo lunar missions and the Apollo 13 accident. Deputy Administrator of NASA (1968-1969). Chair, National Commission on Space (1985-1992).

Park, Archibald B.: Originator of the Landsat concept at the United States Department of Agriculture (USDA).

Parkinson, Bradford W.: Gravity Probe B Coprincipal Investigator at Stanford University (as of 1982). He was Department of Defense Program Director during the period in which the Global Positioning System architecture was created, as well as responsible for its engineering, development, demonstration, and implementation phases. The Bradford W. Parkinson Award honors an outstanding graduate student in the field of global navigation satellite systems.

Parsons, William "Bill" W., Jr. (b. 1956): Space shuttle Program Manager (as of 2003), responsible for the overall management, integration, and operations of the space shuttle program. Director, Stennis Space Center (SSC), Hancock County, Mississippi (2002-2003).

Pecora, William T.: Originator of the Landsat concept at the United States Geological Survey (USGS).

Petersen, Forrest (1922-1990): Rear Admiral, United States Navy (Retired). X-15 Pilot on five flights.

Peterson, Max R.: Project Manager, MErcury Surface, Space ENvironment, GEochemistry, and Ranging (MESSENGER), Johns Hopkins University Applied Physics Laboratory (APL), Baltimore, Maryland.

Petrone, Rocco A. (b. 1926): NASA Associate Administrator (1974-1975). Saturn Project Officer and Director of Launch Operations, Kennedy Space Center (KSC), 1961-1969. Director, Apollo Program Office, (1969-1973). Director, Marshall Space Flight Center (MSFC), Huntsville, Alabama (1973-1974).

Pham, Tuân (b. 1947): Vietnamese Cosmonaut (1979-1980). Soyuz 37 (Salyut 6 Visiting Crew 7) Astronaut-Investigator. First non-European person to fly into space (Soyuz 37). He was selected as part of the sixth international crew for the Intercosmos Program on April 1, 1979.

Phillips, Samuel C. (1921-1990): General, United States Air Force. NASA Director of the Apollo Manned Lunar Landing Program (1964-1969).

Pickering, William H. (1910-2004): Director, Jet Propulsion Laboratory (JPL), Pasadena, California (1954-1976). During his tenure, JPL developed the first U.S. satellite (Explorer 1) and the first successful U.S. cislunar space probe (Pioneer 4). The organization also completed the Mariner flights to Venus and Mars in the early to mid-1960's, the Ranger photographic missions to the Moon in 1964 and 1965, and the Surveyor lunar landings of 1966 and 1967.

Post, Wiley (1898-1935): American aviator. He flew around the world with Harold Charles Gatty, setting a world record in July of 1931. The two men covered 24,903 kilometers in 8 days, 15 hours, and 51 minutes in a Lockheed Vega monoplane, the *Winnie Mae*. Post made the first solo flight around the world in the same plane between July 15 and July 22, 1933, completing a distance of 25,099 kilometers in 7 days, 18 hours, and 49 minutes (a new record for the distance). In principle, the space suits worn by the astronauts are the same as Post envisioned, designed, and successfully test-flew in 1934 to prove his theories about sustained flight at high altitudes.

Ramon, Ilan (1954-2003): Colonel, Israeli Air Force. Israeli Payload Specialist, STS-107. He was killed when the space shuttle *Columbia* (STS-107) disintegrated during orbital reentry into the Earth's atmosphere. In 1997, Colonel Ramon was selected as a Payload Specialist. In July, 1998, he reported for training at the Johnson Space Center (JSC), Houston, Texas. He trained as prime for a payload that included a multispectral camera for recording desert aerosol until *Columbia* launched in 2003.

Rayman, Marc D. (b. 1956): Chief Engineer, Dawn Project, Jet Propulsion Laboratory (JPL), Pasadena, California. Project Manager, Deep Space 1 Hyperextended Mission (2001). Project Manager, Deep Space 1 Extended Mission (2000-2001). The Deep Space Extended Mission included the heroic rescue from the star tracker and the spectacular and flawless encounter with Comet Borrelly. Deputy Mission Manager (1998-2000) and Chief Mission Engineer (1995-2000), Deep Space 1. Asteroid Rayman was named in honor of his dedication and contributions to space exploration. He is the only person to have won the Exceptional Leadership Award and the Exceptional Technical Excellence Award, JPL's highest honors. Among his awards from NASA are the Outstanding Leadership Medal and two Exceptional Achievement Medals.

Reed, R. Dale: NASA engineer who carried the idea of the lifting body concept forward and introduced a conceptual design for a lifting body in the form of a truncated cone.

Reid, Henry J. E. (1895-1968): Engineer in Charge, National Advisory Committee for Aeronautics (NACA)/NASA Langley Research Center (LaRC), Hampton, Virginia (1925-1960). Director, LaRC (1958-1960).

Reiff, Glenn A. (b. 1923): Mariner-Venus Program Manager, Jet Propulsion Laboratory (JPL), Pasadena, California (1967).

Resnik, Judith A. (1949-1986): NASA Astronaut (1978-1986). One of the six women named as NASA's first female astronaut candidates in January, 1978. STS 41-D Mission Specialist. STS 51-L Mission Specialist. She died when the space shuttle *Challenger* was destroyed during launch on mission STS 51-L.

Ride, Sally K. (b. 1951): NASA Astronaut (1978-1987). One of the six women named as NASA's first female astronaut candidates in January, 1978. STS-7 Mission Specialist, STS 41-G Mission Specialist. First American woman in space. Member of the Presidential Commission on the Space Shuttle *Challenger* Accident (1986). Chaired a NASA task force that prepared a report on the future of the civilian space program entitled *Leadership and America's Future in Space* (1986-1987).

Rogers, Russell L. (1928-1967): Lieutenant Colonel, United States Air Force. Pilot Astronaut in the X-20 Dyna-Soar program (1960-1963). Rogers was an Experimental Test Pilot at Edwards Air Force Base when he was selected for the X-20 program in April, 1960. He left the program when it was canceled on December 10, 1963.

Rogers, William P. (1913-2001): Attorney General of the United States (1957-1961). Secretary of State (1969-1973). Chair of the Presidential Commission on the Space Shuttle *Challenger* Accident (1986).

Rushworth, Robert A. (1924-1993): Major General, United States Air Force. X-15 Pilot. He was selected for the X-15 program in 1958 and made his first flight on November 4, 1960. Over the next six years, he made thirty-four flights in the X-15, more than any other pilot. This included a flight to an altitude of 87 kilometers, made on June 27, 1963. This flight qualified Rushworth for astronaut wings. On a later X-15 flight, he was awarded a Distinguished Flying Cross for successfully landing an X-15 after its nose wheel extended at a speed near Mach 5. He made his final X-15 flight on July 1, 1966.

Rutan, Burt (b. 1943): Aircraft designer known for light, strong, unusual-looking, energy-efficient aircraft. He is most famous for his design of the record breaking *Voyager*, which was the first plane to fly around the world without stopping

or refueling, and the suborbital rocket plane SpaceShipOne. Founded Scaled Composites, Inc., in 1982; the company has become one of the world's preeminent aircraft design and prototyping facilities. He was a flight test Project Engineer for the United States Air Force at Edwards Air Force Base, where he worked on nine separate projects, including fighter spin tests and the XC-142 VSTOL transport.

Rutan, Richard "Dick" (b. 1938): Aviator most famous for flying the *Voyager* aircraft around the world nonstop (with the assistance of Jeana Yeager). He and his brother Burt came up with the idea of the *Voyager* aircraft in 1981. This plane design would break the flight distance record of 20,168 kilometers (set by a Boeing B-52 bomber in 1962). Rutan and Yeager took off in *Voyager* on December 14, 1986, from Edwards Air Force Base. After 9 days, 3 minutes, and 44 seconds of flight, they touched down on December 23 with only a few liters of fuel remaining. The 40,211-kilometer trip yielded Rutan and Yeager numerous awards.

Ryan, Robert: Stardust Mission Operations Manager, Jet Propulsion Laboratory (JPL), Pasadena, California. Supervisor of the Flight Mission Control Teams for the Voyager spacecraft, Helios probes, Active Magnetospheric Particle Tracer Explorer (AMPTE), Galileo project, Magellan mission, Ulysses misison, and the Ocean Topography Experiment (TOPEX/Poseidon).

Ryken, John: Bell Aerosystems Company Lunar Landing Research Vehicle (LLRV) Project Manager.

Sagan, Carl (1934-1996): Astronomer, educator, and author. Cornell University professor (1968-1996). Harvard University professor (1960-1968). Cofounder and first President of the Planetary Society who popularized the notion of contact with extraterrestrial intelligences. He played a leading role in NASA's Mariner, Viking, Voyager, and Galileo expeditions to other planets. Received NASA's medals for Exceptional Scientific Achievement and for Distinguished

Public Service (which he received twice), and also received NASA's Apollo Achievement Award. The landing site of the Mars Pathfinder spacecraft was renamed the Carl Sagan Memorial Station in honor of Sagan on July 5, 1997. The marker on the memorial displays a quote from Sagan: "Whatever the reason you're on Mars, I'm glad you're there, and I wish I was with you." Asteroid 2709 Sagan is also named in his honor. Sagan became associated with the catchphrase "billions and billions," although he never actually used that phrase in his book *Cosmos*. Still, his distinctive delivery and frequent use of the word "billions" made him a target of comedic impressions. Sagan took this in good humor, and his final book was entitled *Billions and Billions*. The unit the "Sagan" is used as a joking reference to any count greater than 4,000,000,000.

Sagdeyev, Roald Z. (b. 1932): One of the leading lights of Soviet space science from the 1960's through the 1980's. He was involved in virtually every Soviet lunar and planetary probe during this era, including the highly successful Venera and Vega missions. He also advised Soviet leader Mikhail Sergeyevich Gorbachev on space and arms control at the 1986 Geneva, 1987 Washington, and 1988 Moscow summits.

Sänger, Eugen (1905-1964): Austrian aerospace engineer best known for his contributions to lifting body and ramjet technology. Sänger led a rocket development team in Germany (1936) and conceived a rocket-powered sled, which would launch a bomber with its own rocket engines that would climb to the fringe of space and then "skip" along the upper atmosphere—not actually entering orbit, but able to cover vast distances in a series of suborbital hops. This antipodal bomber was called the *Silbervogel* ("Silver bird") and would have relied on its fuselage creating lift (as a lifting body) to carry it along its suborbital path. Sänger also designed the rocket motors that the space-plane would use, which would need to generate 1 meganewton of thrust. In this design, he was one of the first to suggest using the rocket's fuel as a way of cooling the en-

gine (by circulating it around the rocket nozzle before burning it in the engine). His work on the *Silbervogel* would prove important to the X-15, the X-20 Dyna-Soar, and the space shuttle program.

Scherer, Lee R. (b. 1919): Captain, United States Navy (Retired). Director, Kennedy Space Center (KSC), Florida (1975-1979). Director, Flight Research Center (FRC), Edwards, California, later the Dryden Flight Research Center (DFRC), 1971-1975. Director of Apollo Lunar Exploration (1967-1971). Lunar orbiter Program Manager (1962-1967).

Schiminovich, David: Galaxy Evolution Explorer (GALEX) Science Operations and Data Analysis Manager, Jet Propulsion Laboratory (JPL), Pasadena, California.

Schirra, Walter M., Jr. (b. 1923): Captain, United States Navy (Retired). NASA Astronaut (1959-1970). Mercury-Atlas 8 Pilot, Gemini VI-A Commander, Apollo 7 Commander. Only person to fly all three preshuttle spacecraft. Commanded the first orbital rendezvous of two American spacecraft (Gemini VII and Gemini VI-A). During the Apollo 7 mission, Schirra caught what was perhaps the most famous cold in NASA history. On the advice of the flight surgeon, he took the drug Actifed to relieve his symptoms. Years later, he would become a spokesman for Actifed and would appear in television commercials advertising the product.

Schmitt, Harrison H. "Jack" (b. 1935): Senator (R-New Mexico), 1977-1983. NASA Astronaut (1965-1975) and geologist. Apollo 17 Lunar Module Pilot. Eleventh human and the only scientist to walk on the Moon (Apollo 17). Before joining NASA, Schmitt worked at the U.S. Geological Survey's Astrogeology Center at Flagstaff, Arizona, developing geological field techniques that would be used by the Apollo crews. Following his selection, Schmitt played a key role in training Apollo crews to be geologic observers when they were in lunar orbit and competent geologic field workers when they were on the lunar surface. After each of the landing missions, he participated in the examination and evaluation of the returned lunar samples and helped the crews with the scientific aspects of their mission reports.

Schneider, William C. (b. 1923): Apollo Mission Director and Apollo Program Deputy Director for Missions (1967-1968). Skylab Program Director (1968-1974). Associate Administrator for Space Tracking and Data Systems (1978-1980). Deputy Director of Manned Spaceflight for Mission Operations; served as Mission Director on all Gemini flights after Gemini V.

Schneiderman, Dan: Mariner-Mars Project Manager, Jet Propulsion Laboratory (JPL), Pasadena, California (1971).

Schriever, Bernard A. (b. 1910): General, United States Air Force (Retired). Commander, Air Force Ballistic Missile Division (1954-1959). Commander, Air Force Systems Command (1959-1966). He presided over the development of the Atlas, Thor, and Titan missiles. In developing these missiles, Schriever instituted a systems approach in which the various components of the Atlas and later missiles underwent simultaneous design and testing as part of an overall weapons system. Schriever also introduced the idea of concurrency, allowing components of the missiles to enter production while still in the test phase, speeding development. In 1998, Falcon Air Force Base in Colorado Springs, Colorado, was renamed Schriever Air Force Base in his honor.

Schurmeier, Harris M.: Ranger Project Manager, Voyager 1 Project Manager, Project Manager, Mariner 6/7, Jet Propulsion Laboratory (JPL), Pasadena, California.

Scobee, Francis R. "Dick" (1939-1986): Lieutenant Colonel, United States Air Force (Retired). NASA Astronaut (1978-1986). STS 41-C Pilot, STS 51-L Commander. He died when the space shuttle *Challenger* was destroyed during launch on mission STS 51-L. After his tour of duty in Vietnam, Scobee attended the Aerospace Research Pilot School at Edwards Air Force Base, California. Upon graduation in 1972, he be-

came an Air Force Test Pilot, logging thousands of hours of flight time in dozens of aircraft, including the Boeing 747, the experimental X-24B lifting body, the F-111 Aardvark, and the gigantic C-5 Galaxy.

Scott, David R. (b. 1932): Colonel, United States Air Force (Retired). NASA Astronaut (1963-1972). Gemini VIII Pilot, Apollo 9 Command Module Pilot, Apollo 15 Commander. Seventh human to walk on the Moon (Apollo 15). Drove the first Lunar Roving Vehicle (Apollo 15).

Seamans, Robert C., Jr. (b. 1918): NASA Deputy Administrator (1965-1968). NASA Associate Administrator (1960-1965). National Advisory Committee for Aeronautics (NACA), 1948-1958. Secretary of the Air Force (1969-1973).

Seddon, Margaret Rhea (b. 1947): NASA Astronaut (1978-1977). One of the six women named as NASA's first female astronaut candidates in January, 1978. Mission Specialist, STS 51-D, Mission Specialist, STS-40, and Payload Commander, STS-58. In September, 1996, NASA sent her to Vanderbilt University Medical School in Nashville, Tennessee. There she assisted in the preparation of cardiovascular experiments that flew aboard space shuttle *Columbia* on the STS-90 Neurolab Spacelab flight in April, 1998. Assistant Chief Medical Officer of the Vanderbilt Medical Group in Nashville, Tennessee (as of 1996).

See, Elliot M., Jr. (1927-1966): NASA Astronaut (1962-1966). Original Gemini IX Command Pilot. Died February 28, 1966, in St. Louis, Missouri, in the crash of a T-38 jet.

Shane, Doug: Vice President for Business Development, Director of Flight Operations, and Test Pilot for Scaled Composites (as of 1989). SpaceShipOne Flight Crew.

Shepard, Alan B., Jr. (1923-1998): Rear Admiral, United States Navy (Retired). NASA Astronaut (1959-1974). Mercury-Redstone 3 Pilot, Apollo 14 Commander. Second person in space (Mercury-Redstone 3). He was scheduled to pilot the Mercury-Atlas 10 *Freedom 7-II*, three-day, extended-duration mission in October, 1963, but the MA-10 mission was canceled on June 13, 1963. Shepard was designated as the Command Pilot of the Gemini III mission, and Thomas Stafford was picked as his co-pilot. Virgil I. "Gus" Grissom and John Young took over the mission when Shepard was diagnosed with Meniere's disease and was removed from flight status. Shepard was restored to full flight status in May, 1969, following corrective surgery (using a newly developed method) for Meniere's disease. Fifth human to walk on the Moon (Apollo 14). Chief of the Astronaut Office (1963-1974). During his life he was awarded the Congressional Space Medal of Honor; two NASA Distinguished Service Medals, NASA's Exceptional Achievement Medal, Naval Astronaut Wings, the Navy Distinguished Service Medal, and the Distinguished Flying Cross. He received the Langley Award (the highest award given by the Smithsonian Institution) on May 5, 1964, and has also received the Lambert Trophy, the Iven C. Kincheloe Award, the Cabot Award, the Collier Trophy, the City of New York Gold Medal (1971), and the Achievement Award for 1971. Shepard is the only person to play golf on the Moon (using a converted soil sampler as his club).

Shepherd, William M. (b. 1949): Captain, United States Navy. NASA Astronaut (as of 1984). International Space Station Expedition One ISS Commander. Mission Specialist, STS-27, STS-41, and STS-52.

Shoemaker, Carolyn (b. 1929): Astronomer. Best known for her codiscovery (with David Levy and husband Eugene M. Shoemaker) of Comet Shoemaker-Levy 9. She holds the record for most comets discovered by an individual. She started her astronomical career in 1980, searching for Earth-crossing asteroids and comets at the California Institute of Technology and at Palomar Observatory. In the 1980's and 1990's, Shoemaker used film taken using the wide-field telescope at Palomar and a stereoscope to find objects that moved against the background of fixed stars.

Shoemaker, Eugene M. (1928-1997): Astronomer. Best known for his March, 1993, codiscovery (with his wife Carolyn and colleague David Levy) of a comet that would strike Jupiter sixteen months later. Comet Shoemaker-Levy 9 was only one of the finds that made this husband-and-wife team the leading comet discoverers of their time. They are also credited with discovering more than eight hundred asteroids. The NEAR (Near Earth Asteroid Rendezvous) Shoemaker spacecraft is named in his honor.

Siebold, Pete: Aeronautical Engineer, Experimental Test Pilot, and Flight Test Engineer at Scaled Composites, Inc. On April 8, 2004, Siebold piloted the second powered test flight of SpaceShipOne (flight 13P), which reached a top speed of Mach 1.6 and an altitude of 32.0 kilometers.

Silverstein, Abe (1908-2001): Director, Lewis Research Center (now the Glenn Research Center), Cleveland, Ohio (1961-1970). Key figure in the reorganization of the National Advisory Committee for Aeronautics (NACA) into NASA. Engineer at the NACA Langley Aeronautical Laboratory, Hampton, Virginia (1929-1943). In 1958, Silverstein moved to NASA headquarters, where he initiated and directed efforts leading to the Mercury spaceflights. He later named and laid the groundwork for the Apollo missions.

Slayton, Donald K. "Deke" (1924-1993): NASA Astronaut (1959-1982), Apollo-Soyuz Test Project Docking Module Pilot. Selected as one of the Mercury Seven, he was the only one who did not fly during the Mercury program. Slayton had been scheduled to fly in 1962, on the second orbital flight, but due to an erratic heart rate (idiopathic atrial fibrillation), his place was taken by M. Scott Carpenter. Coordinator of Astronaut Activities (1962-1963). Director of Flight Crew Operations (1963-1972). Manager of the Orbital Flight Test program and first Assistant Director of Flight Crew Operations. After his retirement from NASA, he served as President of Space Services, Inc., a Houston-based company he founded to develop rockets for small commercial payloads. He helped design and build a rocket called the *Conestoga*, which was successfully launched on September 9, 1982.

Smith, Bradford A.: Principal Investigator, Robotic Deep-Space Probes Imaging Experiments, Jet Propulsion Laboratory (JPL), Pasadena, California. Mariner 9 Imaging Team Leader.

Smith, Michael J. (1945-1986): Commander, United States Navy. NASA Astronaut (1980-1986). Pilot, STS 51-L. He died when the space shuttle *Challenger* was destroyed during launch on mission STS 51-L.

Sollazzo, Claudio: European Space Agency (ESA) Huygens Operations Manager at the European Space Operations Centre (ESOC) in Darmstadt, Germany.

Solomon, Sean C.: Director, Department of Terrestrial Magnetism at the Carnegie Institution in Washington, D.C. MErcury Surface, Space ENvironment, GEochemistry, and Ranging (MESSENGER) Principal Investigator.

Southwood, David: European Space Agency (ESA) Science Director in charge of the ESA Science Programme (as of 2001). Head of the ESA Space Science Advisory Committee (1990-1993). Head of the ESA Science Programme Committee (1993-1996).

Spear, Anthony: Venus Radar Mapper (later renamed Magellan) Project Manager, Jet Propulsion Laboratory (JPL), Pasadena, California. JPL Mars Pathfinder Project Manager.

Spitzer, Lyman, Jr. (1914-1997): Director of the astronomy department at Princeton University in Princeton, New Jersey (1947-1979). In 1946, he proposed the development of a large space telescope. He outlined the advantages of greater angular resolution, increased wavelength coverage, and more stability, which led to the development of the Hubble Space Telescope. The Spitzer Space Telescope, formerly the Space Infrared Telescope Facility (SIRTF), is named after him.

Stack, John (1906-1972): Assistant Director, Langley Research Center (LaRC), Hampton, Virginia (1947-1961). He guided much of the research that paved the way for transonic aircraft. In 1947, he shared the Collier Trophy with Chuck Yeager, pilot of the X-1 that broke the sound barrier. He won the award again in 1952 and later won the Wright Brothers Memorial Trophy, among other awards. Director of Aeronautical Research, NASA (1961-1962).

Stafford, Thomas P. (b. 1930): Lieutenant General, United States Air Force (Retired). NASA Astronaut (1962-1975). Gemini VI-A Pilot, Gemini IX-A Command Pilot, Apollo 10 Commander, Apollo-Soyuz Test Project Commander. Flew the first orbital rendezvous of two American spacecraft (Gemini VII/Gemini VI-A). Commanded the first Lunar Module flight in lunar orbit (Apollo 10). Commanded the first joint U.S.-Soviet spaceflight, the Apollo-Soyuz.

Stennis, John C. (1901-1995): Senator (D-Mississippi), 1947-1989. Holds the second-longest congressional tenure in the history of the nation. Member, Mississippi House of Representatives (1928-1947). The John C. Stennis Space Center (SSC), Hancock County, Mississippi, the John C. Stennis National Student Congress, and the aircraft carrier USS *John C. Stennis* are named for him.

Stern, S. Alan: New Horizons Principal Investigator, Southwest Research Institute, Boulder, Colorado.

Stewart, Robert L. (b. 1942): Colonel, United States Army (Retired). NASA Astronaut (1978-1986). STS 41-B Mission Specialist, STS 51-J Mission Specialist. During STS 41-B, Bruce M. McCandless II and he participated in two extravehicular activities to conduct the first flight evaluations of the Manned Maneuvering Unit. These spacewalks were the first untethered operations by humans from a spacecraft in flight.

Stone, Edward C., Jr.: Director, Jet Propulsion Laboratory (JPL), Pasadena, California (1991-2001). Project Scientist for the Voyager Mission.

Stroud, William G. (b. 1923): Chief, Aeronomy and Meteorology Division, Goddard Space Flight

Center (GSFC), Greenbelt, Maryland. Manager of the Television Infrared Observations Satellite (TIROS) project.

Sullivan, Kathryn D. (b. 1951): Commander, U.S. Naval Reserve. NASA Astronaut (1978-1993). One of the six women named as NASA's first female astronaut candidates in January, 1978. Mission Specialist on STS 41-G, STS 31, and STS-45. First American woman to walk in space. Chief Scientist at the National Oceanic and Atmospheric Administration (NOAA), 1993-1996. President and CEO of the Center of Science and Industry (COSI) in Columbus, Ohio (1996-2005). COSI Science Adviser (as of 2005).

Swigert, John L. "Jack," Jr. (1931-1982): NASA Astronaut (1966-1973). Apollo 13 Command Module Pilot. On November 2, 1982, Swigert won the new seat for Colorado's Sixth Congressional District, receiving 64 percent of the popular vote. He died of complications from cancer in Washington, D.C., on December 27, 1982, a week before he would have his begun his term in Congress.

Tamayo-Mendez, Arnaldo (b. 1942): Cuban Cosmonaut (1978-1980). Soyuz 38 (Salyut 6 Visiting Crew 8) Astronaut-Investigator. First space traveler of African descent (Soyuz 38).

Tananbaum, Harvey: Harvard professor. Director, Smithsonian Astrophysical Observatory's Chandra X-Ray Center (CXC). Responsible for overseeing the operation of the Chandra X-Ray Observatory and providing support to scientists at the Obervatory.

Tepper, Morris (b. 1916): Director of NASA Meteorological Programs (1962-1969). Television Infrared Observations Satellite (TIROS) Program Manager and Nimbus Project Scientist at Goddard Space Flight Center (GSFC), Greenbelt, Maryland.

Tereshkova, Valentina (b. 1937): Soviet/Russian Cosmonaut (1961-1997). Member of the Supreme Soviet (1966-1974). Member of the Presidium of the Supreme Soviet (from 1974-1989). Central Committee of the Communist Party (1969-1991). In 1959, Tereshkova joined the Yaroslavl Air Sports Club and became a skilled amateur parachutist. She then volunteered for the Soviet space program and became a Cosmonaut in 1961. She received an Air Force commission and trained for eighteen months before being named pilot of the Vostok 6 mission. On June 16, 1963, Tereshkova became the first woman to be launched into space, making 48 orbits in 70 hours and 50 minutes. After her flight, she studied at the Zhukovski Air Force Academy and graduated as Cosmonaut Engineer in 1969. The same year, the female cosmonaut group was dissolved.

Thompson, Milton O. (b. 1926): Pilot and Astronaut in the X-20 Dyna-Soar program (1960-1963). NASA research pilot, selected as an astronaut for the X-20 Dyna-Soar program in April, 1960. After the Dyna-Soar program was canceled on December 10, 1963, he remained a NASA research pilot and flew the X-15 rocket plane. Later he became Chief Engineer and Director of Research Projects during a long career at the Dryden Flight Research Center (DFRC) in Edwards, California. On August 16, 1963, Thompson became the first person to fly a lifting body, the lightweight M2-F1.

Thompson, Robert F. (b. 1925): Skylab Project Manager, Manned Spacecraft Center (MSC), later the Johnson Space Center (JSC), in Houston, Texas (1967-1970).

Thurman, Sam: Mars Surveyor Flight Operations Manager, Jet Propulsion Laboratory (JPL), Pasadena, California.

Toftoy, Holger N. (1903-1967): Major General, United States Army (Retired). Responsible for bringing the German Rocket Team (under the leadership of Wernher von Braun) to the United States in 1945. Commander of the Redstone Arsenal in Huntsville, Alabama, in 1954. He worked closely with von Braun's teams to develop the Redstone and Jupiter missiles. In the aftermath of Sputnik 1 (1957), he persuaded the Department of Defense to allow the launch of the United States' first Earth-orbiting satellite aboard the Juno 1 missile. The result was Explorer 1's placement into orbit.

Tombaugh, Clyde (1906-1997): Astronomer who discovered Pluto in 1930. During his planet search, Tombaugh photographed 65 percent of the sky and spent seven thousand hours examining about 90 million star images. Besides Pluto, his discoveries included 6 star clusters, 1 cloud of galaxies, 1 comet, and about 775 asteroids. Few astronomers have seen so much of the universe in such minute detail.

Truax, Robert C. (b. 1936): Captain, United States Navy (Retired). Rocket, missile, and space pioneer. Founded Truax Engineering, Inc. in 1966. The company designs low-cost launch vehicles that are privately owned and operated, such as the Volksrocket and Truax launch vehicle. Headed the initial design studies of the Thor missile in 1954.

Truly, Richard H. (b. 1937): Admiral, United States Navy (Retired). NASA Astronaut (1965-1983). Shuttle approach and landing Test Pilot, STS-2 Pilot, and STS-8 Commander. NASA Administrator (1989-1992). He was responsible for getting the shuttle program back into operational mode in the post-*Challenger* era. He was also responsible for the development of the Space Station Freedom program.

Turneaure, John P.: Gravity Probe B Coprincipal Investigator at Stanford University.

Van Allen, James A. (b. 1914): Professor of physics, University of Iowa, 1950-1985. Professor emeritus, 1985-present. His Explorer 1 experiment established the existence of the radiation belts, later named for him, that encircle the Earth. His discovery opened a broad field of research.

Vandenberg, Hoyt S. (1899-1954): General, United States Air Force (Retired). Second Chief of Staff of the Air Force (1948-1952). Director of the Central Intelligence Agency (1945-1947). Vandenberg Air Force Base (originally established in 1941 as the U.S. Army's Camp Cooke) was renamed in his honor in 1958.

Van Hoften, James D. A. "Ox" (b. 1944): NASA Astronaut (1978-1986). STS 41-C Mission Specialist, STS 51-I Mission Specialist. Participated in on-orbit repair of the Solar Maximum satellite using the Manned Maneuvering Unit during STS 41-C. Performed a successful on-orbit rendezvous with the ailing 7,000-kilogram Syncom IV-3 satellite, which was repaired in space and returned to orbit during STS 51-I.

Varghese, Philip: Mars Odyssey Project Manager, Jet Propulsion Laboratory (JPL), Pasadena, California.

Vellinga, Joseph: Stardust Program Manager, Lockheed Martin Astronautics.

Vidal-Madjar, Alfred: French Project Scientist, Far Ultraviolet Spectroscopic Explorer (FUSE), at the Institut d'Astrophysique de Paris (Astrophysical Institute of Paris).

Walker, Joseph A. (1921-1966): Astronaut. He joined the National Advisory Committee for Aeronautics (NACA) in March, 1945, and served as Project Pilot at NASA's Flight Research Center, later Dryden Flight Research Center, (DFRC), Edwards, California. There he worked on pioneering research projects, including the D-558-1, D-558-2, X-1, X-3, X-4, X-5, and the X-15. He also flew programs involving the F-100, F-101, F-102, F-104, and the B-47. Walker made the first NASA X-15 flight on March 25, 1960. He flew the research aircraft twenty-four times and achieved its fastest speed and highest altitude. During a flight on June 27, 1962, he attained a speed of 6,605 kilometers per hour (Mach 5.92). On August 22, 1963 (his last X-15 flight), he reached an altitude of 108 kilometers.

Weaver, Hal: New Horizons Project Scientist, Johns Hopkins University Applied Physics Laboratory (APL), Baltimore, Maryland.

Werner, Michael: Project Scientist, Space Infrared Telescope Facility (SIRTF), Jet Propulsion Laboratory (JPL), Pasadena, California. SIRTF was renamed the Spitzer Space Telescope.

White, Edward H., II (1930-1967): Lieutenant Colonel, United States Air Force. NASA Astronaut

(1962-1967). Gemini IV Pilot, Apollo 1 Senior Pilot. First American to walk in space. Used the first-generation MMU, called the Handheld Maneuvering Unit. White died along with fellow astronauts Virgil I. "Gus" Grissom and Roger B. Chaffee in the Apollo 1 fire at Cape Kennedy, Florida. White Crater lies on the far side of the Moon. It is located just to the south of the huge Apollo impact basin, and to the northeast of Cori Crater. NASA has memorialized the Apollo 1 crew by naming the hills surrounding the Mars Exploration Rover (MER) Spirit's landing site in Gusev Crater in honor of the late astronauts.

White, Robert M. (b. 1924): Major General, United States Air Force (Retired). The Air Force's primary pilot for the X-15 program (1958-1963). In February, 1961, he set an unofficial world speed record of 3,661 kilometers per hour. Over the next eight months, he became the first human to fly an aircraft at Mach 4 and at Mach 5. This amazing record was set on November 9, 1961, when White reached a speed of 6,587 kilometers per hour. This was 150 kilometers per hour faster than the plane had been designed to travel and made White the first human to fly a winged craft six times faster than the speed of sound. On July 17, 1962, he took the X-15 to a record-setting altitude of 96 kilometers, qualifying him for astronaut wings. He also became the first of the tiny handful of "winged astronauts" to earn their wings without using a conventional spacecraft.

Wilkinson, David T. (1935-2002): Physicist and professor, Princeton University, Princeton, New Jersey. Wilkinson Microwave Anisotropy Probe Project Scientist, Princeton University. Wilkinson was a pioneer in cosmic microwave background (CMB) research from the earliest days of its discovery to his work as the project's Instrument Scientist thirty-eight years later. Wilkinson was also a founder of the Cosmic Background Explorer (COBE) satellite, which launched in 1989. The Wilkinson Microwave Anisotropy Probe (WMAP) is named in his honor.

Williams, Walter C. (1919-1995): Project Engineer. Set up flight tests for the X-1, including the first human supersonic flight. Founding Director of the organization that became Dryden Flight Research Center (DFRC). Associate Director, Project Mercury and Associate Director, Space Task Group (1959-1963). Deputy Associate Administrator of the Office of Manned Space Flight (1963-1964). Vice President for Aerospace Corporation (1964-1975). NASA Chief Engineer (1975-1982).

Wood, James (1924-1990): Colonel, United States Air Force (Retired). Pilot Astronaut in the X-20 Dyna-Soar program (1960-1963). Experimental Test Pilot at the Air Force Flight Test Center, Edwards Air Force Base when selected as an X-20 Pilot. After cancellation of the program, he remained in the Air Force, serving as Commander of Test Operations, Edwards Air Force Base.

Yardley, John F. (1925-2001): NASA Associate Administrator for Space Transportation Systems (1974-1981). Responsible for the Get-Away Special concept. Technical Director for Gemini VIII. Manager of Project Mercury at McDonnell Douglas. McDonnell Douglas President (1981-1988).

Yeager, Charles E. "Chuck" (b. 1923): Brigadier General, United States Air Force (Retired). Test Pilot for the Army Air Corps. First person to break the sound barrier when, on October 14, 1947, over dry Rogers Lake in California, he rode the X-1 research aircraft to a speed of 1,065 kilometers per hour. Member of the Presidential Commission on the Space Shuttle *Challenger* Accident (1986).

Yeager, Jeana L. (b. 1952): Aviator, most famous for accompanying Dick Rutan on a nonstop flight around the world in the *Voyager* aircraft (December 14-23, 1986). The flight took 9 days, 3 minutes, and 44 seconds, and covered 40,211 kilometers, more than doubling the old distance record (set by a Boeing B-52 bomber in 1962). In recognition of this achievement, she received the Harmon Trophy.

Young, John W. (b. 1930): NASA Astronaut (1962-2004). Gemini 3 Pilot, Gemini X Command Pilot, Apollo 10 Command Module Pilot, Apollo 16 Commander, STS-1 Commander, STS-9 Commander. Chief of the Space Shuttle Branch of the Astronaut Office (1973-1974). Chief of the Astronaut Office (1974-1987). Special Assistant to the Director for Engineering, Operations, and Safety, Johnson Space Center (JSC), Houston, Texas (1987-1996). Astronaut Science Advisor (ASA) Associate Technical Director (as of 1996). Ninth human to walk on the Moon (Apollo 16). Commander of the first spaceflight of the space shuttle (STS-1). Commander of the first spaceflight of the Spacelab scientific module (STS-9).

Russell R. Tobias and David G. Fisher

Glossary

For a fuller listing of abbreviations and what they stand for, see "List of Abbreviations," which appears in each volume's front matter.

Ablation: The removal of surface material from a body by vaporization, melting, chipping, or another erosive process. Specifically, ablation refers to the intentional removal of material from a nose cone or spacecraft (during high-speed movement through a planetary atmosphere) in order to provide thermal protection to the underlying structure.

Ablative heatshield: A heatshield composed of material that ablates as a spacecraft reenters Earth's atmosphere. The consequent removal of excess heat prevents the spacecraft from burning up.

Ablative material: Special heat-dissipating materials on the surface of a spacecraft that vaporize during reentry.

ABM: *See* Antiballistic missile.

Abort: The unscheduled termination of a mission prior to its completion. Also, to terminate a mission prior to its completion.

Absolute magnitude: The brightness of a star or other celestial body measured at a standard distance of 10 parsecs. *See also* Apparent magnitude, Luminosity, Parsec.

Absolute temperature scale: A temperature scale that sets the lowest possible temperature at zero. Atomic and molecular translational motion ceases at absolute zero, but not vibrational motion (zero-point energy). *See also* Kelvin.

Absorption spectrum: An electromagnetic spectrum that shows dark lines resulting from the passage of the electromagnetic radiation through an absorbing medium, such as the gases found in a star's atmosphere. The absorption lines are characteristic of certain chemical elements and reveal much about the composition of the star's atmosphere. *See also* Electromagnetic spectrum, Emission spectrum, Spectrum.

Achondrite: The rarer of the two main types of stony meteorite, accounting for about 9 percent of all meteorite falls. Achondrites are made of rock that has crystallized from a molten state.

Acquisition: The detection and tracking of an object, signal, satellite, or probe to obtain data or control the path of a spacecraft. *See also* Star tracker.

Acquisition and tracking radar: Radar set that locks onto a strong signal and tracks the object reflecting the signal.

Active experiment: An experiment package carried by a satellite, usually in a canister, which typically has a control circuit, a battery-driven power system, data recording instruments, and environmental control systems. *See also* Passive experiment.

Active satellite: A satellite equipped with onboard electrical power that enables it to transmit signals to Earth or other spacecrafts. Most artificial satellites fit this description. *See also* Passive relay satellite.

Advanced vidicon camera system (AVCS): Spaceborne imaging systems made up of two 800-line cameras with nearly twice the resolution of a normal television camera. Capable of photographing a 3,000-kilometer-wide area with a resolution of 3 kilometers.

Aerobraking: The action of atmospheric drag in slowing down an object that is approaching a planet or some other body with an atmosphere. Also known as atmospheric braking. Where enough atmosphere exists, it can be used to alter the orbit of a spacecraft or decrease a vehicle's velocity prior to landing.

Aerodynamics: The study of the behavior of solid bodies (such as an airplane) moving through gases (such as Earth's atmosphere).

Aerography: The study of land features on Mars.

Aeronautics: The study of aircraft and the flight of these objects in the atmosphere.

Aeronomy: The study of the physics and chemistry of the atmospheres of Earth and other planets.

Aerospace: The space extending from Earth's surface outward to the farthest reaches of the universe.

Agenzia Spaziale Italiana (ASI): *See* Italian Space Agency.

Air lock: A small enclosed area (especially in the space shuttle and space stations) located between the interior of a spacecraft and outer space or another spacecraft into which an astronaut or cargo can pass without depressurizing the spacecraft. Spelled "airlock" in some proper names.

Airglow: A faint glow emitted by Earth that results from interaction between solar radiation and gases in the ionosphere, perceived from space as a halo around the planet. Airglow is known for interfering with Earth-based astronomical observations, making space-based telescopes desirable.

Albedo: The amount of electromagnetic radiation reflected from a nonluminous body, measured from 0 (perfectly absorbing) to 1 (perfectly reflective).

Almaz: The Russian word for "diamond." The Almaz space station was an ambitious, top-secret Soviet project envisioned by Vladimir Chelomei as a piloted, orbiting outpost, equipped with powerful spy cameras, radar, and self-defense weapons. Almaz was approved over the nonmilitary station, Salyut. Leonid Brezhnev, general secretary of the Communist Party of the Soviet Union, wanted to place a space station in orbit before the American Skylab. He ordered Vasili Mishin's OKB-1 to produce a space station in the shortest possible time using the Almaz structural vessel but grafting into it proven Soyuz systems. Three Almaz stations were launched under the guise of Salyut stations. Salyut 2 failed shortly after achieving orbit, but Salyut 3 and 5 both conducted successful manned testing.

Altitude: The distance of an object directly above a surface. Also, the arc or angular distance of a celestial object above or below the horizon.

Amateur radio satellite: A satellite used to relay messages between amateur ham radio operators. These satellites are typically small and simple in design and are often built by universities. Due to their small size and typically low budgets, these satellites are usually launched as secondary payloads, accompanying much larger primary payloads.

Anemometer: An instrument for measuring the force of wind.

Angle of attack: The acute angle between the wing profile (measured along its bottom) and the wing's motion relative to the surrounding air. In the case of a rocket rising through the atmosphere, it is the angle between the long axis of the rocket and the direction of the air flowing past it.

Angstrom: One ten-thousand-millionth of a meter; a unit used to measure electromagnetic wavelengths.

Angular momentum: A property of a rotating body (or a system of rotating bodies) defined as the product of the distribution of the body's mass around the rotational axis (the moment of inertia) and the speed of the body around the axis (the angular velocity). Under no net external torque, angular momentum remains constant; that is, an increase in one of the two factors is compensated by a decrease in the other. Hence, a spinning ice skater will rotate faster as he pulls his arms toward his body; a planet or artificial satellite will move faster in its elliptical orbit as it approaches the point closest to the object around which it is orbiting (periapsis).

Antiballistic missile (ABM): A missile designed to destroy a ballistic missile in flight.

Antimatter: Matter in which atoms are composed of antiparticles; positrons in place of electrons, antiprotons in place of protons, and so forth.

The existence of such antiparticles is accepted, although their configuration as antimatter has yet to be discovered. Antihydrogen, the antimatter counterpart of hydrogen, was first produced in 1995 at the CERN laboratory in Geneva, Switzerland.

Antipodal bomber: Conceived by Austrian engineer Eugen Sänger, an antipodal bomber is a rocket-powered plane that could boost into suborbital space and skip off the atmosphere like a stone skipping over water. When it reached its objective, it would descend by gliding and drop its bomb payload.

Antipode: Point on the surface of a planet exactly 180° opposite a reciprocal point on a line projected through center of body. In Apollo usage, antipode refers to a line positioned from the center of the Moon through the center of Earth and projected to Earth's surface on the opposite side. The antipode crosses the mid-Pacific recovery line along the 165th meridian of longitude once every two hours.

Antisatellite (ASAT): A satellite used to destroy orbiting satellites. This category includes the ASAT interceptors as well as vehicles placed into orbit as test targets.

Aphelion: The point in an object's orbit around the Sun at which it is farthest away from the Sun.

Apoapsis: The point in one object's orbit around another object at which the orbiting object is farthest away from the object being orbited.

Apocynthion: The point in an object's orbit around the Moon at which it is farthest away from the Moon. Named after Cynthia, Roman goddess of the Moon.

Apogee: The point in an object's orbit around Earth at which it is farthest away from Earth.

Apolune: The apocynthion of an artificial satellite.

Apparent magnitude: The brightness of a star or other celestial body as seen from a single point, such as Earth. The brightness is apparent because stars vary in their distance from Earth. *See also* Absolute magnitude, Luminosity.

Apparent motion: The path of movement of a body relative to a fixed point of observation.

Applications Technology Satellite: A satellite designed for developing applications (meteorology, navigation, communication, Earth resources, and the like). Also used as a relay between other satellites and Earth stations.

APU: *See* Auxiliary power unit.

Arc second: Unit of angular measurement which comprises one-sixtieth of an arc minute, or 1/3600 of a degree of arc, or 1/1296000 of a circle.

Arm: *See* Remote Manipulator System, Robot arm.

Array: A system of multiple devices (such as radio aerials or optical telescopes) situated to increase the strength of the data received.

Artificial satellite: A satellite or object sent into orbit around a celestial body. Generally referred to simply as satellites, these spacecraft are usually robotic.

ASAT: *See* Antisatellite.

Ascent stage: The upper of two stages of the Apollo Lunar Module; it carried the crew, their gear and samples, the life-support system, attitude-control jets, a rocket engine, and fuel for liftoff from the Moon, orbit, and docking with the Command and Service Module. *See also* Descent stage.

Asteroid: A small solid body (also known as a planetoid), ranging in size from about 200 meters to 1,000 kilometers in diameter, which orbits the Sun. The solar system is home to thousands of these bodies, most of which exist in a region between the orbits of Mars and Jupiter.

Asteroid belt: The region between Mars and Jupiter (between 2.15 and 3.3 astronomical units from the Sun) where most of the solar system's asteroids have been found.

Asthenosphere: The layer of Earth below the lithosphere.

Astronaut: An American, European, or Canadian, or a person traveling in an American spacecraft, who achieves spaceflight as defined by the Fédération Aéronautique Internationale. *See also* Cosmonaut, Taikonaut.

Astronaut Badge: A military badge of the United States awarded to military aviation pilots who have completed astronaut training with National Aeronautics and Space Administration (NASA) and performed a successful spaceflight. A variation of the Astronaut Badge is issued to civilians who are employed with NASA as specialists on spaceflight missions.

Astronautics: The science and technology of spaceflight, including all aspects of aerodynamics, ballistics, celestial mechanics, physics, and other disciplines as they affect or relate to spaceflight. *See also* Aerodynamics, Aeronautics, Celestial mechanics.

Astronomical unit (AU): The mean distance between the centers of Earth and the Sun: 149,597,870 kilometers (92,955,630 miles). Used for measuring distance within the solar system.

Astronomy: The study of all celestial bodies and phenomena within the universe.

Astronomy satellite: Astronomy satellites are dedicated to the study of stellar objects, ranging from extremely distant nebulae to our own nearby Sun.

Astrophysics: The branch of astronomy dealing with the chemical and physical properties and behaviors of celestial matter and their interactions.

Atmosphere: Any gaseous envelope surrounding a planet or star. Earth's atmosphere consists of five layers: the troposphere, stratosphere, mesosphere, thermosphere (which overlaps the ionosphere), and exosphere.

Atmospheric pressure: The force exerted by Earth's atmosphere, which at sea level is approximately 14.7 pounds per square inch, 101.325 newtons per square meter, or 1 bar. One atmosphere refers to any of these sea-level measures and can be used to refer to atmospheric pressure on other planets: Venus's surface pressure, for example is 90 atmospheres.

Atom: The smallest amount of an element that can exist alone or in combination with other atoms in compounds. Atoms consist of electrons (negatively charged particles), protons (positively charged particles), and neutrons (particles without net charge). The protons and neutrons are located inside a tiny nucleus, whereas the electrons orbit in an expansive cloud surrounding the nucleus. The number of protons (the atomic number) determines the element. The total number of protons and neutrons is the atomic mass number.

Attitude: The orientation of a spacecraft or other body in space relative to a point of reference.

Attitude control system: The combined mechanisms working together to maintain or alter a spacecraft's position relative to its point of reference, including onboard computers, gyroscopes, and star trackers.

ATV: *See* Automated Transfer Vehicle.

AU: *See* Astronomical unit.

Aurora: The colored lights that appear near the poles when charged particles from the Sun become trapped in Earth's magnetic field. The arching, spiraling glows result from the interaction of these particles and atmospheric gases as they follow Earth's magnetic force lines.

Aurora australis: The aurora occurring near Earth's South Pole.

Aurora borealis: The aurora occurring near Earth's North Pole.

Automated Transfer Vehicle (ATV): European Space Agency spacecraft designed to supply the International Space Station with propellant, water, air, payload experiments and the like. In addition, the ATV can reboost the station, restoring its orbit, which shrinks over time as the result of friction caused by its interaction with the atmosphere.

Auxiliary power unit (APU): A secondary, sometimes backup, power-generating system of an air or space vehicle. The auxiliary power unit of the shuttle orbiter is a hydrazine-fueled, turbine-driven power unit that generates mechanical shaft power to drive a hydraulic pump that produces pressure for the orbiter's hydraulic system. The APUs provide power for most vehicle subsystems, including landing gear and brakes, rocket engine gimballing, and moving the shuttle's control surfaces. On aircraft, it is a relatively small, self-contained generator used to start the main engines, usually with compressed air, and to provide electrical power while the aircraft is on the ground. In many aircraft, the APU can also provide electrical power in the air.

AVCS: *See* Advanced vidicon camera system.

Avionics: The electronic devices used onboard an air or spacecraft, or the development, production, or study of those devices.

Axis: The imaginary line around which a celestial body or human-made satellite rotates.

Azimuth: The arc, or angular distance, measured horizontally and moving clockwise between a fixed point (usually true north) and a celestial object. *See also* Altitude.

Backup crew: A group of astronauts/cosmonauts trained to perform the same functions as the primary crew members of a particular space mission (such as the commander, pilot, and flight engineer), in the event that something requires replacement of all or part of the original crew.

Baikonur Cosmodrome: Also called Tyuratam, the world's oldest and largest working space launch facility. It is situated about 200 kilometers to the east of the Aral Sea, on the north bank of the Syr Darya, near the town of Tyuratam, in the south-central part of Kazakhstan. The name Baikonur was chosen to help prevent the West from determining the site's actual location by suggesting that the site was near Baikonur, a mining town about 320 kilometers northeast of the space center. When the Soviet Union collapsed, Kazakhstan agreed to lease the site to Russia.

Ballistic missile: A missile that is not self-guided but rather is aimed and propelled only at the point of launch, following a trajectory determined at launch.

Ballistics: The study of the motion of projectiles in flight, including their trajectories; it is especially important in the launching and course-correction maneuvers of spacecraft.

Band: *See* Frequency, Hertz.

Bar: *See* Atmospheric pressure.

Barbecue maneuver: The deliberately-maintained slow roll (along the long axis) of a spacecraft in orbit so that all exterior surfaces will be evenly heated by the Sun.

Barycenter: The center of mass of a system of two or more bodies.

Basalt: A fine-grained, dark-colored rock of volcanic origin, basalt is the most common extrusive igneous rock on the terrestrial planets and covers about 70 percent of Earth's surface.

Bhangmeter: An optical-flash detector used to detect nuclear explosions from satellites.

Big Bang theory: Accepted by many scientists, this cosmological theory states that the universe evolved from a gigantic explosion of a compressed ball of hot plasma many billions of years ago. The theory holds that matter is still flying outward uniformly from the center of this explosion, and is therefore also referred to as the expanding universe theory. *See also* Steady state theory.

Binary star system: A star system formed of two stars orbiting their combined center of mass. Three types of binaries are visual binaries, which emit radiation in the visible wavelength range and are the most common binaries; spectroscopic binaries, distinguished by their Doppler shifts; and eclipsing binaries, in which one star periodically blocks light from the other as they rotate around each other. It is estimated that more than half of the stars in the galaxy are binaries.

Biosatellite: Derived from the words "biological" and "satellite," a biosatellite is an artificial satellite carrying life forms for the purpose of discovering their reaction to conditions imposed in the space environment.

Biotelemetry: The remote measurement and monitoring of the life functions (such as heart rate) of living beings in space, and the transmissions of such data to the monitoring location (such as Earth).

Black dwarf: A star that has cooled to the point that it no longer emits visible radiation; the end state of a white dwarf star. Black dwarfs should not be confused with brown dwarfs, which are formed when gas contracts to form a star but does not possess enough mass to initiate and sustain hydrogen nuclear fusion. What we now refer to as brown dwarfs were at times called black dwarfs in the 1960's. *See also* White dwarf.

Black hole: A hypothetical celestial body whose existence is predicted by Albert Einstein's general theory of relativity and is accepted by many scientists. In a black hole, matter is so condensed and gravitational forces are so strong that not even light can escape. Black holes are thought to be either the product of a collapsed star (stellar black holes) or the result of the original Big Bang (primordial black holes).

Blueshift: An apparent shortening of electromagnetic wavelengths emitted from a star or other celestial object, indicating movement toward the observer. *See also* Doppler effect.

Boom: A long arm extending outward from a satellite or other spacecraft to hold an instrument or device such as a camera.

Booster: *See* Rocket booster.

Bow shock: The wave created by a planet's magnetic field when it forms an obstacle to the stream of ionized gases flowing outward from the Sun.

Brazilian National Institute for Space Research (INPE): In Portuguese, Instituto Nacional de Pesquisas Espaciais. The INPE manages and coordinates the country's involvement in space-related activities. INPE's mission is to enable Brazilian society to benefit from new developments in space science and technology.

Breccia: A type of rock formed by sharp, angular fragments embedded in fine-grained material such as clay or sand. Among the Moon rocks returned by the Apollo astronauts, breccias were the most common.

Buran: The Russian word for "snowstorm," the Soviet reusable space shuttle similar in design to its American counterpart, but with two important differences: it could be flown automatically and it did not have reusable boosters.

Burn: As a noun, the term used to refer to the firing of a rocket engine, including any burn used during a spaceflight to set the spacecraft on a trajectory toward a planet or other target.

Burnout: The point when combustion ceases in a rocket engine.

C-band: A radio frequency range of 3.9 to 6.2 gigahertz. *See also* Hertz.

Caldera: A very large crater formed by the collapse of the central part of a volcano.

Canadian Space Agency (CSA): In French, L'Agence Spatiale Canadienne (ASC). Established in March, 1989, the Canadian Space Agency (CSA) derives its authority from the Canadian Space Agency Act, sanctioned in December, 1990. Five core functions are carried out by the CSA: Space Programs, Space Technologies, Space Science, Canadian Astronaut Office, and Space Operations.

Canard: A short, stubby, wing-like element affixed to the Apollo launch escape tower to provide the Command Module blunt end forward aerodynamic capture during an abort.

Canopus: The brightest star in the sky after Sirius, visible south of 37° latitude. Canopus is often the target of a spacecraft's star tracker, which uses it as a reference point in steering a course toward the spacecraft's destination.

Capsule communicator (CapCom): A ground-based astronaut who acts as a communications liaison between ground and flight crews during piloted missions.

Cassegrain telescope: The most common type of reflecting telescope, named for its inventor, Guillaume Cassegrain, who proposed the idea in 1672. The telescope contains two mirrors: a concave mirror near its base, which reflects light from the sky onto a convex mirror above it; the convex mirror, in turn, reflects the light back down through a hole in the middle of the concave mirror to the focal point. The Hubble Space Telescope is of Cassegrain design.

CCD: *See* Charge-coupled device.

Celestial mechanics: The branch of physics concerned with the laws that govern the motion (especially the orbits) of artificial and natural celestial bodies.

Celestial sphere: An imaginary sphere surrounding an observer at a fixed point in space (the sphere's center), with a radius extending to infinity, a celestial equator (a belt cutting the sphere into two even halves), and celestial poles (north and south). By reference to these points on the celestial sphere, the observer can describe the position of an object in space.

Celsius scale: A temperature scale, named for its inventor, Anders Celsius (1701-1744), which sets the freezing point of water at zero and the boiling point at 100. Also referred to as the centigrade scale, its increments correspond directly to kelvins. To convert kelvins to degrees Celsius, subtract 273.15. *See also* Kelvin.

Centre National d'Études Spatiales (CNES): *See* French Space Agency.

Centrifugal force: Often described as a fictitious force, it is a reaction to the centripetal force as described in a non-inertial reference frame.

Centrifuge: A device for whirling objects or human beings at high speeds around a vertical axis. A centrifuge exerts an inertial force, which, when countered by centripetal force, aids in the testing of spacecraft hardware or in training astronauts to withstand the forces of launch and re-entry.

Centripetal force: The force always directed to the center of the circle making a motion in a circle possible.

Cepheid variable: A star that has passed its main sequence phase (the greater part of its lifetime) and has entered a transitional phase in its evolution. During this phase the star expands and contracts and at the same time pulsates in brightness. Astronomers find the distance to the star and nearby celestial objects by measuring the star's period of pulsation and extrapolating from that its absolute magnitude. They are then able to compare the absolute magnitude to the apparent magnitude.

Chandrasekhar limit: The maximum possible mass for a white dwarf star, calculated by Subrahmanyan Chandrasekhar in 1931 as approximately 1.4 solar masses (later modified upwards for rapidly rotating white dwarf stars). When the mass exceeds the Chandrasekhar limit, gravity compresses it into a neutron star or black hole.

Charge: A property of matter defined by the excess or deficiency of electrons in comparison to protons. A negative charge results from excess electrons; a positive charge indicates a deficiency of electrons.

Charge-coupled device (CCD): A highly sensitive device, commonly 1 centimeter square, that is sensitive to electromagnetic radiation and contains electrodes and conductor channels overlying an oxide-covered silicon chip, for collecting, storing, and later transferring data to create images of celestial objects and other astronomical phenomena.

Chromosphere: The lower layer of the solar atmosphere (between the photosphere and the solar corona), several thousands of kilometers thick, composed mainly of oxygen, helium, and calcium and is visible only when the photosphere is obscured, as during a solar eclipse. The term also applies to corresponding regions of other stars.

Chronograph: A device for viewing the solar (or another star's) corona. It consists of a solar telescope outfitted with an occulting mechanism to obscure the photosphere (hidden during a solar eclipse) so that the corona is more easily perceived.

Circular orbit: An orbit in which the path described is a circle.

Cislunar: Adjective referring to space between Earth and the Moon.

Closed loop: Automatic control units linked together with a process to form an endless chain.

Closing rate: The speed of approach of two spacecraft preparing for rendezvous.

Coma: *See* Comet.

Comet: A luminous celestial object orbiting the Sun, consisting of a nucleus of water ice and other ices mixed with solid matter, and, as the comet approaches the Sun, a growing coma and tail structure. The coma is a collection of gases and dust particles that evaporate from the nucleus and form a glowing ball around it. The tail structure involves two trailing tails, one plasma and one dust. An anti-tail may also point forward. Comets appear periodically, depending on the parameters of their solar orbits, and vary in size from a few kilometers to thousands of kilometers.

Command and Service Module (CSM): The portion of the Apollo spacecraft consisting of the Command Module and the Service Module.

Command Module (CM): A conical three-person spacecraft that served as the control center and main living area for Apollo missions. The CM was the only part of Apollo built to withstand the heat of reentry. The forward section contained a pair of thrusters for attitude control during reentry, parachutes for landing, and a tunnel for entering the Lunar Module (LM). At the end of the tunnel was an airtight hatch and a removable docking probe used for linking the CM and LM.

Communications satellite: *See* Telecommunications satellite.

Comsat: An abbreviation for "communications satellite." *See also* Telecommunications satellite.

Conjunction: The alignment of two planets or other celestial bodies so that their longitudes on the celestial sphere are the same. Inferior conjunction occurs between two bodies whose orbits are closer to the sun than Earth's; superior conjunction occurs between bodies whose orbits are farther from the Sun than Earth's.

Constellation: A collection of stars that form a pattern as seen from Earth. The stars in these groupings are often quite distant from one another, and are linked primarily by their appearance in the constellations they form (such as the Big Dipper or Ursa Major) against the backdrop of the night sky. Constellations are useful points of reference for astronomers and other stargazers.

Convection: A process that results from the movement of unevenly heated matter in gas or liquid form. In it, hotter matter moves toward and into cooler matter. The resultant circular motion and transfer of heat energy is convection.

Core: The central portion of any celestial body. Often refers to the terrestrial planets but may also

refer to stars. Also, the central part of a launch vehicle, to which strap-on boosters may be added.

Core sample: A sample of rock and soil taken from Earth, the Moon, or another terrestrial body by pressing a cylinder down into the body's surface.

Corona: The outermost portion of the Sun's atmosphere, extending like a halo outward from the Sun's photosphere. The corona consists of extremely hot ionized gases that eventually escape as solar wind. The term is also used to refer to the corresponding region of any star's atmosphere.

Cosmic dust: Tiny solid particles found throughout the universe, also known as interstellar dust. Scientists believe that cosmic dust originated from the primordial universe, the disintegration of comets, the condensation of stellar gases, and other sources.

Cosmic radiation: Particles that are the most energetic known, consisting mainly of protons, along with electrons, positrons, neutrinos, gamma-ray photons, and other nuclei. These particles emanate from a number of sources, both within and beyond the Milky Way, and they bombard atoms in Earth's atmosphere to produce showers of secondary particles such as pions, muons, electrons, and nucleons. If a primary cosmic particle is sufficiently energetic when it hits an atmospheric atom, the secondary particles can pass through matter to reach Earth's surface.

Cosmic ray detector: A device for sensing, measuring, and analyzing the composition of cosmic radiation in an attempt to discover its sources and distribution.

Cosmodrome: The spacecraft and launch vehicle facilities of the Soviet Union and Russia, including assembly, test, and launch facilities. One of the best known is Tyuratam/Baikonur in Kazakhstan.

Cosmology: The study of the origins and evolution of the universe.

Cosmonaut: A Russian, a person from the Soviet Union, or a person traveling in a Russian or Soviet spacecraft who achieves spaceflight as defined by the Fédération Aéronautique Internationale. *See also* Astronaut, Taikonaut.

Countdown: The audible backward counting in fixed units (hours, minutes, seconds) from an arbitrarily chosen starting number to mark the time remaining before an event; also, preparations carried on during such a count. For spaceflight-related events (such as liftoff, satellite release, and docking) items to be checked and functions to be performed are executed in sequence. Usually expressed as "T" (time) "minus" so many minutes and seconds before the event. *See also* Ground-elapsed time, Mission-elapsed time.

Crater: A depression in the surface of a planet or moon caused by the force of a meteorite's impact. Also, the depression that forms at the mouth of a volcano.

Cruise missile: A missile that flies at low altitude by means of an onboard guidance system that senses terrain and identifies the target.

Crust: The outermost layer, or shell, of a moon or planet, such as Earth.

Cryogenic propellant: Liquid propellant (such as liquid oxygen and liquid hydrogen) for rocket engines that must be stored at extremely low temperatures to remain liquid.

Cyclotron radiation: The electromagnetic radiation produced by charged particles as they spiral around magnetic lines of force at extremely high speeds close to that of light.

Data acquisition: The detection, gathering, and storage of data by scientific instruments.

DBS: *See* Direct broadcast satellite.

Deboost: A retrograde maneuver which lowers either perigee or apogee of an orbiting spacecraft. Not to be confused with deorbit.

Declination: Angular measurement of a body above or below a celestial equator, measured north or south along the body's hour circle. Corresponds to Earth's surface latitude.

Deep space: Regions of space beyond the Earth-Moon system.

Deep space probe: A device launched beyond the Earth-Moon system that is designed to investigate other parts of the solar system or beyond. Sometimes called an interplanetary space probe, a reference to spacecraft investigating the planets and the space between them.

Delta (Δ) V: Velocity change.

Density: A measure of the distribution of an object's mass throughout its volume.

Deorbit: To execute maneuvers, such as firing of retrograde rockets, and to reduce orbital velocity in preparation for descent to the surface of a celestial body.

Descent stage: The lower of two stages of the Apollo Lunar Module; it carried a throttleable rocket engine and fuel for landing as well as some scientific equipment and the Lunar Roving Vehicle. *See also* Ascent stage.

Deutschen Zentrum für Luft- und Raumfahrt (DLR): *See* German Aerospace Center.

Dewar: A container, similar to a vacuum bottle, with inner and outer walls. Between these walls, space is evacuated to prevent transfer of heat. Used to store cryogenic fuels at very low temperatures.

Dielectric: Used to characterize any device, substance, or state (insulation materials, a vacuum) that does not readily conduct electricity.

Digital imaging: *See* Imaging.

Direct broadcast satellite (DBS): A telecommunications satellite that is a subset of the domestic communications class, designed to broadcast television signals directly to subscribers' ground-based satellite dishes.

Dirty snowball theory: The model of comets, accepted by most astrophysicists, that considers a comet's nucleus to be composed of a small sphere of ice and rock.

Discovery Program: A NASA program of small probes, which operates under the idea of "faster, better, cheaper." The basic requirements of the Discovery Program are that missions should take less than three years to complete so that they can keep up with technological advancements. *See* Explorer Program.

Diurnal: Occurring daily. The diurnal motion of a planet or other celestial body is its daily path across the sky as seen from a fixed point such as Earth. Diurnal motion depends on the position of the observer.

Dock: To link one spacecraft with another or others while in space. First achieved in in 1966 by Gemini VIII astronauts Neil A. Armstrong and David R. Scott with an Agena target vehicle.

Domestic communications satellite: Domestic communication missions relay domestic transmissions, such as telephone, radio, and television signals, from one point on the Earth's surface to another.

Doppler effect: First described in 1842 by Christian Doppler (1803-1853), the principle that describes the effect, from the perspective of an observer, of an object's movement on the electromagnetic or sound energy that it emits. The wavelength or frequency of this energy appears to increase (shorten) if the energy source is approaching the observer, and the rate (speed) of approach will determine the rate of increase. The opposite is true as the energy source moves away from the observer. Hence, the sound from an ambulance streaking past a motorist seems to rise sharply as it approaches, then fall as it rushes away: The motorist perceives increasingly compressed (shortened) wavelengths of sound as the ambulance approaches, then increasingly stretched (elongated) wavelengths as the ambulance speeds down the road. Electromagnetic (light) energy behaves similarly. Red and infrared rays have longer wavelengths, while the wavelengths of blue and ultraviolet rays are shorter. A source of light that is moving away from an observer will appear more intensely red (called the redshift). If the light source is approaching, it will become more intensely blue (blueshift). These phenomena are measured by observing the spectral lines of the energy source over time. The Doppler effect is fundamental to much of our understanding of the universe and provides strong support for the theory of the expanding universe.

Downlink: Transmissions to Earth from a spacecraft; often used in reference to telecommunications satellites. *See also* Uplink.

Drogue chute: A small parachute designed to pull a larger parachute from stowage or to decrease the velocity of a free-falling spacecraft during reentry.

Early-warning satellite (EW): A spacecraft used to detect the launch of missiles and rockets from the Earth's surface. Information from these satellites is used to quickly detect the launch of offensive missiles and to track the long-term patterns of foreign countries' space programs.

Earth day: Twenty-four hours, or the time required for Earth to complete one rotation on its axis. Scientists measure the other planets' periods of rotation in Earth days.

Earth-orbital probe: An unmanned spacecraft carrying instruments for obtaining information about the near-Earth environment.

Earth resources satellite: This mission class refers to missions that collect information about the Earth's atmosphere, land, and oceans for scientific research and resource management.

Eccentricity: The degree to which an ellipse (or orbital path) departs from circularity. Eccentricity is characterized as high when the ellipse is very elongated.

Eclipse: The obscuring of one celestial body by another. In a lunar eclipse, Earth's shadow obscures the Moon when Earth is situated directly between the Sun and Moon. In a solar eclipse, the Moon is situated between the Sun and Earth in such a way that part or all of the Sun's light is blocked; the total blockage of sunlight (with the exception of the Sun's corona) is called a total eclipse of the Sun.

Ecliptic plane: The plane in which Earth orbits the Sun. From Earth, the ecliptic plane is perceived as the Sun's yearly path through the sky.

Edge of space: *See* Karman Line.

Ejecta: Fractured and/or molten rocky debris thrown out of a crater during an impact event; material discharged by a volcano.

Electromagnetic radiation: Radiation, or a series of waves of energy, consisting of electric and magnetic waves vibrating perpendicularly to each other, or to particles (photons) traveling at the speed of light. Electromagnetic radiation varies in wavelength and frequency as well as in source, which may be thermal or nonthermal. *See also* Electromagnetic spectrum.

Electromagnetic spectrum: The continuum of all possible electromagnetic wavelengths, from the longest (radio waves, which are longer than 0.3 meter), to the shortest (gamma rays, which are shorter than 0.01 nanometer). The shorter the wavelength, the higher the frequency and the greater the energy. Within the electromagnetic spectrum is a range of wavelengths that can be detected by the human eye. This is visible light. Its wavelengths correspond to colors: Red light emits visible radiation with the longest wavelength; violet light emits radiation with the shortest wavelength. None of these types of electromagnetic radiation is discrete; each blends into the surrounding forms. Detection of nonvisible radiation by special instruments (used in such branches of astronomy like infrared astronomy and x-ray astronomy) reveals much about the behavior of celestial bodies and the origins of the universe. Regions of the spectrum (from shortest wavelength to longest) are gamma ray, x ray, ultraviolet, visible, infrared, microwaves, and radio waves.

Electron: An atomic particle that carries a negative charge and a mass about one eighteen-thousandth that of a proton. One or more electrons whirl around the nuclei of all atoms and can also exist independently.

Electronic intelligence satellite (ELINT): A military satellite used to monitor (spy on) the military and domestic electronic communications of other nations. These types of satellites are also known as ferrets. *See also* Ferret.

Electrophoresis: A process for separating cells using a weak electric charge. More easily accomplished in space than on Earth.

Elementary particles: The smallest units of matter or radiation, characterized by electrical charge, mass, and angular momentum. Elementary particles include electrons, neutrinos, quarks, the various mesons, and their corresponding antiparticles (which form antimatter). Photons, the smallest units of electromagnetic radiation, are also considered elementary particles.

ELINT: *See* Electronic intelligence satellite.

Ellipse: An oval-shaped geometric curve formed by a point moving so that the sum of the distances between two points (called foci) around which it moves is always the same. The planets trace out ellipses in their orbits around the Sun.

Elliptical orbit: An orbit that departs from circularity, as most orbits do. A highly elliptical orbit is one whose apoapsis is much greater than its periapsis, resulting in an orbit that traces out an elongated ellipse. The Earth orbits the Sun in a slightly elliptical orbit with the Sun at one focus.

ELV: *See* Expendable launch vehicle.

Emission spectrum: A spectrum showing the array of wavelengths emitted by a thermal source of electromagnetic radiation, such as a star. Atoms in this source subjected to thermal (heat) energy will emit energy as their electrons move from one energy level to another. The wavelengths that characterize these energy jumps correspond to specific elements (such as hydrogen), and appear as a characteristic line on the emission spectrum. The relationships of these lines to one another tell astronomers much about the composition, density, temperature, and other conditions of the energy source.

EMU: *See* Space suit.

End of mission: Space missions that return to Earth officially end when the craft touches the ground or the water. Mission-elapsed time will continue to run until the flight is officially declared to have ended. For the Mercury, Gemini, and Apollo spacecraft, touchdown occurred the moment any part of the spacecraft came in contact with the water. For the space shuttle, touchdown occurs when one of the main landing gear wheels comes in contact with the ground.

Entry corridor: The final flight path of the spacecraft before and during Earth reentry.

Ephemeris: Orbital measurements (such as apogee, perigee, inclination, and period) of one celestial body in relation to another at a given time. In spaceflight, the orbital measurements of a spacecraft relative to the celestial body about which it orbits.

Equatorial orbit: An orbit that follows the equator of the body orbited.

Escape velocity: The velocity at which an object must travel to escape the gravitational attraction of a celestial body. In order for a spacecraft to leave Earth orbit, for example, its engines must exert enough in-orbit thrust to achieve escape velocity.

Eucrite meteorite: A stony meteorite and the oldest known basalt in the solar system. This is the most common class of achondrite meteorite whose parent body is believed to be the asteroid Vesta. Eucrites are basalts—volcanic rocks of magmatic origin—composed primarily of the calcium-poor pigeonite (a monoclinic mineral of the pyroxene group) and the calcium-rich anorthite (a white, grayish, or reddish feldspar occurring in many igneous rocks). Eucrites represent the upper crust of Vesta that solidified on a magma ocean after the core and the mantle had already been formed.

European Space Agency (ESA): The European Space Agency is Europe's gateway to space. ESA was formed in 1975 through the merging of the European Space Research Organization (ESRO) and the European Launcher Development Organization (ELDO).

EVA: *See* Extravehicular activity.

Event horizon: The boundary beyond which an observer cannot see. Also, the boundary beyond which nothing can escape from a black hole, where escape velocity equals the speed of light and thus nothing, not even light, can escape. Therefore, the event horizon is theoretically the spherical delineation of a black hole. *See also* Escape velocity.

Exobiology: The study of the conditions for and potential existence of life beyond Earth.

Exosphere: The outermost layer of Earth's atmosphere.

Expendable launch vehicle (ELV): A launch vehicle not intended for reuse. The Atlas, Delta, Titan, and Saturn launch vehicles fall into this category.

Explorer: A long and ongoing series of small United States scientific spacecraft. Explorer 1, the first successful American satellite, was launched on January 31, 1958, and discovered the Van Allen Belts. Explorer 6 took the first photo of Earth from space. Subprograms of the Explorer series have included ADE (Air Density Explorer), AE (Atmosphere Explorer), DME (Direct Measurement Explorer), EPE (Energetic Particles Explorer), GEOS (Geodetic Earth Orbiting Satellite), IE (Ionosphere Explorer), and IMP (Interplanetary Monitoring Platform). Among the more recent Explorers are IUE (International Ultraviolet Explorer) and COBE (Cosmic Background Explorer).

Explorer Program: The Explorer Program is NASA's continuing level-of-effort program for orbital astronomy and space physics packages. Explorers come in three sizes, defined by budget cap: Medium Explorer (MIDEX), Small Explorer (SMEX), and University-class Explorer (UNEX). *See also* Discovery Program, Medium-class Explorer, Small Explorer, University-class Explorer.

Explosive bolts: Bolts destroyed or severed by a surrounding explosive charge, which can be activated by an electrical impulse.

External Tank: The large tank that stores liquid fuel for the three space shuttle main engines; it separates from the shuttle after all the fuel has been used (about 8.5 minutes after liftoff) and breaks up in the atmosphere.

Extraterrestrial life: *See* Exobiology.

Extravehicular activity (EVA): Popularly known as a spacewalk, any human maneuver taking place partially or fully outside the portion of a spacecraft that houses the astronauts.

Extravehicular mobility unit (EMU): *See* Space suit.

F region: *See* Thermosphere.

Fairing: A piece, part, or structure having a smooth, streamlined outline, used to cover a non-streamlined object or to smooth a junction.

False-color image: An image resembling a photograph, created from data collected by instruments (such as an infrared sensor) aboard a spacecraft and deliberately assigned unnatural colors in order to make nonvisible radiation visible or to highlight distinctions. *See also* Imaging.

Fédération Aéronautique Internationale (FAI): Standard-setting and record-keeping body for aeronautics and astronautics. It is the international governing body for all airborne sports.

Ferret: An electronic intelligence satellite designed to detect hostile electromagnetic radiation. *See also* Electronic intelligence satellite.

Fission (atomic): The breaking of an atomic nucleus into two parts, resulting in a great release of energy.

Flight control system: A system that serves to maintain attitude stability and control during flight.

Flight path: The trajectory of an airborne or spaceborne object relative to a fixed point such as Earth.

Fluid mechanics: The study of the behavior of fluids (gases and liquids) under various conditions, including that of microgravity in spaceflight. Understanding fluid mechanics in space is important to the technology of spaceflight and may also have applications on Earth.

Fluorescence: The property of emitting visible light absorbed from an external source.

Flyby: A close approach to a planet or other celestial object, usually made by a probe for the purpose of gathering data; the maneuver does not include orbit or landing. Also used to refer to a mission that undertakes a flyby.

Focal ratio (f-number): The ratio of (1) the distance between the center of a lens or mirror and its point of focus (focal-length) and (2) the aperture, or diameter, of the mirror lens. The focal ratio of a telescope determines its power of magnification.

Footprint: An area on Earth's surface where a spacecraft is expected to land. Also, an area served by a telecommunications satellite.

Frauenhofer lines: Prominent absorption lines in the Sun's spectrum, first observed by Joseph von Frauenhofer in 1814, indicating the presence of certain elements in the Sun's corona. Also used to refer to such absorption lines in other stars' spectra.

Free-flyer: Any spacecraft capable of solitary flight, and not attached to another for electrical power.

Free-return trajectory: An orbital flight path that allows a spacecraft to reenter Earth's atmosphere safely without assistance.

French Space Agency (CNES): In French, Centre National d'Études Spatiales (the Center for National Space Studies). This state-owned industrial and commercial organization is currently under the joint responsibility of the Ministry of Research and the Ministry of Defense. Created in 1961, CNES is responsible for shaping France's space policy, presenting it to the government, and implementing it.

Frequency: The number of times an event recurs within a specific period of time. Frequency characterizes both sound and electromagnetic radiation and is defined as the ratio of its wavelength to its speed. Frequency is measured in hertz or multiples of hertz.

Fuel cell: A device that joins chemicals to produce electric energy. Different from a battery in that the chemicals are joined in a controlled fashion depending on electrical load. A by-product of fuel cells that use oxygen and hydrogen is potable water. Used on Apollo and space shuttle missions.

Fusion: A thermonuclear reaction in which the nuclei of light elements are joined to form heavier atomic nuclei, releasing energy. The process can be controlled to produce power. It is also the process by which stars form the elements with atomic numbers up to that of iron.

G or g force: Force exerted upon an object by gravity or by reaction to acceleration or deceleration, as in a change of direction: one g is the measure of force required to accelerate a body at the rate of 9.8 meters-per-second/per-second. A person undergoing a 10g acceleration would have an apparent weight ten times his or her actual gravitational weight.

Gabbro: A dark-colored, igneous, crystalline rock.

Gain: The increase in power of a transmitted signal as it is picked up by an antenna.

Galaxy: A collection of stars, other celestial bodies, interstellar gas and dust, and radiation rotating or clustered around a central hub, classified by Edwin Hubble in 1925 as one of four shapes: spiral, elliptical, lenticular, or irregular. Galaxies are thousands of light years across and contain billions of stars, and there are thousands of galaxies in the universe.

Gamma radiation: Electromagnetic radiation with wavelengths less than 0.01 nanometer, the most energetic known in the universe outside cosmic radiation. The ability of gamma rays to penetrate the interstellar matter and radiation of the universe makes them especially valuable to astronomers.

Gamma-ray astronomy: The branch of astronomy that investigates gamma radiation and its sources with detectors sent aloft in satellites such as the Orbiting Solar Observatory 3, SAS2, and COS-B. Among the sources of gamma rays are pulsars and neutron stars.

Gamma-ray bursts: Brief flashes of highly energetic photons (light) that occur at random times and unpredictable locations in the sky.

Gegenschein: A patch of faint light about 20° across, visible from the night side of Earth at a point opposite the Sun in the ecliptic plane, and possibly caused by the reflection of sunlight from a dust tail swept away by the solar wind. *See also* Zodiacal light.

Geiger counter: A device that detects high-energy radiation (including cosmic rays) through a tube that contains gas and an electric current. The radiation causes the gas to ionize, which is transmitted to the current and detected as a sound or a needle jump.

General relativity theory: Fundamental physical theory of gravitation, which corrects and extends Newtonian gravitation, especially at the macroscopic level of stars or planets. General relativity has a unique role among physical theories in the sense that it interprets the gravitational field as a geometric phenomenon. More specifically, it assumes that any object possessing mass will curve the space in which it exists, and that this curvature equals gravity. The theory has held up in every experimental test performed on it since its formulation by Albert Einstein in 1915.

Geocentric orbit: An orbit with Earth as the object orbited.

Geodesy: The science concerned with the size and shape of Earth and its gravitational field.

Geodetic satellite: A satellite used to measure the location of points on the Earth's surface with great accuracy. These measurements are used to determine the exact size and shape of Earth, act as references for mapping, and track movements of Earth's crust.

Geostationary orbit: A type of geosynchronous orbit that is circular and lies in Earth's equatorial plane, at an altitude of approximately 36,000 kilometers. As a result, a satellite in geostationary orbit appears to hover over a fixed point on Earth's surface. *See also* Geosynchronous orbit.

Geosynchronous orbit: A geocentric orbit with a period of 23 hours, 56 minutes, and 4.1 seconds, equal to Earth's rotational period. Such an orbit is also geostationary if it lies in Earth's equatorial plane and is circular. If inclined to the equator, a geosynchronous orbit will appear to trace out a figure eight daily; the size of the figure eight will depend on the angle of inclination. These orbits are used for satellites whose purpose it is to gather data on a particular area of Earth's surface or to transmit signals from one point to another. Many communications satellites are geosynchronous. This type of orbit is sometimes known as the Clarke orbit in honor of Sir Arthur C. Clarke, who first proposed the concept that geostationary satellites would be ideal telecommunications relays.

German Aerospace Center (DLR): In German, Deutschen Zentrum für Luft- und Raumfahrt, or (literally) German Center for Air and Space Travel. DLR manages the country's space activities on behalf of and under the instructions of the German government. The agency plans and carries out German spaceflight programs and activities and represents the interests of the German space community internationally.

GET: *See* Ground-elapsed time.

Gigahertz: *See* Hertz.

Gimbaled motor: A rocket motor mounted on gimbal (a device that has two mutually perpendicular axes of rotation) to correct pitching and yawing.

Glavkosmos or Glavcosmos: The Soviet Union's contact agency for space affairs, based at the Ministry of Machine Building in Moscow. Following the collapse of the Soviet Union in 1991, the interface between Russia's space activities and the outside world became the newly formed Russian Space Agency.

Glide path: The path of descent of an aircraft under no power.

Globular clusters: Spherically shaped congregations of thousands, sometimes millions, of stars, which occur throughout the universe, although more often near elliptical galaxies than spiral galaxies such as the Milky Way. It is believed that globular clusters contain the oldest stars, and because they also contain a variety of stars of different sizes, all occurring at relatively the same distance from Earth, it is possible to learn much from them about the history of stars and the size of the galaxy.

Gravitation: The force of attraction that exists between two bodies, such as Earth and the Moon. In 1687, Sir Isaac Newton described this force as proportional to the distance between them squared. Although gravitation is the weakest of the naturally occurring forces (the others being electromagnetic, and the strong and weak nuclear interactions), it has the broadest range

and is responsible for much celestial movement, including orbital dynamics.

Gravitational constant: The universal constant defined as the force of attraction between two bodies of 1 kilogram in mass separated by 1 meter.

Gravitational field: A force field of attraction exerted around a mass, such as a celestial body.

Gravity assist: A technique, first used with the Mariner 10 probe to Mercury, whereby a spacecraft uses the gravitational and orbital energy of a planet (or moon) to gain energy to achieve a trajectory toward a second destination or to return to Earth.

Great Red Spot: A vast, oval-shaped cloud system occurring at 22° south latitude in Jupiter's atmosphere, rotating counterclockwise. Its name comes from an unknown substance that the convection of the phenomenon pulls to the surface; the substance absorbs violet and ultraviolet radiation and consequently delivers a red hue to the Spot. The Great Red Spot has been observed for more than three centuries.

Greenhouse effect: The heating of a planet's surface and lower atmosphere as a result of trapped infrared radiation. Such radiation becomes trapped when an excess of carbon dioxide in the atmosphere absorbs and remits infrared radiation rather than allowing it to escape. As a result, the atmosphere acts like a greenhouse, heating the planet. The effect on Earth is exacerbated by the burning of fossil fuels, which releases carbon dioxide into the atmosphere.

Grism: Combination of a prism and grating arranged to keep light at a chosen central wavelength undeviated as it passes through the grism. Grisms are normally inserted into a collimated camera beam. The grism then creates a dispersed spectrum centered on the location of the object in the camera field of view.

Ground-elapsed time (GET): *See* Mission-elapsed time.

Ground station: A location on Earth where radio equipment is housed for receiving and sending signals to and from satellites, probes, and other spacecraft.

Ground test: To test, on Earth, craft, devices, and instruments designed for operation in the air or space.

Guidance system: A system that measures and evaluates flight information, correlates it with target data, converts the result into the conditions necessary to achieve the desired flight path, and communicates this data in the form of commands to the flight control system.

Gyroscope: A device that uses a rapidly spinning rotor to assist in stabilization and navigation.

Hard landing: A crash landing. Early probes to the Moon, for example, were designed to take pictures of the lunar surface during free fall before impacting the surface.

Hatch: A tightly sealed door to the outside or to another module of a spacecraft.

Heatshield: A layer of material that protects a space vehicle from overheating, especially upon reentry into Earth's atmosphere. *See also* Ablative heatshield.

Heliocentric orbit: An orbit with the Sun at its center.

Heliopause: The border between the solar system and the surrounding universe, where the solar wind gives way to interstellar matter and winds.

Hertz: The SI unit of frequency. Equals one cycle per second and is named for the German physicist H. R. Hertz (1857-1894) in 1960. Multiples include kilohertz (10^3 hertz), megahertz (10^6 hertz), and gigahertz (10^9 hertz).

High-Earth orbit: Any Earth orbit at a relatively great distance from Earth, such as the geosynchronous orbits of telecommunications satellites.

High-gain antenna: A single-axis, strongly directional antenna that is able to receive or transmit signals at great distances.

Highly elliptical orbit (HEO): This class covers orbits that have large eccentricities (are highly elliptical) with perigees below 3,000 kilometers and apogees above 30,000 kilometers.

Horizon: The line formed where land meets sky, from the perspective of an observer. In astronomy, the horizon also means the circle on the circumference of the celestial sphere that is formed by the intersection of the observer's horizontal plane with the sphere. The particle horizon is the theoretical horizon on the celestial sphere at a distance beyond which particles cannot yet have traveled.

Hubble's law: The principle, articulated in 1929 by Edwin Hubble, that the galaxies are moving away from one another at speeds proportional to their distance: that is, uniformly across time. Hubble deduced this principle from observations of the redshifts in galactic spectra. Along with the discovery of the cosmic microwave background (CMB) radiation by Arno Penzias and Robert Wilson in 1965, Hubble's law forms the basis for the Big Bang theory of the expanding universe. *See also* Big Bang theory, Doppler effect, Redshift.

Hydroxyl-terminated polybutadiene (HTPB): A stable and easily stored synthetic rubber often used in tire manufacturing. HTPB propellant was first used and test flown in 1970 in Aerojet's Astrobee D meteorological sounding rocket. It has been used in the upper stages of the Delta II and Titan IV launch vehicles.

Hypergolic propellant: A form of liquid propellant in which the fuel ignites spontaneously upon contact with an oxidizer, thereby eliminating the need for an ignition system. Since hypergolics remain liquid at normal temperatures, they do not pose the storage problems of cryogenic propellants. Hypergolics are highly toxic and must be handled with extreme care. Hypergolic fuels commonly include hydrazine, monomethyl hydrazine (MMH), and unsymmetrical dimethyl hydrazine (UDMH). The oxidizer is typically nitrogen tetroxide (N_2O_4) or nitric acid (HNO_3).

ICBM: *See* Intercontinental ballistic missile.

IGY: *See* International Geophysical Year.

Imaging: The process of creating a likeness of an object by electronic means.

Impact basin: A large depression in the surface of a planet or a moon, created by the force of meteorite impact.

Inclination: *See* Orbital inclination.

Inertia: The property of an object to resist changes to its state of motion. Being an inherent property of mass, it is present even in the absence of gravity. For example, although a spacecraft may be located far away from any gravitating mass, its inertia must still be overcome in order for it to speed up, slow down, or change direction.

Inertial guidance: Guidance by means of the measurement and integration of acceleration data from on board the spacecraft. The system absorbs and interprets such data to determine speed and position and automatically adjusts the vehicle to a predetermined flight path. Essentially, the system pinpoints where it is and where it is going by analyzing information about where it came from and how it got there. It does not give out any radio frequency signal so it cannot be detected by radar and cannot be jammed.

Inertial Upper Stage (IUS): A two-stage solid-rocket motor used to boost heavy satellites out of low-Earth orbit into a higher orbit. It can be used in conjunction with both the space shuttle and Titan launch vehicles.

Infrared astronomy: The branch of astronomy that examines the infrared emissions of stars and other celestial phenomena. Studying the infrared emissions tells astronomers much about the composition and dynamics of their sources. Because infrared rays cannot readily penetrate most of Earth's atmosphere, knowledge of infrared astronomy has burgeoned through use of the Infrared Astronomical Satellite, the Kuiper Airborne Observatory, and other advances in space age technology.

Infrared radiation: Electromagnetic radiation of wavelengths from 1 to 1,000 micrometers, which occur next to visible light at the red end of the electromagnetic spectrum.

Infrared scanner: An imaging instrument that is sensitive to the infrared spectrum or to heat.

Infrared spectrometer: A spectrometer that takes spectra of infrared radiation emitted by celestial bodies.

Injection: The process of boosting a spacecraft into a calculated trajectory.

Insertion: The process of boosting a spacecraft into an orbit around the Earth or other celestial bodies.

Instituto Nacional de Pesquisas Espacias (INPE): *See* Brazilian National Institute for Space Research.

Instituto Nacional de Tecnica Aeroespacial (INTA): *See* Spanish Space Agency.

Intercontinental ballistic missile (ICBM): A ballistic missile with a range of thousands of kilometers specifically designed to carry a warhead.

Interferometry: A data acquisition technique that uses more than one signal receiver (such as a series of radio telescopes). Signals are combined to form one highly detailed image. *See also* Very long-baseline interferometry.

Intergalactic medium: Matter that exists between galaxies. Although space between galaxies is generally transparent and apparently empty, intergalactic matter must exist, because the combined mass of the galaxies in a galaxy cluster (of which there are many) is much less than that required to exert the gravitational force that forms the cluster. Hence, not only must intergalactic matter exist, but there must be enough of it to account for the missing mass. Postulations include invisible masses (dark matter), such as black holes and other forms of dead stars, as well as the intergalactic gases.

Interkosmos or Intercosmos: The Council for International Cooperation on Space Research at the Academy of Science of the USSR, founded in May, 1966. In September, 1976, Interkosmos voted to conduct flights of international crews from Soviet member states onboard Soviet spacecraft and orbital stations.

Intermediate range ballistic missile (IRBMs): A ballistic missile with a range of approximately 2,800 kilometers.

International Atomic Time (TAI): In Latin, Temps Atomique International. A very accurate and stable time scale, it is a weighted average of the time kept by about 200 cesium atomic clocks in more than fifty national laboratories around the world. It has been available since 1955, and became the international standard on which Coordinated Universal Time (UTC) is based on January 1, 1972. The United States is the single largest contributor to TAI, with clocks maintained at both the National Institute of Standards and Technology in Boulder, Colorado, and the United States Naval Observatory in Washington, D.C.

International Geophysical Year (IGY): The eighteen-month period from July, 1957, to December, 1958, during which many countries cooperated in the study of Earth and the Sun's effect on it. The Space Age began with the launch of Sputnik 1 on October 4, 1957.

Interplanetary space probe: *See* Deep space probe.

Interstellar dust: *See* Cosmic dust.

Interstellar wind: *See* Solar wind, Stellar wind.

Ion: An atom that is not electrically balanced but rather has either more electrons than protons or more protons than electrons. Such atoms are unstable and thus in search of the particles (missing electrons or protons) that will return them to electric balance.

Ionization: The process whereby atoms are made into ions by removal or addition of electrons or protons. Such a process often occurs as a result of excitation of atoms into an energy state in which they lose electrons.

Ionosphere: The ionized layer of gases in Earth's atmosphere that occurs between the thermosphere (below) and the exosphere (above), from approximately fifty to five hundred kilometers above the planet's surface. Within the ionosphere, ionized gases are maintained by the Sun's ultraviolet radiation. The free electrons that result reflect long radio waves that make long-distance radio communication possible. Other planets are known to have ionospheres, including Jupiter, Mars, and Venus.

IRBM: *See* Intermediate range ballistic missile.

Italian Space Agency (ASI): In Italian, Agenzia Spaziale Italiana. Formed in 1988, this government agency identifies, coordinates, and man-

ages Italian space programs and oversees Italy's involvement with ESA.

IUS: *See* Inertial Upper Stage.

Japan Aerospace Exploration Agency (JAXA): JAXA is Japan's unified aerospace agency, formed on October 1, 2003, from the merger of three previously separate organizations: ISAS (the Institute of Space and Astronautical Science), NASDA (the National Space Development Agency of Japan), and NAL (the National Aerospace Laboratory of Japan).

J2000.0 epoch: Used in astronomy, it is precisely Julian date 2451545.0 Terrestrial Time (TT), or January 1, 2000, 12h TT. This is equivalent to January 1, 2000, 11:59:27.816 International Atomic Time (TAI), or January 1, 2000, 11:58:55.816 Coordinated Universal Time (UTC).

Jupiter orbit: The Jupiter orbit class includes all missions that went into orbit around Jupiter or entered the Jovian atmosphere.

K-band: A radio frequency range of about 11 to 15 gigahertz. *See also* Hertz.

Karman Line (or Kármán Line): An internationally designated altitude used to define outer space. According to definitions by the Fédération Aéronautique Internationale (FAI), the Karman Line lies 100 kilometers above Earth's surface. Around this altitude, the Earth's atmosphere becomes negligible (at least for aeronautic purposes) and there is an abrupt increase in atmospheric temperature and interaction with solar radiation. It was named for Theodore von Kármán, a Hungarian aeronautics engineer and physicist who was responsible for many key advances in aerodynamics, particularly in the areas of supersonic and hypersonic airflow characterization.

Kelvin: A unit of temperature on the Kelvin temperature scale, which begins at absolute zero (273.15° Celsius). One unit kelvin is equal to one degree Celsius. The Kelvin scale is particularly suited to scientific (especially astronomical) measurement. *See also* Absolute temperature scale.

Kepler's laws of motion: Three laws of motion discovered by Johannes Kepler and published by him in 1618 and 1619: (1) Each planet moves in an ellipse around the Sun, with the Sun at one of the two foci of that ellipse. (2) A line from the sun to the planet sweeps out equal areas in equal times. (3) The square of the period of a planet's orbit is proportional to the cube of its mean distance from the Sun. *See also* Angular momentum.

Kibo: Japanese for "hope." Also the name of the Japanese Experiment Module (JEM) of the International Space Station.

Kick motor: A rocket motor on a spacecraft designed to boost it, or a payload, from parking orbit into a higher orbit or a different trajectory. Also known as an apogee kick motor, since it is more effective if fired at the spacecraft's orbital apogee.

Kilogram: A metric unit of mass. The equivalent of 1,000 grams or approximately 2.25 pounds.

Kilometer: A metric unit of distance, the equivalent of 1,000 meters or approximately 0.62 mile.

Korabl: The Russian word for "spacecraft." Korabl-Sputniks (spaceship-satellites) were early Soviet spacecraft that served as test vehicles for the Vostok piloted flights. They carried into orbit a variety of animals, including dogs, rats, and mice.

Kosmos: The Russian word for "space." The name refers to a series of satellites that were launched by the Soviet Union and are now being launched by Russia. Any satellite that does not fit into a particular program is designated as a Cosmos satellite.

Kristall: The Russian word for "crystal." Refers to the technological module of the Mir Complex.

Kuiper Belt: An area of the solar system extending from within the orbit of Neptune (at 30 astronomical units, or AU) to 50 AU from the sun, at inclinations consistent with the ecliptic. Objects within the Kuiper Belt are referred to as trans-Neptunian objects, asteroids, or planetoids). The outer boundary of the Kuiper belt is not defined arbitrarily; rather, there appears to be a real and fairly sharp drop off in objects beyond a certain distance. This is sometimes called the Kuiper gap or Kuiper cliff. The cause for this remains a mystery; one hypothesis postulates the existence of an Earth- or Mars-sized object that sweeps away debris.

Kvant (kvant): The Russian word for "quantum." Refers to the Kvant-1 astrophysics module of the Mir Complex and the Kvant-2 scientific and air lock module.

L'Agence Spatiale Canadienne (ASC): *See* Canadian Space Agency.

Lander: A spacecraft or module designed to make a soft landing on the surface of a celestial body: It carries scientific instruments to measure surface conditions.

Laser: Originally an acronym for Light Amplification by Stimulated Emission of Radiation. A beam of infrared, visible, ultraviolet, or shorter-wavelength radiation produced by using electromagnetic radiation to excite the electrons in a suitable material to a higher energy level in their cycles around their atomic nuclei. These electrons are then stimulated, making them jump back down to their normal energy levels. When they do, the electrons emit a stream of coherent radiation: photons with the same wavelength and direction as the originating radiation. This results in a narrow, intense beam of light (or nonvisible radiation), which bounces off a reflector and returns to the propagating material, where the process is repeated, maintaining the laser. Laser technology has a vast range of applications in telecommunications, medicine, and astronomical measurements.

Laser-ranging: A technique that allows scientists at two different Earth stations to determine, very precisely, their distance from each other by bouncing a laser beam off a satellite retroreflector. The time it takes to receive an echo allows each scientist to calculate his distance from the satellite; knowing both distances allows calculation of the distance between the two points on Earth. Over time, these measurements are repeated and changes in the distance between the two Earth locations are noted, providing important information about crustal movements and the likelihood of earthquakes.

Laser reflector: An instrument off which a scientist can bounce or reflect a laser beam in order to measure (usually great) distances.

Latitude: The angular distance from a specified horizontal plane of reference on Earth or the distance north or south of the equatorial plane. In the solar system, the angular distance of a celestial body from the ecliptic plane. *See also* Longitude.

Launch: To boost a body, such as a spacecraft, from a celestial body into space or from one orbit into another orbit or a trajectory. Also, the act of doing so.

Launch complex: The area from which a rocket vehicle is fired, including all the necessary support facilities, such as the launch pad, service structures, umbilical towers, safety equipment, and flame detectors.

Launch control center: A centralized control point for all phases of prelaunch and launch operations. The handover of control to a mission control center occurs at the moment of separation of the space vehicle from all hard ground connections.

Launch Escape System (LES): A mechanism available during the launch of Mercury, Apollo, and Soyuz missions, consisting of a solid-fueled rocket booster set atop the spacecraft. During the early stages of atmospheric flight after launch, the LES could fire to move the spacecraft to safety in the event of an emergency. The first and only use of a launch escape system during a piloted launch was the Soyuz T-10-1 mission. The launch vehicle was destroyed on the launch pad by fire but the escape rocket fired two seconds before the vehicle exploded, saving the crew.

Launch pad: The physical platform from which a spacecraft or launch vehicle is launched. Often called simply a pad. *See also* Launch complex.

Launch site: A location housing a facility designed to handle preparations for launch as well as the launch itself.

Launch vehicle: *See* Expendable launch vehicle.

Launch window: A period of time (usually days or hours) during which conditions, such as weather, orbital position of a rendezvous target spacecraft, and planetary alignment, are in sync for meeting the goals of a particular space mission and launch is therefore possible.

LES: *See* Launch Escape System.

Life-support system: The combined mechanisms that keep an environment capable of sustaining life, including devices that control air pressure, oxygen, temperature, and the like.

Liftoff: The point at which a vertically-rising vehicle leaves its ground-based support structure (a launch platform, the ground, or another vehicle). Rockets, missiles, lunar module ascent stages, vertical takeoff and landing (VTOL) aircraft, and helicopters are examples of vehicles that lift off, rather than take off (like an airplane).

Light-year: The distance that light travels in one year, or approximately 9.5×10^{12} kilometers.

Limb: The outer edge of the apparent (or visible) disk of the Sun, the Moon, a planet, or another celestial body.

Liquid-fueled rocket booster: A rocket booster that uses a liquid propellant. Expendable launch vehicles from Redstone through Titan have used liquid-fueled rockets.

Lithosphere: Earth's crust and top layer of the underlying mantle, about 80 kilometers thick. The term can be used in reference to the solid part of any planet.

Long-range intercontinental ballistic missile (LRICBM): An intercontinental ballistic missile with a range of thousands of kilometers.

Longitude: The angular distance from a specified vertical plane of reference; on Earth, the angular distance east or west of the plane that dissects Earth through the poles at the meridian.

Look angle: Angular limits of vision.

Low-Earth Orbit (LEO): An orbit having an apogee and a perigee below 3,000 kilometers. The large majority of all satellites are in low-Earth orbit.

LRICBM: *See* Long-range intercontinental ballistic missile.

LRVs: *See* Lunar Roving Vehicles.

Luminosity: The brightness of a celestial object.

Lunar day: The time it takes the Moon to complete one rotation on its axis, or approximately 27.33 days.

Lunar Module (LM): The portion of the Apollo spacecraft housed in the Spacecraft Lunar Module Adapter (SLA) of the Saturn V during launch and in which two astronauts could travel to and from the Moon's surface; it was the first piloted spacecraft designed for use exclusively outside Earth's atmosphere. It consisted of the Descent Stage for landing and the Ascent Stage that carried the crew and essential supplies. The Lunar Module (LM) was originally known as the Lunar Excursion Module until the "excursion" part of the name was deemed too frivolous and dropped. In its early life it was referred to occasionally by NASA as the "bug," by its manufacturers as the "LM" ("el-em"), but nearly always by the astronauts as the "lem." *See also* Ascent stage, Command and Service Module, Descent stage.

Lunar orbit: The Lunar orbit class includes all missions that went into orbit around the Moon or impacted the lunar surface.

Lunar Roving Vehicles (LRVs): The Moon vehicles used on Apollo missions 15, 16, and 17 to transport the astronauts several kilometers over the lunar surface. Battery-powered with four wheels and a television camera, these lunar cars enabled the astronauts to transmit their observations to Earth.

Lunar satellite: Lunar missions are designed to gather information about the Earth's Moon for the purpose of scientific research. This class includes orbiters, flybys, hard landers, and soft landers (including piloted missions).

Mach: The ratio of the speed of a moving object to the speed of sound in the surrounding medium.

At Mach 1, the speed of an aircraft equals the speed of sound.

Magellanic Clouds: Refers to the two nearest galaxies outside the Milky Way, which are visible from Earth's Southern Hemisphere. They are the Large Cloud (160,000 light years away) and the Small Cloud (185,000 light-years away). The Magellanic Clouds have been instrumental in establishing an extragalactic distance scale.

Magnetic field: Any force field of attraction created by the mass of a body or the combined masses of multiple bodies. Magnetic fields are responsible for much of the shape of the universe, from the orbits of planets in the solar system to the shapes of galaxies and clusters of galaxies. *See also* Gravitation.

Magnetometer: An instrument that detects disturbances in a magnetic field, within which the body's magnetic lines of force control the movement of ionized particles.

Magnetotail: A tail of nearly parallel lines of magnetic force extending from Earth in the direction away from the Sun.

Magnitude: The brightness of a celestial body expressed numerically. *See also* Absolute magnitude, Apparent magnitude.

Main sequence star: A star, such as the Sun, which produces energy mainly by a hydrogen-to-helium fusion reaction. Most stars spend the greater part of their lifetimes in this state.

Man-rating: Approval for use during a piloted mission. A device that is man-rated is deemed safe for use by or around humans.

Manned Maneuvering Unit (MMU): *See* Space suit.

Mantle: The section of Earth between the lithosphere and the central core.

Mare (pl. maria): A large flat area on the Moon or Mars, so named (after the Latin for "sea") because these areas appear dark and sea-like to the Earth observer.

Maritime satellite: A satellite designed for telecommunications by and for shipping industries. These satellites occupy geostationary orbits over oceans to transmit ship-to-shore communications and data.

Mars orbit: The Mars orbit class includes all missions that went into orbit around Mars or impacted the planet's surface.

Mascon: One of several concentrations of mass located beneath lunar maria, which cause distortion in the orbit of a spacecraft around the Moon.

Maser: An acronym for Microwave Amplification by Stimulated Emission of Radiation. A device similar to a laser in which energy is generated in the same way it is in a laser, but at microwave levels. A maser can exist in nature as a celestial object. Artificial masers are used to amplify weak radio signals. *See also* Laser.

Mass: The amount of matter contained within a body, which determines the amount of gravitation force it exerts. Mass is measured in such units as kilograms and pounds; it differs from weight, however, which is the force exerted on a mass by gravity.

Mass spectrometer: An instrument that identifies the chemical composition of a substance by separating ions by mass and charge.

Materials processing: The manufacture of crystals and other materials in the microgravity environment of a spacecraft, whereby uniform crystal growth and other processes that are difficult to support on Earth can be accomplished for improvements in space technology and for industrial applications.

Matter: A substance that has mass and occupies space, which, along with energy, is responsible for all observable phenomena.

Maunder minimum: Named for E. W. Maunder, who in 1890 discovered a period in the three-hundred-year history of sunspot observations when few sunspots were recorded. Confirmed independently in 1976 by evidence from tree rings, the Maunder minimum covers the years 1645 to 1715, a period also known as the Northern Hemisphere's "Little Ice Age."

Medium-class Explorer (MIDEX): Medium-sized (defined by definition and development costs not exceeding $140 million) spacecraft that carries one or more science instruments. *See also* Explorer, Small Explorer, University-class Explorer.

Medium-Earth orbit (MEO): An orbit having an apogee greater than 3,000 kilometers but less that 30,000 kilometers. Navigation (such as global positioning) and communications (such as Odyssey) missions sometimes used these orbits.

Megahertz: *See* Hertz.

MEO: *See* Medium-Earth orbit.

Mesosphere: The layer of Earth's atmosphere occurring above the stratosphere and below the thermosphere, from about 40 kilometers to 85 kilometers above sea level. This is the coldest layer of the atmosphere.

MET: *See* Mission-elapsed time.

Meteor: A streak of light in Earth's upper atmosphere caused by the burning of a meteoroid.

Meteorite: A meteoroid that does not burn completely and reaches Earth. The three types of meteorites are stony, iron, and stony-iron.

Meteoroid: A particle of interplanetary dust greater than 0.1 millimeter which enters Earth's atmosphere and burns as a result of friction, creating a shooting star.

Meteorological satellite: A satellite that collects data on weather systems for forecast and other analysis. *See also* Weather satellite.

Meter: The metric unit of length, equivalent to approximately 39.37 inches, or a little more than 1 yard.

Metric system: The decimal system of weights and measures, which forms part of the Système International d'Unites. *See also* SI units.

Metric ton: A metric unit of weight equivalent to about one short ton (2,205 pounds). Just as thrust is often measured in pounds in the United States, metric tons are often used as units of thrust in Russia and other countries. *See also* Newton.

Microgravity: Nearly zero gravity. Microgravity exists in a space vehicle because of the minute gravitational forces exerted by objects on one another. The microgravity environment is the ideal environment for certain types of materials processing.

Micrometeorite: A micrometeroid that has contacted Earth's surface.

Micrometeoroid: A meteoroid with a diameter of less than 0.1 millimeter. Because of their size, micrometeoroids rarely burn up but reach Earth's surface as spherules or as cosmic dust particles.

Micropaleontology: The study of microscopic fossils. Discoveries in this field are of potential importance to the study of exobiology as well as to Earth and life sciences.

Microwaves: A form of electromagnetic radiation with wavelengths ranging between 1 millimeter and 30 centimeters, located between infrared and long-wave radio on the electromagnetic spectrum.

MIDEX: *See* Medium-class Explorer.

Military communications satellite: A communications satellite dedicated to relaying signals between elements of a nation's armed forces.

Military Space Forces (VKS): The Russian military space agency. The VKS controls Russia's Plesetsk Cosmodrome launch facility.

Military (unknown/unique) satellite: This category includes all military satellites whose purpose is either unknown (such as still-secret surveillance, ELINT, and EW spacecraft) or unique in nature.

Milky Way: The galaxy in which our solar system is located. The Milky Way is a spiral galaxy that is about 100,000 light-years across. *See also* Galaxy.

Mir: Named for the Russian word for "peace" or "community," the Mir Complex was a large and long-lived Soviet/Russian space station, the first segment of which was launched in February, 1986. Bigger than its predecessors in the Salyut series and composed of several modules, Mir was designed to house more cosmonauts on longer stays than the Salyuts could support. The core of Mir was the base block (the living quarters), equipped with six docking ports to which visiting spacecraft and additional modules could be attached. Mir was gradually expanded by adding laboratory and equipment modules, rearranged for different missions, and upgraded without abandoning the original core

unit. It was almost continuously occupied for thirteen years.

Mission Control Center: A room or building equipped with the means to monitor and control the progress of a spacecraft during all phases of its flight after launch.

Mission duration: The official length of a mission from liftoff to touchdown.

Mission-elapsed time (MET): The time that has elapsed since initial liftoff from Earth, measured in days, hours, minutes, and seconds. Officially, all mission events are based on MET. The moment that MET begins varies among vehicles. For the Saturn V it was the instant the vehicle had traveled three-quarters of an inch vertically. For the space shuttle it is the moment the solid rocket boosters are ignited. The mission officially ends at touchdown, but mission-elapsed time will continue to run until the flight director declares that the flight has ended. *See also* Touchdown.

Mission specialist: An astronaut who has overall responsibility for a mission payload.

MLR: *See* Monodisperse Latex Reactor.

MMU: *See* Space suit.

Molecule: The smallest unit of a compound (as opposed to an element). Formed by a characteristic complex of atoms joined together. The smallest unit of the substance water, for example, is a molecule formed by two hydrogen atoms and one oxygen atom.

Monodisperse Latex Reactor (MLR): A device designed to develop monodisperse, or identically sized, beadlike rubber particles for use in medical and industrial research.

Moon: Any natural satellite orbiting a planet, especially Earth's Moon.

MSS: *See* Multispectral scanner.

Multispectral scanner (MSS): A type of radiometer, used on such satellites as Landsat, which produces detailed false-color images of a planet's surface.

Nanometer: One thousand millionth of a meter; a unit used to express electromagnetic wavelengths.

National Advisory Committee for Aeronautics (NACA): A United States civilian agency for aviation research, chartered in 1915 and operational from 1917 to 1958, when it was absorbed into the newly formed National Aeronautics and Space Administration (NASA). NACA concentrated mainly on laboratory studies at its Langley, Ames, and Lewis Centers, gradually shifting from aerodynamic research to military rocketry as the Cold War made missile development a priority.

National Aeronautics and Space Administration (NASA): A civilian agency of the United States government, formally established on October 1, 1958, under the National Aeronautics and Space Act of 1958. It absorbed the former National Advisory Committee for Aeronautics (NACA), including its 8,000 employees and three major research laboratories—Langley Aeronautical Laboratory, Ames Aeronautical Laboratory, and Lewis Flight Propulsion Laboratory—and two small test facilities. The functions of the organization are to plan, direct, and conduct all American aeronautical and space activities except those that are primarily military. NASA's administrator is a civilian appointed by the president with the advice and consent of the Senate. The administration arranges for the scientific community to take part in planning scientific measurements and observations to be made through the use of aeronautical and space vehicles, and provides for the dissemination of data that results. Under the guidance of the president, the administration participates in the development of programs of international cooperation in space activities.

Navigation satellite: A satellite that provides positional information for any moving object on land, on sea, or in the air, including inner space.

Navsat: An abbreviation for "navigation satellite."

Near-Earth space: Roughly defined as the space environment from the outer reaches of Earth's atmosphere to the path of the Moon's orbit. The area beyond near-Earth space is known as deep space.

Nebula: A celestial body composed of aggregated gas and dust, which may be luminous, reflecting or emitting light under the influence of nearby stars (an emission nebula), or dark, obscuring the light of distant stars and appearing as a silhouette.

Neutral gas analyzer: An instrument that determines the chemical composition of the atmosphere.

Neutrino: An elementary particle that has enormous penetrating power as a result of its lack of electric charge and its almost complete lack of mass. Traveling directly out from the cores of stars as a by-product of nuclear reactions, neutrinos have enormous potential as a source of information on the stars and other astrophysical phenomena.

Neutron: An uncharged particle found in atomic nuclei; its mass is approximately equal to that of a proton.

Neutron stars: The smallest known stars, with diameters of about 20 kilometers and densities matching that of the Sun, consisting of a thin iron shell enclosing a liquid sea of neutrons. The properties of neutron stars are beyond scientists' complete understanding, but they are thought to originate from main sequence stars much larger than the sun, which become supernovae. Rapidly spinning neutron stars are observable as pulsars.

New astronomy: A term used collectively to refer to the areas of astronomy (such as gamma-ray astronomy, infrared astronomy, and x-ray astronomy) investigating electromagnetic emissions by celestial phenomena. The application of space technology and electronics to the development of instruments capable of detecting such data has greatly increased astronomers' understanding of the universe.

Newton: An SI unit of force used to measure thrust.

Northern lights: *See* Aurora, Aurora borealis.

Nose cone: The conically shaped front end of a launch stack, missile, or other spacecraft. The nose cone is a protective shield that improves aerodynamic efficiency.

Nova: A star that emits a sudden radiation of light and quickly (over months or years) returns to its former brightness. *See also* Supernova.

NPO (Nauchno-proizvodstvennoe obedinenie): Soviet/Russian organization for scientific production or research. May have one or more OKBs (OKB is the acronym for the Soviet/Russian Experimental Design Bureau).

NPO Molniya: Soviet/Russian Research and Industrial Corporation. Founded to build the first Russian piloted reusable spacecraft (Buran) in 1976. Molniya is the Russian word for "lightning."

Nuclear detection satellite: A satellite used to detect nuclear explosions on the Earth's surface. Although primarily used to ensure compliance with nuclear treaties, nuclear detection spacecraft have also been used to detect and observe galactic events like supernovae.

Nuclear energy: Energy that is released as a result of interactions between elementary particles and nuclei.

Nuclear reactor: A device, usually located at a nuclear power station, designed to contain nuclear fission reactions during the production of nuclear energy.

Nucleus: The central part of an atom around which electrons rotate. A nucleus can consist of one proton (in a hydrogen atom) or many protons and neutrons (as in an atom of uranium).

O-ring: A ringed rubber gasket, one-quarter inch in width, which acts as a sealant between the bottom and next-to-bottom segments of a shuttle solid-fueled rocket booster. Failure of an O-ring or the insulating putty that surrounds it can allow combustible gases to escape through the joint from within the rocket; this type of failure appears to have been the cause of the *Challenger* accident of January, 1986.

Oblate: Flattened at the poles.

Occultation: The obscuring of one celestial body by another; occurs during a solar eclipse.

OKB (Opytnoe Konstructorskoe Buro): Soviet/Russian Experimental Design Bureau.

OKB-1: Classified Soviet name of Sergei Korolev's design bureau, it later evolved to become the Energia Rocket and Space Corporation (RSC Energia).

OKB-51: Soviet Special Design Bureau that designed the Soviet's first pilotless aircraft.

OKB-52: The classified Soviet name of Vladimir Chelomei's design bureau in Moscow.

OMS: *See* Orbital maneuvering system.

Oort Cloud: A cloud of millions of comets orbiting the Sun between 30,000 and 100,000 astronomical units from the Sun, postulated by Dutch astronomer Jan Oort in 1950.

Opposition: The alignment of Sun, Earth, and a superior planet (one whose orbit is farther from the Sun than Earth's) in a straight line; that is, the superior planet appears in the sky at 180° celestial longitude from the Sun. In this position, the planet is closest to Earth and therefore most easily observed by ground-based instruments.

Optical navigation: Navigation by sight, as opposed to inertial methods, using stars or other visible objects as reference.

Orbit: The path traced out by one celestial or artificial body as it moves around another that exerts greater gravitational force. The distinguishing characteristics of an orbit are called its orbital parameters and include apoapsis, periapsis, inclination to the ecliptic of the body orbited, eccentricity, and period. All orbits trace out an ellipse, of which the body orbited forms at least one of two foci. *See also* Apoapsis, Circular orbit, Eccentricity, Ellipse, Elliptical orbit, Equatorial orbit, Orbital inclination, Parabolic orbit, Parking orbit, Periapsis, Period, Polar orbit, Prograde orbit, Retrograde orbit, Synchronous orbit, Transfer orbit.

Orbital inclination: The angle formed between the orbital plane of a satellite and the equatorial plane of the object orbited.

Orbital maneuvering system (OMS): A system of engines, located on a spacecraft, which provide small amounts of thrust for fine maneuvers in orbit.

Orbiter: A spacecraft intended to orbit, rather than land on, a planet or other celestial body, often used to relay signals from a lander to Earth. *See also* Lander.

Outer space: The region of the Earth's atmosphere above the Karman Line, which defines the edge of space.

Outgassing: The process whereby gases are emitted from solids into a vacuum, referring mainly to the exudation of gases from terrestrial bodies into space, a remnant of the way these bodies were formed.

Oxidizer: In a rocket propellant, a substance such as liquid oxygen or nitrogen tetroxide, which supports combustion of the fuel.

Ozone layer: The thin layer of Earth's atmosphere, located between 12 and 50 kilometers above Earth's surface (in the stratosphere), in which ozone is found in its greatest concentrations. This layer, which absorbs most of the ultraviolet radiation entering the atmosphere, forms a protective blanket around the planet, shielding it from excess radiation.

PAM: *See* Payload Assist Module.

Panspermia: A theory proposed by chemist Svante Arrhenius in 1906, and later modified by Sir Fred Hoyle, which holds that organic molecules (the basis on which life is formed) were transported to Earth by comets. The organic material found in Halley's comet supports the theory, which is further supported by the argument that each process on the road to a biological organism is so improbable that it requires a greater combination of conditions than those possible on Earth alone. The theory is not accepted by many scientists.

Parabolic orbit: An orbit that forms a parabola around the object orbited and hence escapes from the gravitational field of that object with precisely the escape velocity. If the velocity is greater than escape, then the orbit is hyperbolic. For example, comets that leave the solar system and only encounter the Sun once depart either along a parabolic or hyperbolic trajectory.

Parallax: The apparent displacement of a celestial object as seen from two different points on Earth. Knowing this angle allows astronomers to calculate the object's distance from Earth.

Parking orbit: An interim orbit around a celestial body between launch and injection into another orbit or into a trajectory toward another destination.

Parsec: A unit for measuring astronomical distances equivalent to 3.26 light-years.

Particle: *See* Elementary particles.

Passive experiment: An experiment package which requires only exposure to the space environment to perform its investigations.

Passive relay satellite: An early telecommunications satellite, such as Echo, which relayed radio signals from one point to another by bouncing them off its surface.

Payload: Any experiment package, satellite, or other special cargo carried into space by a spacecraft.

Payload Assist Module (PAM): A solid-fueled rocket engine designed to boost satellites into geostationary orbit from the space shuttle's payload bay.

Payload bay: The portion of the space shuttle that carries satellites into space, approximately 18 by 4 meters. Hinged doors at the top of the fuselage open to expose the satellite, which can then be boosted into orbit from the bay by a Payload Assist Module.

Payload specialist: An astronaut who assists a mission specialist in conducting an experiment aboard the space shuttle.

Penumbra: Semidark portion of a shadow in which light is partly cut off, as on the surface of the Moon or Earth away from the Sun where the disc of the Sun is only partly obscured.

Periapsis: The point in one object's orbit around another at which the orbiting object is closest to the object being orbited.

Pericynthion: The point in an object's orbit around the Moon at which it is closest to the Moon.

Perigee: The point in an object's orbit around Earth at which it is closest to Earth.

Perihelion: The point in a solar orbit at which the orbiting object is closest to the Sun.

Perilune: Pericynthion of an artificial satellite.

Period: The time span between repetitions of a cyclic event. An orbital period is the time required for a satellite or moon to make one complete orbit around a planet, a moon, the Sun, or another celestial body.

Photometer: An instrument that measures the brightness of a light source.

Photomultiplier: An instrument for increasing the apparent brightness or strength of a source of light by means of secondary excitation of electrons; effectively, a light (or other radiation) amplifier.

Photon: The smallest theoretical quantity of radiation, visualized as both wavelength energy and as an elementary particle.

Photopolarimeter: An instrument for producing an image of a celestial body (or other light source) by means of polarized light.

Photoreconnaissance: The gathering of information, especially on enemy installations, by means of photography from the air or from space.

Photosphere: The region of the Sun that separates its exterior (the chromosphere and corona) from its interior, forming the boundary between the transparent and opaque gases. The photosphere appears as the bright central disk from Earth, and it is the source of most of the Sun's light.

Photovoltaic cell: A solid-state energy device that converts sunlight into electricity.

Piroda: The Russian word for "nature" and name of the remote sensing module of the Mir Complex.

Pirs: The Russian word for "pier" and name of the Pirs Module (Russian Docking Compartment) of the International Space Station.

Pitch, roll, and yaw: Movements that a spacecraft or aircraft undergo as a result of an external force (like wind) or internal force (like an attitude control engine). Pitch is movement about an axis that is perpendicular to the longitudinal axis and horizontal to a primary body (up and down). Roll is revolution about the longitudinal axis. Yaw is lateral rotation about the vertical axis (side-to-side).

Pitchover: The programmed turn from the vertical that a rocket takes as it describes an arc and points in a direction other than vertical.

Pixel: A small unit arranged with others in a two-dimensional array that contains a discrete portion of an image (as on a television screen) or an electrical charge (as on a charge-coupled device). Together these pixels form an image or other meaningful information.

Planetary satellite: Planetary missions leave the Earth's gravitational field to perform close observations of other planets, asteroids, and comets. These types of missions can range from quick flybys to long-term observations from orbits around the body. Planetary missions have also dropped probes into planets' atmospheres, have soft-landed and hard-landed (crashed) on planets' surfaces, and have flown in formation with comets and asteroids.

Planets: A planet is a nonluminous natural celestial body that orbits the Sun (or another star) and is not categorized as an asteroid or comet. There are nine known planets in the solar system: Mercury, Venus, Earth, Mars, Jupiter, Saturn, Uranus, Neptune, and Pluto.

Plasma: Ionized gas, consisting of roughly equal numbers of free electrons and positive ions. Plasma forms the atmospheres of stars, interstellar and intergalactic matter, nebulae—in fact, most of the matter in the universe. The extremely high excitation of its constituent particles has earned it the label "fourth state of matter," (after the solid, liquid, and gaseous states).

Plasma sheath: The definite outer boundary of Earth's ionosphere, identified by Orbiting Geophysical Observatory 1.

Plate tectonics: The study of continental drift, seafloor spreading, and other dynamics of Earth's lithosphere, which is divided into seven major sections, or plates.

Plesetsk: Cosmodrome founded in 1957 and used to launch spacecraft into different inclination orbits. The cosmodrome includes nine launch pads for launch vehicles, as well as a variety of assembly/testing facilities, telemetry and tracking stations.

Pogo effect: The up-and-down vibrating motion that occurs during the launch of a spacecraft.

Polar orbit: An orbit in which a satellite passes over a planet's or moon's poles.

Polarimeter: An instrument for measuring the degree to which electromagnetic radiation is polarized.

Pole: One of two points where the surface of a planet is intersected by its axis of rotation. In a magnetic field, one of two or more points of concentration of the lines of magnetic force.

Posigrade: Moving in the direction of travel.

Precession: A type of motion that occurs in a rotating body in response to torque: A planet or other rotating body orbiting around a gravitational force, such as the Sun, slowly turns in the direction of its rotation so that, over a long period, each of the planet's poles describes a circle. The fact that precession is exhibited by many planets and moons means that adjustments must be made in the locations to which astronomers look to observe stars and other celestial phenomena. Earth's period of precession is approximately twenty-five thousand years.

Probe: *See* Deep space probe.

Prograde orbit: An orbit that moves in the same direction as the rotation of the body orbited.

Progress: The Russian word for "progress." The Progress is an expendable, autonomous freighter, derived from the Soyuz spacecraft, and launched with the Soyuz launch vehicle. Progress has resupplied Salyut 6, Salyut 7, and Mir, and is currently used for the International Space Station.

Propellant: *See* Cryogenic propellant, Hypergolic propellant, Solid propellant.

Propulsion system: The combined mechanisms that propel a space vehicle, including engines and fuel systems.

Proton: The Russian word for "proton." Refers to the Proton launch vehicle.

Protons: Particles that carry a positive charge. A single proton forms the nucleus of a the hydrogen atom, and protons are found in the nuclei of other chemical elements in combination with neutrons.

Pulsar: A rapidly spinning neutron star that emits a narrow beam of electromagnetic radiation in the form of visible and radio waves (single pulsars), as well as x rays and gamma rays (pulsars occurring in binary star systems). The regular emission of radio waves can be detected on Earth by radio telescopes, and the pulsar's distance from Earth can be detected by the difference in the arrival times of radio waves at different wavelengths.

Quarks: Elementary particles hypothesized to form the known elementary particles (electrons, protons, neutrons, and their antiparticles), characterized by electric charge, flavor, and color. The forces required to break nucleons into their component quarks are so great that quarks do not exist as free particles in nature, although it is thought that neutron stars may consist of a "quark soup" within a solid iron shell.

Quasar: An abbreviation of "quasi-stellar" or "quasi-stellar object." An object that continuously releases a tremendous amount of energy, equivalent to the output of between one million and 100 trillion Suns. This definition includes virtually all kinds of electromagnetic radiation (gamma rays, x rays, ultraviolet, optical, and infrared radiation, microwaves and radio waves), from a very small volume of space to an area about the size of the solar system. As far as is known, all objects satisfying these criteria are located in the nuclei of galaxies, although it is also possible that these objects may be galaxies with black holes at their centers. Discovered in 1963, the first quasar generated considerable excitement among astronomers, and these phenomena continue to be among the most fascinating and mysterious in the universe.

Radar: An acronym for Radio Detection and Ranging. A means of locating and determining the distance to objects by bouncing radio waves off them and measuring the time required to receive the echo.

Radiation: *See* Electromagnetic radiation.

Radio astronomy: The branch of astronomy that examines the radio emissions of celestial objects. Because radio radiation, along with visible radiation, can penetrate Earth's atmosphere, radio receivers have provided much of the data detectable by ground-based, as well as space-based, instruments. Radio emissions also form a significant portion of certain celestial phenomena, such as radio galaxies, quasars, and pulsars.

Radio telescope: A radio receiving aerial, either a dish or a dipole, connected to recording devices, whereby distant radio emissions can be detected.

Radioisotopic thermoelectric generator (RTG): A device for creating power from a radioactive substance, used on spacecraft to supplement the power generated by the solar-energy-collecting solar panels. The RTG is especially important when access to sunlight is weak or nonexistent.

Radiometer: An instrument, used by meteorological and other satellites, that measures Earth's infrared and reflected solar radiance and uses small, selected wavelengths (ultraviolet to microwave) to measure temperature, ozone, and water vapor in the atmosphere. Radiometers are sensitive to one or more wavelength bands in the visible and invisible ranges. If visible wavelengths are used, the satellite can detect cloud vistas from reflected sunlight, resulting in a slightly blurred version of those images photographed directly by astronauts. Using the invisible range, satellites can capture terrestrial radiation, producing images from Earth's radiant energy.

Raketa: The Russian word for "rocket."

RBV: *See* Return beam vidicon.

Real time: Referring to the transmission of signals or other data at the same time that they are used.

Reconnaissance satellite: A satellite that gathers information about enemy military installations.

Red giant: Stars with surface temperatures less than 4,700 kelvins and diameters between 10 and 100 times that of the Sun.

Red Planet: Mars, which appears red when seen through Earth-based telescopes.

Redshift: The apparent lengthening of electromagnetic wavelengths issuing from a celestial object or other source as a result of the object's movement away from the observer. As a result, the spectral lines in the spectra of such an object will shift toward the red end of the electromagnetic spectrum. *See also* Doppler effect.

Reentry: The return of a spacecraft into Earth's atmosphere.

Reflecting telescope: An optical telescope that uses a mirror or mirrors to capture, magnify, and focus light from the object observed. These telescopes, such as the 200-inch reflecting telescope on Palomar Mountain in Southern California, are widely used for Earth-based optical astronomy. *See also* Cassegrain telescope, Refracting telescope.

Refracting telescope: An optical telescope that uses a lens to magnify and focus light from the object observed. Refracting telescopes were used by the earliest astronomers. When reflecting telescopes were perfected in the twentieth century, refracting telescopes became less important in astronomy, although they are still widely used for guided and amateur observations. *See also* Reflecting telescope.

Regolith: A thick layer of rock, broken by the impact of a meteoroid, that overlies the surface of a moon or planet.

Relativity: The physical law, first proposed by Albert Einstein, that states that measurements of time and space are dependent upon the frame of reference in which they are measured. The general theory of relativity applies this law to gravity and mass; the special theory of relativity applies it to the propagation of electric and magnetic phenomena in space and time.

Remote Manipulator System (RMS): The space shuttle's 15-meter-long robot arm. Operated from within the shuttle by an astronaut, the arm duplicates the operator's hand and wrist movements, allowing payloads to be moved and repairs to be made without extravehicular activity. Because the RMS was manufactured by a Canadian aerospace company, it is sometimes called the "Canadarm."

Remote sensing: Acquiring data at a distance by electronic or mechanical means.

Rendezvous: The planned meeting of two spacecraft in orbit and often their maneuvering into proximity of each other in preparation for docking. The first rendezvous in space occurred during the flights of Gemini VII and VI-A on December 15, 1965.

Research satellite: This general class of missions includes the many space science disciplines not falling into any of the other mission categories. A few examples include zero-g materials processing tests, biological studies, and space physics experiments.

Resolution: The degree to which a photographic or other imaging system, or the image produced, clearly distinguishes objects of a certain size. In a photograph with a resolution of 200 meters, for example, the smallest distinguishable objects are 200 meters across.

Restart: Reignition of a rocket engine after it has been inactive during orbit.

Retrofire: The firing of a rocket to slow down or change the orbit of a spacecraft.

Retrograde orbit: An orbit that moves opposite to the rotational direction of the body orbited.

Retroreflector: A device, carried by a satellite, used to reflect laser beams directed at the satellite from Earth. *See also* Laser-ranging.

Retrorocket: A rocket that exerts thrust in the opposite direction of the object's motion to slow down or change the orbit of a spacecraft.

Return beam vidicon (RBV): A camera, used by Earth resources satellites, that takes very high-resolution photographs of the planet's surface from space.

Revolution: One complete orbit of a planet around the Sun or of a natural or artificial satellite around another celestial body.

Rille: A long, narrow valley on the Moon.

RMS: *See* Remote Manipulator System.

Robot arm: A mechanical arm extending from many spacecraft which can be remotely controlled to manipulate instruments and repair equipment. *See also* Remote Manipulator System.

Robotics: The development, construction, and use of computerized machines to replace humans in a variety of tasks requiring precise hand-eye coordination.

Roche limit: Named for Édouard Roche, who discovered it in 1848, the minimum distance from a planet at which a natural satellite can form by accretion: roughly 2.44 times the planet's radius. Within this limit, an existing satellite will be torn apart by gravitational stresses. Saturn's rings, which lie within the planet's Roche limit, may be the remnants of a former moon.

Rocket booster: A propulsion engine used to launch a spacecraft into orbit from Earth or into a different orbit or trajectory from space.

Roll: *See* Pitch, roll, and yaw.

Rollout: The termination of a flight, occurring after touchdown on the landing site and before brakes are set, during which an aircraft or space shuttle rolls to decrease speed.

Rosaviakosmos (RKA): Russian Aviation and Space Agency (now Roscosmos). *See also* Russian Federal Space Agency.

Roscosmos (ross-coss-mose): *See* Russian Federal Space Agency.

Rover: A vehicle for exploring the surface of an extraterrestrial body, it can be piloted, act autonomously, be controlled from a distance, or a combination of the three.

RTG: *See* Radioisotopic thermoelectric generator.

Russian Federal Space Agency (Roscosmos): Formerly the Russian Aviation and Space Agency (RKA), which was created in February, 1992, by a decree issued by the President of the Russian Federation after the breakup of the Soviet Union and the dissolution of the Soviet space program. Renamed the Russian Federal Space Agency in 2004, it is the central body of the federal executive authority and defines the Russian Federation's national policy in the field of space research and exploration. The Agency also performs interdisciplinary coordination of national scientific and application space programs.

S-Band: A radio-frequency band of 1,550 to 5,200 megahertz.

S. P. Korolev Rocket and Space Corporation Energia (RKK Energia): Leader in the Russian rocket and space industry. Established in 1946, it became a pioneer practically in all areas of rocket and space technology. It is the prime contractor for piloted space stations, piloted spacecraft, and space systems built on their basis.

Salyut: The Russian word for "salute," for which the Salyut Space Station is named.

Sample return mission: A mission designated to collect soil and rock samples from another body and return them to Earth.

Satellite: Any body that orbits another of larger mass, usually a planet. Satellites include moons, the small bodies that form planetary rings, and man-made satellites. *See also* Artificial satellites.

Scanning radiometer: An instrument used on meteorological satellites to measure radiation emitted by the atmosphere, especially in the infrared region, to build a picture of atmospheric conditions and temperature that can be used in forecasting.

Scarp: A broken slope or a line of cliffs caused by a fault line or by erosion.

Schwarzschild radius: The radius of a collapsing mass, such as a degenerating star, at which it becomes a black hole—that is, at which its gravitational force will not allow light to escape. The length of this radius depends on the body's mass, and the formula for calculating it was established by Karl Schwarzschild in 1916.

Scientific satellite: A broad term for any satellite dedicated primarily, if not solely, to collecting scientific data, especially data on astrophysical phenomena. Although most satellites can be described as scientific in some sense, the main purpose of a scientific satellite is to broaden

our knowledge of the universe, rather then serve a practical (such as telecommunications), military (reconnaissance), or commercial purpose.

Seismic activity: Any movement in the outer layer of a planet or moon.

Seismometer: A sensitive electronic instrument that measures movements in the outer layer of a planet or moon. The graphs produced by this instrument can be interpreted—on the Richter scale, generally—to determine the magnitude and intensity of seismic activity.

Selenocentric: An orbit having the Moon as its center. A word derived from Selene, the Greek goddess of the Moon.

Selenography: The study of the lunar surface features; the counterpart of geography on Earth.

Selenology: The study of the Moon, analogous to geology on Earth.

Service Module (SM): The aluminum-alloy cylinder portion of the Apollo spacecraft. Its end housed the main engine, which was used to place the spacecraft into lunar orbit and begin the return to Earth. The SM carried the hypergolic (self-igniting) propellants for the main engine, the systems (including fuel cells) used to generate electrical power, and some of the life-support equipment. At four locations on the SM's exterior were clusters of attitude control jets. On the Apollo 15, 16, and 17 missions, the SM also contained a scientific instrument module (SIM) with cameras and other sensors for studying the Moon from orbit. *See also* Command Module.

Service Propulsion System: A large rocket engine that propelled the Apollo Command and Service Module, slowing it as it neared the Moon, and later sending it back on a trajectory toward Earth; also used for mid-course corrections.

Shenzhou: Chinese for "magical vessel." China's first piloted space program. The Shenzhou spacecraft resembles the Russian Soyuz spacecraft, but is larger. Shenzhou 5 carried China's first astronaut, Yang Liwei, on a one-day, fourteen-orbit mission on October 15, 2003.

Shroud: A heat-resistant covering used to protect a spacecraft, payload, or missile, especially during launch.

Shuttle Imaging Radar (SIR): A high-resolution radar system used aboard the space shuttle. It operates at frequencies high enough to penetrate not only Earth's atmosphere but occasionally a few feet beneath its surface.

SI units: The collective units of measurement used in the Système International d'Unites, the system of measurement most widely accepted by scientists. It has seven fundamental or base units: the meter (the base unit of length), kilogram (mass), second (time), ampere (electric current), kelvin (temperature), mole (amount of substance), and candela (luminosity). Other units are derived from these seven base units. They include multiples, fractions, or powers of the base units such as the kilometer (1 meter × 10^3) and the square meter (the unit of area). Further units are derived from combinations of the base units and have their own names: hertz (the unit of frequency, which is cycles per second), newton (force or thrust, measured in kilogram-meters per second squared), pascal (pressure, measured in newtons per square meter), joule (energy, the kilogram-meter), watt (power, measured in joules per second), coulomb (quantity of electricity, measured in ampere-seconds), volt (electric potential, measured in watts per ampere), farad (capacitance, or the ability to store energy, measured in coulombs per volt), and ohm (electrical resistance, measured in volts per ampere). In the United States, the base SI units are used with increasing frequency. Some measures, however, remain more familiarly rendered by English units of measure, even in scientific use: It is common, for example, to refer to rocket thrust in pounds or even metric tons rather than newtons, and atmospheric pressure is often measured in pounds per square inch (or bars and millibars in the centimeter-gram-second system).

SIR: *See* Shuttle Imaging Radar.

Small Explorer (SMEX): A small-sized (defined by definition and development costs not exceeding $71 million) spacecraft that carries one or more science instruments. *See also* Explorer, Medium-class Explorer, University-class Explorer.

SMEX: *See* Small Explorer.

Soft landing: A controlled landing on the surface of a planet or other celestial body designed to minimize damage to the spacecraft and its instrument payload.

Solar array: An assembly of solar cells, as on a solar panel extending from a satellite.

Solar cell: A photovoltaic device that converts solar energy directly into electricity for use in powering a spacecraft.

Solar constant: The amount of solar energy received by a square meter per second on Earth (or one astronomical unit from the Sun), approximately 1,370 watts per square meter per second.

Solar cycle: A period of approximately eleven years during which the number of sunspots visible near the Sun's equator increases to a maximum and then decreases. Other solar activity follows the solar cycle. *See also* Sunspots.

Solar flare: A large arc of charged particles and electromagnetic radiation ejected from the Sun's surface (in the low corona and upper chromosphere) and lasting from a few minutes to several hours. Solar flare activity affects radio transmission on Earth and can produce auroras in Earth's atmosphere.

Solar mass: A unit equivalent to the mass of the Sun, or $1,989 × 10^{33}$ grams. Masses of other stars are sometimes given in solar masses.

Solar orbit: The solar orbit class includes all missions during which the spacecraft has achieved Earth escape velocity and goes into orbit about the Sun. This class includes missions designed to observe the Sun, as well as planetary flyby missions that have entered solar orbit following rendezvous with their targets.

Solar system: The Sun, the planets, asteroids, comets, and other matter that orbit the Sun, and the satellites of those bodies, along with interplanetary space, radiation, and gases. *See also* Planets.

Solar system escape trajectory: The solar system escape trajectory class includes all missions that have achieved solar escape velocity and will thus eventually depart the solar system.

Solar wind: The hot ionized gases, or plasma, that escape the Sun's gravitational field and flow in spirals outward at about 200 to 900 kilometers per second. It consists primarily of free protons, electrons, and alpha particles escaping from the Sun's corona. *See also* Stellar wind.

Solid propellant: Rocket propellant in solid form: cast, extruded, granular, powder, or other.

Solid rocket booster (SRB): One of two rocket boosters, fueled by solid propellants, which assist the main engines of the space shuttle during the first two minutes of ascent. The SRBs then separate from the External Tank, and their descent is slowed by drogue chutes that issue from nose caps. The SRBs splash down and are recovered for later reuse. *See also* External Tank.

Sonar: An acronym for Sound Navigation Ranging. A system for bouncing sonic and supersonic waves off a submerged object in order to determine its distance.

Sounding rocket: A suborbital rocket carrying scientific instruments that take measurements of Earth's atmosphere. Sounding rockets were used before satellites came into prominence.

Sounding sensor: A sonar-like device that probes the atmosphere to detect data about temperature, moisture, and other conditions.

Soyuz: The Russian word for "union," for which the Soyuz spacecraft and launch vehicle are named.

Space adaptation syndrome: *See* Space sickness.

Space Age: The age of space exploration, whose beginning is generally dated from October 4, 1957, the day on which the first artificial satellite, Sputnik 1, was launched into Earth orbit.

Space capsule: A small piloted or robotic spacecraft, such as those used on the Mercury missions, which is pressurized and otherwise environmentally controlled.

Space center: A complex that houses a variety of facilities, including launch facilities, that oversee development of space technology and preparations for or monitoring of space missions.

Space medicine: The study of human health in microgravity and other conditions surrounding space travel, including systems for maintaining health.

Space Race: A term applied primarily to the early years of space exploration (1957 through the 1960's), during the Cold War between the United States and the Soviet Union. Space firsts were a preoccupation of the space programs of both countries, becoming a matter of public concern and national pride, culminating in the race to place the first person on the Moon, a goal achieved in 1969 by the United States. The achievements of both nations in space have been impressive and it is arguably impossible to quantify them in any meaningful fashion.

Space shuttle main engine (SSME): A reusable, liquid-fueled engine that generates 70,000 horsepower. Three SSMEs are clustered in the tail of the U.S. space shuttle, supplying a total thrust of about 5 kilonewtons, somewhat less than one of the F-1 engines of a Saturn 5.

Space sickness: Any health problem experienced as a result of space travel; this varies with individual and spacecraft environment. Specifically, however, space sickness is the nausea and other symptoms associated with adjustment to microgravity. Also known as space adaptation syndrome.

Space stations: Large orbiting structures designed to support piloted operations for extended periods of time (months to years). Space stations have performed both military and civilian objectives.

Space suit: The pressurized garment worn by an astronaut during extravehicular activity, and sometimes within a spacecraft. The space suit worn by space shuttle astronauts outside the spacecraft, called an extravehicular mobility unit, incorporates a portable life-support system, a manned maneuvering unit, a displays and controls module, a liquid cooling and ventilation garment, a urine collection device, and a delivery system for drinking water.

Space tether: *See* Tether.

Spacecraft: Any self-contained, piloted or robotic, space vehicle.

Spaceflight: The Fédération Aéronautique Internationale defines spaceflight as traveling to an altitude 100 kilometers above the surface of the Earth. This definition is followed by all countries except the United States, which maintains the space boundary at 80 kilometers.

Spacewalk: *See* Extravehicular activity.

Spanish Space Agency (INTA): In Spanish, Instituto Nacional de Técnica Aeroespacial, or Spanish National Institute of Aerospace Techniques. INTA is a public research body, specializing in research and development in aerospace technology.

Spectrograph: A type of spectrometer that splits light into its component wavelengths and records the separated wavelengths photographically or by means of a charge-coupled device.

Spectrometer: An instrument that splits electromagnetic radiation into its component wavelengths for viewing or electronic recording of the emitting body's spectrum.

Spectroscope: A device that splits electromagnetic radiation into its component wavelengths, which can then be read by a spectrometer or a spectrograph.

Spectroscopy: The creation and interpretation of spectra using a variety of instruments, including spectrographs and a variety of spectrometers. By analyzing the spectra of celestial bodies, scientists are able to discover much about the composition of those bodies and the chemical reactions taking place in them.

Spectrum: An image that represents the distribution and intensity of electromagnetic radiation from a body. This image can be photographic or a map of lines showing all or selected wavelengths.

Spektr: The Russian word for "spectrum" and name of the Spektr Earth observation (specifi-

cally natural resources and atmospheric studies) module of the Mir Complex.

Spin axis: The line around which a body rotates.

Spin stabilization: The method whereby an artificial satellite is made to spin at a constant rate about a symmetry axis, relying on gyroscopic effects to keep that axis relatively fixed in space.

Spiral galaxy: A galaxy consisting of a bulge of gas and stars at the center around which arms of stars and other celestial bodies, matter, and radiation rotate in a spiral fashion. Spiral galaxies, of which the Milky Way is one, are by far the most common in the universe.

Splashdown: The free-fall landing of a space capsule or other spacecraft in one of Earth's oceans.

Sputnik: The Russian word for "satellite." Name of the world's first artificial satellite.

SRB: *See* Solid rocket booster.

SSME: *See* Space shuttle main engine.

Stage: A self-contained section of a space vehicle, used for propulsion of a spacecraft into orbit and separated from the launch stack after use.

Standby: A piece of equipment available to replace its counterpart on short notice; the act of placing a piece of equipment into a temporary state of non-use to conserve its power, delay results, or otherwise prevent it from functioning.

Star: A large, nearly spherical mass of extremely hot gas bound together by gravity.

Star tracker: An electronic device programmed to detect and lock onto a celestial body, such as the star Canopus, to provide a spacecraft with a fixed point of reference for purposes of navigation.

Steady state theory: A model of the universe, proposed by Hermann Bondi, Thomas Gold, and Fred Hoyle, which posits that the density of matter in the universe remains constant in an expanding universe because it is created at the same rate as that at which old stars die. The theory is less widely subscribed to than the Big Bang theory, because it does not explain the presence of the cosmic microwave background (CMB) radiation. *See also* Big Bang theory.

Stellar wind: The ionized gases, or plasma, that flow out from stars at high speeds, composed mainly of free protons and electrons. *See also* Solar wind.

Stratosphere: The layer of Earth's atmosphere between the troposphere and the mesosphere. It extends from about 15 to 50 kilometers above Earth's surface, roughly coinciding with the ozone layer, which absorbs the Sun's ultraviolet radiation and heats the stratosphere from a low of about −60° Celsius at the bottom to about 0° Celsius at its top. There is no meteorological activity or vertical air movement in this region of the atmosphere.

Suborbital flight: A spaceflight that either completes less than one orbit of Earth or is not intended to reach orbit.

Subsatellite: A satellite carried into orbit by another satellite. Also, a satellite of a moon.

Sunspots: Dark spots that appear on the Sun's surface in cycles, increasing and decreasing in an eleven-year cycle. These dark spots are about 500 kelvins cooler than a surrounding lighter area of the spot, which in turn is another 500 kelvins cooler than the surrounding photosphere. These cooler regions result from magnetic fields.

Superconducting Quantum Interference Device (SQUID): Mechanism used to measure extremely weak signals, such as subtle changes in the human body's electromagnetic energy field. Using a device called a Josephson junction, a SQUID can detect a change of energy as great as 100 billion times weaker than the electromagnetic energy that moves a compass needle. A Josephson junction consists of two superconductors, separated by an insulating layer so thin that electrons can pass through them.

Superconductivity: A phenomenon occurring in certain materials at low temperatures, characterized by the complete absence of electrical resistance and the damping of the interior magnetic field.

Supergiant: The brightest stars in the universe, which also include the largest: Red supergiants are one thousand times the size of the Sun.

Supernova: A nova resulting from the explosion of any star whose mass is 1.4 times the Sun's. *See also* Nova.

Surveillance: Ongoing monitoring of enemy territory or military installations by aircraft or satellites carrying any of a variety of sensors.

Surveillance satellite: Any military spacecraft that provides imagery reconnaissance data. Such missions include visible, infrared, and radar imaging. Information about these satellites (as well as most other military missions) is very often secret. As a result, the information on these vehicles is often sketchy and may be derived from indirect observations.

Swing-by: The close approach of a spacecraft as it passes a planet on a tour of the solar system. *See also* Flyby.

Synchronous orbit: Any orbit whose period equals the rotational period of the object orbited.

Système International d'Unités: *See* SI units.

Taikonaut: A Chinese person or a person traveling in a Chinese spacecraft who achieves spaceflight, as defined by the Fédération Aéronautique Internationale. *Taikong* is a Chinese word that means space or cosmos. In Chinese, astronauts are *yuhangyuan*, or "travelers of the universe." *See also* Astronaut, Cosmonaut.

Technology satellite: This general class of missions includes all missions designed as technical tests of satellite or ground-based systems. Examples include tests of new satellite components, investigations of launch vehicle performance, calibration of ground-based systems (such as radar), and demonstrations of new sensor technologies.

Tectonics: The branch of geology that examines folding and faulting in Earth's crust. *See also* Plate tectonics.

Telecommunications satellite: An artificial satellite dedicated to the receiving and transmission of radio, television, and other communications signals. Such satellites are usually placed in geostationary orbits so that they remain over a fixed point on Earth.

Telemetering: A system for taking measurements within an aerospace vehicle in flight and transmitting them by radio to a ground station.

Telemetry: Real-time transmission of data from a distance via radio signals.

Teleoperations: Manipulation of an orbiter, booster, or instruments in space from Earth via remote control.

Temps Atomique International (TAI): *See* International Atomic Time.

Terminator: The line on a planet or moon that forms the boundary between day and night.

Terrestrial planets: Mercury, Venus, and Mars are called terrestrial planets because they resemble Earth in certain fundamental features (such as density and composition).

Terrestrial Time (TT): A time scale established by the International Astronomical Union (IAU) to serve as the independent argument for apparent geocentric ephemerides. That is, TT is used for the prediction or recording of the positions of celestial bodies as measured by an observer on Earth. It is the successor to Ephemeris Time (ET), but is based on the SI second. Terrestrial Time is effectively equal to International Atomic Time (TAI) plus 32.184 seconds.

Test flight: Experimental operation of an aircraft or spacecraft to determine whether it functions as designed and to identify systems in need of adjustment.

Test range: A site dedicated to the testing of aircraft or spacecraft.

Tether: A chord, cable, or wire connection between a spacecraft and another object in orbit. The earliest tethers were those used as life-lines for astronauts carrying out spacewalks during the pioneering Soviet and American manned orbital missions. Much longer space tethers, however, provide a means of deploying probes to study Earth's outer atmosphere or generating electricity to power a spacecraft or space station.

Thermal mapping: Gathering data from which to construct maps by means of instruments capable of sensing heat.

Thermal tiles: Heat-resistant tiles glued to the aluminum skin of the space shuttle to protect it from overheating upon reentry into Earth's atmosphere. They are favored over ablative material because they are lighter and reusable.

Thermosphere: The highest layer of Earth's atmosphere except for the exosphere, beginning at 85 kilometers above sea level. The oxygen and nitrogen that compose the atmosphere at this level are extremely rarefied and are heated by the Sun's ultraviolet radiation to the point of ionization. As a result, the ionosphere (which lies between 50 and 500 kilometers above sea level) overlaps the thermosphere. This is also the region in which auroras and meteors occur.

Three-axis stabilization: Stabilization of a satellite against pitch, roll, and yaw. *See also* Pitch, roll, and yaw.

Thrust: The force required to propel a vehicle. Refers especially to that force exerted by a rocket engine to launch a space vehicle into orbit or a deep space trajectory. Measured in newtons (SI).

Time dilation: The phenomenon, predicted by Albert Einstein's special theory of relativity, whereby time appears to slow down in a system moving near the speed of light from the vantage point of an observer outside that system.

Topside observation: Electronic scanning of Earth's (or another) atmosphere from above. Used mainly in reference to meteorological satellites.

Touchdown: The moment in time when a spacecraft deliberately makes contact with the ground or water.

Tracking network: A network of tracking stations at different points on the globe that send and receive radio signals to and from spacecraft via large dish antennae that allow continuous communications with spacecraft. Examples are the Spaceflight Tracking and Data Network and the Deep Space Network.

Trajectory: The path traced out by a ballistic missile or by a spacecraft launched from Earth or from orbit toward the Moon or another destination.

Trans-Earth injection: A boost from lunar (or other planetary) orbit which places a spacecraft on a trajectory toward Earth. *See also* Translunar injection.

Transducer: Any device that transforms one type of energy into another, such as a solar cell (which transforms sunlight into electrical power), a thermocouple (which transforms thermal energy into an electric signal), or a Geiger counter (which transforms radioactive into sound energy).

Transfer orbit: The orbit into which a spacecraft is boosted from Earth orbit on its way to orbit around another celestial body. Since the spacecraft does not complete a full revolution of the transfer orbit, but only part of the ellipse, the path it follows describes a trajectory that intersects the final orbit.

Transit: The passage of one celestial body across the face of another or across the observer's meridian.

Translunar injection: The process by which a spacecraft in orbit around Earth is boosted into a trajectory that heads it toward the Moon.

Transponder: A device that receives radio signals and automatically responds to them using the same frequency.

Transport satellite: Transport missions include all spacecraft that are used to ferry people or materials from the Earth to orbit without performing significant research. Missions in this class include space station resupply vehicles and shuttle missions whose primary purpose is to carry satellites into orbit for deployment.

Triangulation: A means of determining the position of an object by calculation from known quantities: The distance between two fixed points and the angles formed between the line described by those points and the line between a third point. Triangulation is the oldest method of determining distances, both on Earth and in space.

Troposphere: The layer of Earth's atmosphere that lies closest to Earth's surface, extending upward to about 8 kilometers. The troposphere is the densest region of the atmosphere and the region in which all meteorological phenomena occur.

Ullage: The volume in a closed tank or container that is not occupied by the stored liquid; the ratio of this volume to the total volume of the tank; also an acceleration to force propellants into the engine pump intake lines before ignition.

Ultraviolet astronomy: The branch of astronomy that examines the ultraviolet emissions of celestial phenomena. Ultraviolet astronomy has developed with the advent of the space age. Because ultraviolet rays are unable to penetrate Earth's atmosphere, instruments carried aloft by satellites such as the International Ultraviolet Explorer have enabled scientists to learn much about celestial bodies, since many of the elements of which they are composed are most evident in spectroscopic measurements.

Ultraviolet radiation: Electromagnetic radiation emitted between the wavelengths of 900 and 3,000 angstroms (between 400 and 2 nanometers). The band between visible violet light and X-radiation on the electromagnetic spectrum.

Ultraviolet spectrometer: An instrument that measures electromagnetic wavelengths in the ultraviolet range, used on satellites such as the International Ultraviolet Explorer and the Extreme Ultraviolet Explorer.

Umbra: Darkest part of a shadow in which light is completely absent, such as the surface of the Moon or Earth away from the Sun where the disc of the Sun is completely obscured.

UNEX: *See* University-class Explorer.

University-class Explorer (UNEX): A very small-sized spacecraft (defined by definition and development costs not exceeding $7.5 million) that carries one or more science instruments. *See also* Explorer, Medium-class Explorer, Small Explorer.

Update pad: Information on spacecraft attitudes, thrust values, event times, or navigational data, voiced up to the crew in standard formats according to the purpose. Examples of this type of information include maneuver updates, navigation checks, landmark tracking, and entry updates.

Uplink: Signals sent up to a satellite from an Earth station. *See also* Downlink.

Vacuum: An area in which absolutely nothing exists. Although a true vacuum never occurs in nature, the behavior of bodies within a vacuum is of concern to physicists studying interplanetary and deep space, in which near-vacuum conditions exist.

Van Allen radiation belts: The two layers of Earth's magnetosphere, discovered by James Van Allen in the late 1950's, in which ionized particles spiral back and forth between Earth's magnetic poles. These zones are of importance because they pose potential hazards to electronic instruments aboard spacecraft.

Variable star: A star whose brightness varies over time as a result of several intrinsic or extrinsic factors. There are thousands of these stars, and they can be categorized into seven classes based on the causes of their variation. *See also* Cepheid variable.

Venus orbit: The Venus orbit class includes all missions that went into orbit around Venus or impacted the planet's surface.

Vernier rocket: A small thruster rocket used in space to make fine corrections to a spacecraft's orientation or trajectory.

Very high orbit (VHO): Any orbit that has a perigee at or above geosynchronous orbit and an apogee above geosynchronous orbit, yet remains in orbit around the Earth (or Earth-Moon system). Orbits in this class are often highly elliptical, with apogees several hundred thousand kilometers in altitude.

Very long-baseline interferometry: A technique used by radio astronomers to increase the sharpness (resolution) of received signals by using several or many radio telescopes at widely spaced locations. Unlike ordinary radio interferometry, in which two or only a few radio telescopes are used, this technique allows for highly accurate mapping of the celestial bodies that emit radio signals. *See also* Interferometry.

Vidicon: A video camera, or a device that converts light into electronic signals that can be transmitted or recorded as pictures.

Volcanism: The dynamic process in which molten material explodes from cracks or other openings in the interior of a planet as it is transferred to the planet's solid surface.

Voskhod: The Russian word for "rising" and the name of the Soviet two-person spacecraft.

Vostok: The Russian word for "east" and the name of the first piloted spacecraft.

Wavelength: A spatial characteristic that in part defines sonic waves and electromagnetic radiation. A wavelength is the length (measured in angstroms or nanometers in the electromagnetic range) between successive crests in a photon's wave pattern as it moves up and down in a direction of propagation. For sonic waves, it is the distance between maximum compression or maximum expansion in the pressure variations of the propagating medium along the direction of travel of the sound.

Weather satellite: Weather satellites monitor the Earth's atmospheric conditions and provide data to help meteorologists predict and understand the Earth's weather patterns. *See also* Meteorological satellite.

Weightlessness: *See* Zero gravity.

White dwarf: A dying star (one that is collapsing in on itself) with a mass 1.4 times that of the Sun or less, and with a radius approximately that of Earth. Such electron-degenerate stars are destined to end their lives as cold, dark spheres, having expended all of their energy but not initially massive enough to end life as neutron stars or black holes.

Wind tunnel: A large tubular structure through which air is forced to flow at high speeds for the purpose of testing the behavior of aircraft and other structures that travel through the atmosphere.

Window: *See* Launch window.

X-band: The range of radio frequencies between 5.2 and 10.9 gigahertz.

X-radiation: Electromagnetic radiation with wavelengths between 0.1 and 10 nanometers; the range lying between gamma and ultraviolet radiation on the electromagnetic spectrum.

X-ray astronomy: The branch of astronomy that examines the x-ray emissions of celestial phenomena. This radiation cannot be studied from the ground, since Earth's atmosphere absorbs most X-radiation. As a result, x-ray astronomy has blossomed with the advent of spacecraft that can carry x-ray telescopes and other detectors into space. The examination of celestial x-ray sources has led, among other things, to the discovery of neutron stars and black holes as members of binary star systems. The Einstein Observatory, launched by NASA in 1978, revealed that nearly all stars emit X-radiation.

Yaw: *See* Pitch, roll, and yaw.

Zarya: The Russian word for "dawn" and name of the Functional Cargo Block of the International Space Station.

Zero gravity: The condition of absolute weightlessness, which occurs in free-fall and is approached in deep space, far from massive bodies. Because all masses exert gravitational force on one another, the condition of zero gravity does not occur in nature. *See also* Microgravity.

Zodiacal light: The glow seen in the west after sunset and in the east before dawn. Caused by sunlight reflecting off microscopic dust particles.

Zvezda: The Russian word for "star" and name of the service module of the International Space Station.

Russell R. Tobias and David G. Fisher

Space Shuttle Missions

Shuttle missions are listed below in chronological order by launch date. Mission numbers do not necessarily reflect chronological order; missions are assigned a Space Transportation System (STS) number early in their development but may experience rescheduled launch dates due to equipment checks, weather, changes in funding, and other exigencies.

Mission	Launch Date	Orbiter
Test flights	1977	Enterprise
STS-1	April 12, 1981	Columbia
STS-2	November 12, 1981	Columbia
STS-3	March 22, 1982	Columbia
STS-4	June 27, 1982	Columbia
STS-5	November 11, 1982	Columbia
STS-6	April 4, 1983	Challenger
STS-7	June 18, 1983	Challenger
STS-8	August 30, 1983	Challenger
STS-9	November 28, 1983	Columbia
STS 41-B	February 3, 1984	Challenger
STS 41-C	April 6, 1984	Challenger
STS 41-D	August 30, 1984	Discovery
STS 41-G	October 5, 1984	Challenger
STS 51-A	November 8, 1984	Discovery
STS 51-C	January 25, 1985	Discovery
STS 51-D	April 12, 1985	Discovery
STS 51-B	April 29, 1985	Challenger
STS 51-G	June 17, 1985	Discovery
STS 51-F	July 29, 1985	Challenger
STS 51-I	August 27, 1985	Discovery
STS 51-J	October 3, 1985	Atlantis
STS 61-A	October 30, 1985	Challenger
STS 61-B	November 26, 1985	Atlantis
STS 61-C	January 12, 1986	Columbia
STS 51-L	January 28, 1986	Challenger
STS-26	September 29, 1988	Discovery
1K1 (USSR)	November 15, 1988	Buran 1.01
STS-27	December 2, 1988	Atlantis
STS-29	March 13, 1989	Discovery
STS-30	May 4, 1989	Atlantis
STS-28	August 8, 1989	Columbia
STS-34	October 18, 1989	Atlantis
STS-33	November 22, 1989	Discovery

Mission	Launch Date	Orbiter
STS-32	January 9, 1990	Columbia
STS-36	February 28, 1990	Atlantis
STS-31	April 24, 1990	Discovery
STS-41	October 6, 1990	Discovery
STS-38	November 15, 1990	Atlantis
STS-35	December 2, 1990	Columbia
STS-37	April 5, 1991	Atlantis
STS-39	April 28, 1991	Discovery
STS-40	June 5, 1991	Columbia
STS-43	August 2, 1991	Atlantis
STS-48	September 12, 1991	Discovery
STS-44	November 24, 1991	Atlantis
STS-42	January 22, 1992	Discovery
STS-45	March 24, 1992	Atlantis
STS-49	May 7, 1992	Endeavour
STS-50	June 25, 1992	Columbia
STS-46	July 31, 1992	Atlantis
STS-47	September 12, 1992	Endeavour
STS-52	October 22, 1992	Columbia
STS-53	December 2, 1992	Discovery
STS-54	January 13, 1993	Endeavour
STS-56	April 8, 1993	Discovery
STS-55	April 26, 1993	Columbia
STS-57	June 21, 1993	Endeavour
STS-51	September 12, 1993	Discovery
STS-58	October 18, 1993	Columbia
STS-61	December 2, 1993	Endeavour
STS-60	February 3, 1994	Discovery
STS-62	March 4, 1994	Columbia
STS-59	April 9, 1994	Endeavour
STS-65	July 8, 1994	Columbia
STS-64	September 9, 1994	Discovery
STS-68	September 30, 1994	Endeavour
STS-66	November 3, 1994	Atlantis

Mission	Launch Date	Orbiter	Mission	Launch Date	Orbiter
STS-63	February 3, 1995	Discovery	STS-96	May 27, 1999	Discovery
STS-67	March 2, 1995	Endeavour	STS-93	July 23, 1999	Columbia
STS-70	June 13, 1995	Discovery	STS-103	December 19, 1999	Discovery
STS-71	June 27, 1995	Atlantis	STS-99	February 11, 2000	Endeavour
STS-69	September 7, 1995	Endeavour	STS-101	May 19, 2000	Atlantis
STS-73	October 20, 1995	Columbia	STS-106	September 8, 2000	Atlantis
STS-74	November 12, 1995	Atlantis	STS-92	October 11, 2000	Discovery
STS-72	January 11, 1996	Endeavour	STS-97	November 30, 2000	Endeavour
STS-75	February 22, 1996	Columbia	STS-98	February 7, 2001	Atlantis
STS-76	March 22, 1996	Atlantis	STS-102	March 8, 2001	Discovery
STS-77	May 19, 1996	Endeavour	STS-100	April 19, 2001	Endeavour
STS-78	June 20, 1996	Columbia	STS-104	July 12, 2001	Atlantis
STS-79	September 16, 1996	Atlantis	STS-105	August 10, 2001	Discovery
STS-80	November 19, 1996	Columbia	STS-108	December 5, 2001	Endeavour
STS-81	January 12, 1997	Atlantis	STS-109	March 1, 2002	Columbia
STS-82	February 11, 1997	Discovery	STS-110	April 8, 2002	Atlantis
STS-83	April 4, 1997	Columbia	STS-111	June 5, 2002	Endeavour
STS-84	May 15, 1997	Atlantis	STS-112	October 7, 2002	Atlantis
STS-94	July 1, 1997	Columbia	STS-113	November 23, 2002	Endeavour
STS-85	August 7, 1997	Discovery	STS-107	January 16, 2003	Columbia
STS-86	September 25, 1997	Atlantis	STS-114	July 23, 2005	Discovery
STS-87	November 19, 1997	Columbia	STS-121	September 9, 2005	Atlantis
STS-89	January 22, 1998	Endeavour	STS-115	February 16, 2006	Atlantis
STS-90	April 17, 1998	Columbia	STS-116	April 23, 2006	Discovery
STS-91	June 2, 1998	Discovery	STS-117	July 13, 2006	Endeavour
STS-95	October 29, 1998	Discovery	STS-118	November 7, 2006	Discovery
STS-88	December 4, 1998	Endeavour	STS-119	December 14, 2006	Endeavour

World Wide Web Pages

The Internet can be a wonderful tool for doing research. Almost any bit of information about nearly any subject is available online. However, two major problems exist with the information found on the World Wide Web. The first has to do with accuracy. Far too many Web sites obtain information from places other than the originator of the material. Many sites do not check the correctness of the information, so that what might have started as a mistyped sentence becomes "fact." The second problem stems from the ability to update the information on a site. This has both good and bad effects. The latest information is always available. What is not readily available is historical information. For example, it is easy to obtain the current space shuttle launch manifest from most of NASA's Web sites, but finding a manifest from just before the 2003 Columbia accident is nearly impossible.

For every site with useful information, there are hundreds of sites filled with useless items. Listed below are Web sites that are current as of 2005, accurate, and authoritative. For major programs, check the sponsoring agency's Web site first. Current mission pages are included where available.

Although every effort has been made to ensure accuracy, Web sites are continually being updated and there may be changes to the site address listed. A subject or keyword search through any of the major search engines will help locate the new address. Please note when entering these Universal Resource Locators (URLs) that proper capitalization and punctuation are required.

ACRIMSAT (Active Cavity Radiometer Irradiance Monitor) Mission, Jet Propulsion Laboratory
http://acrim.jpl.nasa.gov

Aerospace Corporation
http://www.aero.org

Aerospace Dictionary
http://roland.lerc.nasa.gov/~dglover/
    dictionary/content.html

Aerospace Science and Technology Dictionary
http://www.hq.nasa.gov/office/hqlibrary/
    aerospacedictionary/

L'Agence Spatiale Canadienne (ASC)/Canadian Space Agency
http://www.space.gc.ca/asc/index.html

Agenzia Spaziale Italiana (ASI)/Italian Space Agency
http://www.asi.it/sito/english.htm

Air and Space Magazine
http://www.airspacemag.com

Air Force: Launch Information
https://www.patrick.af.mil/launch.htm

Air Force: Space Command
http://www.peterson.af.mil/hqafspc

Amateur Radio on the International Space Station (ARISS)
http://www.arrl.org/ARISS/

Ames Research Center
http://www.nasa.gov/centers/ames

Ames Research Center: History Office
http://history.arc.nasa.gov

Ames Research Center: Multimedia
http://www.nasa.gov/centers/ames/multimedia

Ansari X Prize
http://www.xprize.org/home.php

Apollo: A Retrospective Analysis
http://www.hq.nasa.gov/office/pao/History/
    Apollomon/cover.html

Apollo, Biomedical Results of (NASA SP-368)
http://history.nasa.gov/SP-368/sp368.htm

Apollo by the Numbers: A Statistical Reference (NASA SP-4029)
http://history.nasa.gov/SP-4029/SP-4029.htm

Apollo Expeditions to the Moon (NASA SP-350)
http://www.hq.nasa.gov/office/pao/History/SP-350/cover.html

Apollo Lunar Surface Journal
http://www.hq.nasa.gov/office/pao/History/alsj

Apollo Maniacs: Apollo Spacecraft and Saturn Specifications
http://apollomaniacs.web.infoseek.co.jp/apollo/indexe.htm

Apollo 1
http://www.hq.nasa.gov/office/pao/History/Apollo204

Apollo Over the Moon: A View From Orbit (NASA SP-362)
http://www.hq.nasa.gov/office/pao/History/SP-362/cover.htm

Apollo Press Kits
http://www-lib.ksc.nasa.gov/lib/presskits.html

Apollo Program
http://spaceflight.nasa.gov/history/apollo

Apollo Program: Gallery
http://spaceflight1.nasa.gov/gallery/images/apollo

Apollo Program Summary Report
http://history.nasa.gov/apsr/apsr.htm

Apollo Project Archive
http://www.apolloarchive.com

Apollo Project Archive: Apollo Image Gallery
http://www.apolloarchive.com/apollo_gallery.html

Apollo Project Archive: Lunar Lander Simulator
http://www.apolloarchive.com/lander.html

Apollo Saturn Reference Page
http://www.apollosaturn.com

Apollo-Soyuz Test Project
http://www-pao.ksc.nasa.gov/kscpao/history/astp/astp.html

Apollo-Soyuz Test Project: Gallery
http://spaceflight.nasa.gov/gallery/images/apollo-soyuz

Apollo Spacecraft: A Chronology (NASA SP-4009)
http://www.hq.nasa.gov/office/pao/History/SP-4009/cover.htm

Apollo 13 Review Board, Report of the
http://history.nasa.gov/ap13rb/ap13index.htm

ARISS: Amateur Radio on the International Space Station
http://www.arrl.org/ARISS/

Army Space and Missile Defense Command
http://www.smdc.army.mil/

"Ask Dr. Marc:" NASA Space Place
http://spaceplace.nasa.gov/en/kids/phonedrmarc

Association of Space Explorers
http://www.space-explorers.org

Astronaut Biographies
http://www.jsc.nasa.gov/Bios

Astronaut Biographies, Canadian Astronauts Corps
http://www.space.gc.ca/asc/eng/astronauts/bio.asp

Astronaut Biographies, European Astronaut Corps
http://www.spaceflight.esa.int/file.cfm?filename=astcorps

Astronaut Selection Office
http://nasajobs.nasa.gov/astronauts/

Astronauts Memorial Foundation, The
http://www.amfcse.org/

Astronomy Café, The
http://www.astronomycafe.net

Astronomy Picture of the Day
http://antwrp.gsfc.nasa.gov/apod

Astrophysics Data System
http://adswww.harvard.edu

Atmospheric Laboratory for Applications and
Science (ATLAS)
http://wwwghcc.msfc.nasa.gov/atlas.html

Aviation Week's *AviationNow*
http://www.aviationnow.com/avnow/

Ball Aerospace and Technologies Corp.
http://www.ballaerospace.com/

*"Before This Decade Is Out . . ." Personal Reflections
on the Apollo Program* (NASA SP-4223)
http://history.nasa.gov/SP-4223/sp4223.htm

Boeing Company, The
http://www.boeing.com

Boeing International Space Station: Home
http://www.boeing.com/defense-space/space/
spacestation

Boeing NASA Systems
http://www.boeing.com/defense-space/space/
nasasystems

Boeing Rocketdyne (See Pratt & Whitney:
Space Propulsion)

Boeing Satellite Systems
http://www.boeing.com/defense-space/space/
bss

Brazilian National Institute for Space Research
(INPE)/Instituto Nacional de Pesquisas Espacias
http://www.inpe.br/english

British National Space Centre
http://www.bnsc.gov.uk

Canadian Astronauts Corps
http://www.space.gc.ca/asc/eng/astronauts/
bio.asp

Canadian Space Agency (CSA)/L'Agence
Spatiale Canadienne (ASC)
http://www.space.gc.ca/asc/index.html

Cape Canaveral Air Force Station: Patrick Air
Force Base
https://www.patrick.af.mil/45sw/ccafs/
index.htm

Cassini-Huygens Mission
http://www.nasa.gov/mission_pages/cassini/
main

Centennial of Flight, U.S.
http://www.centennialofflight.gov

Center for Aerospace Information
http://www.sti.nasa.gov

Centre National d'Etudes Spatiales (CNES)/
French Space Agency
http://www.onera.fr/english.html

Challenger *Accident, Report of the Presidential
Commission on the Space Shuttle*
http://science.ksc.nasa.gov/shuttle/missions/
51-l/docs/rogers-commission/table-of-
contents.html

Chandra X-Ray Observatory Center
http://chandra.harvard.edu

*Chariots for Apollo: A History of Manned Lunar
Spacecraft* (NASA SP-4205)
http://www.hq.nasa.gov/office/pao/History/SP-
4205/cover.html

Chronology of Space Exploration, Russian Space
Web
http://www.russianspaceweb.com/
chronology.html

Coalition for Space Exploration
http://www.spacecoalition.com

Columbia Accident Investigation Board
http://caib.nasa.gov

Columbia *Accident Investigation Board, Report of the*
http://www1.nasa.gov/columbia/caib/html

Combined Release and Radiation Effects Satellite (CRRES)
http://www.ball.com/aerospace/crres.html

Compton Gamma Ray Observatory
http://cossc.gsfc.nasa.gov/cossc

Conseil National de Recherches Canada/ National Research Council Canada
http://www.nrc-cnrc.gc.ca/main_e.html

Convair (*see* General Dynamics)

Cosmonaut Biographies
http://www.jsc.nasa.gov/Bios/cosmo.html

Cosmonautics, Tsiolkovsky State Museum of the History of
http://www.informatics.org/museum

Dawn Mission
http://dawn.jpl.nasa.gov

Deep Impact
http://deepimpact.jpl.nasa.gov

Deep Space Network
http://deepspace.jpl.nasa.gov/dsn

Department of Defense, U.S.
http://www.defenselink.mil/

Destination Earth
http://www.earth.nasa.gov

Destination Moon: A History of the Lunar Orbiter Program (NASA TM-3487)
http://www.hq.nasa.gov/office/pao/History/ TM-3487/top.htm

Deutsches Zentrum für Luft- und Raumfahrt (DLR)/German Aerospace Center
http://www.dlr.de/dlr

Dictionary of Technical Terms for Aerospace Use
http://roland.lerc.nasa.gov/~dglover/ dictionary/content.html

Discovery Program
http://discovery.nasa.gov

Dryden Flight Research Center
http://www.nasa.gov/centers/dryden

Dryden Flight Research Center: Biographies
http://www.dfrc.nasa.gov/Newsroom/ Biographies/Pilots

Dryden Flight Research Center: Gallery
http://www.dfrc.nasa.gov/Gallery

Dryden Flight Research Center: History
http://www1.nasa.gov/centers/dryden/history

Dryden Flight Research Center: Test Pilot Biographies
http://www.dfrc.nasa.gov/Newsroom/ Biographies

Earth Observations Photography, Space Shuttle
http://earth.jsc.nasa.gov/sseop/efs/

Earth Observing System
http://eospso.gsfc.nasa.gov

Earth Observing System Data Gateway
http://delenn.gsfc.nasa.gov/~imswww/pub/ imswelcome

Edwards Air Force Base
http://afftc.edwards.af.mil

Encyclopedia Astronautica
http://www.astronautix.com/

Energia: All About the HLLV
http://www.k26.com/buran/index.html

European Astronaut Corps
http://www.spaceflight.esa.int/ file.cfm?filename=astcorps

European Space Agency (ESA)
http://www.esa.int

European Space Agency: Science Programs
http://sci.esa.int

Experimental Communications Satellites
http://roland.lerc.nasa.gov/~dglover/sat/ satcom1.html

Exploration System Mission Directorate
http://exploration.nasa.gov

Explorer Missions
http://nssdc.gsfc.nasa.gov/multi/explorer.html

Far Ultraviolet Spectroscopic Explorer (FUSE)
http://fuse.pha.jhu.edu

Federation of American Scientists
http://www.fas.org/main/home.jsp

Florida Space Authority
http://www.floridaspaceauthority.com

French National Aerospace Research
Establishment/ONERA: Office National
d'Etudes et de Recherches Aerospatiales
http://www.onera.fr/english.html

French Space Agency (CNES)/Centre National
d'Etudes Spatiales
http://www.cnes.fr

FUSE: Far Ultraviolet Spectroscopic Explorer
http://fuse.pha.jhu.edu

Galaxy Evolution Explorer
http://www.galex.caltech.edu

Gamma-ray Large Area Space Telescope
(GLAST)
http://glast.gsfc.nasa.gov

Gemini Program
http://www-pao.ksc.nasa.gov/kscpao/history/
    gemini/gemini.htm

Gemini Program: Gallery
http://spaceflight.nasa.gov/gallery/images/
    gemini

General Dynamics
http://www.gd.com

Geological Survey, U.S.
http://www.usgs.gov

German Space Agency (DLR)/Deutschen
Zentrum für Luft- und Raumfahrt
http://www.dlr.de/dlr

GLAST: Gamma-ray Large Area Space Telescope
http://glast.gsfc.nasa.gov

Glenn Research Center
http://www.nasa.gov/centers/glenn

Glenn Research Center: History
http://grchistory.grc.nasa.gov

Glenn Research Center: Multimedia
http://www.nasa.gov/centers/glenn/multimedia

Global Tropospheric Experiment (GTE)
http://www-gte.larc.nasa.gov

Globalstar
http://www.globalstar.com

Goddard Space Flight Center (GSFC)
http://www.nasa.gov/centers/goddard

Goddard Space Flight Center: High-Energy
Missions
http://heasarc.gsfc.nasa.gov/docs/heasarc/
    missions/alphabet.html

Goddard Space Flight Center: Multimedia
http://www.nasa.gov/centers/goddard/
    multimedia

Government Printing Office: Access
http://www.gpoaccess.gov

Government Printing Office: Online Bookstore
http://bookstore.gpo.gov

Gravity Probe B
http://www.gravityprobeb.com

Great Images in NASA
http://grin.hq.nasa.gov

Grumman (see Northrop Grumman)

Heavens-Above Satellite Observations
http://www.heavens-above.com

High-Energy Missions, Goddard Space Flight
Center
http://heasarc.gsfc.nasa.gov/docs/heasarc/
    missions/alphabet.html

History of Cosmonautics, Tsiolkovsky State
Museum of the
http://www.informatics.org/museum

History Timelines
http://www.hq.nasa.gov/office/pao/History/
    timeline.html

Human Space Flight
http://spaceflight.nasa.gov/home

Human Space Flight: Imagery
http://spaceflight.nasa.gov/gallery

Human Space Flight: International Space
Station
http://spaceflight.nasa.gov/station

Human Space Flight: International Space Station
Assembly
http://spaceflight.nasa.gov/station/assembly

Human Space Flight: International Space Station
Imagery
http://spaceflight1.nasa.gov/gallery/images/
station

Human Space Flight: International Space Station
Reference
http://spaceflight.nasa.gov/station/reference

Human Space Flight: International Space Station
Science
http://spaceflight.nasa.gov/station/science

Human Space Flight: Space Shuttle
http://spaceflight.nasa.gov/shuttle

Human Space Flight: Space Shuttle Gallery
http://spaceflight.nasa.gov/gallery/images/
shuttle

Human Space Flight: Space Shuttle Reference
http://spaceflight.nasa.gov/shuttle/reference

Instituto Nacional de Pesquisas Espacias (INPE)/
Brazilian National Institute for Space Research
http://www.inpe.br/english

Instituto Nacional de Técnica Aeroespacial
(INTA)/Spanish Space Agency
http://www.inpe.br/english

International Space Station
http://www.nasa.gov/mission_pages/station/
main

International Space Station: Assembly
http://spaceflight.nasa.gov/station/assembly

International Space Station: Gallery
http://spaceflight.nasa.gov/gallery/images/
station

International Space Station: History
http://spaceflight.nasa.gov/history/station

International Space Station: Reference
http://spaceflight.nasa.gov/station/reference/
index.html

International Space Station: Science
http://spaceflight.nasa.gov/station/science/
index.html

International Spacecraft and Launch Vehicle
Names: Glossary
http://www.spacecraftnames.info/selection.html

Italian Space Agency (ASI)/Agenzia Spaziale
Italiana
http://www.asi.it/sito/english.htm

Japan Aerospace Exploration Agency (JAXA)
http://www.jaxa.jp/index_e.html

Jerrie Cobb Foundation
http://www.jerrie-cobb.org

Jet Propulsion Laboratory (JPL)
http://www.jpl.nasa.gov/

Jet Propulsion Laboratory: History
http://beacon.jpl.nasa.gov

Jet Propulsion Laboratory: Missions
http://www.jpl.nasa.gov/missions

Jet Propulsion Laboratory: Multimedia
http://www.nasa.gov/centers/jpl/multimedia

John H. Glenn Research Center at Lewis Field
http://www.nasa.gov/centers/glenn

John F. Kennedy Space Center
http://www.nasa.gov/centers/kennedy

John C. Stennis Space Center
http://www.nasa.gov/centers/stennis

Johnson Space Center (JSC)
http://www.nasa.gov/centers/johnson

Johnson Space Center: Digital Image Collection
http://images.jsc.nasa.gov

Johnson Space Center: History
http://www.jsc.nasa.gov/history

Johnson Space Center: History Collection
http://www.jsc.nasa.gov/history/
history_collection/uhcl.htm

Johnson Space Center: Multimedia
http://www.nasa.gov/centers/johnson/
multimedia

Johnson Space Center: White Sands Test Facility
http://www.wstf.nasa.gov

Kennedy Space Center
http://www.nasa.gov/centers/kennedy

Kennedy Space Center: Historical Documents
http://www.nasa.gov/centers/kennedy/about/
history/historydocs.html

Kennedy Space Center: History
http://www.nasa.gov/centers/kennedy/about/
history

Kennedy Space Center: Multimedia
http://www.nasa.gov/centers/kennedy/
multimedia

Kennedy Space Center: News Releases
http://www1.nasa.gov/centers/kennedy/news/
releases

Kennedy Space Center: Reference Material
http://www1.nasa.gov/centers/kennedy/news/
mediaresources/info_publications.html

Kennedy Space Center: Space Shuttle Operations
http://www1.nasa.gov/centers/kennedy/
shuttleoperations

Kennedy Space Center: Space Station Payloads
http://www1.nasa.gov/centers/kennedy/
stationpayloads

Kennedy Space Center Story, The
http://www1.nasa.gov/centers/kennedy/about/
history/story/kscstory.html

Khrunichev State Space Center (Russia)
http://www.intertec.co.at/itc2/partners/
KHRUNICHEV

Konstantin E. Tsiolkovsky State Museum of the
History of Cosmonautics
http://www.informatics.org/museum

Kosmonavtka
http://suzymchale.com/kosmonavtka/
index.html

Langley Research Center (LRC)
http://www.nasa.gov/centers/langley

Langley Research Center: Multimedia
http://www.nasa.gov/centers/langley/
multimedia

Little Joe Series, Project Mercury
http://www.peter.mcquillan.dsl.pipex.com

Lloyd's Satellite Constellations
http://www.ee.surrey.ac.uk/Personal/L.Wood/
constellations

Lockheed Martin
http://www.lockheedmartin.com

Lunar and Planetary Institute
http://www.lpi.usra.edu

Lunar Exploration Timeline
http://nssdc.gsfc.nasa.gov/planetary/lunar/
lunartimeline.html

Lunar Prospector
http://lunar.arc.nasa.gov

Lyndon B. Johnson Space Center
http://www.nasa.gov/centers/johnson

McDonnell Douglas Aerospace (*see* Boeing)

Managing NASA in the Apollo Era (NASA SP-4102)
http://history.nasa.gov/SP-4102/sp4102.htm

Manned Astronautics: Figures and Facts
http://space.kursknet.ru/cosmos/english/
main.sht

Manned Spacecraft Gallery
http://www.lycoming.edu/astr-phy/fisherpdg1

Manned Spaceflight PDF Documents
http://www.geocities.com/bobandrepont/
spacepdf.htm

Mars Exploration Program
http://mars.jpl.nasa.gov/

Mars Exploration Rover Mission
http://marsrovers.jpl.nasa.gov/home

Mars Express
http://mars.jpl.nasa.gov/express/

Mars Global Surveyor
http://mars.jpl.nasa.gov/mgs/

Mars Historical Log of All International Missions
http://mars.jpl.nasa.gov/missions/log/

Mars Odyssey
http://mars.jpl.nasa.gov/odyssey/

Mars Reconnaissance Orbiter
http://mars.jpl.nasa.gov/mro/

Mars Science Laboratory
http://centauri.larc.nasa.gov/msl/

Marshall Space Flight Center
http://www.nasa.gov/centers/marshall

Marshall Space Flight Center: Image Exchange
http://mix.msfc.nasa.gov

Marshall Space Flight Center: Multimedia
http://www.nasa.gov/centers/marshall/
multimedia

Mercury, Project
http://www-pao.ksc.nasa.gov/kscpao/history/
mercury/mercury.htm

Mercury, Project: Gallery
http://spaceflight.nasa.gov/gallery/images/
mercury

Mercury, Project: Little Joe Series
http://www.peter.mcquillan.dsl.pipex.com

Military Space Programs: Federation of American Scientists
http://www.fas.org/spp/military/program/
index.html

Mir Mission Data
http://home.earthlink.net/~cliched/spacecraft/
mirhistory.html

Mir Space Station, Russian Space Web
http://www.russianspaceweb.com/mir.html

MirCorp: The Manned Commercial Space Exploration Company
http://www.mir-corp.com

Missile Defense Agency
http://www.mda.mil/mdalink/html/
mdalink.html

Mission Data for All Missions to Mir
http://home.earthlink.net/~cliched/spacecraft/
mirhistory.html

Mission Patches
http://www.hq.nasa.gov/office/pao/History/
mission_patches.html

Molniya Research & Industrial Corporation
http://www.buran.ru/htm/molniya.htm

Moonport: A History of Apollo Launch Facilities and Operations (NASA SP-4204)
http://www.hq.nasa.gov/office/pao/History/
SP-4204/cover.html

NASA
http://www.nasa.gov

NASA (listing of all NASA-related sites)
http://www.placedirectory.com/nasa.htm

NASA: Astronaut Biographies
http://www.jsc.nasa.gov/Bios

NASA: Astrophysics Data System
http://adswww.harvard.edu

NASA: Biographical and Other Personnel Information
http://www.hq.nasa.gov/office/pao/History/
prsnnl.htm

NASA: Center Directors, Chronological List
http://www.hq.nasa.gov/office/pao/History/
director.html

NASA: Center for Aerospace Information
http://www.sti.nasa.gov/

NASA: Discovery Program
http://discovery.nasa.gov

NASA: Earth Observing System Data Gateway
http://delenn.gsfc.nasa.gov/~imswww/pub/
imswelcome

NASA: Engineers and the Age of Apollo (NASA SP-4104)
http://history.nasa.gov/SP-4104/sp4104.htm

NASA: Experimental Communications
Satellites
http://roland.lerc.nasa.gov/~dglover/sat/
satcom1.html

NASA: Exploration System Mission Directorate
http://exploration.nasa.gov

NASA: Headquarters
http://www.nasa.gov

NASA: History Office
http://history.nasa.gov/

NASA: History Series Publications
http://history.nasa.gov/series95.html

NASA: History Series Publications On-Line
http://www.hq.nasa.gov/office/pao/History/
on-line.html

NASA: History Timelines
http://www.hq.nasa.gov/office/pao/History/
timeline.html

NASA: Human Space Flight
http://spaceflight.nasa.gov/home

NASA: Image Exchange (NIX)
http://nix.nasa.gov

NASA: Information Locator
http://www.sti.nasa.gov/gils/nasa_gils.html

NASA: International Space Station
http://www.nasa.gov/mission_pages/station/
main

NASA: Links to Unmanned Spacecraft and
Satellites
http://roland.lerc.nasa.gov/~dglover/sat/
craft.html

NASA: Mission Patches
http://www.hq.nasa.gov/office/pao/History/
mission_patches.html

NASA: News Highlights
http://www.nasa.gov/news/highlights

NASA: Office of Space Operations
http://www.hq.nasa.gov/osf

NASA: Office of Space Science Missions
http://science.hq.nasa.gov/missions

NASA: Origins of the Universe
http://origins.jpl.nasa.gov

NASA: Personnel
http://www.hq.nasa.gov/office/pao/History/
prsnnl.htm

NASA: Planetary Photojournal
http://photojournal.jpl.nasa.gov

NASA: Satellite Tracking in Real Time
http://science.nasa.gov/realtime

NASA: Science Mission Directorate
http://science.hq.nasa.gov/

NASA: Science Office of Standards and
Technology
http://ssdoo.gsfc.nasa.gov/nost

NASA: Scientific and Technical Information
Server
http://www.sti.nasa.gov

NASA: Shuttle and Rocket Mission Schedule
http://www.nasa.gov/missions/highlights/
schedule.html

NASA: Solar System Mission
http://www.nasa.gov/missions/timeline/
current/solar-system_missions.html

NASA: Space Image Libraries
http://www.okstate.edu/aesp/image.html

NASA: Space Network On-line Information Center
http://msp.gsfc.nasa.gov/tdrss/tdrsshome.html

NASA Space Place, The
http://spaceplace.nasa.gov/en/kids

NASA Space Place: "Ask Dr. Marc"
http://spaceplace.nasa.gov/en/kids/
phonedrmarc

NASA: Space Science
http://spacescience.nasa.gov

NASA: Space Science Data Center
http://nssdc.gsfc.nasa.gov

NASA: Space Science Missions by Phase
http://science.hq.nasa.gov/missions/phase.html

NASA: Space Science Photo Gallery
http://nssdc.gsfc.nasa.gov/photo_gallery

NASA: Space Science Planetary Missions
http://nssdc.gsfc.nasa.gov/planetary/
projects.html

NASA: Space Telerobotics Program
http://ranier.hq.nasa.gov/telerobotics_page/
telerobotics.shtm

NASA: Spacecraft Technical Diagrams and Drawings
http://www.hq.nasa.gov/office/pao/History/
diagrams/diagrams.htm

NASA: Technical Reports Server
http://ntrs.nasa.gov

NASA: Telerobotics Photo Archive
http://ranier.oact.hq.nasa.gov/
telerobotics_page/photos.html

NASA: Television
http://www.nasa.gov/multimedia/nasatv

NASA: Thesaurus
http://www.sti.nasa.gov/thesfrm1.htm

National Air and Space Museum (NOAA)
http://www.nasm.si.edu

National Air and Space Museum: Space History Division
http://www.nasm.si.edu/research/dsh

National Oceanic and Atmospheric Administration
http://www.noaa.gov

National Oceanic and Atmospheric Administration: Library
http://www.lib.noaa.gov

National Oceanic and Atmospheric Administration: Photo Library
http://www.lib.noaa.gov

National Oceanic and Atmospheric Association: National Weather Service
http://www.nws.noaa.gov

National Research Council Canada/Conseil National de Recherches Canada
http://www.nrc-cnrc.gc.ca/main_e.html

National Space Science Data Center
http://nssdc.gsfc.nasa.gov

National Technical Information Service (NTIS)
http://www.ntis.gov

National Weather Service
http://www.nws.noaa.gov

Naval Network and Space Operations Command
http://www.nnsoc.navy.mil

New Horizons Pluto-Kuiper Belt Mission
http://pluto.jhuapl.edu

New Millennium Program
http://nmp.jpl.nasa.gov

Nine Planets, The: A Multimedia Tour of the Solar System
http://seds.lpl.arizona.edu/nineplanets/
nineplanets

NOAA: National Oceanic and Atmospheric Administration
http://www.noaa.gov

NOAA: National Oceanic and Atmospheric
Administration: Library
http://www.lib.noaa.gov

NOAA: National Oceanic and Atmospheric
Administration: Photo Library
http://www.lib.noaa.gov

NOAA: National Weather Service
http://www.nws.noaa.gov

Northrop Grumman Corporation: Space
Technology
http://www.st.northropgrumman.com

NSSDC: NASA Space Science Data Center
http://nssdc.gsfc.nasa.gov

NSSDC: NASA Space Science Photo Gallery
http://nssdc.gsfc.nasa.gov/photo_gallery

NSSDC: NASA Space Science Planetary
Missions
http://nssdc.gsfc.nasa.gov/planetary/
    projects.html

Office National d'Études et de Recherches
Aerospatiales (ONERA)/French National
Aerospace Research Establishment
http://www.onera.fr/english.html

Office of Space Science Missions
http://science.hq.nasa.gov/missions

One NASA
http://www.onenasa.nasa.gov

ONERA-CERT Toulouse Research Center
http://www.cert.fr/index.a.html

ONERA: Office National d'Etudes et de
Recherches Aerospatiales/French National
Aerospace Research Establishment
http://www.onera.fr/english.html

Orbital Sciences Corporation
http://www.orbital.com

Origins of the Universe
http://origins.jpl.nasa.gov

*Partnership: History of the Apollo-Soyuz Test Project,
The*
http://www.hq.nasa.gov/office/pao/History/SP-
    4209/cover.htm

Patrick Air Force Base
https://www.patrick.af.mil

Phoenix Mars Lander 2007
http://phoenix.lpl.arizona.edu/

Planetary Photojournal
http://photojournal.jpl.nasa.gov

Planetary Society, The
http://planetary.org

Planetary Society: Near Earth Objects
http://www.planetary.org/html/neo

Planetary Society: Search for Extraterrestrial
Intelligence
http://www.planetary.org/html/UPDATES/seti/
    index.html

PlanetScapes
http://planetscapes.com

Pratt & Whitney: A United Technologies
Company
http://www.pratt-whitney.com

Pratt & Whitney: Space Propulsion
http://www.pratt-whitney.com/prod_space.asp

Presidential Commission on the Space Shuttle
Challenger *Accident, Report of the*
http://science.ksc.nasa.gov/shuttle/missions/
    51-l/docs/rogers-commission/table-of-
    contents.html

Project Apollo Archive
http://www.apolloarchive.com

Project Apollo Archive: Apollo Image Gallery
http://www.apolloarchive.com/
    apollo_gallery.html

Project Apollo Archive: Lunar Lander Simulator
http://www.apolloarchive.com/lander.html

Project Mercury
http://www-pao.ksc.nasa.gov/kscpao/history/
mercury/mercury.htm

Project Mercury: Gallery
http://spaceflight.nasa.gov/gallery/images/
mercury

Project Mercury: Little Joe Series
http://www.peter.mcquillan.dsl.pipex.com

Radio Amateur Satellite (AMSAT) Corporation,
The
http://www.amsat.org

Rocketdyne (*see* Pratt & Whitney: Space
Propulsion)

Rockwell Automation
http://www.rockwell.com

Rockwell Collins
http://www.rockwellcollins.com

Rockwell International (*see* Rockwell Automation,
Rockwell Collins)

Roscosmos/Russian Federal Space Agency
http://www.federalspace.ru

Russian Designers and Scientists, Russian Space
Web
http://www.russianspaceweb.com/people.html

Russian Federal Space Agency/Roscosmos
http://www.federalspace.ru

Russian Khrunichev State Space Center
http://www.intertec.co.at/itc2/partners/
KHRUNICHEV

Russian Rocketry, Russian Space Web
http://www.russianspaceweb.com/rockets.html

Russian Space Centers, Russian Space Web
http://www.russianspaceweb.com/centers.html

Russian Space Research Institute (IKI)
http://www.iki.rssi.ru/eng/index.htm

Russian Space Science Internet (RSSI)
http://www.rssi.ru

Russian Space Statistics/RKA
http://www.rka-statistics.com

Russian Space Web
http://www.russianspaceweb.com

Russian Space Web: Chronology of Space
Exploration
http://www.russianspaceweb.com/
chronology.html

Russian Space Web: Mir Space Station
http://www.russianspaceweb.com/mir.html

Russian Space Web: Russian Designers and
Scientists
http://www.russianspaceweb.com/people.html

Russian Space Web: Russian Rocketry
http://www.russianspaceweb.com/rockets.html

Russian Space Web: Russian Space Centers
http://www.russianspaceweb.com/centers.html

Russian Space Web: Russian Spacecraft
http://www.russianspaceweb.com/
spacecraft.html

Russian Spacecraft, Russian Space Web
http://www.russianspaceweb.com/
spacecraft.html

SAREX: Space Amateur Radio Experiment
http://www.arrl.org/ARISS/

S. P. Korolev Rocket and Space Corporation
Energia
http://www.energia.ru/english/

Satellite Encyclopedia Online
http://www.tbs-satellite.com/tse/online

Satellite Information, World Data Center System
http://www.ngdc.noaa.gov/wdc

Satellite Tracking in Real Time
http://science.nasa.gov/realtime

Scaled Composites, LLC
http://www.scaled.com

Science Mission Directorate
http://science.hq.nasa.gov/

Scientific and Technical Information Program
http://www.sti.nasa.gov

Search for Extraterrestrial Intelligence (SETI)
Institute
http://www.seti-inst.edu

Shuttle-Mir: History
http://spaceflight.nasa.gov/history/shuttle-mir

Shuttle-Mir: Multimedia
http://spaceflight1.nasa.gov/history/shuttle-
mir/multimedia/multimedia.htm

Shuttle: Press Kits
http://www.shuttlepresskit.com

Skylab Program
http://www-pao.ksc.nasa.gov/kscpao/history/
skylab/skylab.htm

Skylab Program: Gallery
http://spaceflight.nasa.gov/gallery/images/
skylab

Sojourner, Mars Rover
http://mpfwww.jpl.nasa.gov/rover/
sojourner.html

Solar and Heliospheric Observatory (SOHO)
http://sohowww.nascom.nasa.gov

Solar System Exploration
http://sse.jpl.nasa.gov

Space Amateur Radio Experiment (SAREX)
http://www.arrl.org/ARISS/

Space Foundation
http://www.spacefoundation.org

Space Image Libraries
http://www.okstate.edu/aesp/image.html

Space Network On-line Information Center
http://msp.gsfc.nasa.gov/tdrss/tdrsshome.html

Space Place
http://spaceplace.nasa.gov/en/kids

Space Place: "Ask Dr. Marc," NASA
http://spaceplace.nasa.gov/en/kids/
phonedrmarc

Space Research Institute (IKI), Russian
http://www.iki.rssi.ru/eng/index.htm

Space Services Inc.: Memorial Spaceflights
(formerly Celestis)
http://www.memorialspaceflights.com

Space Shuttle: Earth Observations Photography
http://earth.jsc.nasa.gov/sseop/efs/

Space Shuttle: History
http://spaceflight.nasa.gov/shuttle/archives

Space Shuttle: Human Space Flight
http://spaceflight.nasa.gov/shuttle

Space Shuttle: Reference
http://spaceflight.nasa.gov/shuttle/reference

Space Shuttle Gallery: Human Space Flight
http://spaceflight.nasa.gov/gallery/images/
shuttle

Space Station: Science Operations News
http://www.scipoc.msfc.nasa.gov

Space Technology Hall of Fame
http://www.spacetechhalloffame.org

Space Telerobotics Program
http://ranier.hq.nasa.gov/telerobotics_page/
telerobotics.shtm

Space Telescope Science Institute
http://www.stsci.edu

Space Telescope Science Institute: Data Archive
http://archive.stsci.edu

Space Telescope Science Institute: Hubble Space
Telescope
http://www.stsci.edu/hst

Space Telescope Science Institute: James Webb
Space Telescope
http://www.stsci.edu/jwst

SPACECOM: United States Space Command
http://www.peterson.af.mil/hqafspc

SPACEHAB
http://www.spacehab.com

Spaceport Systems International: Commercial Spaceport
http://www.calspace.com

Spanish Space Agency (INTA)/Instituto Nacional de Técnica Aeroespacial
http://www.inta.es/index.asp

Spitzer Space Telescope
http://www.spitzer.caltech.edu/spitzer

Stages to Saturn (SP-4206)
http://history.nasa.gov/SP-4206/contents.htm

Stardust Project
http://stardust.jpl.nasa.gov

Starsem: The Soyuz Company
http://www.starsem.com

Stennis Space Center, John C.
http://www.nasa.gov/centers/stennis

Stennis Space Center: History
http://www.nasa.gov/centers/stennis/about/
    history/history.html

Stennis Space Center: Multimedia
http://www.nasa.gov/centers/stennis/
    multimedia/index.html

Swift Gamma Ray Burst Mission
http://swift.gsfc.nasa.gov/docs/swift/
    swiftsc.html

TDRSS Satellite System
http://msp.gsfc.nasa.gov/tdrss/tdrsshome.html

30th Space Wing
http://www.vandenberg.af.mil

This New Ocean: A History of Project Mercury (SP-4201)
http://www.hq.nasa.gov/office/pao/History/
    SP-4201/toc.htm

Toulouse Research Center (ONERA-CERT)
http://www.cert.fr/index.a.html

TRW, Inc. (*see* Northrop Grumman)

Tsiolkovsky State Museum of the History of Cosmonautics
http://www.informatics.org/museum

U.S. Air Force: Launch Information
https://www.patrick.af.mil/launch.htm

U.S. Air Force: Space Command
http://www.peterson.af.mil/hqafspc

U.S. Army: Space and Missile Defense Command
http://www.smdc.army.mil/

U.S. Centennial of Flight
http://www.centennialofflight.gov

U.S. Geological Survey
http://www.usgs.gov

U.S. Government Printing Office: Access
http://www.gpoaccess.gov

U.S. Government Printing Office: Online Bookstore
http://bookstore.gpo.gov

U.S. Naval Network and Space Operations Command
http://www.nnsoc.navy.mil

U.S. Space Camp
http://www.spacecamp.com

Ulysses
http://ulysses.jpl.nasa.gov

Ulysses, ESA
http://sci.esa.int/ulysses

United Space Alliance
http://www.unitedspacealliance.com

Unmanned Space Project Management: Surveyor and Lunar Orbiter (NASA SP-4901)
http://www.hq.nasa.gov/office/pao/History/
    SP-4901/table.htm

Unmanned Spacecraft and Satellite Links
http://roland.lerc.nasa.gov/~dglover/sat/
    craft.html

Upper Atmosphere Research Satellite (UARS)
http://umpgal.gsfc.nasa.gov/uars-science.html

Vandenberg Air Force Base
http://www.vandenberg.af.mil

Virtual Space Museum (of Cosmonautics)
http://vsm.host.ru/emain.htm

Wallops Island Flight Facility
http://www.wff.nasa.gov

Wallops Island Flight Facility: History
http://www.wff.nasa.gov/about/history.php

Wallops Island Flight Facility: Missions
http://www.wff.nasa.gov/missions/
index_missions.php

Wallops Island Flight Facility: Multimedia
http://www.wff.nasa.gov/multimedia/
index_multimedia.php

What Made Apollo a Success? (NASA SP-287)
http://history.nasa.gov/SP-287/sp287.htm

Where No Man Has Gone Before: A History of Apollo Lunar Exploration Missions (NASA SP-4214)
http://www.hq.nasa.gov/office/pao/History/
SP-4214/cover.html

White Sands Missile Range
http://www.wsmr.army.mil

White Sands Missile Range: History
http://www.wsmr-history.org/History.htm

White Sands Test Facility, Johnson Space Center
http://www.wstf.nasa.gov

Wilkinson Microwave Anisotropy Probe
http://map.gsfc.nasa.gov

World Data Center System for Satellite Information
http://www.ngdc.noaa.gov/wdc

X Prize, Ansari
http://www.xprize.org/home.php

Zarya: Information on Soviet and Russian Spaceflight
http://www.zarya.inf

Russell R. Tobias and David G. Fisher

Indexes

Personages Index

Subject Index

Wolanczyk, Stephan, 982
Wolf, David A., 591, 611, 1138,
    1255, 1481, 1519, 1521, 1541,
    1547-1548
Wolfe, Allen E., 803
Wolfe, John H., 1072, 1080
Women in space, 324-331, 1348,
    1571
Wood, James, 1870, 1957
Worden, Alfred M., 114, 137
Working Group on Interplanetary
    Exploration, 820
World Climate Research Program
    (WCRP), 307, 961
World Magnetic Survey, 1043
World Meteorological
    Organization (WMO), 497,
    676
World Radio Conference (1992),
    970
World War II, 376
World Weather Watch (WWW),
    499
Worldwide Reference System, 709
Wright, Orville and Wilbur, 19
Wright Patterson Air Force Base,
    982
WSF. See Wake Shield Facility

X-1 aircraft, 55, 1265
X-15 aircraft, 982, 1055, 1245,
    1265, 1267, 1276, 1620, 1864-
    1869
X-20 Dyna-Soar, 903, 1870-1874

X-24 aircraft, 983
X-24A aircraft, 577
X-30 aircraft, 1268. See also
    National AeroSpace Plane
X-33 aircraft, 730, 876, 1268
X-34 aircraft, 731, 876, 1268
X-37 aircraft, 731, 876
X-38 aircraft, 577, 747, 1268
X-43 aircraft, 731
X-band, 1997
X Prize, 19-23, 732
X Prize Foundation, 730, 1110
X-radiation, 1997
X-ray astronomy, 240-245, 526-532,
    1570, 1997
X-ray background emission, 1049
X-ray-emitting stars, 242
X-Ray Explorer, 341
X-ray sources, 529, 1023
X-ray telescopes, 240-245
X-Ray Timing Explorer, 256
X rays, 240; solar, 340
XLR-11 rocket engine, 745
XMM-Newton (satellite), 243
XV-15 tilt-rotor vehicle, 16

Y2K problem, 1562
Yang, Liwei, 324, 330
Yankee Clipper (Apollo Control and
    Service Module), 116
Yardley, John F., 461, 490, 893,
    899, 1305, 1621, 1957
Yaw, 925. See also Pitch, roll, and
    yaw

Yeager, Charles E. "Chuck," 1264-
    1265, 1957
Yeager, Jeana L., 1957
"Year 2000" problem. See Y2K
    problem
Yeliseyev, Aleksei, 1226
Yeltsin, Boris, 621
Yeomans, Donald K., 272
Yogi (Martian rock), 860, 862
Yosemite National Park (Landsat
    data), 711
Young, A. Thomas, 1791, 1797
Young, John W., 76, 96, 124, 144,
    150, 152, 159, 427-428, 437,
    468-469, 1305-1306, 1332, 1336,
    1958
Young, Richard E., 411

Zarnecki, J. C., 559
Zarya Control Module, 263, 591,
    599, 616, 622, 627, 632, 1549
Zegrahm Space Voyages, 1110
Zero-g crouch, 1634
Zero gravity, 1997
Zimmerman, Dean, 490
Zodiacal light, 452, 566, 889, 940,
    1049, 1997
Zond missions (Soviet), 819, 1224
Zuber, M. T., 267
Zurek, Richard, 830, 865
Zvezda Service Module, 591, 599,
    627, 632, 1585